T0254263

This book provides a detailed introduction to the principles of Doppler and polarimetric radar, focusing in particular on their use in the analysis of precipitation. The design features and operation of practical radar systems are highlighted throughout the book in order to illustrate important theoretical foundations.

The authors begin by discussing background topics such as electromagnetic scattering, polarization, and wave propagation. They then deal in detail with the engineering aspects of pulsed Doppler polarimetric radar, including the relevant signal theory, spectral estimation techniques, and noise considerations. They close by examining a range of key applications in meteorology and remote sensing.

The book will be of great use to graduate students of electrical engineering and atmospheric science as well as to practitioners involved in the meteorological application of radar systems.

V. N. Bringi received his Ph.D. from Ohio State University and is a Professor of Electrical and Computer Engineering at Colorado State University. He is a pioneer in the field of polarimetric radar meteorology and has published over 100 technical articles.

V. Chandrasekar received his Ph.D. from Colorado State University, where he is a Professor of Electrical and Computer Engineering. He has over twenty years' experience working with various aspects of radar systems. He has also been a visiting professor at The National Research Council of Italy.

POLARIMETRIC DOPPLER WEATHER RADAR

Principles and applications

Professor V. N. Bringi

Professor V. Chandrasekar

CAMBRIDGE
UNIVERSITY PRESS

CAMBRIDGE UNIVERSITY PRESS
Cambridge, New York, Melbourne, Madrid, Cape Town, Singapore, São Paulo

Cambridge University Press
The Edinburgh Building, Cambridge CB2 2RU, UK

Published in the United States of America by Cambridge University Press, New York

www.cambridge.org
Information on this title: www.cambridge.org/9780521623841

First published 2001
This digitally printed first paperback version 2005

A catalogue record for this publication is available from the British Library

ISBN-13 978-0-521-62384-1 hardback
ISBN-10 0-521-62384-7 hardback

ISBN-13 978-0-521-01955-2 paperback
ISBN-10 0-521-01955-9 paperback

With humility and reverence, this work is placed at the feet of Ganesha, the remover of all obstacles.

Contents

4 Dual-polarized wave propagation in precipitation media

5 Doppler radar signal theory and spectral estimation

Preface

Doppler radars are now considered to be an indispensable tool in the measurement and forecasting of atmospheric phenomena. The deployment of WSR–88D radars, the terminal Doppler weather radar (TDWR) at major airports, and wind-profiling radars in the USA can all be considered as major milestones in the operational application of the Doppler principle. Another milestone is the successful deployment of the first precipitation radar in space as part of the Tropical Rainfall Measurement Mission (TRMM). Measurement of the reflectivity and velocity of precipitation particles basically exploits the information contained in the amplitude and phase of the scattered electromagnetic wave. In the last two or three decades, it has become increasingly clear that significant information is also contained in the polarization state of the scattered wave. The physical and experimental basis for the application of radar polarimetry to the study of precipitation is the main subject of this book.

The evaluation of polarimetric measurement options for operational WSR–88D radars is gaining momentum and, if realized, will result in widespread application of polarimetric techniques by a broad segment of the meteorological community. Basic and applied research in polarimetric radar meteorology continues to be strong world-wide, especially in Europe and Japan. We believe that the time has arrived for a detailed treatment of the physical principles underlying coherent polarimetric radar for meteorological applications.

This book is based in part on a graduate class taught by the authors at Colorado State University since the mid-1980s. Our goal has been to present the subject, essentially from first principles, in a self-contained format. Substantial thought has gone into selection and organization of the material in order to provide a good balance between theoretical rigor and practical applications. Examples of radar measurements are frequently given to illustrate the theory and to motivate the reader. Mathematical and physical background normally acquired in an undergraduate program in physics, atmospheric science, or electrical engineering is assumed.

The book is organized into eight chapters and five appendices. Notes are provided at the end of each chapter that provide suggestions for further reading, websites for description of specialized instruments, and examples of data not usually available in the archival journals. The Website for this book is www.engr.colostate.edu/ece/ radar_education where specialized software, homework assignments, etc., are available.

Chapters 1–4 provide the foundation for wave scattering and propagation theory as applied to radar polarimetry. Chapter 1 covers a number of important electromagnetic concepts starting with Maxwell's equations and leading to Rayleigh scattering and dielectric mixing formulas. The bistatic Doppler frequency formula is derived by considering an example of a moving loop illuminated by a plane wave. The time-correlated bistatic cross section of a moving particle is defined, which naturally leads to the Doppler spectrum. The elements of a generic pulsed Doppler radar system are then described.

Chapter 2 deals with the amplitude scattering matrix of spheroids in the Rayleigh–Gans limit. In fact, it is quite remarkable that important polarimetric radar observations of precipitation can, to a large degree, be explained using the spheroidal model. For completeness, the Mie scattering solution is formulated as a boundary value problem, with a review of spherical harmonics provided in Appendix 2.

Chapter 3 provides a detailed description of wave, antenna, and radar polarization. The dual-polarized radar range equation is derived in linear and circular polarization bases. The conventional radar cross section concept is generalized to include the concepts of copolar and cross-polar polarization synthesis and optimal polarizations. Scattering from randomly distributed precipitation particles is treated using the Mueller matrix, which leads to the definition of conventional polarimetric observables such as differential reflectivity, linear depolarization ratio, etc. The polarimetric covariance matrix is defined, which represents complete measurements from a dual-polarized radar. Much of this chapter deals with the structure of the covariance matrix in linear and circular bases including simplifications afforded by symmetry arguments.

Chapter 4 deals with dual-polarized wave propagation in precipitation media and how differential attenuation and differential phase between the two "characteristic" waves can be measured. The structure of the propagation-modified covariance matrix in linear and circular bases is treated in detail. Radar measurements in linear and circular bases are used to illustrate the theory. The hybrid measurement mode, which is currently being explored for the WSR–88D radar system, is explained.

Doppler radar theory, signal statistics, and signal processing form the subject matter of Chapter 5 and are dealt with in a fairly rigorous manner. Although a very brief review of signals and systems theory is provided, the level of treatment assumes prior exposure and familiarity with linear systems theory. The application of Doppler radar to wind retrieval or to the study of storm dynamics is not covered as these topics are dealt with in detail in existing books (Sauvageot 1992; Doviak and Zrnić 1993).

Chapters 6–8, which form nearly 50% of the book, deal with dual-polarized radar systems and applications to meteorology, and would be of most relevance to professionals in the field. Chapter 6 describes how polarization diversity and agility are configured in different radar systems. The topics of antenna performance and system polarization errors are treated in detail. A significant portion of this chapter is devoted to estimation of covariance matrix elements from signal samples under three different pulsing schemes. The practical topic of deriving specific differential phase from range profiles of differential propagation phase is also covered in detail.

Chapter 7 deals with the polarimetric basis for characterizing precipitation and describes methods used to infer hydrometeor types and amounts. Commensurate with the importance of this topic, this chapter occupies nearly one-third of the book. A large number of examples are used to illustrate different methods and approaches. Correction of measured reflectivity and differential reflectivity for attenuation, and differential attenuation due to rain along the propagation path are treated in detail. This chapter ends with a description of fuzzy logic methods applied to the problem of hydrometeor classification, and several examples are provided. With real-time implementation of such methods already in place on some research radars, this topic is both timely and relevant for applications.

Chapter 8 treats the radar rainfall measurement problem from the viewpoint of both physically based and statistical/engineering-based approaches. The error structure of polarimetric-based rain rate algorithms is explained via simulations. The theoretical basis of the area–time integral (ATI) and probability-matched methods (PMM) is covered as well as their polarimetric extensions. A separate section on neural networks and their application to rainfall estimation is included together with examples of performance.

Five appendices are provided, three of which are reviews (electrostatics, spherical harmonics, and the T-matrix method). Appendix 4 derives the transmission matrix, while Appendix 5 details procedures for calculating the variance of the magnitude and phase of correlation functions. It is beyond the scope of this book to give a historical account of polarimetric Doppler radar, for which we defer to *Radar in Meteorology* (American Meteorological Society 1990, Ed. D. Atlas). Since emphasis in this book is primarily on pulsed, dual-polarized Doppler radar, other radar techniques, such as frequency modulated–continuous waveform (FM–CW), pulse compression, or synthetic aperture radar are not covered, nor are radars based on airborne or spaceborne platforms. However, the principles covered in this book will be helpful to the reader wishing to pursue these topics.

Finally, the opportunity to pursue exciting interdisciplinary research, to teach classes in our research area, and to interact with graduate students both in electrical engineering and atmospheric science, have all played a significant role in the writing of this book. Indeed, it is our hope that students will enjoy learning from this book. At the same time, we hope that researchers in the field will find the topics both timely and helpful for their own work.

V. N. Bringi
V. Chandrasekar

Acknowledgments

It is a great pleasure for us to acknowledge the assistance of our colleagues and students. To our students and post-doctoral fellows (Ji Ran, Yoong-Goog Cho, Gwo-Jong Huang, Hongping Liu, Li Liu, John Hubbert, John Beaver, Steve Bolen, Gang Xu, Konrad Gojara, Max Seminario, A. Al-Zaben, and L. Ramaswamy), we are indeed grateful for their assistance given in many forms. A number of our colleagues carefully read draft versions of many of the chapters and provided valuable feedback for which we are grateful (Archibald Hendry, Richard Doviak, Eugene Mueller, Anthony Holt, John Hubbert, Richard Strauch, Eugenio Gorgucci, J. Vivekanandan, and David Brunkow). Many colleagues provided us with original figures which are included in the book, and these have been acknowledged with each figure. The Bureau of Meteorology Research Center, Melbourne, Australia provided C-POL radar data and disdrometer data described in Chapter 7. The 2D-video disdrometer data were provided by Joanneum Research, Graz, Austria. A significant impetus which motivated us in large part to the writing of this book was the relocation of the CHILL radar to Colorado State University in 1990, and now referred to as the CSU–CHILL radar facility operated via a cooperative agreement with the National Science Foundation. The close association that the authors have had with this facility (Eugene Mueller and David Brunkow) and with the Department of Atmospheric Science (Steve Rutledge) has without doubt lead to a better exposition of the subject matter of this book. Another significant factor has been the continued research support from the National Science Foundation (Ron Taylor, Program Manager for Physical Meteorology, now retired) and from the US Army Research Office (Walter "Bud" Flood, retired, and Walter Bach) for our polarimetric radar research, particularly from the early 1980s to the mid-1990s. We are grateful for this continuous period of support during which much of the earlier exciting research results using the NCAR CP-2 radar were obtained (in collaboration with Thomas Seliga and Kultegin Aydin). One of the authors (V.C.) also acknowledges close collaboration with Eugenio Gorgucci and Gianfranco Scarchilli, particularly with reference to polarimetric rain rate algorithms in Chapter 8. Finally, both authors are especially grateful for the support of their respective spouses (Sreedevi Bringi and Nirmala Chandra) during the writing of this book, and to their parents for being a continuous source of inspiration.

Notation

List of symbols

a radius of a sphere or radius of the principal equatorial circle of a spheroid (Fig. 1.8a)

a_n^s expansion coefficients of the scattered electric field (van de Hulst definition), Section 2.4.1

A specific attenuation, eq. (1.133)

A_{dp} specific differential attenuation between the two characteristic waves, eq. (4.7)

$A_{h,v}$ specific attenuation at horizontal (h) or vertical (v) polarization states, eq. (4.76)

A_{sm} asymmetry ratio, eq. (2.99)

ATI   area–time integral

$\vec{A}$ magnetic vector potential, eq. (1.13a)

b semi-major axis length of a prolate spheroid or semi-minor axis length of an oblate spheroid (Fig. 1.8a)

b_n^s expansion coefficients of the scattered electric field (van de Hulst definition), Section 2.4.1

B bandwidth of a radar receiver

$\vec{B}^{i,s}$ magnetic field; superscripts i, s correspond to incident and scattered waves

c speed of light in vacuum

c fractional volume concentration, eq. (1.69)

D diameter of a sphere or characteristic dimension of a spheroid (equi-volume spherical diameter)

D_{max} maximum diameter in a measured size distribution

D_m mass-weighted mean diameter, eq. (7.13)

D_0 median volume diameter, eq. (7.11)

D_z reflectivity-weighted mean diameter, eq. (7.25)

$\hat{e}_{i,r,s}$ unit polarization vector of the incident (i), reflected (r), or scattered (s) waves

$\hat{e}_{h,v}$ unit polarization vector for horizontal (h) and vertical (v) states

$\hat{e}_{R,L}$ unit polarization vector for right-hand (R) and left-hand (L) circular states, eqs. (3.6, 3.8)

$\hat{e}_t$ unit polarization vector of the transmitting antenna

E_0 electric field amplitude

$E_h^{i,r,s}$; $E_v^{i,r,s}$ components of the electric field in the (h, v)-basis, superscript (i, r, s) refers to incident, reflected, or scattered waves, respectively

E_R; E_L components of the electric field in circular basis (R = right-hand component, L = left-hand component)

$E_l^{i,s}$; $E_r^{i,s}$ components of the electric field in the van de Hulst convention (Section 2.4.1); i, s refer to incident and scattered fields, l, r refer to parallel (l) and perpendicular (r) components

$\vec{E}^i$ incident electric field vector

$\vec{E}_T^{in}$ total electric field inside the scatterer, eq. (1.18b)

$\vec{E}^r$ reflected electric field vector

$\vec{E}^s$ scattered electric field vector

$\mathbf{E}^i$; $\mathbf{E}^r$; $\mathbf{E}^s$ 2 × 1 column, or Jones vector, of the incident, reflected, or scattered electric field, eqs. (2.7, 2.12)

f frequency

f ice fraction, eq. (7.87c)

f_c carrier frequency

f_D Doppler frequency

f_{IF} intermediate frequency

f_h; f_v normalized antenna power pattern functions, eq. (6.20)

f_h; f_v forward scattering amplitude at horizontal and vertical polarizations, eq. (8.5)

$f_{\varepsilon h}$; $f_{\varepsilon v}$ cross-polar pattern functions, eq. (6.21)

$f(\theta, \phi)$ normalized antenna pattern function, eq. (5.38)

$f(R_T)$ rain volume fraction above a threshold rain rate, eq. (8.39b)

f_η normalized range–time profile of reflectivity, eq. (5.40)

$f(x)$ probability density function, where x can be voltage, phase, power, etc.

$f_D(D)$ probability density function of drop diameter, eq. (8.2)

$\vec{f}$ vector scattering amplitude, eq. (1.24a)

F dielectric factor, eq. (1.69)

F volume flux of rain, eq. (8.35)

F noise figure, eq. (5.144)

F_e equivalent noise figure, eq. (5.151)

$\vec{F}$ vector radiation amplitude, eq. (3.30)

$g(t)$ impulse response of a receiver filter, eq. (5.51)

G_0 peak boresight gain of an antenna, eq. (5.38)

G_r receiver power gain

$G(\theta, \phi)$ antenna power gain function, eq. (5.38)

$G_{h,v}$ antenna power gain function associated with the h- or v-port of a dual-polarized antenna, eq. (6.19)

G_{0h}; G_{0v} peak boresight gains associated with the h- or v-port of a dual-polarized antenna

G_R; G_L antenna power gain functions associated with the right-hand (R)-port or left-hand (L)-port of a dual-circularly polarized antenna, eq. (3.102)

$G(f)$ frequency-response function of a receiver, eq. (5.50)

$G_n(f)$ normalized frequency response function of a receiver

$h_n^{(2)}$ spherical Hankel function of the second kind, eq. (A2.6)

$h(\tau; t)$ impulse response of the medium, eq. (5.101)

$\hat{h}_{i,r,s}$ unit horizontal polarization vector of incident, reflected, or scattered electric field

$\vec{h}$ effective vector "length" of an antenna, eq. (3.37)

$H(f; t)$ frequency response function of the medium, eq. (5.100)

H_{dr} hail signature function, eq. (7.88)

i $\sqrt{-1}$

$\hat{i}$ unit vector along the direction of the incident wave (Fig. 1.7)

I	in-phase component of a signal	$\lvert K_{p,w}\rvert^2$	dielectric factor of a particle (p) or of water (w), eqs. (3.167, 5.49a)
I	element of the 4×1 Stokes' vector, eq. (3.25a)	K_{dp}	specific differential phase, eq. (4.8)
$\mathbf{I}$	4×1 Stokes' vector, eq. (3.25)	$\mathbf{K}$	covariance matrix of signal samples, eq. (6.73)
$\bar{\bar{\mathbf{I}}}$	identity tensor, or unit diagonal matrix, eq. (2.28b)	l_r	finite bandwidth loss factor, eq. (5.56)
j	$\sqrt{-1}$	l_{sw}	insertion loss factor of a polarization switch, eq. (6.61)
j_n	spherical Bessel function, eq. (A2.6)	l_{wg}	waveguide loss factor, eq. (6.53)
$\mathbf{J}$	coherency matrix, eq. (3.146). Additional subscripts cp and up define completely polarized and completely unpolarized waves, respectively	L	loss factor of a lossy device, eq. (5.152)
		L	linear depolarization ratio; $L = 10^{0.1(\mathrm{LDR})}$, eq. (3.232)
$\mathbf{J}_{10};\mathbf{J}_{01}$	coherency matrix when the transmitted wave is horizontally polarized (subscript "10") or vertically polarized (subscript "01"), eq. (3.181)	$\mathrm{LDR}_{vh};$ LDR_{hv}	linear depolarization ratio for an ensemble of particles, subscript "vh" stands for horizontal transmit/vertical receive, while "hv" stands for vertical transmit/horizontal receive, eq. (3.169)
$\vec{J}_p$	polarization current density, eq. (1.54)		
k	Boltzman constant, 1.38×10^{-23} J K^{-1}, eq. (5.140)	m	refractive index, eq. (1.126)
		$\vec{M}$	vector spherical harmonic, eq. (A2.9a)
k	wave number in a dielectric, eq. (1.8)	$M_{h,v}$	transmitter wave amplitudes exciting the horizontal or vertical, h- or v-, ports of a dual-polarized antenna. $M_h = 1$, $M_v = 0$, represents transmission of a h-polarized wave, while $M_h = 0$, $M_v = 1$, represents a v-polarized wave, eq. (3.59)
k_{eff}	effective wave number (or propagation constant) in a composite material, eq. (1.70)		
$k_{\mathrm{eff}}^{\mathrm{re}};$ $k_{\mathrm{eff}}^{\mathrm{im}}$	real (re) or imaginary (im) component of the effective propagation constant, eq. (1.129)		
$k_{\mathrm{eff}}^{h};$ k_{eff}^{v}	effective propagation constant when the two characteristic waves are polarized horizontally (h) and vertically (v), eq. (4.67)	$M_{R,L}$	as above except for the right-hand (R) or left-hand (L) ports of a dual-polarized antenna, eq. (3.102)
$k_{\mathrm{re}}^{h,v};$ $k_{\mathrm{im}}^{h,v}$	real (re) and imaginary (im) components of k_{eff}^{h} or k_{eff}^{v}, eq. (4.68a,b)	$\mathbf{M}$	4×4 Mueller matrix, eq. (3.28)
		$n; n_c; N$	number of particles per unit volume, eqs. (4.3), (7.26), (1.64)
$\hat{k}_{i,r,s}$	unit vector along direction of the incident (i), reflected (r), or scattered (s) waves	N	number of signal samples

N_0 intercept parameter of an exponential or gamma drop size distribution, eq. (7.12)

$N_0/2$ power spectral density of "white" noise, eq. (5.141)

N_w "intercept" parameter of a normalized gamma drop size distribution, eq. (7.61)

$N(D)$ drop size distribution, eq. (1.128)

$\vec{N}$ vector spherical harmonic, eq. (A2.9b)

p, χ parameters representing differential attenuation and differential phase between the two characteristic waves, eq. (4.13b)

$p_{x,y,z}$ Cartesian components of the dipole moment vector, eq. (1.36a)

$p(t)$ impulse train sampling function, eq. (5.111)

$p(\theta);$ probability density function (pdf) of
$p(\phi)$ the spherical coordinate angles θ or ϕ

$p(\Omega)$ pdf of the symmetry axis in the solid angle interval

$\vec{p}$ vector dipole moment, eq. (1.32c)

P_t transmitter pulse power, eq. (3.33)

$P_t^{h,v}$ transmitter pulse power coupled to the horizontal or vertical, h- or v-ports of a dual-polarized antenna, eqs. (3.57, 3.58)

$P_t^{R,L}$ as above except for a dual-circularly polarized antenna

P_{rad} power radiated by an antenna, eq. (3.31)

P_{rec} received power, eq. (3.130)

$P_{h,v}$ power received by the h-receiver or v-receiver in the hybrid mode, eq. (6.106)

P_{co}^h power received at the copolar (co) h-port of a dual-polarized antenna when horizontal (h) polarization is transmitted, eq. (3.177a)

P_{co}^v power received at the v-port when the transmitted state is vertical (v), eq. (3.177b)

P_{cx} cross-polar (cx) received power, eq. (3.178)

$\bar{P}_o$ averaged received (or signal) power at the receiver output, eq. (6.52a)

$\bar{P}_{ref}$ averaged received (or signal) power at the reference plane, eq. (6.53)

$\bar{P}_{S+N}$ sum of averaged signal power and noise power

P_N noise power, eq. (5.140)

P_s total power scattered by a particle when illuminated by an ideal, linearly polarized plane wave, eq. (1.44)

P_d time-averaged power dissipated within a particle, eq. (1.53)

$\mathbf{P}$ 2×2 ensemble-averaged forward scattering matrix, eq. (4.39)

$P_{hh}; P_{hv};$ elements of the 2×2
$P_{vh}; P_{vv}$ ensemble-averaged forward scattering matrix, eq. (4.40)

$\vec{P}$ volume density of polarization within a dielectric particle

Q quadrature phase component of a signal

Q element of the 4×1 Stokes' vector, eq. (3.25b)

Q_{ext} normalized extinction cross section of a spherical particle, eq. (1.129c)

r spherical coordinate of a point (r, θ, ϕ) or radar range

r axis ratio of a spheroid ($r = b/a$), Fig. 1.8

$\vec{r}; \vec{r}'$ position vectors

$\vec{r}_i$	position vector to a point on the plane of constant phase of a plane wave, Fig. 1.6b	$R_{vh,vh}[l]$	autocorrelation of the vh signal at lag l, eq. (6.76d)
$\hat{r}$	unit vector ($\hat{r} = \vec{r}/r$)	$R_{vv,hv}[l]$	correlation between vv and hv returns at lag l, eq. (6.76e)
r	magnitude of $\vec{r}$	$R_{vv,vv}[l]$	autocorrelation of the vv signal at lag l, eq. (6.76f)
$\bar{r}_m$	mass-weighted mean axis ratio, eq. (7.33)	$R_{v,h}[0]$	correlation between the h and v signal returns at lag 0 in the hybrid
$\bar{r}_z$	reflectivity-weighted mean axis ratio, eq. (7.35)		mode, eq. (6.107)
$\overline{r^2}$	mean-square value of the axis ratio, eq. (3.220a)	$\mathbf{R}$	2×2 rotation matrix, eq. (2.93a)
$\bar{r}$	mean axis ratio, eq. (3.219a)	$\mathbf{R}$	4×4 covariance matrix of the complex signal vector, eq. (5.168d)
$\vec{R}$	vector connecting source and observation points, $\vec{R} = \vec{r} - \vec{r}'$	$\hat{s}$	unit vector along the direction of scattering, Fig. 1.6
R	magnitude of $\vec{R}$	s	distance along the direction of scattering
$\hat{R}$	unit vector, along the direction of $\vec{R}$	$s(t)$	complex signal representation, eq. (1.92)
R	rain rate, eq. (7.64)	$s_{tr}(t)$	transmitted signal, eq. (5.5a)
$\bar{R}$	areal average rain rate, eq. (8.38a)	S_σ	bistatic Doppler power spectrum, eq. (1.113a)
$R; R_v$	autocorrelation function of the received voltage, eqs. (5.80, 5.128)	$S(\omega);$ $S(f)$	Doppler frequency spectrum or power spectral density, eqs. (5.124, 5.125)
R_N	autocorrelation function of "white" noise, eq. (5.158)	$S(v)$	Doppler velocity spectrum, eq. (5.127)
R_{co}	correlation between the copolar received voltages (hh and vv returns), eq. (3.179c)	$S_N(f)$	power spectral density of "white" noise, eq. (5.141)
R_{cx}^h	correlation between copolar and cross-polar signal returns (hh and vh) when the transmitted polarization is horizontal (h), eq. (3.179a)	$S_{11,22}$	principal plane elements of the scattering matrix, eq. (2.89)
		$S_{1,2}$	functions related to the Mie solution, eq. (2.108)
R_{cx}^v	correlation between coplar and cross-polar signal returns (vv and hv) when transmitted state is vertical (v), eq. (3.179b)	$S_{hh}; S_{vh}$	first column of the 2×2 amplitude scattering matrix, eq. (2.6a)
$R_{hh,hh}[l]$	autocorrelation of the hh signal return at lag l, eq. (6.76a)	$S_{hv}; S_{vv}$	second column of the 2×2 amplitude scattering matrix, eq. (2.6a)
$R_{hh,vh}[l]$	correlation between hh and vh returns at lag l, eq. (6.76b)	$S_{RR}; S_{LR}$	first column of the 2×2 amplitude scattering matrix in the circular basis, eq. (3.88a)
$R_{hh,vv}[l]$	correlation between hh and vv returns at lag l, eq. (6.76c)		

$S_{RL}; S_{LL}$	second column of the 2 × 2 amplitude scattering matrix in the circular basis, eq. (3.88)	$\bar{v}$	mean Doppler velocity, eq. (5.127)
S_{BSA}^c	2 × 2 scattering matrix in circular basis, eq. (3.88a)	$v(D)$	terminal velocity of a drop of diameter D, eq. (7.65a)
$S_{BSA};$ S_{FSA}	2 × 2 scattering matrix in back scatter alignment (BSA) or forward scatter alignment (FSA) conventions, eqs. (2.6b, 2.12)	V	volume of sphere or spheroid
		V	element of the 4 × 1 Stokes' vector, eq. (3.25d)
		$V; V_r$	voltage or received voltage
t	time	$V_{h,v}$	received voltage at the h- or v-port of a dual-polarized antenna; also, received voltages in the hybrid mode, eqs. (3.60, 6.106)
t'	retarded time, eq. (1.108b)		
T_0	pulsewidth of a transmitted pulse		
T_0	ambient temperature, eq. (5.145)	$V_{R,L}$	received voltage at the RHC- or LHC-port of a dual-circularly polarized antenna, eqs. (3.105, 3.108a)
T_s	pulse repetition time		
T_e	equivalent noise temperature, eq. (5.149)	V_h^{10}	received voltage at the h-port of an antenna when the transmitted state is horizontal (denoted by superscript "10").
T_D	coherence time of a medium, eq. (5.96)		
T_D	device noise temperature, eq. (5.143)	V_v^{10}	received voltage at the v-port of an antenna when the transmitted state is horizontal
T_N	noise temperature, eq. (5.140)		
$T_{11,22}$	principal plane elements of the transmission matrix, eq. (4.6b)	V_v^{01}	received voltage at the v-port when the transmitted state is vertical (denoted by "01")
$T_{hh}; T_{vh};$ $T_{hv}; T_{vv}$	elements of the 2 × 2 transmission matrix, eqs. (4.19c, 4.42)	V_h^{01}	received voltage at the h-port of an antenna when the transmitted state is vertical
$\mathbf{T}$	transmission matrix in (h, v)-basis, eq. (4.62)		
$\mathbf{T}^c$	transmission matrix in circular basis	$V_h^{11}; V_v^{11}$	received voltage at the h- or v-port when transmitted state (denoted by "11") is slant 45° (or hybrid mode), eq. (4.141)
U	element of the 4 × 1 Stokes' vector, eq. (3.25c)		
U_R	correlation between the RHC and LHC received votage components when the transmitted state is RHC, eq. (4.24), or LHC, eq. (4.25)	$V_h^{1,-1};$ $V_v^{1,-1}$	received voltage at the h- or v-port when the transmitted state ("1, −1") is slant −45°, eq. (4.143)
		$V_{hh}; V_{vh};$ $V_{hv}; V_{vv}$	time samples of the received signal vector, eq. (6.72)
$U_{tr}(t)$	transmitted waveform, eq. (5.5a)	W	rainwater content, eq. (7.11)
$\mathbf{U}$	2 × 2 polarization basis transformation matrix, eq. (3.73a)	$W; W'$	correlation between the RHC and LHC voltage components; prime refers to transmission-modified version, eq. (4.122c)
$v; v_0$	particle velocity or Doppler velocity		
$\vec{v}$	particle velocity vector, eq. (1.108a)		

W	window function, eq. (5.132)	Z'_{dr}	transmission-matrix modified Z_{dr}
$W(t)$	range–time weighting function, eq. (5.52)	α	polarizability of a sphere, eq. (A1.27a)
$W_n(t)$	normalized range–time weighting function, eq. (5.52c)	α	slope of a linear relation between specific attenuation and specific differential phase, Table 7.1
$W(r,\theta,\phi)$	three-dimensional weighting function, eq. (5.60a)		
$W_1; W'_1$	copolar received power ("weak" channel) in a circularly polarized radar; prime refers to transmission-modified version, eq. (4.122a)	$\alpha_{x,y,z}$	elements of the polarizability matrix, eqs. (A1.28, A1.29)
		α_{zb}	polarizability of a spheroid along its symmetry axis, eq. (2.45)
		α	polarizability matrix, eq. (2.28a)
$W_2; W'_2$	cross-polar received power ("strong" channel) in a circularly polarized radar; prime refers to transmission-modified version, eq. (4.122b)	$\bar{\bar{\alpha}}$	polarizability tensor, eq. (2.28b)
		α_{oln}	expansion coefficients related to the Mie solution, eq. (2.104a)
		$\alpha_{hh}, \alpha_{vh},$ α_{hv}, α_{vv}	phase of the elements of the "instantaneous" scattering matrix, eq. (3.174)
$(W/W_2)_+$	complex correlation between the RHC and LHC voltage components when the transmitted polarization state is RHC (subscript "+"), eq. (4.92a)	β	canting angle in the plane of polarization, Fig. 2.10a
		β	angle between the dipole moment and the scattering direction, Fig. 1.9
$(W/W_2)_-$	as above complex correlation between the RHC and LHC voltage components when the transmitted polarization state is LHC (subscript "−"), eq. (4.92b)	β	slope of a linear relation between specific differential attenuation and specific differential phase, Table 7.1
$\mathbf{X}$	polarization error matrix, eq. (6.33)	β_0	mean canting angle
$\mathscr{Z}_{dr}$	differential reflectivity ratio ($\mathscr{Z}_{dr} = 10^{0.1(Z_{dr})}$), eq. (3.231c)	β_{eln}	expansion coefficients related to the Mie solution, eq. (2.104b)
Z	reflectivity factor, eq. (3.166)	β_a^2	gain inequality between the two ports of a dual-polarized antenna, eq. (6.22a)
Z_e	equivalent reflectivity factor, eq. (5.49a)		
$Z_{h,v}$	equivalent reflectivity factor at horizontal or vertical polarization, eq. (6.61c,d)	γ	slope of a linear relation between drop axis ratio and diameter, eq. (7.29a)
Z_0	intrinsic impedance of empty space, eq. (1.29)	$\gamma_{hh}; \gamma_{vh};$ $\gamma_{hv}; \gamma_{vv}$	magnitude of the elements of an "instantaneous" scattering matrix, eq. (3.174)
Z_{dr}	differential reflectivity, eq. (3.168)		
Z_{dp}	difference reflectivity, eq. (7.86a)		
Z'_h	transmission-matrix modified Z_h	$\Gamma(n)$	gamma function

$\Gamma(\Delta f, \Delta t)$ spaced-time/spaced-frequency coherency function, eq. (5.106a)

δ Dirac delta function

δ_{co} scattering differential phase, eq. (7.49)

δ_c scattering differential phase at circular polarization, eq. (3.98)

$\delta_{hh}; \delta_{vh}; \delta_{hv}; \delta_{vv}$ phase angle of the scattering matrix elements, eq. (4.69)

ΔZ_e enhancement in equivalent reflectivity factor in the bright-band relative to the rain below

ε_0 permittivity of empty space

ε_r relative permittivity of a dielectric

$\varepsilon_r'; \varepsilon_r''$ real and imaginary parts of ε_r

ε_{eff} effective permittivity

ε_m radar measurement error, eq. (8.23)

ε_p error due to parameterization of the rain rate, eq. (8.23)

ε_T sum of ε_m and ε_p, eq. (8.23)

$\varepsilon_{h,v}$ complex error terms of an antenna polarization error matrix, eq. (3.63)

$\varepsilon_{R,L}$ complex error terms for circular polarization, eqs. (3.113, 3.114)

η radar reflectivity, eq. (3.167a)

η_c radar reflectivity at circular polarization, eq. (3.202a)

$\eta_{hh,vv}$ copolar radar reflectivities at horizontal and vertical polarizations, eq. (3.164)

η_{vh} cross-polar radar reflectivity; transmit polarization is h and receive polarization is v (subscript "vh"), eq. (3.164c)

η_{hv} cross-polar radar reflectivity; transmit state is v and receive is h

$\bar{\eta}$ resolution-volume averaged radar reflectivity, eq. (5.62)

$\theta_{i,s}$ spherical coordinate angle related to incidence (i) and scattered (s) directions

$\theta_{R,L}$ phase angle of RHC and LHC components

θ_b spherical coordinate angle of the particle symmetry axis, Fig. 2.6a

λ wavelength

$\lambda_{1,2}$ eigenvalues of the two characteristic waves, eq. (4.6a)

$\lambda_{x,y,z}$ depolarization factors of a spheroid, Table A1.1

λ_{zb} depolarizing factor along the spheroid symmetry axis

Λ slope of an exponential drop size distribution or slope parameter of a gamma dsd, eq. (7.12)

$\Lambda_{x,y,z}$ related to $\lambda_{x,y,z}$, Table A1.1

μ parameter of a gamma drop size distribution, eq. (7.12)

μ_0 permeability of empty space

ν parameter of a gamma dsd, note that $\mu = \nu - 1$, eq. (7.27)

ν back scatter amplitude ratio at circular polarization, eq. (3.97a)

$\nu^2; \overline{\nu^2}$ circular depolarization ratio; ensemble-average, eq. (3.202c)

$\vec{\Pi}$ electric Hertz vector, eq. (1.14)

ρ polarization match factor, eq. (3.44)

ρ_0 size parameter of a sphere

ρ_p particle density

ρ_w water density

$\rho_{2,4}$ orientation parameters, eq. (3.201)

ρ_{co} copolar correlation coefficient between the hh and vv return signals, eq. (3.170a)

$\rho[l]$ — correlation coefficient between signal samples at lag l, eq. (5.190)

$\rho_p[l]$ — correlation coefficient for samples of the signal power, eq. (5.190)

$\rho(t_s)$ — normalized autocorrelation function of the received signal, eq. (5.95)

ρ_{cx}^h, ρ_{cx}^v — correlation coefficient between the copolar and cross-polar return signals when the transmitted polarization is horizontal (superscript "h"), eq. (4.78a) or vertical (superscript "v"), eq. (4.78b)

$\rho_{hh,hh}[l]$ — correlation coefficient of the hh return signal at lag l; see in relation to eq. (6.76a)

$\rho_{hh,vh}[l]$ — as above except between hh and vh returns; see eq. (6.76b)

$\rho_{hh,vv}[l]$ — as above except between hh and vv returns; see eq. (6.76c)

$\sigma; \sigma^2$ — standard deviation; variance

σ_m — standard deviation of the mass-weighted mean diameter, eq. (7.50a)

σ_v — standard deviation of velocity distribution, eq. (5.93b)

σ_v — standard deviation of the Doppler velocity spectrum, eq. (5.127)

$\sigma_{a,s,b}$ — absorption (a), scattering (s) and back scatter (b) cross section of a sphere, eqs. (1.56, 1.45, and 1.51), respectively

σ_{bi} — bistatic radar cross section, eq. (1.46b)

σ_{co} — copolar radar cross section, eq. (3.137b)

σ_{cx} — cross-polar radar cross section, eq. (3.137c)

σ_{rt} — generalized bistatic radar cross section, eq. (3.134)

$\sigma_{hh}; \sigma_{vv}$ — copolar radar cross sections at horizontal and vertical polarizations, eq. (8.4)

σ_{ext} — extinction cross section of a sphere, eq. (1.57)

Σ — polarimetric covariance matrix in (h, v)-basis, eq. (3.183)

Σ^c — polarimetric covariance matrix in circular basis, eq. (3.195)

τ — volume of a non-spherical scatterer

τ — range–time

τ — time difference

τ — ellipticity angle of a polarization ellipse, eq. (3.13)

τ — mean canting angle in McCormick and Hendry's definition, eq. (4.33)

ϕ — canting angle in the polarization plane, Oguchi's definition, eq. (4.44)

ϕ — scalar potential function, eq. (1.10a)

$\phi_{i,s}$ — spherical coordinate angle related to directions of incidence (i) and scattering (s)

ϕ_b — spherical coordinate angle related to the symmetry axis of the spheroid, Fig. 2.6

$\Phi_{h,v}$ — phase angles associated with the h- or v-ports of a dual-polarized antenna, eq. (6.19)

Φ_{dp} — differential propagation phase, eq. (4.65b)

χ — characteristic function of a random variable, eq. (5.93a)

$\vec{\chi}$ — vector associated with Rayleigh scattering, eq. (1.48)

$\chi_{i,s,r}$ — complex polarization ratio of an incident (i), scattered (s), or reflected (r) wave, eq. (3.3–3.5)

ψ	angle between the direction of the incident wave and the particle symmetry axis, Fig. 2.10a
ψ	orientation angle of a polarization ellipse, Fig 3.3b
ψ_m	magnetic flux, eq. (1.72)
Ψ_{dp}	differential phase between copolar (hh and vv) received signals, eq. (4.65a)
ω	angular frequency
ω_D	angular Doppler frequency
Ω	solid angle, eq. (2.68)
Ω	frequency axis for discrete time signals
Ω	feature vector, eq. (3.175)

Abbreviations

AGL	above ground level
ATI	area–time integral
BMRC	Bureau of Meteorology Research Center
BPN	back propagation network
BSA	back scatter alignment
CaPE	convective and precipitation/electrification experiment
CDF	cumulative distribution function
CDR	circular depolarization ratio
COHO	coherent oscillator
CW	continuous waves
CSU–CHILL	Colorado State University–University of Chicago/Illinois State Water Survey
DDA	discrete dipole approximation
DFT	discrete Fourier transform
DLR	German Aerospace Research Establishment
dsd	drop size distribution
DSP	digital signal processing
DTFT	discrete time Fourier transform
EDR	elliptical depolarization ratio
EMF	electromotive force
ETL	Environmental Technology Laboratory
FDTD	finite difference–time domain
FFT	fast Fourier transform
FHC	fuzzy logic hydrometeor classifier
FIR	finite impulse response
FSA	forward scatter alignment
HVPS	high volume particle spectrometer
IEEE	Institute of Electrical and Electronic Engineers
IF	intermediate frequency
iid	independent and identically distributed
IIR	infinite impulse response
IWC	ice–water content
LHC	left-hand circular
LNA	low noise amplifier
MBF	membership function
M_EMF	motional EMF
MG	Maxwell-Garnet
NCAR	National Center for Atmospheric Research
NOAA	National Oceanic and Atmospheric Administration
NRC	National Research Council of Canada
OMT	orthomode transducer
ORTT	apparent degree of orientation measured by a circularly polarized radar
pdf	probability density function
PI	particular integral

PMM	probability-matching method		SLDR	slant linear depolarization ratio
PPI	plan position indicator		SNR	signal-to-noise ratio
PPMM	polarimetric probability matching method		STALO	stable local oscillator
			T_EMF	transformer EMF
PRF	pulse repetition frequency		TDWR	terminal Doppler weather radar
PROM	programmable read only memory		TRMM	Tropical Rain Measurement Mission
PRT	pulse repetition time			
RAL	Rutherford Appleton Laboratory		TWT	traveling wave tube
RBF	radial basis function		UTC	universal time coordinates
RHC	right-hand circular		VDH	van de Hulst
RHI	range–height indicator		VSWR	voltage standing wave ratio
RRV	radar resolution volume		WPMM	window probability-matching method
SD	standard deviation		WSR–88D	Weather Service Radar–1988 Doppler
SDSM&T	South Dakota School of Mines and Technology		XPI	cross-polarization isolation

1 Electromagnetic concepts useful for radar applications

The scattering of electromagnetic waves by precipitation particles and their propagation through precipitation media are of fundamental importance in understanding the signal returns from dual-polarized, Doppler weather radars. In this chapter, a number of useful concepts are introduced from first principles for the benefit of those readers who have not had prior exposure to such material. Starting with Maxwell's equations, an integral representation for scattering by a dielectric particle is derived, which leads into Rayleigh scattering by spheres. The Maxwell-Garnet(MG) mixing formula is discussed from an electrostatic perspective (a review of electrostatics is provided in Appendix 1). Faraday's law is used in a simple example to explicitly show the origin of the bistatic Doppler frequency shift. The important concepts of coherent and incoherent addition of waves are illustrated for two- and N-particle cases. The time-correlated bistatic scattering cross section of a single moving sphere is defined, which naturally leads to Doppler spectrum. The transmitting and receiving aspects of a simple Doppler radar system are then explained. This chapter ends with coherent wave propagation through a slab of spherical particles, the concept of an effective propagation constant of precipitation media, and the definition of specific attenuation.

1.1 Review of Maxwell's equations and potentials

Maxwell's time-dependent equations[1] governing the electric $(\vec{E})$ and magnetic $(\vec{B})$ vectors within a material can be written in terms of permittivity of free space (ε_0), permeability of free space (μ_0), and the volume density of polarization $(\vec{P})$. It is assumed here that within the material under consideration, the volume density of free charge is zero, the conductivity is zero (no Ohmic currents), and the volume density of magnetization is zero. Then, $\vec{E}$ and $\vec{B}$ within the interior of the material satisfy,

$$\nabla \times \vec{E} = -\frac{\partial \vec{B}}{\partial t}; \quad \varepsilon_0 \nabla \cdot \vec{E} = -\nabla \cdot \vec{P} \tag{1.1a}$$

$$\frac{1}{\mu_0} \nabla \times \vec{B} = \frac{\partial \vec{P}}{\partial t} + \varepsilon_0 \frac{\partial \vec{E}}{\partial t}; \quad \nabla \cdot \vec{B} = 0 \tag{1.1b}$$

The term $\partial \vec{P}/\partial t$ is identified as the polarization current, which can be thought of as being maintained by external sources, while $\varepsilon_0 \partial \vec{E}/\partial t$ is the free space displacement

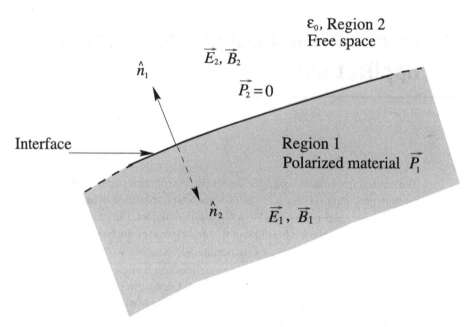

Fig. 1.1. Boundary between a polarized material (Region 1) and free space (Region 2). The unit normals are directed "outward" with respect to the corresponding regions.

current. Fields $\vec{E}$ and $\vec{B}$ are the macroscopic fields in the material, not the "microscopic" or "local" fields. The boundary conditions on the interface between the material and the free space [see Fig. 1.1 and note the directions of the unit normal vectors ($\hat{n}_{1,2}$)] are given as,

$$\hat{n}_1 \times \vec{E}_1 + \hat{n}_2 \times \vec{E}_2 = 0; \quad \varepsilon_0(\hat{n}_1 \cdot \vec{E}_1 + \hat{n}_2 \cdot \vec{E}_2) = -\hat{n}_1 \cdot \vec{P}_1 \tag{1.2a}$$

$$\frac{1}{\mu_0}(\hat{n}_1 \times \vec{B}_1 + \hat{n}_2 \times \vec{B}_2) = 0; \quad \hat{n}_1 \cdot \vec{B}_1 + \hat{n}_2 \cdot \vec{B}_2 = 0 \tag{1.2b}$$

The tangential components of the electric and magnetic fields are continuous across the interface; the normal component of the magnetic vector is continuous; and the normal component of the electric vector is discontinuous by an amount equal to $-\hat{n}_1 \cdot \vec{P}_1 / \varepsilon_0$ or $-\eta_b / \varepsilon_0$, where η_b is the surface density of the bound charge. If the interface is between two materials with different $\vec{P}_1$ and $\vec{P}_2$ on either side, then $\varepsilon_0(\hat{n}_1 \cdot \vec{E}_1 + \hat{n}_2 \cdot \vec{E}_2) = -(\hat{n}_1 \cdot \vec{P}_1 + \hat{n}_2 \cdot \vec{P}_2)$. Note that in Fig. 1.1, $\vec{P}_2$ is taken as zero in free space.

When the time variation of the external sources (e.g. radiating antenna) that maintain the polarization current within the material (or particle) is in sinusoidal steady state at a fixed angular frequency (ω), then Maxwell's equations (1.1) can be transformed by defining,

$$\vec{E}(\vec{r}, t) = \mathrm{Re}\left[\vec{E}^c(\vec{r})e^{j\omega t}\right] \tag{1.3a}$$

$$\vec{B}(\vec{r}, t) = \mathrm{Re}\left[\vec{B}^c(\vec{r})e^{j\omega t}\right] \tag{1.3b}$$

$$\vec{P}(\vec{r}, t) = \mathrm{Re}\!\left[\vec{P}^c(\vec{r})e^{j\omega t}\right] \tag{1.3c}$$

where $\vec{E}^c(\vec{r})$, $\vec{B}^c(\vec{r})$, and $\vec{P}^c(\vec{r})$ are vector-phasors or complex vectors. Substituting (1.3) into (1.1) yields,

$$\nabla \times \vec{E}^c = -j\omega \vec{B}^c; \quad \varepsilon_0 \nabla \cdot \vec{E}^c = -\nabla \cdot \vec{P}^c \tag{1.4a}$$

$$\frac{1}{\mu_0} \nabla \times \vec{B}^c = j\omega \vec{P}^c + j\omega\varepsilon_0 \vec{E}^c; \quad \nabla \cdot \vec{B}^c = 0 \tag{1.4b}$$

The boundary conditions in (1.2) are valid without any change by replacing the real, instantaneous vectors by complex vectors. Note that,

$$\vec{E}^c(\vec{r}) = \vec{E}^c_{\mathrm{real}} + j\vec{E}^c_{\mathrm{im}} \tag{1.5a}$$

$$\vec{E}(\vec{r}, t) = \mathrm{Re}\!\left[\vec{E}^c(\vec{r})e^{j\omega t}\right] \tag{1.5b}$$

Henceforth, the superscript c will be dropped as only the sinusoidal steady state will be considered. For linear materials, a complex relative permittivity is defined ($\varepsilon_r = \varepsilon_r' - j\varepsilon_r''$) as discussed in Appendix 1, where $\vec{P} = \varepsilon_0(\varepsilon_r - 1)\vec{E}$ within the material.

1.1.1 Vector Helmholtz equation

The vector Helmholtz equation for $\vec{E}$ is derived by taking the curl of $\vec{E}$ in (1.4a),

$$\nabla \times \nabla \times \vec{E} = -j\omega \nabla \times \vec{B} \tag{1.6a}$$

$$= -j\omega\mu_0\!\left(j\omega\vec{P} + j\omega\varepsilon_0\vec{E}\right) \tag{1.6b}$$

$$= \omega^2\mu_0\vec{P} + \omega^2\mu_0\varepsilon_0\vec{E} \tag{1.6c}$$

$$= \omega^2\mu_0\vec{P} + k_0^2\vec{E} \tag{1.6d}$$

Thus, the inhomogeneous vector Helmholtz equation for $\vec{E}$ is,

$$\nabla \times \nabla \times \vec{E} - k_0^2\vec{E} = \omega^2\mu_0\vec{P} \tag{1.7a}$$

and a similar equation for $\vec{B}$ is,

$$\nabla \times \nabla \times \vec{B} - k_0^2\vec{B} = j\omega\mu_0\nabla \times \vec{P} \tag{1.7b}$$

where $k_0 = \omega\sqrt{\varepsilon_0\mu_0}$ is the wave number of free space. Here, $\vec{E}$ and $\vec{B}$ are the total fields within the material, and $\vec{P}$ is maintained by external sources. The volume density of polarization or the polarization current, in this context, can be thought of as induced "sources" in (1.7a, b). If the linear relation $\vec{P} = \varepsilon_0(\varepsilon_r - 1)\vec{E}$ is used within the material, then $\vec{E}$ and $\vec{B}$ satisfy the more familiar homogeneous vector Helmholtz equations,

$$\nabla \times \nabla \times \vec{E} - k^2\vec{E} = 0 \tag{1.8a}$$

$$\nabla \times \nabla \times \vec{B} - k^2\vec{B} = 0 \tag{1.8b}$$

where $k = \omega\sqrt{\varepsilon_0\varepsilon_r\mu_0} = k_0\sqrt{\varepsilon_r}$ is the complex wave number of the material.

1.1.2   Scalar, vector and electric Hertz potentials

The $(\vec{E}, \vec{B})$ fields can be expressed in terms of simpler potential functions $(\phi, \vec{A})$, where $\vec{B} = \nabla \times \vec{A}$ and $\vec{E} = -\nabla\phi - j\omega\vec{A}$. These follow from the two Maxwell equations, $\nabla\cdot\vec{B} = 0$ and $\nabla \times \vec{E} + j\omega\vec{B} = 0$. The function ϕ is termed the electric scalar potential, and reduces to the electrostatic potential when $\omega = 0$. The function $\vec{A}$ is termed the magnetic vector potential. These two potentials must also satisfy Maxwell's remaining two equations,

$$\varepsilon_0 \nabla\cdot\vec{E} = -\nabla\cdot\vec{P} \tag{1.9a}$$

$$\frac{1}{\mu_0}\nabla \times \vec{B} = j\omega\vec{P} + j\omega\varepsilon_0\vec{E} \tag{1.9b}$$

Substituting for $\vec{B}$ and $\vec{E}$ in terms of the potentials ϕ and $\vec{A}$ gives,

$$\nabla^2\phi + j\omega\nabla\cdot\vec{A} = \frac{1}{\varepsilon_0}\nabla\cdot\vec{P} \tag{1.10a}$$

$$\nabla \times \nabla \times \vec{A} = j\omega\mu_0\vec{P} - j\omega\mu_0\varepsilon_0\nabla\phi + k_0^2\vec{A} \tag{1.10b}$$

Since $\nabla \times \nabla \times \vec{A} = \nabla(\nabla\cdot\vec{A}) - \nabla^2\vec{A}$,

$$\nabla(\nabla\cdot\vec{A}) - \nabla^2\vec{A} = j\omega\mu_0\vec{P} - j\omega\varepsilon_0\mu_0\nabla\phi + k_0^2\vec{A} \tag{1.11}$$

The magnetic vector potential is not unique as yet, since only its curl is defined as equal to $\vec{B}$. Its divergence can be assigned any convenient value and, in particular, if the Lorentz gauge is used, i.e. $\nabla\cdot\vec{A} = -j\omega\varepsilon_0\mu_0\phi$, then (1.10a, b) become,

$$(\nabla^2 + k_0^2)\phi = \frac{1}{\varepsilon_0}\nabla\cdot\vec{P} \tag{1.12a}$$

$$(\nabla^2 + k_0^2)\vec{A} = -j\omega\mu_0\vec{P} \tag{1.12b}$$

The above are the (inhomogeneous) Helmholtz equations for the potentials $(\phi, \vec{A})$ from which $\vec{E}$ and $\vec{B}$ are derived. In practice, $\vec{E}$ and $\vec{B}$ can be derived from $\vec{A}$ alone using the Lorentz gauge relation,

$$\vec{B} = \nabla \times \vec{A} \tag{1.13a}$$

$$\vec{E} = -\nabla\phi - j\omega\vec{A} \tag{1.13b}$$

$$= \frac{-j}{\omega\varepsilon_0\mu_0}\nabla(\nabla\cdot\vec{A}) - j\omega\vec{A} \tag{1.13c}$$

It is convenient to define the electric Hertz vector $\vec{\Pi} = \vec{A}/j\omega\mu_0\varepsilon_0$, so that $\vec{\Pi}$ satisfies,

$$(\nabla^2 + k_0^2)\vec{\Pi} = \frac{-\vec{P}}{\varepsilon_0} \tag{1.14}$$

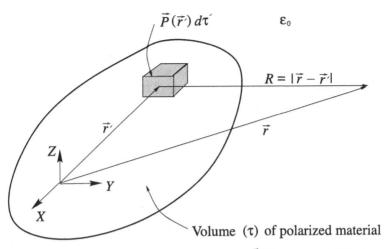

Fig. 1.2. Continuous volume density of polarization $\vec{P}(\vec{r}')$ is defined within the volume τ of polarized material.

and $(\vec{E}, \vec{B})$ are derived from $\vec{\Pi}$ as,

$$\vec{E} = \nabla(\nabla \cdot \vec{\Pi}) + k_0^2 \vec{\Pi} \qquad (1.15a)$$

$$= \nabla \times \nabla \times \vec{\Pi} + (\nabla^2 + k_0^2)\vec{\Pi} \qquad (1.15b)$$

$$= \nabla \times \nabla \times \vec{\Pi} - \frac{\vec{P}}{\varepsilon_0} \qquad (1.15c)$$

$$\vec{B} = j\omega\mu_0\varepsilon_0 \nabla \times \vec{\Pi} \qquad (1.15d)$$

Referring to Fig. 1.2, the particular integral (PI) solution to (1.14) is,

$$\vec{\Pi}(\vec{r}) = \frac{1}{4\pi\varepsilon_0} \int_\tau \vec{P}(\vec{r}') \frac{e^{-jk_0|\vec{r}-\vec{r}'|}}{|\vec{r}-\vec{r}'|} d\tau' \qquad (1.16)$$

where,

$$G_0(\vec{r}/\vec{r}') = \frac{1}{4\pi\varepsilon_0} \frac{e^{-jk_0|\vec{r}-\vec{r}'|}}{|\vec{r}-\vec{r}'|} \qquad (1.17)$$

is the free space Green function for the scalar Helmholtz equation. Note that $\vec{\Pi}$ satisfies no boundary condition (because it is a PI solution) except the radiation condition on a spherical surface at infinity. This radiation condition implies that $\vec{\Pi}$ at large distances (R) must be of the form of outward propagating spherical waves, $\exp(-jk_0R)/R$ (for $\exp(j\omega t)$ time dependence), where $R = |\vec{r} - \vec{r}'|$. One can consider (1.16) as analogous to the electrostatic PI solution given in (A1.5) for the electrostatic potential, where in both cases $\vec{P}$ is maintained by external sources as yet unspecified. When the observation point $\vec{r}$ is outside the volume (τ) of polarized material and in empty space (see Fig. 1.2), then the electric field $\vec{E}(\vec{r})$ from (1.15c) is given as $\nabla \times \nabla \times \vec{\Pi}(\vec{r})$, since $\vec{P}$ is zero in free space.

1.2   Integral representation for scattering by a dielectric particle

Now consider the case of a dielectric particle placed in a known incident field ($\vec{E}^i$) as illustrated in Fig. 1.3a. As discussed in Appendix 1, $\vec{E}^i$ is perturbed by the dielectric particle. What was termed the perturbation field ($\vec{E}^P$), in Appendix 1, in an electrostatic context, is now conventionally termed the scattered field ($\vec{E}^s$). Figure 1.3b illustrates that the "source" for the scattered electric field is the unknown volume density of polarization ($\vec{P}$) within the particle. Inside the particle, $\vec{E}^s$ satisfies $\nabla \times \nabla \times \vec{E}^s - k_0^2 \vec{E}^s = \omega^2 \mu_0 \vec{P}$, and thus $\vec{E}^s$ can be derived from the electric Hertz vector $\vec{\Pi}_s$. Since $\vec{\Pi}_s$ satisfies $(\nabla^2 + k_0^2)\vec{\Pi}_s = -\vec{P}/\varepsilon_0$ inside the particle, the principal integral solution is given by (1.16), and $\vec{E}^s$ outside the particle is given as,

$$\vec{E}^s(\vec{r}) = \nabla \times \nabla \times \frac{1}{4\pi\varepsilon_0} \int_\tau \frac{\vec{P}(\vec{r}')e^{-jk_0|\vec{r}-\vec{r}'|}}{|\vec{r}-\vec{r}'|} \, d\tau' \tag{1.18a}$$

Using the fact that $\vec{P}(\vec{r}') = \varepsilon_0(\varepsilon_r - 1)\vec{E}_T^{in}(\vec{r}')$, where $\vec{E}_T^{in}$ is the total electric field inside the particle, an integral representation for the scattered electric field outside the particle, $\vec{E}^s(\vec{r})$, can be obtained as a volume integral,

$$\vec{E}^s(\vec{r}) = \nabla \times \nabla \times \frac{1}{4\pi} \int_\tau (\varepsilon_r - 1)\vec{E}_T^{in}(\vec{r}')\frac{e^{-jk_0 R}}{R} \, d\tau' \tag{1.18b}$$

where $R = |\vec{r} - \vec{r}'|$, and may be compared with the electrostatic representation for $\vec{E}^P$ in (A1.14b). For homogeneous particles, the term $(\varepsilon_r - 1)$ can be brought outside the integral.

When $r \gg r'$, and recalling that $\vec{r}'$ locates the differential volume element ($d\tau'$) inside the particle, the far-field integral representation for $\vec{E}^s(\vec{r})$ can be derived as follows. First, note that the dual curl operator $[\nabla \times \nabla \times (\cdots)]$ operates on $\vec{r} \equiv (x, y, z)$ only and can be taken inside the integral in (1.18b), Thus,

$$\nabla \times \left[\vec{E}_T^{in}(\vec{r}')\frac{e^{-jk_0 R}}{R} \right] = \nabla \left(\frac{-e^{jk_0 R}}{R} \right) \times \vec{E}_T^{in}(\vec{r}') \tag{1.19a}$$

$$= \left[\frac{1}{R}\nabla(e^{-jk_0 R}) + e^{-jk_0 R}\nabla\left(\frac{1}{R}\right) \right] \times \vec{E}_T^{in}(\vec{r}') \tag{1.19b}$$

$$= \left[\frac{-jk_0 e^{-jk_0 R}\nabla(R)}{R} + \frac{e^{-jk_0 R}(-\hat{R})}{R^2} \right] \times \vec{E}_T^{in}(\vec{r}') \tag{1.19c}$$

$$= \frac{-jk_0(e^{-jk_0 R})\hat{R}}{R}\left(1 + \frac{1}{jk_0 R}\right) \times \vec{E}_T^{in}(\vec{r}') \tag{1.19d}$$

Since $r \gg r'$ and $R = |\vec{r} - \vec{r}'|$, the term $|1/jk_0 R|$ may be neglected, in comparison to unity in (1.19d). Also, the term $(1/R)$ can be approximated as $(1/r)$, while $\hat{R}$ is parallel to $\hat{r}$ (see Fig. 1.4). Hence, (1.19d) reduces to,

$$\nabla \times \left[\vec{E}_T^{in}(\vec{r}')\frac{e^{-jk_0 R}}{R} \right] \approx \frac{(-jk_0)}{r}e^{-jk_0 R}\hat{r} \times \vec{E}_T^{in}(\vec{r}') \tag{1.20}$$

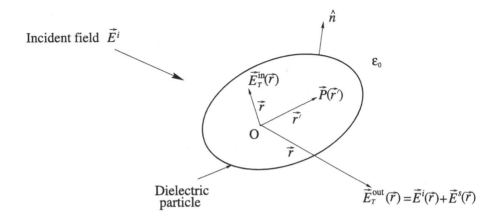

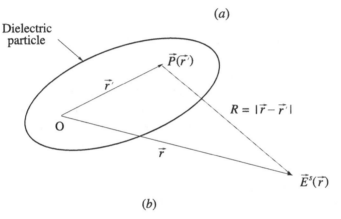

Fig. 1.3. (*a*) Dielectric particle in the presence of an incident electric field $\vec{E}^i$. The particle gets polarized $\vec{P}$ and causes a scattered field component $\vec{E}^s$ outside the particle. The total field is $\vec{E}_T^{in}$ inside the particle. (*b*) Illustration showing that the "source" of the scattered field $\vec{E}^s$ is the volume density of polarization $\vec{P}$ within the particle.

In a similar manner,

$$\nabla \times \nabla \times \left[\vec{E}_T^{in}(\vec{r}\,') \frac{e^{-jk_0 R}}{R} \right] \approx \frac{(-jk_0)(-jk_0)}{r} e^{-jk_0 R} \hat{r} \times \hat{r} \times \vec{E}_T^{in}(\vec{r}\,') \qquad (1.21a)$$

$$= \frac{-k_0^2}{r} e^{-jk_0 R} \hat{r} \times \hat{r} \times \vec{E}_T^{in}(\vec{r}\,') \qquad (1.21b)$$

Substituting (1.21b) into (1.18b),

$$\vec{E}^s(\vec{r}) = \frac{-k_0^2(\varepsilon_r - 1)}{4\pi} \left(\frac{1}{r} \right) \int_\tau e^{-jk_0 R} \hat{r} \times \hat{r} \times \vec{E}_t^{in}(\vec{r}\,')\, d\tau' \qquad (1.22)$$

The final far-field approximation is in the exponential term $\exp(-jk_0 R) = \cos(k_0 R) - j\sin(k_0 R)$, where $R \approx r - \vec{r}\,' \cdot \hat{r}$ (see Fig. 1.4). It is not permissible to approximate

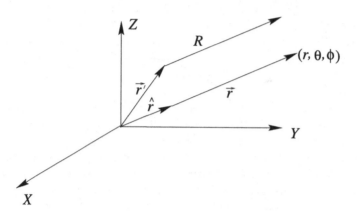

Fig. 1.4. Approximation for R in the far-field, $R \approx r - \vec{r}' \cdot \hat{r}$.

$R \approx r$ in the exponential term since $\theta = k_0 R \approx k_0 r - k_0 \vec{r}' \cdot \hat{r}$, which lies between 0 and 2π. Since k_0 is the wave number of free space ($k_0 = \omega/c = 2\pi/\lambda$, where λ is the wavelength and c the velocity of light in free space, respectively), the correction term $k_0 \vec{r}' \cdot \hat{r} = \vec{r}' \cdot \hat{r}(2\pi/\lambda)$ may become significant if the maximum extent of $\vec{r}'$ is even a small fraction of λ. Since $\vec{r}'$ is a variable of integration and covers all differential volume elements within the particle, the maximum extent of $\vec{r}'$ is of the same order as the maximum dimension of the particle. With this final approximation, (1.22) reduces to,

$$\vec{E}^s(\vec{r}) = \frac{k_0^2}{4\pi}(\varepsilon_r - 1)\frac{e^{-jk_0 r}}{r} \int_\tau \left[\vec{E}_T^{\text{in}}(\vec{r}') - \hat{r}(\hat{r} \cdot \vec{E}_T^{\text{in}})\right] e^{jk_0 \vec{r}' \cdot \hat{r}} \, d\tau' \tag{1.23}$$

where the vector identity $\vec{a} \times \vec{b} \times \vec{c} = \vec{b}(\vec{a} \cdot \vec{c}) - \vec{c}(\vec{a} \cdot \vec{b})$, with $\vec{a} \equiv \hat{r}, \vec{b} \equiv \hat{r}$, and $\vec{c} \equiv \vec{E}_T^{\text{in}}$, is used. The far-field vector scattering amplitude ($\vec{f}$) is defined as,

$$\vec{E}^s = \vec{f} \frac{e^{-jk_0 r}}{r} \tag{1.24a}$$

$$\vec{f} = \frac{k_0^2(\varepsilon_r - 1)}{4\pi} \int_\tau \left[\vec{E}_T^{\text{in}}(\vec{r}') - \hat{r}(\hat{r} \cdot \vec{E}_T^{\text{in}})\right] e^{jk_0 \vec{r}' \cdot \hat{r}} \, d\tau' \tag{1.24b}$$

where $\vec{f} \equiv \vec{f}(\theta, \phi)$, the spherical coordinates (θ, ϕ) referring to the direction of $\vec{r}$ (see Fig. 1.4), i.e. to the scattering direction. From (1.15d) the scattered magnetic vector $\vec{B}^s$ is,

$$\vec{B}^s = j\omega\varepsilon_0\mu_0 \nabla \times \vec{\Pi} \tag{1.25a}$$

$$= j\omega\varepsilon_0\mu_0 \nabla \times \frac{1}{4\pi} \int_\tau \frac{(\varepsilon_r - 1)\vec{E}_T^{\text{in}}(\vec{r}')e^{-jk_0 R}}{R} \, d\tau' \tag{1.25b}$$

Using the far-field approximation,

$$\vec{H}^s = \frac{\vec{B}^s}{\mu_0} = j\omega\varepsilon_0(-jk_0)(\varepsilon_r - 1)\frac{e^{-jk_0r}}{4\pi r}\int_\tau \hat{r} \times \vec{E}_T^{in}(\vec{r}')e^{jk_0\vec{r}'\cdot\hat{r}}\,d\tau' \tag{1.26}$$

where $\vec{H}^s$ is the far-field magnetic intensity vector. Using $k_0 = \omega/c$, $c^2 = 1/\varepsilon_0\mu_0$, and defining the impedance of free space as $Z_0 = \sqrt{\mu_0/\varepsilon_0}$, the far-field relation between $\vec{E}^s$ and $\vec{H}^s$ reduces to the simple form,

$$\vec{E}^s = Z_0(\vec{H}^s \times \hat{r}); \quad \vec{H}^s = Z_0^{-1}(\hat{r} \times \vec{E}^s) \tag{1.27}$$

From both (1.23) and (1.26) it is easy to see that $\hat{r}\cdot\vec{E}^s = \hat{r}\cdot\vec{H}^s = 0$; thus the radial component of the far-field electric and magnetic vectors is zero. Since $\vec{E}^s = \vec{f}(\theta, \phi)\exp(-jk_0r)/r$, it follows that,

$$\vec{H}^s = \frac{1}{Z_0}\left[\hat{r} \times \vec{f}(\theta, \phi)\right]\frac{e^{-jk_0r}}{r} \tag{1.28}$$

$$\left|\frac{\vec{E}^s}{\vec{H}^s}\right| = Z_0 \tag{1.29}$$

Recall that $\vec{E}^s$ and $\vec{H}^s$ are complex vectors or vector-phasors, and so is the vector scattering amplitude $\vec{f}(\theta, \phi)$. The function $rF = |\vec{f}(\theta, \phi)|\exp(-jk_0r)$ satisfies the spherical wave equation,

$$\frac{\partial^2}{\partial r^2}(rF) + k_0^2(rF) = 0 \tag{1.30}$$

as can be verified by direct substitution. The function $F(r, \theta)$, thus describes spherical waves that expand radially with constant velocity of light. The equiphase surfaces are given by $k_0r \equiv$ constant; these are spherical and they are separated by the constant distance $\lambda = 2\pi/k_0$, where λ is the wavelength.

A further simplification of the far-field structure of the scattered wave follows if interest is confined to a relatively small region of space surrounding the point P_0 in Fig. 1.5, whose radial extent is much smaller than r_0 and whose solid angle subtended at the origin is also small. Within this small volume of space, the function $|\vec{f}(\theta, \phi)|r^{-1}$ can be considered constant and equal to $|\vec{f}(\theta_0, \phi_0)|r_0^{-1}$. Consider another point P within this small volume, as also shown in Fig. 1.5. It is clear that the direction to P as described by $\hat{s}$ will be the same as $\hat{r}_0$. However, the phase of the spherical wave at P is approximated as $\exp(-jk_0s)$, while its amplitude is approximated as a constant equal to $K \equiv |\vec{f}(\theta_0, \phi_0)|r_0^{-1}$. Thus within the small volume, the scattered wave characteristics can be written as $F(s) \approx K\exp(-jk_0s)$, which is the so-called "local" plane wave approximation, since the surfaces of the constant phase are given by $k_0s =$ constant, which are plane surfaces orthogonal to $\hat{s}$. The function $F(s)$ satisfies the plane wave equation,

$$\frac{d^2F}{ds^2} + k_0^2F = 0 \tag{1.31}$$

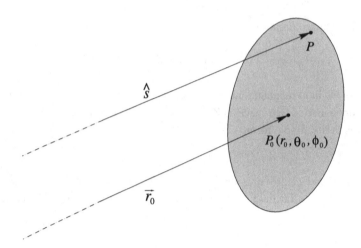

Fig. 1.5. Local plane wave approximation within a small volume centered at P_0.

as can be verified by direct substitution. Thus (1.30) which is the spherical wave equation, is approximated by (1.31), which is the local plane wave equation (King and Prasad 1986).

1.3   Rayleigh scattering by a dielectric sphere

The integral representation for the far-field vector scattering amplitude, $\vec{f}(\theta, \phi)$, given in (1.24b) is fundamental, with the basic unknown being the total electric field in the interior of the dielectric particle. When this internal field is approximated by the corresponding electrostatic solution it is called Rayleigh scattering. In Appendix 1, the case of a dielectric sphere in an uniform incident field is described and extensions are made to the case of prolate/oblate spheroids. For arbitrary-shaped dielectric particles, numerical methods must be used: a surface integral equation technique is described in Van Bladel (1985; Section 3.9) with numerical results given, for example, in Herrick and Senior (1977).

Referring to Fig. 1.6a, the uniform incident field $\vec{E}^i = \hat{z}E_0$ can be used as an electrostatic approximation for an ideal plane wave propagating along the positive Y-axis with amplitude E_0 and linearly polarized along the $\hat{z}$-direction, of the form $\vec{E}^i = \hat{z}E_0 \exp(-jk_0 y)$ (wave polarization will be treated in detail in Chapter 3). As discussed under the electroquasistatic approximation in Appendix 1, this is valid when the sphere diameter is very small compared with the wavelength λ. Substituting from Appendix 1 (A1.24), the electrostatic solution $\vec{E}_T^{in} = \hat{z}3E_0/(\varepsilon_r + 2) = \vec{E}^i(3/\varepsilon_r + 2)$ in (1.24b), and noting that $\vec{E}_T^{in}$ is constant inside the sphere and that $\exp(jk_0\vec{r}'\cdot\hat{r}) \approx 1$, since the maximum extent of $\vec{r}'$ is the sphere radius, results in,

$$\vec{f}(\theta, \phi) = \frac{k_0^2}{4\pi} \frac{(\varepsilon_r - 1)}{(\varepsilon_r + 2)} 3V \left[\vec{E}^i - \hat{r}(\hat{r}\cdot\vec{E}^i) \right] \qquad (1.32a)$$

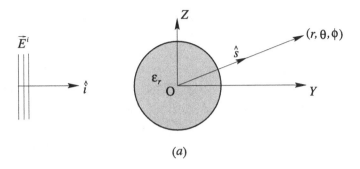

(a)

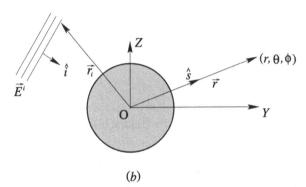

(b)

Fig. 1.6. Dielectric sphere of diameter D in the presence of an ideal plane wave incident field: (a) $\vec{E}^i = \hat{z}E_0 \exp(-jk_0y)$ and (b) $\vec{E}^i = \hat{e}_i E_0 \exp(-jk_0\hat{i}\cdot\vec{r}_i)$. For both (a) and (b), the scattering direction is $\hat{s}$.

where V is the volume of the sphere. Also,

$$\vec{E}^s = \vec{f}(\theta,\phi)\frac{e^{-jk_0r}}{r} \tag{1.32b}$$

Since the dipole moment $\vec{p} = \alpha \vec{E}^i$, and the polarizability $\alpha = 3\varepsilon_0 V(\varepsilon_r - 1)/(\varepsilon_r + 2)$, from (A1.27a),

$$\vec{f}(\theta,\phi) = \frac{k_0^2}{4\pi\varepsilon_0}\left[\vec{p} - \hat{r}(\hat{r}\cdot\vec{p})\right] \tag{1.32c}$$

Equation (1.32c) is fundamental for Rayleigh scattering by a dielectric sphere and, indeed, for any dielectric object whose electrostatic dipole moment can be computed either by separation of variables or by numerical methods. The direction and linear polarization state of the incident plane wave, $\vec{E}^i = \hat{z}E_0 \exp(-jk_0y)$, is implicit in (1.32c), while the scattering direction is $\hat{r} \equiv \hat{s}$ (see Fig. 1.6a). If the direction of the incident plane wave is represented by the unit vector $\hat{i}$, then the vector scattering amplitude can be written as $\vec{f}(\hat{s}, \hat{i})$, where the arguments indicate the directions of scattering and incidence, respectively. The Rayleigh approximation can also be understood in terms of the change in spatial phase of the incident plane wave (spatial phase $\equiv k_0y$)

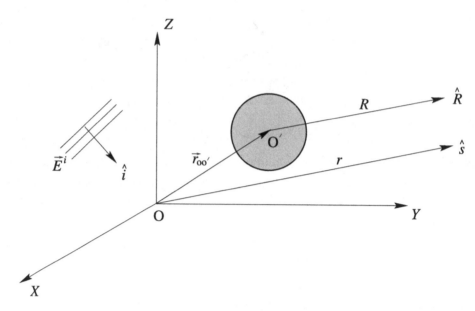

Fig. 1.7. Dielectric sphere with center at O'. The incident field is a plane wave ($\vec{E}^i$).

across the sphere diameter (which must be very small for the approximation to be valid, or $D \ll \lambda$). When calculating the dipole moment ($\vec{p} = \alpha \vec{E}^i$), the incident field is evaluated on the $y = 0$ plane, which can be considered to be the reference plane for the spatial phase term $k_0 y$. When ε_r is complex, then the dipole moment exhibits a time-phase lag with respect to the incident field, as discussed in Appendix 1. Even though $\vec{p} = \alpha \vec{E}^i$ was derived for a particular form for the electrostatic incident field, $\vec{E}^i = \hat{z} E_0$, and associated with a particular incident plane wave, $\vec{E}^i = \hat{z} E_0 \exp(-jk_0 y)$ (see Fig. 1.6a), the equation is valid for a general, linearly polarized plane wave incident field of the form $\vec{E}^i = \hat{e}_i E_0 \exp(-jk_0 \hat{i} \cdot \vec{r}_i)$, where the plane of constant spatial phase is now given by $k_0 \hat{i} \cdot \vec{r}_i = \text{constant}$, $\vec{r}_i \equiv (x, y, z)$ being an arbitrary position vector to any point on this plane (see Fig. 1.6b). The unit vector $\hat{e}_i$ characterizes the linear polarization state and lies in the transverse plane. The spatial phase reference can now be considered to be the plane $\hat{i} \cdot \vec{r}_i = 0$, which contains the origin (or the sphere center). The far-field vector amplitude is still given by (1.32a), with $\vec{E}^i = \hat{e}_i E_0$, or by (1.32c), with $\vec{p} = \alpha \hat{e}_i E_0$.

For later application in Section 1.9, the far-field $\vec{E}^s$ when the sphere center is displaced from the origin needs to be considered (see Fig. 1.7). The spatial phase of the incident field at O' is now $-k_0 \hat{i} \cdot \vec{r}_{OO'}$, while its amplitude is E_0. Thus, (1.32a) can be used to express $\vec{E}^s$ with respect to O' along the scattered direction $\hat{R}$, with $\vec{E}^i(O') = \hat{e}_i E_0 \exp(-jk_0 \hat{i} \cdot \vec{r}_{OO'})$,

$$\vec{E}_s(O') = \frac{k_0^2}{4\pi} \frac{(\varepsilon_r - 1)}{(\varepsilon_r + 2)} 3V \exp(-jk_0 \hat{i} \cdot \vec{r}_{OO'}) [\hat{e}_i E_0 - \hat{R}(\hat{R} \cdot \hat{e}_i E_0)] \frac{e^{-jk_0 R}}{R} \qquad (1.33)$$

Since $R \approx r - \hat{s} \cdot \vec{r}_{OO'}$ and $\hat{R} \approx \hat{s}$ in the far-field, the expression for the scattered field with respect to O becomes,

$$\vec{E}^s(\hat{s}, \hat{i}) = \left\{ \frac{k_0^2}{4\pi} \frac{(\varepsilon_r - 1)}{(\varepsilon_r + 2)} 3V \left[\hat{e}_i E_0 - \hat{s}(\hat{s} \cdot \hat{e}_i) E_0 \right] \right\} \frac{e^{-jk_0 r}}{r} e^{-jk_0(\hat{i}-\hat{s}) \cdot \vec{r}_{OO'}} \tag{1.34}$$

The term in curly brackets is the same as $\vec{f}(\hat{s}, \hat{i})$ in (1.32a), which is the vector scattering amplitude if the center of the sphere were at the origin. Thus,

$$\vec{E}^s = \vec{f}(\hat{s}, \hat{i}) e^{-jk_0(\hat{i}-\hat{s}) \cdot \vec{r}_{OO'}} \frac{e^{-jk_0 r}}{r} \tag{1.35}$$

where the spatial phase term $-k_0(\hat{i} - \hat{s}) \cdot \vec{r}_{OO'}$ accounts for: (i) the spatial phase of the incident wave at O′, and (ii) an adjustment to the spatial phase of the scattered field expressed with respect to O because of the displacement of the sphere center from O to O′. Equation (1.35) will be used later in Section 1.9 to consider the case where the sphere is moving, i.e. where $\vec{r}_{OO'}$ is a function of time.

1.3.1 Extension to Rayleigh scattering by spheroids

Scattering by a prolate/oblate spheroid when the symmetry axis is oriented along the $+Z$-axis (see Fig. 1.8) can be treated using (1.32). This low frequency approximation will also be referred to as Rayleigh–Gans scattering by spheroids. If $\vec{E}^i = \hat{e}_i E_0 \exp(-jk_0 \hat{i} \cdot \vec{r}_i)$, then the Cartesian components of $\vec{p}$ are given by,

$$p_x = \alpha_x E_0 \hat{e}_i \cdot \hat{x}, \quad p_y = \alpha_y E_0 \hat{e}_i \cdot \hat{y}, \quad p_z = \alpha_z E_0 \hat{e}_i \cdot \hat{z} \tag{1.36a}$$

where (see A1.30),

$$\alpha_{x,y,z} = V \varepsilon_0 (\varepsilon_r - 1) \Lambda_{x,y,z} \tag{1.36b}$$

From (1.32b, c),

$$\vec{E}^s = \vec{f}(\hat{s}, \hat{i}) \frac{e^{-jk_0 r}}{r} \tag{1.36c}$$

and

$$\vec{f}(\hat{s}, \hat{i}) = \frac{k_0^2}{4\pi \varepsilon_0} \left[\vec{p} - \hat{s}(\hat{s} \cdot \vec{p}) \right] \tag{1.36d}$$

When $\vec{s} = -\hat{i}$, the scattered direction is along the back scatter or radar direction, since the radar launches the incident wave $\vec{E}^i$. The dependence of $\vec{E}^s$ on the polarization state of the incident wave is clear from (1.36), but will be treated in more detail in Chapter 2, where the amplitude scattering matrix will be formally defined.

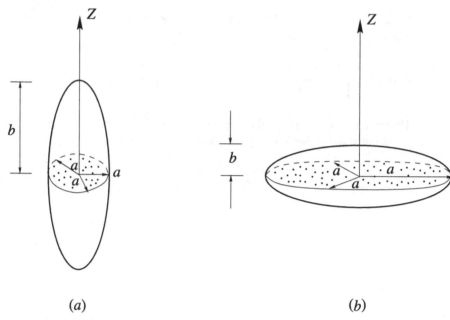

(a)   (b)

Fig. 1.8. (a) A prolate spheroid with symmetry axis along the Z-axis and (b) an oblate spheroid. The principal equatorial cross section in both instances is a circle of radius $= a$, which is shaded.

1.4   Scattering, bistatic, and radar cross sections

Consider two periodic functions $A(t)$ and $B(t)$, where $A(t) = A_0 \cos(\omega t + \theta_A)$, $B(t) = B_0 \cos(\omega t + \theta_B)$. The time-averaged value of $A(t)$ is defined as,

$$\langle A(t) \rangle = \frac{1}{T_p} \int_0^{T_p} A(t)\, dt \tag{1.37}$$

where $T_p = 2\pi/\omega$ is the period. It is clear that,

$$\langle A(t) \rangle = 0 \tag{1.38a}$$

$$\langle A^2(t) \rangle = \frac{A_0^2}{2} \tag{1.38b}$$

$$\langle A(t)B(t) \rangle = \frac{1}{2} A_0 B_0 \cos(\theta_A - \theta_B) \tag{1.38c}$$

Let $\tilde{A}$ and $\tilde{B}$ be phasors defined as,

$$\tilde{A} = A_0 e^{j\theta_A}, \quad \tilde{B} = B_0 e^{j\theta_B} \tag{1.39}$$

It follows directly that,

$$\langle A^2(t) \rangle = \frac{1}{2} \tilde{A}\tilde{A}^*, \quad \langle A(t)B(t) \rangle = \frac{1}{2} \mathrm{Re}(\tilde{A}\tilde{B}^*) \tag{1.40}$$

From elementary circuit analysis, if $\tilde{A}$ and $\tilde{B}$ are identified with the voltage ($\tilde{V}$) and current ($\tilde{I}$), respectively, then $\mathrm{Re}(\tilde{V}\tilde{I}^*)/2$ gives the time-averaged power, while $\tilde{V}\tilde{I}^*/2$ is the complex power.

The complex Poynting vectors of the incident and scattered fields are defined as,

$$\vec{S}_i = \frac{1}{2}\vec{E}^i \times \vec{H}^{i*} \tag{1.41a}$$

$$\vec{S}_s = \frac{1}{2}\vec{E}^s \times \vec{H}^{s*} \tag{1.41b}$$

For $\vec{E}^i = \hat{e}_i E_0 \exp(-jk_0\hat{i}\cdot\vec{r}_i)$, $\vec{H}_i = (\hat{i} \times \vec{E}^i)/Z_0$, from an extension of (1.27), for an ideal plane wave, and

$$\vec{S}_i = \frac{E_0^2}{2Z_0}\hat{i} \tag{1.42a}$$

Similarly, from $\vec{E}^s = \vec{f}\exp(-jk_0r)/r$ and $Z_0\vec{H}_s = (\hat{r} \times \vec{f})\exp(-jk_0r)/r$ in the far-field,

$$\vec{S}_s = \frac{|\vec{f}|^2}{2Z_0}\frac{1}{r^2}\hat{r} \tag{1.42b}$$

Thus, for an ideal incident plane wave and for the far-field scattered spherical wave, the Poynting vector is real and directed along the propagation direction in each case. The real Poynting vector has units of power per unit area (W m^{-2}) and is associated with the time-averaged power density of the incident and scattered fields.

The real transfer function $\mathrm{Re}(T_{i,s})$ (King and Prasad 1986), defined as,

$$\mathrm{Re}(T_{i,s}) = \oint_{\Sigma} \hat{n}\cdot\mathrm{Re}(\vec{S}_{i,s})\,d\Sigma \tag{1.43}$$

where Σ is a closed surface on which $\hat{n}$ is the outward drawn unit normal vector, is associated with the total time-averaged power transfer across the closed surface. For an ideal incident plane wave in free space, it is easily argued that $\mathrm{Re}\,T_i = 0$ for any closed Σ, since the time-averaged power transfer into the surface must be exactly balanced by the time-averaged power transfer out of the surface (conservation of energy). For $\vec{E}^s$ in the far-field, the closed surface (Σ) is usually chosen as a spherical surface of radius r drawn with the center at the origin. Thus,

$$P_s = \mathrm{Re}\,T_s = \frac{1}{2Z_0}\oint_{\Sigma}\hat{r}\cdot\frac{|\vec{f}|^2\hat{r}}{r^2}\,d\Sigma = \frac{1}{2Z_0}\int_{4\pi}|\vec{f}|^2\,d\Omega \tag{1.44}$$

where $d\Sigma = r^2\,d\Omega$, $d\Omega$ being the element of solid angle subtended by $d\Sigma$ at the origin.

$\mathrm{Re}\,T_s$ is the total time-averaged power transfer associated with the scattered field across the spherical surface (Σ) and is denoted as the total scattered power (P_s). The total scattering cross section is defined as,

$$\sigma_s = \int_{4\pi}|\vec{f}|^2\,d\Omega \tag{1.45}$$

which can be interpreted as follows. Since the time-averaged power density of a unit amplitude ($E_0 = 1$ V m^{-1}) incident plane wave is $1/2Z_0$, the product of σ_s (a geometrical cross sectional area) multiplied by $1/2Z_0$ (a power density) can be associated with the power intercepted from the incident wave by the particle, which is then equated to the total scattered power (P_s or Re T_s). Note that the incident plane wave is assumed to be linearly polarized ($\vec{E}^i \equiv \hat{e}_i E_0$).

From (1.44) the power scattered per unit elemental solid angle (or the scattered radiation intensity) is given by,

$$\frac{dP_s}{d\Omega} = \frac{1}{2Z_0}|\vec{f}(\hat{s}, \hat{i})|^2 \tag{1.46a}$$

The bistatic radar cross section is defined as,

$$\sigma_{bi}(\hat{s}, \hat{i}) = 4\pi|\vec{f}(\hat{s}, \hat{i})|^2 \tag{1.46b}$$

which can be interpreted as follows. Since the incident power density is $1/2Z_0$, the product of $\sigma_{bi}(\hat{s}, \hat{i})$ and $1/2Z_0$ gives the power intercepted from the incident wave by the particle. This power is then assumed to be re-radiated isotropically so that the effective scattered radiation intensity is constant and equal to $[\sigma_{bi}/2Z_0](1/4\pi)$, which is then equated to the actual radiation intensity in the $\hat{s}$-direction given in (1.46a), resulting in (1.46b). A generalized bistatic radar cross section accounting for an arbitary (i.e. elliptical) incident polarization state will be defined in Section 3.8. The differential scattering cross section is defined as,

$$\sigma_d(\hat{s}, \hat{i}) = \frac{1}{4\pi}\sigma_{bi}(\hat{s}, \hat{i}) \tag{1.47}$$

and can be similarly interpreted. The radar cross section or back scattering cross section is defined as $\sigma_b(-\hat{i}, \hat{i})$, where the scattering direction is $\hat{s} = -\hat{i}$.

For Rayleigh scattering by a dielectric sphere, the far-field vector amplitude $[\vec{f}(\hat{s}, \hat{i})]$ is given as,

$$\vec{f}(\hat{s}, \hat{i}) = \frac{k_0^2}{4\pi\varepsilon_0}[\vec{p} - \hat{s}(\hat{s}\cdot\vec{p})] = \frac{k_0^2}{4\pi\varepsilon_0}(\vec{\chi}) \tag{1.48}$$

Since $\hat{s}\cdot\vec{\chi} = 0$ and $\hat{s} \times \vec{\chi} = \hat{s} \times \vec{p}$,

$$|\hat{s} \times \vec{\chi}| = |\vec{\chi}| = |\hat{s} \times \vec{p}| = p\sin\beta \tag{1.49}$$

where p is the magnitude of the dipole vector and β is the angle between $\hat{s}$ and $\vec{p}$, see Fig. 1.9. Thus,

$$|\vec{f}(\hat{s}, \hat{i})| = \frac{k_0^2}{4\pi\varepsilon_0}p\sin\beta \tag{1.50a}$$

and

$$\sigma_{bi}(\hat{s}, \hat{i}) = 4\pi\frac{k_0^4}{(4\pi)^2}\frac{1}{\varepsilon_0^2}p^2\sin^2\beta \tag{1.50b}$$

$$= \frac{k_0^4}{4\pi}\left|\frac{3(\varepsilon_r - 1)}{\varepsilon_r + 2}\right|^2 V^2\sin^2\beta \tag{1.50c}$$

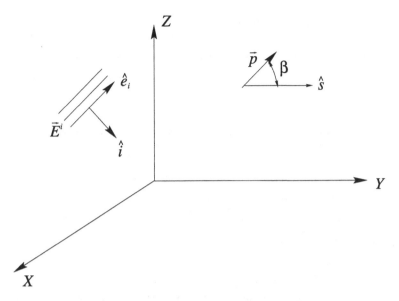

Fig. 1.9. Illustrating the angle β in (1.49) for Rayleigh scattering.

where $\sin^2 \beta = 1 - \cos^2 \beta = 1 - (\hat{s} \cdot \hat{p})^2 = 1 - (\hat{s} \cdot \hat{e}_i)^2$, since $\hat{e}_i$ is parallel to $\vec{p}$. When $\hat{s} = -\hat{i}$, $\sin^2 \beta = 1$, so that the back scattering or radar cross section is,

$$\sigma_b(-\hat{i}, \hat{i}) = \frac{k_0^4}{4\pi} \left| \frac{3(\varepsilon_r - 1)}{\varepsilon_r + 2} \right|^2 V^2 \tag{1.51a}$$

$$= \frac{\pi^5}{\lambda^4} |K|^2 D^6 \tag{1.51b}$$

where D is the sphere diameter, and $|K|^2 = |(\varepsilon_r - 1)/(\varepsilon_r + 2)|^2$ is the dielectric factor. Equation (1.51b) is the classic form used in radar meteorology. It clearly shows Rayleigh scattering as being dependent on λ^{-4}. The size dependence is based on the sixth power of the diameter and it is also weighted by the dielectric factor. Use of (1.50a) in (1.45) gives,

$$\sigma_s = \frac{3}{2\pi} k_0^4 V^2 |K|^2 \tag{1.52}$$

For a sphere, the total scattering cross section (σ_s) and the radar cross section (σ_b) are independent of the state of linear polarization ($\hat{e}_i$) and direction ($\hat{i}$) of the incident wave. The bistatic cross section ($\sigma_{bi}(\hat{s}, \hat{i})$), however, is dependent on $\sin^2 \beta = 1 - (\hat{s} \cdot \hat{e}_i)^2$, which shows that it is dependent on both the state of linear polarization of the incident wave as well as the scattering direction.

1.5   Absorption and extinction cross sections

The time-averaged power dissipated within the dielectric particle is given as,

$$P_d = \frac{1}{2} \mathrm{Re} \int_\tau \vec{E}_T^{\mathrm{in}} \cdot \vec{J}_p^* \, d\tau \tag{1.53}$$

where $\vec{E}_T^{\mathrm{in}}$ is the total electric field inside the particle and $\vec{J}_p$ is the polarization current, which (from Appendix 1) is,

$$\vec{J}_p = j\omega \vec{P} = j\omega\varepsilon_0(\varepsilon_r - 1)\vec{E}_T^{\mathrm{in}} \tag{1.54}$$

Since $\varepsilon_r = \varepsilon_r' - j\varepsilon_r''$, the power loss inside the dielectric is,

$$P_d = \frac{1}{2}\omega\varepsilon_0\varepsilon_r'' \int_\tau |E_T^{\mathrm{in}}|^2 \, d\tau \tag{1.55}$$

Note that τ is the volume of the particle. The absorption cross section (σ_a) is defined as that cross section which when multiplied by the incident power density (of a unit amplitude, linearly polarized incident wave) gives the power loss (P_d) or,

$$\sigma_a\left(\frac{1}{2Z_0}\right) = P_d = \frac{1}{2}\omega\varepsilon_0\varepsilon_r'' \int_\tau |E_T^{\mathrm{in}}|^2 \, d\tau \tag{1.56}$$

Finally, the sum of the total scattering and absorption cross sections is called the extinction cross section,

$$\sigma_{\mathrm{ext}} = \sigma_s + \sigma_a \tag{1.57}$$

The extinction cross section represents the total power loss from the incident wave due to scattering and absorption by the particle.

The optical theorem is stated here (see the proof in the appendix to Chapter 10 of Ishimaru 1991) as,

$$\sigma_{\mathrm{ext}} = \frac{-4\pi}{k_0} \mathrm{Im}\left[\vec{f}(\hat{i}, \hat{i}) \cdot \hat{e}_i\right] \tag{1.58}$$

where $\vec{f}(\hat{i}, \hat{i})$ is the forward scattered vector amplitude and $\vec{E}^i = \hat{e}_i E_0 \exp(-jk_0\hat{i} \cdot \vec{r}_i)$, with $E_0 = 1$ V m^{-1}. In numerical scattering calculations, (1.58) is used to calculate σ_{ext}, and σ_a is derived from $\sigma_a = \sigma_{\mathrm{ext}} - \sigma_s$.

For Rayleigh scattering by a dielectric sphere, (1.56) can be used since $|E_T^{\mathrm{in}}| = 3E_0/(\varepsilon_r + 2)$, which is constant inside the sphere. Thus,

$$\sigma_a = 9k_0 V \frac{\varepsilon_r''}{|\varepsilon_r + 2|^2} \tag{1.59}$$

and the absorption cross section is proportional to the volume of the sphere, i.e. to D^3. For Rayleigh scattering, use of the optical theorem gives $\sigma_{\mathrm{ext}} = \sigma_a$, and thus appears to violate (1.57). Since the sphere diameter $D \ll \lambda$ and $\sigma_s \propto k_0^{-2}(D/\lambda)^6$, while $\sigma_a \propto k_0^{-2}(D/\lambda)^3$, (1.57) reduces to $\sigma_{\text{ext}} = \sigma_a$, which is correct to the order $(D/\lambda)^3$. If the dielectric sphere is lossless ($\varepsilon_r'' = 0$), then Rayleigh scattering and use of the optical theorem gives $\sigma_{\text{ext}} = 0$, which is incorrect since $\sigma_{\text{ext}} = \sigma_s$. This violation of (1.57) occurs because the vector scattering amplitude in the Rayleigh limit is only correct up to the order $(D/\lambda)^3$. This point will be re-visited in Section 2.5, when the low frequency expansion of the Mie coefficients will be considered.

1.6   Clausius–Mosotti equation and Maxwell-Garnet mixing formula

The total macroscopic electric field inside a dielectric ($\vec{E}_T^{\text{in}}$) and the volume density of polarization ($\vec{P}$) are related by $\vec{P} = (\varepsilon - \varepsilon_0)\vec{E}_T^{\text{in}}$. Referring to Appendix 1, see (A1.11), $\vec{E}_T^{\text{in}}$ is not the electric field that polarizes an individual atom or molecule of the material. This field is microscopic in nature and is called the "local" electric field, which differs from $\vec{E}_T^{\text{in}}$. The atomic dipole moment ($\vec{p}_{\text{atom}}$) is then related to $\vec{E}_{\text{local}}$ by $\vec{p}_{\text{atom}} = \alpha_{\text{atom}}\vec{E}_{\text{local}}$, where α_{atom} is the atomic polarizability. In (A1.4), the $\vec{p}_j$ used in defining $\vec{P}$ refers to the atomic dipole moment.

To derive the "local" electric field, the discussion in Section A1.2 can be used to illustrate the basic concept. Figure A1.10 shows a spherical cavity of radius a cut in an uniformly permanently polarized material (with $\vec{P} = P_0\hat{z}$) of infinite extent. It is permissible to consider $\vec{P}$ as the uniform volume density of polarization within a linear, dielectric material of ε placed, for example, between the conducting plates of a parallel plate capacitor. To find the local field that "excites" a particular atom (the jth atom), a spherical cavity is cut whose center is at the location of this atom, and whose radius (a) is small enough so that it does not perturb the macroscopic field ($\vec{E}_T^{\text{in}}$) external to the cavity. However, the radius (a) is "microscopically" large enough to contain a significant number of polarized atoms before the cavity is cut. Note that the process of cutting the cavity implies that all the polarized atoms within the cavity have been removed. From Section A1.2 it can be argued that the macroscopic electric field at the cavity center ($\vec{E}_{\text{center}}$) has jumped from $\vec{E}_T^{\text{in}}$ (no cavity case) to $\vec{E}_T^{\text{in}} + \vec{P}/3\varepsilon_0$ (in the presence of the cavity). Thus,

$$\vec{E}_{\text{center}} = \vec{E}_T^{\text{in}} + \frac{\vec{P}}{3\varepsilon_0} \tag{1.60}$$

The additional term $\vec{P}/3\varepsilon_0$ is due to the surface density of the bound charge (η_b) on the cavity/dielectric interface, across which $\vec{P}$ is discontinuous (zero inside the cavity, and $P_0\hat{z}$ outside the cavity), with the top and bottom hemispheres of the interface being negatively and positively charged, respectively, as shown in Fig. A1.10b. Thus, the additional term $\vec{P}/3\varepsilon_0$ is the electric field contribution to the "local" electric field exciting the jth atom (whose location is at the center of the cavity). This additional electric field component can be viewed as taking into account the effects of all polarized atoms external to the cavity on the jth atom, but considered in a macroscopic manner.

Now, replace all the polarized atoms that were taken out when the cavity was cut, except for the jth atom at the center. These near-polarized atoms can be considered to generate another component of the "local" electric field exciting the jth atom, which is termed $\vec{E}_{near}$. Thus,

$$\vec{E}_{local} = \vec{E}_{center} + \vec{E}_{near} = \vec{E}_T^{in} + \frac{\vec{P}}{3\varepsilon_0} + \vec{E}_{near} \tag{1.61}$$

Lorentz (1952) showed, from symmetry, that for polarized atoms (elemental dipoles) arranged in a simple cubic lattice, $\vec{E}_{near}$ vanishes at any lattice site and, in particular, at the site of the jth atom at the cavity center (see also Jackson 1975, p. 153). If $\vec{E}_{near}$ is zero for this highly symmetric situation, it appears reasonable that $\vec{E}_{near} \approx 0$, even if the polarized "near" atoms are randomly distributed in location. With this approximation, the expression for the "local" electric field exciting the jth atom reduces to,

$$\vec{E}_{local} \approx \vec{E}_T^{in} + \frac{\vec{P}}{3\varepsilon_0} \tag{1.62}$$

If α_{atom} is the atomic polarizability of the jth atom (all atoms are considered to be identical here), then the elemental dipole moment ($\vec{p}_j \equiv \vec{p}$) of the jth atom is related to the "local" electric field by,

$$\vec{p} = \alpha_{atom} \vec{E}_{local} = \alpha_{atom} \left(\vec{E}_T^{in} + \frac{\vec{P}}{3\varepsilon_0} \right) \tag{1.63}$$

If there are N atoms per unit volume, then the volume density of polarization from (A1.4) is $\vec{P} = N\vec{p}$ (from the "microscopic" definition), which can be equated to $(\varepsilon - \varepsilon_0)\vec{E}_T^{in}$ (from the macroscopic definition). Thus,

$$N\alpha_{atom} \left\{ \vec{E}_T^{in} + \frac{\vec{P}}{3\varepsilon_0} \right\} = (\varepsilon - \varepsilon_0) \vec{E}_T^{in} \tag{1.64}$$

Equation (1.64) reduces to the Clausius–Mosotti formula[2] for ε,

$$\frac{\varepsilon}{\varepsilon_0} = 1 + \left(\frac{N\alpha}{\varepsilon_0} \right) \Big/ \left(1 - \frac{N\alpha}{3\varepsilon_0} \right) = \left(1 + \frac{2N\alpha}{3\varepsilon_0} \right) \Big/ \left(1 - \frac{N\alpha}{3\varepsilon_0} \right) \tag{1.65}$$

where $\alpha \equiv \alpha_{atom}$ for convenience. Equation (1.65) can also be written as,

$$\frac{\varepsilon - \varepsilon_0}{\varepsilon + 2\varepsilon_0} = \frac{N\alpha}{3\varepsilon_0} \tag{1.66}$$

The Clausius–Mosotti formula for ε in terms of atomic polarizability (α) and N can be extended to determine the effective permittivity (ε_{eff}) of a two-phase material, with one phase modeled as identical discrete spheres (also known as inclusions) of permittivity ε_i and volume V, which is imbedded in a second phase "matrix" of permittivity ε_m, as illustrated in Fig. 1.10. The dielectric spheres, of which there

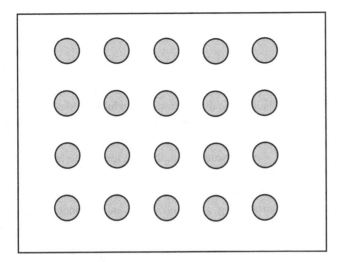

Fig. 1.10. Spherical inclusions of ε_i (shaded areas) imbedded in a matrix material of ε_m.

are assumed to be N per unit volume, are treated as "elemental" dipoles, whose polarizability α is given by eq. (A1.27) generalized to the case of a single sphere ε_i imbedded in a matrix material ε_m,

$$\alpha = \frac{3V\varepsilon_m(\varepsilon_i - \varepsilon_m)}{(\varepsilon_i + 2\varepsilon_m)} \tag{1.67}$$

Equation (1.65) is generalized by recognizing that $\varepsilon/\varepsilon_0$ for the two-phase material is replaced by $\varepsilon_{\text{eff}}/\varepsilon_m$, while $N\alpha/3\varepsilon_0$ is replaced by $N\alpha/3\varepsilon_m$, where α is given in (1.67). Thus,

$$\frac{\varepsilon_{\text{eff}}}{\varepsilon_m} = \left(1 + \frac{2N\alpha}{3\varepsilon_m}\right) \bigg/ \left(1 - \frac{N\alpha}{3\varepsilon_m}\right) \tag{1.68}$$

Defining $c = NV$ as the volume concentration of dielectric spheres (or inclusions) and $F = (\varepsilon_i - \varepsilon_m)/(\varepsilon_i + 2\varepsilon_m)$, (1.68) is simplified as,

$$\frac{\varepsilon_{\text{eff}}}{\varepsilon_m} = \frac{1 + 2cF}{1 - cF} \tag{1.69}$$

which is referred to as the Maxwell-Garnet (1904) mixing formula. Note that (1.69) reduces to $\varepsilon_{\text{eff}} = \varepsilon_m$, when $c = 0$ (i.e. when there are no inclusions), and $\varepsilon_{\text{eff}} = \varepsilon_i$, when $c = 1$. However, it is not permissible to let $c = 1$, since the maximum "packing" for spherical inclusions results in $c \approx 0.63$. When $c \ll 1$, the binomial expansion $(1 - cF)^{-1} \approx 1 + cF$ results in $\varepsilon_{\text{eff}} = \varepsilon_m(1 + 3cF)$. This approximation for a very dilute concentration of spherical inclusions can also be obtained from (1.62), with the term $\vec{P}/3\varepsilon_0$ being neglected, or from the approximation $\vec{E}_{\text{local}} \approx \vec{E}_T^{\text{in}}$. Corresponding to $\varepsilon_{\text{eff}}/\varepsilon_m$ in (1.69), an effective wave number can be defined as,

$$\left(\frac{k_{\text{eff}}}{k_m}\right)^2 = \frac{\varepsilon_{\text{eff}}}{\varepsilon_m} = \frac{1 + 2cF}{1 - cF} \tag{1.70a}$$

and for $c \ll 1$,

$$\frac{k_{\text{eff}}}{k_m} \approx 1 + \frac{3}{2}cF \tag{1.70b}$$

The Maxwell-Garnet formula is dependent upon which component of the two-phase mixture is considered to be the spherical inclusions and which is considered to be the matrix. Following the notation of Meneghini and Liao (1996), the result in (1.69), where the spherical inclusions (ε_i) are imbedded in a matrix (ε_m), can be denoted as $\varepsilon_{\text{MG}(mi)} = \varepsilon_m (1 + 2cF)/(1 - cF)$. Note, again, that c refers to the fractional volume concentration of spherical inclusions, and $F = (\varepsilon_i - \varepsilon_m)/(\varepsilon_i + 2\varepsilon_m)$. For example, if the two-phase material is a mixture of water and ice, then the notation MG_{WI} will refer to $\varepsilon_{\text{MG}(mi)}$ with the matrix being water ($m \equiv W$) and the inclusions being ice ($i \equiv I$). The notation MG_{IW} will refer to water inclusions imbedded in an ice matrix. The Maxwell-Garnet formula does not yield $\text{MG}_{WI} = \text{MG}_{IW}$.

The MG mixing formula can be used to calculate the Rayleigh limit radar and absorption cross sections, see (1.51b, 1.59), of wet or "spongy" hailstones, where ε_r in these equations is replaced by ε_{eff} from the Maxwell-Garnet formula (MG_{IW} or MG_{WI}). The Maxwell-Garnet formula can also be applied to calculate the effective permittivity of snow particles, which may be modeled as a two-phase mixture of ice (I), and air (A). In this case it turns out that $\text{MG}_{IA} \approx MG_{AI}$. Melting snow particles may be modeled as a three-phase mixture of water (W), ice (I), and air (A). The Maxwell-Garnet formula can then be applied in two steps. Following Meneghini and Liao (1996), $\text{MG}_{1;2,3}$ is used to represent both steps in compact notation, where the first step is $\text{MG}_{2,3}$; where material 3 is the inclusion and 2 is the matrix. The effective permittivity from this first step is then taken as the inclusion for the second step, where the matrix is material 1. This is to be distinguished from $\text{MG}_{1,2;3}$, where the first step is $\text{MG}_{1,2}$. The effective permittivity from this step is used as the matrix for the second step, where the inclusion is material 3.

Ellipsoidal- and spheroidal-shaped inclusions have been discussed by Bohren and Battan (1980; 1982) and DeWolf et al. (1990), who generalize the Maxwell-Garnet formulation. A review of dielectric mixing formulas can be found in Sihvola and Kong (1988), who showed that the Maxwell-Garnet formula is a special case of a more general formulation.

1.7   Faraday's law and non-relativistic Doppler shift

The Doppler shift formula is derived in a simple manner in many standard radar texts (e.g. Skolnik 1980). If r is the distance from the radar antenna to any scatterer, the total number of wavelengths (λ) contained in the two-way path between the radar and the scatterer is $2r/\lambda$. Since one wavelength corresponds to 2π radians of spatial phase, the total spatial phase excursion from the radar to the scatterer is $4\pi r/\lambda$. Since the scatterer is moving, the path length ($2r$) and the total spatial phase excursion vary with time. The time derivative of the total spatial phase excursion is the angular Doppler frequency, $\omega_D = (4\pi/\lambda)dr/dt = 4\pi v_0/\lambda$, where v_0 is the velocity of the scatterer. However, this description does not clearly show the two-step nature of the Doppler shift frequency relative to the frequency of the incident wave launched by the radar. This section uses a simple example to illustrate the underlying physics behind both monostatic and bistatic Doppler frequency shifts.

1.7.1 Electromotive force

In order to illustrate the origin of the Doppler shift, consider a simple configuration such as a rectangular loop moving with constant velocity and illuminated by an ideal plane wave. Before this example is considered, some preliminary material related to Faraday's law is reviewed. From (1.1a), Faraday's law in differential form in a fixed frame of reference is,

$$\nabla \times \vec{E} = -\frac{\partial \vec{B}}{\partial t} \tag{1.71}$$

which describes the simple notion that a time-varying magnetic field gives rise to a non-conservative component of the electric field. If $\vec{B} = \nabla \times \vec{A}$, then $\vec{E} = -\nabla\phi - \partial\vec{A}/\partial t$, see (1.13), which implies that the total electric field is composed of conservative $(-\nabla\phi)$ and non-conservative $(\partial\vec{A}/\partial t)$ components. The closed contour integral (or any closed geometrical path in space) of the non-conservative component of the electric field is known as the induced electromotive force (EMF), which is related to the negative time-rate of change of the magnetic flux (ψ_m) across any open surface (Σ) spanning this closed contour (or "loop"). The magnetic flux is defined as,

$$\psi_m = \int_{\text{open } \Sigma} \vec{B}(\vec{r}, t) \cdot \hat{n} \, d\Sigma \tag{1.72}$$

where $\hat{n}$ is a unit normal to Σ. The time-rate of change of ψ_m and the induced EMF are defined as,

$$\frac{d\psi_m}{dt} = \lim_{\Delta t \to 0} \frac{\psi_m(t + \Delta t) - \psi_m(t)}{\Delta t} = \frac{d}{dt} \int_{\text{open } \Sigma} \vec{B}(\vec{r}, t) \cdot \hat{n} \, d\Sigma \tag{1.73a}$$

$$\text{EMF} = -\frac{d\psi_m}{dt} = -\frac{d}{dt} \int_{\Sigma} \vec{B}(\vec{r}, t) \cdot \hat{n} \, d\Sigma \tag{1.73b}$$

Since, in general, both $\vec{B}$ and the open surface (Σ) spanning the closed contour can change with time (for example, due to "loop" movement or rotation), the time derivative d/dt must be viewed as a convective or total derivative,

$$\frac{d}{dt} = \frac{\partial}{\partial t} + \vec{v} \cdot \nabla \tag{1.74a}$$

where $\vec{v}$ is the velocity of the "loop". The total derivative $d\vec{B}/dt$ is,

$$\frac{d\vec{B}}{dt} = \frac{\partial \vec{B}}{\partial t} + (\vec{v} \cdot \nabla)\vec{B} = \frac{\partial \vec{B}}{\partial t} + \nabla \times \vec{B} \times \vec{v} + \vec{v}(\nabla \cdot \vec{B}) \tag{1.74b}$$

Since $\nabla \cdot \vec{B}$ is always zero, see (1.1b), $d\vec{B}/dt = \partial \vec{B}/\partial t + \nabla \times \vec{B} \times \vec{v}$. Thus (1.73a) reduces to,

$$\frac{d\psi_m}{dt} = \frac{d}{dt}\int_\Sigma \vec{B} \cdot \hat{n}\, d\Sigma = \int_\Sigma \frac{\partial \vec{B}}{\partial t} \cdot \hat{n}\, d\Sigma + \int_\Sigma (\nabla \times \vec{B} \times \vec{v}) \cdot \hat{n}\, d\Sigma \tag{1.75a}$$

$$= \int_\Sigma \frac{\partial \vec{B}}{\partial t} \cdot \hat{n}\, d\Sigma + \oint_C (\vec{B} \times \vec{v}) \cdot \vec{dl} \tag{1.75b}$$

The final result for the induced EMF is,

$$\text{EMF} = \frac{-d\psi_m}{dt} = \int_\Sigma \frac{-\partial \vec{B}}{\partial t} \cdot \hat{n}\, d\Sigma + \oint_C (\vec{v} \times \vec{B}) \cdot \vec{dl} \tag{1.76}$$

Note that Stokes' theorem is used to arrive at (1.75b). The first integral on the right-hand side of (1.76) is called the "transformer" EMF, while the second integral is the "motional" EMF.

1.7.2   Example of a moving rectangular loop

Consider an open circuited, rectangular conducting loop moving with uniform velocity $\vec{v}$ in the presence of an incident plane wave with $\vec{B} = \hat{y} B_0 \cos(\omega t - kx)$, as illustrated in Fig. 1.11. If the center of the loop at some time is $(x_c, 0, 0)$, then $x_c = x_0 + v_0 t$, where x_0 is an initial position at $t = 0$. Choosing a clockwise contour C, which coincides with the perimeter of the loop, and the open surface Σ spanning C as the loop surface itself, the unit normal to this surface is $\hat{n} = \hat{y}$. Evaluating the transformer EMF (T_EMF), first, the integrand is,

$$\frac{-\partial \vec{B}}{\partial t} \cdot \hat{n}\, d\Sigma = B_0 \omega \sin(\omega t - kx) h\, dx \tag{1.77}$$

and the integral becomes,

$$\text{T\_EMF} = \omega h B_0 \int_{x=x_0+v_0t-l/2}^{x_0+v_0t+l/2} \sin(\omega t - kx)\, dx \tag{1.78a}$$

$$= \frac{2\omega h B_0}{k} \sin\left(\frac{kl}{2}\right) \sin(\omega t - kv_0 t - kx_0) \tag{1.78b}$$

Note that the total phase of the plane wave $(\omega t - kx)$ changes across the loop dimension along the X-axis, which is the wave propagation direction. Of course, this will happen even if the loop is stationary. However, since the loop is moving, the change in total phase becomes a function of the loop velocity.

The integrand for the motional EMF (M_EMF) is,

$$(\vec{v} \times \vec{B}) \cdot \vec{dl} = v_0 \hat{x} \times [\hat{y} B_0 \cos(\omega t - kx)] \cdot \vec{dl} \tag{1.79a}$$

$$= v_0 [B_0 \cos(\omega t - kx)] \hat{z} \cdot (dx\, \hat{x} + dz\, \hat{z}) \tag{1.79b}$$

$$= v_0 B_0 \cos(\omega t - kx)\, dz \tag{1.79c}$$

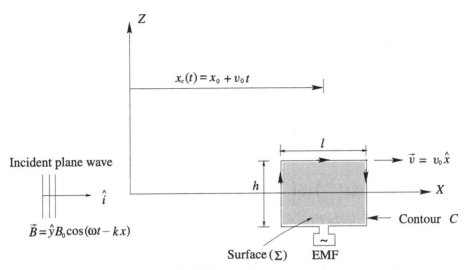

Fig. 1.11. Rectangular loop moving with constant velocity $\vec{v}$ and illuminated by a plane wave with magnetic field $\vec{B}$. Clockwise rotation around the loop is denoted by the contour C, while Σ is the shaded surface. The unit normal to the surface Σ is $\hat{n} = \hat{y}$.

and the integral for M_EMF becomes,

$$\text{M\_EMF} = v_0 B_0 \oint_C \cos(\omega t - kx)\, dz \tag{1.80}$$

Since only the two sections of the rectangular loop parallel to the Z-axis will contribute,

$$\text{M\_EMF} = v_0 B_0 \int_{-h/2}^{h/2} \cos(\omega t - kx)\Big|_{x=x_0+v_0t-l/2}\, dz$$

$$+ v_0 B_0 \int_{h/2}^{-h/2} \cos(\omega t - kx)\Big|_{x=x_0+v_0t+l/2}\, dz \tag{1.81a}$$

$$= v_0 B_0 h \cos\left[\omega t - k\left(x_0 + v_0 t - \frac{l}{2}\right)\right]$$

$$+ v_0 B_0(-h)\cos\left[\omega t - k\left(x_0 + v_0 t + \frac{l}{2}\right)\right] \tag{1.81b}$$

$$= -2v_0 B_0 h \sin\left(\frac{kl}{2}\right)\sin(\omega t - kv_0 t - kx_0) \tag{1.81c}$$

The total EMF is the sum of (1.78b) and (1.81c),

$$\text{EMF} = 2B_0 h \sin\left(\frac{kl}{2}\right)\left(\frac{\omega}{k} - v_0\right)\sin(\omega t - kv_0 t - kx_0) \tag{1.82}$$

Since the wave number of the plane wave is $k = \omega/c = 2\pi f/c = 2\pi/\lambda$, the term $(\omega/k) - v_0 = c - v_0 \approx c$, since v_0 is assumed to be extremely small compared to the velocity of light. Thus, the T_EMF term dominates in (1.82) as far as the amplitude of the EMF is concerned. Now consider the total phase of the sinusoidal term in (1.82), which can be written as,

$$\omega t - k v_0 t - k x_0 = 2\pi f t - \frac{2\pi v_0 t}{\lambda} - k x_0 \tag{1.83a}$$

$$= 2\pi \left(f - \frac{v_0}{\lambda} \right) t - k x_0 \tag{1.83b}$$

The total EMF is dominated by T_EMF, given in (1.78),

$$\text{EMF} \approx 2 B_0 h \left(\frac{\omega}{k} \right) \sin \left(\frac{kl}{2} \right) \sin \left[2\pi \left(f - \frac{v_0}{\lambda} \right) t - k x_0 \right] \tag{1.84a}$$

$$\equiv A \sin(2\pi f' t - k x_0) \tag{1.84b}$$

Thus, the total EMF is sinusoidal with amplitude A, but the frequency (f') is Doppler shifted from the frequency of the incident plane wave, i.e. $f' = f - v_0/\lambda = f - f_D$, where f_D is the Doppler frequency, which equals v_0/λ. The term $k x_0$ may be considered as an initial phase that depends on the position of the loop at time $t = 0$. One inference from examining (1.78) is that the Doppler shift in the EMF is related to the change in phase of the incident plane wave across the moving loop. If the loop were stationary, there would be zero Doppler shift. Since the loop is assumed to move parallel to the direction of propagation of the incident wave, the frequency (f') is smaller than the incident wave frequency (f). If the loop were moving in the opposite direction $(\vec{v} = -v_0 \hat{x})$, then it is clear that the integration limits in (1.78) would be,

$$\text{T\_EMF} = \omega h B_0 \int_{x_0 - v_0 t - l/2}^{x_0 - v_0 t + l/2} \sin(\omega t - k x) \, dx \tag{1.85a}$$

$$= \frac{2\omega h B_0}{k} \sin \left(\frac{kl}{2} \right) \sin(\omega t + k v_0 t - k x_0) \tag{1.85b}$$

In this case the frequency (f') of the EMF is larger than the incident plane wave frequency, i.e. $f' = f + v_0/\lambda$.

Faraday's law was used in this example to explicitly show the origin of the Doppler shifted frequency of the EMF in the moving loop. A simpler derivation for plane wave incidence is based on the total phase of the incident wave at the center of the loop $(x_c, 0, 0)$, which is given by $\omega t - k x_c$, where $x_c = x_0 + v_0 t$ (see Fig. 1.11). Thus,

$$\omega t - k x_c = \omega t - k(x_0 + v_0 t) \tag{1.86a}$$

$$= \omega t - k x_0 - k v_0 t \tag{1.86b}$$

$$= 2\pi \left(f - \frac{v_0}{\lambda} \right) t - k x_0 \tag{1.86c}$$

which is identical to (1.83a). If the loop were accelerating (a) with $x_c(t) = x_0 + v_0 t + a t^2/2$, then the EMF would have a frequency "chirp" with $f' = f - (v_0/\lambda) - (1/2)(at/\lambda)$.

1.7.3   Loop moving at an angle α

A further generalization can be made if the loop is moving at angle α, as shown in Fig. 1.12a. For simplicity, let the center of the loop in the primed coordinate system be at the origin, at $t = 0$, so that $x_c'(t) = |\vec{v}|t$. The transformer EMF is still given by,

$$\text{T\_EMF} = \int_\Sigma -\frac{\partial \vec{B}}{\partial t} \cdot \hat{n}\, d\Sigma \qquad\qquad (1.87a)$$

$$= \int_\Sigma -\frac{\partial \vec{B}}{\partial t} \cdot \hat{y}'\, dx'\, dz' \qquad\qquad (1.87b)$$

Since $\vec{B} = \hat{y} B_0 \cos(\omega t - kx) = \hat{y}' B_0 \cos\left[\omega t - k(x'\cos\alpha - z'\sin\alpha)\right]$, the integrand becomes,

$$-\frac{\partial \vec{B}}{\partial t} \cdot \hat{y}' = \omega B_0 \sin(\omega t - kx'\cos\alpha + kz'\sin\alpha) \qquad\qquad (1.88a)$$

and,

$$\text{T\_EMF} = \omega B_0 \int_{-h/2}^{h/2} \int_{|\vec{v}|t-l/2}^{|\vec{v}|t+l/2} \sin(\omega t - kx'\cos\alpha + kz'\sin\alpha)\, dx'\, dz' \qquad\qquad (1.88b)$$

$$= \frac{4\omega B_0}{k^2 \sin\alpha \cos\alpha} \sin\left(\frac{kh}{2}\sin\alpha\right) \sin\left(\frac{kh}{2}\cos\alpha\right) \sin(\omega t - k|\vec{v}|t\cos\alpha) \qquad (1.88c)$$

$$= A \sin(\omega t - k|\vec{v}|t\cos\alpha) \qquad\qquad (1.88d)$$

where A is the amplitude of the EMF, and the Doppler frequency is $f_D = (|\vec{v}|/\lambda)\cos\alpha$. Since the incident plane wave propagation direction is $\hat{i} = \hat{x}$, the term $k|\vec{v}|\cos\alpha \equiv k\hat{i}\cdot\vec{v}$, and the Doppler frequency is $f_D = \hat{i}\cdot\vec{v}/\lambda$. If the loop velocity vector is orthogonal to the direction of the incident plane wave then $f_D = 0$. The maximum Doppler shift occurs when $\vec{v} = \pm\hat{i}$, or when the loop velocity vector is parallel (or opposite) to the incident wave direction. It can also be deduced that if the incident plane wave is propagating along the $-\hat{x}'$-direction with $\vec{B} = \hat{y}' B_0 \cos(\omega t + kx')$, and the loop velocity vector is $-|\vec{v}|\hat{x}$, then the EMF is still Doppler shifted in frequency by the amount $\hat{i}\cdot\vec{v}/\lambda$ (see Fig. 1.12b).

1.7.4   Re-radiation by the loop

The EMF in the open circuited rectangular loop can be considered as a "source" or generator that can drive an external load. For convenience, let the rectangular loop be short-circuited so that the "load" now becomes the internal resistance of the conducting loop (for simplicity, the self-inductance of the loop is neglected). Thus, an induced current (i) flows in the shorted loop (i = EMF/resistance), and this current will be sinusoidal and Doppler shifted by the same amount as the EMF. The current circulating in the loop can be viewed as an elementary magnetic dipole ($\vec{p}_m$), whose moment is simply the product of the induced current and loop surface area. This magnetic dipole

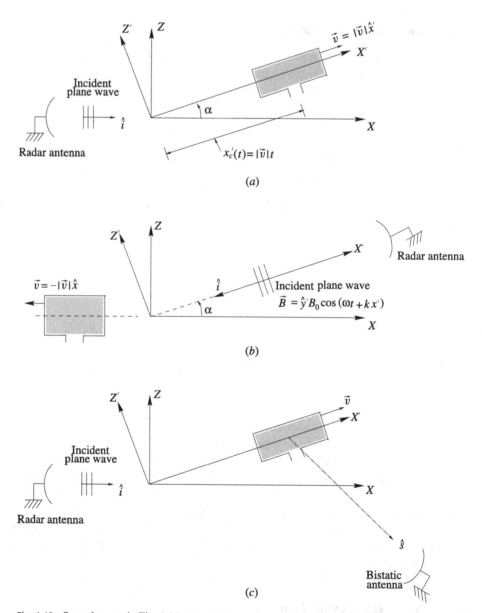

Fig. 1.12. Same loop as in Fig. 1.11, except (a) moving with constant velocity at angle α, (b) with roles of incident plane wave and loop reversed from that illustrated in (a) above, and (c) illustrating the bistatic Doppler shift along the $\hat{s}$-direction.

will radiate an electromagnetic field, which in the far-field will have the form given in (1.32b), with the vector radiation amplitude $(\vec{f})$ dependent on $\vec{p}_m$, much like the role played by $\vec{p}$ in (1.32c). Now consider Fig. 1.12a again, and let the incident plane wave shown there be launched by a radar antenna located at $x \rightarrow -\infty$. Assume that the loop is shorted so that the Doppler shifted magnetic dipole moment is of the form $A \sin(\omega t - k\hat{i} \cdot \vec{v}t)$. This dipole moment will re-radiate an electromagnetic wave, with wave number $k' = 2\pi/\lambda' = (\omega - k\hat{i} \cdot \vec{v}t)/c$ and $\lambda' = \lambda_0/(1 - v/c)$. Since $v/c \ll 1$, $k' \approx k$ and $\lambda' \approx \lambda$. In particular, consider the re-radiated local plane wave along the $\hat{s} = -\hat{x}'$-direction in Fig. 1.12a. This re-radiated plane wave will induce current in the radar antenna which will be Doppler shifted since there is relative motion between the loop and the stationary radar antenna. This relative motion is equivalent to considering the magnetic dipole moment of the loop to be fixed in space and letting the radar antenna move with a velocity $-|\vec{v}|\hat{x}$. This relative motion is illustrated in Fig. 1.12b, where the local plane wave re-radiated by the magnetic dipole is in the $\hat{s} = \hat{i} = -\hat{x}'$-direction and the now "moving" radar antenna is identified with the rectangular loop moving with velocity $-|\vec{v}|\hat{x}$. Thus, the induced current in the radar antenna is Doppler shifted by $\hat{i} \cdot \vec{v}/\lambda$ relative to the magnetic dipole moment. However, the magnetic dipole moment $(\vec{p}_m)$ itself was Doppler shifted relative to the incident wave frequency (f) by an identical amount $\hat{i} \cdot \vec{v}/\lambda$. Thus, the induced current in the radar antenna will be Doppler shifted by $2\hat{i} \cdot \vec{v}/\lambda$ relative to the frequency of the incident wave that it originally launched.

1.7.5 Bistatic Doppler frequency

The bistatic Doppler frequency along the $\hat{s}$-direction can also be derived by referring to the schematic in Fig. 1.12c. The magnetic dipole moment of the loop re-radiates in the far-field along the $\hat{s}$-direction as a local plane wave, and the induced current in the bistatic antenna is Doppler shifted. The relative motion between the loop and the bistatic antenna can be viewed with the magnetic dipole being fixed in space and re-radiating a local plane wave along $\hat{s}$ while the bistatic antenna moves with velocity $-\vec{v}$. Thus, the Doppler frequency is given as $f_D = (-\vec{v}) \cdot \hat{s}/\lambda$ relative to the magnetic dipole moment. Since the magnetic dipole moment is itself Doppler shifted by an amount $\hat{i} \cdot \vec{v}/\lambda$ relative to the frequency of the incident wave, the bistatic Doppler frequency along the $\hat{s}$-direction is given by,

$$f_D = \frac{(\hat{i} - \hat{s}) \cdot \vec{v}}{\lambda} \tag{1.89}$$

When $\hat{s} = -\hat{i}$, which is the radar direction, then f_D reduces to $2\hat{i} \cdot \vec{v}/\lambda$, as before. In the forward direction, where $\hat{s} = \hat{i}$, f_D reduces to zero. In all other re-radiated directions, f_D is given by (1.89). Note that a positive value for f_D implies that $f' = (f - f_D)$ is less than f, while the opposite is true for negative values of f_D.

1.8   Moving dielectric spheres: coherent and incoherent summation

The rectangular loop is now replaced by a moving dielectric sphere, as illustrated in Fig. 1.13. The polarization current $\vec{J}_p$ is analogous to the induced current in the shorted

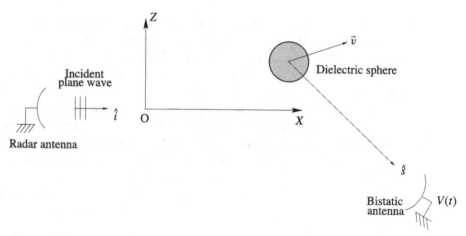

Fig. 1.13. Bistatic Doppler shift along the $\hat{s}$-direction due to a dielectric sphere moving with uniform velocity and illuminated by the incident plane wave.

loop and is thus Doppler shifted in frequency by $\hat{i}\cdot\vec{v}/\lambda$. In the Rayleigh limit, the electric dipole moment $\vec{p}$ is analogous to $\vec{p}_m$ of the shorted loop. The current (or voltage) induced in the bistatic antenna will be sinusoidal and its total phase will be $2\pi(f - f_D)t = \omega t - k(\hat{i} - \hat{s})\cdot\vec{v}t$, where it is assumed that the sphere center is at O at $t = 0$. The angular Doppler frequency is defined as $\omega_D = k(\hat{i} - \hat{s})\cdot\vec{v}$. If the sphere center is at a general initial position $\vec{r}_{in}$ at $t = 0$, then the total phase of the induced current (or voltage) in the bistatic antenna will be $\omega t - k(\hat{i} - \hat{s})\cdot\vec{v}t - k(\hat{i} - \hat{s})\cdot\hat{r}_{in}$, where $\theta = k(\hat{i} - \hat{s})\cdot\vec{r}_{in}$ is the initial phase angle. The voltage can be written, in principle, as,

$$V(t) = A\cos[\omega t - k(\hat{i} - \hat{s})\cdot\vec{v}t - \theta] \equiv \text{Re}\left[Ae^{-j(\omega_D t + \theta)}\right]e^{j\omega t} \tag{1.90}$$

In the radar direction, simply replace $\hat{s} = -\hat{i}$.

   If there is a collection of dielectric spheres and the velocity of the mth sphere is $\vec{v}_m$, the initial phase angle is θ_m, $\omega_{Dm} = k(\hat{i} - \hat{s})\cdot\vec{v}_m$, and the amplitude is A_m, the resultant "phasor" voltage induced in the bistatic antenna is given by superposition as,

$$V = \sum_m A_m e^{-j(\omega_{Dm} t + \theta_m)} \tag{1.91}$$

Note that V, while complex, is not a phasor quantity in the strict sense, since it depends on t. The variation of voltage with time $V(t)$ is considered here to be "slow" compared with the inverse of the angular frequency ω of the incident wave. Hence, $V(t)$ is defined as the envelope of the complex signal $s(t)$,

$$s(t) = V(t)e^{j\omega t} = \left[a(t)e^{j\alpha(t)}\right]e^{j\omega t} \tag{1.92}$$

The function $a(t)$ is known as the envelope magnitude and $\alpha(t)$ as the envelope phase. The real signal, $x(t)$, is obtained from $s(t)$ as,

$$x(t) = \text{Re}[s(t)] = a(t)\cos[\omega t + \alpha(t)] \tag{1.93}$$

The quantities $a(t)\cos\alpha(t)$ and $a(t)\sin\alpha(t)$ are called the in-phase and quadrature phase components, respectively, of the envelope $V(t)$. Doppler radars are configured to measure these two components, as will be discussed in the Section 1.9.2.

For simplicity, consider the case of two dielectric spheres in the summation of (1.91),

$$V = A_1 e^{-j(\omega_1 t + \theta_1)} + A_2 e^{-j(\omega_2 t + \theta_2)} \tag{1.94}$$

where the subscript D on ω_D is dropped for ease of notation. The power is proportional to VV^*, which is,

$$
\begin{aligned}
VV^* &= A_1 A_1^* + A_2 A_2^* + A_1 A_2^* e^{j(\omega_2 - \omega_1)t} e^{j(\theta_2 - \theta_1)} \\
&\quad + A_1^* A_2 e^{j(\omega_1 - \omega_2)t} e^{j(\theta_1 - \theta_2)} \tag{1.95a} \\
&= |A_1|^2 + |A_2|^2 + A_1 A_2^* e^{j\Omega_{21}t} e^{j\theta_{21}} + A_1^* A_2 e^{-j\Omega_{21}t} e^{-j\theta_{21}} \tag{1.95b} \\
&= |A_1|^2 + |A_2|^2 + 2\operatorname{Re}\left(A_1 A_2^* e^{j\Omega_{21}t} e^{j\theta_{21}}\right) \tag{1.95c}
\end{aligned}
$$

where $\Omega_{21} = \omega_2 - \omega_1$, $\theta_{21} = \theta_2 - \theta_1$, and the amplitudes $(A_{1,2})$ are treated as complex. The power fluctuates with time depending on the Doppler "beat" frequency (Ω_{21}). If the two spheres have $\omega_1 = \omega_2$, then

$$VV^* = |A_1|^2 + |A_2|^2 + 2\operatorname{Re}\left(A_1 A_2^* e^{j\theta_{21}}\right) \tag{1.96}$$

In this special case the power does not fluctuate with time, but it does depend on the difference in the initial phase angles $\theta_2 - \theta_1$. Let θ_2 and θ_1 be considered as independent random variables with a uniform probability density function (pdf) in the range $(0, 2\pi)$. Imagine an experiment where each realization of VV^* corresponds to the release of two dielectric spheres with initial phases θ_1 and θ_2. The average of VV^* (angle brackets denote ensemble average here) over a large number of such realizations is given by,

$$
\begin{aligned}
\langle VV^* \rangle &= |A_1|^2 + |A_2|^2 + 2\operatorname{Re}\left(A_1 A_2^* \langle e^{j(\theta_2 - \theta_1)} \rangle\right) \tag{1.97a} \\
&= |A_1|^2 + |A_2|^2 \tag{1.97b}
\end{aligned}
$$

Since $\langle e^{j(\theta_2 - \theta_1)} \rangle = \langle \cos(\theta_2 - \theta_1) \rangle + j\langle \sin(\theta_2 - \theta_1) \rangle$, the average of the sinusoidal functions will be zero. This is a special case of "incoherent" summation, which is valid for two spheres with $\omega_1 = \omega_2$ and initial phases uniformly distributed in $(0, 2\pi)$.

Next, consider any one realization of the release of two dielectric spheres but with $\omega_1 \neq \omega_2$. The power will fluctuate with time as in (1.95c). If a long period time average is considered for VV^*, then the time average of $\exp[j(\omega_2 - \omega_1)t]$ over a long period should vanish so that the time-averaged power again equals the "incoherent" sum in (1.97b).

Now, consider the summation over a large number of spheres, as in (1.91). The power (VV^*) will be,

$$
\begin{aligned}
VV^* &= \sum_m A_m e^{-j(\omega_m t + \theta_m)} \sum_l A_l^* e^{j(\omega_l t + \theta_l)} \tag{1.98a} \\
&= \sum_{m=l} |A_m|^2 + \sum_{m \neq l} \sum A_m A_l^* e^{-j(\omega_m - \omega_l)t} e^{-j(\theta_m - \theta_l)} \tag{1.98b}
\end{aligned}
$$

If the initial phases are independent, identically distributed (iid) random variables with uniform pdf in $(0, 2\pi)$, then the ensemble average of $V V^*$ yields,

$$\langle V V^* \rangle = \sum_m |A_m|^2 \tag{1.99}$$

which is termed the incoherent power sum formula.[3] If any one realization is considered, and a long period time average is considered, then this time average of $V V^*$ also reduces to (1.99).

In the forward scattered direction where $\hat{s} = \hat{i}$, note that ω_m and θ_m are both always zero, from (1.90). Thus (1.98b) reduces to,

$$V V^* = \sum_{m=l} |A_m|^2 + \sum_{m \neq l} \sum A_m A_l^* \qquad \cdot \tag{1.100}$$

Such a summation is referred to as a "coherent" summation, and is valid in the forward scattered direction. However, note that in the forward scattered direction it is not possible to distinguish between the incident wave and the scattered wave via measurements. If m goes from 1 to N (i.e. N spheres), then there are N^2 terms in the coherent summation in (1.100), while there are only N terms in the incoherent summation in (1.99). If the spheres are identical, then the coherent power sum is N times the incoherent power sum.

1.9   Moving dielectric sphere under plane wave incidence

1.9.1   Time-correlated bistatic cross section

The concept of a bistatic cross section of a dielectric sphere under plane wave incidence was introduced in Section 1.4, and can be generalized if the sphere is moving with uniform velocity (Ishimaru 1978). Before introducing this subject it is necessary to briefly consider the subject of retarded time.

Consider scattering of an incident plane wave by a dielectric sphere in the Rayleigh limit, as discussed in Section 1.3. Referring to Fig. 1.6b, where the sphere center is located at the origin, the far-field scattered wave is given by,

$$\vec{E}^s(\vec{r}) = \frac{-k_0^2}{4\pi} \frac{(\varepsilon_r - 1)}{(\varepsilon_r + 2)} 3V \left[\hat{r} \times \hat{r} \times \vec{E}^i(O) \right] \frac{e^{-jk_0 r}}{r} \tag{1.101}$$

where $\vec{E}^i(O)$ is the incident plane wave field evaluated at the origin. For simplicity, let ε_r be real. The time-varying scattered field is written as,

$$\vec{E}^s(t) = \mathrm{Re}\left[\vec{E}^s(\vec{r}) e^{j\omega t} \right] \tag{1.102}$$

and the sinusoidal part is essentially $\cos(\omega t - k_0 r) = \cos[\omega(t - r/c)]$, or $\cos \omega(t - t')$, where $t' = r/c$. The time-varying incident field at the origin is simply $\vec{E}^i(t) =$

$\hat{e}_i E_0 \cos \omega t$, since the spatial phase of the incident field at the origin is zero. The time $t' = r/c$ represents the propagation time needed for a "source disturbance" to travel a distance r at the speed of light. Thus, the total phase of the scattered field $\vec{E}^s$ at time t is related to the phase of the incident wave (or "source") at the origin at an earlier (or retarded) time $(t - t')$.

Now, consider the case where the sphere center is displaced from the origin, as shown in Fig. 1.7. The total phase of the far-field scattered wave at time t is related to the total phase of incident plane wave at the origin O$'$ at the retarded time $(t - R/c)$. The total phase of $\vec{E}^i$ at O$'$ is given by the argument of $\cos(\omega t - k\hat{i} \cdot \vec{r}_{OO'})$. If the sphere is moving with a constant velocity $\vec{v}$, then the location of its center (the vector $\vec{r}_{OO'}$) must be considered at the retarded time $(t - R/c)$. The general equation for $\vec{r}_{OO'}$ is $\vec{r}_{OO'} = \vec{r}_{in} + \vec{v}t$, where $\vec{r}_{in}$ is the location of the sphere center at $t = 0$, which for convenience may be chosen as the origin O.

From (1.33), the far-field scattered wave $\vec{E}^s$ at time t_1 is,

$$\vec{E}^s(t_1) = \frac{-k_0^2}{4\pi} \frac{(\varepsilon_r - 1)}{(\varepsilon_r + 2)} 3V \left[\hat{R}_1 \times \hat{R}_1 \times \vec{E}^i(O) \right] e^{-jk_0\hat{i}\cdot\vec{r}_1(t_1')} \frac{e^{-jk_0 R_1}}{R_1} \tag{1.103}$$

where $t_1' = t_1 - R_1/c$, and $\vec{r}_1(t_1')$ is the location of the sphere center at t_1'. Similarly, $\vec{E}^s$ at time t_2 is given as,

$$\vec{E}^s(t_2) = \frac{-k_0^2}{4\pi} \frac{(\varepsilon_r - 1)}{(\varepsilon_r + 2)} 3V \left[\hat{R}_2 \times \hat{R}_2 \times \vec{E}^i(O) \right] e^{-jk_0\hat{i}\cdot\vec{r}_2(t_2')} \frac{e^{-jk_0 R_2}}{R_2} \tag{1.104}$$

where $\vec{r}_2$ is the location of the sphere center at $t_2' = t_2 - R_2/c$, see Fig. 1.14. Note that $\hat{R}_1 \equiv \hat{R}_2 \equiv \hat{s}$ and $R_1 \approx R_2 \approx r$ in the terms R_1^{-1} and R_2^{-1}. However, $R_1 \approx r - \hat{s} \cdot \vec{r}_1$ and $R_2 \approx r - \hat{s} \cdot \vec{r}_2$ in the exponential terms $\exp(-jk_0 R_1)$ and $\exp(-jk_0 R_2)$. Recognizing that,

$$\vec{f}(\hat{s}, \hat{i}) = \frac{-k_0^2}{4\pi} \frac{(\varepsilon_r - 1)}{(\varepsilon_r + 2)} 3V \left[\hat{s} \times \hat{s} \times \vec{E}^i(O) \right] \tag{1.105}$$

is the vector far-field amplitude for the sphere with its center at O,

$$\vec{E}^s(t_1) = \vec{f}(\hat{s}, \hat{i}) \frac{e^{-jk_0 r}}{r} e^{-jk_0(\hat{i} - \hat{s})\cdot\vec{r}_1(t_1')} \tag{1.106a}$$

$$\vec{E}^s(t_2) = \vec{f}(\hat{s}, \hat{i}) \frac{e^{-jk_0 r}}{r} e^{-jk_0(\hat{i} - \hat{s})\cdot\vec{r}_2(t_2')} \tag{1.106b}$$

Thus, the phase of the far-field scattered wave changes as the sphere moves. The time-correlation between $E^s(t_2)$ and $E^s(t_1)$ is defined as,

$$E^s(t_2) E^{s*}(t_1) = \frac{|\vec{f}(\hat{s}, \hat{i})|^2}{r^2} \exp\left\{ -jk_0(\hat{i} - \hat{s}) \cdot \left[\vec{r}_2(t_2') - \vec{r}_1(t_1') \right] \right\} \tag{1.107}$$

Since,

$$\vec{r}_2(t_2') - \vec{r}_1(t_1') = \vec{v}(t_2' - t_1') \tag{1.108a}$$

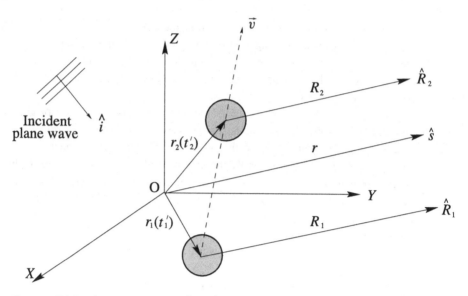

Fig. 1.14. Dielectric sphere located at $\vec{r}_1$ at t_1', and at $\vec{r}_2$ at t_2', where $t_1' = t_1 - (R_1/c)$ and $t_2' = t_2 - (R_2/c)$. The sphere moves at uniform velocity $\vec{v}$ and is illuminated by an incident plane wave. The scattering direction is $\hat{s}$.

and,

$$t_2' - t_1' = t_2 - \frac{R_2}{c} - \left(t_1 - \frac{R_1}{c} \right) \qquad (1.108b)$$

$$\approx t_2 - t_1 \qquad (1.108c)$$

(1.107) reduces to,

$$E^s(t_2)E^{s*}(t_1) = \frac{|\vec{f}(\hat{s}, \hat{i})|^2}{r^2} \exp\left[-jk_0(\hat{i} - \hat{s}) \cdot \vec{v}(t_2 - t_1) \right] \qquad (1.109)$$

Defining $\tau = t_2 - t_1$, the time-correlation between $E^s(t_2)$ and $E^s(t_1)$ becomes,

$$E^s(t_2)E^{s*}(t_1) = \frac{|\vec{f}(\hat{s}, \hat{i})|^2}{r^2} \exp\left[-jk_0(\hat{i} - \hat{s}) \cdot \vec{v}\tau \right] \qquad (1.110)$$

The time-correlated bistatic scattering cross section (Ishimaru 1978) is then defined as,

$$\sigma_{bi}(\hat{s}, \hat{i}, \tau) = \sigma_{bi}(\hat{s}, \hat{i}) \exp\left[-jk_0(\hat{i} - \hat{s}) \cdot \vec{v}\tau \right] \qquad (1.111)$$

In the radar direction $\hat{s} = -\hat{i}$, the time-correlated radar cross section becomes,

$$\sigma_b(-\hat{i}, \hat{i}, \tau) = \sigma_b(-\hat{i}, \hat{i}) \exp\left(-j2k_0\hat{i} \cdot \vec{v}\tau \right) \qquad (1.112a)$$

$$= \sigma_b(-\hat{i}, \hat{i}) \exp\left(-j\frac{4\pi}{\lambda}\hat{i} \cdot \vec{v}\tau \right) \qquad (1.112b)$$

Doppler radars are configured to measure the argument of $\sigma_b(-\hat{i}, \hat{i}, \tau)$ from which the particle velocity component $\hat{i} \cdot \vec{v}$ is obtained. In particular, pulsed Doppler radars measure the argument of $\sigma_b(-\hat{i}, \hat{i}, \tau)$ when $\tau = T_s$, where T_s is the pulse repetition time. Of course, the radar cross section $\sigma(-\hat{i}, \hat{i})$ is also obtained.

The Fourier transform of $\sigma_{bi}(\hat{s}, \hat{i}, \tau)$ is related to the bistatic Doppler power spectrum,

$$S_\sigma(\hat{s}, \hat{i}, \omega'') = \int_{-\infty}^{\infty} \sigma_{bi}(\hat{s}, \hat{i}, \tau) e^{-j\omega''\tau} \, d\tau \qquad (1.113a)$$

$$= \sigma_b(\hat{s}, \hat{i}) \int_{-\infty}^{\infty} \exp\{-j[\omega'' + k_0(\hat{i} - \hat{s}) \cdot \vec{v}]\tau\} \, d\tau \qquad (1.113b)$$

$$= \sigma_b(\hat{s}, \hat{i}) \delta[\omega'' + k_0(\hat{i} - \hat{s}) \cdot \vec{v}] \qquad (1.113c)$$

where δ is the Dirac delta function. The bistatic Doppler spectrum of a single sphere moving with constant velocity $\vec{v}$ is a discrete line at angular frequency $\omega'' = -k_0(\hat{i} - \hat{s}) \cdot \vec{v}$, or $f'' = -(\hat{i} - \hat{s}) \cdot \vec{v}/\lambda$, which is identical to (1.89) except for the negative sign. Hence, f_D defined in (1.89) is $-f''$. Recall that a positive value for f_D will result in the frequency of the induced current in the bistatic antenna being less than the incident wave frequency. The discrete line spectrum corresponding to this case is illustrated in Fig. 1.15a where the line for a single sphere is displaced from the "carrier" or incident wave frequency (f_c). Figure 1.15b shows the line spectrum corresponding to (1.113c), which is the "baseband" version with the "carrier" suppressed. When $\hat{s} = -\hat{i}$, which is in the radar or back scattering direction, $f'' = -2\hat{i} \cdot \vec{v}/\lambda$. A particle which is moving "away" from the radar (i.e. $\hat{i} \cdot \hat{v} > 0$) will have $f'' < 0$, while a particle moving "toward" the radar (i.e. $\hat{i} \cdot \hat{v} < 0$) will have $f'' > 0$, as illustrated in Fig. 1.15b. It should be clear that stationary targets contribute a discrete line at $\omega'' = 0$. Thus, moving targets (with $\hat{i} \cdot \hat{v} \neq 0$) can be distinguished from stationary ones even though their radar cross sections may be several orders of magnitude smaller than for the stationary target. Henceforth, the context will make clear whether the "baseband" Doppler spectrum is referred to, as given in (1.113) and Fig. 1.15b, or whether the spectrum relative to the "carrier" frequency is referred to, as in Fig. 1.15a; see also (1.84) or (1.88d).

1.9.2   Block diagram of a pulsed Doppler radar

A simple block diagram of a pulsed Doppler radar is given in Fig. 1.16, where the transmitting and receiving sub-systems are shown separately for convenience.

The transmitted waveform can be expressed as $s_{tr}(t) = U_{tr}(t) \exp(j2\pi f_0 t)$, where $U_{tr}(t)$ is known as the envelope or modulation function and f_0 is the "carrier" frequency. While a wide variety of modulation types are used in radars, the simplest is the train of rectangular pulses whose pulsewidth (T_0) is typically a few microseconds and whose pulse repetition time (T_s) is a few milliseconds. As mentioned earlier, the phase of the scattered signal from a moving target changes, and to measure this phase change with time (or Doppler frequency), the scattered signal phase must be compared against the

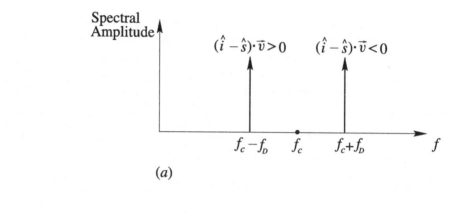

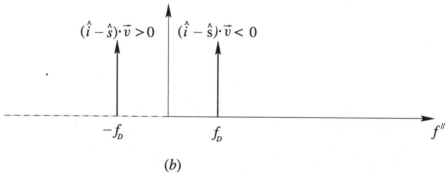

Fig. 1.15. Discrete line spectrum for single sphere moving with uniform velocity $\vec{v}$ as illustrated in Fig. 1.14: (a) spectral lines shown relative to incident wave "carrier" frequency f_c, and (b) "baseband" version corresponding to (1.113c).

phase of the transmitted signal. In coherent systems, the phase of the transmitted signal is extremely stable from pulse-to-pulse (this property is known as phase coherency). Typically, two oscillators, a stable local oscillator (STALO) and a coherent oscillator (COHO), are used as very pure continuous wave (CW) signal sources. The modulator controls the transmitter in pulsed systems, enabling the generation of a train of pulses of specified pulsewidth and pulse repetition time (PRT). The final power amplifier brings the transmitted power to its desired level (pulse powers are typically in the range of hundreds of kilowatts). The final power amplifier in coherent systems is, generally, in the class of linear beam-microwave amplifying tubes, such as Klystrons or traveling wave tubes (TWT). Figure 1.16a is adapted from Skolnik (1980) and shows a generic radar transmitter block diagram. The final "carrier" frequency (f_0) is usually the sum of the STALO (f_s) and COHO (f_c) frequencies. Another class of coherent-on-receive systems is built around transmitters which are power oscillators (e.g. magnetrons), whose phase is unpredictable (or random) and must be measured and tracked from pulse-to-pulse. Thus, the phase of the latest transmitted pulse sets the system phase reference (against which the scattered signal phase is compared).

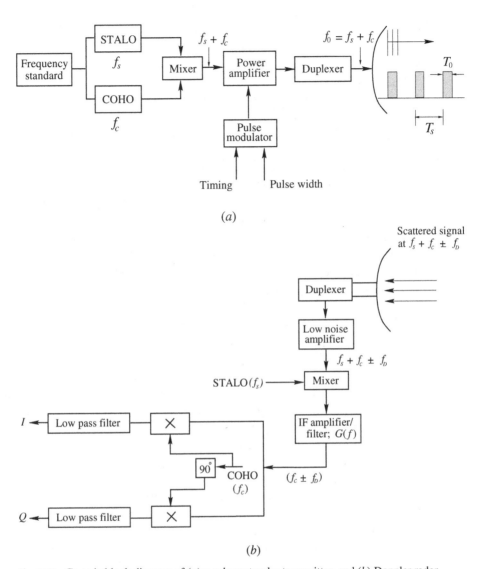

Fig. 1.16. Generic block diagram of (*a*) a coherent radar transmitter, and (*b*) Doppler radar receiver. The coherent oscillator (COHO) frequency, f_c, is the intermediate frequency used in the receiver. The power amplifier is in the class of linear beam-microwave amplifying tubes (klystron or traveling wave tube). STALO, stable local oscillator; f_D, Doppler frequency; IF, intermediate frequency; $G(f)$, frequency-response function of the IF amplifier/filter; I and Q, in-phase and quadrature phase component, respectively.

A simplified block diagram of a Doppler radar receiver is shown in Fig. 1.16*b*. The scattered signal from a moving point target is essentially a scaled replica of the transmitted signal except for a range time delay (t_0) and Doppler frequency shift (f_D), i.e. $s_r(t) = U_{tr}(t - t_0) \exp[j2\pi(f_0 \pm f_D)(t - t_0)]$. In the receiver, the STALO frequency functions as the local oscillator, and the COHO frequency, which is at the intermediate frequency (IF) of the receiver, forms the reference for the phase detector (or in-phase /quadrature phase (I/Q) demodulation reference).

The received signal $s_r(t)$ is first amplified by the low noise amplifier (LNA), and then mixed with the STALO frequency to produce the IF signal just as in a conventional superheterodyne receiver. Recall that the carrier frequency $f_0 = f_s + f_c$. The output signal of the first mixer is, thus, $s_r(t) = U_{tr}(t - t_0) \exp[j2\pi(f_{IF} \pm f_D)(t - t_0)]$. The main amplification and filtering of the received signal is done in the IF portion of the receiver. The frequency response function of the IF part of the receiver is denoted by $G(f)$, which is generally "matched" to the Fourier transform of the complex envelope of $s_r(t)$ or $U_{tr}(t - t_0)$. A useful approximation for pulsed radars is that the IF receiver bandwidth is equal to the reciprocal of the transmitted pulsewidth (T_0^{-1}). For example, a typical pulsewidth of 1 μs will correspond to an IF bandwidth of 1 MHz. If the IF bandwidth is much larger than T_0^{-1}, then additional noise is introduced which lowers the output signal-to-noise (SNR) ratio. If the IF bandwidth is much narrower than T_0^{-1}, then the noise is reduced along with a considerable part of the signal energy. Thus, there is an optimum IF bandwidth for which the output SNR is a maximum (Skolnik 1980).

The output of the IF amplifier is fed into the I/Q demodulator whose function is to produce the in-phase (I) and quadrature phase (Q) components of the envelope of the received signal, which is $U_{tr}(t - t_0) \exp[j2\pi(\pm f_D)(t - t_0)]$. As illustrated in Fig. 1.16b, the COHO reference to the I/Q demodulator is shifted by 90°, and the original and phase-shifted references are compared with $s_r(t)$ and low-pass filtered to form the I and Q "video" outputs. Thus, $I = U_{tr}(t - t_0) \cos[2\pi(f_D)(t - t_0)]$ and $Q = \pm U_{tr}(t - t_0) \sin[2\pi f_D(t - t_0)]$. The envelope of the received signal is $V(t) = I(t) + jQ(t)$. The time-correlated radar cross section $\sigma_b(-\hat{i}, \hat{i}, \tau)$ in (1.112b) is related to the autocorrelation of $V(t)$, i.e. to $[V^*(t)V(t + \tau)]$ with $\tau = mT_s$ (T_s is the PRT); while the radar cross section, $\sigma_b(-\hat{i}, \hat{i})$ is related to $V^*(t)V(t)$. The Fourier transform of the autocorrelation function of $V(t)$ is related to the Doppler power spectrum S_σ in (1.113). Details of Doppler radar theory for random media (such as a precipitation medium) and signal processing methods will be treated in Chapter 5.

1.10   Coherent forward scattering by a slab of dielectric spheres

The propagation of an incident plane wave through a random medium consisting of dielectric particles can be understood by explicit consideration of the following configuration.[4] Consider an incident plane wave $\vec{E}^i = \hat{e}_i E_0 \exp(-jk_0 z)$ propagating along the positive Z-axis, as shown in Fig. 1.17a. A "slab" consisting of identical dielectric spheres between $z = 0$ and $z = d$ is introduced into the path of the incident wave. The presence of the slab perturbs the incident wave and the objective here is to calculate the total electric field at the point P(0, 0, z_0), by superposing the incident wave at P with the scattered wave at P from each sphere. The field "exciting" a particular sphere in the slab (shown at $\vec{r}$ in Fig. 1.17a) is assumed to be the unperturbed incident

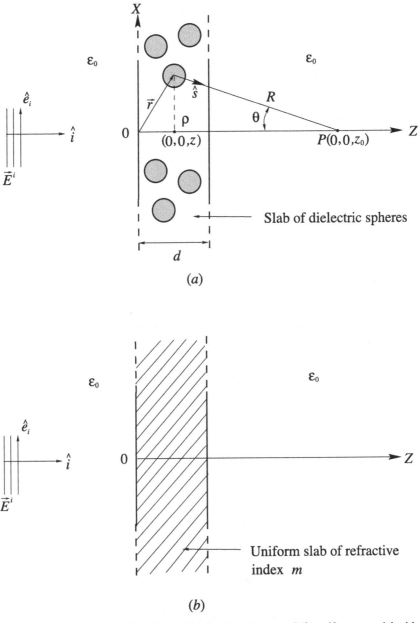

Fig. 1.17. Slab consisting of (a) identical dielectric spheres, and (b) uniform material with refractive index m.

field. This approximation is valid for a sparse distribution of spheres, i.e. the fractional volume concentration (c) of the spheres is very small ($c \ll 1.0$). The derivation here follows Jackson (1975; Section 9.14). The far-field scattered wave at P due to a single sphere at $\vec{r} \equiv (\rho, \phi, z)$ in cylindrical coordinates is given by (1.33). Using $\vec{f}(\hat{s}, \hat{i})$ in

(1.32a), the scattered wave at P due to a single sphere at $\vec{r}$ equals,

$$\vec{f}(\hat{s}, \hat{i}) e^{-jk_0\hat{i}\cdot\hat{r}} \frac{e^{-jk_0R}}{R}, \quad \text{or} \tag{1.114a}$$

$$\vec{f}(\hat{s}, \hat{i}) e^{-jk_0z} \frac{e^{-jk_0R}}{R} \tag{1.114b}$$

Let the number of spheres per unit volume be n. A volume element dV will contain $n\,dV$ spheres. Let dV be the volume element centered at $\vec{r} \equiv (\rho, \phi, z)$. Using the continuum approximation, the electric field $d\vec{E}^s$ due to $n\,dV$ spheres in the volume dV is given by,

$$d\vec{E}^s(P) = \vec{f}(\hat{s}, \hat{i}) \frac{e^{-jk_0R}}{R} e^{-jk_0z} n\,dV \tag{1.115}$$

The total scattered field at P is by superposition,

$$\vec{E}^s(P) = n \int_{\text{slab volume}} \frac{e^{-jk_0R}}{R} \vec{f}(\hat{s}, \hat{i}) e^{-jk_0z}\,dV \tag{1.116}$$

where $dV = \rho\,d\rho\,d\phi\,dz$, $R^2 = \rho^2 + (z_0 - z)^2$, and $R\,dR = \rho\,d\rho$ for a given z. Then $\vec{E}^s(P)$ can be written as,

$$\vec{E}^s(P) = n \int_{\phi=0}^{2\pi} d\phi \int_{z=0}^{d} e^{-jk_0z}\,dz \int_{z_0-z}^{\infty} \frac{e^{-jk_0R}}{R}(R\,dR)\vec{f}(\hat{s}, \hat{i}) \tag{1.117}$$

Note, from Fig. 1.17 that $\mu \equiv \cos\theta = \hat{s}\cdot\hat{i} = (z_0 - z)/R$, so that,

$$\frac{d}{dR} \equiv \frac{d}{d\mu}\frac{d\mu}{dR} = \frac{d}{d\mu}\left[\frac{-(z_0 - z)}{R^2}\right] \tag{1.118}$$

Integrating by parts,

$$\int_{z_0-z}^{\infty} e^{-jk_0R}\vec{f}(\hat{s}, \hat{i})\,dR = \vec{f}(\hat{s}, \hat{i})\frac{e^{-jk_0R}}{(-jk_0)}\bigg|_{R=(z_0-z)}^{R=\infty}$$

$$- \int_{(z_0-z)}^{\infty} \frac{d}{dR}\left[\vec{f}(\hat{s}, \hat{i})\right]\left(\frac{e^{-jk_0R}}{-jk_0}\right)dR \tag{1.119a}$$

$$= \vec{f}(\hat{s}, \hat{i})\frac{e^{-jk_0R}}{(-jk_0)}\bigg|_{R=(z_0-z)}^{\infty}$$

$$- \frac{1}{jk_0}\int_{z_0-z}^{\infty} \frac{(z_0 - z)}{R^2}\frac{d}{d\mu}\left[\vec{f}(\hat{s}, \hat{i})\right]e^{-jk_0R}\,dR \tag{1.119b}$$

Provided $d\vec{f}(\hat{s}, \hat{i})/d\mu$ is well-behaved, the magnitude of the integrand $[(z_0 - z)/k_0R^2]$ $\exp(-jk_0R)$ in the second integral is of the order $1/k_0(z_0 - z)$ times the integrand $\exp(-jk_0R)$ of the first integral in (1.119b). Since the assumption is that the point P is many wavelengths from the slab, the second integral in (1.119b) may be neglected. Hence,

$$\int_{z_0-z}^{\infty} e^{jk_0R} \vec{f}(\hat{s}, \hat{i})\, dR = \vec{f}(\hat{s}, \hat{i}) \frac{e^{-jk_0R}}{(-jk_0)}\Big|_{R=(z_0-z)}^{\infty} \tag{1.120a}$$

$$\approx \vec{f}(\hat{i}, \hat{i}) \frac{e^{-jk_0(z_0-z)}}{(jk_0)} \tag{1.120b}$$

where the oscillating contribution from the upper limit $(R \to \infty)$ is neglected (e.g. it is plausible if n tends to zero as $\rho \to \infty$). From (1.117),

$$\vec{E}^s(P) = n \int_0^{2\pi} d\phi \int_0^d e^{-jk_0z}\, dz \frac{1}{jk_0} \vec{f}(\hat{i}, \hat{i}) e^{-jk_0(z_0-z)} \tag{1.121a}$$

$$= \frac{2\pi nd}{jk_0} \vec{f}(\hat{i}, \hat{i}) e^{-jk_0z_0} \tag{1.121b}$$

The total field at P is,

$$\vec{E}(P) = \vec{E}^i(P) + \vec{E}^s(P) \tag{1.122a}$$

$$= \hat{e}_i E_0 e^{-jk_0z_0} + \frac{2\pi nd}{jk_0} \vec{f}(\hat{i}, \hat{i}) e^{-jk_0z_0} \tag{1.122b}$$

Note that $\vec{f}(\hat{i}, \hat{i})$ is the vector scattering amplitude in the forward scattered direction $\hat{s} = \hat{i}$. From (1.122b), the component of $\vec{E}(P)$ parallel to the incident linear polarization state $(\hat{e}_i)$ with $E_0 = 1 \text{ V m}^{-1}$, is given by,

$$\hat{e}_i \cdot \vec{E}(P) = e^{-jk_0z_0}\left[1 + \frac{2\pi nd}{jk_0} \hat{e}_i \cdot \vec{f}(\hat{i}, \hat{i})\right] \tag{1.123}$$

Equation (1.123) is correct to first-order in the slab thickness (d). The equation also shows that the result of coherent summation in the forward scattered direction, i.e. $|\vec{E}^s|^2$ from (1.121b), is proportional to n^2.

1.10.1 Concept of an effective propagation constant

From the Maxwell-Garnet formula in (1.69), the slab of discrete dielectric spheres in the Rayleigh scattering regime is equivalent to a uniform permittivity of ε_{eff} (or k_{eff} in (1.70a)). This concept can be extended to non-Rayleigh scattering with the proviso that ε_{eff} (or k_{eff}) is no longer given by (1.69, 1.70a). The slab configuration in Fig. 1.17b, when excited by a uniform plane wave at normal incidence, is a standard boundary value problem involving the reflection and transmission of plane waves. An exact solution for the total electric field in the region $z \gg d$ is given in Barber and Hill (1990) as,

$$\vec{E}(z \geq d) = \hat{e}_i \left[\frac{-4me^{-jk_0d(m-1)}}{(m^2+1)(e^{-j2mk_0d}-1) - 2m(e^{-j2mk_0d}+1)}\right] e^{-jk_0z} \tag{1.124}$$

where $m = \sqrt{\varepsilon_{\text{eff}}/\varepsilon_0}$ is the refractive index of the slab. To first-order in d, $\vec{E}(P)$ can be approximated using $\exp(-jk_0d) \approx 1 - jk_0d$, which gives,

$$\vec{E}(P) = \hat{e}_i \left\{ \frac{-4m\,[1 - jk_0d(m-1)]}{(m^2+1)(-j2mk_0d) - 2m(2 - j2mk_0d)} \right\} e^{-jk_0z_0} \tag{1.125a}$$

$$\approx \hat{e}_i\,[1 - jk_0d(m-1)]\,e^{-jk_0z_0} \tag{1.125b}$$

The above approximation, while correct to first-order in slab thickness (d), assumes that $|m-1|$ is small compared with unity (Jackson 1975). Comparing (1.123) with (1.125b) gives,

$$m = 1 + \frac{2\pi n}{k_0^2}\,\hat{e}_i \cdot \vec{f}(\hat{i},\hat{i}) \tag{1.126}$$

Since the effective propagation constant of the slab is $k_{\text{eff}} = k_0 m$,

$$\frac{k_{\text{eff}}}{k_0} = 1 + \frac{2\pi n}{k_0^2}\,\hat{e}_i \cdot \vec{f}(\hat{i},\hat{i}) \tag{1.127}$$

which may be compared with (1.70b), which is valid in the Rayleigh limit with $c \ll 1$. While (1.127) is derived for identical spheres with n spheres per unit volume, it can be generalized to a size distribution of spheres defined by $N(D)$, where $N(D)\,dD$ is the number of spheres per unit volume with sizes in the interval D to $D + dD$. Essentially, the product $n\hat{e}_i \cdot \vec{f}(\hat{i},\hat{i})$ in (1.127) is replaced by,

$$n\hat{e}_i \cdot \vec{f}(\hat{i},\hat{i}) \rightarrow \int_D \hat{e}_i \cdot \vec{f}(\hat{i},\hat{i}; D)N(D)\,dD \tag{1.128}$$

Several additional comments are relevant here. The effective wave number can, in general, be complex, $k_{\text{eff}} = k_{\text{eff}}^{\text{re}} - jk_{\text{eff}}^{\text{im}}$, where,

$$\frac{k_{\text{eff}}^{\text{re}}}{k_0} = 1 + \frac{2\pi n}{k_0^2}\,\text{Re}\big[\hat{e}_i \cdot \vec{f}(\hat{i},\hat{i})\big] \tag{1.129a}$$

$$\frac{k_{\text{eff}}^{\text{im}}}{k_0} = \frac{-2\pi n}{k_0^2}\,\text{Im}\left\{\hat{e}_i \cdot \vec{f}(\hat{i},\hat{i})\right\} \tag{1.129b}$$

$$= \frac{n\sigma_{\text{ext}}}{2k_0} = \frac{3}{8}\frac{cQ_{\text{ext}}}{k_0a} \tag{1.129c}$$

The optical theorem stated in (1.58) has been used to arrive at (1.129c), where $c = (4/3)n\pi a^3$ and Q_{ext} is the normalized extinction cross section $(\sigma_{\text{ext}}/\pi a^2)$, with a being the sphere radius. Using the generalization for a size distribution $N(D)$ given by (1.128),

$$\frac{k_{\text{eff}}^{\text{im}}}{k_0} = \frac{1}{2k_0}\int_D \sigma_{\text{ext}}(D)N(D)\,dD \tag{1.130}$$

The unit for $k_{\text{eff}}^{\text{im}}$ is inverse length. A plane wave propagating along the positive Z-axis in a medium with k_{eff} can be written as,

$$\vec{E} = \hat{e}E_0 \exp(-jk_{\text{eff}}z) \tag{1.131a}$$

$$= \hat{e}E_0 \exp(-jk_{\text{eff}}^{\text{re}}z) \exp(-k_{\text{eff}}^{\text{im}}z) \tag{1.131b}$$

Since the medium is composed of discrete spheres that scatter energy, the electric field above is the coherent field, which suffers attenuation as it propagates in the medium. The specific attenuation (or attenuation per unit distance) is derived from,

$$\frac{|\vec{E}(z)|}{|\vec{E}(0)|} = \exp(-k_{\text{eff}}^{\text{im}}z) \tag{1.132a}$$

or,

$$20\log_{10}\frac{|\vec{E}(z)|}{|\vec{E}(0)|} = -8.686k_{\text{eff}}^{\text{im}}z \tag{1.132b}$$

The specific attenuation (A) in units of decibels per kilometer is defined as,

$$A = 8.686k_{\text{eff}}^{\text{im}}(1000) \tag{1.133a}$$

$$= (8.686 \times 10^3)\frac{n\sigma_{\text{ext}}}{2} \tag{1.133b}$$

$$= (4.343 \times 10^3)n\sigma_{\text{ext}} \tag{1.133c}$$

$$= 4.343 \times 10^3 \int_D \sigma_{\text{ext}}(D)N(D)\,dD \quad \text{dB km}^{-1} \tag{1.133d}$$

where the units for σ_{ext} are m^2 and $N(D)\,dD$ are m^{-3}. Note that the formula for A applies to attenuation of the coherent field amplitude as well as to the coherent power which is proportional to $|E(z)|^2$, since $20\log|\vec{E}| = 10\log|\vec{E}|^2$.

Notes

1. The introductory material in Sections 1.1–1.5 is based on Van Bladel (1985), King and Prasad (1986), and Ishimaru (1991).

2. The Clausius–Mosotti equation is derived, for example, in Jackson (1975, see Section 4.5). It gives the background for understanding dielectric mixture formulas (e.g. Maxwell-Garnet formula). This topic is included here, as polarimetric radar signatures are strongly weighted by the effective dielectric constant of hydrometeor particles. There is extensive literature on dielectric mixing formulas, for example, see the review by Sihvola and Kong (1988).

3. Portions of material in this section are adapted from lecture notes provided by Professor Ramesh Srivastava of the University of Chicago. See also, Atlas (1964).

4. There is extensive literature on coherent wave propagation in a random distribution of discrete scatterers (see, for example, Chapter 6 of Tsang et al. 1985). Uzunoglu and Evans (1978) used

the theory developed by Twersky (1978) to generalize eq. (1.127) as follows,

$$
\frac{k_{\text{eff}}^2}{k_0^2} = \left[1 + \frac{2\pi n}{k_0^2} \langle f(\hat{i}, \hat{i}) \rangle \right]^2 - \left[\frac{2\pi n}{k_0^2} \langle f(-\hat{i}, \hat{i}) \rangle \right]^2
$$

where angle brackets denote ensemble-averaging [see Chapter 3; eqs. (3.160, 3.161). The above formula includes multiple scattering, and first appeared in Waterman and Truell (1961). It reduces to (1.127) if,

$$
\left| 1 + \frac{2\pi n}{k_0^2} \langle f(\hat{i}, \hat{i}) \rangle \right| \gg \left| \frac{2\pi n}{k_0^2} \langle f(-\hat{i}, \hat{i}) \rangle \right|
$$

The above inequality was numerically tested for rain at frequencies ranging from 4 to 100 GHz and was found to be satisfied (Uzunoglu and Evans 1978).

2 Scattering matrix

The scattering matrix of spheroids in the Rayleigh–Gans limit is the principal topic of this chapter. While the assumption of spheroidal shape is somewhat simplistic considering the wide distribution of shapes of natural hydrometeors, it is quite remarkable that important polarimetric radar observations of precipitation can, to a large degree, be explained using the spheroidal model and Rayleigh–Gans scattering. It will also become evident that important concepts gained from Rayleigh–Gans scattering (e.g. the dependence of elevation angle, particle orientation effects), can also be extended to Mie scattering. The scattering matrix is formulated in forward scatter alignment (FSA) and back scatter alignment (BSA) conventions, following van Zyl and Ulaby (1990). Sufficient detail is provided for those readers without prior exposure to this subject matter. Simple hydrometeor orientation models are used to illustrate the behavior of linear polarization radar observables such as differential reflectivity and linear depolarization ratio.

For completeness, the Mie solution for scattering by spherical particles is formulated as a boundary value problem, with vector spherical harmonics and the multipole expansion of the electric field being covered in Appendix 2. Expressions for radar, scattering, and extinction cross sections are given in terms of Mie coefficients. This chapter concludes with a very brief discussion of numerical methods for scattering by non-spherical particles at higher frequencies, and the T-matrix method is briefly described in Appendix 3.

2.1 The forward scatter and back scatter alignment conventions

2.1.1 FSA convention

The scattering matrix is a fundamental descriptor of scattering of an incident plane wave by a single particle. The scattered field components in the far-field are related to the incident field components via a 2×2 amplitude scattering matrix. Since field components are involved, the unit vectors must be carefully defined. A Cartesian coordinate system (XYZ) is introduced, with the origin (O) at the center of the particle. The direction of incidence of the plane wave is specified by the spherical angles θ_i and ϕ_i, or by the unit vector $\hat{k}_i$, as shown in Fig. 2.1a. The unit vectors for the spherical coordinate system form the triplet $\hat{k}_i, \hat{\theta}_i, \hat{\phi}_i$, with $\hat{k}_i = \hat{\theta}_i \times \hat{\phi}_i$. The horizontal ($\hat{h}_i$) and vertical ($\hat{v}_i$) unit vectors are defined as $\hat{v}_i = \hat{\theta}_i$ and $\hat{h}_i = \hat{\phi}_i$, so that the triplet of unit

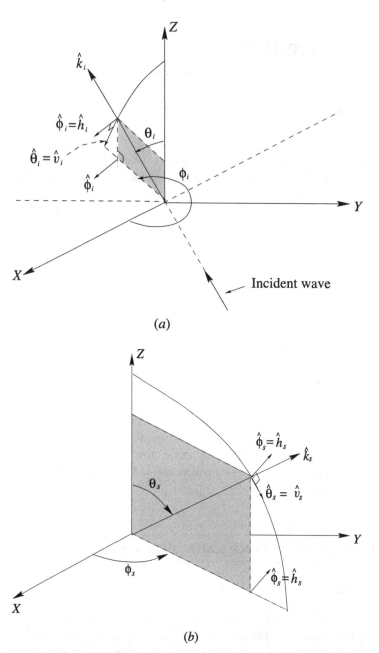

Fig. 2.1. (*a*) Incident wave direction, specified by $\hat{k}_i$, with $\hat{k}_i = \hat{\theta}_i \times \hat{\phi}_i$; and (*b*) the scattered wave direction in the forward scatter alignment (FSA), specified by $\hat{k}_s$, with $\hat{k}_s = \hat{\theta}_s \times \hat{\phi}_s$. For (*a*): the shaded plane is the plane $\phi = \phi_i$; the unit horizontal vector is $\hat{h}_i = \hat{\phi}_i$, and the unit vertical vector is $\hat{v}_i = \hat{\theta}_i$; the triplet of unit vectors for the incident wave is $(\hat{k}_i, \hat{v}_i, \hat{h}_i)$, with $\hat{k}_i = \hat{v}_i \times \hat{h}_i$. For (*b*): the shaded plane is the plane $\phi = \phi_s$; the unit horizontal vector is $\hat{h}_s = \hat{\phi}_s$, and the unit vertical vector is $\hat{v}_s = \hat{\theta}_s$. The triplet of unit vectors is $(\hat{k}_s, \hat{v}_s, \hat{h}_s)$, with $\hat{k}_s = \hat{v}_s \times \hat{h}_s$.

vectors is $(\hat{k}_i, \hat{v}_i, \hat{h}_i)$ with $\hat{k}_i = \hat{v}_i \times \hat{h}_i$. The incident plane wave has the form,

$$\vec{E}^i = \hat{e}_i E_0 \exp(-jk_0 \hat{i} \cdot \vec{r}_i)$$ (2.1)

where $\vec{r}_i$ is the vector from the origin to a point on the plane of constant phase, which is the plane orthogonal to the direction of incidence $(\hat{i})$. Note that k_0 is the wave number of free space $(k_0 = 2\pi/\lambda)$, λ being the wavelength. At the origin, $\vec{E}^i(O) = \hat{e}_i E_0$, which is the plane wave amplitude (the spatial phase reference is at the origin). This vector amplitude is resolved into components as follows,

$$\vec{E}^i(O) = \hat{e}_i E_0 = E_h^i \hat{h}_i + E_v^i \hat{v}_i$$ (2.2)

where $E_h^i = \hat{h}_i \cdot \hat{e}_i E_0$ and $E_v^i = \hat{v}_i \cdot \hat{e}_i E_0$. The vertical and horizontal directions are useful descriptors when $\theta_i \approx 90°$ and the XY-plane is considered to be the Earth's surface. In this case, $\hat{v}_i$ is vertical (or perpendicular) to the Earth's surface and $\hat{h}_i$ is horizontal (or parallel) to this surface. The incident field amplitude can be written, in terms of its two components as a 2×1 column matrix (or Jones vector),

$$\vec{E}^i(O) = \hat{e}_i E_0 \equiv E_0 \begin{bmatrix} \hat{e}_i \cdot \hat{h}_i \\ \hat{e}_i \cdot \hat{v}_i \end{bmatrix} = \begin{bmatrix} E_h^i \\ E_v^i \end{bmatrix}$$ (2.3)

Consider next the far-field scattered wave and let its direction be given by the spherical angles θ_s and ϕ_s, or by the unit vector $\hat{k}_s$, as shown in Fig. 2.1b. The unit vectors for the spherical coordinate system form the triplet $(\hat{k}_s, \hat{\theta}_s, \hat{\phi}_s)$, with $\hat{k}_s = \hat{\theta}_s \times \hat{\phi}_s$. The horizontal and vertical unit vectors in the FSA convention are defined as $\hat{h}_s = \hat{\phi}_s$ and $\hat{v}_s = \hat{\theta}_s$, so that the triplet becomes $(\hat{k}_s, \hat{v}_s, \hat{h}_s)$, with $\hat{k}_s = \hat{v}_s \times \hat{h}_s$. The far-field scattered wave is expressed as (see 1.24a),

$$\vec{E}^s = \frac{\vec{f}(\hat{s}, \hat{i})}{r} \exp(-jk_0 r)$$ (2.4)

The scattered field $(\vec{E}^s)$ is resolved into components as,

$$\vec{E}^s = E_h^s \hat{h}_s + E_v^s \hat{v}_s$$ (2.5)

In (2.4), note that $\hat{s} \equiv \hat{k}_s$ and $\hat{i} \equiv \hat{k}_i$, while r is the magnitude of the position vector $(\vec{r})$. The scattering matrix (or Jones matrix; Jones 1941) in the FSA is defined as,

$$\begin{bmatrix} E_h^s \\ E_v^s \end{bmatrix} = \frac{e^{-jk_0 r}}{r} \begin{bmatrix} S_{hh} & S_{hv} \\ S_{vh} & S_{vv} \end{bmatrix}_{\text{FSA}} \begin{bmatrix} E_h^i \\ E_v^i \end{bmatrix}$$ (2.6a)

The 2×2 scattering matrix is also defined using the symbol $\mathbf{S}_{\text{FSA}}$, where,

$$\mathbf{S}_{\text{FSA}} \equiv \begin{bmatrix} S_{hh} & S_{hv} \\ S_{vh} & S_{vv} \end{bmatrix}_{\text{FSA}}$$ (2.6b)

Equation (2.6a) can be written in compact form as,

$$\mathbf{E}^s = \mathbf{S}_{\text{FSA}} \mathbf{E}^i \left(\frac{e^{-jk_0 r}}{r} \right)$$ (2.7)

To summarize the notation, the arrow on $\vec{E}$ refers to its vector form, while a bold typeface refers to the corresponding 2×1 matrix (or column "vector"). At times it is convenient to drop the spherical wave factor $\exp(-jk_0 r)/r$, so that (2.7) reduces to,

$$\mathbf{E}^s = \mathbf{S}_{\text{FSA}} \mathbf{E}^i \tag{2.8}$$

The context should make it reasonably clear when the spherical wave factor is dropped and when not. Note that the particle is generally located at the origin of the XYZ system, e.g. in the case of a sphere the center is usually at the origin. The elements of $\mathbf{S}_{\text{FSA}}$ are, in

general, functions of both (θ_i, ϕ_i) and (θ_s, ϕ_s). The FSA convention is essentially based on the unit vectors of the spherical coordinate system, which in turn can be related to the unit vectors $(\hat{x}, \hat{y}, \hat{z})$ and incidence angles (θ_i, ϕ_i), or to the scattered angles (θ_s, ϕ_s). For completeness,

$$\hat{\phi}_{i,s} = \hat{h}_{i,s} = -\sin\phi_{i,s}\hat{x} + \cos\phi_{i,s}\hat{y} \tag{2.9a}$$

$$\hat{\theta}_{i,s} = \hat{v}_{i,s} = \cos\phi_{i,s}\cos\theta_{i,s}\hat{x} + \sin\phi_{i,s}\cos\theta_{i,s}\hat{y} - \sin\theta_{i,s}\hat{z} \tag{2.9b}$$

$$\hat{k}_{i,s} = \cos\phi_{i,s}\sin\theta_{i,s}\hat{x} + \sin\phi_{i,s}\sin\theta_{i,s}\hat{y} + \cos\theta_{i,s}\hat{z} \tag{2.9c}$$

In optics, FSA is almost exclusively used (e.g. Azzam and Bashara 1989). The FSA is a "wave-oriented" convention, since the unit vectors $\hat{v}_s$ and $\hat{h}_s$ are always defined with $\hat{k}_s$ aligned with the direction of propagation of the scattered wave. In the back scattered direction, $\hat{k}_s = -\hat{k}_i$, $\theta_s = \pi - \theta_i$, and $\phi_s = \phi_i + \pi$, resulting in $\hat{v}_s = \hat{v}_i$ and $\hat{h}_s = -\hat{h}_i$.

2.1.2   BSA convention

For radar applications the BSA convention is generally used to resolve the components of the scattered field. As will be shown in the next section, this convention leads to a symmetric scattering matrix for particles satisfying reciprocity. Referring to Fig. 2.2 the unit vectors $\hat{h}_r$ and $\hat{v}_r$ are defined as $\hat{h}_r = -\hat{h}_s$ and $\hat{v}_r = \hat{v}_s$. Thus, corresponding to the FSA triplet $(\hat{k}_s, \hat{v}_s, \hat{h}_s)$, with $\hat{k}_s = \hat{v}_s \times \hat{h}_s$, the BSA triplet is $(\hat{k}_r, \hat{v}_r, \hat{h}_r)$, with $\hat{k}_r = -\hat{k}_s = \hat{v}_r \times \hat{h}_r$. The scattered wave can be expressed as $\vec{E}^s = E_h^s \hat{h}_s + E_v^s \hat{v}_s$ in the FSA, or as $\vec{E}^r = E_h^r \hat{h}_r + E_v^r \hat{v}_r$ in BSA. The scattered wave components (E_h^s, E_v^s) in FSA are easily related to the "reflected" wave components (E_h^r, E_v^r) in BSA as,

$$\begin{bmatrix} E_h^r \\ E_v^r \end{bmatrix} = \begin{bmatrix} -1 & 0 \\ 0 & 1 \end{bmatrix} \begin{bmatrix} E_h^s \\ E_v^s \end{bmatrix} \tag{2.10}$$

Substitution of (2.10) in (2.6a) results in,

$$\begin{bmatrix} E_h^r \\ E_v^r \end{bmatrix} = \frac{e^{-jk_0 r}}{r} \begin{bmatrix} -1 & 0 \\ 0 & 1 \end{bmatrix} \begin{bmatrix} S_{hh} & S_{hv} \\ S_{vh} & S_{vv} \end{bmatrix}_{\text{FSA}} \begin{bmatrix} E_h^i \\ E_v^i \end{bmatrix} \tag{2.11}$$

The scattering matrix (or Sinclair matrix; Sinclair 1950) in the BSA is defined as,

$$\mathbf{E}^r = \mathbf{S}_{\text{BSA}} \mathbf{E}^i \tag{2.12}$$

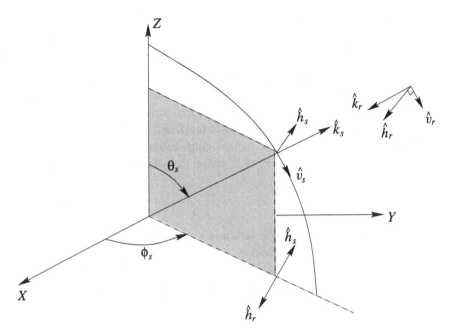

Fig. 2.2. Unit vectors $(\hat{k}_r, \hat{v}_r, \hat{h}_r)$, with $\hat{k}_r = \hat{v}_r \times \hat{h}_r$, for the back scatter alignment (BSA) convention. Note that $\hat{k}_r = -\hat{k}_s$, $\hat{v}_r = \hat{v}_s$, and $\hat{h}_r = -\hat{h}_s$. The scattered wave propagates along the $\hat{k}_s$-direction. The scattered field is resolved as either $\vec{E}^s = E_h^s \hat{h}_s + E_v^s \hat{v}_s$ in FSA, or as $\vec{E}^r = E_h^r \hat{h}_r + E_v^r \hat{v}_r$ in BSA.

where,

$$\mathbf{S}_{BSA} = \begin{bmatrix} -1 & 0 \\ 0 & 1 \end{bmatrix} \mathbf{S}_{FSA} \qquad (2.13)$$

Henceforth, the subscripts FSA or BSA on **S** will denote the alignment convention. The subscript s on the scattered field will also denote use of the FSA convention, while the subscript r on the scattered field will denote the BSA convention.

In the monostatic radar case, where $\hat{k}_s = -\hat{k}_i$, the angles (θ_s, ϕ_s) become $\theta_s = \pi - \theta_i$ and $\phi_s = \phi_i + \pi$. From (2.9) it is clear that $\hat{h}_s = -\hat{h}_i$ and $\hat{v}_s = \hat{v}_i$ in the back scatter direction. Thus in BSA, $\hat{h}_r = \hat{h}_i$, $\hat{v}_r = \hat{v}_i$, and the unit vectors for the incident wave and the "reflected" wave become identical. Also, $\hat{k}_i = \hat{k}_r = -\hat{k}_s$ and the BSA triplet $(\hat{k}_r, \hat{v}_r, \hat{h}_r)$, with $\hat{k}_r = \hat{v}_r \times \hat{h}_r$, becomes identical to the incident wave triplet $(\hat{k}_i, \hat{v}_i, \hat{h}_i)$, with $\hat{k}_i = \hat{v}_i \times \hat{h}_i$. In radar applications, this equivalence condition in BSA is useful, since the polarization state of an antenna, for example, is defined as the polarization of the wave radiated by the antenna, even when it is used as a receiving antenna. The topics of wave and antenna polarization will be treated in more detail in Chapter 3.

In the forward scattered direction, where $\hat{k}_s = \hat{k}_i$, the FSA triplet $(\hat{k}_s, \hat{v}_s, \hat{h}_s)$, becomes identical with the incident wave triplet $(\hat{k}_i, \hat{v}_i, \hat{h}_i)$. For all other scattering directions, the incident and scattered (or reflected) unit vector triplets are not so simply related.

2.2   Reciprocity theorem

The reciprocity theorem is stated here without proof (e.g. see Van Bladel 1985). Following van Zyl and Ulaby (1990), consider an incident plane wave whose vector amplitude at the origin is $\vec{E}^i(O) = E_0 \hat{q}$, as shown in Fig. 2.3$a$. The unit vector $\hat{q}$ describes the polarization state of the incident wave. For the present discussion, let $\hat{q} = \hat{v}_i$ or $\hat{h}_i$, corresponding to vertical or horizontal polarization states. In the far-field, the scattered wave ($\vec{E}^s$) can be written, emphasizing its dependence on the directions of incidence ($\hat{k}_i$) and scattering ($\hat{k}_s$), and on the polarization state of the incident wave ($\hat{q}$), as,

$$\vec{E}^s = \vec{E}^s(\hat{q}; \hat{k}_i, \hat{k}_s) \tag{2.14}$$

Next, consider the situation shown in Fig. 2.3b where the incident wave direction is along $-\hat{k}_s$, with a polarization state described by $\hat{q}'$. The scattering direction is now along $-\hat{k}_i$. Let the scattered wave in this configuration be described by $\vec{E}^s(\hat{q}'; -\hat{k}_s, -\hat{k}_i)$. The reciprocity theorem states that,

$$\hat{q}' \cdot \vec{E}^s(\hat{q}; \hat{k}_i, \hat{k}_s) = \hat{q} \cdot \vec{E}^s(\hat{q}'; -\hat{k}_s, -\hat{k}_i) \tag{2.15}$$

For the monostatic radar case, where $\hat{k}_s = -\hat{k}_i$, let $\hat{q} = \hat{v}_i$ and $\hat{q}' = \hat{h}_i$, which gives,

$$\hat{h}_i \cdot \vec{E}^s(\hat{v}_i; \hat{k}_i, -\hat{k}_i) = \hat{v}_i \cdot \vec{E}^s(\hat{h}_i; \hat{k}_i, -\hat{k}_i) \tag{2.16}$$

From (2.6a),

$$\vec{E}^s(\hat{v}_i; \hat{k}_i, -\hat{k}_i) = (S_{hv}\hat{h}_s + S_{vv}\hat{v}_s)_{\text{FSA}} E_0 \tag{2.17a}$$

$$\vec{E}^s(\hat{h}_i; \hat{k}_i, -\hat{k}_i) = (S_{hh}\hat{h}_s + S_{vh}\hat{v}_s)_{\text{FSA}} E_0 \tag{2.17b}$$

where FSA is used. From (2.16) and using the fact that in the case of monostatic radar, $\hat{h}_i = -\hat{h}_s$ and $\hat{v}_i = \hat{v}_s$,

$$-\hat{h}_s \cdot (S_{hv}\hat{h}_s + S_{vv}\hat{v}_s)_{\text{FSA}} = \hat{v}_s \cdot (S_{hh}\hat{h}_s + S_{vh}\hat{v}_s)_{\text{FSA}} \tag{2.18a}$$

or,

$$(-S_{hv} = S_{vh})_{\text{FSA}} \tag{2.18b}$$

Thus, for monostatic radar and in FSA convention, reciprocity yields the condition,

$$\begin{bmatrix} S_{hh} & S_{hv} \\ S_{vh} & S_{vv} \end{bmatrix}_{\text{FSA}} = \begin{bmatrix} S_{hh} & -S_{vh} \\ S_{vh} & S_{vv} \end{bmatrix}_{\text{FSA}} \tag{2.19}$$

The reciprocity condition given in (2.16) is also valid when $\vec{E}^s$ (now referred to as $\vec{E}^r$) is resolved into components in the BSA. Thus,

$$\vec{E}^r(\hat{v}_i; \hat{k}_i, -\hat{k}_i) = (S_{hv}\hat{h}_r + S_{vv}\hat{v}_r)_{\text{BSA}} E_0 \tag{2.20a}$$

$$\vec{E}^r(\hat{h}_i; \hat{k}_i, -\hat{k}_i) = (S_{hh}\hat{h}_r + S_{vh}\hat{v}_r)_{\text{BSA}} E_0 \tag{2.20b}$$

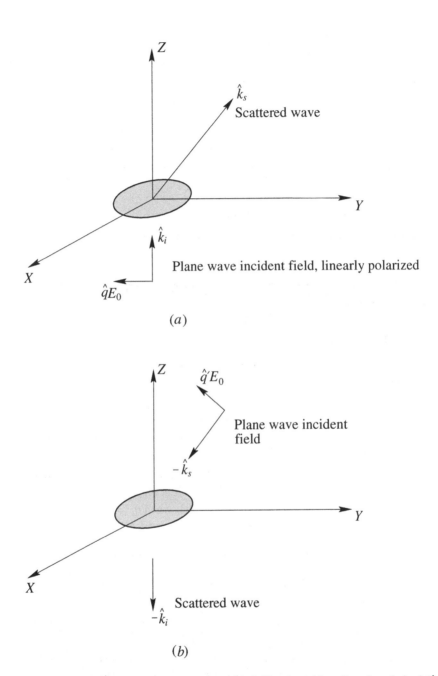

Fig. 2.3. The reciprocity relation. In (a) a particle is illuminated by a linearly polarized $(\hat{q}\,E_0)$ plane wave incident field propagating in the direction $\hat{k}_i$. The scattered wave is along the direction $\hat{k}_s$. In (b) the plane wave incident field is linearly polarized $(\hat{q}'\,E_0)$ and is along the $-\hat{k}_s$-direction. The scattered wave propagates along $-\hat{k}_i$.

Using (2.16) and the fact that in the case of monostatic radar $\hat{h}_i = \hat{h}_r$ and $\hat{v}_i = \hat{v}_r$, gives,

$$\hat{h}_r \cdot (S_{hv}\hat{h}_r + S_{vv}\hat{v}_r)_{\mathrm{BSA}} = \hat{v}_r \cdot (S_{hh}\hat{h}_r + S_{vh}\hat{v}_r)_{\mathrm{BSA}} \tag{2.21a}$$

or,

$$(S_{hv} = S_{vh})_{\mathrm{BSA}} \tag{2.21b}$$

Thus, for the monostatic case, $\mathbf{S}_{\mathrm{BSA}}$ is a symmetric matrix,

$$\begin{bmatrix} S_{hh} & S_{hv} \\ S_{vh} & S_{vv} \end{bmatrix}_{\mathrm{BSA}} = \begin{bmatrix} S_{hh} & S_{vh} \\ S_{vh} & S_{vv} \end{bmatrix}_{\mathrm{BSA}} \tag{2.22}$$

This is one advantage of using the BSA convention for monostatic radar.

2.3   Scattering matrix for sphere and spheroid in the Rayleigh–Gans approximation

2.3.1   Sphere

From (1.32a, b), the far-field scattering of an incident plane wave by a dielectric sphere of volume V is given as,

$$\vec{E}^s = \frac{k_0^2}{4\pi}\left(\frac{\varepsilon_r - 1}{\varepsilon_r + 2}\right) 3V\left[\vec{E}^i - \hat{r}(\hat{r}\cdot\vec{E}^i)\right]\frac{e^{-jk_0 r}}{r} \tag{2.23}$$

Note that the origin is at the center of the sphere, the scattering direction is $\hat{k}_s = \hat{r}$, and $\vec{E}^i$ is the incident wave amplitude at the origin. Let the incident wave direction be along the positive Z-axis, as shown in Fig. 2.4. Let the incident wave polarization state be either $\hat{v}_i$ or $\hat{h}_i$. From (2.9a, b) it is clear that when $\theta_i = \phi_i = 0°$, then $\hat{h}_i = \hat{y}$ and $\hat{v}_i = \hat{x}$. The first column of the scattering matrix in FSA is obtained from (2.6a) upon setting $E_h^i = 1$ and $E_v^i = 0$. In this case,

$$S_{hh} = \frac{k_0^2}{4\pi}\left(\frac{\varepsilon_r - 1}{\varepsilon_r + 2}\right) 3V\hat{\phi}_s \cdot \hat{y} \tag{2.24a}$$

$$S_{vh} = \frac{k_0^2}{4\pi}\left(\frac{\varepsilon_r - 1}{\varepsilon_r + 2}\right) 3V\hat{\theta}_s \cdot \hat{y} \tag{2.24b}$$

Note that $\hat{\phi}_s \cdot \hat{y} = \cos\phi_s$ and $\hat{\theta}_s \cdot \hat{y} = \cos\theta_s \sin\phi_s$. The second column of the scattering matrix is obtained by setting $E_h^i = 0$ and $E_v^i = 1$ in (2.6a). Thus,

$$S_{hv} = \frac{k_0^2}{4\pi}\left(\frac{\varepsilon_r - 1}{\varepsilon_r + 2}\right) 3V\hat{\phi}_s \cdot \hat{x} \tag{2.25a}$$

$$S_{vv} = \frac{k_0^2}{4\pi}\left(\frac{\varepsilon_r - 1}{\varepsilon_r + 2}\right) 3V\hat{\theta}_s \cdot \hat{x} \tag{2.25b}$$

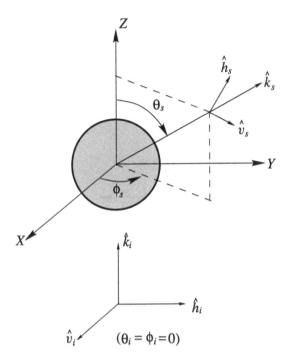

Fig. 2.4. Scattering direction and unit vectors of an incident plane wave in the FSA.

Note that $\hat{\phi}_s \cdot \hat{x} = -\sin\phi_s$ and $\hat{\theta}_s \cdot \hat{x} = \cos\theta_s \cos\phi_s$. From (2.24, 2.25), the explicit expression for the scattering matrix of a dielectric sphere in the Rayleigh limit with an incident plane wave along the positive Z-axis is,

$$\begin{bmatrix} E_h^s \\ E_v^s \end{bmatrix} = \frac{e^{-jk_0 r}}{r} \frac{k_0^2}{4\pi} \left(\frac{\varepsilon_r - 1}{\varepsilon_r + 2}\right) 3V \begin{bmatrix} \cos\phi_s & -\sin\phi_s \\ \cos\theta_s \sin\phi_s & \cos\theta_s \cos\phi_s \end{bmatrix}_{\text{FSA}} \begin{bmatrix} E_h^i \\ E_v^i \end{bmatrix} \quad (2.26)$$

In the back scatter direction, where $\theta_s = \pi - \theta_i = \pi$ and $\phi_s = \phi_i + \pi = \pi$, and suppressing the constant factors,

$$S_{\text{FSA}} = \begin{bmatrix} -1 & 0 \\ 0 & 1 \end{bmatrix} \qquad (2.27a)$$

$$S_{\text{BSA}} = \begin{bmatrix} 1 & 0 \\ 0 & 1 \end{bmatrix} \qquad (2.27b)$$

One advantage of using BSA is that the scattering matrix for a sphere in the back scatter direction assumes the familiar form of the unit matrix.

2.3.2   Spheroid

Consider now the case of oblate or prolate spheroids oriented with their symmetry axes along the positive Z-axis, as shown in Fig. 2.5a, b. Let the incident wave vector be $\hat{k}_i$.

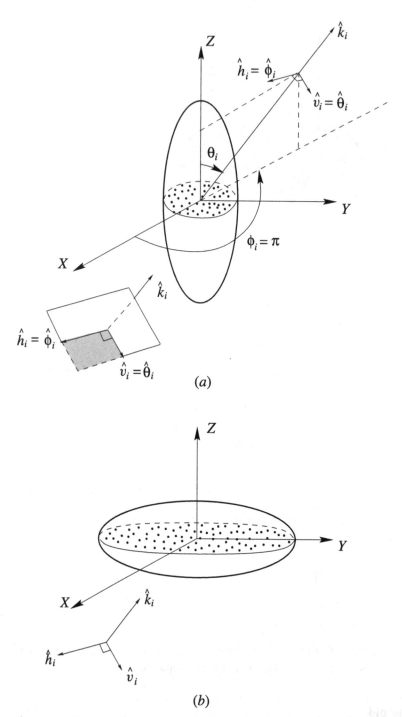

Fig. 2.5. (a) Prolate and (b) oblate spheroids oriented with their symmetry axis along the Z-direction. The incident wave is along $\hat{k}_i$, with $(\theta_i, \phi_i = \pi)$, which corresponds to the negative X-side of the XZ-plane. The spheroids' equatorial planes are shaded.

Note that $\hat{k}_i$ is on the negative X-side of the XZ-plane and, hence, $\phi_i = \pi$. From (2.9), the unit vectors are $\hat{v}_i = -\cos\theta_i\,\hat{x} - \sin\theta_i\,\hat{z}$ and $\hat{h}_i = -\hat{y}$. The incident wave may be considered to be launched by a radar located on the positive X-axis of the XZ-plane and far away from the origin. From (A1.29), let the polarizability $\alpha_x = \alpha_y = \alpha$ (say). The polarizability matrix in (A1.30) can be written as,

$$\boldsymbol{\alpha} = \begin{bmatrix} \alpha & 0 & 0 \\ 0 & \alpha & 0 \\ 0 & 0 & \alpha_z \end{bmatrix} = \begin{bmatrix} \alpha & 0 & 0 \\ 0 & \alpha & 0 \\ 0 & 0 & \alpha \end{bmatrix} + \begin{bmatrix} 0 & 0 & 0 \\ 0 & 0 & 0 \\ 0 & 0 & \alpha_z - \alpha \end{bmatrix} \tag{2.28a}$$

The polarizability tensor ($\bar{\bar{\alpha}}$) can be expressed as,

$$\bar{\bar{\alpha}} = \alpha\bar{\bar{I}} + (\alpha_z - \alpha)\hat{z}\hat{z} \tag{2.28b}$$

where $\bar{\bar{I}}$ is the identity tensor. The double overbar on $\bar{\bar{\alpha}}$ refers to its tensor form, while the corresponding matrix is simply $\boldsymbol{\alpha}$ without the double overbar. From (1.32), the far-field scattered wave along the $\hat{k}_s$-direction can be written as,

$$\vec{E}^s = \frac{k_0^2}{4\pi\varepsilon_0}\left[\vec{p} - \hat{k}_s(\hat{k}_s\cdot\vec{p})\right]\frac{e^{-jk_0r}}{r} \tag{2.29}$$

where $\vec{p} = \bar{\bar{\alpha}}\cdot\vec{E}^i(O)$ and $\vec{E}^i(O)$ is the incident plane wave amplitude at the origin. Thus,

$$\vec{E}^s = \frac{k_0^2}{4\pi\varepsilon_0}\left[\bar{\bar{\alpha}}\cdot\vec{E}^i(O) - \hat{k}_s\left\{\hat{k}_s\cdot[\bar{\bar{\alpha}}\cdot\vec{E}^i(O)]\right\}\right]\frac{e^{-jk_0r}}{r} \tag{2.30}$$

The first column of the scattering matrix in FSA is obtained by setting $E_h^i = 1$ and $E_v^i = 0$ in (2.6a). Thus,

$$(S_{hh})_{\text{FSA}} = \hat{h}_s\cdot\vec{E}^s = \frac{k_0^2}{4\pi\varepsilon_0}\hat{h}_s\cdot\left\{\left[\alpha\bar{\bar{I}} + (\alpha_z - \alpha)\hat{z}\hat{z}\right]\cdot\hat{h}_i\right\} \tag{2.31}$$

The second term in brackets in (2.30) does not contribute to S_{hh}, since $\hat{h}_s\cdot\hat{k}_s = \hat{\theta}_s\cdot\hat{k}_s = 0$. Equation (2.31) simplifies to,

$$(S_{hh})_{\text{FSA}} = \frac{k_0^2}{4\pi\varepsilon_0}\left[\alpha\hat{h}_s\cdot\hat{h}_i + (\alpha_z - \alpha)(\hat{h}_s\cdot\hat{z})(\hat{z}\cdot\hat{h}_i)\right] \tag{2.32}$$

where the properties $\hat{h}_s\cdot\bar{\bar{I}} = \hat{h}_s$ and $\hat{h}_s\cdot(\hat{z}\hat{z})\cdot\hat{h}_i = (\hat{h}_s\cdot\hat{z})(\hat{z}\cdot\hat{h}_i)$ are used (Van Bladel 1985; see Appendix 3 therein). Similarly,

$$(S_{vh})_{\text{FSA}} = \hat{v}_s\cdot\vec{E}^s = \frac{k_0^2}{4\pi\varepsilon_0}\left[\alpha\hat{v}_s\cdot\hat{h}_i + (\alpha_z - \alpha)(\hat{v}_s\cdot\hat{z})(\hat{z}\cdot\hat{h}_i)\right] \tag{2.33}$$

The second column of the scattering matrix in the FSA convention is obtained by setting $E_h^i = 0$ and $E_v^i = 1$ in (2.6a). Thus,

$$(S_{hv})_{\text{FSA}} = \hat{h}_s\cdot\vec{E}^s = \frac{k_0^2}{4\pi\varepsilon_0}\hat{h}_s\cdot\left\{\left[\alpha\bar{\bar{I}} + (\alpha_z - \alpha)\hat{z}\hat{z}\right]\cdot\hat{v}_i\right\} \tag{2.34}$$

$$= \frac{k_0^2}{4\pi\varepsilon_0}\left[\alpha\hat{h}_s\cdot\hat{v}_i + (\alpha_z - \alpha)(\hat{h}_s\cdot\hat{z})(\hat{z}\cdot\hat{v}_i)\right] \tag{2.35}$$

Similarly,

$$(S_{vv})_{\text{FSA}} = \hat{v}_s \cdot \vec{E}^s = \frac{k_0^2}{4\pi\varepsilon_0} \left[\alpha \hat{v}_s \cdot \hat{v}_i + (\alpha_z - \alpha)(\hat{v}_s \cdot \hat{z})(\hat{z} \cdot \hat{v}_i) \right] \tag{2.36}$$

The polarizability terms α and α_z differ for oblate and prolate spheroids through the depolarizing factor λ_z in (A1.22a, b); also see the discussion related to (A1.27)–(A1.30). In radar applications, the back scatter direction $\hat{k}_s = -\hat{k}_i$ is considered with $\theta_s = \pi - \theta_i$ and $\phi_s = \phi_i + \pi = 2\pi$. The unit vectors $\hat{h}_s$ and $\hat{v}_s$ are, from (2.9a, b),

$$\hat{h}_s = \hat{y} = -\hat{h}_i \tag{2.37a}$$

$$\hat{v}_s = -\cos\theta_i \hat{x} - \sin\theta_i \hat{z} = \hat{v}_i \tag{2.37b}$$

A few simple steps yield,

$$\begin{bmatrix} E_h^s \\ E_v^s \end{bmatrix} = \frac{e^{-jk_0 r}}{r} \frac{k_0^2}{4\pi\varepsilon_0} \begin{bmatrix} -\alpha & 0 \\ 0 & \alpha\cos^2\theta_i + \alpha_z\sin^2\theta_i \end{bmatrix}_{\text{FSA}} \begin{bmatrix} E_h^i \\ E_v^i \end{bmatrix} \tag{2.38a}$$

In BSA, use of (2.13) gives,

$$\begin{bmatrix} E_h^r \\ E_v^r \end{bmatrix} = \frac{e^{-jk_0 r}}{r} \frac{k_0^2}{4\pi\varepsilon_0} \begin{bmatrix} \alpha & 0 \\ 0 & \alpha\cos^2\theta_i + \alpha_z\sin^2\theta_i \end{bmatrix}_{\text{BSA}} \begin{bmatrix} E_h^i \\ E_v^i \end{bmatrix} \tag{2.38b}$$

It is clear that alignment of the symmetry axis of the spheroid along the positive Z-axis results in $S_{vh} = S_{hv} = 0$. Two special cases for θ_i are also relevant to radar applications, i.e. $\theta_i = 0°$, or the so-called "vertical" incidence case, and $\theta_i = 90°$, or the so-called "horizontal" incidence case. When $\theta_i = 0°$, S_{BSA} becomes diagonal for both oblate and prolate spheroids,

$$\mathbf{S}_{\text{BSA}}(\theta_i = 0) = \frac{k_0^2 \alpha}{4\pi\varepsilon_0} \begin{bmatrix} 1 & 0 \\ 0 & 1 \end{bmatrix} \tag{2.39a}$$

while for $\theta_i = 90°$,

$$\mathbf{S}_{\text{BSA}}(\theta_i = 90°) = \frac{k_0^2}{4\pi\varepsilon_0} \begin{bmatrix} \alpha & 0 \\ 0 & \alpha_z \end{bmatrix} \tag{2.39b}$$

To summarize, the depolarizing factor λ_z for prolates and oblates is given in (A1.22a, b). The depolarizing factors $\lambda_x = \lambda_y = (1 - \lambda_z)/2$ are simply obtained from the corresponding λ_z. The polarizability elements from (A1.30) are,

$$\alpha_{x,y,z} = V\varepsilon_0(\varepsilon_r - 1)(\Lambda_{x,y,z}) \tag{2.40a}$$

where,

$$\Lambda_{x,y,z} = \frac{1}{(\varepsilon_r - 1)\lambda_{x,y,z} + 1} \tag{2.40b}$$

Note that $\alpha_x = \alpha_y = \alpha$ in (2.38, 2.39). In radar applications the elevation angle of the radar beam is defined as $90° - \theta_i$, so that an elevation angle of $90°$ corresponds to a "vertically" pointing radar beam $(\theta_i = 0°)$, while an elevation angle of $0°$ corresponds to a "horizontally" pointing radar beam $(\theta_i = 90°)$. From (2.38), the amplitude scattering matrix for spheroids oriented as in Fig. 2.5, in the back scatter direction, is a function of the incidence angle (θ_i), the spheroid volume, the relative permittivity ε_r, and the eccentricity (see A1.22a, b). Note that $\alpha_{x,y,z}$ is real whenever ε_r is real. This is a general property of low frequency scattering from dielectric particles of arbitrary shape, that is, the polarizability tensor is real when ε_r is real (e.g. Kleinman 1978). Senior and Weil (1982) have discussed the validity of modeling arbitrary-shaped Rayleigh scatterers by "equivalent" spheroids. For convex scatterers, they showed that an "equivalent" spheroid of the same length-to-width ratio is sufficiently accurate for most practical purposes. However, for concave or multiply-connected shapes that are far removed from spheroids, an equivalent length-to-width approach was determined to be generally inadequate.

2.3.3   Oriented spheroid

The orientation of the spheroid's symmetry axis, given by the angles θ_b and ϕ_b, provides an additional degree of freedom, as illustrated in Fig. 2.6a. Let the incident plane wave be along the direction $\hat{k}_i$, with angles $\theta_i, \phi_i = 180°$. Introduce a $X_b Y_b Z_b$ coordinate system, called the "body" reference frame, which is obtained from the fixed XYZ system by three rotations through the Eulerian angles ϕ_b, θ_b, and ψ_b, as illustrated in Fig. 2.6b. First, rotate by angle ϕ_b about OZ to obtain the $X_1 Y_1 Z$ system. Next, rotate by angle θ_b about OY_1 to obtain the $X_2 Y_1 Z_b$ system. Finally, rotate by ψ_b about OZ_b to obtain the $X_b Y_b Z_b$ system. For particles possessing an axis of symmetry that is lined up with the OZ_b-axis, the third angle (ψ_b) can be set to any convenient value. Here it is set to zero and the orientation of the "body" frame is specified by the angles ϕ_b and θ_b. Euler's transformation relates the unit vectors $\hat{x}_b, \hat{y}_b, \hat{z}_b$ to the unit vectors $\hat{x}, \hat{y}, \hat{z}$, and these are given in Goldstein (1981) as,

$$\begin{bmatrix} \hat{x}_b \\ \hat{y}_b \\ \hat{z}_b \end{bmatrix} = \begin{bmatrix} \cos\theta_b \cos\phi_b & \cos\theta_b \sin\phi_b & -\sin\theta_b \\ -\sin\phi_b & \cos\phi_b & 0 \\ \sin\theta_b \cos\phi_b & \sin\theta_b \sin\phi_b & \cos\theta_b \end{bmatrix} \begin{bmatrix} \hat{x} \\ \hat{y} \\ \hat{z} \end{bmatrix} \tag{2.41}$$

The procedure[1] for determining the elements of the scattering matrix is the same as before, except that the dipole moment $\vec{p}$ is given as,

$$\vec{p} = \bar{\bar{\alpha}} \cdot \vec{E}_b^i(O) = \bar{\bar{\alpha}} \cdot \left(E_{x_b}^i \hat{x}_b + E_{y_b}^i \hat{y}_b + E_{z_b}^i \hat{z}_b \right) \tag{2.42}$$

where $\vec{E}_b^i(O)$ is the incident field vector amplitude at the origin expressed in the $X_b Y_b Z_b$ system. In the XYZ system, the incident field amplitude at the origin is,

$$\vec{E}^i(O) = E_h^i \hat{h}_i + E_v^i \hat{v}_i = E_{x_b}^i \hat{x}_b + E_{y_b}^i \hat{y}_b + E_{z_b}^i \hat{z}_b \tag{2.43}$$

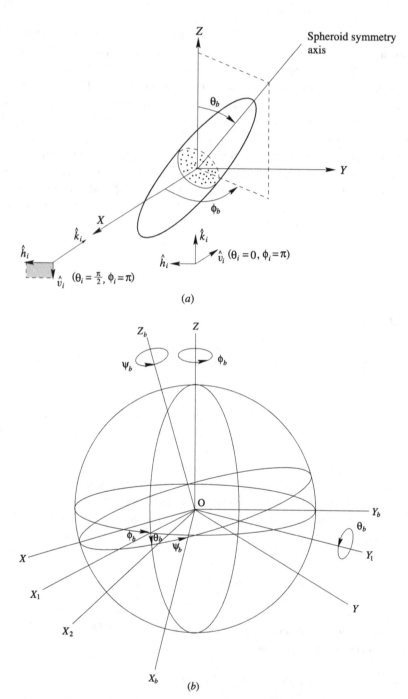

Fig. 2.6. (*a*) Orientation of the symmetry axis of a spheroid is described by the angles (θ_b, ϕ_b). Also shown is the horizontal incidence, for which $\theta_i = \pi/2$; $\phi_i = \pi$, with $\hat{k}_i$ pointed along the $-X$-direction. The unit vectors $\hat{v}_i$, $\hat{h}_i$ satisfy $\hat{v}_i \times \hat{h}_i = \hat{k}_i$. For vertical incidence, $\theta_i = 0$ and $\phi_i = \pi$ as shown. (*b*) Shows the three Eulerian angles of rotation $(\theta_b, \phi_b, \psi_b)$.

and,

$$
\begin{bmatrix} E^i_{x_b} \\ E^i_{y_b} \\ E^i_{z_b} \end{bmatrix} = \begin{bmatrix} \hat{h}_i \cdot \hat{x}_b & \hat{v}_i \cdot \hat{x}_b \\ \hat{h}_i \cdot \hat{y}_b & \hat{v}_i \cdot \hat{y}_b \\ \hat{h}_i \cdot \hat{z}_b & \hat{v}_i \cdot \hat{z}_b \end{bmatrix} \begin{bmatrix} E^i_h \\ E^i_v \end{bmatrix} \tag{2.44}
$$

The general expression for the far-field scattered wave is still given by (2.29), with $\vec{p} = \bar{\bar{\alpha}} \cdot \vec{E}^i_b(\mathrm{O})$ and $\bar{\bar{\alpha}}$ expressed as,

$$
\bar{\bar{\alpha}} = \alpha \bar{\bar{\mathbf{I}}} + (\alpha_{z_b} - \alpha)\hat{z}_b \hat{z}_b \tag{2.45}
$$

Thus,

$$
\vec{E}^s = \frac{k_0^2}{4\pi \varepsilon_0} \left[\bar{\bar{\alpha}} \cdot \vec{E}^i_b(\mathrm{O}) - \hat{k}_s \left\{ \hat{k}_s \cdot \left[\bar{\bar{\alpha}} \cdot \vec{E}^i_b(\mathrm{O}) \right] \right\} \right] \frac{e^{-jk_0 r}}{r} \tag{2.46}
$$

The first column of the scattering matrix is obtained by setting $E^i_h = 1$ and $E^i_v = 0$ in (2.6a). Thus,

$$
(S_{hh})_{\mathrm{FSA}} = \hat{h}_s \cdot \vec{E}^s = \frac{k_0^2}{4\pi \varepsilon_0} \left[\alpha \hat{h}_s \cdot \vec{E}^i_b + (\alpha_{z_b} - \alpha)(\hat{h}_s \cdot \hat{z}_b)(\hat{z}_b \cdot \vec{E}^i_b) \right] \tag{2.47}
$$

Again, the second term in brackets in (2.46) does not contribute to S_{hh}, since $\hat{h}_s \cdot \hat{k}_s = \hat{\theta}_s \cdot \hat{k}_s = 0$. From (2.44), with $E^i_h = 1$ and $E^i_v = 0$,

$$
E^i_{x_b} = \hat{h}_i \cdot \hat{x}_b; \quad E^i_{y_b} = \hat{h}_i \cdot \hat{y}_b; \quad E^i_{z_b} = \hat{h}_i \cdot \hat{z}_b \tag{2.48a}
$$

and,

$$
\vec{E}^i_b = (\hat{h}_i \cdot \hat{x}_b)\hat{x}_b + (\hat{h}_i \cdot \hat{y}_b)\hat{y}_b + (\hat{h}_i \cdot \hat{z}_b)\hat{z}_b \tag{2.48b}
$$

Substituting (2.48b) into (2.47) results in,

$$
(S_{hh})_{\mathrm{FSA}} = \frac{k_0^2}{4\pi \varepsilon_0} \left\{ \alpha \left[(\hat{h}_s \cdot \hat{x}_b)(\hat{h}_i \cdot \hat{x}_b) + (\hat{h}_s \cdot \hat{y}_b)(\hat{h}_i \cdot \hat{y}_b) + (\hat{h}_s \cdot \hat{z}_b)(\hat{h}_i \cdot \hat{z}_b) \right] \right.
$$
$$
\left. + (\alpha_{z_b} - \alpha)(\hat{h}_s \cdot \hat{z}_b)(\hat{h}_i \cdot \hat{z}_b) \right\} \tag{2.49}
$$

The above expression may be compared with (2.32), which is valid when $\theta_b = \phi_b = 0$. Similarly, S_{vh} is obtained by replacing $\hat{h}_s$ with $\hat{v}_s$ in (2.49),

$$
(S_{vh})_{\mathrm{FSA}} = \hat{v}_s \cdot \vec{E}^s = \frac{k_0^2}{4\pi \varepsilon_0} \left[\alpha \hat{v}_s \cdot \vec{E}^i_b + (\alpha_{z_b} - \alpha)(\hat{v}_s \cdot \hat{z}_b)(\hat{z}_b \cdot \vec{E}^i_b) \right]
$$
$$
= \frac{k_0^2}{4\pi \varepsilon_0} \left\{ \alpha \left[(\hat{v}_s \cdot \hat{x}_b)(\hat{h}_i \cdot \hat{x}_b) + (\hat{v}_s \cdot \hat{y}_b)(\hat{h}_i \cdot \hat{y}_b) + (\hat{v}_s \cdot \hat{z}_b)(\hat{h}_i \cdot \hat{z}_b) \right] \right.
$$
$$
\left. + (\alpha_{z_b} - \alpha)(\hat{v}_s \cdot \hat{z}_b)(\hat{h}_i \cdot \hat{z}_b) \right\} \tag{2.50}
$$

The second column of the scattering matrix is obtained by setting $E_h^i = 0$ and $E_v^i = 1$, where now $\vec{E}_b^i = (\hat{v}_i \cdot \hat{x}_b)\hat{x}_b + (\hat{v}_i \cdot \hat{y}_b)\hat{y}_b + (\hat{v}_i \cdot \hat{z}_b)\hat{z}_b$. Similar steps yield,

$$(S_{hv})_{\text{FSA}} = \hat{h}_s \cdot \vec{E}^s = \frac{k_0^2}{4\pi\varepsilon_0}[\alpha\hat{h}_s \cdot \vec{E}_b^i + (\alpha_{z_b} - \alpha)(\hat{h}_s \cdot \hat{z}_b)(\hat{z}_b \cdot \vec{E}_b^i)] \tag{2.51a}$$

$$= \frac{k_0^2}{4\pi\varepsilon_0}\Big\{\alpha\big[(\hat{h}_s \cdot \hat{x}_b)(\hat{v}_i \cdot \hat{x}_b) + (\hat{h}_s \cdot \hat{y}_b)(\hat{v}_i \cdot \hat{y}_b)$$

$$+ (\hat{h}_s \cdot \hat{z}_b)(\hat{v}_i \cdot \hat{z}_b)\big] + (\alpha_{z_b} - \alpha)(\hat{h}_s \cdot \hat{z}_b)(\hat{v}_i \cdot \hat{z}_b)\Big\} \tag{2.51b}$$

Finally, S_{vv} is obtained from (2.51b) by replacing $\hat{h}_s$ with $\hat{v}_s$,

$$(S_{vv})_{\text{FSA}} = \frac{k_0^2}{4\pi\varepsilon_0}\Big\{\alpha\big[(\hat{v}_s \cdot \hat{x}_b)(\hat{v}_i \cdot \hat{x}_b) + (\hat{v}_s \cdot \hat{y}_b)(\hat{v}_i \cdot \hat{y}_b)$$

$$+ (\hat{v}_s \cdot \hat{z}_b)(\hat{v}_i \cdot \hat{z}_b)\big] + (\alpha_{z_b} - \alpha)(\hat{v}_s \cdot \hat{z}_b)(\hat{v}_i \cdot \hat{z}_b)\Big\} \tag{2.52}$$

Consider the case of back scatter in FSA, so that (2.37a, b) applies. Recall that the incident wave direction is the same as in Fig. 2.5a with $(0 \leq \theta_i \leq \pi, \phi_i = \pi)$. Using (2.41) and (2.37a, b) in (2.49–2.52) results in,

$$(S_{hh})_{\text{FSA}} = -\frac{k_0^2}{4\pi\varepsilon_0}\Big[\alpha + (\alpha_{z_b} - \alpha)\sin^2\theta_b\sin^2\phi_b\Big] \tag{2.53a}$$

$$(S_{vh})_{\text{FSA}} = (-S_{hv})_{\text{FSA}}$$

$$= \frac{k_0^2}{4\pi\varepsilon_0}\Bigg[\frac{(\alpha_{z_b} - \alpha)}{2}\Big(\cos\theta_i\sin^2\theta_b\sin 2\phi_b$$

$$+ \sin\theta_i\sin 2\theta_b\sin\phi_b\Big)\Bigg] \tag{2.53b}$$

$$(S_{vv})_{\text{FSA}} = \frac{k_0^2}{4\pi\varepsilon_0}\Bigg[\alpha + (\alpha_{z_b} - \alpha)\Big(\cos^2\theta_i\sin^2\theta_b\cos^2\phi_b$$

$$+ \sin^2\theta_i\cos^2\theta_b + \frac{\sin 2\theta_i\sin 2\theta_b\cos\phi_b}{2}\Big)\Bigg] \tag{2.53c}$$

The scattering matrix elements in BSA are simply obtained from (2.13).

For a single particle, the differential reflectivity (in decibels) is defined as,

$$Z_{\text{dr}} = 10\log_{10}\frac{|S_{hh}|^2}{|S_{vv}|^2} \tag{2.54}$$

and the linear depolarization ratio (LDR$_{vh}$, again in decibels) as,

$$\text{LDR}_{vh} = 10\log_{10}\frac{|S_{vh}|^2}{|S_{hh}|^2} \tag{2.55}$$

The subscript "vh" refers to "transmit horizontal polarization and receive vertical polarization". The radar parameters Z_{dr} and LDR$_{vh}$ can be measured with dual-polarized radar systems, see Chapter 6. They are functions of the incidence angle (θ_i), the particle orientation angles (θ_b, ϕ_b), the relative permittivity (ε_r), and the eccentricity of the spheroid.

2.3.4   Horizontal incidence: $\theta_i = 90°$

Figure 2.6a illustrates the case of "horizontal" incidence of the radar beam or approximately, the case of a low elevation angle of the incident beam (recall that the radar elevation angle = $90° - \theta_i$). The particle orientation is specified by the angle θ_b in the YZ-plane, where $\phi_b = 90°$. The plane of polarization of the incident wave is defined as the plane orthogonal to $\hat{k}_i$, which in this case is the YZ-plane. From (2.53), it follows that,

$$Z_{dr} = 10 \log_{10} \left(\frac{|\alpha \cos^2 \theta_b + \alpha_{z_b} \sin^2 \theta_b|^2}{|\alpha \sin^2 \theta_b + \alpha_{z_b} \cos^2 \theta_b|^2} \right) \tag{2.56}$$

$$LDR_{vh} = 10 \log_{10} \left(\frac{|(\alpha_{z_b} - \alpha) \sin \theta_b \cos \theta_b|^2}{|\alpha \cos^2 \theta_b + \alpha_{z_b} \sin^2 \theta_b|^2} \right) \tag{2.57}$$

From the above, it is clear that both Z_{dr} and LDR_{vh} depend on the orientation angle (θ_b) for a fixed ε_r and spheroid eccentricity.

One application of Z_{dr} is for the detection of columnar or needle-shaped ice crystals that are preferentially oriented ($\theta_b \approx 0°$) in an electrified cloud with a strong vertical electrostatic field, see Fig. 2.7a. In this case, Z_{dr} is proportional to $|\alpha/\alpha_{z_b}|^2$, which for needles (see Table A1.1) reduces to,

$$\left| \frac{\alpha}{\alpha_{z_b}} \right|^2 = \left| \frac{2}{\varepsilon_r + 1} \right|^2 \tag{2.58}$$

Since $\varepsilon_r \geq 1$, the ratio $|\alpha/\alpha_{z_b}|^2 \leq 1$, and Z_{dr} becomes negative (dB). Note also the strong weighting by ε_r. When ε_r is only slightly different from unity (e.g. in very low density snow), the ratio $|\alpha/\alpha_{z_b}|^2$ tends to unity, and Z_{dr} tends to 0 dB even for oriented needles.

A simple orientation model for plates (see also, Table A1.1) falling in air is shown in Fig. 2.7b. Now,

$$Z_{dr} = 10 \log_{10} |\varepsilon_r|^2 \tag{2.59}$$

so that Z_{dr} is positive (in decibels). Again, there is strong weighting by ε_r, and for very low density snow, Z_{dr} tends to 0 dB even for oriented plates. It is obvious that in Fig. 2.7a, b, the LDR_{vh} is $-\infty$ (dB), since $\theta_b = 0$. The rather dramatic weighting of Z_{dr} by ε_r is further illustrated in Fig. 2.8a for oblate spheroids of varying axis ratios: here four values of ε_r, corresponding to low density snowflake, medium density graupel, hail (solid ice), and raindrop are given.

A simple orientation model for oblate raindrops falling in air is also illustrated in Fig. 2.7b. The Z_{dr} for any fixed (b/a) value between zero and unity is,

$$Z_{dr} = 10 \log_{10} \left(\frac{|\alpha|^2}{|\alpha_{z_b}|^2} \right) \tag{2.60a}$$

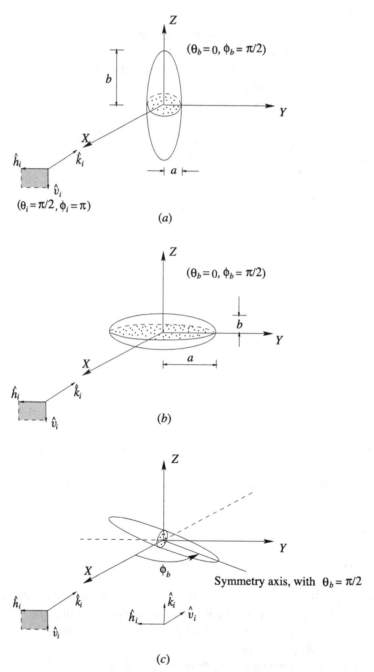

Fig. 2.7. Simple orientation models for (*a*) vertically oriented prolate crystals (e.g. needles); (*b*) oblate raindrops (or plates) falling in air; and (*c*) prolate crystals falling in air, ϕ_b has a uniform probability density function in $(0, 2\pi)$.

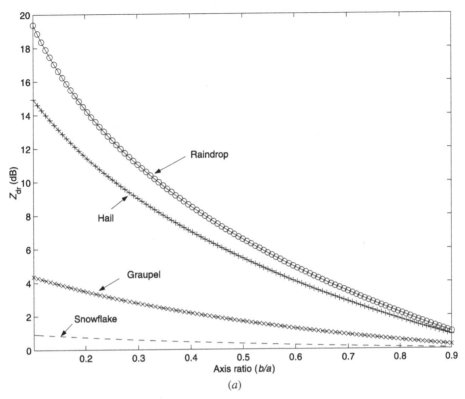

(a)

Fig. 2.8. (*a*) Differential reflectivity (Z_{dr}) of a single particle of oblate shape versus axis ratio. Several particle types with different densities are illustrated to show the strong weighting of Z_{dr} on the dielectric constant for a given (b/a). (*b*) Summary of typical Z_{dr} values of raindrops of various sizes and hail. The black arrows on the hail particle represent the tumbling motions as it falls in a thunderstorm. Adapted from Wakimoto and Bringi (1988). (*c*) Linear depolarization ratio of a single oblate particle with isotropic orientation distribution, see (2.72), versus axis ratio. Particle types are as in Fig. 2.8a. Note strong weighting of LDR with dielectric constant (or density) for a fixed axis ratio.

$$= 10\log_{10}\left(\frac{|\Lambda|^2}{|\Lambda_{z_b}|^2}\right) \tag{2.60b}$$

$$= 10\log_{10}\left[\frac{|1 + \lambda_{z_b}(\varepsilon_r - 1)|^2}{|1 + \lambda(\varepsilon_r - 1)|^2}\right] \tag{2.60c}$$

where λ_{z_b} is given in (A1.22b) and $\lambda = 0.5(1 - \lambda_{z_b})$. Note that here λ is the depolarizing factor, and should not be confused with wavelength. Again, Z_{dr} is positive (in decibels) and is a function of (b/a) for a given ε_r.

   An excellent approximation for raindrops based on curve fitting is given by Jameson (1983), relating Z_{dr} to the axis ratio (b/a) for oblate raindrops, as,

$$10^{0.1(Z_{dr})} \approx (b/a)^{-7/3} \tag{2.61}$$

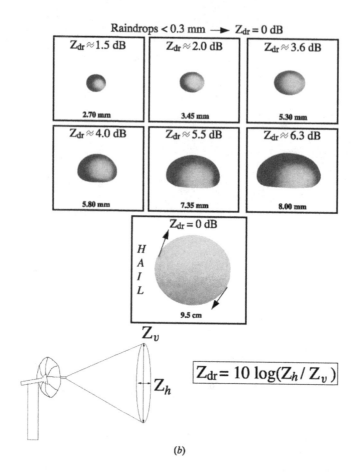

Fig. 2.8. (*cont.*)

Figure 2.8*b* illustrates, more clearly, how Z_{dr} can be used to detect oriented oblate raindrops when the incident beam is at a low elevation angle (i.e. for the case of $\theta_i \approx 90°$). The equilibrium shape of raindrops is nearly oblate, and is primarily governed by a balance between surface tension and gravitational forces (Green 1975). Tiny drops ($D \leq 1$ mm) are spherical with $b/a = 1$, while larger drops have axis ratios that decrease nearly linearly with increasing D. Here, D is the diameter of a spherical drop whose volume equals that of the equivalent oblate spheroid, or $D = 2a(b/a)^{1/3}$. The linear relation $b/a = 1.03 - 0.062D$ (where D is in millimeters) is often used, and thus Z_{dr} can be related to D for single drops. Raindrop shapes and other issues related to using radar polarimetric methods will be discussed in detail in Chapter 7. Figure 2.8*b* also illustrates that $Z_{dr} \rightarrow 0$ dB for hail particles that are nearly spherical in shape (or they appear "spherical" because of tumbling motions).

A simple orientation model for columnar or prolate ice crystals falling in still air is shown in Fig. 2.7*c*, where $\theta_b = 90°$ but ϕ_b is uniformly distributed in the interval

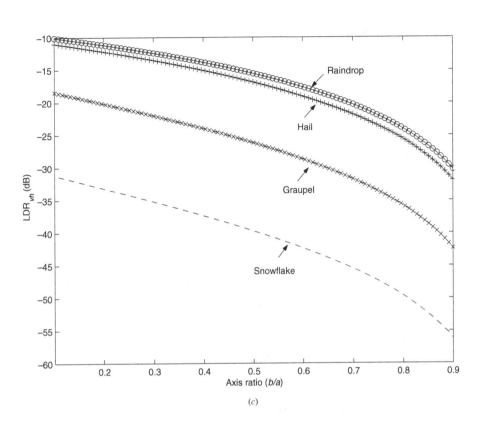

Fig. 2.8. (*cont.*)

$(0, 2\pi)$. From (2.53), and ignoring the multiplication factor, $k_0^2/4\pi\varepsilon_0$ results in,

$$\langle |S_{hh}|^2 \rangle = \langle |\alpha + (\alpha_{z_b} - \alpha)\sin^2\phi_b|^2 \rangle \tag{2.62}$$

$$\langle |S_{vh}|^2 \rangle = 0 \tag{2.63}$$

$$\langle |S_{vv}|^2 \rangle = |\alpha|^2 \tag{2.64}$$

where the angle brackets here denote ensemble averaging with respect to ϕ_b. For Z_{dr} the use of $\langle \sin^2\phi_b \rangle = 0.5$ and $\langle \sin^4\phi_b \rangle = 3/8$ (assuming for simplicity that ε_r is real) gives,

$$Z_{dr} = 10\log_{10}\left(\frac{\langle |S_{hh}|^2 \rangle}{\langle |S_{vv}|^2 \rangle}\right) \tag{2.65a}$$

$$= 10\log_{10}\left[\frac{\alpha^2 + (3/8)(\alpha_{z_b} - \alpha)^2 + \alpha(\alpha_{z_b} - \alpha)}{\alpha^2}\right] \tag{2.65b}$$

In this case Z_{dr} is positive (in decibels) and LDR_{vh} is $-\infty$ (in decibels).

2.3.5   Vertical incidence: $\theta_i = 0°$

The case of $\theta_i = 0°$ is also known as vertical beam incidence. Again, for simplicity let ε_r be real. The scattering matrix elements from (2.53) are,

$$(S_{hh})_{\text{FSA}} = \frac{-k_0^2}{4\pi\varepsilon_0}\left[\alpha + (\alpha_{z_b} - \alpha)\sin^2\theta_b\sin^2\phi_b\right] \tag{2.66a}$$

$$(S_{vh})_{\text{FSA}} = \frac{k_0^2}{4\pi\varepsilon_0}\left[\frac{(\alpha_{z_b} - \alpha)}{2}\sin^2\theta_b\sin 2\phi_b\right] \tag{2.66b}$$

$$(S_{vv})_{\text{FSA}} = \frac{k_0^2}{4\pi\varepsilon_0}\left[\alpha + (\alpha_{z_b} - \alpha)\sin^2\theta_b\cos^2\phi_b\right] \tag{2.66c}$$

For the orientation model shown in Fig. 2.7c, it is easy to see that,

$$Z_{\text{dr}} = 0 \tag{2.67a}$$

$$\text{LDR}_{vh} = 10\log_{10}\left[\frac{\frac{1}{8}(\alpha_{z_b} - \alpha)^2}{\alpha^2 + (3/8)(\alpha_{z_b} - \alpha)^2 + \alpha(\alpha_{z_b} - \alpha)}\right] \tag{2.67b}$$

This condition at vertical incidence is often used to determine any radar system bias error in the measurement of differential reflectivity (see Section 6.3, Fig. 6.22a).

The dependence of LDR_{vh} as a function of θ_i can be used to distinguish between needle-like crystals, oriented as shown in Fig. 2.7c, and plate-like crystals, oriented as in Fig. 2.7b. In the former case, LDR_{vh} decreases with θ_i from the value given in (2.67b) at $\theta_i = 0°$, to $-\infty$ dB at $\theta_i = 90°$. In the latter case, LDR_{vh} is at $-\infty$ dB at any angle θ_i. In practice, the LDR_{vh} lower bound for a well-designed radar system is around -35 dB (see also, Section 6.2). Use of the elevation angle dependence of polarimetric observations in order to distinguish between plate- and column-like ice crystals is discussed further in Sections 3.8.1 and 7.3.

2.3.6   Probability distribution function of the symmetry axis

So far, the angle θ_b has been considered as a deterministic value describing particle orientation. Because the angles θ_b and ϕ_b define the two-dimensional orientation of the spheroid's symmetry axis on a spherical surface, the probability of the symmetry axis pointing within the solid angle $(\Omega, \Omega + d\Omega)$ is expressed as $p(\Omega)\,d\Omega = p(\theta_b, \phi_b)\,d\Omega$, where $d\Omega = \sin\theta_b\,d\theta_b\,d\phi_b$. For example, if the random variables θ_b and ϕ_b are independent, and if ϕ_b is uniform in $(0, 2\pi)$ while θ_b is uniform in $(0, \theta_0)$, then,

$$\int_0^{\theta_0}\int_0^{2\pi} p(\theta_b)p(\phi_b)\,d\Omega = 1 \tag{2.68}$$

which implies that,

$$p(\theta_b) = \frac{1}{1 - \cos\theta_0} \tag{2.69}$$

From (2.66),

$$LDR_{vh} = 10\log_{10}\left[\left|\frac{\frac{1}{8}(\alpha_{z_b} - \alpha)^2\langle\sin^4\theta_b\rangle}{\{\alpha^2 + [(3/8)(\alpha_{z_b} - \alpha)^2]\langle\sin^4\theta_b\rangle + \alpha(\alpha_{z_b} - \alpha)\langle\sin^2\theta_b\rangle\}}\right|\right]$$

(2.70)

It may easily be verified (Russchenberg 1992) that,

$$\langle\sin^2\theta_b\rangle = \frac{1}{1 - \cos\theta_0}\left(\frac{2}{3} - \frac{3}{4}\cos\theta_0 + \frac{1}{12}\cos3\theta_0\right)$$

(2.71a)

$$\langle\sin^4\theta_b\rangle = \frac{1}{1 - \cos\theta_0}\left(\frac{8}{15} - \frac{5}{8}\cos\theta_0 + \frac{5}{48}\cos3\theta_0 - \frac{1}{80}\cos5\theta_0\right)$$

(2.71b)

As a special case, if $\theta_0 = \pi$, implying an isotropic orientation, LDR_{vh} in (2.70) reduces to,

$$LDR_{vh} = 10\log_{10}\left[\frac{(\alpha_{z_b} - \alpha)^2}{(8\alpha^2 + 4\alpha\alpha_{z_b} + 3\alpha_{z_b}^2)}\right]$$

(2.72)

Figure 2.8c illustrates the rather strong weighting of LDR_{vh} by ε_r for the same oblate-shape particle types shown in Fig. 2.8a, i.e. snowflake, graupel, hail, and raindrop. The horizontal axis is the axis ratio (b/a). The case of isotropic orientation (i.e. randomly oriented case) is shown.

Gaussian probability density have been commonly used to describe the orientation of the symmetry axis of a particle. This description is inaccurate, and quickly breaks down for orientation distributions that are not narrow. Perhaps the best way to describe hydrometeor orientation distributions is by the family of Fisher distributions (Mardia 1972). The Fisher density function, where the random variables are θ_b and ϕ_b, is given by,

$$g(\theta_b, \phi_b) = \frac{\kappa\sin\theta_b}{4\pi\sinh(\kappa)}\exp\left\{\kappa\left[\cos\bar{\theta}_b\cos\theta_b + \sin\bar{\theta}_b\sin\theta_b\cos(\phi_b - \bar{\phi}_b)\right]\right\}$$

(2.73)

where $0 \leq \theta_b \leq \pi$; $0 \leq \phi_b \leq 2\pi$ and $\kappa \geq 0$; and $\bar{\theta}_b$ and $\bar{\phi}_b$ are the mean values of θ_b and ϕ_b, respectively, and κ is a parameter that controls the width of the distribution. Note that the multiplicative factor $\sin\theta_b$ in (2.73) comes from the expression for the elemental solid angle $d\Omega$ and is now part of the pdf. For $\kappa = 0$, the Fisher density reduces to the uniform density,

$$g(\theta_b, \phi_b) = \frac{1}{4\pi}\sin\theta_b$$

(2.74)

If the distribution is symmetric about the vertical, then $\bar{\theta}_b = 0°$; and if $\bar{\phi}_b$ can be assumed to be zero, then

$$g(\theta_b, \phi_b) = \frac{\kappa e^{\kappa\cos\theta_b}}{4\pi\sinh(\kappa)}\sin\theta_b$$

(2.75)

The marginal pdfs of θ and ϕ (the subscript "b" is dropped for now) can be written as,

$$p(\theta) = \left[\frac{\kappa}{2\sinh(\kappa)}\right]e^{\kappa\cos\theta}\sin\theta; \quad 0 \leq \theta \leq \pi, \quad \kappa > 0$$

(2.76)

and,

$$p(\phi) = \frac{1}{2\pi}; \quad 0 \le \phi \le 2\pi \tag{2.77}$$

Figure 2.9a shows $p(\theta)$ for selected values of κ, which indicates that the pdf peak shifts towards $\theta = 0°$ as κ increases. Some simple properties of the Fisher distribution can be summarized as follows. The probability that θ lies between two angles (θ_1, θ_2) can be written as,

$$P(\theta_1 < \theta < \theta_2) = \frac{\left(e^{\kappa \cos \theta_1} - e^{\kappa \cos \theta_2}\right)}{\left(e^{\kappa} - e^{-\kappa}\right)} \tag{2.78}$$

When κ is large then,

$$p(\theta > \delta) \approx \exp\left[-\kappa(1 - \cos \delta)\right] \tag{2.79}$$

For large κ, $\kappa\theta^2$ is a chi-square distribution of order two, which is,

$$f(\theta) = \kappa\theta e^{-\frac{1}{2}\kappa\theta^2} \tag{2.80}$$

The Fisher distribution in (2.73) is useful in describing orientation distributions about an arbitrary mean direction in three-dimensional space.

Often, interest is in orientation distributions where the orientation angle is concentrated about $\bar{\theta} = 90°$, but not concentrated in any $\bar{\phi}$-direction. Such orientation distributions can be described by axial distributions (Mardia 1972),

$$g_A(\theta) = \frac{b(\kappa)}{2\pi} e^{-\kappa \cos^2 \theta} \sin \theta \tag{2.81}$$

where,

$$b(\kappa) = \frac{1}{\left(2 \int_0^1 e^{-\kappa t^2} dt\right)} \tag{2.82}$$

The density of the axial distribution has a maximum where $\bar{\theta} = 90°$, and the density is, thus, concentrated around the equator. As κ increases, the distribution becomes more concentrated around $\bar{\theta} = 90°$. Figure 2.9b shows the plot of $g_A(\theta)$ for various values of κ.

2.3.7   General incidence: $0 \le \theta_i \le 90°$

The case of an angle between vertical ($\theta_i = 0°$) and horizontal incidence ($\theta_i = 90°$) is shown in Fig. 2.10a which clearly illustrates the canting angle β in the polarization plane (Holt 1984). The angle ψ is the angle between the incidence direction and the symmetry axis ($\psi = \angle\text{ION}$) in Fig. 2.10a. The line segment QT is the projection of ON

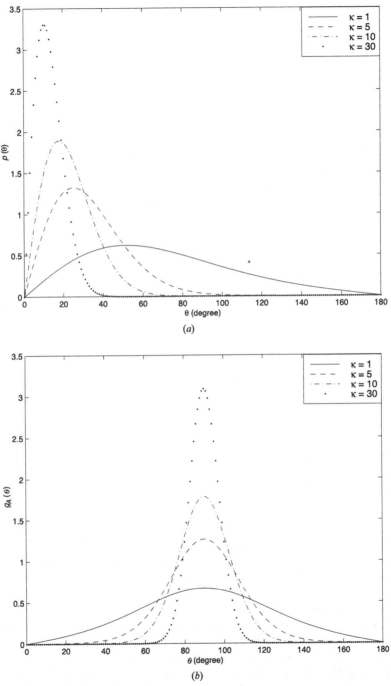

Fig. 2.9. Probability density function for (*a*) the Fisher distribution, see (2.76), and (*b*) the axial distribution, see (2.81), $\kappa = 1, 5, 10, 30$.

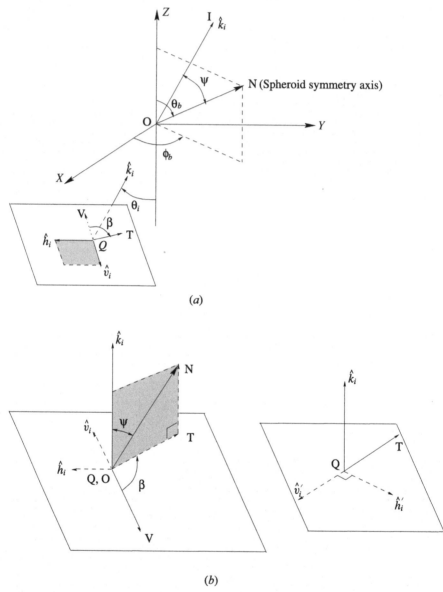

(a)

(b)

Fig. 2.10. (a) The spheroid symmetry axis is oriented along ON with angles θ_b and ϕ_b. The angle between the incidence direction (OI) and ON is ψ. QT is the projection of ON onto the plane of polarization of the incident wave (shown shaded). This plane is orthogonal to $\hat{k}_i$. Positive canting angle (β) in the polarization plane is measured clockwise from QV. Figure adapted from Holt (1984) (© 1984 American Geophysical Union). (b) Illustrates the angles ψ and β from a different perspective view, where $\overline{ON}$ is the spheroid symmetry axis and $\overline{QT}$ is its projection on the plane of polarization. Note that $\overline{ON} \cdot \hat{v}_i = \overline{QT} \cdot \hat{v}_i = -|\overline{ON}| \sin \psi \cos \beta$, while $\overline{ON} \cdot \hat{h}_i = \overline{QT} \cdot \hat{h}_i = -|\overline{ON}| \sin \psi \sin \beta$. The right panel shows the rotated linear basis $(\hat{h}_i', \hat{v}_i')$.

onto the plane of polarization which is the plane orthogonal to $\hat{k}_i$. The canting angle $\beta(= \angle \text{VQT})$ in the polarization plane is the angle measured clockwise between the line segments QV and QT as shown in Fig. 2.10a. Note that QV runs in a direction opposite to $\hat{v}_i$. Recall that $\hat{k}_i$ lies on the $-X$-side of the XZ-plane (where $\phi_i = \pi$), and from (2.9),

$$\hat{k}_i = -\sin \theta_i \hat{x} + \cos \theta_i \hat{z} \tag{2.83a}$$

$$\hat{h}_i = -\hat{y} \tag{2.83b}$$

$$\hat{v}_i = -\cos \theta_i \hat{x} - \sin \theta_i \hat{z} \tag{2.83c}$$

The unit vector along the line segment ON is given by,

$$\hat{\text{ON}} = \cos \phi_b \sin \theta_b \hat{x} + \sin \phi_b \sin \theta_b \hat{y} + \cos \theta_b \hat{z} \tag{2.84}$$

The dot product $\hat{k}_i \cdot \hat{\text{ON}}$ is,

$$\hat{k}_i \cdot \hat{\text{ON}} = \cos \psi = \cos \theta_b \cos \theta_i - \sin \theta_i \cos \phi_b \sin \theta_b \tag{2.85}$$

From Fig. 2.10b it should be clear that $\hat{\text{ON}} \cdot \hat{h}_i = -\sin \psi \sin \beta$ and $\hat{\text{ON}} \cdot \hat{v}_i = -\sin \psi \cos \beta$. Using (2.83) and (2.84), it follows (Holt 1984) that,

$$\cos \beta \sin \psi = \cos \theta_b \sin \theta_i + \sin \theta_b \cos \theta_i \cos \phi_b \tag{2.86a}$$

$$\sin \beta \sin \psi = \sin \theta_b \sin \phi_b \tag{2.86b}$$

The scattering matrix elements in (2.53) are expressed in terms of the three angles θ_i, θ_b, and ϕ_b. Using (2.86), the scattering matrix elements in (2.53) are now expressed in terms of the two angles ψ and β. After some tedious algebra the results are,

$$(S_{hh})_{\text{FSA}} = \frac{-k_0^2}{4\pi \varepsilon_0} \left[\alpha \cos^2 \psi + \sin^2 \psi \left(\alpha_{z_b} \sin^2 \beta + \alpha \cos^2 \beta \right) \right] \tag{2.87a}$$

$$(S_{vh})_{\text{FSA}} = (-S_{hv})_{\text{FSA}} = \frac{k_0^2}{4\pi \varepsilon_0} \left[\frac{(\alpha_{z_b} - \alpha)}{2} \sin^2 \psi \sin 2\beta \right] \tag{2.87b}$$

$$(S_{vv})_{\text{FSA}} = \frac{k_0^2}{4\pi \varepsilon_0} \left[\alpha \cos^2 \psi + \sin^2 \psi \left(\alpha_{z_b} \cos^2 \beta + \alpha \sin^2 \beta \right) \right] \tag{2.87c}$$

From a scattering theory viewpoint, the relative angle ψ between the incidence direction and the particle symmetry axis is more physically meaningful than the angles θ_i and (θ_b, ϕ_b), which are relative to the choice of the XYZ system. From the wave polarization viewpoint, it is the canting angle β in the plane of polarization of the incident wave that is more physically meaningful than the orientation angles (θ_b, ϕ_b) of the symmetry axis. For example, when circular polarization is used one of the radar parameters can be used to measure β, as will be discussed in Chapter 3. However, the particle orientation angles (θ_b, ϕ_b) are microphysically meaningful, especially if the Z-axis is directed along the local vertical direction (i.e. opposite to the direction of gravitational force $\vec{g}$). It should also be clear that the angle β is representative of θ_b only for the case $\theta_i \approx 90°$, i.e for horizontal incidence (a low radar elevation angle). For vertical incidence, $\theta_i \approx 0°$, the polarization plane is the XY-plane and $\beta = \phi_b$. Since ϕ_b is generally assumed to be uniformly distributed in the interval $(0, 2\pi)$, the angle β behaves similarly and is not representative of particle orientation. Several authors, most notably McCormick and Hendry (1975), have assumed that the pdf of β is a symmetric distribution. Oguchi (1983) has assumed that the pdfs of both ψ and β are independent Gaussian. From a physical viewpoint the hydrometeor orientation is specified by the pdfs of θ_b and ϕ_b, and is independent of the angle of incidence. The distribution of ψ and β should follow from (2.85, 2.86).

It is possible to organize (2.87) in such a way as to make the roles of the angles ψ (relevant to scattering) and β (relevant to canting angle) clearer. Before this is done, it is usual in radar applications to use the BSA convention rather than FSA. Thus, using (2.13), (2.87) is expressed in BSA as,

$$(S_{hh})_{\mathrm{BSA}} = \frac{k_0^2}{4\pi\varepsilon_0}\left[\alpha\cos^2\psi + \sin^2\psi\left(\alpha_{z_b}\sin^2\beta + \alpha\cos^2\beta\right)\right] \tag{2.88a}$$

$$(S_{vh})_{\mathrm{BSA}} = (S_{hv})_{\mathrm{BSA}} = \frac{k_0^2}{4\pi\varepsilon_0}\left[\frac{(\alpha_{z_b}-\alpha)}{2}\sin^2\psi\sin 2\beta\right] \tag{2.88b}$$

$$(S_{vv})_{\mathrm{BSA}} = \frac{k_0^2}{4\pi\varepsilon_0}\left[\alpha\cos^2\psi + \sin^2\psi\left(\alpha_{z_b}\cos^2\beta + \alpha\sin^2\beta\right)\right] \tag{2.88c}$$

Next, define $S_{11}(\psi)$ and $S_{22}(\psi)$,

$$S_{11}(\psi) = \frac{k_0^2}{4\pi\varepsilon_0}\left(\alpha\cos^2\psi + \alpha\sin^2\psi\right) \tag{2.89a}$$

$$S_{22}(\psi) = \frac{k_0^2}{4\pi\varepsilon_0}\left(\alpha\cos^2\psi + \alpha_{z_b}\sin^2\psi\right) \tag{2.89b}$$

It follows that,

$$S_{22}(\psi) - S_{11}(\psi) = \frac{k_0^2}{4\pi\varepsilon_0}(\alpha_{z_b}-\alpha)\sin^2\psi \tag{2.90a}$$

$$S_{11}(\psi)\cos^2\beta + S_{22}(\psi)\sin^2\beta = \frac{k_0^2}{4\pi\varepsilon_0}\left[\alpha\cos^2\psi + \sin^2\psi\left(\alpha_{z_b}\sin^2\beta + \alpha\cos^2\beta\right)\right] \tag{2.90b}$$

$$S_{11}(\psi)\sin^2\beta + S_{22}(\psi)\cos^2\beta = \frac{k_0^2}{4\pi\varepsilon_0}\left[\alpha\cos^2\psi + \sin^2\psi\left(\alpha_{z_b}\cos^2\beta + \alpha\sin^2\beta\right)\right] \tag{2.90c}$$

The right-hand side of (2.90) can be readily identified with the right-hand side of (2.88). Thus,

$$(S_{hh})_{\mathrm{BSA}} = \left[S_{11}(\psi)\cos^2\beta + S_{22}(\psi)\sin^2\beta\right] \tag{2.91a}$$

$$(S_{vh})_{\mathrm{BSA}} = (S_{hv})_{\mathrm{BSA}} = \left[\frac{S_{22}(\psi) - S_{11}(\psi)}{2}\sin 2\beta\right] \tag{2.91b}$$

$$(S_{vv})_{BSA} = \left[S_{11}(\psi) \sin^2 \beta + S_{22}(\psi) \cos^2 \beta \right] \tag{2.91c}$$

The incident field amplitude at the origin can be written as,

$$\vec{E}^i(O) = E_h^i \hat{h}_i + E_v^i \hat{v}_i \tag{2.92}$$

If unit vectors $\hat{h}_i'$ and $\hat{v}_i'$ are introduced, which are obtained from $\hat{h}_i$ and $\hat{v}_i$ through rotation by an angle β in the polarization plane, as shown in Fig. 2.11, then,

$$\begin{bmatrix} \hat{h}_i \\ \hat{v}_i \end{bmatrix} = \begin{bmatrix} \cos\beta & \sin\beta \\ -\sin\beta & \cos\beta \end{bmatrix} \begin{bmatrix} \hat{h}_i \\ \hat{v}_i \end{bmatrix}' = \mathbf{R}(\beta) \begin{bmatrix} \hat{h}_i \\ \hat{v}_i \end{bmatrix}' \tag{2.93a}$$

It follows that,

$$\begin{bmatrix} E_h^i \\ E_v^i \end{bmatrix}' = \mathbf{R}(-\beta) \begin{bmatrix} E_h^i \\ E_v^i \end{bmatrix} \tag{2.93b}$$

where $\mathbf{R}(\beta)$ is the rotation matrix and $\mathbf{R}^{-1}(\beta) = \mathbf{R}^t(\beta) = \mathbf{R}(-\beta)$. Note that the superscript $'t'$ represents a matrix transpose. It is easily verified that,

$$\mathbf{R}(\beta) \begin{bmatrix} S_{11}(\psi) & 0 \\ 0 & S_{22}(\psi) \end{bmatrix} \mathbf{R}(-\beta) = \mathbf{S}_{BSA} \tag{2.94}$$

where the elements of $\mathbf{S}_{BSA}$ are as in (2.91). While the above equation is derived for back scattering by spheroids under the Rayleigh–Gans approximation, the form is more generally valid for non-Rayleigh scattering by bodies of revolution (e.g. cones, cylinders).

The roles of the ψ and β angles are now clearer. They enable the $\mathbf{S}_{BSA}$ matrix to be diagonalized using real rotations. The diagonal elements $S_{11}(\psi)$ and $S_{22}(\psi)$ defined in (2.89) are simply the solution to the scattering problem with the particle symmetry axis aligned along the Z-axis (see Fig. 2.7a, b) and interpreting the incidence angle (θ_i) as the relative angle (ψ) between the incidence direction and the symmetry axis, i.e. $\psi = \theta_i$ in this special configuration. From (2.53) with $\theta_b = \phi_b = 0°$,

$$(S_{hh})_{BSA} = \frac{k_0^2}{4\pi\varepsilon_0} \alpha = \frac{k_0^2}{4\pi\varepsilon_0} \alpha(\cos^2\psi + \sin^2\psi) = S_{11}(\psi) \tag{2.95a}$$

$$(S_{vv})_{BSA} = \frac{k_0^2}{4\pi\varepsilon_0} (\alpha\cos^2\psi + \alpha_{zb}\sin^2\psi) = S_{22}(\psi) \tag{2.95b}$$

$$(S_{vh})_{BSA} = 0 \tag{2.95c}$$

It is important to recognize that the above solution does not change if the particle symmetry axis is tilted $(\theta_b = \psi, \phi_b = \pi)$ and the incidence direction is along the positive Z-axis so that the relative angle ψ between the incidence direction and symmetry axis is unchanged $(\psi = \theta_b)$. If the angle ϕ_b now changes to, say, $90°$, keeping the same angle θ_b and incidence direction $\theta_i = 0°$ (see Fig. 2.6a), the relative

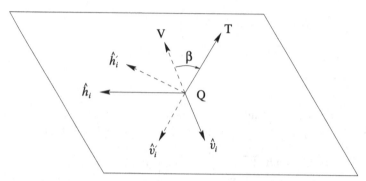

Fig. 2.11. Rotated unit vectors $\hat{h}'_i$ and $\hat{v}'_i$ in the plane of polarization of the incident wave, where β is the canting angle and is a positive angle measured clockwise from the line segment QV.

angle ψ does not change; but to keep the solution the same as in (2.95), the polarization basis $(\hat{h}_i, \hat{v}_i)$ drawn in Fig. 2.6a (see the case of $\theta_i = 0°$) must be rotated by the angle $\beta = 90°$. In this rotated polarization state $(\hat{h}'_i, \hat{v}'_i)$, the solution is the same as in (2.95), i.e. $(S_{hh})' = S_{11}(\psi), (S_{vv})' = S_{22}(\psi)$, and $(S_{vh})' = 0$. Referring now to Figs. 2.10, 2.11 and to (2.94), the process of diagonalization is now clear. For a given θ_i, θ_b, and ϕ_b, the angles ψ and β are computed from (2.86). In the polarization basis $(\hat{h}'_i, \hat{v}'_i)$, which is obtained from $(\hat{h}_i, \hat{v}_i)$ by rotation through the angle β, the solution is the same as in (2.95), i.e. $(S_{hh})' = S_{11}(\psi), (S_{vv})' = S_{22}(\psi)$, and $(S_{vh})' = 0$. Thus,

$$\begin{bmatrix} E^r_h \\ E^r_v \end{bmatrix}' = \frac{e^{-jk_0 r}}{r} \begin{bmatrix} S_{11}(\psi) & 0 \\ 0 & S_{22}(\psi) \end{bmatrix}_{\text{BSA}} \begin{bmatrix} E^i_h \\ E^i_v \end{bmatrix}' \tag{2.96}$$

The scattered (or reflected) field is indicated by the superscript r, since the BSA convention is used. Using (2.93b) to rotate the incident and reflected wave polarization basis from $(\hat{h}'_i, \hat{v}'_i)$ to $(\hat{h}_i, \hat{v}_i)$ yields (2.94). Recall that in the BSA convention, the unit vectors of the incident and reflected waves are identical in the back scattering direction.

The polarization basis $(\hat{h}'_i, \hat{v}'_i)$ may be referred to as the eigenbasis, and $\vec{E}_1 = E^{i'}_h \hat{h}'_i$ and $\vec{E}_2 = E^{i'}_v \hat{v}'_i$ may be referred to as the eigenvectors. For an oblate spheroid it follows on physical grounds that maximal back scattered power will result when the transmitted polarization state is aligned along the spheroid's major axes (or when $\vec{E}_1$ is transmitted), while minimal power will result for alignment along its minor axis (or $\vec{E}_2$). This result can be stated more formally by defining the Graves (1956) power matrix,

$$G = (S_{\text{BSA}})^{t^*} (S_{\text{BSA}}) \tag{2.97}$$

where S_{BSA} is given in (2.94); which, as mentioned earlier, is generally valid for non-Rayleigh scattering by a rotationally symmetric particle. Since G is Hermitian, its eigenvalues are always real. Kostinski and Boerner (1986) showed that these eigenvalues (λ_1, λ_2) equal the maximal and minimal back scattered powers, while the corresponding eigenvectors are the optimal polarization states ($\vec{E}_1$ and $\vec{E}_2$). Substituting (2.94) into (2.97), it follows that,

$$G = \mathbf{R}(\beta)\, G_0 \mathbf{R}(-\beta) \tag{2.98a}$$

where,

$$G_0 = \begin{bmatrix} |S_{11}(\psi)|^2 & 0 \\ 0 & |S_{22}(\psi)|^2 \end{bmatrix} \tag{2.98b}$$

Thus, the Graves matrix (G) of a canted oblate particle can be expressed in terms of the Graves matrix (G_0) based on the "principal" plane elements of the scattering matrix (S_{11} and S_{22}) and the rotation matrix $\mathbf{R}(\beta)$ defined in (2.93a). Since G is obtained from an orthogonal transformation of G_0, the eigenvalues of G are the same as the eigenvalues of G_0, or $\lambda_1 = |S_{11}|^2$ and $\lambda_2 = |S_{22}|^2$. Kwiatkowski et al. (1995) define the asymmetry ratio (A_{sm}) as the ratio of the eigenvalues of G,

$$A_{sm} = \frac{\lambda_1}{\lambda_2} = \frac{|S_{11}(\psi)|^2}{|S_{22}(\psi)|^2} \tag{2.99}$$

The above ratio corresponds to the ratio of maximal to minimal scattered power. It is a generalization of the differential reflectivity (Z_{dr}) defined in (2.54), which depends on the canting angle, whereas A_{sm} does not. The application of the asymmetry ratio to rainfall will be discussed in Section 3.14.4.

In the forward scattered direction ($\theta_s = \theta_i, \phi_s = \phi_i$), FSA must be used and the unit vectors satisfy $\hat{h}_i = \hat{h}_s$ and $\hat{v}_i = \hat{v}_s$. In Chapter 4, the scattering matrix in the forward scattered direction will be needed to study the propagation of coherent waves in a random medium. The scattering matrix in the forward direction is essentially identical with (2.91), and can be derived from the general expressions in (2.49–2.52) and by using $\hat{h}_i = \hat{h}_s = \hat{y}, \hat{v}_i = \hat{v}_s = -\cos\theta_i\hat{x} - \sin\theta_i\hat{y}$. In terms of ψ and β, the forward scattered matrix can be expressed as,

$$\mathbf{S}_{FSA}(\hat{i}, \hat{i}) = \mathbf{R}(\beta) \begin{bmatrix} S_{11}(\hat{i}, \hat{i}; \psi) & 0 \\ 0 & S_{22}(\hat{i}, \hat{i}; \psi) \end{bmatrix} \mathbf{R}(-\beta) \tag{2.100}$$

where the "principal" plane elements are $S_{11}(\hat{i}, \hat{i}; \psi)$ and $S_{22}(\hat{i}, \hat{i}; \psi)$ given in (2.89a, b) in the Rayleigh–Gans limit.

2.4   Mie solution

The Mie solution to scattering of an incident plane wave by dielectric spheres (see Fig. 2.12) is, in essence, a boundary value problem governed by the vector Helmholtz equation.[2] A review of vector spherical harmonics and multipole expansion of the electromagnetic field is provided in Appendix 2. The approach to the Mie solution

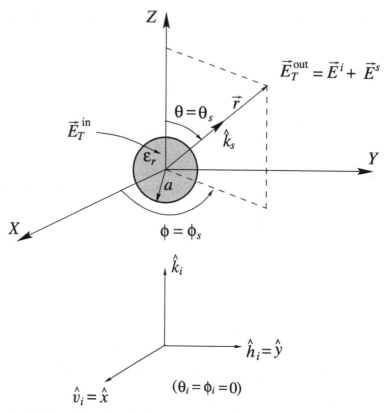

Fig. 2.12. Scattering of an incident plane wave by a dielectric sphere (shaded) of radius a and relative permittivity ε_r. Geometry is shown relevant to the Mie solution.

is similar to the discussion in Appendix 1, which involves the boundary value problem for potentials governed by the Laplace equation for a dielectric sphere in the presence of a uniform, electrostatic incident field (illustrated in Fig. A1.11). In the electrostatic case, the incident potential $\Psi_i = -E_0 r \cos\theta = -E_0 r P_1^0(\cos\theta)$ was an eigenfunction corresponding to a dipole at infinity. In the Mie solution, the incident plane wave $\hat{x} E_0 \exp(-jk_0 z)$ is given by the multipole expansion in (A2.28a). In the electrostatic case, the perturbation potential Ψ_p external to the sphere was of a simple dipole form, i.e. $A_0 \cos\theta / r^2$. In the Mie solution, the corresponding scattered field $\vec{E}^s$ external to the sphere ($r > a$) must be expanded into multipoles as,

$$\vec{E}^s(k_0 r, \theta, \phi) = 2\pi E_0 \sum_{n=1}^{\infty} (-i)^n \gamma_{1n}^{1/2} \left[\alpha_{o1n} \vec{M}_{o1n}(k_0 r, \theta, \phi) + i\beta_{e1n} \vec{N}_{e1n}(k_0 r, \theta, \phi) \right]$$

(2.101)

where α_{o1n} and β_{e1n} are the unknown scattered field expansion coefficients. Note the use of $i = \sqrt{-1}$ here, rather than the more conventional j. In the electrostatic case, the total potential inside the sphere was of the simple form $\Psi_T^{in} = B_0 r \cos\theta$ (i.e. the total electrostatic field inside the sphere ($\vec{E}_T^{in} = -B_0\hat{z}$), was constant). In the Mie solution, the total electric field inside ($r < a$) must be expanded into multipoles,

$$\vec{E}_T^{in}(kr, \theta, \phi) = 2\pi E_0 \sum_{n=1}^{\infty} (-i)^n \gamma_{1n}^{1/2} \left[c_{o1n} \text{ Rg } \vec{M}_{o1n}(kr, \theta, \phi) \right.$$

$$\left. + i d_{e1n} \text{ Rg } \vec{N}_{e1n}(kr, \theta, \phi) \right] \qquad (2.102)$$

where c_{o1n} and d_{e1n} are the unknown internal field expansion coefficients. As explained in Appendix 2, "Rg" refers to the "regular" part of the vector spherical harmonic $\vec{M}$ or $\vec{N}$. Note that the single $m = 1$ mode of the incident field expansion can only excite the $m = 1$ mode in the scattered and total internal electric fields. Also, the incident field polarization ($E_0\hat{x}$) dictates the particular even/odd choices in (2.101, 2.102). By orthogonality, the $m \neq 1$ modes and ($\alpha_{e1n}, \beta_{o1n}$) and ($c_{e1n}, d_{o1n}$) will all be zero. The reverse would be true for the even/odd combinations if the incident polarization state were $E_0\hat{y}$ (see (A2.28b)).

In the electrostatic case, the boundary conditions were particularly simple as given in (A1.16, A1.19) involving continuity of both the potential ($\Psi_t^{in} = \Psi_i + \Psi_p$) and $\varepsilon \partial\Psi/\partial r$ across the spherical surface. In the Mie solution, the general boundary conditions, as stated in (1.2a, b), must be used across the spherical surface at $r = a$,

$$\hat{n}_1 \times (\vec{E}_T^{in}) + \hat{n}_2 \times (\vec{E}^i + \vec{E}^s) = 0 \qquad (2.103a)$$

$$\hat{n}_1 \times (\vec{B}_T^{in}) + \hat{n}_2 \times (\vec{B}^i + \vec{B}^s) = 0 \qquad (2.103b)$$

where $\hat{n}_1 = \hat{r}$, $\hat{n}_2 = -\hat{r}$. The orthogonality of the $\vec{m}_{\sigma mn}(\theta, \phi)$ and $\hat{r} \times \vec{m}_{\sigma mn}(\theta, \phi)$ functions over the surface $r = a$ must be invoked (e.g. see (A2.23)). Once these elements of the Mie solution are grasped the actual steps are tedious but straightforward. The solutions (e.g. Stratton 1941) for the scattered field expansion coefficients (α_{o1n} and β_{e1n}) for the case $\vec{E}^i = \hat{x} E_0$ (see Fig. 2.12) are given as,

$$\alpha_{o1n} = \frac{\rho j_n(\rho)[\rho_0 j_n(\rho_0)]' - \sqrt{\varepsilon_r}\rho_0 j_n(\rho_0)[\rho j_n(\rho)]'}{\sqrt{\varepsilon_r}\rho_0 h_n^{(2)}(\rho_0)[\rho j_n(\rho)]' - \rho j_n(\rho)[\rho_0 h_n^{(2)}(\rho_0)]'} \qquad (2.104a)$$

$$\beta_{e1n} = \frac{\rho_0 j_n(\rho_0)[\rho j_n(\rho)]' - \sqrt{\varepsilon_r}\rho j_n(\rho)[\rho_0 j_n(\rho_0)]'}{\sqrt{\varepsilon_r}\rho j_n(\rho)[\rho_0 h_n^{(2)}(\rho_0)]' - \rho_0 h_n^{(2)}(\rho_0)[\rho j_n(\rho)]'} \qquad (2.104b)$$

where $\rho_0 = k_0 a$; $\rho = \rho_0\sqrt{\varepsilon_r}$; and, as defined in Appendix 2, $[\rho z_n(\rho)]' = d[\rho z_n(\rho)]/d\rho$. When the incident wave polarization is $E_0\hat{y}$, then the corresponding scattered field expansion coefficients obey the simple relation $\alpha_{e1n} = -\alpha_{o1n}$ and $\beta_{o1n} = \beta_{e1n}$.

The far-zone scattered field is of primary interest in radar applications. It is obtained by using (A2.13a, b) in (2.101), which yields,

$$\vec{E}^s(k_0 r, \theta, \phi) = 2\pi E_0 \frac{e^{-i\rho}}{\rho} \sum_{n=1}^{\infty} (-i)^n \gamma_{1n}^{1/2}$$

$$\times \left[\alpha_{o1n}(i)^{n+1} \vec{m}_{o1n}(\theta, \phi) + i(i)^n \beta_{e1n}\hat{r} \times \vec{m}_{e1n}(\theta, \phi) \right] \qquad (2.105a)$$

$$= i2\pi E_0 \frac{e^{-ik_0 r}}{k_0 r} \sum_{n=1}^{\infty} \gamma_{1n}^{1/2} \left[\alpha_{o1n} \vec{m}_{o1n}(\theta, \phi) + \beta_{e1n} \hat{r} \times \vec{m}_{e1n}(\theta, \phi) \right]$$

(2.105b)

Note that $\rho = k_0 r$ now. The angular part of the far-field $\vec{E}^s$ can be explicitly expressed using (A2.10b, c), and is written as,

$$\vec{E}^s(k_0 r, \theta, \phi) = i E_0 \frac{e^{-ik_0 r}}{k_0 r} \sum_{n=1}^{\infty} \frac{2n+1}{n(n+1)} \left\{ \hat{\theta} \left[\alpha_{o1n} \frac{P_n^1(\cos \theta)}{\sin \theta} + \beta_{e1n} \frac{d P_n^1(\cos \theta)}{d\theta} \right] \right.$$

$$\left. \times \cos \phi - \hat{\phi} \left[\alpha_{o1n} \frac{d P_n^1(\cos \theta)}{d\theta} + \beta_{e1n} \frac{P_n^1(\cos \theta)}{\sin \theta} \right] \sin \phi \right\}$$

(2.106)

where $P_n^m(\cos \theta)$ are the associated Legendre functions. Note that the scattered wave direction $(\hat{k}_s)$ is as marked on Fig. 2.12; in the FSA, $\hat{\theta} = \hat{\theta}_s = \hat{v}_s$ and $\hat{\phi} = \hat{\phi}_s = \hat{h}_s$ (refer also to Fig. 2.1b). Since $\theta_i = \phi_i = 0°$, the incident field $\vec{E}^i = E_0 \hat{x} = E_0 \hat{v}_i$. Thus, when $E_0 = 1$ V m^{-1}, the second column of the $\mathbf{S}_{FSA}$ matrix defined in (2.6) is obtained from (2.106) as,

$$(S_{hv})_{FSA} = \frac{-i \sin \phi_s}{k_0} S_1(\theta_s)$$

(2.107a)

$$(S_{vv})_{FSA} = \frac{i \cos \phi_s}{k_0} S_2(\theta_s)$$

(2.107b)

where,

$$S_2(\theta_s) = \sum_{n=1}^{\infty} \frac{2n+1}{n(n+1)} \left[\alpha_{o1n} \frac{P_n^1(\cos \theta_s)}{\sin \theta_s} + \beta_{e1n} \frac{d P_n^1(\cos \theta_s)}{d\theta_s} \right]$$

(2.108a)

$$S_1(\theta_s) = \sum_{n=1}^{\infty} \frac{2n+1}{n(n+1)} \left[\alpha_{o1n} \frac{d P_n^1(\cos \theta_s)}{d\theta_s} + \beta_{e1n} \frac{P_n^1(\cos \theta_s)}{\sin \theta_s} \right]$$

(2.108b)

When $\vec{E}^i = E_0 \hat{y} = E_0 \hat{h}_i$, the corresponding scattered field expansion coefficients are α_{e1n} and β_{o1n}, and they are simply related to α_{o1n} and β_{e1n}, as $\alpha_{e1n} = -\alpha_{o1n}$ and $\beta_{o1n} = \beta_{e1n}$. The far-field $\vec{E}^s$ is now written as (cf. 2.105),

$$\vec{E}^s(k_0 r, \theta, \phi) = i2\pi E_0 \frac{e^{-ik_0 r}}{k_0 r}$$

$$\times \sum_{n=1}^{\infty} \gamma_{1n}^{1/2} \left[\alpha_{e1n} \vec{m}_{e1n}(\theta, \phi) + \beta_{o1n} \hat{r} \times \vec{m}_{o1n}(\theta, \phi) \right]$$

(2.109a)

$$= i E_0 \frac{e^{-ik_0 r}}{k_0 r} \sum_{n=1}^{\infty} \frac{2n+1}{n(n+1)}$$

$$\times \left\{ \hat{\theta} \left[-\alpha_{e1n} \frac{P_n^1(\cos \theta)}{\sin \theta} + \beta_{o1n} \frac{d P_n^1(\cos \theta)}{d\theta} \right] \sin \phi \right.$$

$$+ \hat{\phi} \left[-\alpha_{e1n} \frac{d P_n^1 (\cos \theta)}{d\theta} + \beta_{o1n} \frac{P_n^1 (\cos \theta)}{\sin \theta} \right] \cos \phi \Bigg\} \qquad (2.109b)$$

The first column of the $\mathbf{S}_{\text{FSA}}$ matrix is defined as,

$$(S_{hh})_{\text{FSA}} = \frac{i \cos \phi_s}{k_0} S_1(\theta_s) \qquad (2.110a)$$

$$(S_{vh})_{\text{FSA}} = \frac{i \sin \phi_s}{k_0} S_2(\theta_s) \qquad (2.110b)$$

Finally, the $\mathbf{S}_{\text{FSA}}$ matrix for $\theta_i = \phi_i = 0°$ and, as defined in (2.6a), relates the scattered field components to the incident plane wave field components as,

$$\begin{bmatrix} E_h^s \\ E_v^s \end{bmatrix} = \frac{e^{-ik_0 r}}{r} \begin{bmatrix} \dfrac{i \cos \phi_s}{k_0} S_1(\theta_s) & \dfrac{-i \sin \phi_s}{k_0} S_1(\theta_s) \\ \dfrac{i \sin \phi_s}{k_0} S_2(\theta_s) & \dfrac{i \cos \phi_s}{k_0} S_2(\theta_s) \end{bmatrix}_{\text{FSA}} \begin{bmatrix} E_h^i \\ E_v^i \end{bmatrix} \qquad (2.111)$$

which may be compared with the $\mathbf{S}_{\text{FSA}}$ Rayleigh limit given in (2.26).

2.4.1 van de Hulst convention: scattering and extinction cross sections

For those readers familiar with the van de Hulst (1981) scattering alignment convention, (2.111) can be further simplified. The van de Hulst convention defines the scattering plane ($\phi = \phi_s$) as a reference plane, and unit vectors for $\vec{E}^i$ and $\vec{E}^s$ are defined as being perpendicular (r) or parallel (l) to this plane. Thus, $\vec{E}^i = E_l^i \hat{e}_l^i + E_r^i \hat{e}_r^i$ and $\vec{E}^s = E_l^s \hat{e}_l^s + E_r^s \hat{e}_r^s$, where the unit vectors are as illustrated in Fig. 2.13. The FSA triplet $(\hat{k}_i, \hat{v}_i, \hat{h}_i)$ is now $(\hat{k}_i, \hat{e}_r^i, \hat{e}_l^i)$, while the FSA triplet $(\hat{k}_s, \hat{v}_s, \hat{h}_s)$ is now $(\hat{k}_s, \hat{e}_r^s, \hat{e}_l^s)$. The scattered field expansion coefficients defined by van de Hulst are (a_n^s, b_n^s), which are related to $(\alpha_{o1n}, \beta_{e1n})$ as $a_n^s = -\beta_{e1n}$ and $b_n^s = -\alpha_{o1n}$. Thus, the van de Hulst $S_2^{\text{VDH}}(\theta_s) = -S_2(\theta_s)$ and $S_1^{\text{VDH}}(\theta_s) = -S_1(\theta_s)$. With these definitions in mind, (2.111) can be expressed as,

$$\begin{bmatrix} E_r^s \\ E_l^s \end{bmatrix} = \frac{e^{-ik_0 r}}{ik_0 r} \begin{bmatrix} -1 & 0 \\ 0 & 1 \end{bmatrix} \begin{bmatrix} S_1^{\text{VDH}}(\theta_s) \cos \phi_s & -S_1^{\text{VDH}}(\theta_s) \sin \phi_s \\ S_2^{\text{VDH}}(\theta_s) \sin \phi_s & S_2^{\text{VDH}}(\theta_s) \cos \phi_s \end{bmatrix}$$
$$\begin{bmatrix} -\cos \phi_s & \sin \phi_s \\ \sin \phi_s & \cos \phi_s \end{bmatrix} \begin{bmatrix} E_r^i \\ E_l^i \end{bmatrix} \qquad (2.112a)$$

$$= \frac{e^{-ik_0 r}}{ik_0 r} \begin{bmatrix} S_1^{\text{VDH}}(\theta_s) & 0 \\ 0 & S_2^{\text{VDH}}(\theta_s) \end{bmatrix} \begin{bmatrix} E_r^i \\ E_l^i \end{bmatrix} \qquad (2.112b)$$

A general form for the linear polarization state of an incident plane wave is $E_0 \hat{e}_i = E_0(\cos \varepsilon \hat{h}_i + \sin \varepsilon \hat{v}_i)$. The total scattering cross section of the sphere can be derived using (1.45), which assumes unit amplitude for the linearly polarized incident wave

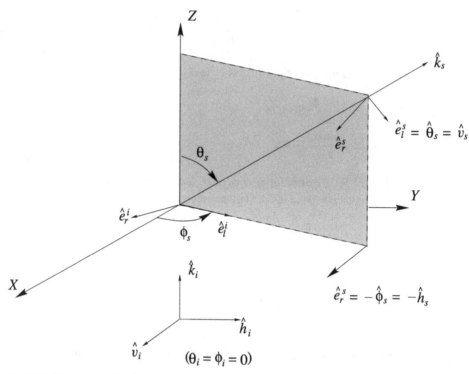

Fig. 2.13. Unit vectors for the van de Hulst convention. The scattering plane is shaded and defined by $\phi = \phi_s$. Note that $\hat{e}_r^i \times \hat{e}_l^i = \hat{k}_i$ and $\hat{e}_r^s \times \hat{e}_l^s = \hat{k}_s$.

$(E_0 = 1 \text{ V m}^{-1})$. Thus,

$$\sigma_s = \int_{4\pi} |\vec{f}|^2 \, d\Omega \tag{2.113a}$$

$$= \frac{1}{k_0^2} \int_0^{2\pi} \int_0^{\pi} \left[S_1^{\text{VDH}}(S_1^{*\text{VDH}})|E_r^i|^2 + S_2^{\text{VDH}}(S_2^{*\text{VDH}})|E_l^i|^2 \right] \sin\theta_s \, d\theta_s \, d\phi_s$$

$$\tag{2.113b}$$

$$= \frac{\pi}{k_0^2} \int_0^{\pi} \left[S_1^{\text{VDH}}(S_1^{*\text{VDH}}) + S_2^{\text{VDH}}(S_2^{*\text{VDH}}) \right] \sin\theta_s \, d\theta_s \tag{2.113c}$$

After some tedious algebra and using the orthogonality of the associated Legendre functions given in Van Bladel (1985; eq. (119) of his Appendix),

$$\sigma_s = \frac{2\pi}{k_0^2} \sum_{n=1}^{\infty} (2n+1) \left(|a_n^s|^2 + |b_n^s|^2 \right) \tag{2.114}$$

Note that the total scattering cross section of a sphere is independent of the linear polarization state of the incident plane wave. While wave polarization will be considered in detail in Chapter 3, it will suffice here to mention that $\tan \varepsilon$ controls the amplitude ratio of the two orthogonal components of $\vec{E}^i$. From symmetry, it can also be deduced that σ_s should also be independent of the incidence direction ($\hat{k}_i$).

The extinction cross section from the optical theorem is given in (1.58) as,

$$\sigma_{ext} = \frac{-4\pi}{k_0} \, \text{Im}\left[\vec{f}(\hat{i}, \hat{i}) \cdot \hat{e}_i\right] \tag{2.115}$$

From (2.111) and using the fact that $\theta_s = \theta_i = 0°$, $\phi_i = \phi_s = 0°$, and that, in the forward scattered direction, $\hat{v}_s = \hat{v}_i$, $\hat{h}_s = \hat{h}_i$,

$$\hat{e}_i \cdot \vec{f}(\hat{i}, \hat{i}) = \frac{i}{k_0}\left[\hat{h}_i S_1(0) \cos\varepsilon + \hat{v}_i S_2(0) \sin\varepsilon\right] \cdot \left[\hat{v}_i \sin\varepsilon + \hat{h}_i \cos\varepsilon\right] \tag{2.116a}$$

$$= \frac{i}{k_0}\left[S_1(0) \cos^2\varepsilon + S_2(0) \sin^2\varepsilon\right] \tag{2.116b}$$

Using eqs. (98, 105, 112) from Appendix 4 of Van Bladel (1985) it can be shown that for $\theta_s = 0°$,

$$\frac{d P_n^1(\cos\theta_s)}{d\theta_s} = \frac{P_n^1(\cos\theta_s)}{\sin\theta_s} = \frac{n(n+1)}{2} \tag{2.117}$$

Substituting (2.108a, b) into (2.116b), and using the above relation yields,

$$\sigma_{ext} = -\frac{2\pi}{k_0^2} \sum_{n=1}^{\infty} (2n+1) \, \text{Re}(\alpha_{o1n} + \beta_{e1n}) \tag{2.118}$$

$$= \frac{2\pi}{k_0^2} \sum_{n=1}^{\infty} (2n+1) \, \text{Re}(a_n^s + b_n^s) \tag{2.119}$$

The extinction cross section is also independent of the incident linear polarization state and, from symmetry, will also be independent of the direction of incidence ($\hat{k}_i$).

2.4.2   Bistatic radar cross section

From (1.46b), the bistatic radar cross section is,

$$\sigma_{bi}(\hat{s}, \hat{i}) = 4\pi |\vec{f}(\hat{s}, \hat{i})|^2 \tag{2.120}$$

and will depend on the linear polarization state of the incident wave. One application of the use of bistatic Doppler principles in meteorology is where a primary transmitting radar is used as the radiation source, and a second bistatic antenna is used to detect the bistatic Doppler shift frequency.[3] For example, Fig. 2.14 illustrates a vertically pointing primary transmitting radar, illuminating a raindrop falling with velocity $\vec{v}$. For simplicity, a bistatic antenna is located in the $\phi_s = 90°$ plane (YZ-plane) and along the scattering angle θ_s. From (1.111), the time-correlated bistatic scattering cross section is,

$$\sigma_{bi}(\hat{s}, \hat{i}, \tau) = \sigma_{bi}(\hat{s}, \hat{i}) \exp\left[-jk_0(\hat{i} - \hat{s}) \cdot \vec{v}\tau\right] \tag{2.121a}$$

$$= 4\pi |\vec{f}(\hat{s}, \hat{i})|^2 \exp\left[-jk_0(\hat{i} - \hat{s}) \cdot \vec{v}\tau\right] \tag{2.121b}$$

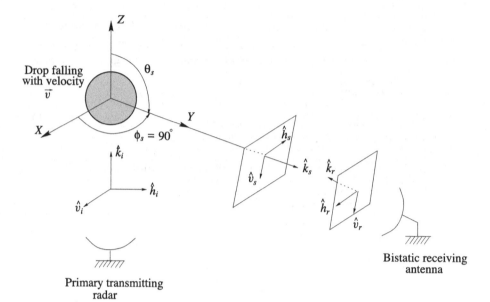

Fig. 2.14. Geometry illustrating bistatic scattering from a drop, with fall velocity $\vec{v}$, when illuminated by a vertical incidence beam from the primary transmitting radar. The scattering direction is specified by $(\theta_s, \phi_s = 90°)$. Note BSA unit vectors $(\hat{k}_r, \hat{v}_r, \hat{h}_r)$ at the bistatic receiving antenna.

From (2.111) with $\phi_s = 90°$,

$$\vec{f}(\hat{s}, \hat{i}) = \frac{-i}{k_0} S_1(\theta_s) E_v^i \hat{h}_s + \frac{i}{k_0} S_2(\theta_s) E_h^i \hat{v}_s \qquad (2.122)$$

where $E_h^i = E_0 \cos \varepsilon$ and $E_v^i = E_0 \sin \varepsilon$. If the primary radar transmits $E_h^i = E_0$ and $E_v^i = 0$, and if the polarization state of the bistatic antenna is horizontal ($\hat{h}_r = -\hat{h}_s$), then no voltage will be received at the bistatic receiver; as a result no bistatic Doppler shift will be detected. The same holds true if $E_h^i = 0$, $E_v^i = E_0$, and the bistatic antenna is vertically polarized ($\hat{v}_r = \hat{v}_s$). However, if $E_h^i = E_v^i = E_0/\sqrt{2}$ (or $\varepsilon = 45°$), then a vertically (or horizontally) polarized bistatic antenna can be used to detect the bistatic Doppler shift. In Chapter 3, see eq. (3.134), the bistatic radar cross section taking into account the polarization states of both the primary transmitting radar and the bistatic antenna, will be formally defined.

2.4.3 Radar cross section

The back scatter or radar cross section is defined as,

$$\sigma_b(-\hat{i}, \hat{i}) = 4\pi |\vec{f}(-\hat{i}, \hat{i})|^2 \qquad (2.123)$$

and for a sphere is independent of the linear polarization state of the incident wave.

From (2.111), with $\phi_s = \phi_i + \pi = \pi$ and $\theta_s = \pi - \theta_i = \pi$,

$$\vec{f}(-\hat{i}, \hat{i}) = \frac{-i}{k_0} \left[S_1(\pi) E_h^i \hat{h}_s + S_2(\pi) E_v^i \hat{v}_s \right] \tag{2.124}$$

Again, using Appendix 4 of Van Bladel (1985) with $\theta_s = \pi$,

$$\frac{dP_n^1}{d\theta_s}(\cos\theta_s) = (-1)^n \frac{n(n+1)}{2} = -\frac{P_n^1(\cos\theta_s)}{\sin\theta_s} \tag{2.125}$$

Using the above in (2.108) and substituting into (2.124) yields,

$$\sigma_b(-\hat{i}, \hat{i}) = \frac{\pi}{k_0^2} \left| \sum_{n=1}^{\infty} (-1)^n (2n+1)(\alpha_{o1n} - \beta_{e1n}) \right|^2 \tag{2.126a}$$

$$= \frac{\pi}{k_0^2} \left| \sum_{n=1}^{\infty} (-1)^n (2n+1)(a_n^s - b_n^s) \right|^2 \tag{2.126b}$$

The expressions for σ_s, $\sigma_{bi}(\hat{s}, \hat{i})$, and $\sigma_b(-\hat{i}, \hat{i})$ from the Mie solution may be compared with the corresponding Rayleigh limit solutions in (1.52), (1.50c), and (1.51b). From the review in Appendix 2 of vector spherical harmonics and dipole radiation, the coefficients β_{e11} and β_{o11} are related to induced electric dipoles in the sphere along the $\hat{x}$- and $\hat{y}$-directions (see A2.20, A2.21), while α_{e11} and α_{o11} are related to induced magnetic dipoles in the sphere along these directions. The coefficients corresponding to $n = 2$ are related to induced quadrupole moments in the sphere. Fast numerical methods (e.g. Wiscombe 1980; Barber and Hill 1990) are now available that can compute the values of α_{o1n} and β_{e1n} given in (2.104a, b). The number of multipoles in the summation in (2.126) is of the order of k_0a, and based on extensive calculations for spheres it was found that,

$$n_{\max} = k_0a + 4.05(k_0a)^{1/3} + 2 \tag{2.127}$$

2.5   Mie coefficients in powers of k_0a: low frequency approximation

The scattered field expansion coefficients (or simply the Mie coefficients) $\alpha_{o1n}(= -b_n^s)$ and $\beta_{e1n}(= -a_n^s)$ in (2.104) take particularly simple forms when k_0a and $\sqrt{\varepsilon_r}k_0a$ are small compared with unity. Since $k_0a = a(2\pi/\lambda) = (\omega/c)a$, this condition can be achieved either by decreasing the frequency for a particle of fixed radius a, or by decreasing the radius at a fixed frequency. The parameter $\rho_0 = k_0a$ is often referred to as the "size" parameter (the subscript on ρ may be dropped for convenience). The small argument approximations for the spherical Bessel and Hankel functions are given in Stratton (1941) as,

$$j_n(\rho) = 2^n \rho^n \frac{n!}{(2n+1)!} + O(\rho^{n+2}) \tag{2.128a}$$

$$h_n^{(2)}(\rho) = \frac{i2n!}{2^n n! \rho^{n+1}} + O\left(\frac{1}{\rho^{n-1}}\right) \tag{2.128b}$$

where $O(\ )$ refers to the "order of". The small argument approximations for the terms $[\rho j_n(\rho)]'$ and $[\rho h_n^{(2)}(\rho)]'$ are easily derived as,

$$[\rho j_n(\rho)]' = \rho j_{n-1}(\rho) - n j_n(\rho) = \frac{\rho^n 2^n (n+1)!}{(2n+1)!} + O(\rho^{n+2}) \tag{2.129a}$$

$$[\rho h_n^{(2)}(\rho)]' = \rho h_{n-1}^{(2)} - n h_n^{(2)}(\rho) = \frac{-i(2n)!}{2^n (n-1)! \rho^{n+1}} + O\left(\frac{1}{\rho^{n-1}}\right) \tag{2.129b}$$

Using the above small argument approximations in (2.104a, b) it is possible to express the lowest order coefficients $(a_1^s, b_1^s, a_2^s,$ etc.) in powers of $k_0 a$. For example, van de Hulst (1981) has given expressions for a_1^s, b_1^s, and a_2^s which are correct up to order of $(k_0 a)^6$, $(k_0 a)^5$, and $(k_0 a)^5$, respectively, as,

$$a_1^s = i A \rho^3 (1 + T \rho^2 - i A \rho^3) \tag{2.130a}$$

$$b_1^s = i A U \rho^5 \tag{2.130b}$$

$$a_2^s = i A W \rho^5 \tag{2.130c}$$

where $\rho = k_0 a$, and A, T, U and W are defined as,

$$A = \frac{2}{3}\left(\frac{\varepsilon_r - 1}{\varepsilon_r + 2}\right); \quad T = \frac{3}{5}\frac{(\varepsilon_r - 2)}{(\varepsilon_r + 2)} \tag{2.131a}$$

$$U = \frac{1}{30}(\varepsilon_r + 2); \quad W = \frac{1}{10}\frac{(\varepsilon_r + 2)}{(2\varepsilon_r + 3)} \tag{2.131b}$$

The total scattering cross section from (2.114), correct up to order ρ^3, involves only $a_1^s \approx i A \rho^3$,

$$\sigma_s = \frac{2\pi}{k_0^2}(3)|a_1^s|^2 \tag{2.132a}$$

$$= \frac{8\pi}{3}(k_0)^4 \left|\frac{\varepsilon_r - 1}{\varepsilon_r + 2}\right|^2 a^6 \tag{2.132b}$$

which is essentially the same as (1.52). The radar cross section $\sigma_b(-\hat{i}, \hat{i})$ from (2.126b), correct up to order ρ^3, also only involves the expansion coefficients $a_1^s \approx i A \rho^3$,

$$\sigma_b(-\hat{i}, \hat{i}) = \frac{\pi}{k_0^2}|(-1)(3)a_1^s|^2 \tag{2.133a}$$

$$= \frac{\pi}{k_0^2}(9)|A|^2 \rho^6 \tag{2.133b}$$

$$= \frac{4\pi}{k_0^2}\left|\frac{\varepsilon_r - 1}{\varepsilon_r + 2}\right|^2 (k_0 a)^6 \tag{2.133c}$$

$$= \frac{\pi^5}{\lambda^4}\left|\frac{\varepsilon_r - 1}{\varepsilon_r + 2}\right|^2 D^6 \tag{2.133d}$$

which is identical to (1.51b). Recall that $D = 2a$ is the diameter of the sphere. It should be clear by now that $a_1^s = -\beta_{e11} \approx i A \rho^3$ is the coefficient related to the $\hat{x}$-directed induced dipole in the sphere. Finally, consider the extinction cross section given in (2.119). It is convenient to consider the lossless (ε_r is real) and lossy (ε_r is complex) cases separately. If ε_r is complex, then for σ_{ext} to be correct up to the order ρ^3 involves only $a_1^s \approx i A \rho^3$,

$$\sigma_{\text{ext}} = \frac{2\pi}{k_0^2} (3) \operatorname{Re}(a_1^s) \tag{2.134a}$$

$$= \frac{4\pi}{k_0^2} \rho^3 \operatorname{Re}\left(i \frac{\varepsilon_r - 1}{\varepsilon_r + 2} \right) \tag{2.134b}$$

$$= 9 k_0 V \frac{\varepsilon_r''}{|\varepsilon_r + 2|^2} \tag{2.134c}$$

which is identical to the absorption cross section in (1.59). Recall that V is the volume of the sphere. Thus, when ε_r is complex, and up to the order ρ^3, the extinction cross section equals the absorption cross section. As discussed at the end of Section 1.5, the leading term in the total scattering cross section is proportional to ρ^6, and thus the general result ($\sigma_{\text{ext}} = \sigma_a + \sigma_s$), when applied to order ρ^3, is not violated. If ε_r is real, then σ_{ext} from (2.119) is,

$$\sigma_{\text{ext}} = \frac{2\pi}{k_0^2} (3) A^2 \rho^6 \tag{2.135}$$

which is essentially the same as σ_s in (2.132b).

In radar applications to scattering by typical hydrometeors,[4] it is useful to estimate the size parameter (ρ) at which the simpler solution based on a_1^s, b_1^s, and a_2^s given in (2.130a, b, c) starts to deviate significantly from the Mie solution. Figure 2.15 compares the back scatter cross section versus ρ using the Mie solution and the low frequency approximation in (2.130). Figure 2.16 shows similar comparisons for the extinction cross section.

2.6 Numerical scattering methods for non-spherical particles

It is beyond the scope of this book to provide a comprehensive review of numerical scattering methods applicable to non-spherical hydrometeors.[5] However, a brief overview of the T-matrix method is provided in Appendix 3. Excellent reviews of methods developed and used in the 1970s/early-1980s by the electromagnetic propagation community investigating attenuation and cross-polarization of waves by hydrometeors along the Earth–satellite propagation path can be found in Oguchi (1981) or Holt (1982). More recent summaries are included in the OPEX *Reference Handbook on Radar*,[5] which compares several numerical schemes such as point matching (Morrison and Cross 1974), the Fredholm integral method (Holt and Santoso 1972) and the T-matrix method (Barber and Hill 1990). Because of the large increase in computational power made available in the mid-1980s/early-1990s, the discrete dipole approximation (DDA)

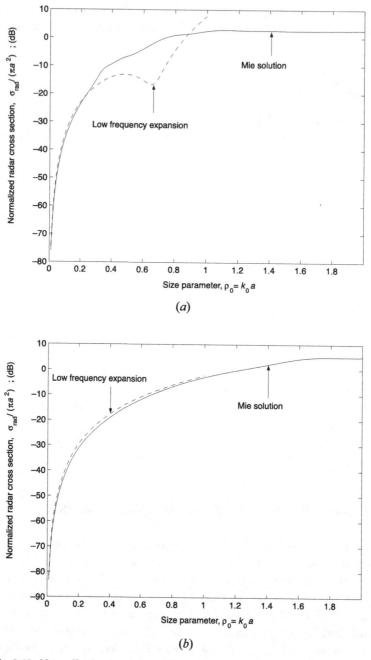

Fig. 2.15. Normalized radar cross section versus size parameter using the Mie solution and the low frequency expansion coefficients in (2.131): (*a*) water spheres, $\varepsilon_r = 80 - j25$; (*b*) ice spheres, $\varepsilon_r = 3.16 - j0.02$. Note that for a 4-mm-diameter sphere, the size parameter is 0.42 at a wavelength of 3 cm.

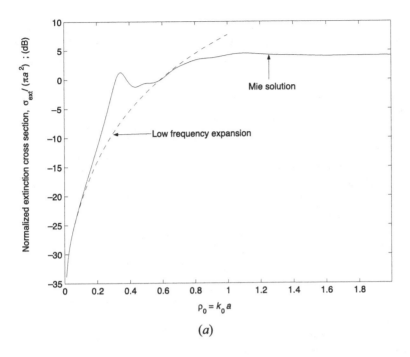

(a)

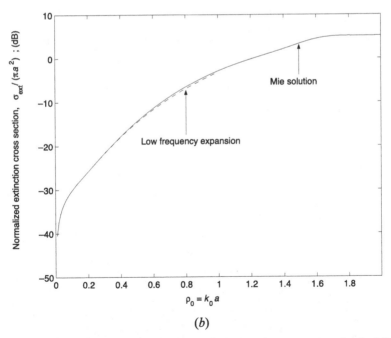

(b)

Fig. 2.16. Normalized extinction cross section versus size parameter using the Mie solution and the low frequency expansion coefficients in (2.131): (*a*) water, $\varepsilon_r = 80 - j25$; (*b*) ice, $\varepsilon_r = 3.16 - j0.02$.

and finite difference–time domain (FDTD) methods have proven very successful for computing scattering by complex shaped ice crystals in the 90–220 GHz frequency range (Goedecke and O'Brien 1988; Bohren and Singham 1991; Tang and Aydin 1995).

Notes

1. Scattering by spheroidal hydrometeors in the Rayleigh–Gans limit was treated by Atlas et al. (1953). Mathur and Mueller (1956) used the low frequency expansion method of Stevenson (1953) to treat scattering by oblate raindrops. Low frequency methods can be used for arbitrary-shaped particles and extensive literature exists on this topic. The approach used in Section 2.3.3 is based on Senior and Sarabandi (1990). Kerker (1969) and Bohren and Huffman (1983) are also very useful references for scattering theory.

2. Extensive literature exists on Mie scattering by spheres (e.g. van de Hulst 1981; Bohren and Huffman 1983). Barber and Hill (1990) provide computer programs for the Mie solution with detailed documentation and is highly recommended.

3. Bistatic radar for meteorology is described by Doviak (1972). More recently, Wurman (1994) has provided a simplified description of the bistatic Doppler principle to measure the wind field vector.

4. Because of easy availability of computer programs for the Mie solution only sample calculations are illustrated in Figs. 2.15 and 2.16.

5. Computer programs for scattering by bodies of revolution using the T-matrix method are provided by Barber and Hill (1990), and this reference includes detailed documentation and examples. The T-matrix method was originally developed by Waterman (e.g. Waterman 1971) and a good description is provided by Barber and Yeh (1975). The OLYMPUS Propagation Experiments (OPEX) web site at,

   http://www.estec.esa.nl/xewww/cost255/opex.htm

   contains the *Reference Handbook on Radar*.

3 Wave, antenna, and radar polarization

The subject of polarization for completely and partially polarized waves has been extensively treated throughout the literature (e.g. Mott 1992). This chapter presents a classical approach to wave and antenna polarization for readers who have not been previously exposed to this topic. The first goal is to derive the voltage form of the dual-polarized radar range equation for scattering by a single particle in both the linear and circular bases. This is followed by a generalization of the usual (single-polarized) power form of the radar range equation to include the concepts of copolar and cross-polar response surfaces and characteristic polarizations.

For scattering by a large number of distributed particles, the ensemble-averaged Mueller matrix is formally defined and conventional radar observables, such as reflectivity, differential reflectivity, the linear depolarization ratio, and the copolar correlation coefficient, are defined in terms of Mueller matrix elements. The time-averaged polarimetric covariance matrix is formally defined from the concept of an "instantaneous" back scatter matrix and the associated "feature" vector. Dual-polarized radars which are configured to measure the three power elements and three (complex) correlation elements of the covariance matrix offer a complete characterization of scattering from precipitation particles. The relation between the covariance matrices in the linear (h/v) and circular bases is examined in detail, including simplifications afforded by symmetry arguments. A number of examples of radar measurements are presented to illustrate the theory. This chapter ends with a discussion of a possible framework for hydrometeor classification using observables in both linear and circular bases (a different hydrometeor classification approach based on fuzzy logic will be described in Chapter 7).

3.1 Polarization state of a plane wave

A general expression for an incident plane wave with real amplitude E_0, propagating along the $\hat{k}_i$-direction, and with unit vectors $(\hat{h}_i, \hat{v}_i)$ defined such that $\hat{k}_i = \hat{v}_i \times \hat{h}_i$ (as discussed in Section 2.1) is given as,

$$\vec{E}^i(\vec{r}) = E_0 \hat{e}_i \exp(-jk_0\hat{k}_i \cdot \vec{r}) \tag{3.1a}$$
$$= (E_h^i \hat{h}_i + E_v^i \hat{v}_i) \exp(-jk_0\hat{k}_i \cdot \vec{r}) \tag{3.1b}$$

where the complex unit vector $\hat{e}_i$ defines the polarization state of the plane wave. The components E_h^i and E_v^i are, in general, complex and the electric field $\vec{E}^i(O)$ can be written as,

$$\vec{E}^i(O) = E_0 \hat{e}_i = |E_h^i| e^{j\theta_h} \hat{h}_i + |E_v^i| e^{j\theta_v} \hat{v}_i \qquad (3.2a)$$

$$= e^{j\theta_h} \left[|E_h^i| \hat{h}_i + |E_v^i| e^{j(\theta_v - \theta_h)} \hat{v}_i \right] \qquad (3.2b)$$

Since $|\vec{E}^i(O)| = E_0$, the unit polarization vector is expressed as,

$$\hat{e}_i = e^{j\theta_h} \left(\frac{|E_h^i|}{E_0} \hat{h}_i + \frac{|E_v^i|}{E_0} e^{j\delta} \hat{v}_i \right) = e^{j\theta_h} \left[(\cos \varepsilon) \hat{h}_i + (\sin \varepsilon) e^{j\delta} \hat{v}_i \right] \qquad (3.2c)$$

where $\delta = \theta_v - \theta_h$, and $\tan \varepsilon = |E_v^i|/|E_h^i|$. The complex polarization ratio (χ_i, or simply the polarization ratio) is defined as,

$$\chi_i = \frac{E_v^i}{E_h^i} = (\tan \varepsilon) e^{j\delta} \qquad (3.3)$$

and it contains information on the amplitude ratio and phase difference between E_v^i and E_h^i. The absolute phase θ_h can be set to zero. The definition of the unit polarization vector in (3.2c) is seen to involve the orthogonal real, unit vector basis ($\hat{h}_i, \hat{v}_i$). Also, the definition of χ_i in (3.3), as the ratio of vertical to horizontal components, implicitly assumes that the triplet ($\hat{k}_i, \hat{v}_i, \hat{h}_i$) satisfies $\hat{k}_i = \hat{v}_i \times \hat{h}_i$, i.e. the forward scatter alignment (FSA) convention.

As discussed in Section 1.2, the far-field scattered wave ($\vec{E}^s$) can be considered as satisfying approximate plane wave behavior within a small volume of space at large distances from the particle. In the FSA convention, the complex polarization ratio of the scattered wave is defined as,

$$\chi_s = \frac{E_v^s}{E_h^s} \qquad (3.4)$$

where the triplet ($\hat{k}_s, \hat{v}_s, \hat{h}_s$) satisfies $\hat{k}_s = \hat{v}_s \times \hat{h}_s$. In the back scatter alignment (BSA) convention, the complex polarization ratio of $\vec{E}^r$ is defined as,

$$\chi_r = \frac{E_v^r}{E_h^r} = \frac{E_v^s}{-E_h^s} = -\chi_s \qquad (3.5)$$

since $\hat{h}_r = -\hat{h}_s$, $\hat{v}_r = \hat{v}_s$, and $\hat{k}_r = -\hat{k}_s$ (see Fig. 2.2).

From (3.2c), a unit complex polarization vector can be defined with $\varepsilon = 45°$ and $\delta = 90°$, as

$$\hat{e}_R = \frac{1}{\sqrt{2}} (\hat{h}_i + j\hat{v}_i) \qquad (3.6)$$

where the subscript R signifies the right-hand circular state.

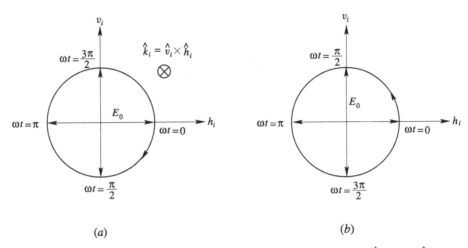

Fig. 3.1. Circular polarization state of a wave propagating along the direction $\hat{k}_i = \hat{v}_i \times \hat{h}_i$. E_0 is the (real) amplitude: (*a*) right-hand sense, (*b*) left-hand sense. The reference direction is horizontal (along the h_i axis).

In the time-domain representation of $\vec{E}^i$ ($\equiv E_0\hat{e}_R$), the horizontal and vertical components of $\vec{E}^i$ are,

$$\begin{bmatrix} E_h^i(t) \\ E_v^i(t) \end{bmatrix} = \frac{E_0}{\sqrt{2}} \begin{bmatrix} \cos \omega t \\ \cos(\omega t + 90°) \end{bmatrix} = \frac{E_0}{\sqrt{2}} \begin{bmatrix} \cos \omega t \\ -\sin \omega t \end{bmatrix} \tag{3.7}$$

where E_0 is a real amplitude. As illustrated in Fig. 3.1*a*, the tip of the resultant electric field vector traces a clockwise rotation around a circle as ωt increases from 0 to 2π. Since the wave is assumed to propagate along $\hat{k}_i$, with $\hat{k}_i = \hat{v}_i \times \hat{h}_i$, the Institute of Electrical and Electronic Engineers (IEEE) convention defines this clockwise rotation (as ωt changes from 0 to 2π), viewed by an observer looking in the $\hat{k}_i$-direction as a right-hand circularly (RHC) polarized wave. The complex polarization ratio $\chi_i = j$ for a RHC state.

Again, from (3.2c), a unit left-hand circular (LHC) polarization vector is defined with $\varepsilon = 45°$ and $\delta = -90°$ as,

$$\hat{e}_L = \frac{1}{\sqrt{2}}(\hat{h}_i - j\hat{v}_i) \tag{3.8}$$

and the corresponding circle is illustrated in Fig. 3.1*b*. The tip of the resultant electric field traces a counter-clockwise rotation as ωt increases from 0 to 2π, and this defines a left-hand circularly polarized state. The complex polarization ratio is now $\chi_i = -j$. Note that when $\omega t = 0$ and with $E_0 = 1$ V m^{-1}, the resultant electric field in either the RHC or LHC state lines up along the horizontal (h_i) axis, which is also usually chosen as a reference direction. Thus, the unit circular polarization basis vectors, as defined in (3.6) and (3.8), line up along this reference direction at $\omega t = 0$. Some authors, notably McCormick and Hendry (1975), use the vertical direction as a reference, resulting in $\hat{e}_{R,\text{MH}} = (\hat{v}_i - j\hat{h}_i)/\sqrt{2}$ and $\hat{e}_{L,\text{MH}} = (\hat{v}_i + j\hat{h}_i)/\sqrt{2}$.

From (3.6) and (3.8), the relation between unit circular polarization vectors $(\hat{e}_R, \hat{e}_L)$ and the unit vectors $(\hat{h}_i, \hat{v}_i)$ is expressed via a basis transformation matrix,

$$\begin{bmatrix} \hat{e}_R \\ \hat{e}_L \end{bmatrix} = \frac{1}{\sqrt{2}} \begin{bmatrix} 1 & j \\ 1 & -j \end{bmatrix} \begin{bmatrix} \hat{h}_i \\ \hat{v}_i \end{bmatrix} \tag{3.9}$$

3.1.1   Polarization ellipse

When the electric field $\vec{E}^i(O)$ is expressed in the complex circular basis as,

$$\vec{E}^i(O) = E_R \hat{e}_R + E_L \hat{e}_L \tag{3.10}$$

the components E_R and E_L (the superscript i is dropped for now on the circular components) can, in general, be complex and $\vec{E}^i(O)$ can be expressed as,

$$\vec{E}^i(O) = |E_R| e^{j\theta_R} \hat{e}_R + |E_L| e^{j\theta_L} \hat{e}_L \tag{3.11}$$

The circle corresponding to $|E_R| e^{j\theta_R}$ is illustrated in Fig. 3.2a, where the time-domain components corresponding to (3.7) are now,

$$\begin{bmatrix} E_h(t) \\ E_v(t) \end{bmatrix} = \frac{|E_R|}{\sqrt{2}} \begin{bmatrix} \cos(\omega t + \theta_R) \\ -\sin(\omega t + \theta_R) \end{bmatrix} \tag{3.12}$$

and the role of the "initial" phase angle θ_R is clearer. If θ_R is positive, the tip of the electric field at $\omega t = 0$ shifts clockwise. If θ_R is negative, then the shift is counter-clockwise from the horizontal reference direction. Figure 3.2b illustrates the case for the circle corresponding to $|E_L| e^{j\theta_L}$. For both circular components, the tip of the electric field rotates at a constant angular velocity (ω). A superposition of the two circular components, as defined in (3.11) with different amplitudes given by $|E_R|, |E_L|$ and different initial phase angles θ_R, θ_L, gives rise to the general elliptical polarization state, where the tip of the resultant electric field traces an ellipse as ωt increases from 0 to 2π. Of course, it is more conventional to use (3.2), involving the real basis vectors $(\hat{h}_i, \hat{v}_i)$, to mathematically derive the parameters of the ellipse, but the use of circular basis vectors is more appealing from a geometric viewpoint.

Figure 3.3a shows the superposition of the two circular components as represented in (3.11), assuming $|E_R| > |E_L|$ and positive "initial" phase angles θ_R and θ_L, with $\theta_L > \theta_R$. Note that $\alpha = 2\pi - (\theta_L + \theta_R)$. Since the two circular "phasors" rotate at the same angular speed ω, they will line up after rotating in the opposite sense (starting from their respective initial phase angles) through the angle $\alpha/2$. This will locate the major axis of the ellipse. The phasors will oppose after rotating in the opposite sense through the angle $(\alpha - \pi)/2$, and this will locate the minor axis of the ellipse. It is conventional to define the orientation angle of the ellipse (ψ) as the angle between the major axis and the horizontal reference direction (h_i). A positive angle is measured counter-clockwise from this reference direction. From Fig. 3.3a, the orientation angle $\psi = \theta_L + \alpha/2 - \pi = (\theta_L - \theta_R)/2$, with ψ in the range $-\pi/2$ to $\pi/2$. Figure 3.3b shows the resultant ellipse, with the "ellipticity" angle τ defined as,

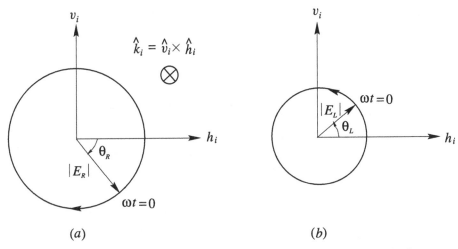

(a)   (b)

Fig. 3.2. Illustrates the initial phase angles θ_R and θ_L measured from the horizontal reference direction for a wave propagating along the direction $\hat{k}_i = \hat{v}_i \times \hat{h}_i$. (a) Right-hand sense, with amplitude $|E_R|$ and with positive θ_R measured clockwise from the horizontal reference; (b) left-hand sense, with amplitude $|E_L|$ and with positive θ_L measured counter-clockwise from the horizontal reference.

$$\tan\tau = \frac{|E_R| - |E_L|}{|E_R| + |E_L|}; \quad \frac{-\pi}{4} \leq \tau \leq \frac{\pi}{4} \tag{3.13}$$

A little thought will show that the sense of rotation of the tip of the electric field will be clockwise when $|E_R| > |E_L|$ and counter-clockwise when $|E_L| > |E_R|$. It should be clear that $\tau = \pm\pi/4$ yields circular polarization states, while $\tau = 0°$ yields linear polarization states oriented at an angle ψ from the reference direction.

Further insight is gained by rotating the axes by the angle ψ as shown in Fig. 3.3b such that the unit vectors $(\hat{h}'_i, \hat{v}'_i)$ are now aligned along the major and minor axes of the ellipse. Thus,

$$\begin{bmatrix} \hat{h}_i \\ \hat{v}_i \end{bmatrix} = \begin{bmatrix} \cos\psi & -\sin\psi \\ \sin\psi & \cos\psi \end{bmatrix} \begin{bmatrix} \hat{h}'_i \\ \hat{v}'_i \end{bmatrix} \tag{3.14}$$

Starting from (3.11),

$$\vec{E}^i(O) = E_R \hat{e}_R + E_L \hat{e}_L \tag{3.15a}$$

$$= \frac{E_R}{\sqrt{2}}(\hat{h}_i + j\hat{v}_i) + \frac{E_L}{\sqrt{2}}(\hat{h}_i - j\hat{v}_i) \tag{3.15b}$$

$$= \frac{\hat{h}'_i}{\sqrt{2}}\left(E_R e^{j\psi} + E_L e^{-j\psi}\right) + j\frac{\hat{v}'_i}{\sqrt{2}}\left(E_R e^{j\psi} - E_L e^{-j\psi}\right) \tag{3.15c}$$

Using $\psi = (\theta_L - \theta_R)/2$, $E_R = |E_R|e^{j\theta_R}$, and $E_L = |E_L|e^{j\theta_L}$ in the above yields,

$$\vec{E}^i(O) = e^{j(\theta_R + \theta_L)/2}\left[\hat{h}'_i\left(\frac{|E_R| + |E_L|}{\sqrt{2}}\right) + j\hat{v}'_i\left(\frac{|E_R| - |E_L|}{\sqrt{2}}\right)\right] \tag{3.16}$$

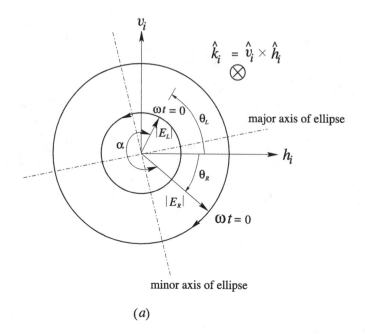

$$(a)$$

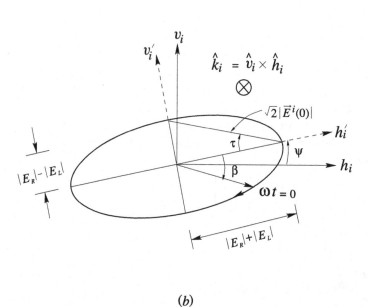

$$(b)$$

Fig. 3.3. (a) The right- and left-hand circular components, $|E_R|e^{j\theta_R}$ and $|E_L|e^{j\theta_L}$, respectively, for a wave propagating along the direction $\hat{k}_i = \hat{v}_i \times \hat{h}_i$. (b) Resultant elliptical polarization with right-hand sense. Ellipticity angle is τ; and the orientation angle (ψ) is measured counter-clockwise from the horizontal reference direction.

Defining $\beta = (\theta_R + \theta_L)/2$, the time-domain representation of $\vec{E}^i(O)$ is,

$$\vec{E}^i(t) = \hat{h}'_i\left[\left(\frac{|E_R| + |E_L|}{\sqrt{2}}\right)\cos(\omega t + \beta)\right] + \hat{v}'_i\left[\left(\frac{|E_R| - |E_L|}{\sqrt{2}}\right)\cos\left(\omega t + \beta + \frac{\pi}{2}\right)\right]$$

$$(3.17)$$

The role of the angle β becomes more clear as it locates the tip of the electric field vector relative to the h'_i-axis at $\omega t = 0$. Since the magnitude of $\vec{E}^i(O)$ is given as,

$$|\vec{E}^i(O)| = \sqrt{|E_R|^2 + |E_L|^2}$$

$$(3.18)$$

it is possible to define,

$$\cos\tau = \frac{|E_R| + |E_L|}{\sqrt{2}|\vec{E}^i(O)|}$$

$$(3.19a)$$

$$\sin\tau = \frac{|E_R| - |E_L|}{\sqrt{2}|\vec{E}^i(O)|}$$

$$(3.19b)$$

Substituting in (3.16) gives,

$$\vec{E}^i(O) = |\vec{E}^i(O)|e^{j\beta}\left(\hat{h}'_i\cos\tau + j\hat{v}'_i\sin\tau\right)$$

$$(3.20)$$

and using (3.14) gives,

$$\vec{E}^i(O) = |\vec{E}^i(O)|e^{j\beta}\left[(\cos\tau\cos\psi - j\sin\tau\sin\psi)\hat{h}_i + (\cos\tau\sin\psi + j\cos\psi\sin\tau)\hat{v}_i\right]$$

$$(3.21)$$

Since $\vec{E}^i(O) = E_h^i\hat{h}_i + E_v^i\hat{v}_i$, the components are related as,

$$\begin{bmatrix} E_h^i \\ E_v^i \end{bmatrix} = |\vec{E}^i(O)|e^{j\beta}\begin{bmatrix} \cos\psi & -\sin\psi \\ \sin\psi & \cos\psi \end{bmatrix}\begin{bmatrix} \cos\tau \\ j\sin\tau \end{bmatrix}$$

$$(3.22a)$$

and the complex polarization ratio is,

$$\chi_i = \frac{\cos\tau\sin\psi + j\cos\psi\sin\tau}{\cos\tau\cos\psi - j\sin\tau\sin\psi} = \frac{\tan\psi + j\tan\tau}{1 - j\tan\psi\tan\tau}$$

$$(3.22b)$$

It is also evident from (3.2c) that,

$$e^{j\theta_h}\begin{bmatrix} \cos\varepsilon \\ (\sin\varepsilon)e^{j\delta} \end{bmatrix} = e^{j\beta}\begin{bmatrix} \cos\psi & -\sin\psi \\ \sin\psi & \cos\psi \end{bmatrix}\begin{bmatrix} \cos\tau \\ j\sin\tau \end{bmatrix}$$

$$(3.23)$$

which relates the angles $(\theta_h, \varepsilon, \delta)$ to (β, ψ, τ). The "absolute" phases θ_h and β play no essential role in describing the polarization state of the wave. The unit polarization vector $\hat{e}_i$ can be written as,

$$\hat{e}_i = \left[(\cos\varepsilon)\hat{h}_i + (\sin\varepsilon)e^{j\delta}\hat{v}_i\right]$$

$$(3.24a)$$

$$= \left[(\cos\tau\cos\psi - j\sin\tau\sin\psi)\hat{h}_i + (\cos\tau\sin\psi + j\cos\psi\sin\tau)\hat{v}_i\right]$$

$$(3.24b)$$

The unit polarization vector orthogonal to $\hat{e}_i$ is simply obtained by setting $\psi \rightarrow \psi + \pi/2$ and $\tau \rightarrow -\tau$ in (3.24b), i.e. the major axis is orthogonal and the sense of rotation is opposite to $\hat{e}_i$. While it is possible to develop a general elliptical, orthogonal basis to express $\vec{E}^i(O)$, it is sufficient for the purposes of the treatment here to use the real linear basis $(\hat{h}_i, \hat{v}_i)$ or the complex circular basis $(\hat{e}_R, \hat{e}_L)$.

3.1.2 Stokes' vector

The elements of the Stokes' vector corresponding to a plane wave with electric field $\vec{E}^i(O) = E^i_h \hat{h}_i + E^i_v \hat{v}_i$, are defined as,

$$I = |E^i_h|^2 + |E^i_v|^2 = |\vec{E}^i(O)|^2 \tag{3.25a}$$

$$Q = |E^i_h|^2 - |E^i_v|^2 = I \cos 2\psi \cos 2\tau = I \cos 2\varepsilon \tag{3.25b}$$

$$U = 2\operatorname{Re}(E^{i*}_h E^i_v) = I \sin 2\psi \cos 2\tau = I \sin 2\varepsilon \cos \delta \tag{3.25c}$$

$$V = 2\operatorname{Im}(E^{i*}_h E^i_v) = I \sin 2\tau = I \sin 2\varepsilon \sin \delta \tag{3.25d}$$

where the representation in terms of the geometrical ellipse angles (ψ, τ) is based on (3.22a). The element I corresponds to the total power of the wave. The remaining three parameters (Q, U, V) are generally normalized by I and represent the polarization state of the wave. Figure 3.4 shows the Poincaré sphere, and each point on the sphere corresponds to a state of polarization given by the Cartesian coordinates of the point (Q, U, V), with 2ψ being the conventional azimuth angle and $90° - 2\tau$ being the conventional polar angle of the spherical coordinate system. The north pole $(\tau = \pi/4)$ and the south pole $(\tau = -\pi/4)$ correspond to the right- and left-hand circular states, respectively. The equator $(\tau = 0°)$ corresponds to a linear state oriented at angle ψ. The top (bottom) hemisphere corresponds to a right-hand (left-hand) elliptical state.[1]

The scattering matrix relating the far-field scattered wave $\vec{E}^s$ to the incident plane wave $\vec{E}^i$ given in (2.7), i.e. $\mathbf{E}^s = \mathbf{S}_{\text{FSA}} \mathbf{E}^i \exp(-jk_0 r)/r$, can be transformed into a relationship between the Stokes' vectors of the scattered and incident fields involving the Mueller matrix. From (3.25), it is clear that the incident Stokes' vector can be expressed as,

$$\begin{bmatrix} I_i \\ Q_i \\ U_i \\ V_i \end{bmatrix} = \begin{bmatrix} 1 & 1 & 0 & 0 \\ 1 & -1 & 0 & 0 \\ 0 & 0 & 1 & 1 \\ 0 & 0 & -j & j \end{bmatrix} \begin{bmatrix} E^i_h E^{i*}_h \\ E^i_v E^{i*}_v \\ E^{i*}_h E^i_v \\ E^i_h E^{i*}_v \end{bmatrix} \tag{3.26}$$

where the 4×4 matrix, with constant elements, defined above can be denoted by $\mathbf{R}$. In FSA, the scattered Stokes' vector is also related to (E^s_h, E^s_v) in the same way as in

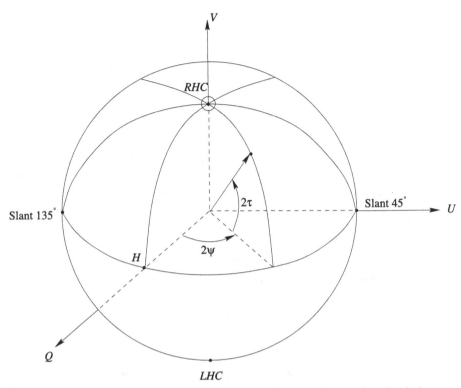

Fig. 3.4. The Poincaré sphere and representative polarization states (the northern hemisphere represents a right-handed sense).

(3.26),

$$
\begin{bmatrix} I_s \\ Q_s \\ U_s \\ V_s \end{bmatrix} = [\mathbf{R}] \begin{bmatrix} E_h^s E_h^{s*} \\ E_v^s E_v^{s*} \\ E_h^{s*} E_v^s \\ E_h^s E_v^{s*} \end{bmatrix} \tag{3.27}
$$

Using the above, (2.7) is transformed as,

$$
\begin{bmatrix} I_s \\ Q_s \\ U_s \\ V_s \end{bmatrix} = \frac{1}{r^2} [\mathbf{M}]_{FSA} \begin{bmatrix} I_i \\ Q_i \\ U_i \\ V_i \end{bmatrix} \tag{3.28}
$$

where $[\mathbf{M}]$ is the 4×4 Mueller matrix in FSA, which relates the Stokes' vectors of the scattered spherical wave and the incident plane wave. In compact notation,

$$
\mathbf{I}_s = \frac{1}{r^2} \mathbf{M} \mathbf{I}_i \tag{3.29}
$$

where it is conventional to drop the FSA descriptor on $\mathbf{M}$. The elements of $\mathbf{M}$ are given in terms of $\mathbf{S}_{FSA}$ elements in Mott (1992); see also (3.163a).

3.2   Basics of antenna radiation and reception

Radiation by antennas, in principle, can be treated in a similar way to scattering of an incident plane wave by a dielectric particle. In Section 1.2, the far-field scattered wave was expressed as an integral over the volume of the particle and involving the total internal electric field or, equivalently, the polarization current ($\vec{J}_p$) at each point within the particle. The incident plane wave was interpreted as "maintaining" the polarization current within the particle. In the case of antenna radiation, the radar transmitter is connected to the antenna and it "maintains" the field distribution in the plane of the aperture. The far-field radiated electric field can then be expressed as an integral over the aperture plane. The far-field radiated electric field is a spherical wave of the form,

$$\vec{E}(r_a, \theta_a, \phi_a) = \vec{F}(\theta_a, \phi_a) \frac{e^{-jk_0 r_a}}{r_a} \tag{3.30}$$

where $\vec{F}(\theta_a, \phi_a)$ is the vector radiation amplitude, see Fig. 3.5. The local plane wave approximation, as discussed in Section 1.2, also holds when interest is confined to a small volume of space at great distance from the antenna. Similar to the concept of total scattered power defined in (1.44), the total radiated power can be expressed as,

$$P_{\text{rad}} = \frac{1}{2Z_0} \oint_{4\pi} |\vec{F}(\theta_a, \phi_a)|^2 \, d\Omega \tag{3.31}$$

Assuming, for simplicity, that the antenna is lossless, this radiated power must equal the time-averaged power supplied by the radar transmitter (P_t). The power radiated per unit solid angle in the direction (θ_a, ϕ_a) is defined as the radiation intensity $U(\theta_a, \phi_a)$,

$$U(\theta_a, \phi_a) = \frac{d P_{\text{rad}}}{d\Omega} = \frac{1}{2Z_0} |\vec{F}(\theta_a, \phi_a)|^2 \tag{3.32}$$

If the transmitted power were radiated isotropically, the radiation intensity would be constant and equal to $P_t/4\pi$. The antenna gain in the direction (θ_a, ϕ_a) is defined as the ratio of the antenna radiation intensity $U(\theta_a, \phi_a)$ to $P_t/4\pi$,

$$G(\theta_a, \phi_a) = \frac{U(\theta_a, \phi_a)}{(P_t/4\pi)} = \frac{4\pi}{P_t} \frac{1}{2Z_0} |\vec{F}(\theta_a, \phi_a)|^2 \tag{3.33a}$$

$$= \frac{2\pi}{P_t Z_0} |\vec{F}(\theta_a, \phi_a)|^2 \tag{3.33b}$$

The above equation can be written as,

$$|\vec{F}(\theta_a, \phi_a)| = \sqrt{\frac{P_t Z_0}{2\pi}} \sqrt{G(\theta_a, \phi_a)} \tag{3.34}$$

From the discussion in Section 3.1 it should be clear that the vector radiation amplitude $\vec{F}(\theta, \phi)$ defines the polarization state of the radiated wave under the local plane wave approximation. Thus, the unit polarization vector of the radiated wave is,

$$\hat{e} = \frac{\vec{F}(\theta_a, \phi_a)}{|\vec{F}(\theta_a, \phi_a)|} \tag{3.35a}$$

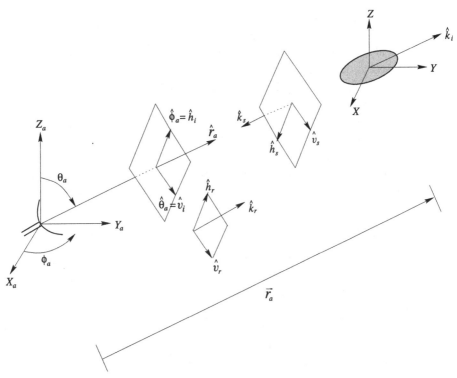

Fig. 3.5. Antenna coordinate system $X_a Y_a Z_a$ and direction of radiation (θ_a, ϕ_a). A particle is located at the origin of the XYZ system. The incident wave unit vectors are $(\hat{k}_i, \hat{v}_i, \hat{h}_i) \equiv (\hat{r}_a, \hat{\theta}_a, \hat{\phi}_a)$. In FSA convention, the unit vectors representing the back scattered wave are $(\hat{k}_s, \hat{v}_s, \hat{h}_s)$; in BSA convention, the corresponding unit vectors are $(\hat{k}_r, \hat{v}_r, \hat{h}_r) \equiv (\hat{k}_i, \hat{v}_i, \hat{h}_i)$.

and,

$$\vec{F}(\theta_a, \phi_a) = \hat{e}|\vec{F}(\theta_a, \phi_a)| = \hat{e}\sqrt{\frac{P_t Z_0}{2\pi}}\sqrt{G(\theta_a, \phi_a)} \tag{3.35b}$$

Hence, the radiated electric field in (3.30) may be expressed in terms of the transmitter power, antenna gain, and the unit polarization vector as,

$$\vec{E}(r_a, \theta_a, \phi_a) = \hat{e}\sqrt{\frac{P_t Z_0}{2\pi}}\sqrt{G(\theta_a, \phi_a)}\frac{e^{-jk_0 r_a}}{r_a} \tag{3.36}$$

An effective antenna "length" vector can be defined as,

$$\vec{h}(\theta_a, \phi_a) = \frac{\lambda}{\sqrt{4\pi}}\hat{e}\sqrt{G(\theta_a, \phi_a)} \tag{3.37}$$

Note that the unit polarization vector of the antenna and the effective length vector are defined when the antenna is considered to be a radiator. Generally, an antenna coordinate system is introduced with the $X_a Y_a$-plane being parallel to the Earth's surface and (θ_a, ϕ_a) being the conventional spherical coordinate angles, as illustrated in Fig. 3.5. The spatial phase reference for the radiated wave expressed in (3.36) is now at the origin of the antenna coordinate system. The unit vector triplet $(\hat{r}_a, \hat{\theta}_a, \hat{\phi}_a)$ is used to define the radiation direction as well as the basis for expressing the unit polarization vector $\hat{e}$, which lies in the plane of polarization. For clarity, Fig. 3.5 also shows a particle located at the origin of an XYZ coordinate system, which is identical to the $X_a Y_a Z_a$ system except for translation of the origin. It should be clear that the unit vector triplet $(\hat{k}_i, \hat{\theta}_a, \hat{\phi}_a)$ is the same as the incident wave unit vector triplet $(\hat{k}_i, \hat{v}_i, \hat{h}_i)$, with $\hat{k}_i = \hat{v}_i \times \hat{h}_i$ as defined in Chapter 2 (see Fig. 2.1a) for the forward scatter alignment (FSA) convention. Thus, the unit polarization vector $\hat{e}$ ($\equiv \hat{e}_i$), of the wave radiated by the antenna can generally be expressed as in (3.24a, b) using the $(\hat{h}_i, \hat{v}_i)$ basis.

3.2.1 Voltage equation

In Fig. 3.5, let the electric field radiated by the antenna be the incident wave, which in turn is scattered by the particle. Consider the back scattered electric field along the $\hat{k}_s = -\hat{k}_i$ direction, with $\theta_s = \pi - \theta_i$ and $\phi_s = \phi_i + \pi$ using the FSA convention. As shown in Section 2.1, the unit vector triplet $(\hat{k}_s, \hat{v}_s, \hat{h}_s)$, with $\hat{k}_s = \hat{v}_s \times \hat{h}_s$, is simply related to $(\hat{k}_i, \hat{v}_i, \hat{h}_i)$ by $\hat{k}_s = -\hat{k}_i$, $\hat{v}_s = \hat{v}_i$, and $\hat{h}_s = -\hat{h}_i$, as illustrated in Fig. 3.5. It was also shown in Section 2.1 that the unit vector triplet $(\hat{k}_r, \hat{v}_r, \hat{h}_r)$ appropriate for the BSA convention coincides with $(\hat{k}_i, \hat{v}_i, \hat{h}_i)$ or $(\hat{r}_a, \hat{\theta}_a, \hat{\phi}_a)$ along the back scattering direction, as illustrated in Fig. 3.5. The scattered field can be expressed as $\vec{E}^s = E_h^s \hat{h}_s + E_v^s \hat{v}_s$ in FSA, or as $\vec{E}^r = E_h^r \hat{h}_r + E_v^r \hat{v}_r$ in BSA, where $\vec{E}^s = S_{FSA} \vec{E}^i$ or $\vec{E}^r = S_{BSA} \vec{E}^i$ and the spherical wave factor has been dropped for convenience, see (2.8).

When the antenna, characterized by its effective length vector $\vec{h}$, is used as a receiver, the open circuit voltage (V) across its terminal (or "port"), due to the scattered field $\vec{E}^r$, expressed in the BSA convention (Sinclair 1950; Kennaugh 1952) is,

$$V = \frac{1}{\sqrt{2Z_0}} \vec{h} \cdot \vec{E}^r \tag{3.38}$$

where Z_0 is the intrinsic impedance of free space and $1/\sqrt{2Z_0}$ is a normalization factor such that $|V|^2$ will have units of power (W). The above equation is referred to as the voltage equation. This and the scattering matrix equation, defined by $\vec{E}^r = S_{BSA} \vec{E}^i$, form the two basic equations governing radar polarimetry. Note that even though $\hat{h}$ and $\vec{E}^r$ are complex vectors, the voltage equation is not written as a conventional inner product (e.g. $\vec{A} \cdot \vec{B}^*$). It is also emphasized that both $\vec{h}$ and $\vec{E}^r$ are expressed in the antenna coordinate system or BSA convention as illustrated in Fig. 3.5, noting that $(\hat{r}_a, \hat{\theta}_a, \hat{\phi}_a) \equiv (\hat{k}_i, \hat{v}_i, \hat{h}_i) \equiv (\hat{k}_r, \hat{v}_r, \hat{h}_r)$.

The voltage equation can also be expressed in mixed matrix/vector notation (Poelman and Guy 1984) as,

$$V = \frac{1}{\sqrt{2Z_0}} \vec{h} \cdot \begin{bmatrix} -1 & 0 \\ 0 & 1 \end{bmatrix} \vec{E}^s \tag{3.39}$$

where (2.10) has been used to relate $\vec{E}^r$ and $\vec{E}^s$. It is emphasized that $\vec{E}^s$ is expressed in FSA convention in (3.39). In terms of the ellipse angles (ψ, τ), the use of this convention makes it easier to interpret the received power geometrically. Using (3.37), the voltage equation is expressed as,

$$V = \frac{\lambda}{\sqrt{8\pi Z_0}} \sqrt{G(\theta_a, \phi_a)} \hat{e}_i \cdot \begin{bmatrix} -1 & 0 \\ 0 & 1 \end{bmatrix} \vec{E}^s \qquad (3.40)$$

where $\hat{e}_i$ is the complex unit vector characterizing the antenna polarization state and the subscript i refers to the incident wave launched by the antenna. Assume that the antenna is configured to launch a general elliptically polarized wave characterized by (ψ_i, τ_i), as in (3.24b),

$$\hat{e}_i = \begin{bmatrix} \cos \psi_i & -\sin \psi_i \\ \sin \psi_i & \cos \psi_i \end{bmatrix} \begin{bmatrix} \cos \tau_i \\ j \sin \tau_i \end{bmatrix} \qquad (3.41)$$

Let the scattered field $\vec{E}^s$ in FSA convention also be elliptically polarized with parameters (ψ_s, τ_s),

$$\vec{E}^s = |\vec{E}^s| \begin{bmatrix} \cos \psi_s & -\sin \psi_s \\ \sin \psi_s & \cos \psi_s \end{bmatrix} \begin{bmatrix} \cos \tau_s \\ j \sin \tau_s \end{bmatrix} \qquad (3.42)$$

The voltage equation can now be written in terms of the geometrical parameters of the antenna and scattered wave ellipses (noting that the dot product implies use of transposed matrices, i.e. $\vec{A} \cdot \vec{B} \equiv \mathbf{A}^t \mathbf{B}$) as,

$$V = |\vec{E}^s| \frac{\lambda}{\sqrt{8\pi Z_0}} \sqrt{G(\theta_a, \phi_a)} [\cos \tau_i \; j \sin \tau_i] \begin{bmatrix} \cos \psi_i & \sin \psi_i \\ -\sin \psi_i & \cos \psi_i \end{bmatrix} \begin{bmatrix} -1 & 0 \\ 0 & 1 \end{bmatrix}$$
$$\begin{bmatrix} \cos \psi_s & -\sin \psi_s \\ \sin \psi_s & \cos \psi_s \end{bmatrix} \begin{bmatrix} \cos \tau_s \\ j \sin \tau_s \end{bmatrix} \qquad (3.43)$$

Assuming a right-hand sense, these two ellipses are illustrated in Fig. 3.6. Note that the orientation angle ψ_i is measured from the horizontal reference direction $\hat{h}_i$, while for ψ_s the horizontal reference direction is $\hat{h}_s$. A positive ψ angle is measured, in each case, starting at the corresponding reference direction and rotating toward the respective v axis. The received power at the antenna port is proportional to the magnitude of the square of the voltage ($|V|^2$). Normalizing $|V|^2$ by the factor $|\vec{E}^s|^2 \lambda^2 G / 8\pi Z_0$ gives an expression for the polarization match factor (ρ) or the polarization efficiency of the receiving antenna. After some tedious algebra, a compact form for ρ emerges (Huynen 1970),

$$\rho = \frac{|V|^2 8\pi Z_0}{|\vec{E}^s|^2 \lambda^2 G} = \frac{1}{2} [1 + \sin 2\tau_i \sin 2\tau_s + \cos 2\tau_i \cos 2\tau_s \cos 2(\psi_i + \psi_s)] \qquad (3.44)$$

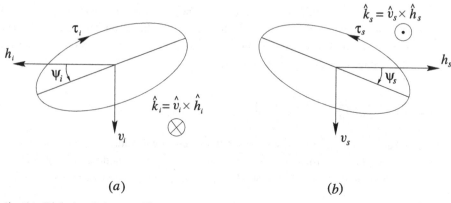

(a) (b)

Fig. 3.6. Right-handed sense ellipses for: (a) the incident wave, where the direction of propagation is $\hat{k}_i = \hat{v}_i \times \hat{h}_i$; and (b) the back scattered wave, which is expressed in FSA convention, where the direction of propagation is $\hat{k}_s = \hat{v}_s \times \hat{h}_s$.

The parameter ρ is also called the "wave-to-antenna" coupling factor and lies between zero and unity. The polarization match factor is unity when $\tau_i = \tau_s$ and $\psi_i = -\psi_s$. This condition implies that when the two ellipses are viewed from behind the antenna, they have the same shape and opposite sense of rotation and their major axes are aligned. The polarization match factor is zero when $\tau_i = -\tau_s$ and $\psi_i + \psi_s = \pi/2$. This condition implies that the two ellipses have the same sense of rotation and the same shape, but their major axes are orthogonal when viewed from behind the antenna.

3.2.2 Polarization efficiency in terms of χ

An expression for ρ can also be derived using the complex polarization ratios of the antenna and the scattered wave (Mott 1992). The complex polarization ratio of the incident wave launched by the antenna is defined as,

$$\chi_i = \frac{E_v^i}{E_h^i} = \frac{\tan \psi_i + j \tan \tau_i}{1 - j \tan \psi_i \tan \tau_i} \tag{3.45}$$

where $\vec{E}^i = E_h^i \hat{h}_i + E_v^i \hat{v}_i$ is the incident electric field radiated by the antenna, see (3.1b, 3.22b), and for now the spherical wave factor is dropped. The complex polarization ratio of the scattered field in FSA convention is,

$$\chi_s = \frac{E_v^s}{E_h^s} = \frac{\tan \psi_s + j \tan \tau_s}{1 - j \tan \psi_s \tan \tau_s} \tag{3.46}$$

while in BSA convention,

$$\chi_r = \frac{E_v^r}{E_h^r} = \frac{E_v^s}{-E_h^s} = -\chi_s \tag{3.47}$$

The unit polarization vector $\hat{e}_i$ can be expressed as,

$$\hat{e}_i = \frac{E_h^i(\hat{h}_i + \chi_i \hat{v}_i)}{|E_h^i|(1 + \chi_i \chi_i^*)^{1/2}} \tag{3.48}$$

where for a complex vector, $|\vec{A}| = (\vec{A} \cdot \vec{A}^*)^{1/2}$. Similarly, the scattered field is expressed as,

$$\vec{E}^s = E_h^s(\hat{h}_s + \chi_s \hat{v}_s) \tag{3.49a}$$

$$|\vec{E}^s| = |E_h^s|(1 + \chi_s \chi_s^*) \tag{3.49b}$$

Using (3.40), the voltage equation is,

$$V = \frac{\lambda}{\sqrt{8\pi Z_0}} \sqrt{G(\theta_a, \phi_a)} \frac{E_h^i E_h^s}{|E_h^i|(1 + \chi_i \chi_i^*)^{1/2}} [1 \ \chi_i] \begin{bmatrix} -1 & 0 \\ 0 & 1 \end{bmatrix} \begin{bmatrix} 1 \\ \chi_s \end{bmatrix} \tag{3.50}$$

The polarization match factor corresponding to (3.44) can be expressed as,

$$\rho = \frac{|V|^2 8\pi Z_0}{|E^s|^2 \lambda^2 G} = \frac{(1 - \chi_i \chi_s)(1 - \chi_i^* \chi_s^*)}{(1 + \chi_i \chi_i^*)(1 + \chi_s \chi_s^*)} \tag{3.51a}$$

$$= \frac{(1 + \chi_i \chi_r)(1 + \chi_i^* \chi_r^*)}{(1 + \chi_i \chi_i^*)(1 + \chi_r \chi_r^*)} \tag{3.51b}$$

where (3.47) is also used. It is clear that ρ equals unity when $\chi_s = -\chi_i^*$, or when $\chi_r = \chi_i^*$. Substituting $\tau_s = \tau_i$ and $\psi_s = -\psi_i$ in (3.46) results in the match condition $\chi_s = -\chi_i^*$, i.e.

$$\chi_s = \frac{\tan \psi_s + j \tan \tau_s}{1 - j \tan \psi_s \tan \tau_s} = \frac{\tan(-\psi_i) + j \tan \tau_i}{1 - j \tan(-\psi_i) \tan \tau_i} \tag{3.52a}$$

$$= \frac{-(\tan \psi_i - j \tan \tau_i)}{(1 + j \tan \psi_i \tan \tau_i)} \tag{3.52b}$$

$$= -\chi_i^* \tag{3.52c}$$

From (3.51), the polarization match factor (ρ) is zero when $\chi_s = 1/\chi_i$, or when $\chi_r = -1/\chi_i$. In terms of the ellipse parameters, the condition $\chi_s = 1/\chi_i$ corresponds to $\tau_s = -\tau_i$ and $\psi_s = \pi/2 - \psi_i$, which can be confirmed by direct substitution,

$$\chi_s = \frac{\tan \psi_s + j \tan \tau_s}{1 - j \tan \psi_s \tan \tau_s} = \frac{\tan(\pi/2 - \psi_i) - j \tan \tau_i}{1 + j \tan[(\pi/2) - \psi_i] \tan \tau_i} \tag{3.53a}$$

$$= \frac{1}{\chi_i} \tag{3.53b}$$

Under this condition of $\rho = 0$, the scattered wave is said to be "cross-polarized" with respect to the antenna.

Some special cases are considered next. If the antenna is horizontally polarized ($\tau_i = 0°$, $\psi_i = 0°$) and if the back scattered wave ($\vec{E}^s$) is linearly polarized ($\tau_s = 0°$) at an orientation angle ψ_s, then from (3.44),

$$\rho = \frac{1}{2}(1 + \cos 2\psi_s) \tag{3.54}$$

and the polarization match factor is unity when $\psi_s = 0°$, or when the back scattered wave is also horizontally polarized. The match factor is zero when $\psi_s = 90°$, or when $\vec{E}^s$ is vertically polarized.

If the antenna is slant linear polarized ($\tau_i = 0°$, $\psi_i = 45°$) and if $\vec{E}^s$ is linearly polarized ($\tau_s = 0°$) at an orientation angle ψ_s then,

$$\rho = \frac{1}{2}\left[1 + \cos 2(45° + \psi_s)\right] \tag{3.55}$$

The match factor is unity when $\psi_s = -45°$ (see Fig. 3.5) or, when viewed from behind the antenna, the orientation of the scattered wave lines up with the antenna's polarization orientation angle. Of course, $\rho = 0$ when $\psi_s = 45°$ or, when viewed from behind the antenna, the orientation of $\vec{E}^s$ is orthogonal to the antenna's polarization orientation angle.

If the antenna is right-hand circularly polarized ($\tau_i = \pi/4$, $\psi_i \equiv$ arbitrary) and if $\vec{E}^s$ is also right-hand circularly polarized ($\tau_s = \pi/4$, $\psi_s \equiv$ arbitrary) then

$$\rho = \frac{1}{2}(1 + \sin 2\tau_i \sin 2\tau_s) = 1 \tag{3.56}$$

If $\vec{E}^s$ is left-hand circularly polarized ($\tau_s = -\pi/4$, $\psi_s \equiv$ arbitrary) then it is clear that $\rho = 0$.

3.3   Dual-polarized antennas: linear polarization basis

A dual-polarized antenna will have two "ports", designated as horizontal (or h-port) and vertical (or v-port) polarization ports as illustrated in Fig. 3.7a. Let M_h and M_v be the wave amplitudes at the input plane, where $|M_h| \propto \sqrt{P_t^h}$ and $|M_v| \propto \sqrt{P_t^v}$; and $P_t^{h,v}$ is the transmitter power. When only the h-port of the dual-polarized antenna is excited, the radiated electric field can be written following (3.36) as,

$$\vec{E}^i = \sqrt{\frac{P_t^h Z_0}{2\pi}}\sqrt{G_h}\hat{h}_i\frac{e^{-jk_0r_a}}{r_a} \tag{3.57}$$

where G_h is the gain associated with the h-port of the antenna; P_t^h is the transmitter power; and $\hat{h}_i$ is the unit polarization vector, which for a well-designed antenna should be constant across the main radiation lobe of the antenna. For generality, a phase term $\exp(j\Phi_h)$ associated with the h-port should be introduced in (3.57), but is omitted here

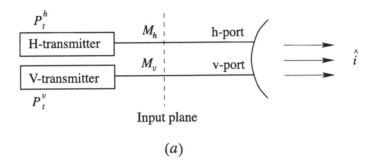

Input plane

(a)

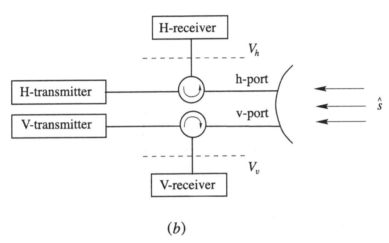

(b)

Fig. 3.7. (a) Dual-polarized antenna with two ports fed by two separate transmitters. The input plane wave amplitudes are M_h and M_v, which are proportional to $\sqrt{P_t^h}$ and $\sqrt{P_t^v}$. (b) Illustrates receive wave amplitudes V_h and V_v at receiver input planes. Circulators route signals into the two receivers.

(it will be considered later in Section 6.2, on antenna performance characteristics). The normalized power gain pattern associated with the h-port is defined as $f_h = G_h/G_{0h}$, where G_{0h} is the peak boresight gain. Similarly, when only the v-port of the antenna is excited, the radiated electric field can be expressed as,

$$\vec{E}^i = \sqrt{\frac{P_t^v Z_0}{2\pi}} \sqrt{G_v}\,\hat{v}_i \frac{e^{-jk_0 r_a}}{r_a}$$

(3.58)

where G_v is the gain associated with the v-port and P_t^v is the transmitter power. Again, a phase term $\exp(j\Phi_v)$ associated with the v-port is omitted in (3.58). By superposition, when both ports are excited, the radiated electric field ($\vec{E}^i = E_h^i \hat{h}_i + E_v^i \hat{v}_i$) can be expressed in matrix form as,

$$\begin{bmatrix} E_h^i \\ E_v^i \end{bmatrix} = \sqrt{\frac{Z_0}{2\pi}} \begin{bmatrix} \sqrt{G_h} & 0 \\ 0 & \sqrt{G_v} \end{bmatrix} \begin{bmatrix} M_h \\ M_v \end{bmatrix} \frac{e^{-jk_0 r_a}}{r_a}$$

(3.59)

Reception of the reflected wave ($\vec{E}^r$) in BSA by the dual-polarized antenna can be handled using (3.38), where the voltage induced in the h-port of the antenna is given by,

$$V_h = \frac{\lambda}{\sqrt{8\pi Z_0}} \sqrt{G_h} \hat{h}_i \cdot \vec{E}^r \tag{3.60a}$$

and the voltage induced in the v-port is given by,

$$V_v = \frac{\lambda}{\sqrt{8\pi Z_0}} \sqrt{G_v} \hat{v}_i \cdot \vec{E}^r \tag{3.60b}$$

In matrix form the voltage equation is,

$$\begin{bmatrix} V_h \\ V_v \end{bmatrix} = \frac{\lambda}{\sqrt{8\pi Z_0}} \begin{bmatrix} \sqrt{G_h} & 0 \\ 0 & \sqrt{G_v} \end{bmatrix} \begin{bmatrix} E_h^r \\ E_v^r \end{bmatrix} \tag{3.61}$$

with $\vec{E}^r = E_h^r \hat{h}_r + E_v^r \hat{v}_r$. See Fig. 3.7b for an illustration where the received wave amplitudes are assumed to be given at the receiver inputs.

An ideal antenna will have $G_h = G_v$. Gain inequality between the two ports of a non-ideal antenna can be described by the ratio G_h/G_v. In addition, when only the h-port is excited, for example, the unit polarization vector of the radiated field can be represented as $\hat{e}_h = i_h \hat{h}_i + \varepsilon_h \hat{v}_i$, where ε_h is a complex error term that accounts for deviation of polarization purity across the antenna beam. Similarly, when the v-port is excited, the unit polarization vector can be written as $\hat{e}_v = \varepsilon_v \hat{h}_i + i_v \hat{v}_i$. It follows that (3.60a, b) are modified as,

$$V_h = \frac{\lambda}{\sqrt{8\pi Z_0}} \sqrt{G_h} \hat{e}_h \cdot \vec{E}^r \tag{3.62a}$$

$$V_v = \frac{\lambda}{\sqrt{8\pi Z_0}} \sqrt{G_v} \hat{e}_v \cdot \vec{E}^r \tag{3.62b}$$

while (3.61) is modified as,

$$\begin{bmatrix} V_h \\ V_v \end{bmatrix} = \frac{\lambda}{\sqrt{8\pi Z_0}} \begin{bmatrix} \sqrt{G_h} & 0 \\ 0 & \sqrt{G_v} \end{bmatrix} \begin{bmatrix} i_h & \varepsilon_h \\ \varepsilon_v & i_v \end{bmatrix} \begin{bmatrix} E_h^r \\ E_v^r \end{bmatrix} \tag{3.63}$$

Note that $i_{h,v}^2 + |\varepsilon_{h,v}|^2 = 1$ to maintain $\hat{e}_{h,v}$ as unit vectors. Using (2.10), the voltage equation is derived, where the scattered field $\vec{E}^s$ is in FSA convention,

$$\begin{bmatrix} V_h \\ V_v \end{bmatrix} = \frac{\lambda}{\sqrt{8\pi Z_0}} \begin{bmatrix} \sqrt{G_h} & 0 \\ 0 & \sqrt{G_v} \end{bmatrix} \begin{bmatrix} i_h & \varepsilon_h \\ \varepsilon_v & i_v \end{bmatrix} \begin{bmatrix} -1 & 0 \\ 0 & 1 \end{bmatrix} \begin{bmatrix} E_h^s \\ E_v^s \end{bmatrix} \tag{3.64}$$

The topic of antenna polarization errors will be considered in more detail in Section 6.2.

3.4   Radar range equation for a single particle: linear polarization basis

Referring to Fig. 3.5, let $\vec{r}_a$ be the vector drawn from the origin of the antenna coordinate system to the origin of the XYZ system (i.e. to the center of the particle). Then, (3.59) gives the electric field incident on the particle, with the spatial phase referenced to the antenna. In BSA, the reflected field is expressed as,

$$\mathbf{E}^r = \frac{e^{-jk_0 r}}{r}S_{BSA}\mathbf{E}^i \tag{3.65}$$

where the spatial phase is now referenced to the center of the particle. With $\bar{E}^i$ as given in (3.59), and along the back scattered direction (with $r = r_a$), (3.65) can be written as,

$$\begin{bmatrix} E_h^r \\ E_v^r \end{bmatrix} = \sqrt{\frac{Z_0}{2\pi}}\frac{e^{-jk_0 r}}{r}\begin{bmatrix} S_{hh} & S_{hv} \\ S_{vh} & S_{vv} \end{bmatrix}_{BSA}\begin{bmatrix} \sqrt{G_h} & 0 \\ 0 & \sqrt{G_v} \end{bmatrix}\begin{bmatrix} M_h \\ M_v \end{bmatrix}\frac{e^{-jk_0 r}}{r} \tag{3.66}$$

Substituting (3.66) into (3.61) gives the voltage form of the dual-polarized radar range equation, relating the received wave amplitudes (V_h, V_v) to the input plane wave amplitudes (M_h, M_v) in terms of antenna characteristics, range to the particle, and the particle scattering matrix as,

$$\begin{bmatrix} V_h \\ V_v \end{bmatrix} = \frac{\lambda}{4\pi r^2}e^{-j2k_0 r}\begin{bmatrix} \sqrt{G_h} & 0 \\ 0 & \sqrt{G_v} \end{bmatrix}\begin{bmatrix} S_{hh} & S_{hv} \\ S_{vh} & S_{vv} \end{bmatrix}_{BSA}$$
$$\begin{bmatrix} \sqrt{G_h} & 0 \\ 0 & \sqrt{G_v} \end{bmatrix}\begin{bmatrix} M_h \\ M_v \end{bmatrix} \tag{3.67}$$

For an ideal antenna, $G_h = G_v = G$, so that (3.67) simplifies to,

$$\begin{bmatrix} V_h \\ V_v \end{bmatrix} = \frac{\lambda G}{4\pi r^2}e^{-j2k_0 r}\begin{bmatrix} S_{hh} & S_{hv} \\ S_{vh} & S_{vv} \end{bmatrix}_{BSA}\begin{bmatrix} M_h \\ M_v \end{bmatrix} \tag{3.68a}$$

The use of (2.10) results in,

$$\begin{bmatrix} V_h \\ V_v \end{bmatrix} = \frac{\lambda G}{4\pi r^2}e^{-j2k_0 r}\begin{bmatrix} -1 & 0 \\ 0 & 1 \end{bmatrix}\begin{bmatrix} S_{hh} & S_{hv} \\ S_{vh} & S_{vv} \end{bmatrix}_{FSA}\begin{bmatrix} M_h \\ M_v \end{bmatrix} \tag{3.68b}$$

which is valid when the scattering matrix is expressed in the FSA convention.

The transmitter powers are implicit in the input plane wave amplitudes ($|M_{h,v}| \propto \sqrt{P_{h,v}}$). It should be clear that when the input plane wave amplitude is $(1\ 0)^t$, then the received wave amplitudes are proportional to the first column of the scattering matrix, i.e. $V_h^{10} \propto S_{hh}$ and $V_v^{10} \propto S_{vh}$. When the input amplitude is $(0\ 1)^t$, then $V_h^{01} \propto S_{hv}$ and $V_v^{01} \propto S_{vv}$. Referring to Fig. 3.7a, this means that the H-transmitter and V-transmitter can be "fired" alternately. Figure 3.7b also implies that the H-receiver will alternately receive signals proportional to S_{hh} and S_{hv}, while the V-receiver will alternately receive signals proportional to S_{vh} and S_{vv}.

The conventional form of the single particle radar range equation involving the received power (P_{co}^h) and radar cross section (σ_{hh}) is obtained, for example, from (3.68a) with input $(M_h\ 0)^t$ as,

$$V_h = \frac{\lambda G}{4\pi r^2} e^{-j2k_0 r} S_{hh} M_h \tag{3.69}$$

and,

$$P_{co}^h = |V_h|^2 = \frac{\lambda^2 G^2}{(4\pi)^2 r^4} |S_{hh}|^2 P_t^h \tag{3.70a}$$

$$= \frac{\lambda^2 G^2}{(4\pi)^3 r^4} 4\pi |S_{hh}|^2 P_t^h \tag{3.70b}$$

$$= \frac{\lambda^2 G^2}{(4\pi)^3 r^4} \sigma_{hh} P_t^h \tag{3.70c}$$

where $\sigma_{hh} = 4\pi |S_{hh}|^2$, P_t^h is the H-transmitter power, and P_{co}^h is the received power in the H-receiver. The subscript "co" stands for copolar.

3.5   Change of polarization basis: linear to circular basis

In this section, the scattering matrix is transformed from linear basis to circular basis. Consider first the change of the polarization basis from linear $(\hat{h}_i, \hat{v}_i)$ to circular $(\hat{e}_R^i, \hat{e}_L^i)$ as described in (3.9), which is repeated here,

$$\begin{bmatrix} \hat{e}_R^i \\ \hat{e}_L^i \end{bmatrix} = \frac{1}{\sqrt{2}} \begin{bmatrix} 1 & j \\ 1 & -j \end{bmatrix} \begin{bmatrix} \hat{h}_i \\ \hat{v}_i \end{bmatrix} \tag{3.71}$$

Let the incident electric field $\vec{E}^i = E_h^i \hat{h}_i + E_v^i \hat{v}_i$ be expressed in circular basis as $\vec{E}^i = E_R^i \hat{e}_R^i + E_L^i \hat{e}_L^i$. It follows from (3.71) that the components (E_h^i, E_v^i) and (E_R^i, E_L^i) are related as,

$$\begin{bmatrix} E_h^i \\ E_v^i \end{bmatrix} = \frac{1}{\sqrt{2}} \begin{bmatrix} 1 & 1 \\ j & -j \end{bmatrix} \begin{bmatrix} E_R^i \\ E_L^i \end{bmatrix} \tag{3.72a}$$

$$\begin{bmatrix} E_R^i \\ E_L^i \end{bmatrix} = \frac{1}{\sqrt{2}} \begin{bmatrix} 1 & -j \\ 1 & j \end{bmatrix} \begin{bmatrix} E_h^i \\ E_v^i \end{bmatrix} \tag{3.72b}$$

Define the transformation matrix $U(RL \rightarrow hv)$ as,[2]

$$U(RL \rightarrow hv) = \frac{1}{\sqrt{2}} \begin{bmatrix} 1 & 1 \\ j & -j \end{bmatrix} \tag{3.73a}$$

and $U(hv \rightarrow RL)$ as,

$$U(hv \rightarrow RL) = \frac{1}{\sqrt{2}} \begin{bmatrix} 1 & -j \\ 1 & j \end{bmatrix} \tag{3.73b}$$

where the matrix U is unitary $(U^{-1} = U^{t*}, |\det U| = 1)$.

3.5.1   FSA scattering matrix in circular basis

In FSA, the scattering matrix relates the incident field $\vec{E}^i$ to the scattered field $\vec{E}^s$ as,

$$\begin{bmatrix} E_h^s \\ E_v^s \end{bmatrix} = \begin{bmatrix} S_{hh} & S_{hv} \\ S_{vh} & S_{vv} \end{bmatrix}_{\text{FSA}} \begin{bmatrix} E_h^i \\ E_v^i \end{bmatrix} \tag{3.74}$$

where the spherical wave factor has been dropped. Refer again to Fig. 3.5, where the back scatter direction and the unit vector triplets $(\hat{k}_i, \hat{v}_i, \hat{h}_i)$ and $(\hat{k}_s, \hat{v}_s, \hat{h}_s)$ are illustrated. Let the scattered field be expressed as $\vec{E}^s = E_h^s \hat{h}_s + E_v^s \hat{v}_s$ in the linear basis and as $\vec{E}^s = E_R^s \hat{e}_R^s + E_L^s \hat{e}_L^s$ in the circular basis. In the FSA convention, the change of basis from linear $(\hat{h}_s, \hat{v}_s)$ to circular $(\hat{e}_R^s, \hat{e}_L^s)$ is the same as given in (3.71),

$$\begin{bmatrix} \hat{e}_R^s \\ \hat{e}_L^s \end{bmatrix} = \frac{1}{\sqrt{2}} \begin{bmatrix} 1 & j \\ 1 & -j \end{bmatrix} \begin{bmatrix} \hat{h}_s \\ \hat{v}_s \end{bmatrix} \tag{3.75}$$

However, note that the horizontal reference direction is now along the $\hat{h}_s$-direction, while in (3.71) the reference direction is $\hat{h}_i$, which equals $-\hat{h}_s$. From (3.75), the components transform as given in (3.72a, b),

$$\begin{bmatrix} E_h^s \\ E_v^s \end{bmatrix} = \mathbf{U}(RL \to hv) \begin{bmatrix} E_R^s \\ E_L^s \end{bmatrix} \tag{3.76a}$$

$$\begin{bmatrix} E_R^s \\ E_L^s \end{bmatrix} = \mathbf{U}(hv \to RL) \begin{bmatrix} E_h^s \\ E_v^s \end{bmatrix} \tag{3.76b}$$

Substituting (3.72a) and (3.76a) into (3.74) results in,

$$\mathbf{U}(RL \to hv) \begin{bmatrix} E_R^s \\ E_L^s \end{bmatrix} = \begin{bmatrix} S_{hh} & S_{hv} \\ S_{vh} & S_{vv} \end{bmatrix}_{\text{FSA}} \mathbf{U}(RL \to hv) \begin{bmatrix} E_R^i \\ E_L^i \end{bmatrix} \tag{3.77}$$

or,

$$\begin{bmatrix} E_R^s \\ E_L^s \end{bmatrix} = [\mathbf{U}(RL \to hv)]^{-1} \mathbf{S}_{\text{FSA}} \mathbf{U}(RL \to hv) \begin{bmatrix} E_R^i \\ E_L^i \end{bmatrix} \tag{3.78}$$

The FSA scattering matrix in the circular basis is defined as,

$$\mathbf{S}_{\text{FSA}}^c = [\mathbf{U}(RL \to hv)]^{-1} \mathbf{S}_{\text{FSA}} \mathbf{U}(RL \to hv) \tag{3.79}$$

3.5.2   BSA scattering matrix in circular basis

The BSA scattering matrix in the circular basis can be derived as follows. First, note from (3.5) that the polarization ratio $\chi_r = -\chi_s$. Since the polarization ratio corresponding to $\hat{e}_R^s$ is j, it follows that the polarization ratio corresponding to $\hat{e}_R^r$ should be $-j$, or,

$$\hat{e}_R^r = \frac{1}{\sqrt{2}}(\hat{h}_r - j\hat{v}_r) \tag{3.80}$$

As a quick check, the time-domain representation of $\hat{e}_R^r$ is of the form $(\hat{h}_r \cos \omega t + \hat{v}_r \sin \omega t)$, which at $\omega t = 0$ lines up along the $\hat{h}_r$-direction, and with increasing ωt, the tip of the electric field rotates clockwise when viewed from the particle (see Fig. 3.5). Similarly, the unit LHC vector $\hat{e}_L^r$ is given as $\hat{e}_L^r = (\hat{h}_r + j\hat{v}_r)/\sqrt{2}$, and the tip of the electric field again lines up along the $\hat{h}_r$-direction at $\omega t = 0$ and rotates counter-clockwise when viewed from the particle. Hence, its polarization ratio $\chi_r = j$ (whereas $\chi_s = -j$). Thus, the transformation from a linear $(\hat{h}_r, \hat{v}_r)$ to a circular $(\hat{e}_R^r, \hat{e}_L^r)$ basis is given as,

$$\begin{bmatrix} \hat{e}_R^r \\ \hat{e}_L^r \end{bmatrix} = \frac{1}{\sqrt{2}} \begin{bmatrix} 1 & -j \\ 1 & j \end{bmatrix} \begin{bmatrix} \hat{h}_r \\ \hat{v}_r \end{bmatrix} \tag{3.81}$$

Let the "reflected" field $\vec{E}^r$ be expressed as $E_h^r \hat{h}_r + E_v^r \hat{v}_r$, or as $E_R^r \hat{e}_R^r + E_L^r \hat{e}_L^r$. From (3.81) it follows that the components are related, as,

$$\begin{bmatrix} E_h^r \\ E_v^r \end{bmatrix} = \frac{1}{\sqrt{2}} \begin{bmatrix} 1 & 1 \\ -j & j \end{bmatrix} \begin{bmatrix} E_R^r \\ E_L^r \end{bmatrix} \tag{3.82a}$$

$$= [U(RL \rightarrow hv)]^* \begin{bmatrix} E_R^r \\ E_L^r \end{bmatrix} \tag{3.82b}$$

The conjugation on $U(RL \rightarrow hv)$ which relates the "reflected" field components in the linear and circular bases in the BSA convention, must be contrasted with (3.76a), which is valid for the FSA convention.

The BSA scattering matrix in the circular basis (S_{BSA}^c) now follows. Starting from the linear polarization basis,

$$\begin{bmatrix} E_h^r \\ E_v^r \end{bmatrix} = \begin{bmatrix} S_{hh} & S_{hv} \\ S_{vh} & S_{vv} \end{bmatrix}_{BSA} \begin{bmatrix} E_h^i \\ E_v^i \end{bmatrix} \tag{3.83}$$

and using (3.82b) and (3.72a) gives,

$$[U(RL \rightarrow hv)]^* \begin{bmatrix} E_R^r \\ E_L^r \end{bmatrix} = \begin{bmatrix} S_{hh} & S_{hv} \\ S_{vh} & S_{vv} \end{bmatrix}_{BSA} U(RL \rightarrow hv) \begin{bmatrix} E_R^i \\ E_L^i \end{bmatrix} \tag{3.84}$$

or,

$$\begin{bmatrix} E_R^r \\ E_L^r \end{bmatrix} = [U(RL \rightarrow hv)]^t S_{BSA} U(RL \rightarrow hv) \begin{bmatrix} E_R^i \\ E_L^i \end{bmatrix} \tag{3.85}$$

$$= (S_{BSA}^c) \begin{bmatrix} E_R^i \\ E_L^i \end{bmatrix} \tag{3.86}$$

The BSA scattering matrix in the circular basis is thus,

$$\mathbf{S}_{BSA}^c = [\mathbf{U}(RL \rightarrow hv)]^t \mathbf{S}_{BSA} \mathbf{U}(RL \rightarrow hv) \tag{3.87}$$

From (3.87) and using reciprocity ($S_{hv} = S_{vh}$)$_{BSA}$ along the back scattering direction, the explicit elements of $\mathbf{S}_{BSA}^c$ are written as,

$$\mathbf{S}_{BSA}^c = \begin{bmatrix} S_{RR} & S_{RL} \\ S_{LR} & S_{LL} \end{bmatrix}_{BSA} \tag{3.88a}$$

$$= \frac{1}{2} \begin{bmatrix} S_{hh} - S_{vv} + j2S_{hv} & S_{hh} + S_{vv} \\ S_{hh} + S_{vv} & S_{hh} - S_{vv} - j2S_{hv} \end{bmatrix}_{BSA} \tag{3.88b}$$

$$= \frac{S_{hh} + S_{vv}}{2} \begin{bmatrix} \dfrac{S_{hh} - S_{vv} + j2S_{hv}}{S_{hh} + S_{vv}} & 1 \\ 1 & \dfrac{S_{hh} - S_{vv} - j2S_{hv}}{S_{hh} + S_{vv}} \end{bmatrix}_{BSA} \tag{3.88c}$$

Note that reciprocity in the circular basis corresponds to $S_{RL} = S_{LR}$, but in general no corresponding condition results for S_{RR} and S_{LL}. For a spherical particle, $S_{hh} = S_{vv}$ and $S_{hv} = 0$, so that,

$$\mathbf{S}_{BSA}^c = S_{hh} \begin{bmatrix} 0 & 1 \\ 1 & 0 \end{bmatrix} \tag{3.89}$$

Thus, when a right-hand circularly polarized wave is incident on the particle ($E_R^i = 1$, $E_L^i = 0$), the back scattered wave is left-hand circularly polarized, i.e. the polarization sense is opposite to the transmitted sense. This principle is used by single channel, circularly polarized target detection radars to reduce rain "clutter" by spherical drops. The target (e.g. aircraft) itself, is generally characterized by $|S_{hh}|^2 \neq |S_{vv}|^2$ and $|S_{hv}|^2 \neq 0$, with $|S_{RR}|^2$ and $|S_{LR}|^2$ being approximately of the same order. A single channel, circularly polarized radar, i.e one capable of measuring the $|S_{RR}|^2$ term, for example, will be sensitive only to the target return, the spherical drop return being zero.

3.5.3 Radar observables in circular basis

A simple example, based on a spheroid in the Rayleigh limit (taken from Section 2.3), will show the theoretical advantage of using circular polarization (McCormick and Hendry 1975). Refer to Fig. 2.6a, and let $\theta_i = 90°$, $\phi_b = 90°$. The elements of $\mathbf{S}_{BSA}^c$ (dropping the factor $k_0^2/4\pi\varepsilon_0$) can be written as,

$$S_{RR} = \frac{1}{2}(\alpha - \alpha_{z_b})e^{-j2\theta_b} \tag{3.90a}$$

$$S_{RL} = S_{LR} = \frac{1}{2}(\alpha + \alpha_{z_b}) \tag{3.90b}$$

$$S_{LL} = \frac{1}{2}(\alpha - \alpha_{z_b})e^{j2\theta_b} \tag{3.90c}$$

Recall that (θ_b, ϕ_b) define the orientation of the spheroid's symmetry axis, see (2.41), and that α and α_{z_b} are the polarizability elements defined by (2.45). The circular depolarization ratio (CDR) of the particle (measured in dB) is defined as,

$$\text{CDR} = 10 \log_{10} \frac{|S_{RR}|^2}{|S_{LR}|^2} \tag{3.91a}$$

$$= 10 \log_{10} \frac{|\alpha - \alpha_{z_b}|^2}{|\alpha + \alpha_{z_b}|^2} \tag{3.91b}$$

which is independent of θ_b, and for a fixed ε_r is proportional to the spheroid eccentricity only. The CDR may be compared with the linear depolarization ratio (LDR$_{vh}$), given in (2.57); the latter does depend on θ_b. Thus, if the intent is to estimate particle "shape" independent of its "orientation" for the case illustrated in Fig. 2.6a, then CDR is preferred over LDR$_{vh}$. It should also be clear from (3.90) that the orientation angle θ_b can be estimated as follows,

$$-4\theta_b = \arg[S_{RR} S_{RL}^*] - \arg[S_{LL} S_{RL}^*] \tag{3.92}$$

It should be emphasized that Rayleigh scattering by spheroids for the geometry illustrated in Fig. 2.6a, with $\theta_i = 90°$, has been assumed.

For the more general case illustrated in Fig. 2.10a, where $0 < \theta_i < 90°$, and in terms of the canting angle β, in the polarization plane, and the angle ψ, between the incident direction and the spheroid's symmetry axis, the elements of $\mathbf{S}_{BSA}^c$ from (2.91) are,

$$S_{RR} = \frac{1}{2} [S_{11}(\psi) - S_{22}(\psi)] e^{-j2\beta} \tag{3.93a}$$

$$S_{RL} = S_{LR} = \frac{1}{2} [S_{11}(\psi) + S_{22}(\psi)] \tag{3.93b}$$

$$S_{LL} = \frac{1}{2} [S_{11}(\psi) - S_{22}(\psi)] e^{j2\beta} \tag{3.93c}$$

where $S_{11}(\psi)$ and $S_{22}(\psi)$ are defined in (2.89a,b). Using (2.94) and (3.87), a more compact form for $\mathbf{S}_{BSA}^c$ is,

$$\mathbf{S}_{BSA}^c = \mathbf{U}' \mathbf{R}(\beta) \begin{bmatrix} S_{11}(\psi) & 0 \\ 0 & S_{22}(\psi) \end{bmatrix} \mathbf{R}(-\beta) \mathbf{U} \tag{3.94}$$

where $\mathbf{U} \equiv \mathbf{U}(RL \rightarrow hv)$. The form of (3.94) is more generally valid for non-Rayleigh scattering by bodies of revolution. The CDR is,

$$10 \log_{10} \frac{|S_{11}(\psi) - S_{22}(\psi)|^2}{|S_{11}(\psi) + S_{22}(\psi)|^2} \tag{3.95}$$

which may be compared with the asymmetry ratio (A_{sm}) defined in (2.99). The canting angle in the polarization plane can be derived as,

$$-4\beta = \arg[S_{RR} S_{LR}^*] - \arg[S_{LL} S_{RL}^*] \tag{3.96}$$

The parameter ν, i.e. the back scatter amplitude ratio at circular polarization, is defined for Rayleigh scattering and real ε_r as,

$$\nu = \frac{S_{11}(\psi) - S_{22}(\psi)}{S_{11}(\psi) + S_{22}(\psi)} \tag{3.97a}$$

$$= \frac{(\alpha - \alpha_{z_b}) \sin^2(\psi)}{(\alpha_{z_b} - \alpha) \sin^2 \psi + 2\alpha} \tag{3.97b}$$

When $\psi = 90°$, $\nu = (\alpha - \alpha_{z_b})/(\alpha + \alpha_{z_b})$. In the Rayleigh limit and when ε_r is real, the polarizabilities are real. As a result, ν is positive for oblate spheroids ($\alpha > \alpha_{z_b}$) and negative for prolates ($\alpha < \alpha_{z_b}$). The above convention is opposite to that of McCormick and Hendry (1975), who assume that ν is positive for prolates and negative for oblates. For non-Rayleigh scattering, or when ε_r is complex, the parameter ν is complex and written as,

$$\nu = |\nu| e^{j\delta_c} = \frac{S_{11}(\psi) - S_{22}(\psi)}{S_{11}(\psi) + S_{22}(\psi)} \tag{3.98}$$

The CDR $= 10 \log_{10} |\nu|^2$, while δ_c is referred to as the scattering differential phase in the circular basis. Note that CDR is now dependent on the angle ψ. Recall the discussion related to (2.95), where $S_{11}(\psi)$ and $S_{22}(\psi)$ are simply solutions to the scattering problem for the case where the symmetry axis is aligned along the Z-axis and ψ is interpreted to be the incident angle θ_i (see Fig. 2.7a, b). Thus, for a fixed ε_r, CDR can now be related to the "apparent" shape of the spheroid. For example, when $\psi = 0°$, the "apparent" shape is circular ($S_{11} = S_{22}$); while $\psi = 90°$ results in a measure of the eccentricity of the principal elliptical cross section of the spheroid. It should be clear that when $\psi = 0°$, the angle β is not meaningful, since $S_{RR} = 0$. In most practical situations, vertical incidence ($\theta_i = 0°$) of the radar beam, as illustrated in Fig. 2.7c, gives $\psi = 90°$ and $\beta = \phi_b$. Since ϕ_b is uniformly distributed in $(0, 2\pi)$ in this orientation model, the angle β does not give meaningful "orientation" information. Only for the case of near horizontal incidence ($\theta_i \approx 90°$) of the radar beam (illustrated in Fig. 2.6a) is the angle β a good measure of the orientation angle θ_b.

3.6   Radar range equation: circular basis

The voltage form of the radar range equation for a single particle in the circular basis, corresponding to (3.67) for the linear basis, can be obtained by starting with (3.86),

$$\begin{bmatrix} E_R^r \\ E_L^r \end{bmatrix} = S_{BSA}^c \begin{bmatrix} E_R^i \\ E_L^i \end{bmatrix} \frac{e^{-jk_0 r}}{r} \tag{3.99}$$

where the spherical wave factor has been re-introduced.

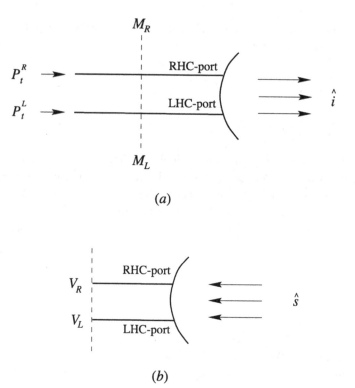

$$(a)$$

$$(b)$$

Fig. 3.8. (a) Dual-circularly polarized antenna as a radiator, and (b) back scattered wave and received voltages (V_R and V_L).

A dual-circularly polarized antenna will have two "ports", designated RHC-port and LHC-port, as illustrated in Fig. 3.8a. Following McCormick (1981), when the RHC-port only is excited with power (P_t^R), the radiated electric field $\vec{E}^i$ corresponding to (3.57) can be written as,

$$\vec{E}^i = \sqrt{\frac{P_t^R Z_0}{2\pi}} \sqrt{G_R} \hat{e}_R^i \frac{e^{-jk_0 r}}{r} \tag{3.100}$$

where G_R is the gain associated with the RHC-port and $\hat{e}_R^i$ is the unit RHC vector. Similarly, when only the LHC-port is excited with power (P_t^L), the radiated electric field is,

$$\vec{E}^i = \sqrt{\frac{P_t^L Z_0}{2\pi}} \sqrt{G_L} \hat{e}_L^i \frac{e^{-jk_0 r}}{r} \tag{3.101}$$

The gain inequality between ports can be represented by the ratio G_R/G_L. The phase inequality between ports can be introduced by the term $\exp[j(\Phi_R - \Phi_L)]$, but is omitted here. The vector $\hat{e}_L^i$ represents the unit LHC vector. When both ports are excited and the input wave amplitudes are $|M_R| \propto \sqrt{P_t^R}$ and $|M_L| \propto \sqrt{P_t^L}$, the radiated field

$(\vec{E}^i = E_R^i \hat{e}_R^i + E_L^i \hat{e}_L^i)$ can be written in matrix form as,

$$
\begin{bmatrix} E_R^i \\ E_L^i \end{bmatrix} = \sqrt{\frac{Z_0}{2\pi}} \begin{bmatrix} \sqrt{G_R} & 0 \\ 0 & \sqrt{G_L} \end{bmatrix} \begin{bmatrix} M_R \\ M_L \end{bmatrix} \frac{e^{-jk_0 r}}{r} \tag{3.102}
$$

Reception of the "reflected" wave $\vec{E}^r$ by a dual-circularly polarized antenna, as illustrated in Fig. 3.8b, can be handled using (3.38) which is invariant under a change of basis as shown by the following (Mieras 1986). The effective antenna length vector transforms according to (3.72a) which is written in compact notation as,

$$
\mathbf{h} = \mathbf{U}(RL \rightarrow hv)\mathbf{h}' \tag{3.103}
$$

where $\mathbf{h} = [h_h \ h_v]^t$ are the components in the linear basis, $\mathbf{h}' = [h_R \ h_L]^t$ are the components in the circular basis, and $\mathbf{U}(RL \rightarrow hv)$ is as given in (3.73a). From (3.82b), the "reflected" wave transforms as, $\mathbf{E}^r = [\mathbf{U}(RL \rightarrow hv)]^* \mathbf{E}'^r$, where $\mathbf{E}^r = [E_h^r \ E_v^r]^t$ and $\mathbf{E}'^r = [E_R'^r \ E_L'^r]^t$. The voltage equation (3.38), thus transforms as follows, using mixed matrix/vector notation with $\vec{a} \cdot \vec{b} \equiv (\mathbf{a}^t)\mathbf{b}$,

$$
V = \frac{1}{\sqrt{2Z_0}} \vec{h} \cdot \vec{E}^r \tag{3.104a}
$$

$$
= \frac{1}{\sqrt{2Z_0}} [\mathbf{h}]^t [\mathbf{E}^r] \tag{3.104b}
$$

$$
= \frac{1}{\sqrt{2Z_0}} [\mathbf{U}\mathbf{h}']^t [\mathbf{U}^* \mathbf{E}'^r] \tag{3.104c}
$$

$$
= \frac{1}{\sqrt{2Z_0}} [\mathbf{h}']^t [\mathbf{U}^t \mathbf{U}^*][\mathbf{E}'^r] = \frac{1}{\sqrt{2Z_0}} [\mathbf{h}']^t [\mathbf{E}'^r] = \frac{1}{\sqrt{2Z_0}} \vec{h}' \cdot \vec{E}'^r \tag{3.104d}
$$

The unitary matrix $\mathbf{U}$ satisfies $[\mathbf{U}^{-1}]^t = \mathbf{U}^*$.

From (3.104d), the voltage induced in the RHC-port by $\vec{E}^r$ given in the circular basis is,

$$
V_R = \frac{\lambda}{\sqrt{8\pi Z_0}} \sqrt{G_R} \hat{e}_R^i \cdot (E_R^r \hat{e}_R^r + E_L^r \hat{e}_L^r) \tag{3.105}
$$

since from (3.37) and (3.100), the effective length vector of the antenna corresponding to excitation of the RHC-port is,

$$
\vec{h} = \frac{\lambda}{\sqrt{4\pi}} \hat{e}_R^i \sqrt{G_R} \tag{3.106}
$$

Using the fact that $\hat{e}_R^i \cdot \hat{e}_R^r = 1$, while $\hat{e}_R^i \cdot \hat{e}_L^r = 0$, which can be verified easily using (3.71) and (3.81), and recognizing that $\hat{h}_i \cdot \hat{h}_r = \hat{v}_i \cdot \hat{v}_r = 1$, (3.105) reduces to,

$$
V_R = \frac{\lambda}{\sqrt{8\pi Z_0}} \sqrt{G_R} E_R^r \tag{3.107}
$$

Similarly, the voltage induced in the LHC-port is given by,

$$V_L = \frac{\lambda}{\sqrt{8\pi Z_0}}\sqrt{G_L}\hat{e}_L^i \cdot (E_R^r \hat{e}_R^r + E_L^r \hat{e}_L^r) \tag{3.108a}$$

$$= \frac{\lambda}{\sqrt{8\pi Z_0}}\sqrt{G_L}E_L^r \tag{3.108b}$$

Combining (3.107) and (3.108b) in matrix form gives,

$$\begin{bmatrix} V_R \\ V_L \end{bmatrix} = \frac{\lambda}{\sqrt{8\pi Z_0}}\begin{bmatrix} \sqrt{G_R} & 0 \\ 0 & \sqrt{G_L} \end{bmatrix}\begin{bmatrix} E_R^r \\ E_L^r \end{bmatrix} \tag{3.109}$$

Using (3.99) and (3.102) in (3.109) gives the voltage form of the radar range equation for a single particle in the circular basis,

$$\begin{bmatrix} V_R \\ V_L \end{bmatrix} = \frac{\lambda}{4\pi r^2}e^{-j2k_0 r}\begin{bmatrix} \sqrt{G_R} & 0 \\ 0 & \sqrt{G_L} \end{bmatrix}(S_{BSA}^c)\begin{bmatrix} \sqrt{G_R} & 0 \\ 0 & \sqrt{G_L} \end{bmatrix}\begin{bmatrix} M_R \\ M_L \end{bmatrix} \tag{3.110}$$

For an ideal antenna with $G_R = G_L$, the equation simplifies to,

$$\begin{bmatrix} V_R \\ V_L \end{bmatrix} = \frac{\lambda G}{4\pi r^2}e^{-j2k_0 r}[S_{BSA}^c]\begin{bmatrix} M_R \\ M_L \end{bmatrix} \tag{3.111}$$

which may be compared with (3.68a).

Non-ideal circular polarized antennas will radiate a small fraction of the undesired polarization state. For example, when the RHC-port is excited the effective length vector corresponding to that port given in (3.106) will be modified as,

$$\vec{h} = \frac{\lambda}{\sqrt{4\pi}}(i_R\hat{e}_R^i + \varepsilon_R\hat{e}_L^i)\sqrt{G_R} \tag{3.112}$$

where ε_R is a small complex error term and $i_R^2 + |\varepsilon_R|^2 = 1$. It follows that (3.107) will be modified as,

$$V_R = \frac{\lambda}{\sqrt{8\pi Z_0}}\sqrt{G_R}(i_R E_R^r + \varepsilon_R E_L^r) \tag{3.113}$$

Similarly, (3.108b) will be modified as,

$$V_L = \frac{\lambda}{\sqrt{8\pi Z_0}}\sqrt{G_L}(\varepsilon_L E_R^r + i_L E_L^r) \tag{3.114}$$

where ε_L is a small complex error term and $i_L^2 + |\varepsilon_L|^2 = 1$. Finally, (3.110) will be modified as,

$$\begin{bmatrix} V_R \\ V_L \end{bmatrix} = \frac{\lambda}{4\pi r^2}e^{-j2k_0 r}\begin{bmatrix} \sqrt{G_R} & 0 \\ 0 & \sqrt{G_L} \end{bmatrix}\begin{bmatrix} i_R & \varepsilon_R \\ \varepsilon_L & i_L \end{bmatrix}[S_{BSA}^c]$$

$$\begin{bmatrix} i_R & \varepsilon_L \\ \varepsilon_R & i_L \end{bmatrix}\begin{bmatrix} \sqrt{G_R} & 0 \\ 0 & \sqrt{G_L} \end{bmatrix}\begin{bmatrix} M_R \\ M_L \end{bmatrix} \tag{3.115}$$

3.7   Bilinear form of the voltage equation

The voltage equation in (3.38) when combined with the definition of the scattering matrix ($\vec{E}^r = S_{BSA}\vec{E}^i$) results in the bilinear form of the voltage equation (Kennaugh 1952),

$$V = \frac{1}{\sqrt{2Z_0}}\vec{h}\cdot\vec{E}^r \tag{3.116a}$$

$$= \frac{1}{\sqrt{2Z_0}}\vec{h}\cdot[S_{BSA}\vec{E}^i] \tag{3.116b}$$

In matrix notation, the dot product $\vec{a}\cdot\vec{b} \equiv \mathbf{a}^t\mathbf{b}$, and (3.116) is written as,

$$V = \frac{1}{\sqrt{2Z_0}}[\mathbf{h}^t][\mathbf{E}^r] \tag{3.117a}$$

$$= \frac{1}{\sqrt{2Z_0}}[\mathbf{h}^t][S_{BSA}][\mathbf{E}^i] \tag{3.117b}$$

The bilinear form of the voltage equation is invariant to a change of polarization basis. The general polarization basis is the elliptical basis, where the incident electric field is expressed as $\vec{E}^i = E_1^i\hat{e}_1^i + E_2^i\hat{e}_2^i$; where $\hat{e}_1^i$ and $\hat{e}_2^i$ are unit complex elliptical polarization vectors, which are orthogonal, i.e. $\hat{e}_1^i\cdot\hat{e}_2^{i*} = 0$. Let the transformation matrix $\mathbf{U}$ relate the components E_1^i, E_2^i to the components in the linear polarization basis E_h^i, E_v^i by,

$$\begin{bmatrix} E_h^i \\ E_v^i \end{bmatrix} = \mathbf{U}(e_1e_2 \to hv)\begin{bmatrix} E_1^i \\ E_2^i \end{bmatrix} \tag{3.118a}$$

or,

$$\mathbf{E}^i = \mathbf{U}\mathbf{E}^{i'} \tag{3.118b}$$

The general form of the unitary matrix $\mathbf{U}$ can be found, for example, in Agrawal and Boerner (1989) or in Hubbert (1994),

$$\mathbf{U}(e_1e_2 \to hv) = \frac{1}{\sqrt{1+\chi\chi^*}}\begin{bmatrix} e^{j\psi_1} & -\chi^*e^{j\psi_2} \\ \chi e^{j\psi_1} & e^{j\psi_2} \end{bmatrix} \tag{3.119}$$

where χ is the complex polarization ratio defined in (3.3) or (3.22b). The unit vectors $\hat{e}_1$ and $\hat{e}_2$ are related to $\hat{h}_i$ and $\hat{v}_i$ by the transpose of $\mathbf{U}$,

$$\begin{bmatrix} \hat{e}_1 \\ \hat{e}_2 \end{bmatrix} = \frac{1}{\sqrt{1+\chi\chi^*}}\begin{bmatrix} e^{j\psi_1} & \chi e^{j\psi_1} \\ -\chi^*e^{j\psi_2} & e^{j\psi_2} \end{bmatrix}\begin{bmatrix} \hat{h}_i \\ \hat{v}_i \end{bmatrix} \tag{3.120}$$

The phases ψ_1 and ψ_2 must be chosen carefully for consistency in radar applications whenever differential phases between the orthogonal components of the electric field need to be compared in different bases. The simplest values for ψ_1 and ψ_2, are, namely,

$\psi_1 = \psi_2 = 0°$. With this choice, and when $\hat{e}_1, \hat{e}_2$ correspond to $\hat{e}_R, \hat{e}_L$, i.e. $\chi = j$ in (3.120), the result is,

$$\begin{bmatrix} \hat{e}_R \\ \hat{e}_L \end{bmatrix} = \frac{1}{\sqrt{2}} \begin{bmatrix} 1 & j \\ j & 1 \end{bmatrix} \begin{bmatrix} \hat{h}_i \\ \hat{v}_i \end{bmatrix} \tag{3.121}$$

which is different from (3.9) and from McCormick and Hendry (1975). The circular base vector $\hat{e}_R$ in (3.121) lines up along the horizontal direction at $\omega t = 0$, while $\hat{e}_L$ lines up along the vertical direction. These different conventions can be a source of confusion, and unless the convention is explicitly stated the reader must figure out which convention is being used before comparing results from different authors.

The components of the antenna effective length vector ($\vec{h}$) will also transform as in (3.118),

$$\begin{bmatrix} h_h \\ h_v \end{bmatrix} = \mathbf{U}(e_1 e_2 \rightarrow hv) \begin{bmatrix} h_1 \\ h_2 \end{bmatrix} \tag{3.122a}$$

or,

$$\mathbf{h} = \mathbf{U}\mathbf{h}' \tag{3.122b}$$

Note the use of a prime to represent fields in the orthogonal elliptical basis. Substituting (3.118b) and (3.122b) into (3.117b) results in,

$$V = \frac{1}{\sqrt{2Z_0}} [\mathbf{U}\mathbf{h}']^t [\mathbf{S}_{\text{BSA}}][\mathbf{U}\mathbf{E}^{i'}] \tag{3.123a}$$

$$= \frac{1}{\sqrt{2Z_0}} [\mathbf{h}']^t [\mathbf{U}^t \mathbf{S}_{\text{BSA}} \mathbf{U}][\mathbf{E}^{i'}] \tag{3.123b}$$

Defining $\mathbf{S}'_{\text{BSA}} = \mathbf{U}^t \mathbf{S}_{\text{BSA}} \mathbf{U}$, which transforms the BSA scattering matrix from a linear basis to an elliptical basis, it is clear that the bilinear voltage equation is,

$$V = \frac{1}{\sqrt{2Z_0}} [\mathbf{h}']^t [\mathbf{S}'_{\text{BSA}}][\mathbf{E}^{i'}] \tag{3.124}$$

which is invariant to a change of polarization basis.

As a corollary, the voltage equation in (3.117a) can be used to determine how the "reflected" field $\vec{E}^r$ in BSA transforms under the change of basis defined in (3.118a). Starting from,

$$V = \frac{1}{\sqrt{2Z_0}} [\mathbf{h}^t][\mathbf{E}^r] \tag{3.125a}$$

$$= \frac{1}{\sqrt{2Z_0}} [\mathbf{U}\mathbf{h}']^t [\mathbf{E}^r] \tag{3.125b}$$

$$= \frac{1}{\sqrt{2Z_0}} [\mathbf{h}']^t [\mathbf{U}^t][\mathbf{E}^r] \tag{3.125c}$$

Defining $\mathbf{E}^{r'} = \mathbf{U}^t\mathbf{E}^r$, gives,

$$V = \frac{1}{\sqrt{2Z_0}}[\mathbf{h}']^t[\mathbf{E}^{r'}] \tag{3.126}$$

Since $\mathbf{U}$ is unitary,

$$\mathbf{E}^r = [\mathbf{U}^t]^{-1}[\mathbf{E}^{r'}] = [\mathbf{U}^*][\mathbf{E}^{r'}] \tag{3.127}$$

Thus, while the incident field and the antenna length vector transform as $\mathbf{E}^i = \mathbf{U}\mathbf{E}^{i'}$ and $\mathbf{h} = \mathbf{U}\mathbf{h}'$, the "reflected" field in the BSA transforms as $\mathbf{E}^r = \mathbf{U}^*\mathbf{E}^{r'}$ (Mieras 1986). Note that (3.82) described the same transformation for $\mathbf{E}^r$ for the special case of the circular basis. When the scattered field is described in FSA, it should be clear that it transforms, as $\mathbf{E}^s = \mathbf{U}\mathbf{E}^{s'}$.

3.8   Polarization synthesis and characteristic polarizations

In Section 1.4, the bistatic radar cross section of a particle was defined as,

$$\sigma_{\text{bi}}(\hat{s}, \hat{i}) = 4\pi|\vec{f}(\hat{s}, \hat{i})|^2 \tag{3.128}$$

and the back scatter or radar cross section as,

$$\sigma_b(-\hat{i}, \hat{i}) = 4\pi|\vec{f}(-\hat{i}, \hat{i})|^2 \tag{3.129}$$

It was assumed that a linearly polarized wave was incident on the particle, $\vec{E}^i = \hat{e}_i E_0$, where $\hat{e}_i$ is a real, unit polarization vector. Also, the cross sections were defined independent of any receiving antenna.

The bilinear form of the voltage equation in (3.116b) allows for a more general definition of the bistatic radar cross section, taking into account the polarization characteristics of separate transmitting and receiving antennas as illustrated in Fig. 3.9. The back scattering case ($\hat{s} = -\hat{i}$) simply implies co-located antennas. The received power (P_{rec}) at the receiving antenna is proportional to $|V|^2$, where V is given by (3.116b), or,

$$P_{\text{rec}} = \frac{1}{2Z_0}|\vec{h}\cdot(\mathbf{S}_{\text{BSA}}\vec{E}^i)|^2 \tag{3.130}$$

To make the notation more explicit, the subscript r will refer to the receiving antenna ($\vec{h} \equiv \vec{h}_r$), while the subscript t will refer to the transmitting antenna ($\vec{E}^i \equiv \vec{E}^t$). Thus,

$$\vec{h}_r = \frac{\lambda}{\sqrt{4\pi}}\hat{e}_r\sqrt{G_r} \tag{3.131}$$

where $\hat{e}_r$ is the complex polarization unit vector characterizing the receiving antenna (when it is used as a radiator). Also, from (3.36), the electric field radiated by the transmitting antenna ($\vec{E}^i \equiv \vec{E}^t$) is given as,

$$\vec{E}^t = \hat{e}_t\sqrt{\frac{P_t Z_0}{2\pi}}\sqrt{G_t} \tag{3.132}$$

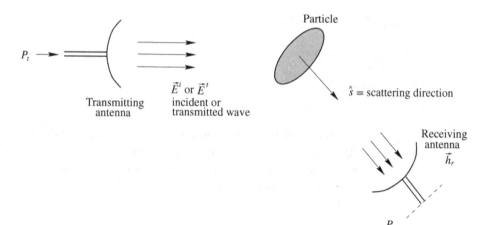

Fig. 3.9. Illustrates general definition of bistatic radar cross section of a particle taking into account the polarization characteristics of transmitting and receiving antennas. Note that $\vec{h}_r$ is the effective vector "length" of the receiving antenna when it is assumed to be a radiator.

where $\hat{e}_t$ is the complex polarization unit vector of the transmitting antenna, and the spherical wave factor has been dropped since it is not relevant to this discussion. Thus (3.130) can be written as,

$$P_{rec} = \frac{\lambda^2 G_r G_t P_t}{(4\pi)^2} \left| \hat{e}_r \cdot (\mathbf{S}_{BSA} \hat{e}_t) \right|^2 \qquad (3.133a)$$

$$= \frac{\lambda^2 G_r G_t P_t}{(4\pi)^3} 4\pi \left| \hat{e}_r \cdot (\mathbf{S}_{BSA} \hat{e}_t) \right|^2 \qquad (3.133b)$$

The bistatic radar cross section (σ_{rt}) accounting for the polarization characteristics of both the transmitting and receiving antennas is defined (Kennaugh 1952; van Zyl and Ulaby 1990) as,

$$\sigma_{rt} = 4\pi \left| \hat{e}_r \cdot (\mathbf{S}_{BSA} \hat{e}_t) \right|^2 \qquad (3.134)$$

In the particularly important case of back scattering by a single particle located at a range r, the power equation in (3.133b), including the r^{-4} range dependence, is,

$$P_{rec} = \frac{\lambda^2 G_r G_t P_t}{(4\pi)^3 r^4} \sigma_{rt} \qquad \bullet \qquad (3.135)$$

which is a generalized form of the conventional radar range equation (compare with (3.70c)).

Now $\hat{e}_t$ can be expressed in terms of geometrical ellipse parameters, as in (3.24b), which in matrix form is,

$$\hat{e}_t \equiv \begin{bmatrix} \cos \psi_t & -\sin \psi_t \\ \sin \psi_t & \cos \psi_t \end{bmatrix} \begin{bmatrix} \cos \tau_t \\ j \sin \tau_t \end{bmatrix} \qquad (3.136)$$

The copolar polarization synthesis equation is defined by σ_{rt}, with $\hat{e}_r = \hat{e}_t$ in (3.134), ψ_t varying from $-\pi/2$ to $+\pi/2$, and τ_t varying from $-\pi/4$ to $+\pi/4$. Thus, a copolar response surface is generated from the polarization synthesis equation for a given particle scattering matrix ($\mathbf{S}_{BSA}$). Similarly, a cross-polar response surface is generated when $\hat{e}_r$ is set orthogonal to $\hat{e}_t$. Recall that the unit ellipse orthogonal to $\hat{e}_t$ is obtained by setting $\psi_t \rightarrow \psi_t + \pi/2$ and $\tau_t \rightarrow -\tau_t$ in (3.136).

Consider the scattering matrix of a sphere, which (excluding constant factors) is,

$$\mathbf{S}_{BSA} = \begin{bmatrix} 1 & 0 \\ 0 & 1 \end{bmatrix} \tag{3.137a}$$

Substituting (3.137a) into (3.134) gives the copolar scattering cross section (σ_{co}) and the cross-polar scattering cross section (σ_{cx}) as,

$$\sigma_{co} = 4\pi (2\cos^2 \tau_t - 1)^2 \tag{3.137b}$$

$$\sigma_{cx} = 4\pi (4\cos^2 \tau_t \sin^2 \tau_t) \tag{3.137c}$$

Figure 3.10 shows the classic copolar (a) and cross-polar (b) polarization response surfaces for a sphere.[3] Note that the maximum copolar response occurs at linear polarization (ellipticity angle, $\tau_t = 0°$), while the copolar response is zero when $\tau_t = \pm\pi/4$, i.e. at circular polarization. In contrast, maximum cross-polar response occurs at $\tau_t = \pm\pi/4$, while the cross-polarized response is zero at linear polarization ($\tau_t = 0°$).

3.8.1   Elliptical depolarization ratio

From (3.137b, c), it is easy to see that for a sphere $\sigma_{co} = \sigma_{cx}$, when $\tau_t = \pm22.5°$. When $|\tau_t| < 22.5°$ then $\sigma_{cx} < \sigma_{co}$, while for $22.5 \le |\tau_t| \le 45°$ the reverse is true, i.e. $\sigma_{cx} > \sigma_{co}$. The concept of a depolarization ratio, introduced earlier as linear, see (2.55), and circular, see (3.95), depolarization ratios was based on extreme values of $\tau_t = 0°$ and $\tau_t = \pm45°$. The corresponding depolarization ratio definitions were $\mathrm{LDR}_{vh} = 10\log_{10}(\sigma_{cx}/\sigma_{co})$ and $\mathrm{CDR} = 10\log_{10}(\sigma_{co}/\sigma_{cx})$, respectively, and for a sphere both values tend to $-\infty$ dB. In the presence of antenna polarization errors, the lower bound for well-designed systems will be around -35 dB (see Section 6.2). Ice particles are generally weakly depolarizing with the result that σ_{cx} (for LDR_{vh}) or σ_{co} (for CDR) fall close to the receiver noise floor and cannot be detected. Matrosov (1991) has proposed transmitting an elliptically polarized wave ($\tau_t \approx 40°$) in order to increase σ_{co}, primarily for ice cloud particle-type classification. The corresponding elliptical depolarization ratio (EDR) is defined as,

$$\mathrm{EDR} = 10\log_{10}\left(\frac{\sigma_{co}}{\sigma_{cx}}\right) \tag{3.138}$$

The US National Oceanic and Atmospheric Administration Environmental Technology Laboratory (NOAA/ETL) operates a 35-GHz cloud-sensing Doppler radar (see Fig. 6.20

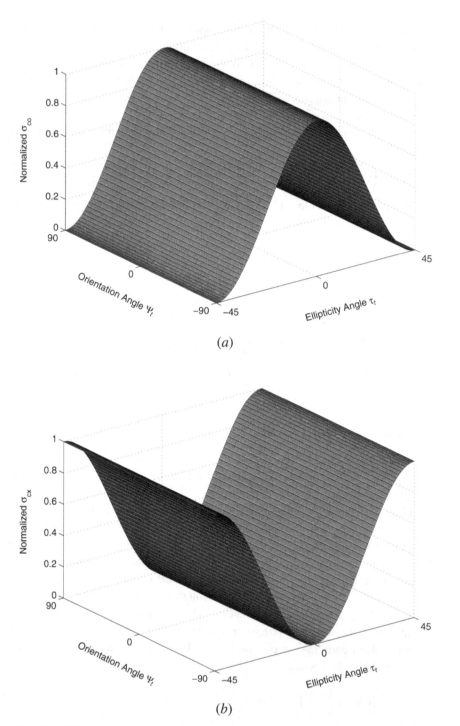

Fig. 3.10. Polarization response surfaces for a sphere: (*a*) copolar, and (*b*) cross-polar [refer to (3.137b, c)].

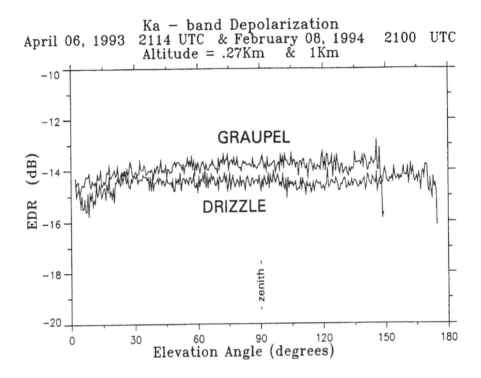

Fig. 3.11. RHI scans through drizzle (2114 UTC 6 April 1993) and graupel (2100 UTC, February 8, 1994). The elliptical depolarization ratio (EDR) is defined in (3.138). The transmitted ellipse has $\tau_t = 39.75°$, see also (3.137). From Reinking et al. (1997).

for a picture of the antenna), which transmits an elliptically polarized wave with $\tau_t = 39.75°$: a two-receiver system is used to measure σ_{co} and σ_{cx}. For spherical particles, the EDR from (3.137b, c), assuming $\tau_t = 39.75°$, is -14.64 dB, and this is expected from tiny spherical raindrops in drizzle. For spheres, S_{BSA} is not dependent on the elevation angle (the elevation angle $= 90° - \theta_i$, where θ_i is the angle of incidence). Figure 3.11 shows EDR versus the elevation angle in drizzle using the NOAA/ETL 35-GHz radar system (Reinking et al. 1997). The average EDR from the data is -14.4 ± 0.3 dB, which is in excellent agreement with theoretical predictions, and thus drizzle forms a reference signature against which data from other types of ice particles can be compared. For example, Fig. 3.11 also shows EDR versus elevation angle for graupel (i.e. for a compact snow pellet). These particles are typically conical in shape but can also be irregular. Because of the expected wide variety of shapes and canting angles, the EDR tends to be slightly higher than for drizzle (on average by about 1.2 dB). Figure 3.12 shows a rather dramatic example from planar crystals that can be modeled as oblate in shape (having an axis ratio $b/a \approx 0.1$) and with $\theta_b \approx 0°$ (see Fig. 2.7b). Thus, at horizontal incidence ($\theta_i = 90°$, or elevation angle $= 0°$) they appear highly

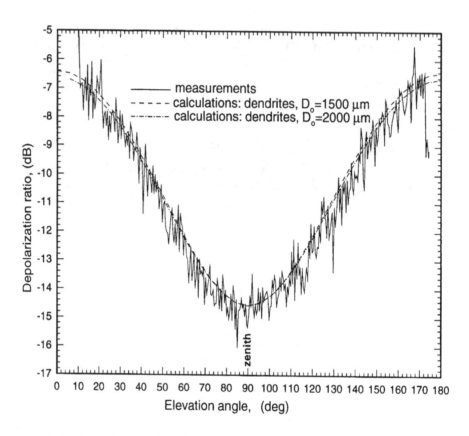

Fig. 3.12. RHI scans from dendrites (1-km above ground level at 2057 UTC, March 11, 1993):
(———) measured curve; (– – –) dendrites with median sizes of 1500 μm
($b/a \approx 0.038$, $\rho \approx 0.51$ g cm^{-3}); and (— · —) dendrites with median sizes of 2000 μm
($b/a \approx 0.035$, $\rho \approx 0.45$ g cm^{-3}). From Reinking et al. (1997).

oblate and oriented, while at vertical incidence ($\theta_i = 0°$ or elevation angle $= 90°$) they
appear circular. A maximum EDR occurs for an elevation angle of $0°$ (or $180°$) and
a minimum EDR occurs at an elevation angle of $90°$. The swing in EDR measured
from Fig. 3.12 is approximately 8 dB. Theoretical calculations using Rayleigh–Gans
approximation for oblate spheroids ($b/a \approx 0.035$, density ≈ 0.5 g cm^{-3}) are also
shown, which form an excellent fit to the data [other examples of ice particle EDRs can
be found in Reinking et al. (1997); see also, Section 7.3].

3.8.2   Copolar and cross-polar nulls for a single particle

Kennaugh first introduced the important theoretical concept of characteristic null
polarization states of a single target, which correspond to those transmit states (ψ_t, τ_t)
that produce zero copolar or cross-polar responses, i.e. copolar or cross-polar nulls.[4]

The reports published by Kennaugh have been compiled and edited by Moffatt and Garbacz (1984). These characteristic polarizations can be considered as reference polarizations for a fixed particle, and serve to establish a target-oriented system of polarization response (Huynen 1970; McCormick and Hendry 1985). For a single target with a symmetric scattering matrix as expressed by $\mathbf{S}_{BSA}$, there are, in general, two cross-polar nulls and two copolar nulls. The derivation of the transmit polarization states that give rise to copolar or cross-polar nulls for a given scattering matrix $\mathbf{S}_{BSA}$ is facilitated by using the complex polarization ratio χ_t of the transmitted (or incident) wave defined in (3.45). The derivation can be found in Agrawal and Boerner (1989) or Hubbert (1994), for example, and only the final results are given here. The transmit polarization ratios that give rise to copolar nulls are given as,

$$\chi_t^{co} = \left(\frac{-2S_{hv} \pm \sqrt{4S_{hv}^2 - 4S_{vv}S_{hh}}}{2S_{vv}} \right)_{BSA} \tag{3.139}$$

The transmit polarization ratios that give rise to cross-polar nulls are given as,

$$\chi_t^{cr} = \left[\frac{-(|S_{hh}|^2 - |S_{vv}|^2) \pm \sqrt{(|S_{hh}|^2 - |S_{vv}|^2)^2 + 4|S_{hv}S_{hh}^* + S_{vv}S_{hv}^*|^2}}{2(S_{hv}S_{hh}^* + S_{vv}S_{hv}^*)} \right]_{BSA} \tag{3.140}$$

Note that $[S_{hv} = S_{vh}]_{BSA}$ in the monostatic radar case. If $\mathbf{S}_{BSA}$ has the form given in (2.94), which is valid for non-Rayleigh scattering by bodies of revolution, then solution of (3.140) gives $\chi_1^{cr} = \cot \beta$ and $\chi_2^{cr} = -\tan \beta$. Recall from the geometry in Fig. 2.10b that the rotated linear unit vector $\hat{v}_i'$ lies in the plane containing both the symmetry axis and the incidence direction, while $\hat{h}_i'$ is perpendicular to this plane (this plane is shaded in Fig. 2.10b). For bodies of revolution, the transmission of $\vec{E}^{i'} = \hat{h}_i'$ or $\hat{v}_i'$ will give two cross-polar nulls, as is also evident from examining (2.96). Since $\hat{v}_i' = \hat{v}_i \cos \beta + \hat{h}_i \sin \beta$, while $\hat{h}_i' = -\hat{v}_i \sin \beta + \hat{h}_i \cos \beta$, the corresponding polarization ratios are $\cot \beta$ and $-\tan \beta$, respectively.

Bodies of revolution are a special case of symmetry for all incidence directions, i.e. a plane of symmetry containing the incidence direction always exists. Targets that exhibit symmetry about a plane containing a restrictive set of incidence directions (but which are not bodies of revolution) can also have an $\mathbf{S}_{BSA}$ of the form given in (2.94), e.g. a flat, rectangular conducting plate at normal incidence, and, hence, possess cross-polar null polarization ratios that are real. A general condition for a target's χ_t^{cr} to be real is obtained from (3.140) as,

$$\text{Im} \left(S_{hv}S_{hh}^* + S_{vv}S_{hv}^* \right) = 0 \tag{3.141}$$

A corollary to condition (3.141) is that $|S_{RR}|^2 = |S_{LL}|^2$ for the $\mathbf{S}_{BSA}^c$ defined in (3.88a). Condition (3.141) is also satisfied by arbitrary-shaped dielectric particles (with real ε_r) in the low frequency scattering case, since the polarizability tensor is real and, therefore, the scattering matrix elements are also real.

3.9   Partially polarized waves: coherency matrix and Stokes' vector

So far only monochromatic waves of the form,

$$\vec{E}^i(t) = \text{Re}\left(\vec{E}^i e^{j\omega t}\right) \tag{3.142a}$$

$$= \hat{h}_i |E_h^i| \cos(\omega t + \theta_h^i) + \hat{v}_i |E_v^i| \cos(\omega t + \theta_v^i) \tag{3.142b}$$

have been considered. Such waves are completely polarized, with the tip of the electric field tracing an ellipse (as ωt varies from 0 to 2π) of constant shape and orientation angle. Generally, the wave radiated by a radar system, here termed the incident wave, can be considered as completely polarized. In addition, only a single target or particle that is fixed in space has been considered so far, having a scattering matrix defined by $\mathbf{E}^s = \mathbf{S}_{\text{FSA}} \mathbf{E}^i$, so that the scattered wave is also completely polarized.

In Section 1.8, the concepts of coherent and incoherent summation were introduced by considering the resultant scattered signal from a collection of moving dielectric spheres with the mth sphere having a velocity $\vec{v}_m$ and corresponding angular Doppler shift frequency $\omega_m = k(\hat{i} - \hat{s}) \cdot \vec{v}_m$, an initial phase angle θ_m, and amplitude A_m. The complex signal was defined, see (1.92), as,

$$s(t) = \left[\sum_m A_m e^{-j(\omega_m t + \theta_m)}\right] e^{j\omega t} = \left[\gamma(t) e^{j\alpha(t)}\right] e^{j\omega t} \tag{3.143}$$

which can be considered as a quasi-monochromatic wave when $\gamma(t)$ and $\alpha(t)$ vary slowly relative to $1/\omega$.

A similar extension can be made to the two orthogonal components of the scattered electric field, $\vec{E}^s = E_h^s \hat{h}_s + E_v^s \hat{v}_s$, from a collection of moving particles. These orthogonal components may be expressed as signals of the form,

$$E_h^s(t) = \left[\sum_m |S_{hh}^m| e^{j\delta_{hh}^m} e^{-j(\omega_m t + \theta_m)}\right] e^{j\omega t} = \left[\gamma_{hh}(t) e^{j\alpha_{hh}(t)}\right] e^{j\omega t} \tag{3.144a}$$

$$E_v^s(t) = \left[\sum_m |S_{vh}^m| e^{j\delta_{vh}^m} e^{-j(\omega_m t + \theta_m)}\right] e^{j\omega t} = \left[\gamma_{vh}(t) e^{j\alpha_{vh}(t)}\right] e^{j\omega t} \tag{3.144b}$$

where $S_{hh} = |S_{hh}| e^{j\delta_{hh}}$ and $S_{vh} = |S_{vh}| e^{j\delta_{vh}}$, and it is assumed that the angular Doppler frequency and initial phases are independent of the polarization state. Note that in (3.144), the incident field has components $E_h^i = 1$, $E_v^i = 0$. For each component of $\vec{E}^s$, the envelope magnitude and phase (γ and α) are assumed to vary slowly with time (with bandwidth $\Delta\omega \ll \omega$), i.e. they are considered to be quasi-monochromatic waves. Dropping the $\exp(j\omega t)$ time factor in (3.144), the components of $\vec{E}^s$ can be written as,

$$E_h^s(t) = \sum_m E_h^{s(m)} = \gamma_{hh}(t) e^{j\alpha_{hh}(t)} \tag{3.145a}$$

$$E_v^s(t) = \sum_m E_v^{s(m)} = \gamma_{vh}(t) e^{j\alpha_{vh}(t)} \tag{3.145b}$$

The tip of the scattered electric field ($\vec{E}^s$) can be thought of as tracing an ellipse whose amplitude, ellipticity, and orientation angle vary with time. Such a wave is considered to be partially polarized.

3.9.1   Coherency matrix

The coherency matrix of a partially polarized scattered wave is defined (Born and Wolf 1975) as,

$$\mathbf{J} = \langle \mathbf{E}^s [\mathbf{E}^s]^{t*} \rangle = \begin{bmatrix} \langle E_h^s E_h^{s*} \rangle & \langle E_h^s E_v^{s*} \rangle \\ \langle E_v^s E_h^{s*} \rangle & \langle E_v^s E_v^{s*} \rangle \end{bmatrix} \tag{3.146}$$

where angle brackets denote time averaging. The coherency matrix can be transformed to a new polarization basis (e.g. a circular basis) by using the unitary matrix $\mathbf{U}$ defined in (3.73). If $\mathbf{J}'$ is the coherency matrix in the circular basis, and $\mathbf{E}' = \mathbf{U}(hv \rightarrow RL)\mathbf{E}$, then

$$\mathbf{J}' = \langle \mathbf{E}'[\mathbf{E}']^{t*} \rangle \tag{3.147a}$$

$$= \langle [\mathbf{UE}][\mathbf{UE}]^{t*} \rangle \tag{3.147b}$$

$$= \mathbf{U} \langle \mathbf{EE}^{t*} \rangle \mathbf{U}^{-1} \tag{3.147c}$$

$$= \mathbf{UJU}^{-1} \tag{3.147d}$$

The complex correlation coefficient between the two orthogonal components of $\vec{E}^s$ is defined as,

$$\mu = \frac{\langle E_h^s E_v^{s*} \rangle}{\sqrt{\langle |E_h^s|^2 \rangle \langle |E_v^s|^2 \rangle}} \tag{3.148}$$

and $|\mu| \leq 1$ (from the Schwartz inequality). The determinant of $\mathbf{J}$ can be written as,

$$\det(\mathbf{J}) = \langle |E_h^s|^2 \rangle \langle |E_v^s|^2 \rangle \left(1 - |\mu|^2 \right) \tag{3.149}$$

Waves that are completely unpolarized are characterized by $\mu = 0$ and $\langle |E_h^s|^2 \rangle = \langle |E_v^s|^2 \rangle$, e.g. natural light. Completely polarized waves are characterized by $|\mu| = 1$, resulting in $\det(\mathbf{J}) = 0$. A partially polarized wave can be considered as a sum of a completely polarized (cp) wave and a completely unpolarized (up) wave (Born and Wolf 1975),

$$\mathbf{J} = \mathbf{J}_{cp} + \mathbf{J}_{up} \tag{3.150a}$$

or,

$$\begin{bmatrix} J_{11} & J_{12} \\ J_{21} & J_{22} \end{bmatrix} = \begin{bmatrix} \langle |E_h^{cp}|^2 \rangle & \langle E_h^{cp} E_v^{cp*} \rangle \\ \langle E_v^{cp} E_h^{cp*} \rangle & \langle |E_v^{cp}|^2 \rangle \end{bmatrix} + \begin{bmatrix} \langle |E_h^{up}|^2 & 0 \\ 0 & \langle |E_h^{up}|^2 \rangle \end{bmatrix}$$

$$\tag{3.150b}$$

The power contained in the completely polarized part is proportional to the sum of the power contained in its two orthogonal components,

$$P_{cp} = \frac{1}{2Z_0} \left\{ \langle |E_h^{cp}|^2 \rangle + \langle |E_v^{cp}|^2 \rangle \right\} \tag{3.151}$$

The total power of the partially polarized wave is,

$$P_{TOT} = \frac{1}{2Z_0}(J_{11} + J_{22}) = P_{cp} + P_{up} \tag{3.152}$$

The degree of polarization (d) of the partially polarized wave is defined as the ratio of P_{cp}/P_{TOT}, which can be written as,

$$d = \sqrt{1 - \frac{4\det(\mathbf{J})}{(J_{11} + J_{22})^2}} \tag{3.153}$$

For completely polarized waves it should be clear that $d = 1$ from $\det(\mathbf{J}) = 0$, while for completely unpolarized waves, $d = 0$. For partially polarized waves, $0 < d < 1$.

The coherency matrix description of a partially polarized wave closely follows what a dual-polarized radar system can extract from $\vec{E}^s$, as illustrated in Fig. 3.7b, i.e. the power in each channel and the complex interchannel correlation coefficient. Note that two coherency matrices result from the two transmission states, ($E_h^i = 1$, $E_v^i = 0$) and ($E_h^i = 0$, $E_v^i = 1$), for example, when the H-transmitter and V-transmitter in Fig. 3.7 are "fired" alternately. However, as shown by McCormick (1989), while these two coherency matrices (obtained from any pair of orthogonally transmitted polarizations) are sufficient for the case of completely polarized back scatter, they are not sufficient for the case of partially polarized back scatter. This point will become clearer when the average Mueller and covariance matrices are defined in Section 3.9.2.

3.9.2   Stokes' vector of partially polarized waves

An alternate description of partially polarized waves is based on the Stokes' vector, originally introduced for studies of partially polarized light. The Stokes' vector of a completely polarized wave was given in (3.25), which defined a vector with four elements, $[I \ Q \ U \ V]^t$, in terms of the geometrical ellipse parameters (ψ, τ) and the total power of the wave. For a partially polarized scattered wave with components E_h^s and E_v^s, the Stokes' vector is defined as,

$$\begin{bmatrix} I_s \\ Q_s \\ U_s \\ V_s \end{bmatrix} = \begin{bmatrix} \langle |E_h^s|^2 \rangle + \langle |E_v^s|^2 \rangle \\ \langle |E_h^s|^2 \rangle - \langle |E_v^s|^2 \rangle \\ 2\operatorname{Re}\langle E_h^{s*} E_v^s \rangle \\ 2\operatorname{Im}\langle E_h^{s*} E_v^s \rangle \end{bmatrix} \tag{3.154a}$$

$$= \begin{bmatrix} 1 & 0 & 0 & 1 \\ 1 & 0 & 0 & -1 \\ 0 & 1 & 1 & 0 \\ 0 & j & -j & 0 \end{bmatrix} \begin{bmatrix} J_{11} \\ J_{12} \\ J_{21} \\ J_{22} \end{bmatrix} \tag{3.154b}$$

For a completely unpolarized wave $I = 2J_{11}$ and $Q = U = V = 0$, based on $J_{11} = J_{22}$ and $\mu = 0$. For a completely polarized wave, it follows from (3.25) that $I^2 = Q^2 + U^2 + V^2$, which also follows from $\det \mathbf{J} = 0$. The Stokes' vector of a partially polarized wave can be written as the sum of the Stokes' vectors of the completely polarized and completely unpolarized parts,

$$
\begin{bmatrix} I \\ Q \\ U \\ V \end{bmatrix} = \begin{bmatrix} I_{cp} \\ Q \\ U \\ V \end{bmatrix} + \begin{bmatrix} I - I_{cp} \\ 0 \\ 0 \\ 0 \end{bmatrix} \tag{3.155}
$$

and the degree of the polarization is $d = I_{cp}/I$. It should be clear that only the completely polarized part can be represented on the Poincaré sphere shown in Fig. 3.4.

The Stokes' vector elements of independent waves propagating in the same direction can be added (Born and Wolf 1975). From (3.144, 3.145) and using the concept of the initial phases being identical and independently distributed (iid) random variables with uniform pdf in $(0, 2\pi)$ (see also, Section 1.8), the Stokes' vector elements of the resultant scattered wave, $\vec{E}^s = E_h^s(t)\hat{h}_s + E_v^s(t)\hat{v}_s$, can be expressed as,

$$
I_s = \sum_m I_s^m = \sum_m \left[\langle |E_h^{s(m)}|^2 \rangle + \langle |E_v^{s(m)}|^2 \rangle \right] \tag{3.156}
$$

and, similarly, $Q_s = \sum Q_s^m$, $U_s = \sum_m U_s^m$, and $V_s = \sum V_s^m$; where I_s^m, Q_s^m, U_s^m, and V_s^m are the Stokes' vector elements of the scattered field, $\vec{E}^{s(m)} = E_h^{s(m)}\hat{h}_s + E_v^{s(m)}\hat{v}_s$, from the mth particle. Equation (3.156) essentially follows from the incoherent power sum formula stated in (1.99).

3.10 Ensemble-averaged Mueller matrix

The Mueller matrix of a single particle was defined in (3.28) as a 4×4 matrix relating the scattered field Stokes' vector to the incident field Stokes' vector. In the case of radar scattering from a distribution of particles, the model described by (3.144), the incident field launched by the radar is completely polarized but the scattered field is partially polarized. If $\mathbf{I}_i$ is the incident Stokes' vector corresponding to the incident vector radiation amplitude in (3.35b), and if each scatterer in the radar resolution volume (RRV; to be formally defined in Chapter 5) is illuminated by $\mathbf{I}_i \equiv \mathbf{I}_i^m$, then for the mth scatterer at the range r_m,

$$
\mathbf{I}_s^m = \frac{1}{r_m^4} \mathbf{M}^m \mathbf{I}_i^m \tag{3.157}
$$

Note that the range dependence here is r_m^{-4}, rather than r_m^{-2} in (3.29), because here the incident wave is a spherical wave as defined in (3.30), rather than an ideal plane wave.

From (3.156), the resultant scattered Stokes' vector is,

$$\mathbf{I}_s = \sum_m \frac{1}{r_m^4} \mathbf{M}^m \mathbf{I}_i^m \tag{3.158}$$

The summation is over all particles within the radar resolution volume. Since the number of particles in the RRV is large, and if it is assumed that the spatial distribution of particles in the RRV is homogeneous, then the summation in (3.158) can be replaced by an integral over the RRV,

$$\mathbf{I}_s = \langle n\mathbf{M} \rangle \int_{RRV} \frac{\mathbf{I}_i}{r^4} dV \tag{3.159}$$

where n is the number concentration of the particles (m^{-3}), dV is the volume element, and r is the range to dV. The angle brackets on $n\mathbf{M}$ now correspond to an ensemble average over particle states. In going from (3.158) to (3.159), the strong law of large numbers (Papoulis 1965) has been used, which states that the summation of individual Mueller matrices in (3.158) converges to the ensemble-averaged Mueller matrix with probability 1. If $s_1, s_2, \ldots, s_N$ are particle states, e.g. shape, orientation, thermodynamic phase, etc., and if $p(s_1, s_2, \ldots, s_N)$ is the joint probability density function, then the ensemble average of any function (f) over particle states can be defined as,

$$\langle f \rangle = \int_{s_1} \cdots \int_{s_N} f(s_1, s_2, \ldots, s_N) p(s_1, s_2, \ldots, s_N) \, ds_1 \, ds_2 \cdots ds_N \tag{3.160}$$

In addition, let D represent a characteristic dimension of the particles, and let $N(D) \, dD$ be the number of particles per unit volume with characteristic dimension between D and $D + dD$. Thus, $N(D)$ represents the usual size distribution function. The ensemble average, $\langle n\mathbf{M} \rangle$, can be generalized to include averaging over the size distribution as,

$$\langle n\mathbf{M} \rangle = \int_D N(D) \int_{s_1} \cdots \int_{s_N} \mathbf{M}(D; s_1, s_2, \ldots, s_N) p(s_1, s_2, \ldots, s_N) \, ds_1 \, ds_2 \cdots ds_N \, dD \tag{3.161}$$

The notation $\langle n\mathbf{M} \rangle$ refers to the ensemble-averaged Mueller matrix. The integral in (3.159) is related to a three-dimensional weighting function, to be defined in Chapter 5.

The scattering matrix of a single, fixed particle in the back scatter direction satisfies reciprocity and has five independent real elements as seen from,

$$\begin{bmatrix} S_{hh} & S_{hv} \\ S_{vh} & S_{vv} \end{bmatrix}_{FSA} = e^{j\delta_{hh}} \begin{bmatrix} |S_{hh}| & |S_{hv}| e^{j(\delta_{hv} - \delta_{hh})} \\ |S_{vh}| e^{j(\delta_{vh} - \delta_{hh})} & |S_{vv}| e^{j(\delta_{vv} - \delta_{hh})} \end{bmatrix}_{FSA} \tag{3.162}$$

The absolute phase (δ_{hh}) of the S_{hh} element is not relevant in practice. The above form of the matrix is referred to as the relative phase scattering matrix.

The elements of $\mathbf{M}$ in terms of the elements of $\mathbf{S}_{FSA}$ for a single, fixed particle are given, for example, in Mott (1992). The ensemble-averaged Mueller matrix, $\langle n\mathbf{M} \rangle$, given by (3.161), follows in terms of the ensemble average of the particular combination of $\mathbf{S}_{FSA}$ matrix elements given below,

$$\langle n\mathbf{M} \rangle = \left\langle n \begin{bmatrix} \frac{1}{2}(E_1 + E_2 + E_3 + E_4) & \frac{1}{2}(E_1 - E_2 - E_3 + E_4) \\ \frac{1}{2}(E_1 - E_2 + E_3 - E_4) & \frac{1}{2}(E_1 + E_2 - E_3 - E_4) \\ \mathrm{Re}(S_{hh}S_{vh}^* + S_{hv}S_{vv}^*) & \mathrm{Re}(S_{hh}S_{vh}^* - S_{hv}S_{vv}^*) \\ -\mathrm{Im}(S_{hh}S_{vh}^* + S_{hv}S_{vv}^*) & -\mathrm{Im}(S_{hh}S_{vh}^* - S_{hv}S_{vv}^*) \end{bmatrix} \right.$$

$$\left. \begin{matrix} \mathrm{Re}(S_{hh}S_{hv}^* + S_{vh}S_{vv}^*) & \mathrm{Im}(S_{hh}S_{hv}^* + S_{vh}S_{vv}^*) \\ \mathrm{Re}(S_{hh}S_{hv}^* - S_{vh}S_{vv}^*) & \mathrm{Im}(S_{hh}S_{hv}^* - S_{vh}S_{vv}^*) \\ \mathrm{Re}(S_{hh}S_{vv}^* + S_{hv}S_{vh}^*) & \mathrm{Im}(S_{hh}S_{vv}^* - S_{hv}S_{vh}^*) \\ -\mathrm{Im}(S_{hh}S_{vv}^* + S_{hv}S_{vh}^*) & \mathrm{Re}(S_{hh}S_{vv}^* - S_{hv}S_{vh}^*) \end{matrix} \right]_{\mathrm{FSA}} \right\rangle \qquad (3.163a)$$

where,

$$\begin{bmatrix} E_1 \\ E_2 \\ E_3 \\ E_4 \end{bmatrix} = \begin{bmatrix} |S_{hh}|^2 \\ |S_{vv}|^2 \\ |S_{hv}|^2 \\ |S_{vh}|^2 \end{bmatrix}_{\mathrm{FSA}} \qquad (3.163b)$$

3.10.1 Radar observables in linear basis

Several radar observables can now be defined. The copolar back scattering cross sections per unit volume ($m^2\ m^{-3}$) at horizontal and vertical polarizations are η_{hh} and η_{vv},

$$\eta_{hh} = \langle n4\pi |S_{hh}|^2 \rangle = \left\langle 4\pi n \frac{1}{2}(M_{11} + M_{12} + M_{21} + M_{22}) \right\rangle \qquad (3.164a)$$

$$\eta_{vv} = \langle n4\pi |S_{vv}|^2 \rangle = \left\langle 4\pi n \frac{1}{2}(M_{11} - M_{12} - M_{21} + M_{22}) \right\rangle \qquad (3.164b)$$

The cross-polar back scattering cross section per unit volume ($m^2\ m^{-3}$) is defined as,

$$\eta_{vh} = \langle n4\pi |S_{vh}|^2 \rangle = \left\langle 4\pi n \frac{1}{2}(M_{11} - M_{12} + M_{21} - M_{22}) \right\rangle \qquad (3.164c)$$

From reciprocity, $\eta_{vh} = \eta_{hv}$; see Tragl (1990) for a formal proof.[5] Note that the subscript vh refers to horizontal transmit polarization state and vertical receive state and vice versa.

The back scattering cross section per unit volume (η) is commonly referred to as the radar reflectivity. If the particles are spherical, then $\eta = \langle n4\pi |S|^2 \rangle$, where $S = S_{hh} = S_{vv}$. Since the radar cross section $\sigma_b = 4\pi |S|^2$, the reflectivity $\eta = \langle n\sigma_b \rangle$. In the

Rayleigh limit, σ_b is given by (1.51b) and, hence,

$$\eta_{\text{Ray}} = \frac{\pi^5}{\lambda^4}\left|\frac{\varepsilon_r - 1}{\varepsilon_r + 2}\right|^2 \langle nD^6 \rangle \tag{3.165a}$$

$$= \frac{\pi^5}{\lambda^4}|K_p|^2 \int_D D^6 N(D)\,dD \tag{3.165b}$$

$$= \frac{\pi^5}{\lambda^4}|K_p|^2 Z \tag{3.165c}$$

where $|K_p|^2 = |(\varepsilon_r - 1)/(\varepsilon_r + 2)|^2$ is the dielectric factor, and the reflectivity factor (Z) is defined as,

$$Z = \int_D D^6 N(D)\,dD \tag{3.166}$$

The reflectivity factor is commonly expressed in units of $mm^6\ m^{-3}$, i.e. in the above integral D is in millimeters and $N(D)\,dD$ is the number of spheres per cubic meter in the diameter interval $(D, D + dD)$. Since precipitation particles can vary in diameter over many orders of magnitude (e.g. cloud droplets $\approx 50\ \mu m$ to raindrops $\approx 5\ mm$), the logarithmic transformation $10\log_{10}(Z)$ is used and its units are in decibels of Z relative to $1\ mm^6\ m^{-3}$ (which corresponds to $0\ dBZ$). In the case of Mie scattering it is conventional to define the reflectivity factor (Z) as,

$$Z = \frac{\lambda^4}{\pi^5|K_p|^2}\eta \tag{3.167a}$$

$$= \frac{\lambda^4}{\pi^5|K_p|^2}\langle n\sigma_b \rangle \tag{3.167b}$$

$$= \frac{\lambda^4}{\pi^5|K_p|^2}\int_D \sigma_b(D)N(D)\,dD \tag{3.167c}$$

The single particle differential reflectivity and linear depolarization ratios defined earlier in (2.54, 2.55) (both measured in dB) can be generalized as,

$$Z_{\text{dr}} = 10\log_{10}\left(\frac{\eta_{hh}}{\eta_{vv}}\right) = 10\log_{10}\left(\frac{\langle 4\pi n|S_{hh}|^2 \rangle}{\langle 4\pi n|S_{vv}|^2 \rangle}\right) \tag{3.168}$$

$$\text{LDR}_{vh} = 10\log_{10}\left(\frac{\eta_{vh}}{\eta_{hh}}\right) = 10\log_{10}\left(\frac{\langle 4\pi n|S_{vh}|^2 \rangle}{\langle 4\pi n|S_{hh}|^2 \rangle}\right) \tag{3.169}$$

The magnitude of the copolar correlation coefficient is defined as,

$$|\rho_{\text{co}}| = \frac{|\langle n(S_{hh}S_{vv}^*)_{\text{FSA}} \rangle|}{\sqrt{\langle n|S_{hh}|^2 \rangle \langle n|S_{vv}|^2 \rangle}} \tag{3.170a}$$

$$= \frac{\sqrt{\{[\langle n(M_{33} + M_{44})\rangle]^2 + [\langle n(M_{34} - M_{43})\rangle]^2\}}}{\sqrt{\langle n(M_{11} + M_{12} + M_{21} + M_{22})\rangle \langle n(M_{11} - M_{12} - M_{21} + M_{22})\rangle}} \tag{3.170b}$$

Note that $M_{11}, M_{12}, \ldots$, etc., refer to the elements of $\mathbf{M}$ in the square brackets in (3.163a). The joint probability density function of particle states, defined in (3.160), is usually simplified by assuming independence between the states, e.g. independence between shape and orientation distributions. For spheroids in the Rayleigh limit, the elements of $\mathbf{S}_{\text{FSA}}$ are given in (2.53) or (2.87). It shows that $\langle n\mathbf{M} \rangle$ will depend, in general, on the angle of incidence (θ_i), the probability density function $p(\theta_b, \phi_b)$ describing the orientation of the symmetry axis, the probability density function of the spheroid axis ratio, the distribution of characteristic size (for spheroids the equivalent spherical diameter is a characteristic size), and on the dielectric constant. Some simple orientation models were considered in Section 2.3, and illustrated in Fig. 2.7. Since θ_i is known, and if ε_r can be assumed for the expected precipitation type, and if each of the three distributions of size, axis ratio, and orientation can be considered to have at most two unknown parameters (typically, the mean and variance), then there are six unknowns to be estimated from $\langle n\mathbf{M} \rangle$. This is a formidable estimation problem even under the simplistic assumptions of Rayleigh–Gans scattering by spheroids. Thus, maximum use must be made of a priori knowledge and self-consistent principles to make the problem more tractable, as will be discussed in Chapter 7.

3.11   Time-averaged Mueller and covariance matrices

3.11.1 Mueller matrix

It is also possible to arrive at the concept of a time-averaged Mueller matrix in terms of the "instantaneous" scattering matrices (Mott 1992). For this concept to be meaningful, the radar observation time must be small compared with the coherency time of the medium defined in Chapter 5, eq. (5.96). It is sufficient to mention here that modern dual-polarized radar systems[6] are capable of switching the transmitted polarization between orthogonal states (e.g. between $E_h^i = 1$, $E_v^i = 0$ and $E_h^i = 0$, $E_v^i = 1$) at time intervals of a few milliseconds, which is much smaller than the coherency time of the precipitation medium at typical microwave radar frequencies.

Consider again the model described in (3.144, 3.145) which describes the scattered field components when the excitation is $E_h^i = 1$, $E_v^i = 0$, i.e. the first column of the scattering matrix has the elements $\gamma_{hh}(t)\exp[j\alpha_{hh}(t)]$ and $\gamma_{vh}(t)\exp[j\alpha_{vh}(t)]$. Similarly, when the excitation is $E_h^i = 0$, $E_v^i = 1$, the second column of the matrix will have the elements $\gamma_{hv}(t)\exp[j\alpha_{hv}(t)]$ and $\gamma_{vv}(t)\exp[j\alpha_{vv}(t)]$. In practice, the second column will be separated in time from the first column by a few milliseconds. Here it is assumed that they are available at the same time. The "instantaneous" back scatter matrix can be written as,

$$\mathbf{S}_{\text{FSA}}(t) = \begin{bmatrix} \gamma_{hh}(t)e^{j\alpha_{hh}(t)} & \gamma_{hv}(t)e^{j\alpha_{hv}(t)} \\ \gamma_{vh}(t)e^{j\alpha_{vh}(t)} & \gamma_{vv}(t)e^{j\alpha_{vv}(t)} \end{bmatrix}_{\text{FSA}} \tag{3.171}$$

and,

$$\begin{bmatrix} E_h^s(t) \\ E_v^s(t) \end{bmatrix} = \mathbf{S}_{\mathrm{FSA}}(t) \begin{bmatrix} E_h^i \\ E_v^i \end{bmatrix} \tag{3.172}$$

Equation (3.172) can be transformed into a relation between the scattered Stokes' vector and the incident Stokes' vector involving the time-averaged (denoted by the use of angle brackets here and in Section 3.11.2) Mueller matrix,

$$\mathbf{I}_s = \frac{1}{r^2} \langle \mathbf{M} \rangle \mathbf{I}_i \tag{3.173}$$

where the elements of $\mathbf{M}$ are exactly the same as (3.163a), except that the complex envelope signals $\left(\gamma_{hh} e^{j\alpha_{hh}}, \gamma_{hv} e^{j\alpha_{hv}}, \gamma_{vh} e^{j\alpha_{vh}}, \gamma_{vv} e^{j\alpha_{vv}} \right)$ are used instead of $(S_{hh}, S_{hv}, S_{vh}, S_{vv})$.

3.11.2 Polarimetric covariance matrix

An alternative formulation to the time-averaged Mueller matrix which is suitable for radar applications is based on the polarimetric covariance matrix. The "instantaneous" back scatter matrix in the BSA is reciprocal and can be written as,

$$\mathbf{S}_{\mathrm{BSA}}(t) = \begin{bmatrix} \gamma_{hh}(t) e^{j\alpha_{hh}(t)} & \gamma_{hv}(t) e^{j\alpha_{hv}(t)} \\ \gamma_{hv}(t) e^{j\alpha_{hv}(t)} & \gamma_{vv}(t) e^{j\alpha_{vv}(t)} \end{bmatrix}_{\mathrm{BSA}} \tag{3.174}$$

where the model described by (3.144) is assumed, except in BSA convention. A feature "vector" is introduced, which is defined as,

$$\mathbf{\Omega} = \left[\gamma_{hh}(t) e^{j\alpha_{hh}(t)} \quad \sqrt{2} \gamma_{hv}(t) e^{j\alpha_{hv}(t)} \quad \gamma_{vv}(t) e^{j\alpha_{vv}(t)} \right]_{\mathrm{BSA}}^t \tag{3.175}$$

A factor of $\sqrt{2}$ is introduced to satisfy norm conservation, i.e. $|\mathbf{\Omega}^{t*}\mathbf{\Omega}| = |\gamma_{hh}|^2 + 2|\gamma_{hv}|^2 + |\gamma_{vv}|^2 = \mathrm{span}[\mathbf{S}_{\mathrm{BSA}}]$. Note that $\mathrm{span}[\mathbf{S}_{\mathrm{BSA}}]$ is invariant to a change of polarization basis. The polarization covariance matrix corresponding to this feature vector is defined as $\mathbf{\Sigma}_{\mathrm{BSA}} = \langle \mathbf{\Omega}\mathbf{\Omega}^{*t} \rangle$ (Tragl 1990), or,

$$\mathbf{\Sigma}_{\mathrm{BSA}} = \left\langle \begin{bmatrix} \gamma_{hh}(t) e^{j\alpha_{hh}(t)} \\ \sqrt{2} \gamma_{hv}(t) e^{j\alpha_{hv}(t)} \\ \gamma_{vv}(t) e^{j\alpha_{vv}(t)} \end{bmatrix} \begin{bmatrix} \gamma_{hh}(t) e^{-j\alpha_{hh}(t)} & \sqrt{2} \gamma_{hv}(t) e^{-j\alpha_{hv}(t)} & \gamma_{vv}(t) e^{-j\alpha_{vv}(t)} \end{bmatrix} \right\rangle \tag{3.176a}$$

$$= \begin{bmatrix} \langle \gamma_{hh}^2 \rangle & \sqrt{2} \langle \gamma_{hh}\gamma_{hv} e^{j(\alpha_{hh}-\alpha_{hv})} \rangle & \langle \gamma_{hh}\gamma_{vv} e^{j(\alpha_{hh}-\alpha_{vv})} \rangle \\ \sqrt{2} \langle \gamma_{hv}\gamma_{hh} e^{j(\alpha_{hv}-\alpha_{hh})} \rangle & 2\langle \gamma_{hv}^2 \rangle & \sqrt{2} \langle \gamma_{hv}\gamma_{vv} e^{j(\alpha_{hv}-\alpha_{vv})} \rangle \\ \langle \gamma_{vv}\gamma_{hh} e^{j(\alpha_{vv}-\alpha_{hh})} \rangle & \sqrt{2} \langle \gamma_{vv}\gamma_{hv} e^{j(\alpha_{vv}-\alpha_{hv})} \rangle & \langle \gamma_{vv}^2 \rangle \end{bmatrix} \tag{3.176b}$$

There are nine (real) independent terms in $\mathbf{\Sigma}$, i.e. the three power terms along the diagonal and three complex correlation terms. The matrix $\mathbf{\Sigma}_{\mathrm{BSA}}$ is Hermitian symmetric by definition, and is written as time averages of the complex envelope magnitude and phase of the resultant signals given in (3.144, 3.145), which express scattering by a collection of moving particles constituting the precipitation medium. A dual-polarized radar as illustrated in Fig. 3.7a, b is configured to measure Σ_{BSA} if the H- and V-transmitters are "fired" alternately with a pulse repetition time (PRT) of a few milliseconds. The time-averaged back scattered copolar powers are defined as,

$$P_{co}^h = C\langle \gamma_{hh}^2 \rangle \tag{3.177a}$$

$$P_{co}^v = C\langle \gamma_{vv}^2 \rangle \tag{3.177b}$$

and the time-averaged cross-polar power is defined as,

$$P_{cx} = C\langle \gamma_{hv}^2 \rangle \tag{3.178}$$

where C is a constant. The three correlation terms are defined as,

$$R_{cx}^h = C\left\langle \gamma_{hh}\gamma_{vh}e^{j(\alpha_{hh}-\alpha_{vh})} \right\rangle \tag{3.179a}$$

$$R_{cx}^v = C\left\langle \gamma_{vv}\gamma_{hv}e^{j(\alpha_{vv}-\alpha_{hv})} \right\rangle \tag{3.179b}$$

$$R_{co} = C\left\langle \gamma_{hh}\gamma_{vv}e^{j(\alpha_{hh}-\alpha_{vv})} \right\rangle \tag{3.179c}$$

Note that the superscript (h or v) on P_{co} and R_{cx} now denotes the transmitted polarization state. With this notation Σ_{BSA} can be written (dropping the constant C) as,

$$\Sigma_{BSA} = \begin{bmatrix} P_{co}^h & \sqrt{2}R_{cx}^h & R_{co} \\ \sqrt{2}(R_{cx}^h)^* & 2P_{cx} & \sqrt{2}(R_{cx}^v)^* \\ (R_{co})^* & \sqrt{2}R_{cx}^v & P_{co}^v \end{bmatrix} \tag{3.180}$$

The term R_{co} involves correlating copolar signals $\gamma_{hh}\exp(j\alpha_{hh})$ and $\gamma_{vv}\exp(j\alpha_{vv})$, which are typically separated in time by one PRT, and special processing methods are needed to estimate it as will be discussed in Section 6.4. The remaining correlation terms (R_{cx}^h and R_{cx}^v) are based on the elements of either the first or second column of the matrix in (3.174) which are available at the same time. The matrix Σ_{BSA} is a complete polarimetric characterization of a random distribution of reciprocal particles (reciprocal means that the reciprocity condition is satisfied). It will now be clear why the pair of coherency matrices (obtained from transmitting $E_h^i = 1, E_v^i = 0$ and $E_h^i = 0, E_v^i = 1$) are not complete for partially polarized back scatter. In the notation of (3.180), the two coherency matrices can be written as,

$$\mathbf{J}_{10} = \begin{bmatrix} P_{co}^h & R_{cx}^h \\ (R_{cx}^h)^* & P_{cx} \end{bmatrix} \tag{3.181a}$$

$$\mathbf{J}_{01} = \begin{bmatrix} P_{cx} & (R_{cx}^v)^* \\ (R_{cx}^v) & P_{co}^v \end{bmatrix} \tag{3.181b}$$

where the subscript 10 corresponds to ($E_h^i = 1$, $E_v^i = 0$), while 01 stands for ($E_h^i = 0$, $E_v^i = 1$) and (3.146) is used. Thus there are seven (real) independent terms and the missing term is R_{co} (McCormick 1989). The early circularly polarized radar systems (McCormick et al. 1972; McCormick and Hendry 1975) were designed to measure the coherency matrix when the transmitted polarization was fixed at one circular state, e.g. right-hand circular ($E_R^i = 1$, $E_L^i = 0$ in (3.86)). Such radar systems (with fixed radiated polarization) are referred to as possessing polarization "diversity", according to the definition of McCormick (1989). The term polarization "agility" refers to the capability of the radar system to change the radiated polarization state on a pulse-to-pulse basis, typically between two orthogonal states. In radars possessing diversity but not agility, the radiated polarization can generally be changed "slowly" (by an order of seconds rather that milliseconds). As shown by McCormick (1989), a complete set of polarimetric measurements is obtained with polarization diversity systems by measuring the pair of coherency matrices with ($E_R^i = 1$, $E_L^i = 0$ and $E_R^i = 0$, $E_L^i = 1$) and, in addition, measuring a coherency matrix with a slant $45°$ linear transmission (linear polarization oriented at $45°$, i.e. $\tau = 0°$ and $\phi = 45°$ in (3.21)). However, as will be seen later in Chapter 6, accuracy considerations make polarization agility highly desirable for measurement of some parameters.

3.11.3 Radar observables in linear basis

The ergodicity principle is now used to define radar observables in terms of the elements of the time-averaged covariance matrix. The differential reflectivity (dB) is,

$$Z_{dr} = 10 \log_{10} \left(\frac{P_{co}^h}{P_{co}^v} \right) = 10 \log_{10} \left(\frac{\langle \gamma_{hh}^2 \rangle}{\langle \gamma_{vv}^2 \rangle} \right) \tag{3.182a}$$

The linear depolarization ratio (dB) is,

$$LDR_{vh} = 10 \log_{10} \left(\frac{P_{cx}}{P_{co}^h} \right) = 10 \log_{10} \left(\frac{\langle \gamma_{vh}^2 \rangle}{\langle \gamma_{vv}^2 \rangle} \right) \tag{3.182b}$$

The magnitude of the copolar correlation coefficient is,

$$|\rho_{co}| = \frac{|R_{co}^*|}{\sqrt{P_{co}^h P_{co}^v}} = \frac{|\langle \gamma_{hh} \gamma_{vv} e^{j(\alpha_{vv} - \alpha_{hh})} \rangle|}{\sqrt{\langle \gamma_{hh}^2 \rangle \langle \gamma_{vv}^2 \rangle}} \tag{3.182c}$$

As mentioned earlier, the R_{co} term involves correlating copolar signals $\gamma_{hh} \exp(j\alpha_{hh})$ and $\gamma_{vv} \exp(j\alpha_{vv})$, which are typically separated in time by one pulse repetition time, and special processing methods must be used to estimate $|\rho_{co}|$ because of the mean Doppler shift that occurs (see Section 6.4).

More generally, the time-averaged Σ_{BSA} in (3.176b) can be expressed using the ergodicity principle, as an ensemble-averaged covariance matrix defined as,

$$\Sigma_{BSA} = \left\langle n \begin{bmatrix} |S_{hh}|^2 & \sqrt{2}(S_{hh}S_{hv}^*) & (S_{hh}S_{vv}^*) \\ \sqrt{2}(S_{hv}S_{hh}^*) & 2|S_{hv}|^2 & \sqrt{2}(S_{hv}S_{vv}^*) \\ (S_{vv}S_{hh}^*) & \sqrt{2}(S_{vv}S_{hv}^*) & |S_{vv}|^2 \end{bmatrix}_{BSA} \right\rangle \quad (3.183)$$

where the angle brackets denote ensemble averaging, for example,

$$\langle n|S_{hh}|^2\rangle = \int_D N(D) \int_{s_1} \cdots \int_{s_N} |S_{hh}(D; s_1, s_2, \ldots, s_N)|^2 p(s_1, s_2, \ldots, s_N)$$
$$\times ds_1\, ds_2 \ldots ds_N\, dD \quad (3.184)$$

Note that corresponding to (3.175), $\Omega = [S_{hh}\ \sqrt{2}S_{hv}\ S_{vv}]_{BSA}^t$ is now the feature vector.[7]

It is useful here to construct model Σ_{BSA} matrices for the case of oblate raindrops by averaging over $N(D)$ and over the orientation distribution described by (θ_b, ϕ_b). Here, D is the volume-equivalent spherical diameter of the oblate drop. The assumed model is as follows:

- Drop size distribution

$$N(D) = N_o \exp\left(-3.67\frac{D}{D_o}\right)$$

$$N_o = 8000\ mm^{-1}\ m^{-3}$$

$$D_o = 2.5\ mm$$

- Orientation distribution

  Fisher distribution, see (2.73)
  $$\bar\theta_b = \bar\phi_b = 0°; \quad \kappa = 100$$

- Incidence direction

  $\theta_i = 90°$ (case of horizontal incidence, see Fig. 2.7b)

The normalized covariance matrix (the elements are normalized by $\langle n|S_{hh}|^2\rangle$ in (3.183)) has been computed by Hubbert and Bringi (1996) for the above model,

$$\Sigma_{BSA} = \begin{bmatrix} 1.0 & 0.0 & 0.618e^{-j5°} \\ 0.0 & 0.0036 & 0.0 \\ 0.618e^{j5°} & 0.0 & 0.412 \end{bmatrix} \quad (3.185)$$

The terms Z_{dr}, LDR_{vh}, and $|\rho_{co}|$ from the above are, respectively, 3.85 dB, -27.45 dB, and 0.963. The back scatter differential phase (δ_{co}) is defined here in BSA as $\delta_{co} = \arg\langle n S_{hh}^* S_{vv}\rangle$ which is $5°$ in this example.

The effect of changing the orientation distribution parameters to $\bar{\theta}_b = 20°$ and $\bar{\phi}_b = 90°$ in the Fisher pdf results in,

$$
\Sigma_{\text{BSA}} = \begin{bmatrix}
1.0 & 0.181e^{-j171°} & 0.696e^{-j3.7°} \\
0.181e^{j171°} & 0.043 & 0.117e^{j166°} \\
0.696e^{j3.7°} & 0.117e^{-j166°} & 0.507
\end{bmatrix} \tag{3.186}
$$

The main change is in the zero elements of (3.185), which are now non-zero, and in the significant increase in LDR_{vh} to -16.7 dB.

3.12 Some implications of symmetry in scattering

3.12.1 "Mirror" reflection symmetry

Precipitation ensembles as a whole can exhibit "mirror" reflection symmetry about a plane. For example, spheroids with a symmetrical canting angle distribution about a mean canting angle β_0 possess this symmetry about the shaded plane in Fig. 2.10b, with $\beta = \beta_0$. Figure 3.13 illustrates this symmetry for identical prolates for the special case of $\psi = 90°$ (i.e. the angle between the direction of incidence and the particle symmetry axis is $90°$). Because of the symmetrical pdf of β about β_0, the correlation terms $R_{\text{cx}}^{h'} = \langle S_{h'h'} S_{v'h'}^* \rangle$ and $R_{\text{cx}}^{v'} = \langle S_{v'v'} S_{h'v'}^* \rangle$ will now be zero, which is shown as follows. If β' is the canting angle relative to β_0, then from (2.94) and using the rotation matrix $\mathbf{R}(\beta')$ and letting $S_{11,22}(\psi = 90°; \beta_0) \equiv S_{11}, S_{22}$, results in,

$$
S_{h'h'} = S_{11} \cos^2 \beta' + S_{22} \sin^2 \beta' \tag{3.187a}
$$
$$
S_{h'v'} = S_{v'h'} = (S_{22} - S_{11}) \sin \beta' \cos \beta' \tag{3.187b}
$$
$$
S_{v'v'} = S_{11} \sin^2 \beta' + S_{22} \cos^2 \beta' \tag{3.187c}
$$

The correlation term $R_{\text{cx}}^{h'}$ can be expressed as,

$$
R_{\text{cx}}^{h'} = \langle S_{h'h'} S_{v'h'}^* \rangle = \langle [S_{11}(S_{22}^* - S_{11}^*) \sin \beta' \cos^3 \beta' + S_{22}(S_{22}^* - S_{11}^*) \sin^3 \beta' \cos \beta'] \rangle \tag{3.188a}
$$
$$
= S_{11}(S_{22}^* - S_{11}^*)\langle \sin \beta' \cos^3 \beta' \rangle + S_{22}(S_{22}^* - S_{11}^*)\langle \sin^3 \beta' \cos \beta' \rangle \tag{3.188b}
$$

If $p(\beta')$ is the symmetric pdf about the angle β_0, then using $\sin \beta' \cos^3 \beta' = (\sin 2\beta')/4 + (\sin 4\beta')/8$, results in,

$$
\langle \sin \beta' \cos^3 \beta' \rangle = \int_{-\pi/2}^{+\pi/2} \left(\frac{1}{4} \sin 2\beta' + \frac{1}{8} \sin 4\beta' \right) p(\beta') \, d\beta' = 0 \tag{3.189}
$$

Similarly, $\langle \sin^3 \beta' \cos \beta' \rangle = 0$, and as a result $R_{\text{cx}}^{h'} = R_{\text{cx}}^{v'} = 0$. This condition is also true if the angle $\psi \neq 90°$.

Nghiem et al. (1992) have defined "mirror" reflection symmetry in a general manner which makes no assumptions about the scattering medium. Referring to Fig. 3.14,

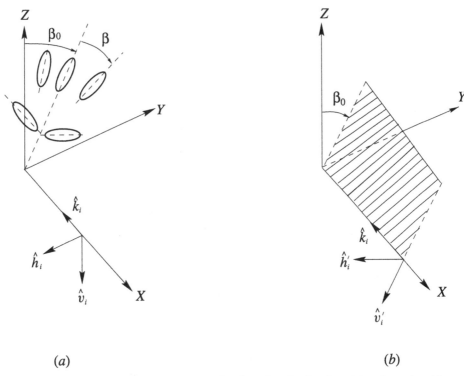

(a)   (b)

Fig. 3.13. "Mirror" reflection symmetry about the plane $\beta = \beta_0$; showing: (a) prolate spheroids with symmetry axes in the YZ-plane, with symmetric pdf about β_0 and where the direction of incidence is $\hat{k}_i$; (b) the symmetry plane (shaded). The $(\hat{h}'_i, \hat{v}'_i)$ basis is obtained by rotating $(\hat{h}_i, \hat{v}_i)$ by the angle β_0.

reflection symmetry about a plane containing the incidence direction is defined with reference to the rotated basis vectors $(\hat{h}'', \hat{v}'')$ and $(\hat{h}''', \hat{v}''')$, where $(\hat{h}'', \hat{v}'')$ corresponds to clockwise rotation by an angle α, while $(\hat{h}''', \hat{v}''')$ corresponds to counter-clockwise rotation by the angle $\alpha + 90°$. If the condition $R_{cx}^{h''} = R_{cx}^{v'''}$ defining reflection symmetry holds, then Nghiem et al. (1992) have shown that $R_{cx}^{v'} = R_{cx}^{h'} = 0$. Thus, in the $(\hat{h}', \hat{v}')$ basis, the term Σ'_{BSA} defined in (3.183) has four zeros,

$$\Sigma'_{BSA} = \begin{bmatrix} \langle n|S_{h'h'}|^2\rangle & 0 & \langle nS_{h'h'}S^*_{v'v'}\rangle \\ 0 & \langle n2|S_{h'v'}|^2\rangle & 0 \\ \langle nS_{v'v'}S^*_{h'h'}\rangle & 0 & \langle n|S_{v'v'}|^2\rangle \end{bmatrix} \tag{3.190}$$

The above covariance matrix has only five (real) independent elements, compared with the nine (real) elements of the general covariance matrix in (3.183).

3.12.2 Azimuthal symmetry/polarization plane isotropy

Nghiem et al. (1992) have also considered other forms of symmetry, such as rotational symmetry and azimuthal symmetry. Azimuthal symmetry is most obvious when

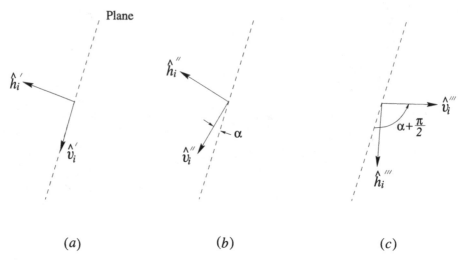

Fig. 3.14. General definition of "mirror" reflection symmetry about a plane containing the incidence direction $\hat{k}_i = \hat{v}'_i \times \hat{h}'_i$: (a) original basis $(\hat{h}'_i, \hat{v}'_i)$; (b) basis rotated clockwise by α; (c) basis rotated counter-clockwise by $[\alpha + (\pi/2)]$. Mirror reflection symmetry about the plane is defined by $R^{h''}_{cx} = R^{v'''}_{cx}$.

precipitation is observed when the angle of incidence $\theta_i = 0°$ (or vertically pointing antenna with elevation angle of $90°$). A simple orientation model for prolate ice crystals is shown in Fig. 2.7c, with $\theta_b = 90°$ and ϕ_b being uniformly distributed in $(0, 2\pi)$. This geometry corresponds to $\psi = 90°$ and $\beta = \phi_b$, and thus, in the plane of polarization, the canting angle may be considered to be uniformly distributed.

Using (2.91), the Σ_{BSA} matrix elements for spheroids can be constructed as follows (the notation is from Tang and Aydin 1995). Defining,

$$P = |S_{11}|^2 + |S_{22}|^2 \tag{3.191a}$$

$$Q = |S_{11}|^2 - |S_{22}|^2 \tag{3.191b}$$

$$R + jI = (S_{11}S^*_{22}) \tag{3.191c}$$

$$\cos 4\beta = c_4 \tag{3.191d}$$

$$\cos 2\beta = c_2 \tag{3.191e}$$

$$\sin 2\beta = s_2 \tag{3.191f}$$

$$\sin 4\beta = s_4 \tag{3.191g}$$

and using (3.183) results in,

$$\langle n|S_{hh}|^2\rangle = \left\langle n\left[\frac{P(3+c_4)}{8} + \frac{Qc_2}{2} + \frac{R(1-c_4)}{4}\right]\right\rangle \tag{3.192a}$$

$$\langle n|S_{vv}|^2\rangle = \left\langle n\left[\frac{P(3+c_4)}{8} - \frac{Qc_2}{2} + \frac{R(1-c_4)}{4}\right]\right\rangle \tag{3.192b}$$

$$\langle n|S_{hv}|^2\rangle = \left\langle n\left[\frac{P(1-c_4)}{8} - \frac{R(1-c_4)}{4}\right]\right\rangle \tag{3.192c}$$

$$\langle n S_{hh} S_{vv}^*\rangle = \left\langle n\left[\frac{P(1-c_4)}{8} + \frac{R(3+c_4)}{4} + jIc_2\right]\right\rangle \tag{3.192d}$$

$$\langle n S_{hh} S_{hv}^*\rangle = \left\langle n\left[\frac{-Q s_2}{4} - \frac{P s_4}{8} + \frac{R s_4}{4} + j\frac{I s_2}{2}\right]\right\rangle \tag{3.192e}$$

$$\langle n S_{hv} S_{vv}^*\rangle = \left\langle n\left[\frac{-Q s_2}{4} + \frac{P s_4}{8} - \frac{R s_4}{4} - j\frac{I s_2}{2}\right]\right\rangle \tag{3.192f}$$

If the size, shape, and orientation distributions are independent, and if β is uniformly distributed in $(-\pi/2, \pi/2)$, then the average trigonometric functions vanish, and the covariance matrix for azimuthal symmetry reduces to,

$$\Sigma_{\mathrm{BSA}} = \frac{\pi}{4}\begin{bmatrix} \left\langle n\left(\frac{3P}{2}+R\right)\right\rangle & 0 & \left\langle n\left(\frac{P}{2}+3R\right)\right\rangle \\ 0 & 2\left\langle n\left(\frac{P}{2}-R\right)\right\rangle & 0 \\ \left\langle n\left(\frac{P}{2}+3R\right)\right\rangle & 0 & \left\langle n\left(\frac{3P}{2}+R\right)\right\rangle \end{bmatrix} \tag{3.193}$$

where the angle brackets now denote ensemble averaging, for example, over size and shape. Note that there are only two independent parameters, P and R. Tang and Aydin (1995) have given the equivalent form of the Mueller matrix corresponding to (3.193). Under azimuthal symmetry it is clear that,

$$Z_{\mathrm{dr}} = 0 \quad (\mathrm{dB}) \tag{3.194a}$$

$$\mathrm{LDR} = \mathrm{LDR}_{vh} = \mathrm{LDR}_{hv} = 10\log_{10}\left[\frac{\langle n\,(P/2-R)\rangle}{\langle n\,(3P/2+R)\rangle}\right] \tag{3.194b}$$

$$|\rho_{\mathrm{co}}| = \frac{\langle n\,(P/2+3R)\rangle}{\langle n\,(3P/2+R)\rangle} \tag{3.194c}$$

Also, it follows that $|\rho_{\mathrm{co}}| = 1 - 2 \times 10^{0.1(\mathrm{LDR})}$ in the special case considered here for spheroids. However, it is more generally valid for particles possessing rotational symmetry (e.g. cones) and for non-Rayleigh scattering. Note, also, that ψ need not be 90°, since S_{11} and S_{22} can be evaluated at any ψ, as in (2.89). This implies that θ_i need not be zero or 90°, as long as the canting angle in the polarization plane is uniformly distributed. In this case, the term azimuthal symmetry is not appropriate and instead will be referred to as polarization plane isotropy. Figure 3.15 compares the direct measurement of $|\rho_{\mathrm{co}}|$ with that calculated from $1 - 2 \times 10^{0.1(\mathrm{LDR})}$ at vertical incidence in a uniform precipitation event with a melting layer (near a range of 0.3 km) using the 95-GHz airborne dual-polarized radar[8] described by Galloway et al. (1997). The excellent agreement indicates that azimuthal symmetry is satisfied in this case.

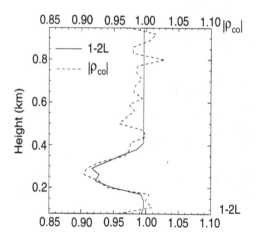

Fig. 3.15. Plot of $|\rho_{\text{co}}|$ and $1 - 2L$ demonstrating azimuthal symmetry, see (3.194), at vertical incidence. Note that LDR $= 10 \log_{10}(L)$. Radar data from the 95-GHz cloud radar on board the University of Wyoming King Air aircraft. From Galloway et al. (1997).

3.13   Covariance matrix in circular basis

Using the feature vector $\Omega = [S_{RR} \; \sqrt{2}S_{RL} \; S_{LL}]^t_{\text{BSA}}$, and noting that $S_{RL} = S_{LR}$ by reciprocity, the covariance matrix is expressed as,

$$\Sigma^c_{\text{BSA}} = \left\langle n \begin{bmatrix} |S_{RR}|^2 & \sqrt{2}(S_{RR}S^*_{RL}) & (S_{RR}S^*_{LL}) \\ \sqrt{2}(S^*_{RR}S_{RL}) & 2|S_{RL}|^2 & \sqrt{2}(S_{RL}S^*_{LL}) \\ (S^*_{RR}S_{LL}) & \sqrt{2}(S^*_{RL}S_{LL}) & |S_{LL}|^2 \end{bmatrix} \right\rangle \tag{3.195}$$

Using (3.88b), these elements can be expressed in terms of the elements of the covariance matrix in the linear $(\hat{h}, \hat{v})$ basis as,

$$\langle n|S_{RR}|^2 \rangle = \left\langle \frac{n}{4} \left[|S_{hh}|^2 + |S_{vv}|^2 + 4|S_{hv}|^2 - 2\,\text{Re}(S_{hh}S^*_{vv}) + 4\,\text{Im}(S_{hh} - S_{vv})S^*_{hv} \right] \right\rangle$$
$$\tag{3.196a}$$

$$\langle n(S_{RR}S^*_{RL}) \rangle = \left\langle \frac{n}{4} \left[|S_{hh}|^2 - |S_{vv}|^2 + j2\,\text{Im}(S_{hh}S^*_{vv}) + j2S_{hv}(S^*_{hh} + S^*_{vv}) \right] \right\rangle \tag{3.196b}$$

$$\langle n(S_{RR}S^*_{LL}) \rangle = \left\langle \frac{n}{4} \left[|S_{hh}|^2 + |S_{vv}|^2 - 4|S_{hv}|^2 - 2\,\text{Re}(S_{hh}S^*_{vv}) + j4\,\text{Re}(S_{hh} - S_{vv})S^*_{hv} \right] \right\rangle$$
$$\tag{3.196c}$$

$$\langle n|S_{RL}|^2 \rangle = \left\langle \frac{n}{4} \left[|S_{hh}|^2 + |S_{vv}|^2 + 2\,\text{Re}(S_{hh}S^*_{vv}) \right] \right\rangle \tag{3.196d}$$

$$\langle n|S_{LL}|^2 \rangle = \left\langle \frac{n}{4} \left[|S_{hh}|^2 + |S_{vv}|^2 + 4|S_{hv}|^2 - 2\,\text{Re}(S_{hh}S^*_{vv}) - 4\,\text{Im}(S_{hh} - S_{vv})S^*_{hv} \right] \right\rangle$$
$$\tag{3.196e}$$

$$\langle n(S_{LL}S^*_{RL}) \rangle = \left\langle \frac{n}{4} \left[|S_{hh}|^2 - |S_{vv}|^2 + j2\,\text{Im}(S_{hh}S^*_{vv}) - j2S_{hv}(S^*_{hh} + S^*_{vv}) \right] \right\rangle \tag{3.196f}$$

Use is made of the simple relations $z_1 z_2^* - z_1^* z_2 = j2\,\text{Im}(z_1 z_2^*) = -j2\,\text{Im}(z_1^* z_2)$ and $z_1 z_2^* + z_1^* z_2 = 2\,\text{Re}(z_1 z_2^*) = 2\,\text{Re}(z_1^* z_2)$, where z_1, z_2 are complex.

Since single particles, which exhibit symmetry about a plane containing the incidence direction and the axis of symmetry, satisfy the condition $\text{Im}(S_{hh}^* S_{hv} + S_{vv} S_{hv}^*) = 0$, as discussed in Section 3.8, it follows that $\langle n|S_{RR}|^2\rangle = \langle n|S_{LL}|^2\rangle$. This result occurs because of the symmetry of each particle in the ensemble, independently of how they are distributed with size, orientation, etc.

3.13.1 Azimuthal symmetry/polarization plane isotropy

In the case of azimuthal symmetry, the circular basis covariance matrix, using (3.192), reduces to the simple form,

$$\Sigma_{\text{BSA}}^c = \frac{\pi}{4}\begin{bmatrix} \langle n(P-2R)\rangle & 0 & 0 \\ 0 & 2\langle n(P+2R)\rangle & 0 \\ 0 & 0 & \langle n(P-2R)\rangle \end{bmatrix} \tag{3.197}$$

where P and R are defined in (3.191) and the angle brackets now denote ensemble averaging over size, shape, etc. The circular depolarization ratio (CDR) in this case is,

$$\text{CDR} = 10\log_{10}\left[\frac{\langle n|S_{RR}|^2\rangle}{\langle n|S_{RL}|^2\rangle}\right] \tag{3.198a}$$

$$= 10\log_{10}\left[\frac{\langle n(P-2R)\rangle}{\langle n(P+2R)\rangle}\right] \tag{3.198b}$$

$$= 10\log_{10}\left[\frac{\langle n|S_{11}-S_{22}|^2\rangle}{\langle n|S_{11}+S_{22}|^2\rangle}\right] \tag{3.198c}$$

3.13.2 "Mirror" reflection symmetry: $\beta_0 = 0°$

Consider now the case of "mirror" reflection symmetry, with the plane of symmetry being the XZ-plane ($\beta_0 = 0°$ case), as illustrated in Fig. 3.13. If $p(\beta)$ is the canting angle pdf, which is assumed to be symmetric about $\beta = 0°$, then from (3.192e, f), it follows that $\langle n S_{hh} S_{hv}^*\rangle = \langle n S_{hv} S_{vv}^*\rangle = 0$, since averages of $\sin 2\beta$ and $\sin 4\beta$ vanish. It follows from (3.196b, f) that $\langle n S_{LL} S_{RL}^*\rangle = \langle n S_{RR} S_{RL}^*\rangle$, and (3.196) reduces to,

$$\langle n|S_{RR}|^2\rangle = \left\langle \frac{n}{4}(P-2R)\right\rangle = \left\langle \frac{n}{4}|S_{11}-S_{22}|^2\right\rangle \tag{3.199a}$$

$$\langle n(S_{RR}S_{RL}^*)\rangle = \left\langle \frac{n}{4}(Q+j2I)\cos 2\beta\right\rangle = \left\langle \frac{n}{4}(S_{11}-S_{22})(S_{11}^*+S_{22}^*)\cos 2\beta\right\rangle \tag{3.199b}$$

$$\langle n(S_{RR}S_{LL}^*)\rangle = \left\langle \frac{n}{4}(P-2R)\cos 4\beta\right\rangle = \left\langle \frac{n}{4}|S_{11}-S_{22}|^2\cos 4\beta\right\rangle \tag{3.199c}$$

$$\langle n|S_{RL}|^2\rangle = \left\langle \frac{n}{4}(P+2R)\right\rangle = \left\langle \frac{n}{4}|S_{11}+S_{22}|^2\right\rangle \tag{3.199d}$$

The above results are valid even if $\psi \neq 90°$, as long as "mirror" reflection symmetry is about the plane $\beta = 0°$, since S_{11} and S_{22} can be replaced by $S_{11}(\psi)$ and $S_{22}(\psi)$. The covariance matrix under this symmetry can be expressed as,

$$\Sigma_{BSA}^c =$$
$$\begin{bmatrix} \langle \frac{n}{4}|S_{11} - S_{22}|^2 \rangle & \sqrt{2}\langle \frac{n}{4}(S_{11} - S_{22})(S_{11}^* + S_{22}^*) \cos 2\beta \rangle & \langle \frac{n}{4}|S_{11} - S_{22}|^2 \cos 4\beta \rangle \\ - & 2\langle \frac{n}{4}|S_{11} + S_{22}|^2 \rangle & - \\ - & \sqrt{2}\langle \frac{n}{4}(S_{11} - S_{22})(S_{11}^* + S_{22}^*) \cos 2\beta \rangle & \langle \frac{n}{4}|S_{11} - S_{22}|^2 \rangle \end{bmatrix}$$

$$(3.200)$$

The elements left blank in the above matrix can be obtained from the Hermitian property. Hendry et al. (1987) define two "orientation" parameters related to averages of $\cos 2\beta$ and $\cos 4\beta$ over the symmetrical pdf $p(\beta)$ as,

$$\rho_2 = \int_{-\pi/2}^{\pi/2} \cos(2\beta) p(\beta) \, d\beta \qquad (3.201a)$$

$$\rho_4 = \int_{-\pi/2}^{\pi/2} \cos(4\beta) p(\beta) \, d\beta \qquad (3.201b)$$

The assumption is made that the canting angle distribution is independent of other particle state distributions such as shape or size. The back scatter cross section per unit volume at circular polarization is defined as,

$$\langle \eta_c \rangle = \langle n4\pi \frac{1}{4}|S_{11} + S_{22}|^2 \rangle \qquad (3.202a)$$

The elements of the matrix in (3.200) can be expressed as,

$$\langle \frac{n}{4}|S_{11} - S_{22}|^2 \rangle = \langle \frac{n}{4}\frac{|S_{11} - S_{22}|^2}{|S_{11} + S_{22}|^2}|S_{11} + S_{22}|^2 \rangle \qquad (3.202b)$$

$$= \langle \frac{\eta_c}{4\pi}|v|^2 \rangle \qquad (3.202c)$$

where $|v|^2 = |S_{11} - S_{22}|^2/|S_{11} + S_{22}|^2$. Also,

$$\langle \frac{n}{4}(S_{11} - S_{22})(S_{11}^* + S_{22}^*) \cos 2\beta \rangle = \langle \frac{n}{4}\frac{(S_{11} - S_{22})(S_{11} + S_{22})(S_{11}^* + S_{22}^*)}{(S_{11} + S_{22})} \rangle \langle \cos 2\beta \rangle \qquad (3.203a)$$

$$= \langle \frac{\eta_c}{4\pi}v \rangle \rho_2 \qquad (3.203b)$$

where $v = (S_{11} - S_{22})/(S_{11} + S_{22})$; see also (3.98). Also,

$$\langle \frac{n}{4}|S_{11} - S_{22}|^2 \cos 4\beta \rangle = \langle \frac{n}{4}\frac{|S_{11} - S_{22}|^2}{|S_{11} + S_{22}|^2}|S_{11} + S_{22}|^2 \rangle \rho_4 \qquad (3.204a)$$

$$= \langle \frac{\eta_c}{4\pi}|v|^2 \rangle \rho_4 \qquad (3.204b)$$

Thus, (3.200) can be written in compact form as,

$$
\Sigma_{BSA}^c = \frac{1}{4\pi}
\begin{bmatrix}
\langle \eta_c|v|^2 \rangle & \sqrt{2}\rho_2\langle \eta_c v \rangle & \langle \eta_c|v|^2 \rangle \rho_4 \\
\sqrt{2}\rho_2\langle \eta_c v^* \rangle & 2\langle \eta_c \rangle & \sqrt{2}\rho_2\langle \eta_c v^* \rangle \\
\langle \eta_c|v|^2 \rangle \rho_4 & \sqrt{2}\rho_2\langle \eta_c v \rangle & \langle \eta_c|v|^2 \rangle
\end{bmatrix}
\tag{3.205}
$$

3.13.3 "Mirror" reflection symmetry: $\beta_0 \neq 0°$

Finally, consider the case of the "mirror" reflection symmetry plane tilted at β_0, as illustrated in Fig. 3.13b. The intent is to modify the covariance matrix in (3.200) which is valid for $\beta_0 = 0°$ to the case of $\beta_0 \neq 0°$. The circular base vectors $(\hat{e}'_R, \hat{e}'_L)$ corresponding to $(\hat{h}'_i, \hat{v}'_i)$ and the circular base vectors $(\hat{e}_R, \hat{e}_L)$ corresponding to $(\hat{h}_i, \hat{v}_i)$ are related by,

$$
\begin{bmatrix} \hat{e}_R \\ \hat{e}_L \end{bmatrix} =
\begin{bmatrix} e^{-j\beta_0} & 0 \\ 0 & e^{j\beta_0} \end{bmatrix}
\begin{bmatrix} \hat{e}'_R \\ \hat{e}'_L \end{bmatrix}
\tag{3.206}
$$

Note that the horizontal reference direction for $(\hat{e}_R, \hat{e}_L)$ is along $\hat{h}_i$. It is easily verified that $\sqrt{2}\hat{e}_R = (\hat{h}'_i + j\hat{v}'_i)\exp(-j\beta_0)$ has a time-varying representation, as $\sqrt{2}\hat{e}_R \equiv \hat{h}'_i \cos(\omega t - \beta_0) + \hat{v}'_i \cos[\omega t - \beta_0 + (\pi/2)]$, or at $\omega t = 0$, $\sqrt{2}\hat{e}_R \equiv \hat{h}'_i \cos\beta_0 + \hat{v}'_i \cos[(\pi/2) - \beta_0] = \hat{h}'_i \cos\beta_0 + \hat{v}'_i \sin\beta_0 = \hat{h}_i$. From (3.206) it follows that the circular components of the incident field are related by,

$$
\begin{bmatrix} E^i_R \\ E^i_L \end{bmatrix}' =
\begin{bmatrix} e^{-j\beta_0} & 0 \\ 0 & e^{j\beta_0} \end{bmatrix}
\begin{bmatrix} E^i_R \\ E^i_L \end{bmatrix}
\tag{3.207}
$$

From Section 3.7, the single particle scattering matrix (in BSA) is related in the two bases as,

$$
\begin{bmatrix} S_{RR} & S_{RL} \\ S_{LR} & S_{LL} \end{bmatrix} =
\begin{bmatrix} e^{-j\beta_0} & 0 \\ 0 & e^{j\beta_0} \end{bmatrix}
\begin{bmatrix} S'_{RR} & S'_{RL} \\ S'_{LR} & S'_{LL} \end{bmatrix}
\begin{bmatrix} e^{-j\beta_0} & 0 \\ 0 & e^{j\beta_0} \end{bmatrix}
\tag{3.208a}
$$

$$
= \begin{bmatrix} S'_{RR}e^{-j2\beta_0} & S'_{RL} \\ S'_{LR} & S'_{LL}e^{j2\beta_0} \end{bmatrix}
\tag{3.208b}
$$

The feature vector is $\Omega = [S'_{RR}e^{-j2\beta_0} \ \sqrt{2}S'_{RL} \ S'_{LL}e^{j2\beta_0}]^t$, and the covariance matrix is,

$$
\Sigma_{BSA}^c = \left\langle n
\begin{bmatrix}
|S'_{RR}|^2 & \sqrt{2}(S'_{RR}S'^*_{RL})e^{-j2\beta_0} & (S'_{RR}S'^*_{LL})e^{-j4\beta_0} \\
\sqrt{2}(S'^*_{RR}S'_{RL})e^{j2\beta_0} & 2|S'_{RL}|^2 & \sqrt{2}(S'_{RL}S'^*_{LL})e^{-j2\beta_0} \\
(S'^*_{RR}S'_{LL})e^{j4\beta_0} & \sqrt{2}(S'^*_{RL}S'_{LL})e^{j2\beta_0} & |S'_{LL}|^2
\end{bmatrix}
\right\rangle
\tag{3.209}
$$

From (3.187), where β' is the canting angle relative to β_0 (see Fig. 3.13) and $p(\beta')$ is the symmetric pdf about β_0, it should be clear that the steps used to arrive at (3.200) can be repeated by replacing β with β' resulting in,

$$
\Sigma_{BSA}^c =
\begin{bmatrix}
\left\langle \frac{n}{4}|S_{11} - S_{22}|^2 \right\rangle & \sqrt{2}\left\langle \frac{n}{4}(S_{11} - S_{22})(S_{11}^* + S_{22}^*)\cos 2\beta' \right\rangle e^{-j2\beta_0} \\
- & 2\left\langle \frac{n}{4}|S_{11} + S_{22}|^2 \right\rangle \\
- & \sqrt{2}\left\langle \frac{n}{4}(S_{11} - S_{22})(S_{11}^* + S_{22}^*)\cos 2\beta' \right\rangle e^{j2\beta_0} \\
\end{bmatrix}
$$

$$
\left\langle \frac{n}{4}|S_{11} - S_{22}|^2 \cos 4\beta' \right\rangle e^{-j4\beta_0}
$$

$$
-
$$

$$
\left\langle \frac{n}{4}|S_{11} - S_{22}|^2 \right\rangle
$$

(3.210)

The "orientation" parameters ρ_2 and ρ_4 are, as before,

$$
\rho_2 = \int_{-\pi/2}^{+\pi/2} \cos(2\beta')p(\beta')\,d\beta' \tag{3.211a}
$$

$$
\rho_4 = \int_{-\pi/2}^{+\pi/2} \cos(4\beta')p(\beta')\,d\beta' \tag{3.211b}
$$

The covariance matrix in (3.205) is thus modified to,

$$
\Sigma_{BSA}^c = \frac{1}{4\pi}
\begin{bmatrix}
\langle \eta_c |v|^2 \rangle & \sqrt{2}\rho_2 \langle \eta_c v \rangle e^{-j2\beta_0} & \rho_4 \langle \eta_c |v|^2 \rangle e^{-j4\beta_0} \\
\sqrt{2}\rho_2 \langle \eta_c v^* \rangle e^{j2\beta_0} & 2\langle \eta_c \rangle & \sqrt{2}\rho_2 \langle \eta_c v^* \rangle e^{-j2\beta_0} \\
\rho_4 \langle \eta_c |v|^2 \rangle e^{j4\beta_0} & \sqrt{2}\rho_2 \langle \eta_c v \rangle e^{j2\beta_0} & \langle \eta_c |v|^2 \rangle \\
\end{bmatrix}
$$

(3.212)

One advantage of the circular basis is that the angle β_0 can be directly obtained from,

$$
-4\beta_0 = \arg\left(\langle S_{RR}S_{LR}^* \rangle\right) - \arg\left(\langle S_{LL}S_{RL}^* \rangle\right) \tag{3.213}
$$

which is a generalization of (3.96).

3.13.4 Radar observables in circular basis

It is useful to define the following ratios, from McCormick and Hendry (1975),

$$
\overline{v^2} = \frac{\langle \eta_c |v|^2 \rangle}{\langle \eta_c \rangle} = \frac{\langle |S_{RR}|^2 \rangle}{\langle |S_{RL}|^2 \rangle} \tag{3.214a}
$$

where $\overline{v^2}$ is the circular depolarization ratio, and,

$$
\bar{v} = \frac{\langle \eta_c v \rangle}{\langle \eta_c \rangle} \tag{3.214b}
$$

The orientation parameter "ORTT" is defined as,

$$\text{ORTT} = \frac{|\langle S_{RR} S_{RL}^* \rangle|}{\sqrt{\langle |S_{RR}|^2 \rangle \langle |S_{RL}|^2 \rangle}} \tag{3.214c}$$

from which,

$$\text{ORTT} = \rho_2 \frac{|\bar{v}|}{(\overline{v^2})^{1/2}} \tag{3.214d}$$

The ratio $|\bar{v}|/(\overline{v^2})^{1/2}$ is termed the "shape" factor. Thus, ORTT is a product of the "shape" factor and ρ_2. It is desirable to estimate them separately for the purpose of hydrometeor classification. A direct measurement of ρ_2 does not appear possible, but ρ_4 can be obtained as,

$$\rho_4 = \frac{|\langle S_{RR} S_{LL}^* \rangle|}{\sqrt{\langle |S_{RR}|^2 \rangle \langle |S_{LL}|^2 \rangle}} \tag{3.215}$$

if the radar system is polarization "agile" (capable of transmitting alternate RHC and LHC states from pulse-to-pulse). Since correlating signals separated by one PRT is involved, special processing methods (see Section 6.4.3) must be used to correct for the mean Doppler shift that occurs (similar to estimation of $|\rho_{co}|$ in (3.182c)). For narrow Gaussian distributions, the relation $\rho_4 = (\rho_2)^4$ can be used (Hendry et al. 1987). Hendry and coworkers also show numerically that the ratio ρ_4/ρ_2 is not very different between Gaussian, rectangular, and triangular distributions.

3.13.5 Rotating linear basis

If the radar is not polarization "agile", then the measurement of ρ_4 involves additional measurements in the linear polarization basis (Hendry et al. 1987; McCormick 1989). For example, Hendry et al. (1987) show that measurements made in the "rotating" linear basis $(\hat{h}_i, \hat{v}_i)$ as a function of α varying from 0 to 180° can yield ρ_4 directly (see Fig. 3.14b for a definition of α),

$$\frac{1 + \rho_4}{1 - \rho_4} = \frac{\max[P_{cx}(\alpha)]}{\min[P_{cx}(\alpha)]} = \frac{\max(\langle n|S_{hv}|^2 \rangle)}{\min(\langle n|S_{hv}|^2 \rangle)} \tag{3.216}$$

The above equation will be derived in Section 6.1.2. Figure 3.16a shows plots of $P_{co}^v(\alpha)$, $P_{cx}(\alpha)$, and linear depolarization ratio LDR(α) in heavy rain at low elevation angles at 10 GHz. Note that the marks "V" and "H" along the abscissa correspond to transmission of nominal vertical and horizontal polarization states. In the case of "mirror" reflection symmetry, min $P_{cx}(\alpha)$ will occur when $\alpha = \beta_0$, or $\alpha = \beta_0 + (\pi/2)$. For the rain data, it can be concluded that $\beta_0 \approx 0$. The large swing between max $P_{cx}(\alpha)$ and min $P_{cx}(\alpha)$ of around 14 dB, is characteristic of a highly oriented medium, such as oblate raindrops ($\rho_4 = 0.914$; $\rho_2 = 0.979$ assuming a Gaussian pdf) with symmetry axes near-vertical (i.e. the orientation model illustrated in Fig. 2.7b).

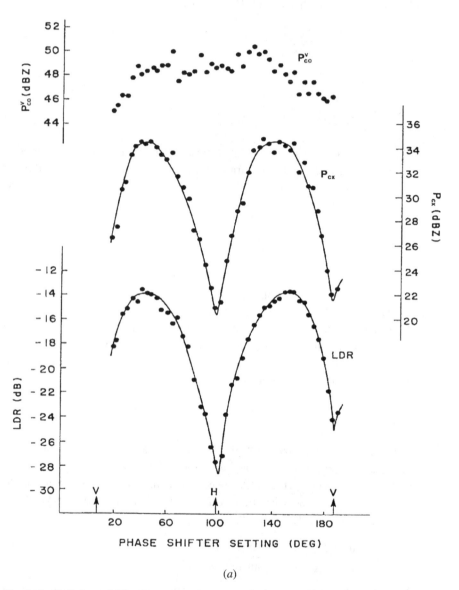

Fig. 3.16. Variations of P_{co}^v, P_{cx}, and LDR during (a) heavy rain at Ottawa, Ontario, July 6, 1984, as the direction of the transmitted linear polarization was varied; and (b) moderately heavy snow at Ottawa, Ontario on March 4, 1985. Note that the abscissa also corresponds to α in (3.216). (a, b) From Hendry et al. (1987).

The back scattering cross section per unit volume which is proportional to $P_{co}^v(\alpha)$ is not very sensitive to α in the range $\pm 45°$. From (3.214d), the "shape" factor was determined to be 0.95, which is very close to unity. As a contrast, Fig. 3.16b shows similar data in

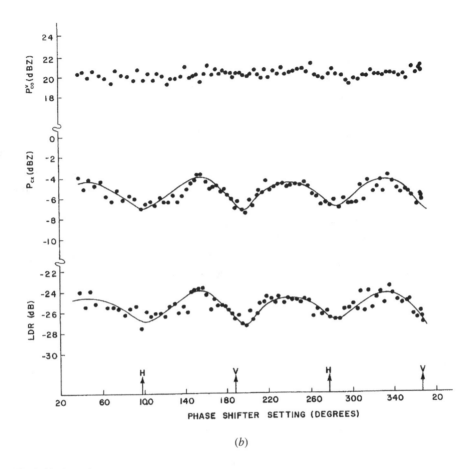

(b)

Fig. 3.16. (*cont.*)

moderately heavy snow. The swing in $P_{cx}(\alpha)$ is much less now (around 3 dB) and ρ_4 is low (0.28; $\rho_2 = 0.73$ assuming a Gaussian pdf). The "shape" factor was determined to be low, at around 0.39, and β_0 was estimated at $10°$. However, large values of β_0 were found throughout the storm, with β_0 reaching $20°$ on occasion.

If the particles are modeled as Rayleigh–Gans spheroids, the "shape" factor can be related to the variance in the axis ratio distribution. The variance in v is defined as,

$$\sigma_v^2 = \overline{v^2} - (\bar{v})^2 \qquad (3.217)$$

For oblate raindrops, Jameson (1983) showed through numerical curve fitting that,

$$v \approx 0.7\left(1 - \frac{b}{a}\right) \qquad (3.218a)$$

$$= 0.7(1 - r); \quad 0.6 \le r \le 1 \qquad (3.218b)$$

where the axis ratio $r = b/a$; b being the dimension along the spheroids's symmetry axis, while a is the radius of the equatorial circle (see Fig. 2.7a, b). Substituting (3.218) into (3.217), it follows that,

$$\sigma_v^2 = (0.7)^2 \left[\overline{r^2} - (\bar{r})^2\right] \tag{3.219a}$$

$$= (0.7)^2 \sigma_r^2 = \overline{v^2}\left[1 - \frac{(\bar{v})^2}{\overline{v^2}}\right] \tag{3.219b}$$

where σ_r^2 is the variance of the axis ratio distribution. Thus, deviations of the "shape" factor from unity are suggestive of a larger variance in axis ratios. Since the "shape" factor for rain is very close to unity in the data in Fig. 3.16a, it may be concluded that the variance in axis ratios (e.g. due to drop oscillations) is small, while the opposite is true for the snow data.

At this point it is useful to reiterate the assumptions behind the form of the circular basis covariance matrix given in (3.212). The ensemble as a whole must possess "mirror" reflection symmetry about the plane $\beta = \beta_0$, the particle canting angle distribution must be independent of other particle state distributions such as size, shape, etc., and each particle must have a plane of symmetry as discussed in Section 3.8 (e.g. bodies of revolution automatically
satisfy this condition for all incidence angles).

3.13.6 Covariance matrix for a two-component mixture of hydrometeor types

Precipitation often occurs as a mixture of different hydrometeor types, e.g. raindrops mixed with hailstones, or snow aggregates mixed with ice crystals. The covariance (or Mueller) matrix of the mixture is a superposition of the individual covariance (or Mueller) matrices of the separate components. In some cases one component of the mixture (e.g. raindrops) may exhibit "mirror" reflection symmetry, while the other component (e.g. hail) may exhibit polarization plane isotropy.[9] In such a case, it is still possible to estimate β_0 using (3.213). To show this, simply combine the covariance matrices given by (3.197) and (3.212) to yield,

$$\Sigma_{BSA}^c =$$

$$\frac{1}{4\pi}\begin{bmatrix} \langle\eta_{c_1}|v_1|^2\rangle + \langle\eta_{c_2}|v_2|^2\rangle & \sqrt{2}\rho_2\langle\eta_{c_2}v_2\rangle e^{-j2\beta_0} & \rho_4\langle\eta_{c_2}|v_2|^2\rangle e^{-j4\beta_0} \\ - & 2[\langle\eta_{c_1}\rangle + \langle\eta_{c_2}\rangle] & - \\ - & - & \langle\eta_{c_1}|v_1|^2\rangle + \langle\eta_{c_2}|v_2|^2\rangle \end{bmatrix} \tag{3.220}$$

where the subscripts 1 and 2 stand for the azimuthal (hail) and "mirror" symmetric (rain) covariance matrix elements, respectively. It follows immediately that β_0 is given by (3.213), since the non-diagonal elements of the "mirror" symmetric matrix are unaltered. From (3.214a), the circular depolarization ratio of the mixture is,

$$\overline{v_m^2} = \frac{\langle\eta_{c_1}|v_1|^2\rangle + \langle\eta_{c_2}|v_2|^2\rangle}{\langle\eta_{c_1}\rangle + \langle\eta_{c_2}\rangle} \tag{3.221}$$

The back scatter cross section per unit volume of the mixture is,

$$\langle \eta_{cm} \rangle = \langle \eta_{c_1} \rangle + \langle \eta_{c_2} \rangle \tag{3.222}$$

and typically, the η_{c_1} of hail will exceed the η_{c_2} of rain and $\overline{v_m^2} \approx \overline{v_1^2}$. From (3.214d), the ORTT of the mixture can be expressed as,

$$\text{ORTT}_m = \frac{\rho_2 |\overline{v_2}|}{(\overline{v_m^2})^{1/2} \sqrt{1 + (\langle \eta_{c_1} \rangle)/(\langle \eta_{c_2} \rangle)}} \tag{3.223}$$

and thus, the "shape" factor of component 2 cannot be isolated as a product of ρ_2 and $|\overline{v_2}|/(\overline{v_2^2})^{1/2}$. From (3.215),

$$\frac{|\langle S_{RR} S_{LL}^* \rangle|}{\sqrt{\langle |S_{RR}|^2 \rangle \langle |S_{LL}|^2 \rangle}} = \frac{\rho_4}{\left[1 + (\langle \eta_{c_1} |v_1|^2 \rangle / \langle \eta_{c_2} |v_2|^2 \rangle)\right]} \tag{3.224}$$

and a direct estimation of ρ_4 is not possible.

3.14   Relation between linear and circular radar observables

3.14.1   Azimuthal symmetry/polarization plane isotropy

Consider first the case of azimuthal symmetry, with Σ_{BSA} given by (3.193) in linear $(\hat{h}, \hat{v})$ basis, and Σ_{BSA}^c given in circular basis by (3.197). Recall that azimuthal symmetry need not necessarily imply that the elevation angle of the radar is $90°$ (i.e. that it is a vertically pointing beam). It also holds for any angle θ_i, as illustrated in Fig. 2.10a, as long as azimuthal symmetry is valid in the plane of polarization, e.g. when $p(\beta)$ is uniformly distributed in $(-\pi/2, \pi/2)$. The term azimuthal symmetry will be replaced by polarization plane isotropy in this case. Thus, in (3.193) or (3.197), S_{11} and S_{22} are replaced by $S_{11}(\psi)$ and $S_{22}(\psi)$, where ψ is the angle between the incident direction and the symmetry axis (Fig. 2.10a). The only polarimetric observable is either $|\rho_{co}|$ or LDR in the linear basis, and CDR in the circular basis. It is easy to see that, for azimuthal symmetry (Tang and Aydin 1995),

$$10^{0.1(\text{CDR})} = \frac{2 \times 10^{0.1(\text{LDR})}}{1 - 10^{0.1(\text{LDR})}} \tag{3.225a}$$

$$|\rho_{co}| = 1 - 2 \times 10^{0.1(\text{LDR})} \tag{3.225b}$$

3.14.2   "Mirror" reflection symmetry: $\beta_0 \neq 0°$

Next, consider the case of "mirror" reflection symmetry, with the symmetry plane defined by $\beta = \beta_0$. Starting from (3.192a), the term $\langle n|S_{hh}|^2 \rangle$ can be expressed in terms of the elements of the circular covariance matrix in (3.205) as follows:

$$\langle n|S_{hh}|^2 \rangle = \left\langle n \left[\frac{P(3 + c_4)}{8} + \frac{Q c_2}{2} + \frac{R(1 - c_4)}{4} \right] \right\rangle \tag{3.226a}$$

$$= \left\langle \frac{n}{4} \left[P + 2R + 2Qc_2 + \frac{1}{2}(P - 2R)(c_4 + 1) \right] \right\rangle \tag{3.226b}$$

$$= \left\langle \frac{n}{4} \left[|S_{11} + S_{22}|^2 + 2\,\mathrm{Re}\{(S_{11} - S_{22})(S_{11}^* + S_{22}^*)\} \cos 2\beta \right. \right.$$
$$\left. \left. + \frac{1}{2}|S_{11} - S_{22}|^2(\cos 4\beta + 1) \right] \right\rangle \tag{3.226c}$$

$$= \frac{1}{4\pi} \left[\langle \eta_c \rangle + 2\rho_2 \cos 2\beta_0 \,\mathrm{Re}\langle \eta_c v \rangle + \frac{1}{2}(1 + \rho_4 \cos 4\beta_0)\langle \eta_c |v|^2 \rangle \right] \tag{3.226d}$$

In going from (3.226b) to (3.226c, d) use is made of (3.199) and (3.202–3.204). Also, since the pdf of β is assumed to be symmetric about β_0, it follows that $\langle \cos 2\beta \rangle = \rho_2 \cos 2\beta_0$, $\langle \cos 4\beta \rangle = \rho_4 \cos 4\beta_0$, $\langle \sin 2\beta \rangle = \rho_2 \sin 2\beta_0$ and $\langle \sin 4\beta \rangle = \rho_4 \sin 4\beta_0$, where ρ_2 and ρ_4 are defined in (3.211). Similarly,

$$\langle n|S_{vv}|^2 \rangle = \frac{1}{4\pi} \left[\langle \eta_c \rangle - 2\rho_2 \cos 2\beta_0 \,\mathrm{Re}\langle \eta_c v \rangle + \frac{1}{2}(1 + \rho_4 \cos 4\beta_0)\langle \eta_c |v|^2 \rangle \right] \tag{3.227a}$$

$$\langle n|S_{hv}|^2 \rangle = \frac{1}{4\pi} \left[\frac{1}{2}(1 - \rho_4 \cos 4\beta_0)\langle \eta_c |v|^2 \rangle \right] \tag{3.227b}$$

$$\langle n S_{hh} S_{vv}^* \rangle = \frac{1}{4\pi} \left(\langle \eta_c \rangle - \frac{(1 + \rho_4 \cos 4\beta_0)}{2} \langle \eta_c |v|^2 \rangle + j2\rho_2 \cos 2\beta_0 \,\mathrm{Im}\langle \eta_c v \rangle \right) \tag{3.227c}$$

$$\langle n S_{hh} S_{hv}^* \rangle = -\frac{1}{4\pi} \left(\rho_2 \sin 2\beta_0 \langle \eta_c v^* \rangle + \frac{\langle \eta_c |v|^2 \rangle}{2} \rho_4 \sin 4\beta_0 \right) \tag{3.227d}$$

$$\langle n S_{vv} S_{hv}^* \rangle = -\frac{1}{4\pi} \left[\rho_2 \sin 2\beta_0 \langle \eta_c v^* \rangle - \frac{\langle \eta_c |v|^2 \rangle}{2} \rho_4 \sin 4\beta_0 \right] \tag{3.227e}$$

Thus,

$$Z_{dr} = 10 \log_{10} \left[\frac{1 + 2\rho_2 \cos 2\beta_0 \,\mathrm{Re}(\bar{v}) + \frac{1}{2}(1 + \rho_4 \cos 4\beta_0)\overline{v^2}}{1 - 2\rho_2 \cos 2\beta_0 \,\mathrm{Re}(\bar{v}) + \frac{1}{2}(1 + \rho_4 \cos 4\beta_0)\overline{v^2}} \right] \tag{3.228a}$$

$$\mathrm{LDR}_{vh} = 10 \log_{10} \left[\frac{\frac{1}{2}(1 - \rho_4 \cos 4\beta_0)\overline{v^2}}{1 + 2\rho_2 \cos 2\beta_0 \,\mathrm{Re}(\bar{v}) + \frac{1}{2}(1 + \rho_4 \cos 4\beta_0)\overline{v^2}} \right] \tag{3.228b}$$

It is clearly seen from the above that the conventional linear[10] observables Z_{dr} and LDR_{vh} are dependent, in a complicated way, upon the "orientation" parameters β_0, ρ_2 and ρ_4 and upon the "shape"-related parameters $\bar{v}$ and $\overline{v^2}$. Equation (3.228a) is essentially identical to eq. (32) of Hendry et al. (1987), accounting for their opposite convention in defining v for oblates/prolates, see (3.98). Note that $\langle n S_{hh} S_{hv}^* \rangle = \langle n S_{vv} S_{hv}^* \rangle = 0$, when $\beta_0 = 0°$.

3.14.3 Case of $\beta_0 = 0°$

It is possible to derive an expression for the ratio $(1 + \rho_4)/(1 - \rho_4)$, given earlier in (3.216), in terms of linear radar observables for the case of $\beta_0 = 0°$. Following Jameson and Davé (1988), the terms $(1 + \rho_4)$ and $(1 - \rho_4)$ can be obtained from (3.227b, c) as,

$$\frac{1}{2}(1 + \rho_4) = \frac{\langle \eta_c \rangle - \mathrm{Re}\langle 4\pi (n S_{hh} S_{vv}^*) \rangle}{\langle \eta_c |v^2| \rangle} \tag{3.229a}$$

$$\frac{1}{2}(1 - \rho_4) = \frac{\langle 4\pi n |S_{hv}|^2 \rangle}{\langle \eta_c |v^2| \rangle} \tag{3.229b}$$

Thus,

$$\frac{1 + \rho_4}{1 - \rho_4} = \frac{\langle \eta_c \rangle - \mathrm{Re}\langle 4\pi n S_{hh} S_{vv}^* \rangle}{\langle 4\pi n |S_{hv}|^2 \rangle} \tag{3.230a}$$

$$= \left(\frac{\langle \eta_c \rangle}{\langle 4\pi n |S_{hh}|^2 \rangle} - \frac{\mathrm{Re}\langle 4\pi n S_{hh} S_{vv}^* \rangle}{\langle 4\pi n |S_{hh}|^2 \rangle} \right) \Big/ 10^{0.1(\mathrm{LDR}_{vh})} \tag{3.230b}$$

From (3.196d),

$$\langle \eta_c \rangle = \langle 4\pi n |S_{RL}|^2 \rangle = \left\langle 4\pi \frac{n}{4} \left[|S_{hh}|^2 + |S_{vv}|^2 + 2\,\mathrm{Re}(S_{hh} S_{vv}^*) \right] \right\rangle \tag{3.231a}$$

$$= \left\langle 4\pi \frac{n}{4} |S_{hh}|^2 \right\rangle \left(1 + \frac{\langle n |S_{vv}|^2 \rangle}{\langle n |S_{hh}|^2 \rangle} + \frac{2\,\mathrm{Re}\langle n S_{hh} S_{vv}^* \rangle}{\langle n |S_{hh}|^2 \rangle} \right) \tag{3.231b}$$

or,

$$\frac{\langle \eta_c \rangle}{\langle 4\pi n |S_{hh}|^2 \rangle} = \frac{1}{4\mathcal{Z}_{dr}} \left[\mathcal{Z}_{dr} + 1 + 2\sqrt{\mathcal{Z}_{dr}}\,\mathrm{Re}(|\rho_{co}|e^{-j\delta_{co}}) \right] \tag{3.231c}$$

where $\mathcal{Z}_{dr} = (\langle n |S_{hh}|^2 \rangle)/(\langle n |S_{vv}|^2 \rangle)$ is the differential reflectivity expressed as a ratio $(\mathcal{Z}_{dr} = 10^{0.1(Z_{dr})})$, and,

$$|\rho_{co}|e^{-j\delta_{co}} = \frac{\langle n S_{hh} S_{vv}^* \rangle}{\sqrt{\langle n |S_{hh}|^2 \rangle \langle n |S_{vv}|^2 \rangle}} \tag{3.231d}$$

Note that the back scatter differential phase, δ_{co}, is defined in BSA as $\arg\langle n S_{hh}^* S_{vv} \rangle$. Using (3.231c, d) in (3.230b) leads to,

$$\frac{1 + \rho_4}{1 - \rho_4} = \frac{\mathcal{Z}_{dr} + 1 - 2\sqrt{\mathcal{Z}_{dr}}\,\mathrm{Re}(|\rho_{co}|e^{-j\delta_{co}})}{4\mathcal{Z}_{dr}L} \tag{3.232}$$

where L is the linear depolarization ratio $[L = 10^{0.1(\mathrm{LDR}_{vh})}]$. The above equation is only valid when $\beta_0 = 0°$, whereas (3.216) is valid for "mirror" reflection symmetry about the plane $\beta = \beta_0$. Jameson and Davé (1988) define their "canting" parameter (γ) as,

$$\gamma = \frac{1}{4} \left(\frac{1 - \rho_4}{1 + \rho_4} \right) \tag{3.233}$$

which lies in the range $0 \leq \gamma \leq 0.25$, the upper limiting value representing "isotropic" distribution of canting angles. Equation (3.232) can be written for Rayleigh–Gans scatterers ($\delta_{co} = 0$) as,

$$10^{0.1(\text{LDR}_{vh})} = L = \left(\frac{1 + \mathcal{Z}_{dr} - 2\sqrt{\mathcal{Z}_{dr}}|\rho_{co}|}{\mathcal{Z}_{dr}} \right) \gamma \tag{3.234}$$

which shows that the linear depolarization ratio (LDR_{vh}) cannot, by itself, be used to separate "shape" effects from canting angle effects (the equivalent expression in terms of circular observables is given in (3.228b)). As the canting angle distribution becomes "isotropic", $L \rightarrow 2\gamma(1 - |\rho_{co}|)$, and the upper limit representing $L \rightarrow (1 - |\rho_{co}|)/2$, which is identical to (3.225b). It will be shown later in Section 6.5 that measurement accuracies in $|\rho_{co}|$ limit the practical application of (3.234) as far as a direct estimation of γ is concerned using the standard linear polarization observables (L, $\mathcal{Z}_{dr}$, and $|\rho_{co}|$). Figure 3.17 shows a plot of $|\rho_{co}|$ versus Z_{dr} based on (3.234). The LDR_{vh} is fixed at -20 dB (or $L = 0.01$), and the γ parameter (or ρ_4) is varied assuming a Gaussian pdf with zero mean canting angle and standard deviation σ_β, see (3.201b). The term ρ_4 may be approximated by $\exp(-8\sigma_\beta^2)$ for narrow Gaussian distributions. Figure 3.17 is valid for Rayleigh–Gans particles possessing an axis of symmetry with a mean canting angle of zero. The figure shows that a wide range of possible combinations of σ_β (or ρ_4), $|\rho_{co}|$, and Z_{dr} are capable of producing an LDR_{vh} of -20 dB, for example.

3.14.4 Asymmetry ratio

The asymmetry ratio for a single canted oblate particle was defined in (2.99) as a generalization of differential reflectivity and was shown to be independent of the canting angle. This concept can be extended to an ensemble of oblate raindrops with Gaussian canting angle distribution (with a mean β_0 and a standard deviation of σ_β).

   Before this extension can be made, the relative average scattering matrix (or the rms scattering matrix) of an ensemble of particles needs to be defined. While such a definition is not necessarily unique, the one proposed by Chandra et al. (1987) is appealing on physical grounds. Using the notation of Section 3.11, the relative average scattering matrix is defined as,

$$\mathbf{S}_{AR} = \begin{bmatrix} \sqrt{\langle n|S_{hh}|^2 \rangle} & \sqrt{\langle n|S_{hv}|^2 \rangle}\exp(j\arg\langle nS_{hv}S_{vv}^* \rangle) \\ \sqrt{\langle n|S_{vh}|^2 \rangle}\exp(j\arg\langle nS_{vh}S_{hh}^* \rangle) & \sqrt{\langle n|S_{vv}|^2 \rangle}\exp(j\arg\langle nS_{vv}S_{hh}^* \rangle) \end{bmatrix} \tag{3.235}$$

Consider an ensemble of raindrops satisfying "mirror" reflection symmetry about $\beta_0 = 0°$. In the Rayleigh–Gans limit, $\mathbf{S}_{AR}$ reduces to,

$$\mathbf{S}_{AR} = \begin{bmatrix} \sqrt{\langle n|S_{hh}|^2 \rangle} & \sqrt{\langle n|S_{hv}|^2 \rangle} \\ \sqrt{\langle n|S_{vh}|^2 \rangle} & \sqrt{\langle n|S_{vv}|^2 \rangle} \end{bmatrix} \tag{3.236a}$$

$$\equiv \begin{bmatrix} \sqrt{P_{co}^h} & \sqrt{P_{cx}} \\ \sqrt{P_{cx}} & \sqrt{P_{co}^v} \end{bmatrix} \tag{3.236b}$$

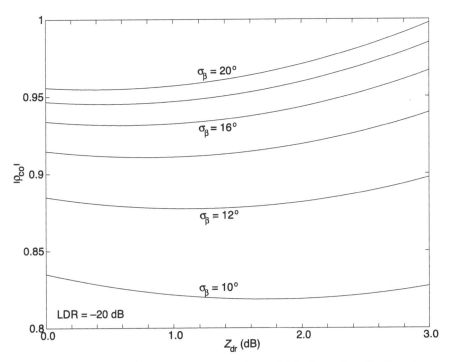

Fig. 3.17. Plot of $|\rho_{co}|$ versus Z_{dr} for various σ_β for Rayleigh–Gans spheroids. Note that LDR $= -20$ dB. Also, $\rho_4 = \exp(-8\sigma_\beta^2)$ for narrow Gaussian distribution of the canting angle. Mean canting angle is zero.

where (3.236b) follows from ergodicity, and P_{co}^h, P_{co}^v, and P_{cx} are the average received powers measured by a dual-polarized radar as defined in (3.177, 3.178).

The asymmetry ratio (A_{sm}) defined in (2.99) for a single particle is now extended to an ensemble of raindrops as the ratio of the eigenvalues of $(S_{AR})^{t*}S_{AR}$. Figure 3.18 shows A_{sm} and $\mathfrak{z}_{dr}$ (expressed in decibels) as a function of σ_β for a rain model (see Section 3.11.3) with an exponential size distribution, $N(D)$, a Gaussian canting angle distribution with $\beta_0 = 0°$ and variable σ_β, and axis ratio versus D relation valid for equilibrium raindrop shapes, see (7.3). Figure 3.18 shows results for two rain rates using this model (rain rate is defined in (7.64)), and it is clear that the asymmetry ratio is less sensitive to increasing σ_β as compared to Z_{dr} (see Kwiatkowski et al. 1995). The behavior of A_{sm} with σ_β is similar to the circular depolarization ratio (CDR), which is independent of both the mean and standard deviation of the canting angle distribution, see (3.212, 3.214a). The asymmetry ratio can be calculated from data collected with a dual-linear polarized radar (i.e. using measurements of P_{co}^h, P_{co}^v, and P_{cx}) and calculating the ratio of the eigenvalues of $(S_{AR})^{t*}S_{AR}$ with S_{AR} in (3.236b). Huang et al. (1997) show that a scatter plot of $10\log_{10}(A_{sm}) - Z_{dr}$ versus LDR$_{vh}$ obtained from radar measurements in rain can be used to set bounds on σ_β. Figure 3.19 shows data collected with the CSU–CHILL radar in a convective rain event. Also

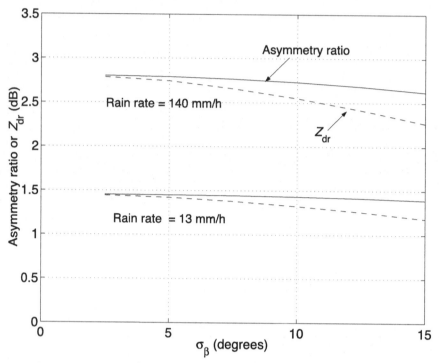

Fig. 3.18. Scattering simulations of A_{sm} and Z_{dr} (both in decibels) versus σ_β for a rain model assuming a Gaussian canting angle distribution with a mean of 0° and a standard deviation of σ_β. Curves are shown for an exponential drop size distribution assuming equilibrium shapes, see (7.3), for high and low rain rates. Calculations are performed at a frequency of 3 GHz (S-band).

plotted in the figure are the results from scattering simulations (same rain model as used in Fig. 3.18) assuming a σ_β of 5, 7, and 10°. It is noted that most of the data points fall between a σ_β of 5 and 10°, with 7° being a reasonable average, which is consistent with σ_β deduced from ρ_4, using the rotating linear polarization data shown in Fig. 3.16a. Here, ρ_4 was deduced to be 0.914 and for a Gaussian canting angle pdf, $\rho_4 = \exp(-8\sigma_\beta^2)$ resulting in $\sigma_\beta = 6°$.

3.14.5 Framework for classification of hydrometeor types

It should be reiterated that the discussion in this section relating circular to linear observables is only valid for the following assumptions:

- "mirror" reflection symmetry occurs,

- the particle canting angle distribution must be independent of other particle state distributions such as size, shape, etc., and

- each particle must have a plane of symmetry as discussed in Section 3.8 (e.g. bodies of revolution).

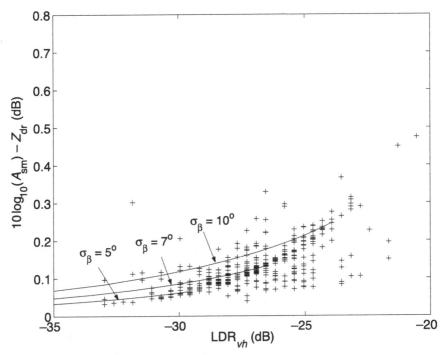

Fig. 3.19. Radar data obtained with the CSU–CHILL radar in a convective rain cell in Colorado: (+) data from range resolution volumes along the beam, as the antenna scans the rain cell; (——) data from scattering simulations for $\sigma_\beta = 5, 7$, and $10°$.

Raindrops and pristine ice crystals are generally assumed to approximate these conditions. However, precipitation is often a mixture of different hydrometeor types, and some hydrometeors do not satisfy the listed symmetry conditions (e.g. wet snow or hailstones). Self-consistency of the linear and circular observables offers a method of validating these assumptions and has been used by Hendry et al. (1987) for data acquired in rain, snow, and the "bright-band" at 10 GHz. However, there are very few radars capable of both linear and circular polarization observations. Separation of hydrometeor "shape" effects from canting effects is important via separate estimation of the shape factor $|\bar{v}|/(\overline{v^2})^{1/2}$ and the orientation parameter ρ_2 (obtainable from ρ_4 and assuming a canting angle model). The direct measurement of ρ_4 using rotating linear basis, see (3.216), was done by Hendry et al. (1987), but it is "slow" and suitable only for steady precipitation. Pulse-to-pulse switching between RHC and LHC states may permit a direct estimate of ρ_4, see (3.215), but has not been tested yet (in principle, the procedure would be similar to that used for $|\rho_{co}|$ in the $(\hat{h}, \hat{v})$ basis, see Section 6.4.3). As alluded to earlier, the retrieval of ρ_4 using (3.232) may not be accurate enough in practice; moreover it assumes $\beta_0 = 0°$.

   A potentially useful framework for hydrometeor-type classification is likely to involve both linear and circular basis measurements using polarization agility. One of the

Table 3.1. Comparison of circular and linear radar observables

	\multicolumn{4}{c}{Data from Hendry et al. (1987)}				\multicolumn{3}{c}{Data from Doviak and Zrnić (1993)}		
	CDR (dB)	$\bar{v}/(\overline{v^2})^{1/2}$	ρ_2	Z_{dr} (dB)	Z_{dr} (dB)	LDR_{vh} (dB)	$|\rho_{co}|$
Rain	−14	0.95	0.98	2.87	0.5–4	−27 to −34	>0.97
Snow	−22	0.38	0.73	0.35	0–0.5	<−34	>0.99
Bright-band	−9.5	0.35	0.65	1.11	0–3.0	−13 to −18	0.8–0.95

earliest attempts at hydrometeor classification using circular polarization observables (η_c, ORTT and CDR; see Section 3.13.2) was presented by Hendry and Antar (1984). A five-dimensional measurement space, which includes the back scatter cross section at horizontal polarization (η_{hh}); the circular depolarization ratio (CDR); the shape factor $|\bar{v}|/(\overline{v^2})^{1/2}$; the orientation parameter ρ_2; and conventional Z_{dr}; may offer a method for hydrometeor classification. Classification methods based strictly on linear polarization measurements (η_{hh}, Z_{dr}, LDR_{vh}, $|\rho_{co}|$) are discussed in Chapter 8 of Doviak and Zrnić (1993), see their Table 8.1. Table 3.1 adapted from Hendry et al. (1987), shows their data at 10 GHz for rain, snow, and the "bright-band". Table 3.1 also shows the range of values at 3 GHz expected for Z_{dr}, LDR_{vh}, and $|\rho_{co}|$ from Table 8.1 of Doviak and Zrnić (1993), in similar precipitation, to illustrate the comparison. In the next chapter, propagation effects, such as differential attenuation and the differential phase (in rain), will be shown to complicate the interpretation of the circular polarization observables, and to a lesser extent the linear polarization observables. A more detailed discussion of hydrometeor classification is presented in Chapter 7.

Notes

1. The definition here is consistent with Deschamps and Mast (1973). However, the IEEE Standard defines the north (south) pole of the Poincaré sphere to be LHC (RHC). This is because $\tan \tau$ is defined as $(|E_L| - |E_R|)/(|E_L| + |E_R|)$ as opposed to (3.13).

2. The notation, $U(RL \rightarrow hv)$, is adopted from Boerner et al. (1992). As given in eq. (3.72a), the term $RL \rightarrow hv$ refers to transformation of

$$\begin{bmatrix} E_R \\ E_L \end{bmatrix} \text{ to } \begin{bmatrix} E_h \\ E_v \end{bmatrix}$$

3. The copolar and cross-polar response surfaces for a variety of single scatterers can be found, for example, in Agrawal and Boerner (1989), Hubbert (1994), or van Zyl and Ulaby (1990). Similar surfaces for distributed scatterers are given in van Zyl et al. (1987), Tragl (1990), or Hubbert and Bringi (1996).

4. The theory of optimal polarizations for distributed scatterers has been treated, for example, by McCormick and Hendry (1985), van Zyl et al. (1987), or Boerner et al. (1991).

5. Tragl (1990) shows more generally that transmission of orthogonal polarizations yields the same cross-polar received power. Here, cross-polar means the receive state orthogonal to that transmitted. For example, if χ is the polarization ratio, see (3.3), of the transmitted signal, then let the cross-polar received power from reciprocal random scatterers be represented by $P_{cx}(\chi)$. From (3.45), the transmit state orthogonal to χ is $-1/\chi^*$. Tragl shows that $P_{cx}(-1/\chi^*) = P_{cx}(\chi)$.

6. The German DLR C-band radar appears to have been the first meteorological radar system designed to measure the time-series of "instantaneous" scattering matrices (Schroth et al. 1988). The "instantaneous" scattering matrix of military targets and clutter has been measured by defense radars, for example, see Poelman and Guy (1984) and Mott (1992).

7. Tragl (1990) shows how the covariance matrix in (3.183) transforms under a change of polarization basis. He also develops a theory of optimal polarizations.

8. The 95-GHz radar operated on the University of Wyoming King Air aircraft is described at

   http://www-das.uwyo.edu/wcr

9. Hubbert and Bringi (1996) have modeled the case of oriented rain mixed with hail in (h, v) basis and have determined the optimal polarizations. They show that the mean canting angle of oriented raindrops can be estimated in a mixture with hail using the characteristic polarizations.

10. McCormick (1979) gave the first approximation for Z_{dr} in terms of circular polarization observables, which can be obtained from the general relation by assuming $\rho_4 \approx 1$ and $\beta_0 \approx 0$. Expressions for Z_{dr} in terms of circular observables were given later by Bebbington et al. (1987) and Jameson and Davé (1988). These later expressions accounted for differential propagation phase in rain (see also, Section 4.6).

4 Dual-polarized wave propagation in precipitation media

Attenuation caused by hydrometeors, along the propagation path from the radar to the resolution volume of interest, has been studied since the beginnings of radar history (Ryde 1946; see also, the historical review by Atlas and Ulbrich 1990). The expression for specific attenuation was derived in (1.133d), using fairly simple concepts. The specific attenuation is, in essence, the extinction cross section of the particles weighted by the size distribution. In the microwave frequency band (<35 GHz), absorption by water along the propagation path (cloud water, raindrops, or melting hydrometeors) is the principal cause for attenuation (absorption by gaseous constituents in the atmosphere is excluded). In particular, attenuation caused by rain along the propagation path must be estimated so that the intrinsic reflectivity from a given resolution volume can be retrieved. Even at a relatively low radar frequency of 3 GHz (S-band range), attenuation effects can be significant when the propagation path passes through multiple rain cells of high intensity. With a single polarized radar, it is not possible to measure the cumulative attenuation along a propagation path since no reference signal is generally available against which to compare the attenuated signal (an exception is satellite-based radar for which the ocean surface forms a reference, e.g. the Tropical Rain Measurement Mission (TRMM) precipitation radar; see Meneghini and Kozu 1990). One significant advantage of dual-polarized radar is the additional measurement of a differential propagation phase, which can be used to "correct" for cumulative attenuation due to rain along the path ("correction" algorithms will be described in Chapter 7).

This chapter deals primarily with differential attenuation and the differential phase caused by oriented hydrometeors (such as raindrops or ice crystals) along the propagation path.[1] The differential propagation phase between horizontal and vertical polarizations due to oriented raindrops along the propagation path, is of fundamental importance both in rainfall estimation (see Chapter 8) and in "correction" of attenuation effects (see Chapter 7). A systematic theoretical and experimental radar study of the propagation of circularly polarized waves in the atmosphere was conducted at the National Research Council (NRC) of Canada (e.g. see Seliga et al. 1990 for a historical review). This chapter will describe, in part, the theoretical approach adopted by the NRC, as well as selected experimental results.

Propagation effects have an impact on the choice of transmit and receive polarization states. A classic example is the change-over of the transmit polarization state of the WSR–88D radars. Circular polarization was initially chosen for the WSR–88D without provision for measuring the full coherency matrix. As shown in this chapter, depolarization caused by a differential propagation phase in rain can seriously impact the measurement of reflectivity unless the full coherency matrix is measured. A subsequent change to horizontal polarization essentially resolved the problem for the WSR–88D system. Current evaluation of polarimetric techniques for WSR–88D radars gives renewed emphasis on a proper understanding of dual-polarized wave propagation through precipitation (Doviak et al. 2000).

4.1   Coherent wave propagation

In Section 1.10, the propagation of a linearly polarized wave in a medium composed of a sparse distribution of identical dielectric spheres was considered: the pertinent results may be summarized as follows. An effective wave number, k_{eff}, for the medium can be defined, which in general is complex. The propagation of a linearly polarized plane wave in such a medium (for example, along the positive Z-axis) can be described by,

$$\vec{E} = \hat{e} E_0 \exp(-jk_{eff}z) \tag{4.1}$$

$$\frac{d\vec{E}}{dz} = (-jk_{eff})\vec{E} \tag{4.2}$$

$$k_{eff} = k_0 + \frac{2\pi n}{k_0} \hat{e} \cdot \vec{f}(\hat{i}, \hat{i}) \tag{4.3}$$

where $\vec{f}(\hat{i}, \hat{i})$ is the forward scattered vector amplitude and n is the number of the spheres per unit volume. The essential approximation here is that each scatterer is "excited" by the unperturbed plane wave incident field (valid for sparse concentrations, $c = nV \ll 1$, where V is the sphere volume).

The differential equation in (4.2) can be written in matrix form (if $\vec{E} = E_h \hat{h} + E_v \hat{v}$), and assuming spherical particles, as,

$$\frac{d}{dz} \begin{bmatrix} E_h \\ E_v \end{bmatrix} = \begin{bmatrix} -jk_{eff} & 0 \\ 0 & -jk_{eff} \end{bmatrix} \begin{bmatrix} E_h \\ E_v \end{bmatrix} \tag{4.4}$$

An FSA convention is assumed with $\hat{v} \times \hat{h} = \hat{k}$, where $\hat{k}$ is the forward scattered direction ($\hat{k} = \hat{i}$). An extension of (4.4) can be made to identical, oriented spheroids with number density n, as illustrated in Fig. 4.1, assuming that the symmetry axes are aligned along the $\hat{v}$-direction (i.e. the canting angle in the polarization plane is zero),

$$\frac{d}{dz} \begin{bmatrix} E_h \\ E_v \end{bmatrix} = \begin{bmatrix} -jk_{eff}^h & 0 \\ 0 & -jk_{eff}^v \end{bmatrix} \begin{bmatrix} E_h \\ E_v \end{bmatrix} \tag{4.5a}$$

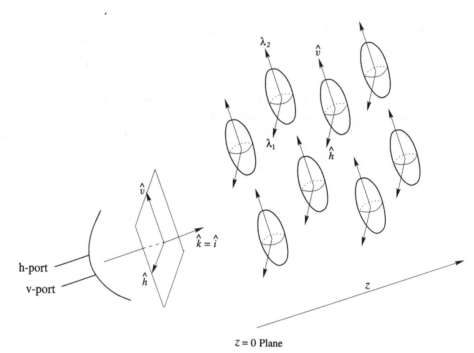

$z = 0$ Plane

Fig. 4.1. A propagation medium, z kilometers in depth, consisting of oriented prolate spheroids whose axes of symmetry are along $\hat{v}$. The two characteristic waves are horizontally and vertically polarized with eigenvalues $\lambda_1 = -jk_{\text{eff}}^h$, $\lambda_2 = -jk_{\text{eff}}^v$.

$$= \left[\begin{array}{cc} -j\left\{k_0 + \dfrac{2\pi n}{k_0}\hat{h}\cdot\vec{f}(\hat{i},\hat{i})\right\} & 0 \\ 0 & -j\left\{k_0 + \dfrac{2\pi n}{k_0}\hat{v}\cdot\vec{f}(\hat{i},\hat{i})\right\} \end{array} \right] \left[\begin{array}{c} E_h \\ E_v \end{array} \right]$$

(4.5b)

Note that, for oriented spheroids with symmetry axes along $\hat{v}$, the differential equation in (4.5) is uncoupled, and $\lambda_1 = -jk_{\text{eff}}^h$ and $\lambda_2 = -jk_{\text{eff}}^v$ can be considered as eigenvalues of the medium. The plane wave solution to (4.5a) is simply,

$$\left[\begin{array}{c} E_h(z) \\ E_v(z) \end{array} \right] = \left[\begin{array}{cc} e^{\lambda_1 z} & 0 \\ 0 & e^{\lambda_2 z} \end{array} \right] \left[\begin{array}{c} E_h(0) \\ E_v(0) \end{array} \right]$$

(4.6a)

$$= \left[\begin{array}{cc} T_{11} & 0 \\ 0 & T_{22} \end{array} \right] \left[\begin{array}{c} E_h(0) \\ E_v(0) \end{array} \right]$$

(4.6b)

where T_{11}, T_{22} are elements of the transmission matrix (**T**); and $E_h(0), E_v(0)$ are components at the reference plane $z = 0$. The two solutions given by (4.6), which are uncoupled, are also called the "characteristic" plane wave solutions for the medium. The specific differential attenuation (dB km^{-1}) between the two "characteristic" plane waves, $\{20\log_{10}[|E_h(z)/E_v(z)|]\}$, follows from (1.132),

$$A_{\mathrm{dp}} = -(8.686 \times 10^3)\,\mathrm{Re}(\lambda_1 - \lambda_2) \tag{4.7a}$$

$$= -(8.686 \times 10^3)\frac{2\pi n}{k_0}\,\mathrm{Im}\left[\hat{h}\cdot\vec{f}(\hat{i},\hat{i}) - \hat{v}\cdot\vec{f}(\hat{i},\hat{i})\right] \tag{4.7b}$$

The specific differential phase (rad km^{-1}) between the two "characteristic" waves is defined as,

$$K_{\mathrm{dp}} = -(10^3)\,\mathrm{Im}(\lambda_1 - \lambda_2) \tag{4.8a}$$

$$= 10^3 \left(\frac{2\pi n}{k_0}\right)\mathrm{Re}[\hat{h}\cdot\vec{f}(\hat{i},\hat{i}) - \hat{v}\cdot\vec{f}(\hat{i},\hat{i})] \tag{4.8b}$$

In the above two equations k_0 is measured in m^{-1}, the scattering amplitude in m, and the number density (n) in m^{-3}. If Rayleigh–Gans scattering from identical, vertically oriented prolate spheroids (with real ε_r) is considered, then from (2.100) and (2.89a, b), valid for the case of forward scattering (i.e. $\beta = 0°$, $\psi = 90°$),

$$\hat{h}\cdot\vec{f}(\hat{i},\hat{i}) = \frac{k_0^2}{4\pi\varepsilon_0}\alpha \tag{4.9a}$$

$$\hat{v}\cdot\vec{f}(\hat{i},\hat{i}) = \frac{k_0^2}{4\pi\varepsilon_0}\alpha_{z_b} \tag{4.9b}$$

Thus A_{dp} is zero, while,

$$K_{\mathrm{dp}} = 10^3\frac{nk_0}{2\varepsilon_0}(\alpha - \alpha_{z_b}) \tag{4.10}$$

Since for prolates $\alpha < \alpha_{z_b}$, the specific differential propagation phase is negative, or equivalently, the phase velocity of the characteristic wave corresponding to vertical polarization is "slowed down" relative to the horizontally polarized characteristic wave. The opposite would be true for oriented oblate spheroids, e.g. raindrops.

4.1.1 Oriented particles ($\beta = 0°$ case): circular polarization

If circularly polarized components are considered, then (4.6) is transformed using (3.76a),

$$\begin{bmatrix} E_R(z) \\ E_L(z) \end{bmatrix} = [\mathbf{U}(RL \to hv)]^{-1} \begin{bmatrix} e^{\lambda_1 z} & 0 \\ 0 & e^{\lambda_2 z} \end{bmatrix} \mathbf{U}(RL \to hv) \begin{bmatrix} E_R(0) \\ E_L(0) \end{bmatrix} \tag{4.11a}$$

$$= \frac{(e^{\lambda_1 z} + e^{\lambda_2 z})}{2}$$

$$\times \begin{bmatrix} 1 & \tanh[(\lambda_1 - \lambda_2)(z/2)] \\ \tanh[(\lambda_1 - \lambda_2)(z/2)] & 1 \end{bmatrix} \begin{bmatrix} E_R(0) \\ E_L(0) \end{bmatrix} \tag{4.11b}$$

The term $[(\lambda_1 - \lambda_2)/2] z$ can be expressed as,

$$(\lambda_1 - \lambda_2)\frac{z}{2} = -j\left(\frac{k_{\text{eff}}^h - k_{\text{eff}}^v}{2}\right) z \tag{4.12a}$$

$$= -j\left(\frac{K_{dp}z}{2}\right) - \left(\frac{A_{dp}}{8.686}\frac{z}{2}\right) \tag{4.12b}$$

where the propagation distance (z) is in km. The hyperbolic tangent function in (4.11b) can be expressed as,

$$\tanh\left[\frac{(\lambda_1 - \lambda_2)z}{2}\right] = -\tanh\left[j\left(\frac{K_{dp}z}{2}\right) + \frac{A_{dp}z}{(8.686)2}\right] \tag{4.13a}$$

$$= -pe^{jx} \tag{4.13b}$$

Also,

$$\frac{e^{\lambda_1 z} + e^{\lambda_2 z}}{2} = e^{(\lambda_1 + \lambda_2)z/2} \cosh\left[(\lambda_1 - \lambda_2)\frac{z}{2}\right] \tag{4.14a}$$

$$= e^{(\lambda_1 + \lambda_2)z/2} \frac{1}{\sqrt{1 - (pe^{jx})^2}} \tag{4.14b}$$

Thus, in the notation of McCormick and Hendry (1973), (4.11) reduces to,

$$\begin{bmatrix} E_R(z) \\ E_L(z) \end{bmatrix} = \frac{e^{-j(k_{\text{eff}}^h + k_{\text{eff}}^v)z/2}}{\sqrt{1 - (pe^{jx})^2}} \begin{bmatrix} 1 & -pe^{jx} \\ -pe^{jx} & 1 \end{bmatrix} \begin{bmatrix} E_R(0) \\ E_L(0) \end{bmatrix} \tag{4.15}$$

where $(k_{\text{eff}}^h + k_{\text{eff}}^v)/2$ may be regarded as a "mean" propagation constant. The negative sign on pe^{jx} is due to the horizontal reference direction chosen here for the unit, circular base vectors, as opposed to McCormick and Hendry (1973) who chose the vertical reference direction.

It is instructive to consider the case where $A_{dp} = 0$ (e.g. Rayleigh scattering with real ε_r),

$$pe^{jx} = \tanh\left[j\left(\frac{K_{dp}z}{2}\right)\right] \tag{4.16a}$$

$$= j \tan\left(\frac{K_{dp}z}{2}\right) \tag{4.16b}$$

so that $\chi = \pi/2$ and $p = \tan(K_{dp}z/2)$. If the initial state is right-hand circular at $z = 0$, i.e. $E_R(0) = 1$ and $E_L(0) = 0$ in (4.15), then after a propagation distance of z kilometers, the wave polarization is elliptical with,

$$E_R(z) = \frac{1}{\sqrt{1 + p^2}} \tag{4.17a}$$

$$E_L(z) = \frac{1}{\sqrt{1 + p^2}} pe^{j\pi/2}e^{j\pi} \tag{4.17b}$$

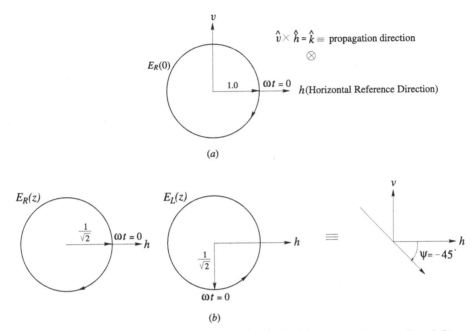

Fig. 4.2. (a) RHC wave represented by $E_R(0) = 1$ at the initial reference plane $z = 0$; and (b) after propagating a distance z such that $(K_{dp}z/2) = 45°$, the superposition of $E_R(z) = 1/\sqrt{2}$ and $E_L(z) = (1/\sqrt{2})e^{j(3\pi/2)}$ results in a linear polarization oriented at $-45°$.

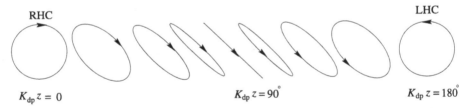

Fig. 4.3. Change in polarization state of an initial RHC wave due to a differential propagation phase $(K_{dp}z)$ varying from 0 to $180°$.

When $p = 1$ or $K_{dp}z = 90°$ (e.g. if $K_{dp} = 2°$ km^{-1} and $z = 45$ km), the polarization state is linear and oriented at $\psi = -45°$ from the horizontal reference direction as sketched in Fig. 4.2. Note that $\psi = (\theta_L - \theta_R)/2$, in general (see Section 3.1). When $K_{dp}z = 180°$, then $E_R(z) = 0$ and $|E_L(z)| = 1$, and the polarization state is LHC (completely reversed from the initial RHC state at $z = 0$). Figure 4.3 illustrates the change in polarization state of an initial RHC wave at $z = 0$ with increasing $(K_{dp}z)$ from zero to $180°$. As will be seen later (see Fig. 4.8), it is not unusual for $K_{dp}z$ to reach between 60 and $100°$ in very intense rain propagation paths at 3-GHz frequency (S-band).

4.1.2   Uniformly canted particles: linear polarization

The transmission matrix in (4.6) can be generalized to the case where the symmetry axis of the identical spheroids is uniformly canted in the polarization plane by an angle β, as illustrated in Fig. 4.4. Note that a positive β angle is measured clockwise from the vertical direction. The angle ψ between the incident direction and the symmetry axis is assumed to be $90°$. Using the rotation matrix, $\mathbf{R}(\beta)$, defined in (2.93a) and repeated here,

$$\mathbf{R}(\beta) = \begin{bmatrix} \cos\beta & \sin\beta \\ -\sin\beta & \cos\beta \end{bmatrix} \tag{4.18}$$

it should be clear that the new transmission matrix is,

$$\begin{bmatrix} E_h(z) \\ E_v(z) \end{bmatrix} = \mathbf{R}(\beta) \begin{bmatrix} e^{\lambda_1 z} & 0 \\ 0 & e^{\lambda_2 z} \end{bmatrix} \mathbf{R}(-\beta) \begin{bmatrix} E_h(0) \\ E_v(0) \end{bmatrix} \tag{4.19a}$$

$$= \begin{bmatrix} (e^{\lambda_1 z}\cos^2\beta + e^{\lambda_2 z}\sin^2\beta) & (e^{\lambda_2 z} - e^{\lambda_1 z})(\sin 2\beta/2) \\ (e^{\lambda_2 z} - e^{\lambda_1 z})(\sin 2\beta/2) & (e^{\lambda_1 z}\sin^2\beta + e^{\lambda_2 z}\cos^2\beta) \end{bmatrix} \begin{bmatrix} E_h(0) \\ E_v(0) \end{bmatrix} \tag{4.19b}$$

$$= \begin{bmatrix} T_{hh} & T_{hv} \\ T_{vh} & T_{vv} \end{bmatrix} \begin{bmatrix} E_h(0) \\ E_v(0) \end{bmatrix} \tag{4.19c}$$

where λ_1 and λ_2 are the eigenvalues of the medium defined in (4.5),

$$\lambda_1 = -jk_{\text{eff}}^h = -j\left[k_0 + \frac{2\pi n}{k_0}\hat{h}\cdot\vec{f}(\hat{i}, \hat{i}; \beta = 0°, \psi = 90°)\right] \tag{4.20a}$$

$$\lambda_2 = -jk_{\text{eff}}^v = -j\left[k_0 + \frac{2\pi n}{k_0}\hat{v}\cdot\vec{f}(\hat{i}, \hat{i}; \beta = 0°, \psi = 90°)\right] \tag{4.20b}$$

Note that the forward scattered amplitude in (4.20) is evaluated for the case of $\beta = 0°$ and $\psi = 90°$ as in Fig. 4.1. The effect of the canting angle is of course taken into account, as in (4.19b). It is clear from (4.19c) that "cross-coupling" exists, and that an initially vertical polarization state $(E_h(0) = 0, E_v(0) = 1)$ after propagating a distance z into the medium, will become "depolarized", i.e. the initial vertical polarization state is altered due to generation of a small horizontally polarized component proportional to T_{hv}. From (4.19), the ratio (T_{hv}/T_{vv}) can be expressed as,

$$\frac{T_{hv}}{T_{vv}} = \frac{pe^{j\chi}\sin 2\beta}{1 + pe^{j\chi}\cos 2\beta} \tag{4.21a}$$

and,

$$\left|\frac{T_{hv}}{T_{vv}}\right| = \frac{p\sin 2\beta}{\sqrt{1 + 2p\cos\chi\cos 2\beta + p^2\cos^2 2\beta}} \tag{4.21b}$$

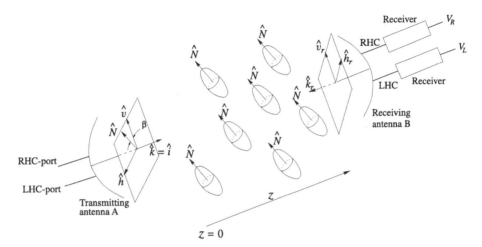

Fig. 4.4. Oriented spheroids, with symmetry axes along $\hat{N}$. The canting angle in the polarization plane is β.

where $pe^{j\chi}$ is defined in (4.13b). The cross-polarization isolation (XPI) is defined as $\mathrm{XPI}_v = 20\log(|T_{vv}|/|T_{hv}|)$, and is used in communication link design. For small β and zero A_{dp} (or $\chi = \pi/2$), the ratio becomes approximately $2\beta \sin(K_{\mathrm{dp}}z/2)$. When $K_{\mathrm{dp}}z = 90°$, for example, then $|T_{hv}/T_{vv}| = \sqrt{2}\beta$. If β is 5°, then the "depolarization" is around -18 dB, i.e. the magnitude of the horizontally polarized electric field component is 18-dB down relative to the vertically polarized field component (assuming that at $z = 0$ the wave was purely vertical). Raindrops are nearly oblate spheroidal in shape and are highly oriented with small β and thus, this approximation is a good measure of the propagation-induced "depolarization" caused by a differential propagation phase. Ice crystals oriented by in-cloud electrostatic fields can also cause "depolarization" due to a differential propagation phase, as was experimentally measured on satellite–Earth propagation links (see the review by Oguchi 1983). At long wavelengths, the K_{dp} due to ice crystals is typically quite small ($\leq 0.5°$ km^{-1}); however, the angle β can be large, and if the oriented crystals extend over a long path (tens of kilometers) then the "depolarization" can become significant.

4.1.3   Uniformly canted particles: circular polarization

It is instructive to transform (4.19b) to a circular basis as,

$$\begin{bmatrix} E_R(z) \\ E_L(z) \end{bmatrix} = [\mathbf{U}(RL \to hv)]^{-1}\mathbf{R}(\beta)\begin{bmatrix} e^{\lambda_1 z} & 0 \\ 0 & e^{\lambda_2 z} \end{bmatrix}\mathbf{R}(-\beta)\mathbf{U}(RL \to hv)\begin{bmatrix} E_R(0) \\ E_L(0) \end{bmatrix}$$
(4.22a)

$$= \frac{e^{(\lambda_1+\lambda_2)z/2}}{\sqrt{1-(pe^{j\chi})^2}}\begin{bmatrix} 1 & -pe^{j(\chi+2\beta)} \\ -pe^{j(\chi-2\beta)} & 1 \end{bmatrix}\begin{bmatrix} E_R(0) \\ E_L(0) \end{bmatrix}$$
(4.22b)

where $pe^{j\chi}$ is defined in (4.13b). By comparing (4.22b) with the case for $\beta = 0°$ in (4.15), it is seen that the canting angle (β) enters the transmission matrix as a phase adjustment to χ in the off-diagonal elements. Note that a positive β is measured clockwise from the vertical (see Fig. 2.10), whereas McCormick and Hendry (1973) define their equivalent angle (τ) as positive when measured counter-clockwise from the vertical. An initially right-hand circularly polarized wave ($E_R(0) = 1, E_L(0) = 0$) will get "depolarized" due to propagation, so that after a distance of z kilometers,

$$20 \log_{10} \left| \frac{E_L(z)}{E_R(z)} \right| = 20 \log_{10}(p) \tag{4.23}$$

which is independent of the angle β. The correlation between $E_R(z)$ and $E_L(z)$ defined as $U_R = E_R^* E_L / |E_R|^2$, can be used to estimate ($\chi - 2\beta$),

$$U_R(z) = \frac{E_R^*(z) E_L(z)}{|E_R(z)|^2} = p e^{j(\chi - 2\beta + \pi)} \tag{4.24}$$

If the initial polarization is left-hand circular ($E_R(0) = 0, E_L(0) = 1$), then define $U_L = E_R E_L^* / |E_L|^2$ so that

$$U_L(z) = p e^{j(\chi + 2\beta + \pi)} \tag{4.25}$$

Thus, the angle β can be determined from,

$$\beta = \frac{1}{4} \left[\arg U_L(z) - \arg U_R(z) \right] \tag{4.26}$$

McCormick and Hendry (1973) have reported on a method to measure the differential attenuation and differential phase shift in a one-way propagation path through rain, using a circularly polarized transmitter and receiver configuration (Fig. 4.4). Equations (3.113, 3.114) are now used to relate the voltages (V_R, V_L), at the receiver antenna ports, and the circular components of the electric field at the receiving antenna, i.e. $E_R^r(z)$ and $E_L^r(z)$ as expressed in the BSA convention,

$$V_R = \frac{\lambda}{\sqrt{8\pi Z_0}} \sqrt{G_b} \left[E_R^r(z) + \varepsilon_{R,b} E_L^r(z) \right] \tag{4.27a}$$

$$V_L = \frac{\lambda}{\sqrt{8\pi Z_0}} \sqrt{G_b} \left[\varepsilon_{L,b} E_R^r(z) + E_L^r(z) \right] \tag{4.27b}$$

where $G_R = G_L = G_b$ and, for well-designed antennas, $i_R \approx i_L \approx 1.0$, with small error terms ($\varepsilon_{R,b}$ and $\varepsilon_{L,b}$) representing deviations from polarization purity. The incident wave launched by the transmitting antenna when its RHC-port is excited can be expressed as,

$$E_R^i = \sqrt{\frac{Z_0}{2\pi}} \sqrt{G_a} \frac{e^{-jk_0 r}}{r} \tag{4.28a}$$

$$E_L^i = \sqrt{\frac{Z_0}{2\pi}} \sqrt{G_a} \varepsilon_{R,a} \frac{e^{-jk_0 r}}{r} \tag{4.28b}$$

where $\varepsilon_{R,a}$ is the small error term that represents a non-ideal antenna and the fraction of undesired LHC state that is radiated along with the desired RHC state. In the absence of any transmission medium between the two antennas, and expressing FSA circular components (E_R^i, E_L^i) at the receiving antenna $(r \equiv z)$ in BSA convention, see (3.81, 3.75), results in,

$$V_R = K'(1 + \varepsilon_{R,b}\varepsilon_{R,a}) \approx K' \tag{4.29a}$$

$$V_L = K'(\varepsilon_{L,b} + \varepsilon_{R,a}) \tag{4.29b}$$

where,

$$K' = -\frac{\lambda}{4\pi}\sqrt{G_a G_b}\frac{e^{-jk_0 z}}{z} \tag{4.29c}$$

The error term of $O(|\varepsilon|^2)$ in (4.29a) is neglected. Thus, in the absence of precipitation, the correlation between V_R and V_L is,

$$U_R(\text{no precip}) = \frac{V_R^* V_L}{|V_R|^2} = \varepsilon_{L,b} + \varepsilon_{R,a} \tag{4.30}$$

which gives a measure of the sum of the complex error terms of the transmitting and receiving antennas. In the presence of precipitation uniformly filling the space between the transmitter and receiver,

$$V_R \approx K \tag{4.31a}$$

$$V_L \approx K\left[\varepsilon_{L,b} + pe^{j(\chi - 2\beta + \pi)} + \varepsilon_{R,a}\right] \tag{4.31b}$$

where K is now,

$$K = \frac{-\lambda}{4\pi}\sqrt{G_a G_b}\frac{e^{(\lambda_1 + \lambda_2)z/2}}{z\sqrt{1 - (pe^{j\chi})^2}} \tag{4.31c}$$

In deriving (4.31a), error terms of the order $O(|\varepsilon|)$ and $O(|\varepsilon|^2)$ are neglected when compared to unity. In (4.31b), error terms of the order $O(|\varepsilon|^2)$ are neglected. The correlation, U_R is now,

$$U_R = \varepsilon_{L,b} + pe^{j(\chi - 2\beta + \pi)} + \varepsilon_{R,a} \tag{4.32}$$

Thus, with the occurrence of precipitation along the transmission path, the deviation of U_R in the complex plane from that given in (4.30), gives a measure of the quantity $p\exp[j(\chi - 2\beta + \pi)]$. Reverting to the definition in McCormick and Hendry (1973) where the vertical direction is chosen as a reference for the circular base vectors, which gets rid of $e^{j\pi}$ in (4.32), and recognizing that their positive canting angle (τ) is measured counter-clockwise from the vertical (or $\tau \equiv -\beta$), yields,

$$U_R = \varepsilon_{L,b} + pe^{j(\chi + 2\tau)} + \varepsilon_{R,a} \tag{4.33}$$

In particular, if the medium has zero A_{dp}, then $\chi = \pi/2$, and, if in addition $\tau = 0°$, then,

$$U_R = \varepsilon_{L,b} + \tan\left(\frac{K_{\text{dp}}z}{2}\right)e^{j\pi/2} + \varepsilon_{R,a} \tag{4.34}$$

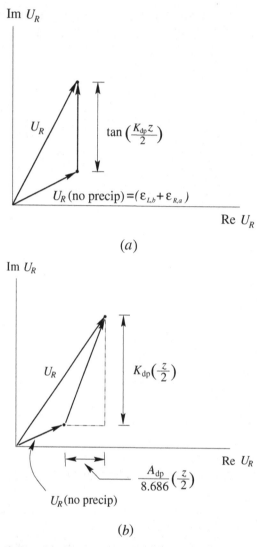

Fig. 4.5. Idealized complex plane plot of U_R: (a) for the case of a differential phase only along a propagation path of length z, and (b) using the approximation
$$\tanh^{-1}\left[(U_R) - U_R(\text{no precip})\right] \approx U_R - U_R(\text{no precip}).$$

An idealized complex plane plot of U_R is shown in Fig. 4.5a for the case of a pure differential phase shift (e.g. raindrops at long wavelength). As will be shown later in Chapter 8, the rain rate is related to K_{dp}. Thus, as the rain rate increases, the complex plane plot of U_R will increase parallel to the imaginary axis. Note that $K_{dp} > 0$ for oblate raindrops. As the frequency increases, the differential attenuation (A_{dp}) becomes non-zero, and U_R then has a component parallel to the real axis. Estimates of K_{dp} and A_{dp} are possible if it is assumed that $\tau = 0°$. From (4.33) and using (4.13b), it is easily established that,

$$(\lambda_2 - \lambda_1)\frac{z}{2} = \tanh^{-1}\left[U_R - U_R(\text{no precip})\right] \tag{4.35a}$$

$$= \left(jK_{dp} + \frac{A_{dp}}{8.686}\right)\frac{z}{2} \tag{4.35b}$$

Note that z is the distance of the propagation path illustrated in Fig. 4.4, and it is assumed that precipitation uniformly fills the entire path from transmitter to receiver. Using the approximation $\tanh^{-1} x \approx x$ for $|x| \ll 1$, typically valid for short paths, it follows that,

$$A_{dp} = (8.686)\frac{2}{z}\text{Re}\left[U_R - U_R(\text{no precip})\right] \qquad \text{dB km}^{-1} \tag{4.36a}$$

$$K_{dp} = \frac{2}{z}\text{Im}\left[U_R - U_R(\text{no precip})\right] \qquad \text{rad km}^{-1} \tag{4.36b}$$

where z is in kilometers. Figure 4.5b illustrates this measurement principle, while Fig. 4.6 shows data for a summer rain shower along a $z = 0.85$ km path at a frequency of 16.5 GHz. Note that the transmitted radiation was LHC for this data set and therefore U_L is plotted in Fig. 4.6. Scales for differential attenuation and differential phase shift are marked alongside the axes. The data points are shown at 5-s intervals, even though the data rate was in fact higher than 1 s^{-1}. The rain rate from a tipping bucket gage is also given at each datum point.

The importance of non-zero τ ($\equiv -\beta$), which is the mean canting angle, can be assessed by transmitting right-hand and left-hand circular polarization, alternately. Similarly to (4.26), it is easy to establish that,

$$4\tau = \arg\left[U_R - U_R(\text{no precip})\right] - \arg\left[U_L - U_L(\text{no precip})\right] \tag{4.37}$$

4.2   Oguchi's solution

A general description of coherent wave propagation in a sparsely distributed discrete scattering medium (Ishimaru and Cheung 1980) was used by Oguchi (1983) to analyze propagation-induced depolarization by precipitation. A review of coherent wave propagation through precipitation media can be found in Olsen (1982). Again, there are two eigenvalues and two associated characteristic waves that propagate without any depolarization. Let the scattering matrix in the forward direction in FSA convention, see (2.100), be ensemble averaged, see (3.161), over the particle states (which now includes, for example, size, shape, and orientation) and be represented as,

$$\langle nS_{FSA}(\hat{i}, \hat{i})\rangle = \begin{bmatrix} \langle nS_{hh}(\hat{i}, \hat{i})\rangle & \langle nS_{hv}(\hat{i}, \hat{i})\rangle \\ \langle nS_{vh}(\hat{i}, \hat{i})\rangle & \langle nS_{vv}(\hat{i}, \hat{i})\rangle \end{bmatrix}_{FSA} \tag{4.38}$$

Following Ishimaru and Cheung (1980), let the matrix $\mathbf{P}$ be defined as,

$$\mathbf{P} = -j\frac{2\pi}{k_0}\langle nS_{FSA}(\hat{i}, \hat{i})\rangle \tag{4.39}$$

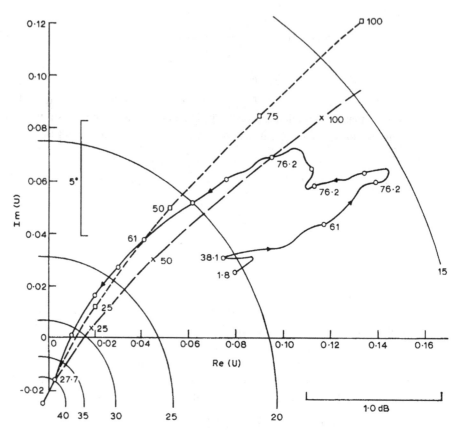

Fig. 4.6. Observation of U_L plotted in a complex plane at 16.5 GHz in a summer shower shown along a 0.85-km path. Note that LHC was transmitted. Data points are open circles connected by a solid line and are shown at 5-s intervals. Rain rate (mm h^{-1}) measured by a gage along the path, is noted alongside each datum point. Scales of differential attenuation and differential phase are indicated alongside the Re(U_L) and Im(U_L) axes for a 1-km distance. The two dashed lines are from theoretical calculations at 18 and 19 GHz. The solid circles represent constant values of $10 \log_{10}(p^2)$, where $p^2 = |U_L - U_L(\text{no precip})|^2$, see also Fig. 4.5. The $U_L(\text{no precip})$ point is located at the bottom left of the diagram near $(0, -0.02)$. It represents the sum of the antenna error terms $\varepsilon_{L,a} + \varepsilon_{R,b}$ and also locates the center of the circles. The 20 dB circle, for example, defines points where the power in the RHC component is 20 dB below that in the transmitted LHC component. From McCormick and Hendry (1973). (© *Electron. Lett.* 1973, IEE.)

The differential equation governing the coherent propagation of an electric field, $\vec{E}(z) = E_h(z)\hat{h} + E_v(z)\hat{v}$, along the positive Z-axis (selected only for convenience), is obtained as a generalization of (4.5b),

$$\frac{d}{dz}\begin{bmatrix} E_h(z) \\ E_v(z) \end{bmatrix} = \begin{bmatrix} -jk_0 + P_{hh} & P_{hv} \\ P_{vh} & -jk_0 + P_{vv} \end{bmatrix}\begin{bmatrix} E_h(z) \\ E_v(z) \end{bmatrix} \qquad (4.40)$$

where the elements of $\mathbf{P}$ are defined in (4.39). The off-diagonal elements do not have the term $-jk_0$, since the unperturbed incident field does not generate any cross-polar component. Equation (4.40) is a coupled set of two differential equations that can be solved by seeking solutions of the form $\exp(\lambda z)$, where $\lambda = -jk_{\mathrm{eff}}$ and k_{eff} is the effective propagation constant. There are, in general, two eigenvalues and two eigenvectors. The eigenvalues are given by,

$$\lambda_1 = -j\left[k_0 + \frac{j}{2}(P_{hh} + P_{vv} + \gamma)\right] \tag{4.41a}$$

$$\lambda_2 = -j\left[k_0 + \frac{j}{2}(P_{hh} + P_{vv} - \gamma)\right] \tag{4.41b}$$

where,

$$\gamma = \sqrt{(P_{hh} - P_{vv})^2 + 4P_{hv}P_{vh}} \tag{4.41c}$$

The sign of the radical has been chosen so that in the case of very small cross-coupling $(P_{hv}, P_{vh} \to 0)$, the sign of γ is the same as the sign of $(P_{hh} - P_{vv})$. Thus, $\lambda_1 = -jk_{\mathrm{eff}}^h$ and $\lambda_2 = -jk_{\mathrm{eff}}^v$, will be close to horizontal and vertical polarizations, respectively, when $P_{hv}, P_{vh} \to 0$. The solution to (4.40) is given in Appendix 4, and is written as,

$$\left[\begin{array}{c} E_h(z) \\ E_v(z) \end{array}\right] = \left[\begin{array}{cc} T_{hh} & T_{hv} \\ T_{vh} & T_{vv} \end{array}\right]\left[\begin{array}{c} E_h(0) \\ E_v(0) \end{array}\right] \tag{4.42}$$

where $\mathbf{T}$ is the transmission matrix with elements given as,

$$T_{hh} = e^{\lambda_1 z}\cos^2\phi + e^{\lambda_2 z}\sin^2\phi \tag{4.43a}$$
$$T_{vv} = e^{\lambda_1 z}\sin^2\phi + e^{\lambda_2 z}\cos^2\phi \tag{4.43b}$$
$$T_{vh} = T_{hv} = (e^{\lambda_1 z} - e^{\lambda_2 z})\frac{\sin 2\phi}{2} \tag{4.43c}$$

where,

$$\phi = \frac{1}{2}\tan^{-1}\left(\frac{2P_{vh}}{P_{hh} - P_{vv}}\right) \tag{4.44}$$

and $\lambda_{1,2}$ are given in (4.41a, b). The angle ϕ can, in general, be complex, since the elements of $\mathbf{P}$ are in terms of $\langle nS_{\mathrm{FSA}}(\hat{i}, \hat{i})\rangle$,

$$\phi = \frac{1}{2}\tan^{-1}\left[\frac{2n\langle S_{vh}(\hat{i}, \hat{i})\rangle}{\langle nS_{hh}(\hat{i}, \hat{i})\rangle - \langle nS_{vv}(\hat{i}, \hat{i})\rangle}\right]_{\mathrm{FSA}} \tag{4.45}$$

and the scattering matrix elements can be complex.

A particularly simple interpretation of the angle ϕ involves the case of identical oriented spheroids of number density n, as illustrated in Fig. 4.4, with the angle β having a Gaussian pdf with mean β_0 and standard deviation σ_β. Note that the angle

ψ between the incident direction and the symmetry axis is assumed to be 90°. If the "principal plane" forward scattering matrix elements in FSA for a single particle are defined, see also (2.100), as,

$$S_{11}(\hat{i}, \hat{i}) = S_{hh}(\hat{i}, \hat{i}; \beta = 0) \tag{4.46a}$$

$$S_{22}(\hat{i}, \hat{i}) = S_{vv}(\hat{i}, \hat{i}; \beta = 0) \tag{4.46b}$$

then,

$$\mathbf{S}(\hat{i}, \hat{i}; \beta) = \mathbf{R}(\beta) \begin{bmatrix} S_{11} & 0 \\ 0 & S_{22} \end{bmatrix} \mathbf{R}(-\beta) \tag{4.47}$$

where $\mathbf{R}(\beta)$ is defined in (2.93a), and the ensemble average in (4.39) becomes,

$$\mathbf{P} = -j\frac{2\pi n}{k_0} \int_{-\infty}^{\infty} \mathbf{S}(\hat{i}, \hat{i}; \beta) \frac{1}{\sqrt{2\pi}\sigma_\beta} e^{-(\beta-\beta_0)^2/2\sigma_\beta^2} \, d\beta \tag{4.48}$$

Using the relations,

$$\int_{-\infty}^{\infty} \sin^2\beta \frac{1}{\sqrt{2\pi}\sigma_\beta} e^{-(\beta-\beta_0)^2/2\sigma_\beta^2} \, d\beta = \frac{1}{2}\left(1 - e^{-2\sigma_\beta^2} \cos 2\beta_0\right) \tag{4.49a}$$

$$\int_{-\infty}^{\infty} \cos^2\beta \frac{1}{\sqrt{2\pi}\sigma_\beta} e^{-(\beta-\beta_0)^2/2\sigma_\beta^2} \, d\beta = \frac{1}{2}\left(1 + e^{-2\sigma_\beta^2} \cos 2\beta_0\right) \tag{4.49b}$$

$$\int_{-\infty}^{\infty} \frac{\sin 2\beta}{2} \frac{1}{\sqrt{2\pi}\sigma_\beta} e^{-(\beta-\beta_0)^2/2\sigma_\beta^2} \, d\beta = \frac{1}{2}e^{-2\sigma_\beta^2} \sin 2\beta_0 \tag{4.49c}$$

it is straightforward to show that,

$$\lambda_1 - \lambda_2 = -j\frac{2\pi}{k_0}n(S_{11} - S_{22})e^{-2\sigma_\beta^2} \tag{4.50a}$$

$$\phi = -\beta_0 \tag{4.50b}$$

Thus, the angle ϕ is identical to the mean of the Gaussian pdf. The negative sign is a result of the angle β_0 being defined here as positive when measured in a clockwise sense from the $\hat{v}$-direction (see also, Fig. 2.10) when viewed along the direction of propagation, $\hat{k} = \hat{v} \times \hat{h}$. However, the angle ϕ is defined as being positive when measured in a counter-clockwise sense from $\hat{v}$ (Oguchi 1983). The integration limits for β in (4.49) should be $-\pi/2$ and $\pi/2$, which are the physical limits of the canting angle for spheroids. Substituting (4.50b) in (4.43a–c) yields the transmission matrix elements, which are of the same form as a constant β in (4.19b), except that β in (4.19b) is to be replaced by β_0. Of course, the eigenvalues in the constant β case (4.20a, b), will be different from the Gaussian case, e.g. $(\lambda_1 - \lambda_2)$ from (4.20a, b) is different from (4.50a) by the factor $\exp(-2\sigma_\beta^2)$. Thus, the effect of having a "spread" in the canting angle (β) is to reduce the specific differential attenuation (A_{dp}) and specific differential phase (K_{dp}) between the two "characteristic" waves by the factor $\exp(-2\sigma_\beta^2)$ relative to A_{dp} and K_{dp} for the constant β case.

The angle ϕ continues to be equal to $-\beta_0$ if the pdf of β is symmetric about β_0 (i.e. not necessarily Gaussian). Oguchi (1983) has given the result for $\lambda_1 - \lambda_2$ when β is Gaussian and, in addition, ψ is also a distributed Gaussian with mean ψ_0 and standard deviation σ_ψ,

$$\lambda_1 - \lambda_2 = -j\frac{2\pi}{k_0}n(S_{11} - S_{22})\frac{(1 + e^{-2\sigma_\psi^2}\cos 2\psi_0)}{2}e^{-2\sigma_\beta^2} \tag{4.51a}$$

$$\phi = -\beta_0 \tag{4.51b}$$

More generally, if the pdf of β is symmetric (about β_0), then the medium exhibits "mirror" reflection symmetry, as discussed in Section 3.12. This condition yields $\phi = -\beta_0$, and the characteristic polarizations of the medium become linear and orthogonal.

The ensemble average in (4.39) can include a mixture of different particle types, such as raindrops and hailstones, or small pristine ice crystals with larger snow particles, with each component of the mixture having different size, shape, and orientation distributions. For example, consider the simple case of identical highly oriented oblate raindrops (number density n_1) with $\beta_{01} = 0$ and $\sigma_{\beta_1} \approx 5$–$10°$, mixed with "tumbling" oblate hailstones (number density n_2) with $\beta_{02} = 0$ and large σ_{β_2}. The ensemble average in (4.39) now becomes,

$$\mathbf{P} = \sum_{k=1}^{2} -j\frac{2\pi n_k}{k_0}\langle \mathbf{S}_k(\hat{i}, \hat{i}; \beta_k)\rangle \tag{4.52}$$

where $k = 1$ represents raindrops and $k = 2$ represents hailstones. It should be clear now that (4.50a) will become,

$$\lambda_1 - \lambda_2 = \sum_{k=1}^{2} -j\frac{2\pi n_k}{k_0}(S_{11} - S_{22})_k e^{-2\sigma_{\beta k}^2} \tag{4.53a}$$

$$\approx -j\frac{2\pi n_1}{k_0}(S_{11} - S_{22})_1 e^{-2\sigma_{\beta_1}^2} \tag{4.53b}$$

where the last approximation follows from the fact that $\exp(-2\sigma_{\beta_2}^2) \ll \exp(-2\sigma_{\beta_1}^2)$. From (4.8), the specific differential phase (rad km^{-1}), in particular, is given for the rain/hail mixture as,

$$K_{dp} = 10^3\frac{(2\pi n_1)}{k_0}\text{Re}(S_{11} - S_{22})_1 e^{-2\sigma_{\beta_1}^2} \tag{4.54a}$$

$$\approx 10^3\frac{(2\pi n_1)}{k_0}\text{Re}(S_{11} - S_{22})_1 \tag{4.54b}$$

Note that n_1 is in m^{-3}, k_0 is in m^{-1} and $S_{11,22}$ is in m. Techniques for the radar measurement of K_{dp} will be discussed later in Chapter 6, but it suffices here to mention that in a rain/hail mixture, as modeled here, K_{dp} is only proportional to the highly oriented rain component of the mixture. Since K_{dp} in rain will be shown in Chapter 7 to be proportional to rain rate (or rainwater content), an important radar-based measurement for estimating the rain rate when precipitation is mixed with hailstones is available.

4.3   Radar range equation with transmission matrix: linear polarization basis

The transmission matrix is also sometimes referred to as the propagation matrix. The radar range equation for a single particle described by $\mathbf{S}_{BSA}$ was derived in Section 3.4, and will now be extended to the case where a uniform precipitation medium exists between the radar and the particle. This uniform medium will be described by the transmission matrix defined in (4.42). The first-order multiple scattering approximation defined by Ishimaru (1978) will be used to derive the radar range equation with propagation effect included, as illustrated in Fig. 4.7. Under this approximation, the transmitted wave launched by the radar suffers depolarization along the propagation path between the radar and the scattering resolution volume; the polarization state of the back scattered wave is then transformed by $\mathbf{E}^r = \mathbf{S}_{BSA}\mathbf{E}^i$; and finally the back scattered wave suffers depolarization on the return path to the radar. The scattering resolution volume will be defined later in Section 5.3, but it is sufficient here to mention that for pulsed radars, the volume will extend from $c(t - T_0)/2$ to $ct/2$, in range where T_0 is the pulsewidth and c is the light velocity. For the present discussion, a single particle described by $\mathbf{S}_{BSA}$ will be fixed within this scattering volume.

Start first with (3.59), which gives an expression for the field incident on a particle at range r_a in the absence of any propagation medium and is repeated here for the case of an ideal antenna as,

$$\begin{bmatrix} E_h^i \\ E_v^i \end{bmatrix} = \sqrt{\frac{Z_0 G}{2\pi}} \frac{e^{-jk_0 r_a}}{r_a} \begin{bmatrix} M_h \\ M_v \end{bmatrix} \qquad (4.55)$$

In the presence of a uniform precipitation medium with transmission matrix, $\mathbf{T}$, given in (4.42), the spherical wave phase term $\exp(-jk_0 r_a)$ is replaced by $\mathbf{T}$,

$$\begin{bmatrix} E_h^i \\ E_v^i \end{bmatrix} = \sqrt{\frac{Z_0 G}{2\pi}} \frac{1}{r} \begin{bmatrix} T_{hh} & T_{hv} \\ T_{vh} & T_{vv} \end{bmatrix} \begin{bmatrix} M_h \\ M_v \end{bmatrix} \qquad (4.56)$$

where $r \equiv r_a \equiv z$ is the propagation path distance (or range) to the fixed particle. In going from (4.55) to (4.56) an additional extension has been made in applying the plane wave transmission solution of (4.42) to spherical waves. The spatial phase reference is at the transmitting antenna in (4.55, 4.56).

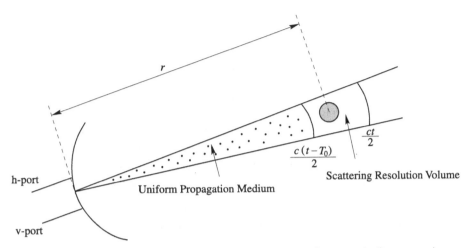

Fig. 4.7. In a pulsed radar with pulsewidth T_0, the voltage at any time t results from scattering within the resolution volume as indicated. In this idealized illustration, a single spherical particle is located within the resolution volume.

The next step is to consider back scattering from the single, fixed particle within the radar resolution volume, as illustrated in Fig. 4.7. The BSA convention is adopted, with $\mathbf{E}^r = \mathbf{S}_{\mathrm{BSA}}\mathbf{E}^i$. Thus, the electric field back scattered by the particle is given as,

$$
\begin{bmatrix} E_h^r \\ E_v^r \end{bmatrix} = \frac{e^{-jk_0 r}}{r} \begin{bmatrix} S_{hh} & S_{hv} \\ S_{vh} & S_{vv} \end{bmatrix}_{\mathrm{BSA}} \begin{bmatrix} E_h^i \\ E_v^i \end{bmatrix}
\tag{4.57}
$$

where the spherical wave factor expresses "outgoing" radiation from the particle in empty space, and the spatial phase reference is now located at the origin of the coordinate system centered on the particle. However, the medium between the particle and the radar is described by the same transmission matrix as in (4.42), so that the spherical wave phase term $\exp(-jk_0 r)$ is again replaced by $\mathbf{T}$,

$$
\begin{bmatrix} E_h^r \\ E_v^r \end{bmatrix} = \frac{1}{r} \begin{bmatrix} T_{hh} & T_{hv} \\ T_{vh} & T_{vv} \end{bmatrix} \begin{bmatrix} S_{hh} & S_{hv} \\ S_{vh} & S_{vv} \end{bmatrix}_{\mathrm{BSA}} \begin{bmatrix} E_h^i \\ E_v^i \end{bmatrix}
\tag{4.58}
$$

Substituting (4.56) into (4.58) yields,

$$
\begin{bmatrix} E_h^r \\ E_v^r \end{bmatrix} = \sqrt{\frac{Z_0 G}{2\pi}} \frac{1}{r^2} [\mathbf{T}][\mathbf{S}_{\mathrm{BSA}}][\mathbf{T}] \begin{bmatrix} M_h \\ M_v \end{bmatrix}
\tag{4.59}
$$

In essence then, the only extension made in going from the case of free space to that of a uniform precipitation medium is to replace the spherical wave factor $\exp(-jk_0 r)/r$ by $(\mathbf{T}/r)$. This is done twice to reflect the two-way propagation: once for the spherical wave radiated by the antenna, and once for the spherical wave radiated by the particle. The spatial phase reference in (4.59) is at the origin of the Cartesian system centered on the particle that is used to describe the back scatter matrix. The last step is to derive the voltages at the two ports of the antenna, using (3.61) for an ideal antenna,

$$\begin{bmatrix} V_h \\ V_v \end{bmatrix} = \frac{\lambda G}{4\pi r^2}[\mathbf{T}][\mathbf{S}_{\text{BSA}}][\mathbf{T}]\begin{bmatrix} M_h \\ M_v \end{bmatrix} \tag{4.60}$$

In one sense, the transmission matrix acts to "distort" the estimation of $\mathbf{S}_{\text{BSA}}$ using the measured voltages. However, there is information contained in estimating a key parameter of $\mathbf{T}$, i.e. the differential phase between the "characteristic" waves can be determined by appropriate processing of the received voltages using a polarization "agile" radar capable of transmitting alternate horizontally ($M_h = 1, M_v = 0$) and vertically ($M_h = 0, M_v = 1$) polarized pulses (see Chapter 6).

4.3.1 Uniform oblate raindrops (zero canting angle)

Consider a transmission matrix with $\phi = 0°$, e.g. a uniform rain medium of oblate drops with zero canting angle, and let $\mathbf{S}_{\text{BSA}}$ represent a fixed, spherical particle. Then it follows quite simply, that under the excitation $M_h = 1, M_v = 0$, the voltage at the h-port is,

$$V_h^{10} = \frac{\lambda G}{4\pi r^2}e^{2\lambda_1 r}S \tag{4.61}$$

while for the excitation ($M_h = 0, M_v = 1$), the voltage at the v-port is,

$$V_v^{01} = \frac{\lambda G}{4\pi r^2}e^{2\lambda_2 r}S \tag{4.62}$$

where $S = S_{hh} = S_{vv}$ is the scattering matrix element for a spherical particle. Note the use of superscripts "10", to represent $M_h = 1, M_v = 0$, and "01", to represent $M_h = 0, M_v = 1$. The argument of $\left[(V_h^{10})^*(V_v^{01})\right]$ is,

$$\arg\left[(V_h^{10})^*V_v^{01}\right] = \arg\left[e^{2(\lambda_2-\lambda_1)r}\right] \tag{4.63a}$$

$$= 2K_{\text{dp}}r \tag{4.63b}$$

where (4.8) has been used. Note that K_{dp} is in units of rad km^{-1}, while r is in km. Equation (4.63b) shows that a simple algorithm for K_{dp} can be constructed in this special case. It is reiterated that K_{dp}, as defined here, is positive for horizontally oriented oblate particles and negative for vertically oriented prolates. If the scattering matrix of the fixed particle is given as,

$$\mathbf{S}_{\text{BSA}} = \begin{bmatrix} S_{hh} & 0 \\ 0 & S_{vv} \end{bmatrix}_{\text{BSA}} = e^{j\delta_{hh}}\begin{bmatrix} |S_{hh}| & 0 \\ 0 & |S_{vv}|e^{j(\delta_{vv}-\delta_{hh})} \end{bmatrix}_{\text{BSA}} \tag{4.64}$$

then (4.63) becomes,

$$\arg\left[(V_h^{10})^*(V_v^{01})\right] = \Psi_{\text{dp}} = \arg\left[e^{2(\lambda_2-\lambda_1)r}\right] + \arg(S_{hh}^*S_{vv})_{\text{BSA}} \tag{4.65a}$$

$$= 2K_{\text{dp}}r + \delta_{\text{co}} = \Phi_{\text{dp}} + \delta_{\text{co}} \tag{4.65b}$$

where $\delta_{co} = \arg(S_{hh}^* S_{vv})$, is the back scatter differential phase.[2] Note the definition of Ψ_{dp}, which equals the sum of the differential propagation phase (Φ_{dp}) and δ_{co}. For Rayleigh scattering and real ε_r, there is no differential phase on scattering. Non-zero values of δ_{co} are generally indicative of Mie scattering effects (i.e. higher order induced multipoles). The two-way differential propagation phase is defined as $\Phi_{dp} = 2K_{dp}r$.

Figure 4.8a shows measurements of Ψ_{dp} versus range in heavy rain at 3-GHz frequency using the CSU–CHILL radar. Since raindrops at 3 GHz (or $\lambda = 10$ cm) are Rayleigh–Gans scatterers, the $\delta_{co} \approx 0°$. Figure 4.8a thus shows the differential propagation phase (Φ_{dp}) in heavy rain starting from around $-60°$ (which is the differential phase of the system) and increasing to $150°$, over a range of 30 km. The dashed line is a "filtered" version of Φ_{dp}, from which K_{dp} is computed as $K_{dp} = [\Phi_{dp}(r_2) - \Phi_{dp}(r_1)]/[2(r_2 - r_1)]$ (see Section 6.6). The resultant K_{dp} as a function of range is plotted in Fig. 4.8b. Rather high values of K_{dp} ($\approx 7°$ km^{-1}) are noted near the 22-km range.

4.3.2   General form of S_{BSA} with a diagonal transmission matrix

It is useful to consider the general form for S_{BSA}, and a diagonal transmission matrix. Equation (4.60) then reduces to,

$$
\begin{bmatrix} V_h \\ V_v \end{bmatrix} = \frac{\lambda G}{4\pi r^2} \begin{bmatrix} S_{hh}e^{2\lambda_1 r} & S_{hv}e^{(\lambda_1+\lambda_2)r} \\ S_{vh}e^{(\lambda_1+\lambda_2)r} & S_{vv}e^{2\lambda_2 r} \end{bmatrix} \begin{bmatrix} M_h \\ M_v \end{bmatrix} \tag{4.66a}
$$

$$
= \frac{\lambda G}{4\pi r^2} \begin{bmatrix} |S_{hh}|e^{j\delta_{hh}}e^{-j2k_{\text{eff}}^h r} & |S_{hv}|e^{j\delta_{hv}}e^{-j(k_{\text{eff}}^h+k_{\text{eff}}^v)r} \\ |S_{vh}|e^{j\delta_{vh}}e^{-j(k_{\text{eff}}^h+k_{\text{eff}}^v)r} & |S_{vv}|e^{j\delta_{vv}}e^{-j2k_{\text{eff}}^v r} \end{bmatrix} \begin{bmatrix} M_h \\ M_v \end{bmatrix} \tag{4.66b}
$$

where $\lambda_1 = -jk_{\text{eff}}^h$, $\lambda_2 = -jk_{\text{eff}}^v$. Note that,

$$
k_{\text{eff}}^h = k_{\text{re}}^h - jk_{\text{im}}^h = k_0 + \frac{2\pi}{k_0}\langle f_{hh}(\hat{i},\hat{i})\rangle \tag{4.67a}
$$

$$
k_{\text{eff}}^v = k_{\text{re}}^v - jk_{\text{im}}^v = k_0 + \frac{2\pi}{k_0}\langle f_{vv}(\hat{i},\hat{i})\rangle \tag{4.67b}
$$

and,

$$
k_{\text{re}}^{h,v} = k_0 + \frac{2\pi}{k_0}\,\text{Re}\langle f_{hh,vv}(\hat{i},\hat{i})\rangle = k_0 + k_{p,\text{re}}^{h,v} \tag{4.68a}
$$

$$
k_{\text{im}}^{h,v} = -\frac{2\pi}{k_0}\,\text{Im}\langle f_{hh,vv}(\hat{i},\hat{i})\rangle = k_{p,\text{im}}^{h,v} \tag{4.68b}
$$

where k_p is the "perturbation" component of the wave number due to the presence of the medium. The notation $f_{hh,vv}(\hat{i},\hat{i})$ is used here to represent the forward scattered matrix elements to distinguish them from the back scattered matrix elements. Thus,

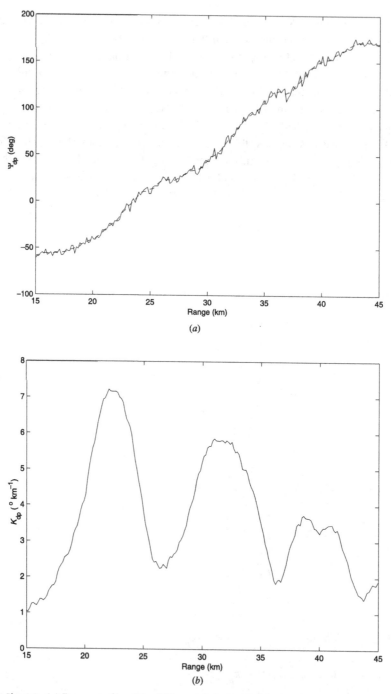

Fig. 4.8. (*a*) Range profile of the differential phase (Ψ_{dp}) at 3 GHz in a convective storm at a low elevation angle using the CSU–CHILL radar. Dashed line is a filtered version (see Section 6.6). Since $\delta_{co} \approx 0°$, the measured $\Psi_{dp} \approx \Phi_{dp}$, which is the differential propagation phase. (*b*) Specific differential phase (or K_{dp}) obtained from the filtered Ψ_{dp}.

(4.66b) becomes,

$$\begin{bmatrix} V_h \\ V_v \end{bmatrix} = \frac{\lambda G}{4\pi r^2} e^{-j2k_0 r}$$

$$\begin{bmatrix} |S_{hh}|e^{-2k_{im}^h r} e^{j\delta_{hh}} e^{-j2k_{p,re}^h r} & |S_{hv}|e^{-(k_{im}^h+k_{im}^v)r} e^{j\delta_{hv}} e^{-j(k_{p,re}^h+k_{p,re}^v)r} \\ |S_{vh}|e^{-(k_{im}^h+k_{im}^v)r} e^{j\delta_{vh}} e^{-j(k_{p,re}^h+k_{p,re}^v)r} & |S_{vv}|e^{-2k_{im}^v r} e^{j\delta_{vv}} e^{-j2k_{p,re}^v r} \end{bmatrix} \begin{bmatrix} M_h \\ M_v \end{bmatrix}$$

(4.69)

The above form is sometimes useful since it shows the attenuation and phase terms due to the "perturbation" part of the effective wave numbers of the medium as separate from the scattering matrix phase terms as well as the empty space spatial phase term $(-j2k_0 r)$. Note that r is the propagation path length in Fig. 4.7, and is essentially the same as the range to the fixed particle within the scattering resolution volume, since the range extent of the volume $cT_0/2$ is generally very small compared to r (i.e. a few hundred meters for typical pulsewidths versus tens of kilometers).

The differential reflectivity (Z_{dr}) of a single particle, in the absence of a propagation medium, was defined in (2.54). From (4.69), Z_{dr} inclusive of the effects of a diagonal transmission matrix, is,

$$Z_{dr} = 10\log_{10} \frac{|V_h^{10}|^2}{|V_v^{01}|^2} \tag{4.70a}$$

$$= 10\log_{10} \frac{|S_{hh}|^2}{|S_{vv}|^2} + 10\log_{10} e^{-4(k_{im}^h - k_{im}^v)r} \tag{4.70b}$$

$$= Z_{dr}^{int} - 40(k_{im}^h - k_{im}^v)r(\log_{10} e) \tag{4.70c}$$

$$= Z_{dr}^{int} - 2A_{dp}r \tag{4.70d}$$

where A_{dp} is defined in (4.7) and is the specific differential attenuation in dB km^{-1}, and r is in kilometers. The superscript "int" on Z_{dr} stands for the "intrinsic" value of Z_{dr} in the absence of any propagation effects. Thus, the intrinsic differential reflectivity of the particle is reduced by the net differential attenuation caused by the medium. Figure 4.9 shows observations of differential reflectivity as a function of range in heavy rain at a frequency of 5.5 GHz. Note the strong effects of differential attenuation, a characteristic feature at high frequencies ($f \gtrsim 5$ GHz) primarily due to differential absorption. Methods for correcting the measured differential reflectivity for differential attenuation effects will be considered in Section 7.4.

The linear depolarization ratio (LDR$_{vh}$) for a single particle was defined in (2.55) in the absence of propagation effects. In the presence of a diagonal transmission matrix,

$$LDR_{vh} = 10\log_{10} \frac{|V_v^{10}|^2}{|V_h^{10}|^2} \tag{4.71a}$$

$$= 10\log_{10} \frac{|S_{vh}|^2}{|S_{hh}|^2} + 10\log_{10} e^{2(k_{im}^h - k_{im}^v)r} \tag{4.71b}$$

$$= LDR_{vh}^{int} + A_{dp}r \tag{4.71c}$$

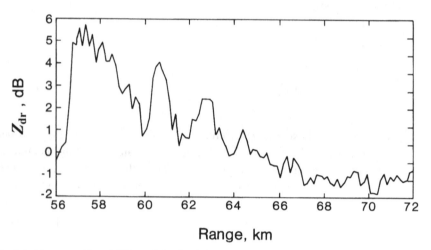

Fig. 4.9. Range profile of differential reflectivity in heavy rain at a frequency of 5.5 GHz to illustrate the strong effects of differential attenuation beyond 63 km, see (4.70d). Data from the German DLR radar located near Munich (Meischner et al. 1991b).

Thus LDR_{vh} is increased by differential attenuation due to the propagation medium. However, $\arg\left[(V_h^{10})^*(V_v^{01})\right]$ is unaffected by differential attenuation which is a desirable property for the estimation of K_{dp}.

Equation (4.69) can be written in slightly different form as,

$$
\begin{bmatrix} V_h \\ V_v \end{bmatrix} = \frac{\lambda G}{4\pi r^2} e^{-j2k_0 r}
$$

$$
\begin{bmatrix} |S_{hh}| e^{-2k_{im}^h r} e^{j\delta_{hh}} e^{-j2\Phi_{hh}} & |S_{hv}| e^{-(k_{im}^h + k_{im}^v)r} e^{j\delta_{hv}} e^{-j(\Phi_{hh}+\Phi_{vv})} \\ |S_{vh}| e^{-(k_{im}^h + k_{im}^v)r} e^{j\delta_{vh}} e^{-j(\Phi_{hh}+\Phi_{vv})} & |S_{vv}| e^{-2k_{im}^v r} e^{j\delta_{vv}} e^{-j2\Phi_{vv}} \end{bmatrix} \begin{bmatrix} M_h \\ M_v \end{bmatrix}
$$

$$(4.72)$$

where $\Phi_{hh} = k_{p,re}^h r$, $\Phi_{vv} = k_{p,re}^v r$, and $\Phi_{dp} = 2(\Phi_{hh} - \Phi_{vv})$. The second column of the above matrix [or the matrix in (4.69)] can be used to define the "second" linear depolarization ratio as,

$$
\text{LDR}_{hv} = 10 \log_{10} \frac{|V_h^{01}|^2}{|V_v^{01}|^2} = 10 \log_{10} \frac{|S_{hv}|^2}{|S_{vv}|^2} + 10 \log_{10} e^{-2(k_{im}^h - k_{im}^v)r} \tag{4.73a}
$$

$$
= \text{LDR}_{hv}^{int} - A_{dp} r \tag{4.73b}
$$

It follows that $\text{LDR}_{hv} - \text{LDR}_{vh} = Z_{dr}$ for reciprocal scattering (i.e. $S_{hv} = S_{vh}$).

The power in the h-channel is defined as,

$$
P_{co}^h = |V_h^{10}|^2 \tag{4.74}
$$

$$
= \frac{\lambda^2 G^2}{16\pi^2 r^4} |S_{hh}|^2 e^{-4k_{im}^h r} = P_{co}^{h,int} e^{-4k_{im}^h r} \tag{4.75}
$$

The specific attenuation (in dB km^{-1}) at h-polarization is defined as in (1.133a),

$$A_h = 8.686 \times 10^3 \, k_{\text{im}}^h \tag{4.76}$$

and, the term $10 \log_{10}\left[\exp(-4k_{\text{im}}^h r)\right] = -2A_h r$, where r is in km. Thus, the intrinsic power in the h-channel is attenuated (in decibels) by $2A_h r$ due to the propagation medium (methods to correct for attenuation are discussed in Section 7.4).

The complex correlation coefficient between the diagonal elements of the matrix in (4.72) is defined as,

$$\frac{(V_h^{10})^*(V_v^{01})}{\sqrt{|V_h^{10}|^2|V_v^{01}|^2}} = \frac{|S_{hh}||S_{vv}|e^{j(\delta_{vv}-\delta_{hh})}}{\sqrt{|S_{hh}|^2|S_{vv}|^2}}e^{j\Phi_{\text{dp}}} \tag{4.77a}$$

$$= |\rho_{\text{co}}|e^{j(\delta_{\text{co}}+\Phi_{\text{dp}})} \tag{4.77b}$$

$$= |\rho_{\text{co}}|e^{j(\Psi_{\text{dp}})} \tag{4.77c}$$

Note that the above correlation coefficient is unaffected by differential attenuation. Two other complex correlation coefficients can be defined from (4.72),

$$(\rho_{\text{cx}}^h) = \frac{(V_h^{10})(V_v^{10})^*}{\sqrt{|V_h^{10}|^2|V_v^{10}|^2}} = \frac{|S_{hh}||S_{vh}|e^{j(\delta_{hh}-\delta_{vh})}e^{-j\Phi_{\text{dp}}/2}}{\sqrt{|S_{hh}|^2|S_{vh}|^2}} \tag{4.78a}$$

$$(\rho_{\text{cx}}^v) = \frac{(V_v^{01})(V_h^{01})^*}{\sqrt{|V_v^{01}|^2|V_h^{01}|^2}} = \frac{|S_{vv}||S_{hv}|e^{j(\delta_{vv}-\delta_{hv})}}{\sqrt{|S_{vv}|^2|S_{hv}|^2}}e^{j\Phi_{\text{dp}}/2} \tag{4.78b}$$

The subscript "cx" on ρ stands for "cross", similar to R_{cx} in (3.179), while the superscript h or v refers to the transmitted polarization state. It is clear that the argument of any one of the complex correlation coefficients can be used to estimate Φ_{dp}. However, practical considerations, i.e. stronger copolar signal strengths, make it desirable to use (4.77). It is reiterated that the entire discussion from (4.61) onwards until (4.78) is based on a diagonal transmission matrix.

Equation (4.60) is valid for a homogeneous propagation path, but it can be generalized for an inhomogeneous path by segmenting the path into a cascade of homogeneous transmission matrices $[\mathbf{T}_1], [\mathbf{T}_2], \ldots, [\mathbf{T}_n]$ as illustrated in Fig. 4.10,

$$\begin{bmatrix} V_h \\ V_v \end{bmatrix} = \frac{\lambda G}{4\pi r^2}[\mathbf{T}_1][\mathbf{T}_2]\cdots[\mathbf{T}_n][\mathbf{S}_{\text{BSA}}][\mathbf{T}_n][\mathbf{T}_{n-1}]\cdots[\mathbf{T}_1] \tag{4.79}$$

When the transmission matrix is not diagonal, and for simplicity if $\mathbf{S}_{\text{BSA}}$ corresponds to a fixed spherical particle, (4.60) reduces to,

$$\begin{bmatrix} V_h \\ V_v \end{bmatrix} = \frac{\lambda G}{4\pi r^2}S\begin{bmatrix} T_{hh}^2 + T_{hv}^2 & T_{hv}(T_{hh} + T_{vv}) \\ T_{hv}(T_{hh} + T_{vv}) & T_{hv}^2 + T_{vv}^2 \end{bmatrix}\begin{bmatrix} M_h \\ M_v \end{bmatrix} \tag{4.80}$$

It is clear that Z_{dr} and LDR_{vh}, defined in (4.70a) and (4.71a), will now be "distorted" relative to their intrinsic values (0 and $-\infty$ dB, respectively) for a sphere, and the

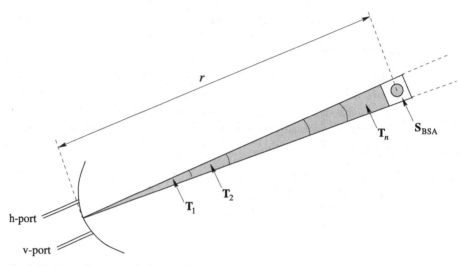

Fig. 4.10. Cascading transmission matrices along an inhomogeneous propagation path. The resolution volume at range r is represented by $\mathbf{S}_{BSA}$ which refers to the scattering matrix of a single particle.

"distortion" will depend on the angle ϕ in (4.45) as well as the differential propagation phase (Φ_{dp}) between the "characteristic" waves (Holt 1984; Sachidananda and Zrnić 1985). It is convenient to transform (4.60) to a circular basis as was done earlier for a one-way propagation path, see (4.22a), to treat the case of a non-diagonal transmission matrix.

4.4   Radar range equation with transmission matrix: circular polarization basis

The circular basis equivalent of (4.56) is obtained using (4.22a),

$$\begin{bmatrix} E_R^i \\ E_L^i \end{bmatrix} = \sqrt{\frac{Z_0 G}{2\pi}} \frac{1}{r} [U(RL \to hv)]^{-1} \begin{bmatrix} T_{hh} & T_{hv} \\ T_{vh} & T_{vv} \end{bmatrix} U(RL \to hv) \begin{bmatrix} M_R \\ M_L \end{bmatrix} \quad (4.81)$$

Now consider (4.58); using (3.82b) and (3.72a) results in,

$$[U(RL \to hv)]^* \begin{bmatrix} E_R^r \\ E_L^r \end{bmatrix} = \frac{1}{r}[T][\mathbf{S}_{BSA}]U(RL \to hv) \begin{bmatrix} E_R^i \\ E_L^i \end{bmatrix} \quad (4.82)$$

or,

$$\begin{bmatrix} E_R^r \\ E_L^r \end{bmatrix} = \frac{1}{r}[U(RL \to hv)]^t \mathbf{T}\mathbf{S}_{BSA} U(RL \to hv) \begin{bmatrix} E_R^i \\ E_L^i \end{bmatrix} \quad (4.83a)$$

$$= \frac{1}{r} U^t T [U^t]^{-1} U^t \mathbf{S}_{BSA} U \begin{bmatrix} E_R^i \\ E_L^i \end{bmatrix} \quad (4.83b)$$

$$= \frac{1}{r}[\mathbf{U}^t\mathbf{T}\mathbf{U}^*][\mathbf{S}^c_{\text{BSA}}]\begin{bmatrix} E^i_R \\ E^i_L \end{bmatrix} \tag{4.83c}$$

where (3.87) is used, i.e. $\mathbf{S}^c_{\text{BSA}} = \mathbf{U}^t\mathbf{S}_{\text{BSA}}\mathbf{U}$. Also, let $\mathbf{U}(RL \to hv) \equiv \mathbf{U}$. Substituting (4.81) into (4.83c) results in,

$$\begin{bmatrix} E^r_R \\ E^r_L \end{bmatrix} = \sqrt{\frac{Z_0 G}{2\pi}}\frac{1}{r^2}[\mathbf{U}^t\mathbf{T}\mathbf{U}^*][\mathbf{S}^c_{\text{BSA}}][\mathbf{U}^{-1}\mathbf{T}\mathbf{U}]\begin{bmatrix} M_R \\ M_L \end{bmatrix} \tag{4.84}$$

which is the circular basis equivalent of (4.59). The final step is to use (3.109) and assuming an ideal, dual-polarized antenna, the voltages in the RHC- and LHC-ports are,

$$\begin{bmatrix} V_R \\ V_L \end{bmatrix} = \frac{\lambda G}{4\pi r^2}[\mathbf{U}^t\mathbf{T}\mathbf{U}^*][\mathbf{S}^c_{\text{BSA}}][\mathbf{U}^{-1}\mathbf{T}\mathbf{U}]\begin{bmatrix} M_R \\ M_L \end{bmatrix} \tag{4.85}$$

The above equation is the single particle radar range equation inclusive of the transmission matrix. It is convenient to define the transmission matrix in the circular basis as $\mathbf{T}^c = [\mathbf{U}^{-1}][\mathbf{T}][\mathbf{U}]$. Also $[\mathbf{T}^c]^t = \mathbf{U}^t\mathbf{T}\mathbf{U}^*$, since $\mathbf{T}$ is symmetric and $\mathbf{U}$ is a unitary matrix ($\mathbf{U}^{-1} = [\mathbf{U}^*]^t$). Thus, (4.85) may be written in compact form as (McCormick and Hendry 1975),

$$\begin{bmatrix} V_R \\ V_L \end{bmatrix} = \frac{\lambda G}{4\pi r^2}[\mathbf{T}^c]^t[\mathbf{S}^c_{\text{BSA}}][\mathbf{T}^c]\begin{bmatrix} M_R \\ M_L \end{bmatrix} \tag{4.86}$$

Note that (4.43) gives the elements of $\mathbf{T}$ in terms of the eigenvalues $\lambda_{1,2}$ and the angle ϕ, which may be expressed as,

$$\mathbf{T} = \mathbf{R}(-\phi)\begin{bmatrix} e^{\lambda_1 r} & 0 \\ 0 & e^{\lambda_2 r} \end{bmatrix}\mathbf{R}(\phi) \tag{4.87}$$

where $\mathbf{R}(\phi)$ is the rotation matrix, see (2.93a). Hence, the matrix $\mathbf{T}^c$ can be identified with the constant canting model transmission matrix in (4.19a), except that $\phi \equiv -\beta$ (the negative sign is due to opposite ways of measuring the angle from the vertical direction). The circular basis $\mathbf{T}^c$ can be written as,

$$\mathbf{T}^c = \frac{1}{2}\begin{bmatrix} e^{\lambda_1 r} + e^{\lambda_2 r} & (e^{\lambda_1 r} - e^{\lambda_2 r})e^{-j2\phi} \\ (e^{\lambda_1 r} - e^{\lambda_2 r})e^{j2\phi} & e^{\lambda_1 r} + e^{\lambda_2 r} \end{bmatrix} \tag{4.88a}$$

$$= e^{(\lambda_1 + \lambda_2)(r/2)}\begin{bmatrix} \cosh[(\lambda_1 - \lambda_2)(r/2)] & e^{-j2\phi}\sinh[(\lambda_1 - \lambda_2)(r/2)] \\ e^{j2\phi}\sinh[(\lambda_1 - \lambda_2)(r/2)] & \cosh[(\lambda_1 - \lambda_2)(r/2)] \end{bmatrix} \tag{4.88b}$$

$$= \frac{e^{(\lambda_1 + \lambda_2)(r/2)}}{\sqrt{1 - (pe^{j\chi})^2}}\begin{bmatrix} 1 & -pe^{j(\chi - 2\phi)} \\ -pe^{j(\chi + 2\phi)} & 1 \end{bmatrix} \tag{4.88c}$$

where $pe^{j\chi}$ is defined as $\tanh[(\lambda_2 - \lambda_1)r/2]$, with $\lambda_{1,2}$ given in (4.41). Using (4.88c) in (4.86) and, for simplicity, assuming $\mathbf{S}^c_{\text{BSA}}$ corresponds to a sphere as given in (3.89), results in,

$$\begin{bmatrix} V_R \\ V_L \end{bmatrix} = \frac{\lambda G}{4\pi r^2} \frac{e^{(\lambda_1+\lambda_2)r} S}{[1 - (pe^{j\chi})^2]} \begin{bmatrix} -2pe^{j(\chi+2\phi)} & 1 + p^2 e^{j2\chi} \\ 1 + p^2 e^{j2\chi} & -2pe^{j(\chi-2\phi)} \end{bmatrix} \begin{bmatrix} M_R \\ M_L \end{bmatrix} \quad (4.89a)$$

$$= \frac{\lambda G}{4\pi r^2} e^{(\lambda_1+\lambda_2)r} S \begin{bmatrix} -e^{j2\phi} \sinh[(\lambda_2 - \lambda_1)r] & \cosh[(\lambda_2 - \lambda_1)r] \\ \cosh[(\lambda_2 - \lambda_1)r] & -e^{-j2\phi} \sinh[(\lambda_2 - \lambda_1)r] \end{bmatrix}$$

$$(4.89b)$$

As noted in relation to (4.15), the negative sign on $-e^{j2\phi} \sinh[(\lambda_2 - \lambda_1)r]$ is due to the choice of the horizontal reference for the unit circular base vectors made here, as opposed to the vertical reference in McCormick and Hendry (1975). Reverting to their choice of reference direction, (4.89b) becomes,

$$\begin{bmatrix} V_R \\ V_L \end{bmatrix} = \frac{\lambda G}{4\pi r^2} e^{(\lambda_1+\lambda_2)r} S$$

$$\times \begin{bmatrix} e^{j2\phi} \sinh[(\lambda_2 - \lambda_1)r] & \cosh[(\lambda_2 - \lambda_1)r] \\ \cosh[(\lambda_2 - \lambda_1)r] & e^{-j2\phi} \sinh[(\lambda_2 - \lambda_1)r] \end{bmatrix} \begin{bmatrix} M_R \\ M_L \end{bmatrix}$$

$$(4.90)$$

As alluded to in the introduction, the WSR–88D radars were designed, and the first ones constructed for the US National Weather Service, as circularly polarized radars transmitting, for example, right-hand circular polarization ($M_R = 1, M_L = 0$), but receiving only V_L in (4.90) via a single receiver. The decision not to use a dual-channel receiver system, i.e. not measuring both V_R and V_L, proved to be a major design flaw, as can easily be seen from (4.90). With $M_R = 1, M_L = 0$, it is clear that,

$$V_L = \frac{\lambda G}{4\pi r^2} e^{(\lambda_1+\lambda_2)r} S \cosh[(\lambda_2 - \lambda_1)r] \quad (4.91a)$$

$$= \frac{\lambda G}{4\pi r^2} e^{(\lambda_1+\lambda_2)r} S \cosh\left(\frac{A_{dp}r}{8.686} + jK_{dp}r\right) \quad (4.91b)$$

$$\approx \frac{\lambda G}{4\pi r^2} e^{(\lambda_1+\lambda_2)r} S \cos(K_{dp}r) \quad (4.91c)$$

where the assumption $A_{dp} \approx 0$ has been made in going from (4.91b) to (4.91c). Since uniform heavy rain at 3 GHz introduces a substantial differential propagation phase, $\Phi_{dp} = 2K_{dp}r$ (e.g. Fig. 4.8), it is clear that $V_L = 0$ whenever $K_{dp}r$ is an odd multiple of $n\pi/2$ ($n = 1, 3, 5, \ldots$) or, equivalently, whenever Φ_{dp} is an odd multiple of $n\pi$. For example, in Fig. 4.8, Φ_{dp} increases by $210°$ over a range of 30 km, so that $|V_L|^2$ from a range resolution volume located at 45 km would be reduced (or "attenuated") by a factor of around 15 due to the differential propagation phase alone. Once this was recognized, the WSR–88D systems were changed to operate at horizontal polarization.

The observable defined by McCormick and Hendry (1975) as $(W/W_2)_+$, where "+" refers to $M_R = 1, M_L = 0$, i.e. transmission of RHC polarization, is defined as,

$$\left(\frac{W}{W_2}\right)_+ = \frac{V_R^{10}(V_L^{10})^*}{|V_L^{10}|^2} = e^{j2\phi} \tanh[(\lambda_2 - \lambda_1)r] \quad (4.92a)$$

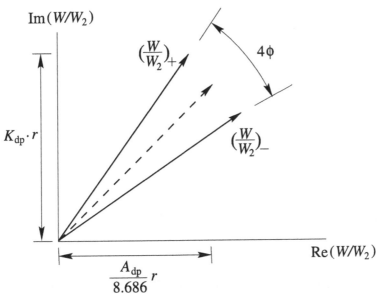

Fig. 4.11. Idealized complex plane plot of (W/W_2) corresponding to (4.94). The dashed line corresponds to $\phi = 0°$. The range r is from the radar to the spherical target in Fig. 4.7. The propagation medium is uniform throughout, from the radar to the spherical target. The approximation $\tanh^{-1}(W/W_2) \approx W/W_2$ is used to estimate A_{dp} and K_{dp}.

The observable $(W/W_2)_-$, where "$-$" refers to $M_R = 0$, $M_L = 1$, i.e. transmission of LHC polarization, is defined as,

$$\left(\frac{W}{W_2}\right)_- = \frac{V_L^{01}(V_R^{01})^*}{|V_R^{01}|^2} = e^{-j2\phi}\tanh[(\lambda_2 - \lambda_1)r] \qquad (4.92b)$$

Thus,

$$4\phi = \arg\left(\frac{W}{W_2}\right)_+ - \arg\left(\frac{W}{W_2}\right)_- \qquad (4.93)$$

It is reiterated that $\mathbf{S}_{BSA}^c$ in (4.86) has been chosen to correspond to a sphere, for simplicity, in order to focus solely on propagation effects, introduced by the uniform precipitation medium, between the radar and the sphere at range r, as illustrated in Fig. 4.7. From (4.92),

$$(\lambda_2 - \lambda_1)r = \left(\frac{A_{dp}}{8.686} + jK_{dp}\right)r = \tanh^{-1}\left[e^{-j2\phi}\left(\frac{W}{W_2}\right)_+\right] \qquad (4.94a)$$

$$= \tanh^{-1}\left[e^{j2\phi}\left(\frac{W}{W_2}\right)_-\right] \qquad (4.94b)$$

An idealized complex plane plot of $(W/W_2)_\pm$ is illustrated in Fig. 4.11, where the dashed line refers to (W/W_2) when $\phi = 0°$. The angle ϕ can be estimated via (4.93),

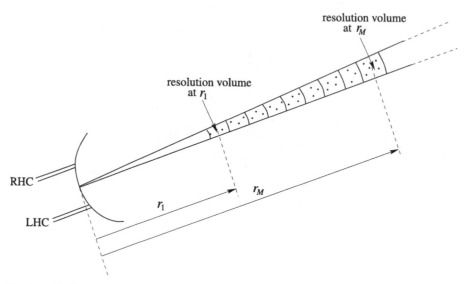

Fig. 4.12. Uniform precipitation medium between ranges r_1 and r_M. Resolution volumes are typically spaced $cT_0/2$-apart, where T_0 is the pulsewidth.

and then A_{dp} and K_{dp} can be estimated using (4.94). If $|W/W_2| \ll 1$, then using the approximation $\tanh^{-1}(W/W_2) \approx W/W_2$ yields the graphical values for A_{dp} and K_{dp} illustrated in Fig. 4.11.

Pulsed radars, as will be discussed in more detail in Chapter 5, are capable of measuring V_R and V_L as a function of r ($= ct/2$) due to their "range gating" capability, i.e. data are available at consecutive resolution volumes, typically spaced $cT_0/2$-apart along the range axis. If the precipitation medium is uniform between ranges r_1 and r_M (as illustrated in Fig. 4.12), then using the approximation $\tanh^{-1} x \approx x$, (4.94) becomes,

$$\left[\frac{W}{W_2}(r_M) \right]_{\pm} - \left[\frac{W}{W_2}(r_1) \right]_{\pm} \approx e^{\pm j2\phi} \left(\frac{A_{dp}}{8.686} + jK_{dp} \right) (r_M - r_1) \tag{4.95}$$

Figure 4.13 shows observations from Hendry and Antar (1984) in heavy rain at a frequency of 10 GHz. The "gate" numbers are marked from 2 to 8, which cover a range interval from 29.6–32.6 km. Instead of starting from the origin, as in the idealized illustration in Fig. 4.11, the complex plane plot starts at different locations for $(W/W_2)_+$ and $(W/W_2)_-$. Since the medium between the radar and the first range "gate" is empty space, the (W/W_2) there is obtained from (4.86) with no transmission matrix,

$$\begin{bmatrix} V_R(r_1) \\ V_L(r_1) \end{bmatrix} = \frac{\lambda G}{4\pi r_1^2} S^c_{BSA}(r_1) \begin{bmatrix} M_R \\ M_L \end{bmatrix} \tag{4.96a}$$

$$= \frac{\lambda G}{4\pi r_1^2} \begin{bmatrix} S_{RR}(r_1) & S_{RL}(r_1) \\ S_{LR}(r_1) & S_{LL}(r_1) \end{bmatrix} \begin{bmatrix} M_R \\ M_L \end{bmatrix} \tag{4.96b}$$

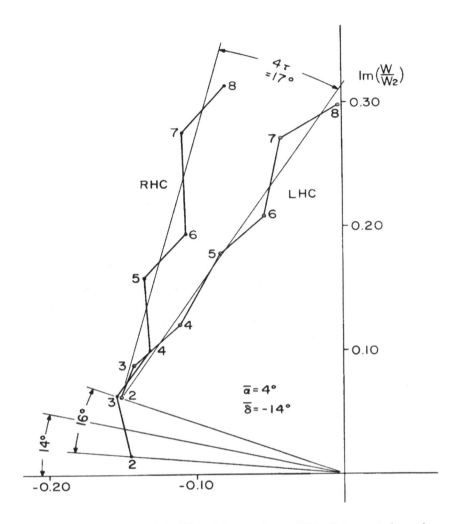

Fig. 4.13. Complex plane plot of $(W/W_2)_\pm$ in heavy rain at 10 GHz. Data are at a low radar elevation angle of 1.3°, with the range 29.6–32.6 km (gates 2–8 as marked). The angle $4\phi = 17°$ (see also, Fig. 4.11). The mean canting angle at gate 2 is $-4\bar{\beta} = \arg(W/W_2)_+ - \arg(W/W_2)_- = 16°$. The corresponding mean scattering differential phase is $\delta_c = \frac{1}{2}[\arg(W/W_2)_+ + \arg(W/W_2)_-] - 180° = -14°$. From Hendry and Antar (1984). In their convention $\tau \equiv \phi$ and $\bar{\alpha} \equiv -\bar{\beta}$. (© 1984 American Geophysical Union.)

and,

$$\left[\frac{W}{W_2}(r_1)\right]_+ = \frac{V_R^{10}(V_L^{10})^*}{|V_L^{10}|^2} = \frac{S_{RR}(r_1)S_{LR}^*(r_1)}{|S_{LR}(r_1)|^2} \tag{4.97a}$$

$$\left[\frac{W}{W_2}(r_1)\right]_- = \frac{V_L^{01}(V_R^{01})^*}{|V_R^{01}|^2} = \frac{S_{LL}(r_1)S_{RL}^*(r_1)}{|S_{RL}(r_1)|^2} \tag{4.97b}$$

The elements of $\mathbf{S}^c_{BSA}$ are given in (3.93) for a spheroid in the Rayleigh–Gans limit, but the form is valid even for non-Rayleigh scattering. From (3.96),

$$-4\beta = \arg[S_{RR}(r_1)S^*_{LR}(r_1)] - \arg[S_{LL}(r_1)S^*_{RL}(r_1)] \tag{4.98}$$

where β is the canting angle in the polarization plane. From (3.93) and (3.98),

$$\delta_c = \frac{1}{2}\{\arg[S_{RR}(r_1)S^*_{LR}(r_1)] + \arg[(S_{LL}(r_1)S^*_{RL}(r_1)]\} \tag{4.99}$$

where δ_c is the scattering differential phase for circular polarization. Rayleigh limit expressions for $S_{11,22}$ are given in (2.89). If ε_r is real, then $\delta_c = 0°$ for oblates and $180°$ for prolates, since the polarizabilities α and α_{zb} in (2.89) are real. For plates, $\delta_c = \arg[(\varepsilon_r - 1)/(\varepsilon_r + 1)]$ and for needles, $\delta_c = \arg[(\varepsilon_r - 1)/(\varepsilon_r + 3)] + 180°$. Since $\varepsilon_r = \varepsilon'_r - j\varepsilon''_r$ for $\exp(j\omega t)$ time dependence, the δ_c for plates will be slightly negative. For oblate raindrops in the Rayleigh limit $\delta_c \approx -2°$ and is nearly independent of eccentricity. For non-Rayleigh scattering by oblate drops, δ_c will vary from -2 to $-90°$ depending on both eccentricity and on the frequency (e.g. see Table 3 of McCormick and Hendry 1975). A distinguishing property of δ_c is that significant deviation from zero is indicative of the importance of non-Rayleigh scattering. It is related to $\delta_{co} = \arg(S_{vv}S^*_{hh})_{BSA}$, defined earlier in (4.65).

Thus, the displacement of the initial values of $(W/W_2)_\pm$ from the origin is related to back scattering from the first resolution volume. From Fig. 4.13 and eq. (4.93) it is clear that $4\phi = 17°$, as marked, using straight line fits to $(W/W_2)_\pm$ between gates 2 and 8. The slantwise progression of (W/W_2) with increasing range is indicative of both differential attenuation and a differential phase. For greater accuracy, it is better to estimate A_{dp} and K_{dp} (after estimating ϕ) without invoking the approximation $\tanh^{-1} x \approx x$ used in (4.95). Using (4.94), and identifying "gate" 8 with r_7 and "gate" 2 with r_1 in Fig. 4.13, results in an estimate of mean A_{dp} and K_{dp} over the range increment $(r_7 - r_1 = 3 \text{ km})$,

$$(\lambda_2 - \lambda_1) = \frac{A_{dp}}{8.686} + jK_{dp}$$

$$= \frac{1}{(r_7 - r_1)}\left\{\tanh^{-1}\left[e^{-j2\phi}\frac{W}{W_2}(r_7)\right]_+ - \tanh^{-1}\left[e^{-j2\phi}\frac{W}{W_2}(r_1)\right]_+\right\} \tag{4.100a}$$

$$= \frac{1}{(r_7 - r_1)}\left\{\tanh^{-1}\left[e^{j2\phi}\frac{W}{W_2}(r_7)\right]_- - \tanh^{-1}\left[e^{j2\phi}\frac{W}{W_2}(r_1)\right]_-\right\} \tag{4.100b}$$

The back scattering effect at gate 1 can be deduced from (4.98), which estimates the canting angle β ($\approx -4°$), which is marked in Fig. 4.13. The convention for positive β is indicated in Fig. 2.10, i.e. positive β is measured clockwise from the line segment QV. Thus, the estimate for ϕ ($= 4.25°$) over the propagation path from (4.93) is consistent with β ($= -4°$) obtained for the first range gate via (4.98). This is reasonable since the raindrops in the first scattering resolution volume are presumably similarly oriented to the drops along the propagation path. From (4.99), the scattering differential phase is

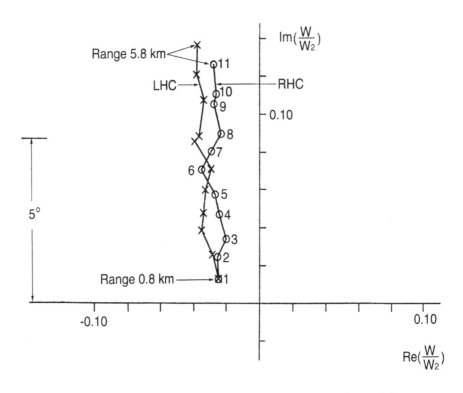

Fig. 4.14. Complex plane plot of $(W/W_2)_\pm$ at an elevation angle of $17°$ through heavy snow at 16.5 GHz. Range gate numbers are marked, with the range to first gate being 0.8 km and to the last being 5.8 km. A scale indicating a differential phase of $5°$ for a range increment of 1 km is included (see also, Fig. 4.11). Figure courtesy of A. Hendry.

estimated as $\delta_c \approx -14°$, indicating non-Rayleigh scattering by oblate drops in heavy rain at a frequency of 10 GHz.

A characteristic feature of the (W/W_2) plot in a non-attenuating medium is its vertical excursion parallel to the imaginary axis. Figure 4.14, from Hendry et al. (1981), shows observations of (W/W_2) at 16.5 GHz in heavy snow. The similarity of the RHC and LHC plots shows that $\phi \approx 0°$. Using (4.95) between range gates 1 and 11 (a distance of 5.0 km) gives $K_{dp} \approx 1.3° \ \text{km}^{-1}$. Since large snowflakes are of low density and are irregular in shape, it is unlikely that they contribute to the differential propagation phase. However, small ice crystals (e.g. plates or needles) fall in air with approximate orientation as illustrated in Fig. 2.7b, c, and they are of higher density. In a mixture of large snowflakes and tiny crystals, K_{dp} is mostly due to the tiny, oriented crystals; much like the K_{dp} in a rain/hail mixture which is mainly due to smaller, highly oriented raindrops. The excursion of (W/W_2) with increasing range parallel to the positive imaginary axis in Fig. 4.14 yields a positive K_{dp}, implying plate-like (or oblate) crystal orientation as in Fig. 2.7b, or prolate-like crystal orientation

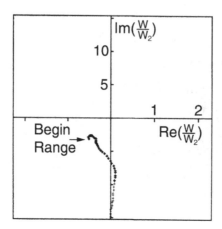

Fig. 4.15. Circularly polarized radar data at 10 GHz illustrating the downward vertical excursion in W/W_2 due to vertically oriented ice crystals, e.g. needles. The total differential phase (Φ_{dp}) is 15° over 9 km. This radar is operated by the New Mexico Institute of Mines and Technology, Socorro, NM. Figure courtesy of Paul Krehbiel.

in Fig. 2.7c. These orientation models will give positive K_{dp} at low radar elevation angles ($\theta_i \approx 90°$). If the prolate crystals are oriented "vertically" as in Fig. 2.7a, e.g. due to vertical in-cloud electric fields, then K_{dp} will be negative, and the $(W/.W_2)$ plot will have an excursion parallel to the negative imaginary axis with increasing range. Figure 4.15 shows an example of a (W/W_2) plot through a region of ice crystals aligned vertically by electric fields in the upper part of a storm.[3] Such "vertical" alignment of crystals can be used to detect the build up of in-cloud electric fields. The application of (W/W_2) plots to detect "vertically" oriented crystals and then subsequent re-alignment to "horizontal" orientation following a lightning discharge was pioneered by McCormick and Hendry (1979a). Subsequent studies using circular polarization by Metcalf (1993; 1995) at 3-GHz frequency and by Krehbiel et al. (1996) at 10 GHz have generally validated the Canadian results (Hendry and McCormick 1976; Hendry and Antar 1982). Caylor and Chandrasekar (1996) used a 3 GHz linearly polarized radar[4] to detect changes in K_{dp} from slightly negative to slightly positive values as the "vertically" oriented crystals re-aligned to conventional "horizontal" orientation following lightning discharge. Figure 4.16 shows direct measurement of K_{dp} before and after a lightning discharge.

4.5   Transmission-modified covariance matrix

So far the radar range equation was derived for a single, fixed particle within the scattering volume so that the effects of the transmission matrix could be clearly identified. However, the scattering volume itself is filled with precipitation. If the

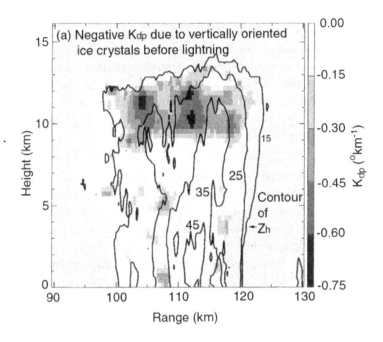

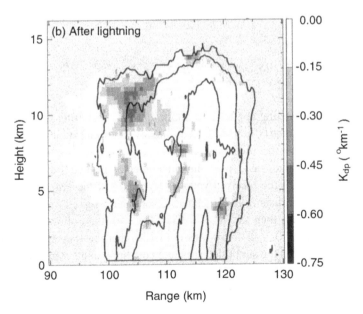

Fig. 4.16. Direct measurement of K_{dp} at 3 GHz using the NCAR/CP-2 radar in Florida illustrating changes in K_{dp} from: (*a*) negative values due to "vertically" oriented crystals, to (*b*) more conventional near-zero values following lightning discharge. From Caylor and Chandrasekar (1996).

precipitation is homogeneous, then the physical properties of the particles do not vary with range and hence the scattering volume is similar to the transmission medium. In the non-homogeneous case, the precipitation in the scattering volume can be different from the transmission medium (e.g. transition from rain to hail in a thunderstorm).

The radar range equation in (4.72) was derived for a diagonal transmission matrix, and for a general scattering matrix for a single, fixed particle. It can be written as,

$$
\begin{bmatrix} V_h \\ V_v \end{bmatrix} = \frac{\lambda G}{4\pi r^2} e^{-j2k_0 r} \begin{bmatrix} S'_{hh} & S'_{hv} \\ S'_{vh} & S'_{vv} \end{bmatrix}_{\text{BSA}} \begin{bmatrix} M_h \\ M_v \end{bmatrix}
$$
(4.101)

where S'_{BSA} can be considered to be the transmission-modified scattering matrix with elements as given in (4.72). To develop the more general case of the scattering volume being filled with randomly distributed precipitation particles, the concept of an "instantaneous" scattering matrix and a resulting ensemble-averaged covariance matrix using the ergodic principle, as discussed in Section 3.11, is invoked. Thus, S'_{BSA} can be considered to be the "instantaneous" transmission-modified scattering matrix with, feature vector $\Omega = [S'_{hh} \ \sqrt{2}S'_{hv} \ S'_{vv}]^t$, and Σ'_{BSA} the ensemble-averaged covariance matrix, to be with feature vector $\langle \Omega'(\Omega'^*)^t \rangle$. The elements of Σ'_{BSA} are

$$\langle n|S'_{hh}|^2 \rangle = \langle n|S_{hh}|^2 \rangle e^{-4k^h_{\text{im}}r}$$
(4.102a)

$$\sqrt{2}\langle nS'_{hh}S'^*_{hv} \rangle = \sqrt{2}\langle n|S_{hh}||S_{hv}|e^{j(\delta_{hh}-\delta_{hv})} \rangle e^{-2k^h_{\text{im}}r} e^{-(k^h_{\text{im}}+k^v_{\text{im}})r} e^{-j(\Phi_{dp}/2)}$$
(4.102b)

$$\langle nS'_{hh}S'^*_{vv} \rangle = \langle n|S_{hh}||S_{vv}|e^{j(\delta_{hh}-\delta_{vv})} \rangle e^{-2k^h_{\text{im}}r} e^{-2k^v_{\text{im}}r} e^{-j\Phi_{dp}}$$
(4.102c)

$$2\langle n|S'_{hv}|^2 \rangle = 2\langle n|S_{hv}|^2 \rangle e^{-2(k^h_{\text{im}}+k^v_{\text{im}})r}$$
(4.102d)

$$\sqrt{2}\langle nS'_{hv}S'^*_{vv} \rangle = \sqrt{2}\langle n|S_{hv}||S_{vv}|e^{j(\delta_{hv}-\delta_{vv})} \rangle e^{-2k^v_{\text{im}}r} e^{-(k^h_{\text{im}}+k^v_{\text{im}})r} e^{-j(\Phi_{dp}/2)}$$
(4.102e)

$$\langle n|S'_{vv}|^2 \rangle = \langle n|S_{vv}|^2 \rangle e^{-4k^v_{\text{im}}r}$$
(4.102f)

Note that ensemble averaging applies only to the intrinsic scattering matrix elements, since the transmission matrix elements have already been ensemble averaged as in (4.39) or (4.68). Thus, the transmission medium is assumed to be "steady" and represented by its average value over the time interval needed for estimating the second-order moments of the received voltages (typically of the order of 50–100 ms for conventional weather radars). If the scattering volume possesses "mirror" reflection symmetry about $\beta_0 = 0°$, then $\langle S_{hh}S^*_{hv} \rangle = \langle S_{vv}S^*_{hv} \rangle = 0$ [this follows from (3.227d, 3.227e) by setting $\beta_0 = 0°$]. It follows from (4.102) that the transmission-modified covariance matrix also has the simplified structure corresponding to "mirror" reflection symmetry,

$$
\Sigma'_{\text{BSA}} = e^{-2(k^h_{\text{im}}+k^v_{\text{im}})r}
\begin{bmatrix}
\langle n|S_{hh}|^2 \rangle e^{-2(k^h_{\text{im}}-k^v_{\text{im}})r} & 0 & \langle n|S_{hh}||S_{vv}|e^{j(\delta_{hh}-\delta_{vv})} \rangle e^{-j\Phi_{dp}} \\
0 & 2\langle n|S_{hv}|^2 \rangle & 0 \\
\langle n|S_{hh}||S_{vv}|e^{j(\delta_{vv}-\delta_{hh})} \rangle e^{j\Phi_{dp}} & 0 & \langle n|S_{vv}|^2 \rangle e^{2(k^h_{\text{im}}-k^v_{\text{im}})r}
\end{bmatrix}
$$
(4.103)

The differential reflectivity in this case is a generalization of (4.70) and is expressed as,

$$Z'_{dr} = 10\log_{10}\left(\frac{\langle n|S_{hh}|^2\rangle}{\langle n|S_{vv}|^2\rangle}\right) + 10\log_{10}e^{-4(k^h_{im}-k^v_{im})r} \tag{4.104a}$$

$$= Z^{int}_{dr} - 2A_{dp}r \tag{4.104b}$$

The linear depolarization ratio (LDR'_{vh}) from (4.103) is a generalization of (4.71),

$$LDR'_{vh} = 10\log_{10}\left(\frac{\langle n|S_{vh}|^2\rangle}{\langle n|S_{hh}|^2\rangle}\right) + 10\log_{10}e^{2(k^h_{im}-k^v_{im})r} \tag{4.105a}$$

$$= LDR^{int}_{vh} + A_{dp}r \tag{4.105b}$$

The copolar correlation coefficient is a generalization of (4.77),

$$|\rho_{co}|e^{j(\delta_{co}+\Phi_{dp})} = \frac{\langle n|S_{hh}||S_{vv}|e^{j(\delta_{vv}-\delta_{hh})}\rangle}{\sqrt{\langle n|S_{hh}|^2\rangle\langle n|S_{vv}|^2\rangle}}e^{j\Phi_{dp}} \tag{4.106}$$

At long radar wavelengths ($\lambda \approx 10$ cm) the specific differential attenuation due to rain is very small and the total differential attenuation corresponding to $\exp[2(k^h_{im} - k^v_{im})r]$ can be neglected. Also, $\delta_{co} \approx 0°$ for Rayleigh scatterers. Thus, (4.103) has the simplified structure,

$$\Sigma'_{BSA} = e^{-2(k^h_{im}+k^v_{im})r}\langle n|S_{hh}|^2\rangle \begin{bmatrix} 1 & 0 & \dfrac{|\rho_{co}|}{\sqrt{\mathcal{Z}_{dr}}}e^{-j\Phi_{dp}} \\ 0 & 2L & 0 \\ \dfrac{|\rho_{co}|}{\sqrt{\mathcal{Z}_{dr}}}e^{j\Phi_{dp}} & 0 & \dfrac{1}{\mathcal{Z}_{dr}} \end{bmatrix} \tag{4.107}$$

where the notation from Section 3.14 is used, i.e. $L = 10^{0.1(LDR_{vh})}$, $\mathcal{Z}_{dr} = 10^{0.1(Z_{dr})}$, and $|\rho_{co}|$ is defined in (3.231d). The effect of the transmission matrix is to impose a pure differential phase via Φ_{dp}. Also, L, $\mathcal{Z}_{dr}$ and $|\rho_{co}|$, which are representative of the scattering volume, are unaltered by the transmission matrix. To reiterate, the assumptions used here are: (i) a diagonal transmission matrix, which imposes a pure differential phase only; (ii) "mirror" reflection symmetry about $\beta_0 = 0°$ for the scattering volume; and (iii) Rayleigh scattering.

4.5.1 Transmission-modified covariance matrix: circular basis

A similar procedure is followed for deriving the transmission-modified covariance matrix in circular basis assuming "mirror" reflection symmetry about $\beta = \beta_0$. The ensemble-averaged transmission matrix in the circular basis is given by (4.88), with $(\lambda_1 - \lambda_2)$ and ϕ given in (4.50a, b). Starting from (4.86),

$$\begin{bmatrix} V_R \\ V_L \end{bmatrix} = \frac{\lambda G}{4\pi r^2}[T^c]^t[S^c_{BSA}][T^c]\begin{bmatrix} M_R \\ M_L \end{bmatrix} \tag{4.108a}$$

$$= \frac{\lambda G}{4\pi r^2} [\mathbf{T}^c]^t \begin{bmatrix} S_{RR} & S_{RL} \\ S_{LR} & S_{LL} \end{bmatrix}_{\text{BSA}} [\mathbf{T}^c] \begin{bmatrix} M_R \\ M_L \end{bmatrix} \tag{4.108b}$$

$$= \frac{\lambda G}{4\pi r^2} e^{(\lambda_1 + \lambda_2)r} \begin{bmatrix} S'_{RR} & S'_{RL} \\ S'_{LR} & S'_{LL} \end{bmatrix}_{\text{BSA}} \begin{bmatrix} M_R \\ M_L \end{bmatrix} \tag{4.108c}$$

Let $\mathbf{\Omega}' = [S'_{RR} \ \sqrt{2} S'_{RL} \ S'_{LL}]^t$ be the feature vector corresponding to the "instantaneous" transmission-modified scattering matrix in the circular basis. The elements of $\mathbf{S}^{'c}_{\text{BSA}}$ are obtained by substituting (4.88b) into (4.108b),

$$S'_{RR} = e^{j2\phi} \left[S_{RL} \sinh 2u + \frac{1}{2} \left(S_{RR} e^{-j2\phi} - S_{LL} e^{j2\phi} \right) \right.$$
$$\left. + \frac{\cosh 2u}{2} \left(S_{RR} e^{-j2\phi} + S_{LL} e^{j2\phi} \right) \right] \tag{4.109a}$$

$$S'_{RL} = S'_{LR} = S_{RL} \cosh 2u + \frac{\sinh 2u}{2} \left(S_{RR} e^{-j2\phi} + S_{LL} e^{j2\phi} \right) \tag{4.109b}$$

$$S'_{LL} = e^{-j2\phi} \left[S_{RL} \sinh 2u - \frac{1}{2} \left(S_{RR} e^{-j2\phi} - S_{LL} e^{j2\phi} \right) \right.$$
$$\left. + \frac{\cosh 2u}{2} \left(S_{RR} e^{-j2\phi} + S_{LL} e^{j2\phi} \right) \right] \tag{4.109c}$$

where $2u = (\lambda_1 - \lambda_2)r$, r being the range to the scattering volume. The ensemble-averaged covariance matrix follows from $\mathbf{\Sigma}^{'c}_{\text{BSA}} = \langle \mathbf{\Omega}' (\mathbf{\Omega}'^*)^t \rangle$. The unmodified covariance matrix, or the "intrinsic" covariance matrix, of the scattering volume is assumed to be given by (3.212), with "mirror" reflection symmetry about $\beta = \beta_0$. The elements of $\mathbf{\Sigma}^{'c}_{\text{BSA}}$ follow from (4.109) after straightforward but tedious algebra,

$$\langle n|S'_{RR}|^2 \rangle = \langle n|S'_{LL}|^2 \rangle$$
$$= \left(\langle \eta_c \rangle | \sinh 2u|^2 \right) + \left[\langle \eta_c |v^2| \rangle \frac{(1 - \rho_4)}{2} \right] + \left[\langle \eta_c |v^2| \rangle | \cosh 2u|^2 \frac{(1 + \rho_4)}{2} \right]$$
$$+ \left[2\rho_2 \operatorname{Re} \left(\sinh 2u \cosh 2u^* \langle \eta_c v^* \rangle \right) \right] \tag{4.110a}$$

$$\langle n|S'_{RL}|^2 \rangle = \langle n|S'_{LR}|^2 \rangle$$
$$= \left(\langle \eta_c \rangle | \cosh 2u|^2 \right) + \left[| \sinh 2u|^2 \langle \eta_c |v^2| \rangle \frac{(1 + \rho_4)}{2} \right]$$
$$+ \left[2\rho_2 \operatorname{Re} \left(\cosh 2u \sinh 2u^* \langle \eta_c v^* \rangle \right) \right] \tag{4.110b}$$

$$\langle n S'_{RR} S'^*_{RL} \rangle = e^{j2\phi} \left[\langle \eta_c \rangle \cosh 2u^* \sinh 2u + \rho_2 \left(\langle \eta_c v \rangle | \cosh 2u|^2 + \langle \eta_c v^* \rangle | \sinh 2u|^2 \right) \right.$$
$$\left. + \frac{(1 + \rho_4)}{2} \langle \eta_c |v|^2 \rangle \sinh 2u^* \cosh 2u \right] \tag{4.110c}$$

$$\langle n S'_{LL} S'^*_{RL} \rangle = e^{-j4\phi} \langle n S'_{RR} S'^*_{RL} \rangle \tag{4.110d}$$

$$\langle n S'_{RR} S'^*_{LL} \rangle = e^{j4\phi} \left[\langle \eta_c \rangle \left(|\sinh 2u|^2 + \frac{|\cosh 2u|^2}{2} - \frac{1}{2} \right) \right.$$
$$\left. + \frac{\rho_4}{2} \langle \eta_c |v|^2 \rangle \left(1 + |\cosh 2u|^2 \right) + 2\rho_2 \operatorname{Re}\left(\sinh 2u \cosh 2u^* \langle \eta_c v^* \rangle \right) \right]$$

$$(4.110e)$$

Note that $\phi = -\beta_0$; ρ_2, ρ_4, $\langle \eta_c \rangle$, $\langle \eta_c |v|^2 \rangle$, and $\langle \eta_c v \rangle$ are "intrinsic" values representing the scattering volume and are defined in (3.201)–(3.204). The transmission-modified covariance matrix is seen to be considerably more complicated than the unmodified version given in (3.212).

The observables $(W/W_2)_+$ and $(W/W_2)_-$, defined earlier in (4.92), can now be generalized,

$$\left(\frac{W'}{W'_2} \right)_+ = \frac{\langle n S'_{RR} S'^*_{RL} \rangle}{\langle n |S'_{RL}|^2 \rangle}$$
$$= e^{j2\phi} \left[\tanh 2u + \rho_2 \left(\bar{v} + \overline{v^*} |\tanh 2u|^2 \right) + \frac{(1 + \rho_4)}{2} \overline{v^2} \tanh 2u^* \right]$$

$$(4.111)$$

where the approximation $\langle n |S'_{RL}|^2 \rangle \approx \langle \eta_c \rangle |\cosh 2u|^2$ has been used, and $\bar{v}$ and $\overline{v^2}$ are defined in (3.214a, b). Also,

$$\left(\frac{W'}{W'_2} \right)_- = \frac{\langle n S'_{LL} S'^*_{LR} \rangle}{\langle n |S'_{LR}|^2 \rangle} = e^{-j4\phi} \left(\frac{W'}{W'_2} \right)_+$$

$$(4.112)$$

The mean canting angle of the propagation medium is obtained from,

$$4\phi = -4\beta_0 = \arg\left(\frac{W'}{W'_2} \right)_+ - \arg\left(\frac{W'}{W'_2} \right)_-$$

$$(4.113)$$

which is a significant advantage of using circular polarization, but the radiated polarization must be "switched" from RHC to LHC preferably on a pulse-to-pulse basis, i.e. polarization agility on transmission.

If the precipitation is homogeneous between ranges r_1 and r_m then using (4.111), (4.95) can be generalized as,

$$\left[\frac{W'}{W'_2}(r_M) \right]_\pm - \left[\frac{W'}{W'_2}(r_1) \right]_\pm = e^{\pm j2\phi} \left[\tanh[(\lambda_1 - \lambda_2)r_M] - \tanh[(\lambda_1 - \lambda_2)r_1] \right.$$
$$+ \rho_2 \overline{v^*} \left\{ |\tanh[(\lambda_1 - \lambda_2)r_M]|^2 - |\tanh[(\lambda_1 - \lambda_2)r_1]|^2 \right\}$$
$$\left. + \frac{(1 + \rho_4)}{2} \overline{v^2} \left\{ \tanh[(\lambda_1 - \lambda_2)^* r_M] - \tanh[(\lambda_1 - \lambda_2)^* r_1] \right\} \right]$$

$$(4.114)$$

However, the extraction of A_{dp} and K_{dp} from the complex plane plot of (W'/W'_2), as described by McCormick and Hendry (1975), involves further assumptions; in particular, the terms in (4.114) involving $\overline{v^*}$ and $\overline{v^2}$ must be negligible compared with $\{\tanh[(\lambda_1 - \lambda_2)]r_M - \tanh[(\lambda_1 - \lambda_2)]r_1\}$. An alternative method of calculating Φ_{dp} from circular observables is given in Section 4.6. Finally, if the approximation $\tanh x \approx x$ holds, then (4.114) reduces to,

$$\left[\frac{W'}{W'_2}(r_M)\right]_{\pm} - \left[\frac{W'}{W'_2}(r_1)\right]_{\pm} = e^{\pm j2\phi}(\lambda_1 - \lambda_2)(r_M - r_1) \tag{4.115a}$$

$$= -e^{\pm j2\phi}\left(\frac{A_{dp}}{8.686} + jK_{dp}\right)(r_M - r_1) \tag{4.115b}$$

The negative sign on the right-hand side of (4.115b) is a result of the horizontal reference direction for the circular basis vectors, as opposed to the choice of vertical reference direction made by McCormick and Hendry (1975). The accuracy of (4.115) will improve if $(r_M - r_1)$ is made as small as possible. Typical research weather radars provide range samples spaced a few hundred meters apart. The advantage of circular polarization, in addition to estimating ϕ, is that A_{dp} and K_{dp} can be independently estimated from the complex plane plot of (W'/W'_2) along uniform range sections of precipitation, and several examples were provided earlier in Section 4.4.

The other circular observables, such as circular depolarization ratio (CDR) and ORTT, defined in (3.214a, c) are adversely affected by the transmission medium, as is obvious from (4.110). From (4.110a, b), the transmission-modified CDR is,

$$10^{0.1(\text{CDR}')} = \frac{\langle n|S'_{RR}|^2\rangle}{\langle n|S'_{RL}|^2\rangle} \approx |\tanh 2u|^2 + \overline{v^2} + 2\,\text{Re}\left[\tanh(2u)\overline{v^*}\right] \tag{4.116}$$

where the approximation $\langle n|S'_{RL}|^2\rangle \approx \langle \eta_c\rangle|\cosh 2u|^2$ has been used, along with ρ_2 and ρ_4 being close to unity, e.g. as in a highly oriented rain medium. If in addition, Rayleigh scattering is valid (i.e. $\bar{v}$ is real) and $2u = (\lambda_1 - \lambda_2)r = -jK_{dp}r$, which implies that the propagation medium imposes a pure differential phase only, then (4.116) reduces to,

$$10^{0.1(\text{CDR}')} = \frac{\langle n|S'_{RR}|^2\rangle}{\langle n|S'_{RL}|^2\rangle} \approx \overline{v^2} + \tan^2(K_{dp}r) \tag{4.117}$$

Thus, the intrinsic CDR (which is related to $\overline{v^2}$) is increased by $\tan^2(K_{dp}r) = \tan^2(\Phi_{dp}/2)$. Figure 4.8 showed an example of Φ_{dp} in heavy rain at 3 GHz. For example, if $\Phi_{dp} = 120°$ and the intrinsic CDR is assumed to be -15 dB (or $\overline{v^2} = 10^{-1.5}$), then CDR' would be 4.8 dB.

A characteristic feature of the transmission-modified CDR and ORTT observables is that they will exhibit an increasing trend with Φ_{dp} (up to $\Phi_{dp} = 180°$), and thus "correction" procedures must be used to retrieve the intrinsic values (e.g. Bebbington et al. 1987; Jameson and Davé 1988). Figure 4.17a, b shows profiles of CDR' and ORTT' as a function of range, together with their "corrected" values. These data were acquired by the Alberta Research Council's S-band circularly polarized radar system[5] (Torlaschi and Holt 1993). Figure 4.17c shows the Φ_{dp} recovered from the circularly

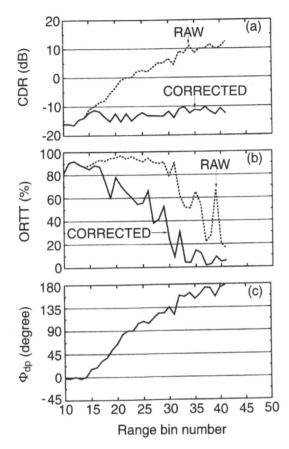

Fig. 4.17. Range profile of radar observations at 3 GHz from the Alberta Research Council's circularly polarized radar, Penhold, Alberta. (*a*) Measured and "corrected" values of the circular depolarization ratio (CDR); (*b*) similar results for the correlation coefficient, (ORTT); and (*c*) differential propagation phase (Φ_{dp}) using (4.127). From Torlaschi and Holt (1993, ©1993 IEEE).

polarized observations, using a method described in Section 4.6 (Holt 1988). Note how CDR' increases rapidly with Φ_{dp} in the range 10–30 km approximately in accord with (4.117).

4.6 Relation between linear and circular radar observables in the presence of propagation effects

In Section 3.14, the relation between circular and linear polarization radar observables was formulated in the absence of any propagation effects. In this section, the extraction of some linear radar observables from circular radar observables will be considered within the context of the scattering volume and the transmission medium, both of which are assumed to satisfy "mirror" reflection symmetry about $\beta_0 = 0°$. Simplifications are possible if the transmission medium imposes a pure differential phase shift ($\lambda_1 - \lambda_2 \approx -jK_{dp}$) and if Rayleigh scattering is assumed. In Section 4.5, the elements of the transmission-modified covariance matrix were derived in the circular basis. In this section, these same elements will be derived in terms of a transmission-modified covariance matrix in the linear basis. The methodology is similar to the earlier derivation, i.e. construct: (i) "instantaneous" transmission-modified scattering matrix, (ii) a feature vector ($\boldsymbol{\Omega}$), and (iii) the ensemble-averaged covariance matrix $\boldsymbol{\Sigma} = \langle\boldsymbol{\Omega}\boldsymbol{\Omega}^{*t}\rangle$, using ergodic principles. Note that ensemble averaging only applies to the intrinsic scattering terms.

Starting with (4.85) and using the relation $\mathbf{S}^c_{\text{BSA}} = \mathbf{U}^t\mathbf{S}_{\text{BSA}}\mathbf{U}$, see (3.87),

$$\begin{bmatrix} V_R \\ V_L \end{bmatrix} = \frac{\lambda G}{4\pi r^2}[\mathbf{U}^t\mathbf{T}\mathbf{U}^*][\mathbf{U}^t\mathbf{S}_{\text{BSA}}\mathbf{U}][\mathbf{U}^{-1}\mathbf{T}\mathbf{U}]\begin{bmatrix} M_R \\ M_L \end{bmatrix} \tag{4.118a}$$

$$= \frac{\lambda G}{4\pi r^2}\mathbf{U}^t\mathbf{T}\mathbf{S}_{\text{BSA}}\mathbf{T}\mathbf{U}\begin{bmatrix} M_R \\ M_L \end{bmatrix} \tag{4.118b}$$

where the transmission matrix ($\mathbf{T}$) is given in (4.87) and simplified for the case of ϕ ($= -\beta_0$) being zero, and $\mathbf{U} \equiv \mathbf{U}(RL \to hv)$ is given in (3.73a). With these substitutions, (4.118b) reduces to,

$$\begin{bmatrix} V_R \\ V_L \end{bmatrix} = \frac{\lambda G}{4\pi r^2}\frac{1}{2}$$

$$\begin{bmatrix} S_{hh}e^{2\lambda_1 r} + j2S_{hv}e^{(\lambda_1+\lambda_2)r} - S_{vv}e^{2\lambda_2 r} & S_{hh}e^{2\lambda_1 r} + S_{vv}e^{2\lambda_2 r} \\ S_{hh}e^{2\lambda_1 r} + S_{vv}e^{2\lambda_2 r} & S_{hh}e^{2\lambda_1 r} - j2S_{hv}e^{(\lambda_1+\lambda_2)r} - S_{vv}e^{2\lambda_2 r} \end{bmatrix}\begin{bmatrix} M_R \\ M_L \end{bmatrix} \tag{4.119a}$$

$$= \frac{\lambda G}{4\pi r^2}\begin{bmatrix} S'_{RR} & S'_{RL} \\ S'_{RL} & S'_{LL} \end{bmatrix}\begin{bmatrix} M_R \\ M_L \end{bmatrix} \tag{4.119b}$$

The feature vector is $\boldsymbol{\Omega}' = [S'_{RR}\ \sqrt{2}S'_{RL}\ S'_{LL}]^t$ and $\boldsymbol{\Sigma}'^c_{\text{BSA}} = \langle\boldsymbol{\Omega}'(\boldsymbol{\Omega}'^*)^t\rangle$. It will be convenient to express $\lambda_{1,2}$ as in (4.67, 4.68),

$$\lambda_1 = -jk^h_{\text{eff}} = -jk_0 - jk^h_{p,\text{re}} - k^h_{\text{im}} \tag{4.120a}$$

$$\lambda_2 = -jk^v_{\text{eff}} = -jk_0 - jk^v_{p,\text{re}} - k^v_{\text{im}} \tag{4.120b}$$

with $\Phi_{dp} = 2(k^h_{p,\text{re}} - k^v_{p,\text{re}})r$. The elements of $\boldsymbol{\Sigma}'^c_{\text{BSA}}$ can be expressed as,

$$\langle n|S'_{RR}|^2\rangle = \frac{e^{-2(k^h_{\text{im}}+k^v_{\text{im}})r}}{4}\left[\langle n|S_{hh}|^2\rangle e^{-2(k^h_{\text{im}}-k^v_{\text{im}})r} + 4\langle n|S_{hv}|^2\rangle\right.$$

$$\left. + \langle n|S_{vv}|^2\rangle e^{2(k^h_{\text{im}}-k^v_{\text{im}})r} - 2\,\text{Re}(\langle nS_{hh}S^*_{vv}\rangle e^{-j\Phi_{dp}})\right] \tag{4.121a}$$

$$2\langle n|S'_{RL}|^2\rangle = \frac{e^{-2(k^h_{\text{im}}+k^v_{\text{im}})r}}{2}\left[\langle n|S_{hh}|^2\rangle e^{-2(k^h_{\text{im}}-k^v_{\text{im}})r}\right.$$

$$+ \langle n|S_{vv}|^2\rangle e^{2(k_{im}^h - k_{im}^v)r} + 2\,\mathrm{Re}\big(\langle n S_{hh} S_{vv}^*\rangle e^{-j\Phi_{dp}}\big)\Big] \tag{4.121b}$$

$$\sqrt{2}\langle n S_{RR}' S_{RL}'^*\rangle = \frac{\sqrt{2}e^{-2(k_{im}^h + k_{im}^v)r}}{4}\Big[\langle n|S_{hh}|^2\rangle e^{-2(k_{im}^h - k_{im}^v)r}$$
$$- \langle n|S_{vv}|^2\rangle e^{2(k_{im}^h - k_{im}^v)r} + j2\,\mathrm{Im}\big(\langle n S_{hh} S_{vv}^*\rangle e^{-j\Phi_{dp}}\big)\Big] \tag{4.121c}$$

$$\langle n|S_{RR}' S_{LL}'^*\rangle = \frac{e^{-2(k_{im}^h + k_{im}^v)r}}{4}\Big[\langle n|S_{hh}|^2\rangle e^{-2(k_{im}^h - k_{im}^v)r} - 4\langle n|S_{hv}|^2\rangle$$
$$+ \langle n|S_{vv}|^2\rangle e^{2(k_{im}^h - k_{im}^v)r} - 2\,\mathrm{Re}\big(\langle n S_{hh} S_{vv}^*\rangle e^{-j\Phi_{dp}}\big)\Big] \tag{4.121d}$$

Symmetry properties of the intrinsic covariance matrix in $(\hat{h}, \hat{v})$ basis have been used as described in Section 3.12. It also follows that $\langle n|S_{RR}'|^2\rangle = \langle n|S_{LL}'|^2\rangle$ and $\langle n S_{RR}' S_{RL}'^*\rangle = \langle n S_{LL}' S_{RL}'^*\rangle$. Note how the elements of $\mathbf{\Sigma}_{BSA}'$ in (4.103) are incorporated into $\mathbf{\Sigma}_{BSA}'^{c}$ given above. If the transmission medium imposes a pure differential phase only (i.e. $k_{im}^h = k_{im}^v = 0$), then the transmission-modified covariance matrix elements in the circular basis are affected by a pure differential phase shift via the complex term $\langle n S_{hh} S_{vv}^*\rangle \exp(-j\Phi_{dp})$ only. Holt (1988) has derived a procedure to extract $\mathcal{F}_{dr}$ (or Z_{dr}) and Φ_{dp} from the elements of the transmission-modified circular covariance matrix given in (4.121). Defining,

$$W_1' = \langle n|S_{RR}'|^2\rangle \tag{4.122a}$$
$$W_2' = \langle n|S_{RL}'|^2\rangle \tag{4.122b}$$
$$W' = \langle n S_{RR}' S_{RL}'^*\rangle \tag{4.122c}$$

it follows from (4.121a, b) that,

$$\frac{W_2' - W_1'}{2} = \frac{e^{-2(k_{im}^h + k_{im}^v)r}}{2}\Big[\mathrm{Re}\big(\langle n S_{hh} S_{vv}^*\rangle e^{-j\Phi_{dp}}\big) - \langle n|S_{hv}|^2\rangle\Big] \tag{4.123a}$$

$$\frac{W_2' + W_1'}{2} = \frac{e^{-2(k_{im}^h + k_{im}^v)r}}{4}\Big[\langle n|S_{hh}|^2\rangle e^{-2(k_{im}^h - k_{im}^v)r}$$
$$+ \langle n|S_{vv}|^2\rangle e^{2(k_{im}^h - k_{im}^v)r} + 2\langle n|S_{hv}|^2\rangle\Big] \tag{4.123b}$$

From (4.121c) it follows that,

$$j\,\mathrm{Im}\,W' = j\frac{e^{-2(k_{im}^h + k_{im}^v)r}}{2}\,\mathrm{Im}\big(\langle n S_{hh} S_{vv}^*\rangle e^{-j\Phi_{dp}}\big) \tag{4.124a}$$

$$\mathrm{Re}\,W' = \frac{e^{-2(k_{im}^h + k_{im}^v)r}}{4}\Big[\langle n|S_{hh}|^2\rangle e^{-2(k_{im}^h - k_{im}^v)r} - \langle n|S_{vv}|^2\rangle e^{2(k_{im}^h - k_{im}^v)r}\Big] \tag{4.124b}$$

Combining (4.123a) and (4.124b) and noting that $\delta_{co} = \arg\langle n S_{hh}^* S_{vv}\rangle$ gives,

$$\frac{W_2' - W_1'}{2} - j\,\mathrm{Im}\,W' = \frac{e^{-2(k_{im}^h + k_{im}^v)r}\langle n|S_{hh}|^2\rangle}{2}\left[\frac{|\rho_{co}|e^{j(\delta_{co} + \Phi_{dp})}}{\sqrt{\mathcal{F}_{dr}}} - L\right] \tag{4.125}$$

while combining (4.123b) with (4.124b) gives,

$$\frac{W_2' + W_1'}{2} + \text{Re } W' = \frac{e^{-2(k_{im}^h + k_{im}^v)r}}{2} \left[\langle n|S_{hh}|^2 \rangle e^{-2(k_{im}^h - k_{im}^v)r} + \langle n|S_{hv}|^2 \rangle \right] \qquad (4.126a)$$

$$\frac{W_2' + W_1'}{2} - \text{Re } W' = \frac{e^{-2(k_{im}^h + k_{im}^v)r}}{2} \left[\langle n|S_{vv}|^2 \rangle e^{2(k_{im}^h - k_{im}^v)r} + \langle n|S_{hv}|^2 \rangle \right] \qquad (4.126b)$$

From (4.125), if the approximation $L \ll |\rho_{co}|/\sqrt{\mathcal{Z}_{dr}}$ holds (see Table 3.1), which is an excellent approximation at frequencies ≤ 10 GHz, then,

$$\arg\left(\frac{W_2' - W_1'}{2} - j \text{ Im } W' \right) = \delta_{co} + \Phi_{dp} = \Psi_{dp} \qquad (4.127)$$

For Rayleigh scattering, $\delta_{co} \approx 0°$, so that Φ_{dp} can be derived from W_1', W_2', and W'. From (4.126a, b),

$$\frac{W_2' + W_1' + 2 \text{ Re } W'}{W_2' + W_1' - 2 \text{ Re } W'} = \mathcal{Z}_{dr} e^{-4(k_{im}^h - k_{im}^v)r} \left[\frac{1 + Le^{2(k_{im}^h - k_{im}^v)r}}{1 + L\mathcal{Z}_{dr} e^{-2(k_{im}^h - k_{im}^v)r}} \right] \qquad (4.128)$$

While L and $L\mathcal{Z}_{dr}$ are typically much smaller than unity, the term $L \exp[2(k_{im}^h - k_{im}^v)r]$ in the numerator can increase with range. In decibels, it can be written as $10 \log_{10} L + A_{dp}r$, see (4.71b, c). At frequencies around 3 GHz, and for propagation through a rain medium, A_{dp} is small, and for normal propagation paths, the term $L \exp[2(k_{im}^h - k_{im}^v)r]$ can be neglected in comparison to unity. Thus, (4.128) reduces to,

$$\mathcal{Z}_{dr} = \left(\frac{W_2' + W_1' + 2 \text{ Re } W'}{W_2' + W_1' - 2 \text{ Re } W'} \right) e^{4(k_{im}^h - k_{im}^v)r} \qquad (4.129a)$$

or,

$$Z_{dr} = 10 \log_{10} \left(\frac{W_2' + W_1' + 2Re W'}{W_2' + W_1' - 2Re W'} \right) + 2A_{dp}r \qquad (4.129b)$$

For propagation through rain media at microwave frequencies, Holt (1988) and Bringi et al. (1990) have shown that A_{dp} can be linearly related to K_{dp}, and this provides for "correction" of Z_{dr} due to differential attenuation (see also, Section 7.4).

Some general comments are appropriate here. In the absence of media with significant propagation effects, e.g. rain or aligned ice crystals, the form of the circular covariance matrix in (3.212) is valid for "mirror" reflection symmetry about $\beta = \beta_0$, and the measurements can be used to separately estimate β_0, the circular back scatter cross section per unit volume ($\langle \eta_c \rangle$), the "shape" factor $[|\bar{\nu}|/(\bar{\nu^2})^{1/2}]$, the orientation parameter (ρ_2), and the differential scattering phase (δ_c). These are indeed microphysically important measurements and can be used for hydrometeor-type classification. The theory is well-developed due to the pioneering work of McCormick, Hendry, and their coworkers at the National Research Council of Canada, who obtained experimental data (though not complete in the sense of measuring all elements of the covariance matrix) at 3, 10, and 16.5 GHz. On the other hand, the elements of the covariance matrix in the $(\hat{h}, \hat{v})$ basis, e.g. (3.227), are not simply related to the intrinsic scattering properties such as mean canting angle β_0, or the "shape" and orientation factors. While in principle, microphysically important circular parameters can be derived from the linear covariance matrix elements, practical system considerations, such as deviation from the assumption of "mirror" reflection symmetry (as well as individual particle symmetry) and complications due to a mixture of hydrometeor types, can lead to unacceptable ambiguities.

In the presence of significant propagation effects, the transmission-modified circular covariance matrix is not so simply related to the intrinsic hydrometeor scattering properties, and "correction" for the effects of differential attenuation and the differential phase caused by the transmission-matrix becomes complicated. However, when the propagation path is uniform (i.e. intrinsic scattering properties are constant along the path), the complex plane plot of $(W'/W'_2)_\pm$ yields the mean tilt angle (ϕ) of the propagation medium, as well as separate estimates of A_{dp} and K_{dp} (subject to assumptions regarding symmetry and neglecting certain cross-products of $\bar{v}, \overline{v^2}$). While the validity of these assumptions is not entirely known in oriented media such as rain and ice crystals, impressive experimental data have been reported on estimation of A_{dp}, K_{dp}, and shifts in the mean tilt angle of ice crystals due to the build-up and decay of in-cloud electric fields.

The transmission-modified covariance matrix in the linear basis $(\hat{h}, \hat{v})$ for the case of $\phi = 0$ or 90°, preserves the structure of the unmodified covariance matrix and, in particular, the intrinsic scattering parameters such as Z_{dr} and LDR are simply modified by differential attenuation, while the copolar correlation coefficient is simply modified by the differential propagation phase. Simple algorithms permit the direct estimation of Φ_{dp}, but separate estimation of A_{dp} is not possible. "Correction" of Z_{dr} due to differential attenuation is discussed in Section 7.4. In highly oriented media such as rain $(\phi \approx 0°)$ or ice crystals $(\phi \approx 0$ or 90°), the simple and direct measurement of Φ_{dp} has proven to be a significant advantage for the linear $(\hat{h}, \hat{v})$ basis.[6] Also, at frequencies near 3 GHz, the "correction" of Z_{dr} and LDR due to differential attenuation in rain is generally negligible, and approximate corrections based on the linear relation $A_{dp} = \alpha K_{dp}$ suffice in practice. Thus, even though the intrinsic scattering parameters, Z_{dr}, LDR, and $|\rho_{co}|$ bear a complicated relation to the hydrometeor "shape" and orientation factors, the addition of K_{dp} has proven advantageous in the quantitative measurement of rainfall, the detection of ice crystal orientation (from $\phi \approx 0-90°$ via changes in the sign of K_{dp}), and improved prospects for hydrometeor classification using the five-parameter set $(\eta_{hh}, Z_{dr}, \text{LDR}, |\rho_{co}|, \text{and } K_{dp})$. These have become active research areas (hydrometeor classification is discussed in Chapter 7, while rain rate algorithms are discussed in Chapter 8).

If the mean tilt angle (ϕ) of the propagation medium is significantly different from zero or 90°, then the linear basis must be rotated by this angle to maintain the simple structure of the transmission-modified covariance matrix. In the case of "mirror" reflection symmetry about the mean tilt angle, the transmission-modified covariance matrix has four zero elements corresponding to correlation terms such as $\langle n S_{h'h'} S_{h'v'}^* \rangle = 0$. Very little experimental data exist using rotated linear bases (Hendry et al. 1987; Antar and Hendry 1987). They found normalized correlation values of $\langle n S_{h'h'} S_{h'v'}^* \rangle$, which reached minima of around 0.4 in rain and ≤ 0.1 in snow, and a "bright" band with corresponding mean tilt angles of near 0, 20, and 0°, respectively. In theory, an accurate measurement of the transmission-modified covariance matrix in the linear $(\hat{h}, \hat{v})$ basis can be transformed into the "characteristic" linear basis in which the correlation terms corresponding to $\langle n S_{h'h'} S_{h'v'}^* \rangle$ attain a minimum value, from which the mean tilt angle may be determined (Hubbert and Bringi 1996). In this "characteristic" basis, the transmission-modified covariance matrix will retain a simplified structure, as in (4.103). For Rayleigh scattering and negligible differential attenuation, the matrix is further simplified, as in (4.107). The Φ'_{dp} in this basis is easily obtained as $\arg\langle n S_{hh}'^* S_{vv}' \rangle$, and this permits the unmodified covariance matrix to be estimated in the "characteristic" basis. The Z'_{dr} and LDR' in this basis follow directly as they are unaffected by the pure differential phase. Several options are now available to estimate the circular parameters. For example, transformation of the unmodified covariance matrix from the "characteristic" basis to one that is rotated by 45° from this basis will yield estimates of ρ_4 via (3.216). Of course, in theory, transformation to a circular basis is also possible, so that the "shape" factor $\bar{v}/(\overline{v^2})^{1/2}$ can be estimated. The practical application of this procedure is likely to be limited by weak signals in the cross-polar channel (i.e. by a low signal-to-noise ratio), and by system polarization errors when the mean tilt angle is close to 0 or 90° (system polarization errors are discussed in Section 6.2).

4.7   Measurements in a "hybrid" basis

The application of polarimetric radar measurements in the linear $(\hat{h}, \hat{v})$ basis to improving rain rate estimates, and in hail detection using Z_{dr} and Φ_{dp}, can be considered as a major advancement in radar meteorology. The physical principles and experimental results will be discussed in more detail in Chapters 7 and 8. It is likely that the next major upgrade to the US National Weather Service radars (WSR–88D) will involve modification for polarimetric capability, principally for measuring Z_{dr}, Ψ_{dp}, and $|\rho_{co}|$ (Doviak et al. 2000). The conventional measurement technique involves polarization agility, i.e. the ability to change the radiated polarization state from horizontal (h) to vertical (v) on alternate pulses (Seliga and Bringi 1976). This involves high power switching (via electronically controlled ferrite circulator switches, or mechanical waveguide switches) if a single transmitter is used, as will be discussed in detail in Section 6.1.1. Since the primary operational application is towards rainfall measurement and hail detection, and because the rain medium appears to satisfy "mirror" reflection symmetry about $\phi = 0°$, leading to a diagonal transmission matrix, while the hail medium appears to satisfy polarization plane isotropy, it seems reasonable to expect that radiation of fixed linear polarization oriented at 45°, with simultaneous reception of the h and v polarized back scattered signal components, will be sufficient to measure Z_{dr}, Ψ_{dp}, and $|\rho_{co}|$. The slant 45° scheme was also originally suggested by Seliga and Bringi (1976) for the measurement of Z_{dr}. Further, it can be generalized that radiation of any fixed polarization ellipse with a unit polarization vector $\hat{e}_i = \cos\varepsilon\,\hat{h}_i + \sin\varepsilon\,e^{j\theta}\,\hat{v}_i$, with $\varepsilon = 45°$ and arbitrary θ, can also be used. These polarization states correspond to points along the two circles of longitude, with $2\psi = \pm\pi/2$, on the Poincaré sphere, illustrated in Fig. 3.4. These polarization states have equal power in the h and v components, with an arbitrary (but fixed) phase difference in θ. The radiated electric field is modified by the transmission and scattering matrices, as $\mathbf{T}\mathbf{S}_{BSA}\mathbf{T}$, see (4.59). The radar measures the coherency matrix $\mathbf{J}$, see (3.146), corresponding to a received electric field with components (E_h^r, E_v^r),

$$\mathbf{J} = \langle \mathbf{E}^r [\mathbf{E}^r]^{t*}\rangle = \begin{bmatrix} \langle E_h^r E_h^{r*}\rangle & \langle E_h^r E_v^{r*}\rangle \\ \langle E_v^r E_h^{r*}\rangle & \langle E_v^r E_v^{r*}\rangle \end{bmatrix} \tag{4.130}$$

The Z_{dr}, $|\rho_{co}|$, and Ψ_{dp} are estimated as,

$$Z_{dr}^{hy} = 10\log_{10}\frac{\langle|E_h^r|^2\rangle}{\langle|E_v^r|^2\rangle} \tag{4.131a}$$

$$|\rho_{co}^{hy}| = \frac{|\langle E_h^{r*} E_v^r\rangle|}{\sqrt{\langle|E_h^r|^2\rangle\langle|E_v^r|^2\rangle}} \tag{4.131b}$$

$$\Psi_{dp}^{hy} = \arg\left(\langle E_h^{r*} E_v^r\rangle\right) \tag{4.131c}$$

Because the received electric field is not resolved into components that are copolar and cross-polar to the radiated polarization state, this measurement scheme is referred to as a "hybrid" (superscript "hy" as in (4.131)) basis[7] as opposed to the classical orthogonal basis. Note, however, that the coherency matrix $\mathbf{J}$ can be transformed to other bases via $\mathbf{J}' = \mathbf{U}\mathbf{J}\mathbf{U}^{-1}$, see (3.147d), so that orthogonal basis retrieval is not strictly excluded.

Measurements in the "hybrid" basis will, in principle, lead to no error if the transmission and scattering matrices are diagonal. From (4.59),

$$\begin{bmatrix} E_h^r \\ E_v^r \end{bmatrix} = \sqrt{\frac{Z_0 G}{2\pi}}\frac{1}{r^2}\mathbf{T}\mathbf{S}_{BSA}\mathbf{T}\begin{bmatrix} M_h \\ M_v \end{bmatrix} \tag{4.132a}$$

$$= \sqrt{\frac{Z_0 G}{2\pi}}\frac{1}{r^2}\begin{bmatrix} e^{\lambda_1 r} & 0 \\ 0 & e^{\lambda_2 r} \end{bmatrix}\begin{bmatrix} S_{hh} & 0 \\ 0 & S_{vv} \end{bmatrix}\begin{bmatrix} e^{\lambda_1 r} & 0 \\ 0 & e^{\lambda_2 r} \end{bmatrix}\begin{bmatrix} \sqrt{P_t}\cos\varepsilon \\ \sqrt{P_t}\sin\varepsilon\, e^{j\theta} \end{bmatrix} \tag{4.132b}$$

and,

$$Z_{dr}^{hy} = 10\log_{10}\frac{\langle|S_{hh}|^2\rangle}{\langle|S_{vv}|^2\rangle} + 10\log_{10}e^{-4(k_{im}^h - k_{im}^v)r} + 10\log_{10}(\cot^2\varepsilon) \tag{4.133a}$$

$$= Z_{dr} - 2A_{dp}r + 10\log_{10}(\cot^2\varepsilon) \tag{4.133b}$$

$$\Psi_{dp}^{hy} = \arg\langle S_{hh}^* S_{vv}\rangle + 2(k_{p,re}^h - k_{p,re}^v)r + \theta \qquad (4.133c)$$

$$= \delta_{co} + \Phi_{dp} + \theta \qquad (4.133d)$$

$$|\rho_{co}^{hy}| = \frac{|\langle S_{hh}^* S_{vv}\rangle|}{\sqrt{\langle|S_{hh}|^2\rangle\langle|S_{vv}|^2\rangle}} \qquad (4.133e)$$

When $\varepsilon = 45°$, the above are identical to (4.70d), (4.65b), and (4.77), respectively, except for the addition of a constant system phase θ in the Ψ_{dp} relation. In practice, Z_{dr} bias errors can occur if ε is not exactly equal to $45°$. The accuracy in Ψ_{dp} will depend on the stability of θ, which should be very high since only passive microwave devices are involved.

4.7.1 Errors caused by non-diagonal S_{BSA}

The hybrid scheme will result in errors if either the transmission or scattering matrices are non-diagonal. Large values of L (or LDR_{vh}) will cause scattering "depolarization", which leads to errors in the hybrid estimators. Starting from (4.132a),

$$\begin{bmatrix} E_h^r \\ E_r^v \end{bmatrix} = \sqrt{\frac{Z_0 G}{2\pi}}\frac{1}{r^2}\begin{bmatrix} e^{\lambda_1 r} & 0 \\ 0 & e^{\lambda_2 r} \end{bmatrix}\begin{bmatrix} S_{hh} & S_{hv} \\ S_{vh} & S_{vv} \end{bmatrix}_{BSA}$$
$$\begin{bmatrix} e^{\lambda_1 r} & 0 \\ 0 & e^{\lambda_2 r} \end{bmatrix}\begin{bmatrix} \sqrt{P_t}\cos\varepsilon \\ \sqrt{P_t}\sin\varepsilon e^{j\theta} \end{bmatrix}$$

$$(4.134a)$$

$$= \sqrt{\frac{Z_0 G}{2\pi}}\frac{1}{r^2}\begin{bmatrix} S_{hh}e^{2\lambda_1 r} & S_{hv}e^{(\lambda_1+\lambda_2)r} \\ S_{hv}e^{(\lambda_1+\lambda_2)r} & S_{vv}e^{2\lambda_2 r} \end{bmatrix}\begin{bmatrix} \sqrt{P_t}\cos\varepsilon \\ \sqrt{P_t}\sin\varepsilon e^{j\theta} \end{bmatrix} \qquad (4.134b)$$

The elements of the coherency matrix are,

$$\langle|E_h^r|^2\rangle = \frac{Z_0 G P_t}{2\pi}\frac{1}{r^4}e^{-2(k_{im}^h+k_{im}^v)r}\left[\langle n|S_{hh}|^2\rangle e^{-2(k_{im}^h-k_{im}^v)r}\cos^2\varepsilon + \langle n|S_{hv}|^2\rangle\sin^2\varepsilon\right]$$

$$(4.135a)$$

$$\langle|E_v^r|^2\rangle = \frac{Z_0 G P_t}{2\pi}\frac{1}{r^4}e^{-2(k_{im}^h+k_{im}^v)r}\left[\langle n|S_{hv}|^2\rangle\cos^2\varepsilon + \langle n|S_{vv}|^2\rangle e^{2(k_{im}^h-k_{im}^v)r}\sin^2\varepsilon\right]$$

$$(4.135b)$$

$$\langle E_h^{r*} E_v^r\rangle = \frac{Z_0 G P_t}{2\pi}\frac{1}{r^4}\frac{\sin 2\varepsilon}{2}e^{-2(k_{im}^h+k_{im}^v)r}\left[\langle n S_{hh}^* S_{vv}\rangle e^{j\theta}e^{j\Phi_{dp}} + \langle n|S_{hv}|^2\rangle e^{-j\theta}\right]$$

$$(4.135c)$$

The main assumption made in deriving the above is that the hydrometeors in the scattering volume satisfy "mirror" reflection symmetry about $\beta_0 = 0°$, i.e. $\langle n S_{hh}S_{hv}^*\rangle = \langle n S_{vv}S_{hv}^*\rangle = 0$. Now, let $\varepsilon = 45°$. The hybrid estimator for differential reflectivity is,

$$Z_{dr}^{hy} = Z_{dr}e^{-4(k_{im}^h-k_{im}^v)r}\left[\frac{1 + Le^{2(k_{im}^h-k_{im}^v)r}}{1 + LZ_{dr}e^{-2(k_{im}^h-k_{im}^v)r}}\right] \qquad (4.136)$$

Note that $Z_{dr}^{hy} = 10 \log_{10}(\chi_{dr}^{hy})$. The hybrid estimator for Ψ_{dp} is,

$$\Psi_{dp}^{hy} = \arg\langle (E_h^r)^* E_v^r \rangle \tag{4.137a}$$

$$= \arg\left[\langle S_{hh}^* S_{vv} \rangle e^{j(\theta + \Phi_{dp})} + \langle n|S_{hv}|^2 \rangle e^{-j\theta} \right] \tag{4.137b}$$

$$= \arg\left[\frac{|\rho_{co}| e^{j(\delta_{co} + \theta + \Phi_{dp})}}{\sqrt{\chi_{dr}}} + Le^{-j\theta} \right] \tag{4.137c}$$

and the hybrid estimator for $|\rho_{co}|$ is,

$$|\rho_{co}^{hy}| = |\rho_{co}| \frac{|1 + (L\sqrt{\chi_{dr}}/|\rho_{co}|)e^{-j(\delta_{co} + 2\theta + \Phi_{dp})}|}{\sqrt{1 + L}\sqrt{1 + L\chi_{dr}}} \tag{4.138}$$

Note that the hybrid estimators for Z_{dr} and Ψ_{dp} have essentially the same dependency on L and on the propagation factors as the estimators using the circular observables W_1', W_2', and W'; see (4.125, 4.128) and compare with (4.136, 4.137c). This close resemblance follows from the fact that slant 45° linear polarization and circular polarization states are similar as far as propagation through a medium with $\phi = 0°$ is concerned. The scattering "depolarization" error in the hybrid estimates will be small if $L\sqrt{\chi_{dr}}/|\rho_{co}| \ll 1$. At the WSR–88D frequency of 3 GHz, the maximum values of LDR$_{vh}$ are around -15 dB in the "bright-band" or in large hail near the surface, while $\chi_{dr} \approx 1$ and $|\rho_{co}| \approx 0.8$. The maximum error in Z_{dr}^{hy} in such cases should not exceed a few tenths of a decibel. The errors in Ψ_{dp}^{hy} and $|\rho_{co}^{hy}|$ will also depend on $\Phi_{dp} + 2\theta$ assuming Rayleigh scattering ($\delta_{co} \approx 0°$). Calculations (Doviak et al. 2000) show that the maximum error in Ψ_{dp}^{hy} should not exceed $\pm 3°$, while the maximum error in $|\rho_{co}^{hy}|$ could be as high as 6%. The estimated $|\rho_{co}^{hy}|$ is always less than $|\rho_{co}|$, which will tend to accentuate the signatures in the "bright-band" or in large hail (refer to Sections 7.2.2 and 7.3 for a detailed discussion of polarimetric variables in hailstorms and the "bright-band").

4.7.2 Error in Z_{dr} due to a non-zero mean canting angle

The assumption of a diagonal transmission matrix for the rain medium needs to be fully evaluated ($\phi = 0°$ assumption). While ϕ is expected to be very close to 0° based on theoretical considerations, even small deviations from this assumption (for example, $\phi = 5°$) can lead to large errors in the hybrid scheme, especially for Z_{dr}, and to a lesser extent for Ψ_{dp} and $|\rho_{co}|$ (Sachidananda and Zrnić 1985; Doviak et al. 2000). Assuming the model described by (4.50), where $\phi \equiv -\beta_0$ and $\lambda_1 - \lambda_2 = -j(2\pi n/k_0)(S_{11} - S_{22}) \exp(-2\sigma_\beta^2)$, Equation (4.80) can be used to evaluate the error in Z_{dr} to first-order in ϕ. Letting $M_{h,v} = 1$ in (4.80) for the hybrid scheme ($\varepsilon = 0°, \theta = 0°$), the received voltages are,

$$V_h^{11} \approx \frac{\lambda G}{4\pi r^2} S \sqrt{P_t} \left[T_{hh}^2 + T_{hv}(T_{hh} + T_{vv}) \right] \tag{4.139a}$$

$$V_v^{11} \approx \frac{\lambda G}{4\pi r^2} S \sqrt{P_t} \left[T_{hv}(T_{hh} + T_{vv}) + T_{vv}^2 \right] \tag{4.139b}$$

where T_{hv}^2 is neglected compared with T_{hh}^2 or T_{vv}^2, since ϕ is small. The superscript "11" refers to $M_h = 1$, $M_v = 1$, i.e. slant $45°$ polarization. From (4.43) and using the approximation $\cos 2\phi \approx 1$, $\sin 2\phi \approx 2\phi$,

$$T_{hh} \approx e^{\lambda_1 r} \tag{4.140a}$$

$$T_{vv} \approx e^{\lambda_2 r} \tag{4.140b}$$

$$T_{hv} \approx (e^{\lambda_1 r} - e^{\lambda_2 r})(2\phi) \tag{4.140c}$$

Using the above in (4.139),

$$V_h^{11} = \frac{\lambda G}{4\pi r^2} S \sqrt{P_t} \left[e^{2\lambda_1 r}(1 + \phi) - \phi e^{2\lambda_2 r} \right] \tag{4.141a}$$

$$V_v^{11} = \frac{\lambda G}{4\pi r^2} S \sqrt{P_t} \left[e^{2\lambda_2 r}(1 - \phi) + \phi e^{2\lambda_1 r} \right] \tag{4.141b}$$

If the rain medium imposes a pure differential phase shift, the hybrid estimate reduces to,

$$\mathcal{Z}_{dr}^{hy} = \frac{|V_h^{11}|^2}{|V_v^{11}|^2} \tag{4.142a}$$

$$= \frac{1 + 2\phi(1 - \cos \Phi_{dp})}{1 - 2\phi(1 - \cos \Phi_{dp})} \tag{4.142b}$$

which is an approximation correct to first-order in ϕ. Note that the scattering volume has a single, fixed, spherical particle whose intrinsic $\mathcal{Z}_{dr} = 1$. Thus, the error in $\mathcal{Z}_{dr}^{hy}$ will depend on ϕ and Φ_{dp}, being a maximum when $\Phi_{dp} = 180°$. For example, if $\phi = 5°$ with $\Phi_{dp} = 180°$, $\mathcal{Z}_{dr}^{hy} \approx 2$ and the error in the hybrid estimator is 3 dB. To keep the error within 0.2 dB, ϕ should be $\leq 0.33°$. Calculations by Doviak et al. (2000) show that the maximum error in Φ_{dp}^{hy} is around $2°$, which is not significant. The error in $|\rho_{co}^{hy}|$ was found to be a maximum at $\Phi_{dp} = 180°$ and was typically less than 1%, which is also not significant.

For greater accuracy, Sachidananda and Zrnić (1985) proposed radiating alternate pulses of slant $\pm 45°$ linear polarization to reduce the error caused by non-zero ϕ. Setting $(M_h = 1, M_v = -1)$ in (4.80) and using the same assumptions,

$$V_h^{1,-1} = \frac{\lambda G}{4\pi r^2} S \sqrt{P_t} \left[e^{2\lambda_1 r}(1 - \phi) + \phi e^{2\lambda_2 r} \right] \tag{4.143a}$$

$$V_v^{1,-1} = \frac{\lambda G}{4\pi r^2} S \sqrt{P_t} \left[-e^{2\lambda_2 r}(1 + \phi) + \phi e^{2\lambda_1 r} \right] \tag{4.143b}$$

A new estimator for $\breve{z}_{dr}$ is defined, based on using the signal returns from the alternate $\pm 45°$ radiated polarizations as,

$$\breve{z}_{dr}^{hy} = \frac{|V_h^{11}|^2 + |V_h^{1,-1}|^2}{|V_v^{11}|^2 + |V_v^{1,-1}|^2} \tag{4.144a}$$

$$\approx \frac{1 + 2\phi - 2\phi \cos \Phi_{dp} + 1 - 2\phi + 2\phi \cos \Phi_{dp}}{1 - 2\phi + 2\phi \cos \Phi_{dp} + 1 + 2\phi - 2\phi \cos \Phi_{dp}} \tag{4.144b}$$

$$= 1 \tag{4.144c}$$

Thus, to first-order in ϕ, there is no error in the slant $\pm 45°$ hybrid scheme as far as $\breve{z}_{dr}^{hy}$ is concerned.

Notes

1. Considerable theoretical and experimental work done by the propagation community for communications applications is also applicable to radar meteorology (see, for example, the review by Oguchi 1983). A wealth of information is available at the Web site of the OLYMPUS Propagation Experimenters (OPEX) Results, in particular, a set of reference handbooks, which can be downloaded from,

   http://www.estec.esa.nl/xewww/cost255/opex.htm

2. There does not appear to be a standard definition of δ_{co}. The definition here is based on the $\exp(jwt)$ time convention, and is consistent with $\Psi_{dp} = \Phi_{dp} + \delta_{co}$. For example, positive values of δ_{co} will cause a local increase in the otherwise monotonic increasing range profile of the differential propagation phase in rain.

3. These radar data are from the X-band circularly polarized radar operated by the New Mexico Institute of Mining and Technology, Socorro, New Mexico. System specifications are available in Bringi and Hendry (1990). This radar has recently been modified for "hybrid" basis measurement (see Section 4.7).

4. These data were acquired with the NCAR/CP-2 radar in Central Florida in the summer of 1991 as part of the Convective and Precipitation/Electrification (CaPE) experiment.

5. The Alberta Research Council's S-band circularly polarized radar was designed for hail detection studies. Early on, Humphries (1974) recognized the importance of Φ_{dp} in rain in modifying the measurement of CDR and ORTT. However, it was much later that Bebbington et al. (1987) and Holt (1988), at the University of Essex, UK, showed how Z_{dr} and Φ_{dp} could be derived from the circular measurements.

6. Mueller (1984) developed the Φ_{dp} algorithm based on signal returns from a polarization agile/single receiver radar operating in (h, v) basis. This algorithm must be considered as a significant advancement in radar meteorology. The specific differential phase (K_{dp}), which is derived from the range profile of Φ_{dp}, is related to rain rate, and thus the Mueller algorithm is significant because of the importance of K_{dp} in the radar measurement of rainfall. Seliga and Bringi (1978) originally proposed that K_{dp} could lead to improved accuracy in the measurement of rainfall when combined with Z_{dr} (see Chapter 8). The accuracy of the Mueller algorithm was evaluated by Sachidananda and Zrnić (1986) and the first experimental

Φ_{dp} data with a polarization agile/single receiver system were reported by Sachidananda and Zrnić (1987).

7. The term "hybrid" basis was proposed by Chandra et al. (1999). Examples of data collected with the C-band DLR radar may be found in this reference. For examples of data collected with the CSU–CHILL S-band radar in this mode, refer to Brunkow et al. (1997), Holt et al. (1999), or Beaver et al. (1999). For examples of data collected in this mode at X-band with the New Mexico radar, including a geometric description based on the Poincaré sphere, see Scott et al. (2001).

5 Doppler radar signal theory and spectral estimation

The operational deployment of WSR–88D radars throughout the USA can be considered as a major milestone in the application of the Doppler principle to radar meteorology. Throughout the world, the operational use of Doppler radars is considered to be an indispensable tool for monitoring the development of hazardous storms. Doppler radar theory and techniques in meteorology have reached a level of maturity where even the non-specialist can begin to use and apply the data with little formal training. This chapter attempts to provide the interested reader with a fairly rigorous approach to Doppler radar principles.[1] Following a very brief review of signal and system theory, the received voltage from a random distribution of precipitation particles is formulated for an arbitrary transmitted waveform. The expression for mean power is formulated in terms of the intrinsic reflectivity and a three-dimensional weighting function. The range–time autocorrelation function of the received voltage is formulated in terms of the range–time profile of the time-correlated scattering cross section of the particles and of the time correlation of the transmitted waveform. Next, the sample–time autocorrelation function is derived and the concept of coherency time of the precipitation medium is introduced, which dictates the pulse repetition time for coherent phase measurement. The concept of a spaced-time–spaced-frequency coherency function is introduced, to illustrate how frequency diversity can be used to obtain nearly uncorrelated voltage samples from the same resolution volume.

Following a brief coverage of sampling, discrete time processing and power spectral density, the statistical properties of the received signal are described, in particular, the joint distribution of phases and factors related to the accuracy of the mean Doppler velocity estimate. The joint distribution of dual-polarized signals is also described, together with factors that relate to the accuracy of differential reflectivity and differential phase estimates.

The Doppler spectrum and mean velocity estimates are formulated from sampled signals, including the classic pulse-pair estimate . The chapter concludes with a description of several modern spectral estimation techniques, together with examples.

5.1   Review of signals and systems

A brief review of signals and systems is provided which will be useful in the study of Doppler radar theory for media with randomly distributed scatterers such

211

as precipitation.[2] Radar signals (either transmitted or received) come under a class described as "narrow band signals", for which the Fourier transform is non-negligible only in a finite band of frequencies. Such signals can be represented as,

$$x(t) = a(t) \cos[2\pi f_0 t + \alpha(t)] \tag{5.1}$$

where f_0 is the "carrier" frequency. The functions $a(t)$ and $\alpha(t)$ represent the amplitude and phase modulation of the signal, respectively, and they are assumed to vary "slowly" compared with f_0^{-1} (they are low frequency or baseband signals). Equation (5.1) can be written as,

$$x(t) = a(t) \cos \alpha(t) \cos 2\pi f_0 t - a(t) \sin \alpha(t) \sin 2\pi f_0 t \tag{5.2a}$$
$$= I(t) \cos 2\pi f_0 t - Q(t) \sin 2\pi f_0 t \tag{5.2b}$$

where $I(t)$ and $Q(t)$ are the in-phase and quadrature phase components of the modulation, respectively. The complex signal $s(t)$ corresponding to the real signal $x(t)$ is defined as,

$$s(t) = [I(t) + jQ(t)]e^{j2\pi f_0 t} \tag{5.3a}$$
$$= a(t)e^{j\alpha(t)}e^{j2\pi f_0 t} \tag{5.3b}$$

It follows that $x(t) = \text{Re}[s(t)]$. If the carrier frequency term, $\exp(j2\pi f_0 t)$, in (5.3b) is suppressed, then the complex envelope of $s(t)$ is defined as $V(t)$,

$$V(t) = I(t) + jQ(t) = a(t)e^{j\alpha(t)} \tag{5.4}$$

To distinguish between transmitted and received signals, the transmitted signal is written as,

$$s_{tr}(t) = U_{tr}(t)e^{j2\pi f_0 t} \tag{5.5a}$$

while the received signal is written as,

$$s_r(t) = V_r(t)e^{j2\pi f_0 t} \tag{5.5b}$$

where $U_{tr}(t)$ and $V_r(t)$ are the complex envelopes. For simplicity, $V_r(t)$ will generally be referred to as the received voltage, and $U_{tr}(t)$ will be referred to as the transmitted waveform.

5.1.1   Fourier transform

The Fourier transform of a signal $v(t)$ is given by the integral,

$$V(f) = \int_{-\infty}^{\infty} v(t)e^{-j2\pi ft} \, dt \tag{5.6a}$$

with the inverse transform relation,

$$v(t) = \int_{-\infty}^{\infty} V(f)e^{j2\pi ft}\, df \tag{5.6b}$$

The signal $v(t)$ must satisfy certain conditions for the Fourier transform to exist, and one of the important conditions is that the function $v(t)$ must be absolutely integrable,

$$\int_{-\infty}^{\infty} |v(t)|\, dt < \infty \tag{5.6c}$$

Typically, signals that are physically realized, such as radar transmit and receive waveforms, have Fourier transforms. Signals of finite energy can be expressed as a continuous sum of sinusoids with frequencies in the interval $-\infty$ to ∞. Thus, the Fourier transform provides a frequency domain decomposition of the signal $v(t)$ with $V(f)$ giving the relative amplitude of the various frequency components. For deterministic signals, $V(f)$ is commonly referred to as the spectrum.

5.1.2   Some useful properties of the Fourier transform

Let $v(t)$ and $V(f)$ be Fourier transform pairs, indicated as $v(t) \Longleftrightarrow V(f)$.

(i) Linearity and superposition
Let $v_1(t) \Longleftrightarrow V_1(f)$ and $v_2(t) \Longleftrightarrow V_2(f)$, then for constants c_1 and c_2,

$$c_1 v_1(t) + c_2 v_2(t) \Longleftrightarrow c_1 V_1(f) + c_2 V_2(f) \tag{5.7}$$

This property indicates that the linear combination of two signals results in linear combinations of their spectra, which is useful in analyzing signals in the presence of noise.

(ii) Time shifting property
The Fourier transform of the time shifted signal $v(t - t_0)$ is given by $V(f)\exp(-j2\pi f t_0)$. The time shifting property indicates that if a function is shifted in time by t_0 then the equivalent effect in the frequency domain is multiplication of $V(f)$ by $\exp(-j2\pi f t_0)$.

(iii) Frequency shifting
If $v(t) \Longleftrightarrow V(f)$ then,

$$e^{j2\pi f_0 t} v(t) \Longleftrightarrow V(f - f_0) \tag{5.8}$$

Thus, multiplication of $v(t)$ by $\exp(j2\pi f_0 t)$ is equivalent to shifting the spectrum by f_0. This is the "modulation" property, which results in shifting of the spectra, and can be used, for example, to study the spectrum of a radar transmitted waveform. An example

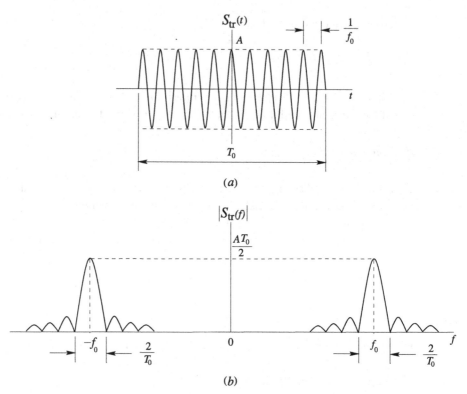

Fig. 5.1. (*a*) Radar transmitted signal shown as a sinusoid of frequency (f_0) modulated by a rectangular pulse of duration T_0. (*b*) Magnitude of the Fourier transform (or spectrum) of the transmitted signal.

of a radar transmitted signal is a high frequency sinusoid modulated by a rectangular pulse (see Fig. 5.1*a*), which can be expressed as,

$$s_{tr}(t) = U_{tr}(t) \cos(2\pi f_0 t) \tag{5.9a}$$

where the rectangular pulse,

$$U_{tr}(t) = \begin{cases} A; & -\dfrac{T_0}{2} < t < \dfrac{T_0}{2} \\ 0; & |t| \geq \dfrac{T_0}{2} \end{cases} \tag{5.9b}$$

and T_0 is the pulsewidth.   Cos $(2\pi f_0 t)$ can be expressed as $[\exp(j2\pi f_0 t) + \exp(-j2\pi f_0 t)]/2$. Applying the frequency shifting property,

$$S_{tr}(f) = \frac{1}{2}[U_{tr}(f - f_0) + U_{tr}(f + f_0)] \tag{5.10}$$

where $U_{tr}(f)$ is the Fourier transform of the rectangular pulse. It can be evaluated as,

$$U_{tr}(f) = A \int_{-T_0/2}^{T_0/2} \exp(-j2\pi f t)\, dt = AT_0 \left(\frac{\sin \pi f T_0}{\pi f T_0} \right) \tag{5.11}$$

$U_{tr}(f)$ can be expressed using the sinc function notation as,

$$U_{tr}(f) = AT_0 \, \text{sinc}(fT_0) \tag{5.12a}$$

where,

$$\text{sinc}(x) \equiv \left(\frac{\sin \pi x}{\pi x} \right) \tag{5.12b}$$

Substituting (5.12) into (5.10), the Fourier transform of the transmitted signal (see Fig. 5.1b) can be written as,

$$S_{tr}(f) = \frac{AT_0}{2} \{ \text{sinc}[(f - f_0)T_0] + \text{sinc}[(f + f_0)T_0] \} \tag{5.13}$$

5.1.3 Transmission of signals through linear systems

In the time domain, a linear system, such as a filter (see Fig. 5.2), is described in terms of its impulse response, which is the response of the system to a Dirac delta function input, $\delta(t)$. Let the response of the system to an impulse be $h(t)$. Then the response of the system, $y(t)$, to an arbitrary input $x(t)$ can be written as,

$$y(t) = \int_{-\infty}^{\infty} x(\tau)h(t - \tau) \, d\tau \tag{5.14a}$$

where the above equation can be considered as a composition of the response to various impulse functions weighted by $x(t)$. The form of the above integral is called the convolution integral. The convolution integral is commutative,

$$y(t) = \int_{-\infty}^{\infty} h(\tau)x(t - \tau) \, d\tau \tag{5.14b}$$

$$y(t) = x(t) \star h(t) = h(t) \star x(t) \tag{5.14c}$$

where the symbol $\star$ indicates convolution. The Fourier transform of the output signal of a linear system can be obtained from the property of the Fourier transform of the convolution of two signals. Let $v_1(t) \Longleftrightarrow V_1(f)$ and $v_2(t) \Longleftrightarrow V_2(f)$ be two Fourier transform pairs. Then,

$$v_1(t) \star v_2(t) \Longleftrightarrow V_1(f)V_2(f) \tag{5.15}$$

Thus, convolution in the time domain results in a product in the frequency domain. The converse is also true, i.e.

$$v_1(t)v_2(t) \Longleftrightarrow V_1(f) \star V_2(f) \tag{5.16}$$

Therefore, convolution in the frequency domain results in a product in the time domain. Since the output signal of a linear system is obtained as a convolution of the impulse response with the input signal, in the frequency domain this can be expressed as,

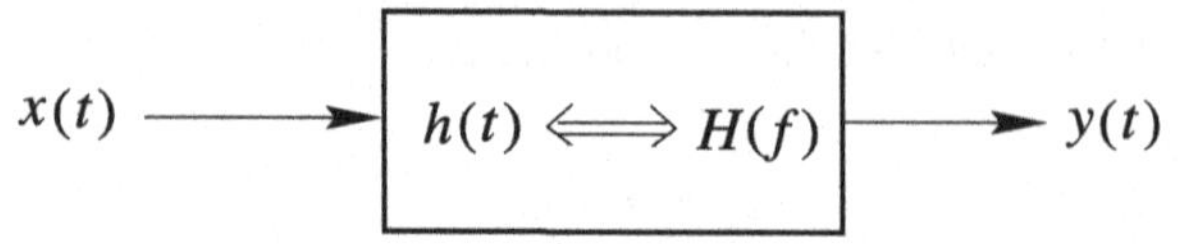

$$x(t) \longrightarrow \boxed{h(t) \Longleftrightarrow H(f)} \longrightarrow y(t)$$

Fig. 5.2. A linear time invariant system is characterized by its impulse response $h(t)$, or by the frequency response function $H(f)$. The input signal is $x(t)$, and the output signal is $y(t)$.

$$Y(f) = X(f)H(f) \tag{5.17}$$

where $y(t) \Longleftrightarrow Y(f)$, $x(t) \Longleftrightarrow X(f)$, and $h(t) \Longleftrightarrow H(f)$. $H(f)$ is referred to as the frequency response or transfer function of the filter. $H(f)$ for a given f_0 can be interpreted as the output when the input signal $x(t) = \exp(j2\pi f_0 t)$.

5.1.4   Gaussian-shaped pulse and filters

Let $g(t)$ be a Gaussian-shaped pulse described as,

$$g(t) = \frac{1}{\sigma\sqrt{2\pi}} e^{-t^2/2\sigma^2} \tag{5.18}$$

where the corresponding Fourier transform $[G(f)]$ can be obtained as,

$$G(f) = e^{-4\pi^2 f^2/2(1/\sigma^2)} = e^{-f^2/2\sigma_f^2} \tag{5.19}$$

Thus, the Fourier transform of a Gaussian-shaped pulse also has a Gaussian shape. From (5.19), $\sigma_f = 1/(2\pi\sigma)$, indicating that if $g(t)$ is wide in the time domain then $G(f)$ is narrow in the frequency domain. $G(f)$ can be used to approximate the shape of the transfer function of a radar receiver. In reality the transfer function is not exactly Gaussian in shape. Nevertheless, the Gaussian shape provides an analytical basis to compute parameters related to receiver filters. The transfer function of a Gaussian-shaped filter can be written as,

$$|G(f)| = G_0 e^{-f^2/2\sigma_f^2} \tag{5.20}$$

where G_0 is the maximum gain of the filter. The transfer function is, in general, characterized by both magnitude and phase functions, but only the magnitude is considered in (5.20). Also, the power loss due to the filter will involve $|G(f)|^2$, which is shown in Fig. 5.3. The 6-dB power bandwidth of this filter is defined as the width between frequencies (f_6) at which $|G(f)|^2$ is 6-dB below its peak (which occurs at $f = 0$) or,

$$\frac{G_0^2}{4} = G_0^2 \exp\left(\frac{-f_6^2}{\sigma_f^2}\right) \tag{5.21a}$$

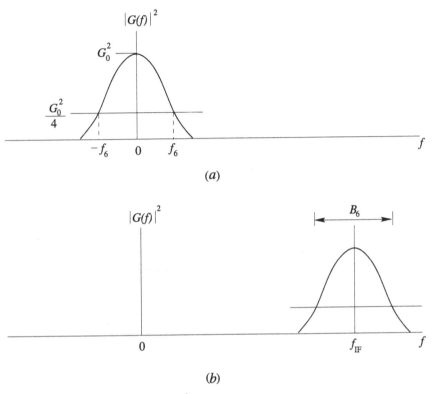

Fig. 5.3. The Gaussian form of $|G(f)|^2$ and the definition of the 6-dB bandwidth, $B_6 = 2f_6$: (a) baseband version, (b) centered on the intermediate frequency (f_{IF}).

From the above,

$$\sigma_f^2 = \frac{f_6^2}{\ln 4} \tag{5.21b}$$

The 6-dB power bandwidth B_6 is given by $2f_6$. However, note that it is conventional in systems theory to define bandwidth as the 3-dB power bandwidth.

5.2   Received signal from precipitation

The received signal due to a point scatterer moving at a uniform velocity is a scaled replica of the transmitted waveform, $U_{\text{tr}}(t)$, except shifted in time by the range–time delay (t_0) and shifted in frequency by the Doppler shift ($f_D = -2\hat{i}\cdot\vec{v}/\lambda$; see Section 1.9). The received voltage at the antenna port for continuous wave (CW) illumination is given by, see (3.69),

$$V_r(t) = \frac{\lambda G\sqrt{P_t}}{4\pi r^2}(S)e^{-j2k_0r} \tag{5.22}$$

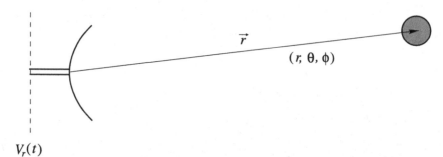

$V_r(t)$

Fig. 5.4. A single sphere is located at $\vec{r}$ with respect to an origin at the radar. As the sphere moves, r and the phase $2k_0 r$ of V_r change with time. Note that $k_0 = 2\pi/\lambda$.

where S is the scattering matrix element of an (assumed) spherical particle ($S = S_{hh} = S_{vv}$) and a single-polarized configuration is assumed for simplicity. The geometry is illustrated in Fig. 5.4, and the functional dependence of r on t can be written as $r(t)$. Using (5.5b), the received signal $s_r(t)$ can be expressed as,

$$s_r(t) = V_r(t)e^{j2\pi f_0 t} \tag{5.23a}$$

$$= \frac{\lambda G \sqrt{P_t}}{4\pi r^2}(S)e^{-j2\pi f_0 (2r/c)}e^{j2\pi f_0 t} \tag{5.23b}$$

$$= \frac{\lambda G \sqrt{P_t}}{4\pi r^2}(S)e^{-j2\pi f_0 \tau}e^{j2\pi f_0 t} \tag{5.23c}$$

$$= Ae^{-j2\pi f_0 \tau}e^{j2\pi f_0 t} \tag{5.23d}$$

where $\tau = 2r/c$, and $k_0 = 2\pi/\lambda = 2\pi f_0/c$, with c being the velocity of light. The factor $\lambda G \sqrt{P_t} S/4\pi r^2$ is, for convenience, denoted as an amplitude A. In compact notation,

$$s_r(t) = Ae^{j2\pi f_0 (t-\tau)} \tag{5.24}$$

Since the transmitted signal, see (5.5a), is a continuous wave at the frequency f_0, then $s_{tr}(t) = \exp(j2\pi f_0 t)$ and (5.24) can be expressed as,

$$s_r(t) = As_{tr}(t - \tau) \tag{5.25}$$

The term $\exp(-j2\pi f_0 \tau)$ can be expressed as,

$$e^{-j2\pi f_0 \tau} = e^{-j(4\pi r/\lambda)} = e^{-j\theta} \tag{5.26}$$

The functional dependence of r on t results in θ varying with t. Thus, the phase θ of the scattered wave from the particle changes with its movement relative to the radar, and the time-rate of change of θ is related to the Doppler frequency shift.

Let the transmitted signal be of the general form $s_{tr}(t) = U_{tr}(t)\exp(j2\pi f_0 t)$. From (5.25) it follows that the received signal $s_r(t)$ can be written as,

$$s_r(t) = AU_{tr}(t - \tau)e^{j2\pi f_0 (t-\tau)} \tag{5.27a}$$

$$= Ae^{-j2\pi f_0 \tau}U_{tr}(t - \tau)e^{j2\pi f_0 t} \tag{5.27b}$$

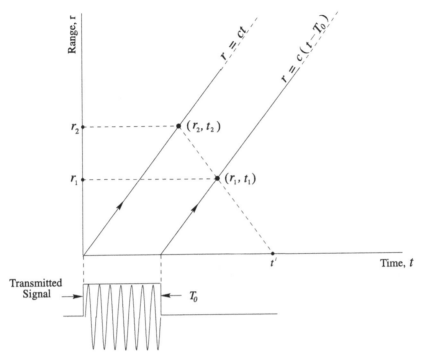

Fig. 5.5. Idealized representation of scattering by two particles in the range–time plane. In this plane, the two particles are located at (r_1, t_1) and (r_2, t_2). The transmitted signal is a rectangular pulse of width T_0, shown below the time axis. The leading and trailing edges of the transmitted signal are on the two characteristic lines defined by $r = ct$ and $r = c(t - T_0)$. The scattered signal at the radar at t' contains contributions from the two particles located along the characteristic line whose slope equals $-c$, i.e. the line $r - r_1 = -c(t - t_1)$. Note that c is the velocity of light.

From (5.5b), the received voltage $V_r(t)$ is,

$$V_r(t) = A e^{-j2\pi f_0 \tau} U_{\mathrm{tr}}(t - \tau) \tag{5.28}$$

The above expression is valid for a single moving particle.

In general, the scattering amplitude can be time-varying and the functional dependence can be expressed as $A(\tau; t)$, as shown by the following example. Consider an idealized situation where two particles are located at radial ranges r_1 and r_2 with time-varying scattering amplitudes (see Fig. 5.5). A rectangular transmitted waveform with pulsewidth T_0 is shown below the horizontal time axis. Note that the ordinate represents radial range. The propagation[3] of the leading edge of the transmitted pulse can be represented by the characteristic line $r = ct$, while the trailing edge is represented by the line $r = c(t - T_0)$. The reflected signal at the radar at time t' contains contributions from the two particles located along the characteristic line with slope equal to $-c$, i.e. the line whose equation is $r - r_1 = -c(t - t_1)$, as shown in Fig. 5.5. Thus, the reflected signal at the radar at t' is made up of a reflected contribution from the scatterer

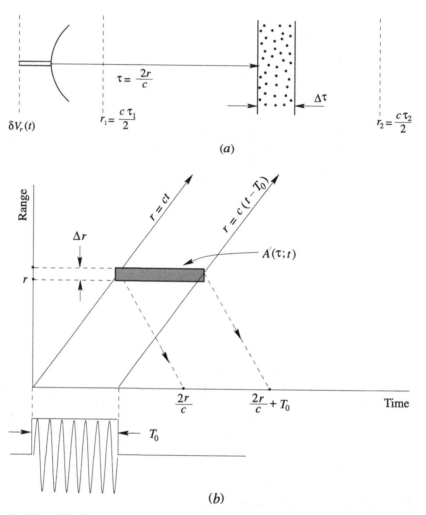

Fig. 5.6. (*a*) The incremental voltage $\delta V_r(t)$ due to scattering from particles located within a shell extending from $(\tau, \tau + \Delta\tau)$ or $(r, r + \Delta r)$. Superposition is used to treat the case when particles extend from r_1 to r_2. (*b*) Range–time diagram for a shell of particles between r and $r + \Delta r$. The function $A'(\tau; t)$ represents the scattering from the shell per unit increment of Δr. The transmitted signal is a rectangular pulse of width T_0.

located at r_2, whose scattering amplitude is evaluated at $t_2 = t' - r_2/c$, and the scatterer located at r_1, whose scattering amplitude is evaluated at $t_1 = t' - r_1/c$. Therefore, the reflected signal amplitude can have the fundamental form $A(\tau; t)$ or $A(r; t)$, where $\tau = 2r/c$ defines the range–time.

In general, precipitation is composed of a large number of hydrometeors extending over a large range with widely different scattering amplitudes and moving with different velocities. Consider now the received voltage due to scattering by particles within a shell extending from r to $r + \Delta r$ (or τ to $\tau + \Delta\tau$), as illustrated in Fig. 5.6a. The

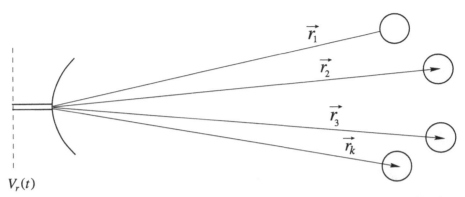

$V_r(t)$

Fig. 5.7. A random collection of particles whose "instantaneous" locations are described by the set of vectors $\vec{r}_k$ ($k = 1, 2, \ldots$) relative to the radar. The phase angles are $\theta_k = 2\pi f_0 \tau_k = (4\pi/\lambda)r_k$. For a random distribution of particles, θ_k are uniformly distributed in $(-\pi, \pi)$. The scattering amplitude A_k has, in general, the functional dependence $A_k(\tau_k; t)$. The received voltage $V_r(t)$ for a general transmitted waveform is given by (5.32).

corresponding range–time diagram for a rectangular transmitted pulse of width T_0 is illustrated in Fig. 5.6b. Now define $A'(\tau; t)$ as the scattering amplitude of a shell of particles per unit range increment of $\Delta\tau$ (or Δr). The received voltage increment from this shell follows from (5.28) as,

$$\delta V_r(t) = A'(\tau; t)e^{-j2\pi f_0 \tau} U_{tr}(t - \tau)\Delta\tau \tag{5.29}$$

If the scattering medium extends from r_1 to r_2, then,

$$V_r(t) = \int_{\tau_1}^{\tau_2} A'(\tau; t)e^{-j2\pi f_0 \tau} U_{tr}(t - \tau)\, d\tau \tag{5.30}$$

Formally, the lower and upper limits can be extended from zero to ∞ so that the general form of the received voltage for an arbitrary transmitted waveform (not necessarily a rectangular pulse) is given by,

$$V_r(t) = \int_0^\infty A'(\tau; t)e^{-j2\pi f_0 \tau} U_{tr}(t - \tau)\, d\tau \tag{5.31}$$

From (5.28) the received voltage can also be expressed as a discrete sum of the contribution from individual particles in the medium as,

$$V_r(t) = \sum_k A_k(\tau_k; t)e^{-j2\pi f_0 \tau_k} U_{tr}(t - \tau_k) \tag{5.32}$$

where A_k is the scattering amplitude of the kth particle, and $\tau_k = 2r_k/c$ (see Fig. 5.7). For any particular frequency, $V_r(t)$ for a given t can be interpreted as a sum of elemental phasors (see Fig. 5.8). Consider now $V_r(t + \Delta t)$. During Δt, the particles move and their corresponding r_k (or $\tau_k = 2r_k/c$ and $\theta_k = (4\pi/\lambda)r_k$) will change. If Δt is very small, the elemental phasors at t and $t + \Delta t$ will be similarly aligned. As Δt increases,

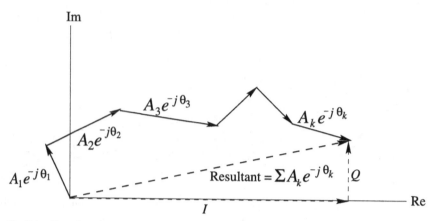

Fig. 5.8. Complex plane plot of the "instantaneous" sum of elemental phasors represented by eq. (5.32) for $V_r(t)$. The phase angle θ_k is assumed to be uniformly distributed in $(-\pi, \pi)$. The resultant phasor has components $I + jQ$, where I is referred to as the in-phase component and Q as the quadrature-phase component.

the elemental phasors will be less aligned. If Δt is sufficiently large, such that Δr_k is a significant fraction of the wavelength, then $V_r(t + \Delta t)$ will be uncorrelated with $V_r(t)$. This separation time (Δt) at which this correlation decreases to a specified value is called the coherence time of the medium (and will be discussed further in Section 5.5; see also, Marshall and Hitschfeld 1953).

5.3   Mean power of the received signal

One of the key radar measurements is the mean power corresponding to the received voltage, $V_r(t)$, which can be related to the back scatter cross section per unit volume (or the reflectivity) of the precipitation. At a given instant of time, $V_r(t)$ is a random variable due to the random locations and random scattering amplitudes of the particles, and therefore $V_r(t)$ is a random process. In essence, the mean power is the variance of $V_r(t)$. Using (5.32), general expressions for $V_r(t_1)$ and $V_r(t_2)$ can be written,

$$V_r(t_1) = \sum_k A_k(\tau_k; t_1)e^{-j\theta_k} U_{\mathrm{tr}}(t_1 - \tau_k) \tag{5.33a}$$

$$V_r(t_2) = \sum_m A_m(\tau_m; t_2)e^{-j\theta_m} U_{\mathrm{tr}}(t_2 - \tau_m) \tag{5.33b}$$

Let $t_1 = t_2 = t$ and form the product,

$$V_r^*(t)V_r(t) = \sum_{k=m} |A_k(\tau_k; t)|^2 |U_{\mathrm{tr}}(t - \tau_k)|^2$$

$$+ \sum_k \sum_{k \neq m} A_k A_m^* e^{j(\theta_m - \theta_k)} U_{\mathrm{tr}}(t - \tau_k) U_{\mathrm{tr}}^*(t - \tau_m) \tag{5.34}$$

The mean received power, $\bar{P}_r(t)$, is the expected value of $V_r^*(t)V_r(t)$, which is denoted as $\langle|V_r(t)|^2\rangle$, where angle brackets indicate ensemble average. Because the received voltage is normalized as in (3.38), the magnitude square of the voltage will have units of power (or watts). Here, it is assumed that the received power is referenced to the antenna port. The expectation of the double-sum in (5.34) will vanish because $\langle\exp[j(\theta_m - \theta_k)]\rangle = 0$, since $\theta_{k,m}$ are independent and identically distributed (iid) random variables, which are uniformly distributed in $(-\pi, \pi)$. Since the positions of the hydrometeors are independent of the scattering amplitude it can be assumed that the amplitudes A_k and phases θ_k are independent. Also, recall that $A \equiv \lambda G\sqrt{P_t}S/4\pi r^2$. Thus, $\langle|V_r(t)|^2\rangle$ can be expressed as,

$$\bar{P}_r(t) = \langle|V_r(t)|^2\rangle = \sum_k \langle|A_k(\tau_k; t)|^2\rangle|U_{tr}(t - \tau_k)|^2 \tag{5.35a}$$

$$= \frac{\lambda^2 P_t}{(4\pi)^3} \sum_k \left\langle \frac{G_k^2 4\pi |S_k|^2}{r_k^4} \right\rangle |U_{tr}(t - \tau_k)|^2 \tag{5.35b}$$

In going from (5.35a) to (5.35b), the time-dependence of $A(\tau; t)$ is dropped, because in weather radar applications it is reasonable to assume that $\langle|A(\tau; t)|^2\rangle$ is stationary over the pulse duration. Thus, the mean received power is related to the incoherent sum (see also, Section 1.8) of the radar cross sections of all the scatterers in the medium, except that it is appropriately weighted by the antenna gain function and the transmit waveform. Note that $\sigma_b = 4\pi|S|^2$, see (2.123).

Let $\eta(r, \theta, \phi)$ be the average radar cross section per unit volume (or reflectivity), defined such that,

$$\eta(r, \theta, \phi)\Delta V = \sum_k \langle 4\pi|S|^2\rangle \tag{5.36}$$

where the summation is over all scatterers in an elemental volume ΔV. The mean received power can then be expressed as an integral of $\eta(r, \theta, \phi)$ over three-dimensional space, but appropriately weighted by the antenna gain function and the transmit waveform,

$$\bar{P}_r(t) = \frac{\lambda^2 P_t}{(4\pi)^3} \int_V \frac{G^2(\theta, \phi)}{r^4} \eta(r, \theta, \phi)|U_{tr}(t - \tau)|^2 \, dV \tag{5.37}$$

The antenna power gain function can be expressed as,

$$G(\theta, \phi) = G_0 f(\theta, \phi) \tag{5.38}$$

where $f(\theta, \phi)$ is the normalized power pattern function and G_0 is the peak power gain (see Fig. 5.9). Substituting (5.38) into (5.37) and replacing $\tau = 2r/c$, and expressing $dV = r^2 d\Omega \, dr$, where $d\Omega$ is the elemental solid angle, results in,

$$\bar{P}_r(t) = \frac{\lambda^2 P_t G_0^2}{(4\pi)^3} \int_0^{2\pi} \int_0^{\pi} f^2(\theta, \phi) \int_0^{\infty} \frac{\eta(r, \theta, \phi)}{r^2} \left|U_{tr}\left(t - \frac{2r}{c}\right)\right|^2 dr \, d\Omega \tag{5.39}$$

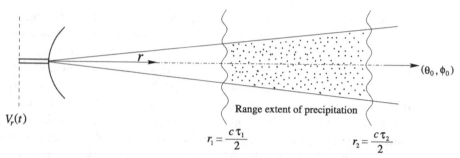

Fig. 5.9. Geometry relevant to derivation of the weather radar range equation. The antenna boresight direction is assumed to be (θ_0, ϕ_0) with respect to a spherical coordinate system at the antenna. The power gain function of the antenna is $G(\theta, \phi) = G_0 f(\theta, \phi)$. The precipitation medium extends from r_1 to r_2 in range (or from τ_1 to τ_2). The range-axis is related to the τ-axis via $r = c\tau/2$.

Defining $f_\eta(\tau, \theta, \phi) = \eta(r, \theta, \phi)/r^2$ evaluated at $r = c\tau/2$, as the range-normalized radar cross section per unit volume (see Fig. 5.10a), (5.39) can be written as,

$$\bar{P}_r(t) = \frac{\lambda^2 P_t G_0^2}{(4\pi)^3} \int_0^{2\pi} \int_0^\pi f^2(\theta, \phi) \int_0^\infty f_\eta(\tau, \theta, \phi) |U_{tr}(t - \tau)|^2 \left(\frac{c}{2}\right) d\tau \, d\Omega \quad (5.40)$$

Note that weighting of $f_\eta(\tau, \theta, \phi)$ by the antenna gain function and the transmit waveform restricts the volume of the scatterers in three-dimensional space that contributes to $\bar{P}_r(t)$. The range weighting is, in essence, a convolution of $f_\eta(\tau)$ with $|U_{tr}(\tau)|^2$, whereas the cross-range weighting is provided by the square of the antenna power pattern. Because the range weighting is provided by the transmit waveform, $U_{tr}(t)$ itself can be termed the range-weighting function [or $W(t)$]. The expression in (5.40) gives the mean received power (referenced to the antenna port) due to an arbitrary transmitted waveform. If the transmitted waveform is a rectangular pulse of width T_0,

$$U_{tr}(t) = 1; \quad 0 < t < T_0 \quad (5.41a)$$
$$= 0; \quad \text{otherwise} \quad (5.41b)$$

then,

$$U_{tr}(t - \tau) = 1; \quad t - T_0 < \tau < t \quad (5.42a)$$
$$= 0; \quad \text{otherwise} \quad (5.42b)$$

and (5.40) reduces to,

$$\bar{P}_r(t) = \langle |V_r(t)|^2 \rangle = \frac{\lambda^2 P_t G_0^2}{(4\pi)^3} \int\int f^2(\theta, \phi) \int_{t-T_0}^t f_\eta(\tau, \theta, \phi) \left(\frac{c}{2}\right) d\tau \, d\Omega \quad (5.43)$$

The uniform range–time weighting of f_η by the rectangular pulse is illustrated in Fig. 5.10a. If $f_\eta(\tau, \theta, \phi)$ is constant over the interval $t - T_0$ to t, then,

$$\left(\frac{c}{2}\right) \int_{t-T_0}^t f_\eta(\tau, \theta, \phi) \, d\tau \approx \left(\frac{cT_0}{2}\right) f_\eta(t_0, \theta, \phi) = \left(\frac{cT_0}{2}\right) \frac{\eta(r_0, \theta, \phi)}{r_0^2} \quad (5.44)$$

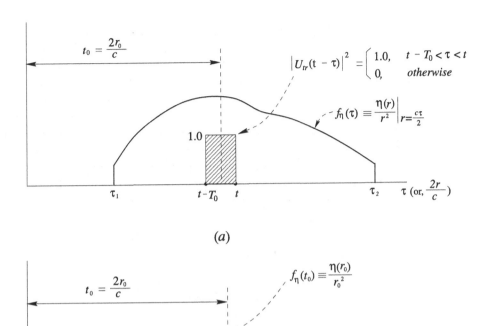

Fig. 5.10. Range–time profile of f_η, where $f_\eta \equiv \eta(r)/r^2$. (a) The transmitted waveform is a rectangular pulse of width T_0. The diagram illustrates the uniform range–time weighting of f_η by the rectangular pulse. The mean power of the received signal at time t is assigned the range $r_0 = ct_0/2$, i.e. to the center of the transmitted pulse. (b) The range–time weighting function is $|W(t - \tau)|^2$, and is assumed here to be symmetric in shape with a peak at $t_0 = 2r_0/c$.

where $t_0 = 2r_0/c$ as shown in Fig. 5.10a. The mean power is assigned to the range r_0. Equation (5.43) reduces to,

$$\bar{P}_r(r_0) = \left(\frac{cT_0}{2}\right)\left[\frac{\lambda^2 P_t G_0^2}{(4\pi)^3 r_0^2}\right]\left[\iint f^2(\theta, \phi)\eta(r_0, \theta, \phi)\,d\Omega\right] \tag{5.45}$$

Equation (5.45) gives the important result that the received power at the antenna port at time t due to a rectangular transmitted pulse is related to the integration of reflectivity about a range $r_0 = ct_0/2$, with the limits of integration along a range given by $r_0 - \Delta r/2$ to $r_0 + \Delta r/2$, where $\Delta r = cT_0/2$. In other words, when a pulse of duration T_0 is transmitted, the signal power received at time t is due to precipitation particles in the resolution volume located at a distance r from the radar given by $ct/2$. However, if a sequence of pulses are transmitted, which are spaced T_s apart, the observations of $V_r(t)$, $V_r(t+T_s)$, $V_r(t+2T_s)$, ..., $V_r(t+mT_s)$ provide temporal samples of the received signal from a resolution volume located $ct/2$ from the radar. In addition, since the transmit pulses are radiated T_s apart, the maximum unambiguous range is $r_{max} = cT_s/2$. If $r > r_{max}$, the corresponding time at which $V_r(t)$ has to be sampled will be such that $t > T_s$. If $t > T_s$, then the range will be ambiguous because the back scatter can be from $r = ct/2$, or from $r' = c(t - T_s)/2$ due to the next pulse. The range–time diagram (Fig. 5.11) illustrates the situation, assuming point scatterers at ranges r' and r''. The received voltage at time t is due to contributions along the characteristic line from the two scatterers as shown. Therefore, for unambiguous operation[4] all scatterers must be within the maximum range r_{max}.

The main radiation lobe of pencil-beam antennas (typical for weather radars) can be approximated by a Gaussian function,

$$f(\theta - \theta_0, \phi - \phi_0) = \exp\left\{-4\ln(2)\left[\frac{(\theta - \theta_0)^2}{\theta_1^2} + \frac{(\phi - \phi_0)^2}{\phi_1^2}\right]\right\} \tag{5.46}$$

where θ_1 and ϕ_1 are the conventional 3-dB beamwidths and (θ_0, ϕ_0) is the boresight direction. For narrow beam antennas, Probert-Jones (1962) derived the approximate relation,

$$\iint f^2(\theta - \theta_0, \phi - \phi_0)\,d\Omega \approx \frac{\pi\theta_1\phi_1}{8\ln 2} \tag{5.47}$$

Figure 5.12 illustrates an actual antenna pattern for the CSU–CHILL antenna, along with a Gaussian fit to the main lobe, using θ_1 from the measured pattern. When (5.47) is used in (5.45) and assuming the reflectivity is homogeneous in the resolution volume, the weather radar range equation becomes,

$$\bar{P}_r(r_0) = \left(\frac{cT_0}{2}\right)\left[\frac{\lambda^2 P_t G_0^2}{(4\pi)^3}\right]\left(\frac{\pi\theta_1\phi_1}{8\ln 2}\right)\frac{\eta(r_0)}{r_0^2} \tag{5.48}$$

The SI system of units is appropriate here. Note that P_t is the transmitter pulse power (in watts) and $\eta(r_0)$ is the back scatter cross section per unit volume (m^2 m^{-3}). The velocity of light in vacuum is c. It is conventional in radar meteorology to express η, in (5.48), in terms of the equivalent reflectivity factor (Z_e). Since the dielectric factor $|K_p|^2 = |(\varepsilon_r - 1)/(\varepsilon_r + 2)|^2$ of the particles within the resolution volume is generally not known, see (3.167), it is conventional to set the dielectric factor assuming the ε_r of water (i.e. $|K_p|^2 = |K_w|^2$ in (3.167), where the subscript w stands for water). Thus, Z_e is defined as,

$$Z_e = \frac{\lambda^4}{\pi^5 |K_w|^2}\eta \tag{5.49a}$$

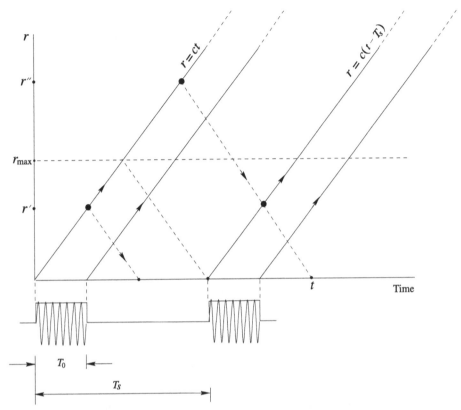

Fig. 5.11. Idealized illustration showing that the maximum unambiguous range is $r_{max} = cT_s/2$. The received voltage at t is due to contributions from the scatterer at range r'', due to the first pulse incident on it, and from the scatterer at range r', due to the second pulse.

and the radar range equation can be expressed as,

$$\bar{P}_r(r_0) = \left(\frac{cT_0}{2}\right)\left[\frac{P_t G_0^2}{\lambda^2(4\pi)^3}\right]\left[\frac{\pi\theta_1\phi_1}{8\ln 2}\right]\frac{\pi^5|K_w|^2 Z_e(r_0)}{r_0^2} \tag{5.49b}$$

The received power is thus related to a meteorological significant quantity referred to as the equivalent reflectivity factor (Z_e) which is conventionally expressed in $mm^6\ m^{-3}$.

Since the above mean power is referenced to the antenna port, the effects of the receiver amplifier/filter have not been accounted for as yet. Typically, the filter is centered at the IF frequency (see Fig. 1.16b), but may also be at the baseband frequency. From Fig. 5.13 it should be clear that at the output of the receiver,

$$V_o(f) = G(f)V_r(f) \tag{5.50}$$

The frequency-response function $G(f)$ can be expressed as $\sqrt{G_r}G_n(f)$, where $G_n(f)$ is the normalized frequency response of the receiver. The gain (G_r) reflects the entire power gain of the receiver.

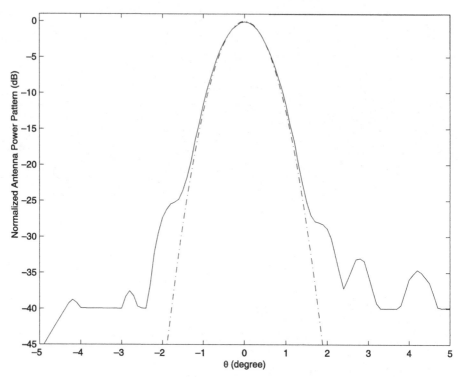

Fig. 5.12. Illustrating a Gaussian fit (— · —) to the main lobe of the measured (——) CSU–CHILL antenna pattern. The 3-dB beamwidth, $\theta_1 = 0.97°$, is taken from the measured pattern.

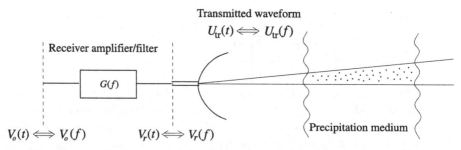

Fig. 5.13. The transmitted waveform is $U_{tr} \Longleftrightarrow U_{tr}(f)$. The received signal at the antenna port is $V_r(t) \Longleftrightarrow V_r(f)$. The frequency response function of the receiver amplifier/filter is $G(f)$. At the output of the receiver amplifier/filter, $V_o(f) = G(f)V_r(f)$.

Consider the ideal case when the transmitted waveform, $U_{tr}(t)$, is a rectangular pulse of width T_0, and the receiver filter is of infinite bandwidth. The received voltage at t is a superposition of back scatter from all particles located in the shaded region shown in

Fig. 5.10a. If the receiver filter has finite bandwidth, then,

$$V_o(t) = V_r(t) \star g(t) \tag{5.51a}$$

$$= \int_0^\infty V_r(\tau)g(t - \tau)\,d\tau \tag{5.51b}$$

where $g(t)$ is the impulse response of the receiver filter. Convolution implies that $V_o(t)$ is the weighted-average of $V_r(\tau)$ with the weighting function being $g(t - \tau)$. It should be clear from Fig. 5.14 that at t', V_o is also composed of contributions from particles outside the segment AB but with non-uniform weighting. It can be argued that this weighting is equivalent to an elongated transmitted waveform whose envelope $W(t)$ is the weighted-average of $U_{\text{tr}}(t)$, with the weighting function being $g(-t)$ or,

$$W(t) = \int_0^\infty U_{\text{tr}}(\tau)g(t - \tau)\,d\tau \tag{5.52a}$$

$$\equiv U_{\text{tr}}(t) \star g(t) \tag{5.52b}$$

Thus, in the presence of a finite receiver bandwidth and a rectangular transmitted waveform, the range–time weighting function is $W(t)$. This concept can be extended to any arbitrary transmitted waveform. The normalized range–time weighting function can be defined as,

$$W_n(t) = U_{\text{tr}}(t) \star g_n(t) \tag{5.52c}$$

where $g_n(t) \Longleftrightarrow G_n(f)$.

In the presence of a finite receiver bandwidth, the mean power at the output of the filter can be obtained from (5.40) by replacing $|U_{\text{tr}}(t - \tau)|^2$ with $|W(t - \tau)|^2$,

$$\bar{P}_o(t) = \langle|V_o(t)|^2\rangle = \frac{\lambda^2 P_t G_0^2}{(4\pi)^3} \iint f^2(\theta, \phi) \int_0^\infty f_\eta(\tau, \theta, \phi)|W(t - \tau)|^2\left(\frac{c}{2}\right) d\tau\, d\Omega \tag{5.53}$$

Figure 5.10b illustrates the range-weighting of the reflectivity profile (f_η) assuming a symmetric shape for $|W(t)|^2$. For the signal received at t, the scatterers located at $t_0 = 2r_0/c$ are given maximum weight as illustrated. Also, the mean power at t is assigned the range $r_0 = ct_0/2$.

In the presence of a finite receiver bandwidth, the resolution in range is determined by $|W(t)|^2$, which is related to the convolution of $g(t)$ and $U_{\text{tr}}(t)$. Typically, the pulsewidth and the width of $g(t)$ are selected to be compatible, and both will determine the actual range resolution.

There is also a loss in received power[5] due to the finite bandwidth of the filter because some of the spectral components in $V_r(f)$ will be eliminated. The ratio of mean power at the input of the filter to the mean power at the output of the filter is,

$$\frac{\bar{P}_r(t)}{\bar{P}_o(t)} = \frac{\displaystyle\int_0^\infty f_\eta(\tau)|U_{\text{tr}}(t - \tau)|^2(c/2)\,d\tau}{\displaystyle\int_0^\infty f_\eta(\tau)|W(t - \tau)|^2(c/2)\,d\tau} \tag{5.54}$$

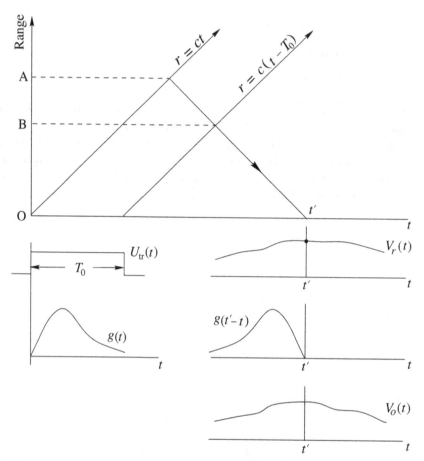

Fig. 5.14. Range–time diagram illustrating the location of contributing scatterers to $V_r(t)$ and $V_o(t)$. $U_{tr}(t)$ is a rectangular transmit pulse of width T_0. At t', $V_r(t')$ is due to scatterers located in the range segment AB. In the presence of a finite bandwidth of the receiver, $V_o(t') = \int_0^\infty V_r(t)g(t'-t)\,dt$. Note that $V_o(t')$ is obtained as the area under the product of $V_r(t)$ and $g(t'-t)$. Since the product $g(t'-t)V_r(t)$ involves $V_r(t)$ when $t \leq t'$, it is clear that the contributing scatterers are located outside the segment AB, i.e. along the segment OA.

where (5.40, 5.53) are used, and with no loss of generality it is assumed that $f_\eta(\tau,\theta,\phi) \equiv f_\eta(\tau)$. If $f_\eta(\tau)$, which is the range profile of reflectivity, is uniform over the range-weighting function, and if $U_{tr}(t)$ is a rectangular pulse of width T_0, then,

$$\frac{\bar{P}_r(t)}{\bar{P}_o(t)} = \frac{cT_0/2}{G_r \int_0^\infty |W_n(\tau)|^2 (c/2)\,d\tau} \tag{5.55a}$$

$$= \frac{cT_0/2}{G_r \int_0^\infty |W_n(r)|^2\,dr} \tag{5.55b}$$

Note that $W_n(r)$ is the normalized range-weighting function. The finite bandwidth loss factor (l_r) can be defined by comparing $\bar{P}_o(t)$ to the output power assuming an infinite bandwidth receiver with the same receiver gain G_r (Doviak and Zrnić 1993),

$$l_r = \frac{cT_0/2}{\displaystyle\int_0^\infty |W_n(r)|^2\, dr} \tag{5.56}$$

where $l_r \geq 1$. Because of finite bandwidth, the function $W_n(r)$ becomes wider than $cT_0/2$ in range, and the net effect is a power loss. Since $\bar{P}_o(t) = \bar{P}_r(t)G_r/l_r$, the weather radar range equation derived earlier, see (5.48), is modified to include the finite bandwidth loss factor and the receiver power gain (G_r) as,

$$\bar{P}_o(r_0) = \left(\frac{cT_0}{2}\right)\left(\frac{G_r}{l_r}\right)\left[\frac{\lambda^2 P_t G_0^2}{(4\pi)^3}\right]\left[\frac{\pi\theta_1\phi_1}{8\ln 2}\right]\frac{\eta(r_0)}{r_0^2} \tag{5.57a}$$

$$= \frac{C\eta}{r_0^2} \tag{5.57b}$$

where C is a radar system constant. Note that $\bar{P}_o(r_0)$ is referenced to the receiver output. The term η is called the radar reflectivity (in units of $m^2\,m^{-3}$), or the back scatter cross section per unit volume.

Equation (5.53) expresses the mean output power, $\bar{P}_o(t)$, as a three-dimensional integral of $f_\eta(r,\theta,\phi)$. However, the antenna gain function and the range-weighting function restrict the volume in space that contributes to $\bar{P}_o(t)$. Let $\eta(r,\theta,\phi)$ be locally uniform around (r_0,θ_0,ϕ_0). An "effective" volume around (r_0,θ_0,ϕ_0) can be defined such that scatterers in this volume dominate the contribution to $\bar{P}_o(t)$. This concept can be formalized by defining a resolution volume (V_6; Doviak and Zrnić 1993) in space, such that its angular extent is defined by the 6-dB beamwidth of $f^2(\theta,\phi)$, and its radial extent by the 6-dB width of $|W_n(r)|^2$.

If the intrinsic radar reflectivity η is not uniform either along range or orthogonal to it (i.e. cross-beam directions which are along θ,ϕ), then (5.53) can be expressed as,

$$\bar{P}_o(t) = \frac{\lambda^2 P_t G_0^2 G_r}{(4\pi)^3}\left(\frac{cT_0}{2l_r}\right)\iiint \frac{\eta(r,\theta,\phi)}{r^4}f^2(\theta,\phi)|W_n(r)|^2\, dV \tag{5.58}$$

where $dV = r^2\, d\Omega\, dr$ is the elemental volume. Combining the range-weighting due to $|W_n(r)|^2$ and angular-weighting due to $f^2(\theta,\phi)$, a normalized three-dimensional weighting function $W(r,\theta,\phi)$ can be defined such that,

$$\bar{P}_o(t) = \frac{\lambda^2 P_t G_0^2 G_r}{(4\pi)^3}\left(\frac{cT_0}{2l_r}\right)\left[\iint f^2(\theta,\phi)\,d\Omega\right]\iiint \frac{\eta(r,\theta,\phi)}{r^2}W(r,\theta,\phi)\, dV \tag{5.59}$$

where,

$$W(r,\theta,\phi) = \left[\frac{f^2(\theta,\phi)}{\displaystyle\iint f^2(\theta,\phi)\,d\Omega}\right]\left[\frac{|W_n(r)|^2}{r^2}\right] \tag{5.60a}$$

and,

$$\iiint W(r, \theta, \phi) \, dV = 1 \tag{5.60b}$$

With this definition, $\bar{P}_o(t)$ can be expressed as,

$$\bar{P}_o(t) = C \iiint \frac{\eta(r, \theta, \phi)}{r^2} W(r, \theta, \phi) \, dV \tag{5.61a}$$

where C is a radar system constant,

$$C = \left(\frac{cT_0}{2}\right) \left(\frac{G_r}{l_r}\right) \left[\frac{\lambda^2 P_t G_0^2}{(4\pi)^3}\right] \left[\iint f^2(\theta, \phi) \, d\Omega\right] \tag{5.61b}$$

Note that C above is the same as defined in (5.57) for a Gaussian-shaped antenna pattern. The averaged reflectivity, $\bar{\eta}(r_0, \theta_0, \phi_0)$, is defined as,

$$\bar{\eta}(r_0, \theta_0, \phi_0) = \frac{\bar{P}_o(r_0)}{C} \tag{5.62a}$$

$$= \iiint \frac{\eta(r, \theta, \phi)}{r^2} W(r, \theta, \phi) \, dV \tag{5.62b}$$

It is convenient to express $W(r, \theta, \phi)$ about the boresight direction (θ_0, ϕ_0) as $W(r, \theta - \theta_0, \phi - \phi_0)$. Equation (5.62b) can then be re-expressed as,

$$\bar{\eta}(r_0, \theta_0, \phi_0) = \iiint \frac{\eta(r, \theta, \phi)}{r^2} W(r, \theta - \theta_0, \phi - \phi_0) \, dV \tag{5.63}$$

The above equation implies a two-dimensional convolution over a spherical surface of given radius r between $\eta(r, \theta, \phi)$ and $W(r, \theta - \theta_0, \phi - \phi_0)$. It is obvious from (5.60a) that, for a given r, $W(r, \theta - \theta_0, \phi - \phi_0)$ is the weighting due to the square of the antenna pattern function. The above convolution implies "smoothing" of the cross-beam reflectivity profile by the main lobe of the antenna and is referred to as "beam smoothing". Equation (5.63) is also written in a form convenient for studying errors induced by the close-in sidelobes of the antenna as the beam "scans" the reflectivity profile. Precipitation is often characterized by strong gradients of reflectivity which can span many orders of magnitude over short distances (up to 30–40 dB km^{-1}). When the main lobe of the antenna (along the θ_0, ϕ_0-direction) is pointed at weak reflectivity and the sidelobes are illuminated by strong reflectivity, errors will "leak" into the reflectivity estimates at (θ_0, ϕ_0). It is difficult to "correct" for these errors, and the best approach is to use well-designed antennas that reduce the close-in sidelobes as much as practical. Typical center-fed parabolic reflector antennas used in weather radars have close-in sidelobes around 25 dB (or better) below the main lobe peak (the two-way sidelobes are 50 dB or better below the main lobe peak). Offset-feed reflector antennas allow for significantly lower sidelobe levels. Antenna performance characteristics will be discussed in Section 6.2.

5.4   Coherency matrix measurements

In the previous section, the weather radar range equation was derived for the single polarized case. The extension to the dual-polarized case is fairly direct. Assume that the transmitted waveform is a horizontally polarized rectangular pulse (with pulse power $P_t^h = P_t$), and that two identical receivers are used for reception of the back scattered horizontal and vertical polarized signals as illustrated in Fig. 5.15. Also, assume an ideal antenna with matched gains at the two ports ($G_h = G_v = G$). Let the scattering matrix, S_{BSA}, of any particle be defined, e.g. see (3.68).

The received voltage at the h-port of the antenna can be written (see (5.33)) as,

$$V_h^{10}(t) = \sum_k \frac{\lambda G_k \sqrt{P_t}}{4\pi r_k^2} S_{hh}^k e^{-j\theta_k} U_{tr}\left(t - \frac{2r_k}{c}\right) \tag{5.64}$$

where S_{hh}^k is the scattering matrix element. The superscript "10" on V_h refers to transmission of a horizontally polarized pulse [$M_h = 1, M_v = 0$ in (3.68)]. Similarly, the received voltage at the v-port of the antenna is,

$$V_v^{10}(t) = \sum_k \frac{\lambda G_k \sqrt{P_t}}{4\pi r_k^2} S_{vh}^k e^{-j\theta_k} U_{tr}\left(t - \frac{2r_k}{c}\right) \tag{5.65}$$

Following the steps used in Section 5.3 exactly, the mean power at the h-port of the antenna can be expressed as,

$$\langle |V_h^{10}(t)|^2\rangle = \frac{C\eta_{hh}}{r_0^2} \tag{5.66}$$

where C is defined in (5.57) except G_r and l_r are unity here, and $\eta_{hh} = \langle n\sigma_{hh}\rangle = 4\pi\langle n|S_{hh}|^2\rangle$ is the horizontally polarized reflectivity. Similarly,

$$\langle |V_v^{10}(t)|^2\rangle = \frac{C\eta_{vh}}{r_0^2} \tag{5.67}$$

where $\eta_{vh} = \langle n\sigma_{vh}\rangle = 4\pi\langle n|S_{vh}|^2\rangle$ is the cross-polarized reflectivity. These two mean powers are two elements of the coherency matrix J_{10}, see (3.181a), earlier referred to as $P_{co}^h = \langle |V_h^{10}|^2\rangle$ and $P_{cx} = \langle |V_v^{10}|^2\rangle$.

The complex cross-correlation is defined as,

$$R_{cx}^h = \langle V_h^{10}(t) \left[V_v^{10}(t)\right]^*\rangle \tag{5.68}$$

which forms the off-diagonal element of J_{10}. Note that this complex cross-correlation is based on the received voltages at the same time, t. Similar to (5.34), the complex cross-correlation can be expressed as,

$$\langle V_h^{10}(t)[V_v^{10}(t)]^*\rangle = \left\langle \left|\sum_{k=m} \frac{\lambda^2 G_k^2 P_t}{(4\pi)^2 r_k^2} S_{hh}^k (S_{vh}^m)^* \left|U_{tr}\left(t - \frac{2r_k}{c}\right)\right|^2\right|\right\rangle$$
$$+ \left\langle \sum_{k\neq m}\sum \frac{\lambda^2 G_k G_m P_t}{(4\pi)^2 r_k^2 r_m^2} S_{hh}^k (S_{vh}^m)^* e^{j(\theta_m - \theta_k)} U_{tr}\left(t - \frac{2r_k}{c}\right) U_{tr}^*\left(t - \frac{2r_m}{c}\right)\right\rangle \tag{5.69}$$

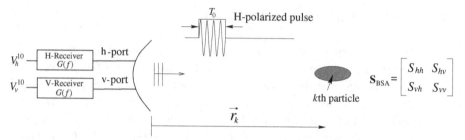

Fig. 5.15. Ideal dual-polarized radar system with two matched receivers. The transmitted waveform is rectangular with pulse power P_t^h and is horizontally polarized. The transmitter is not shown. The two receivers are assumed to be identical.

Again, the expectation over the double-sum will vanish because θ_k and θ_m are iid random variables uniform in $(-\pi, \pi)$, and it is assumed that the scattering amplitudes and phases are independent. Thus, (5.69) reduces to,

$$R_{cx}^h = \langle V_h^{10}(t)[V_v^{10}(t)]^* \rangle = \frac{C}{r_0^2} 4\pi \langle n S_{hh} S_{vh}^* \rangle \tag{5.70}$$

The linear depolarization ratio is defined as,

$$\mathrm{LDR}_{vh} = 10 \log_{10} \left[\frac{\langle |V_v^{10}(t)|^2 \rangle}{\langle |V_h^{10}(t)|^2 \rangle} \right] \tag{5.71a}$$

$$= 10 \log_{10} \left(\frac{\eta_{vh}}{\eta_{hh}} \right) \tag{5.71b}$$

Note that the subscript "vh" on LDR stands for horizontal transmit state and vertical receive state.

The elements of the coherency matrix $\mathbf{J}_{01}$ corresponding to transmission of a vertically polarized pulse with pulse power $P_t^v = P_t$ can be similarly expressed as,

$$\langle |V_h^{01}(t)|^2 \rangle = \frac{C \eta_{hv}}{r_0^2} \tag{5.72a}$$

$$\langle |V_v^{01}(t)|^2 \rangle = \frac{C \eta_{vv}}{r_0^2} \tag{5.72b}$$

$$\langle V_h^{01}(t)[V_v^{01}(t)]^* \rangle = \frac{C}{r_0^2} 4\pi \langle n S_{hv} S_{vv}^* \rangle \tag{5.72c}$$

$$\mathrm{LDR}_{hv} = 10 \log_{10} \left(\frac{\eta_{hv}}{\eta_{vv}} \right) \tag{5.72d}$$

Note that the superscript "01" corresponds to transmission of a vertically polarized pulse ($M_h = 0$, $M_v = 1$ in (3.68)). Also, the subscript "hv" on LDR stands for vertical transmit state and horizontal receive state. From reciprocity $\eta_{hv} = \eta_{vh}$. The combined measurement of $\mathbf{J}_{10}$ and $\mathbf{J}_{01}$ gives three real power terms corresponding to η_{hh}, η_{vv}, and η_{hv}, and two complex terms corresponding to $\langle n S_{hh} S_{vh}^* \rangle$ and $\langle n S_{hv} S_{vv}^* \rangle$.

5.5   Autocorrelation of the received signal

5.5.1   Range–time autocorrelation

General expressions for the received voltages, $V_r(t_1)$ and $V_r(t_2)$, at the antenna port given by (5.33) can be used to formulate the autocorrelation function, $\langle V_r^*(t)V_r(t + \Delta t)\rangle$. For a rectangular transmitted pulse of width T_0, the range–time autocorrelation function will be shown in this section to be triangular in shape for $|\Delta t| \leq T_0$. An extension will be made for the shape of the autocorrelation function at the output of the receiver which will account for the finite bandwidth of the receiver.

From (5.33), the product $V_r^*(t_1)V_r(t_2)$ is expressed as,

$$V_r^*(t_1)V_r(t_2) = \sum_{k=m} A_k^*(\tau_k; t_1)A_k(\tau_k; t_2)e^{-j[\theta_k(t_2)-\theta_k(t_1)]}U_{\text{tr}}^*(t_1 - \tau_k)U_{\text{tr}}(t_2 - \tau_k)$$

$$+ \sum_{k \neq m}\sum A_k^*(\tau_k; t_1)A_m(\tau_m; t_2)e^{j\theta_k(t_1)}e^{-j\theta_m(t_2)}U_{\text{tr}}^*(t_1 - \tau_k)U_{\text{tr}}(t_2 - \tau_m)$$

$$(5.73)$$

Note from Fig. 5.7 that $\theta_k(t) = (4\pi/\lambda)r_k(t) = (4\pi/\lambda)\hat{r}\cdot\vec{r}_k(t)$. Hence,

$$\theta_k(t_2) - \theta_k(t_1) = \frac{4\pi}{\lambda}\hat{r}\cdot[\vec{r}_k(t_2) - \vec{r}_k(t_1)] \tag{5.74a}$$

$$= \frac{4\pi}{\lambda}\hat{r}\cdot\vec{v}_k\,\Delta t \tag{5.74b}$$

where $\vec{v}_k$ is the velocity of the kth scatterer and $\Delta t = t_2 - t_1$. Strictly speaking, Δt equals $t_2' - t_1'$, where $t_{1,2}'$ are the retarded times discussed in Section 1.9. From (1.108c), $t_2' - t_1' \approx t_2 - t_1$ and, henceforth, these time differences will not be distinguished. Note that the expectation of $\exp[j\theta_k(t_1)]\exp[-j\theta_m(t_2)]$ will vanish because $\theta_{k,m}$ are iid random variables uniformly distributed in $(-\pi, \pi)$. Thus, the expectation of $V_r^*(t_1)V_r(t_2)$ reduces to,

$$\langle V_r^*(t_1)V_r(t_2)\rangle = \sum_k \langle A_k^*(\tau_k; t_1)A_k(\tau_k; t_2)\rangle\langle e^{-j(4\pi/\lambda)\hat{r}\cdot\vec{v}_k\Delta t}\rangle$$

$$\times U_{\text{tr}}^*(t_1 - \tau_k)U_{\text{tr}}(t_2 - \tau_k) \tag{5.75}$$

Invoking the same principle used in going from (5.36) to (5.37), the discrete sum over all scatterers is now replaced by a volume integral,

$$\langle V_r^*(t_1)V_r(t_2)\rangle = \int_V \langle A^*(\tau; t_1)A(\tau; t_2)\rangle\langle e^{-j(4\pi/\lambda)\hat{r}\cdot\vec{v}\Delta t}\rangle$$

$$\times U_{\text{tr}}^*\left(t_1 - \frac{2r}{c}\right)U_{\text{tr}}\left(t_2 - \frac{2r}{c}\right)dV \tag{5.76}$$

where $\tau = 2r/c$. It is implicit that $\vec{v}$ is a function of $\tau = 2r/c$.

Invoking stationarity,

$$\langle A^*(\tau; t_1)A(\tau; t_2)\rangle = R_A(\tau; \Delta t) \tag{5.77a}$$

$$= \frac{\lambda^2 G^2 P_t}{(4\pi)^3 r^4}\eta(r; \Delta t) \tag{5.77b}$$

$$= \frac{\lambda^2 G_0^2 f^2(\theta, \phi) P_t}{(4\pi)^3 r^4}\eta(r; \Delta t) \tag{5.77c}$$

where $R_A(\tau; \Delta t)$ is the autocorrelation function of A, and $\eta(r; \Delta t)$ is the autocorrelation of the scattering amplitude per unit volume. When $\Delta t = 0$, $\eta(r; 0)$ is the back scatter cross section per unit volume defined earlier as η in (5.36). Changing particle characteristics such as drop vibration, for example, can cause the ratio $\eta(r; \Delta t)/\eta(r)$ to decrease from unity. An associated coherence time (T_A) can be defined where this ratio falls to $1/e$. For precipitation particles, T_A is expected to be of the order of many tens of milliseconds. Similar to (5.40), (5.76) can be expressed as,

$$\langle V_r^*(t_1)V_r(t_2)\rangle = \frac{\lambda^2 P_t G_0^2}{(4\pi)^3} \iint f^2(\theta, \phi)\,d\Omega \int_0^\infty \frac{\eta(r; \Delta t)}{r^2}$$
$$\times \langle e^{-j(4\pi/\lambda)\hat{r}\cdot\vec{v}\Delta t}\rangle U_{tr}^*\left(t_1 - \frac{2r}{c}\right)U_{tr}\left(t_2 - \frac{2r}{c}\right)dr \tag{5.78a}$$

$$= \left(\frac{c}{2}\right)\left[\frac{\lambda^2 P_t G_0^2}{(4\pi)^3}\right]\left[\frac{\pi\theta_1\phi_1}{8\ln 2}\right]\int_0^\infty f_\eta(\tau; \Delta t)U_{tr}^*(t_1 - \tau)U_{tr}(t_2 - \tau)\,d\tau \tag{5.78b}$$

where $\eta(r; \Delta t)\langle\exp[-j(4\pi/\lambda)\hat{r}\cdot\vec{v}\Delta t]\rangle$ can be viewed as a generalized time-correlated scattering cross section per unit volume, see also (1.112), and $f_\eta(\tau; \Delta t)$ is given by,

$$f_\eta(\tau; \Delta t) = \frac{\eta(r; \Delta t)}{r^2}\langle e^{-j(4\pi/\lambda)\hat{r}\cdot\vec{v}\Delta t}\rangle\bigg|_{r=c\tau/2} \tag{5.79}$$

Equation (5.78b) can be written as,

$$R_v(t_1, t_2) = C'\int_0^\infty f_\eta(\tau; \Delta t)U_{tr}^*(t_1 - \tau)U_{tr}(t_2 - \tau)\,d\tau \tag{5.80}$$

where $R_v(t_1, t_2) = \langle V_r^*(t_1)V_r(t_2)\rangle$ and $C' = (c/2)(\lambda^2 P_t G_0^2\pi\theta_1\phi_1)/(4\pi)^3 8\ln 2$. If $t_1 = t_2 = t$, then $R_v(t, t)$ is,

$$R_v(t, t) = C'\int_0^\infty f_\eta(\tau; 0)U_{tr}^*(t - \tau)U_{tr}(t - \tau)\,d\tau \tag{5.81}$$

which is similar to (5.40). Equation (5.80) clearly shows how the autocorrelation function depends on the range–time profile of the time-correlated scattering cross section of the particles and on the time correlation of the transmitted waveform (Ishimaru 1978).

Consider now the case of a rectangular transmitted pulse of width T_0, and let $t_1 = t$ and $t_2 = t + \Delta t$, where Δt is of the order of T_0 (but $\ll T_s$). Equation (5.80) can be expressed as,

$$R_v(t, t + \Delta t) = C' \int_0^\infty f_\eta(\tau; \Delta t) U_{\mathrm{tr}}^*(t - \tau) U_{\mathrm{tr}}(t + \Delta t - \tau) \, d\tau \qquad (5.82)$$

For precipitation particles, when Δt is of the order of the pulse width (microseconds), $f_\eta(\tau; \Delta t) \approx f_\eta(\tau; 0)$. This approximation implies that the time correlation of the transmitted waveform dominates the range–time autocorrelation function of the received signal. When long duration transmit waveforms are used as in pulse-compression techniques, this approximation is no longer valid. Thus,

$$R_v(t, t + \Delta t) \approx C' f_\eta(t; 0) \int_{t-T_0+\Delta t}^t U_{\mathrm{tr}}^*(t - \tau) U_{\mathrm{tr}}(t + \Delta t - \tau) \, d\tau \qquad (5.83)$$

where the integration limits are written for positive Δt (see Fig. 5.16). If Δt is negative, then the limits must be replaced by $(t - T_0)$ to $(t - |\Delta t|)$. These limits follow simply from the overlapping regions of the two pulses $U_{\mathrm{tr}}(t - \tau)$ and $U_{\mathrm{tr}}(t + \Delta t - \tau)$. From Fig. 5.16, the overlapping region is $T_0 - |\Delta t|$ for $|\Delta t| \leq T_0$ and corresponds to the well-known result that the correlation between two rectangular pulses of width T_0 is a triangle of width $2T_0$. Thus,

$$R_v(t, t + \Delta t) = C' \frac{\eta(r)}{r^2} T_0 \left(1 - \frac{|\Delta t|}{T_0} \right); \quad |\Delta t| \leq T_0 \qquad (5.84)$$

When $\Delta t = 0$, $R_v(t, t) = \bar{P}_r(t)$ and,

$$R_v(t, t + \Delta t) = \bar{P}_r(t) \left(1 - \frac{|\Delta t|}{T_0} \right); \quad |\Delta t| \leq T_0 \qquad (5.85a)$$

The above result for a rectangular transmitted waveform can be generalized to an arbitrary $U_{\mathrm{tr}}(t)$ as,

$$R_v(t, t + \Delta t) = \bar{P}_r(t) R_u(\Delta t) \qquad (5.85b)$$

where $R_u(\Delta t)$ is the autocorrelation function of $U_{\mathrm{tr}}(t)$.

The autocorrelation function of the received signal at the output of the receiver can be related to that at the input through the impulse response of the filter. From linear system theory (e.g. Papoulis 1965), the result is,

$$\langle V_o^*(t) V_o(t + \Delta t) \rangle = [\langle V_r^*(t) V_r(t + \Delta t) \rangle] \star g^*(-\Delta t) \star g(\Delta t) \qquad (5.86)$$

where $g(t)$ is the impulse response of the receiver (the frequency-response function is $G(f)$), and "$\star$" implies a convolution operation. If $G(f)$ is a Gaussian function, see (5.20), then (5.86) can be solved numerically for various values of $B_6 T_0$ (Doviak and Zrnić 1993; see their Fig. 4.10). The right-hand side of (5.86) can be described, in

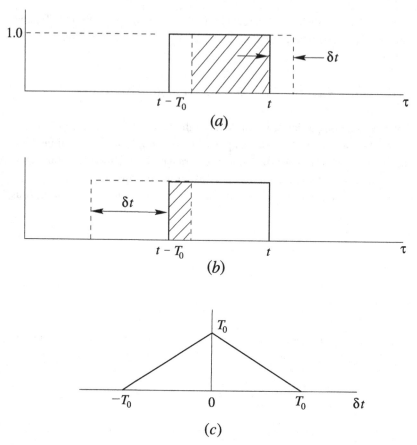

Fig. 5.16. (*a*) The rectangular pulse $U_{tr}(t - \tau)$ is shown as a solid line, while $U_{tr}(t + \delta t - \tau)$ is shown as a dashed line for positive δt. The overlapping region is shaded and its area equals $T_0 - \delta t$. (*b*) The δt is now negative, i.e. the shaded pulse is "shifted' to the left of the solid pulse. The overlap region is shaded and its area equals $T_0 - |\delta t|$. (*c*) The triangular shape is the result of the convolution integral in (5.83) between two rectangular pulses.

essence, as a convolution between a Gaussian function and a triangular function. When $B_6 T_0 \geq 4.0$, the numerical results indicate that the autocorrelation at the receiver output is controlled by the triangular function, i.e. by the time correlation of the rectangular pulse as in Fig. 5.16. When $B_6 T_0 < 0.5$, it is controlled by B_6 and the autocorrelation magnitude falls off "slowly" with increase in Δt.

5.5.2   Sample–time autocorrelation due to transmit pulse train

The radar range equation and the range–time autocorrelation function were formulated for the case of a single transmit pulse. Consider now a periodic pulse train, spaced T_s apart, as illustrated in Fig. 5.17, and assume that the maximum range extent of the precipitation medium is less than $cT_s/2$ (see also, Fig. 5.11). Figure 5.17 illustrates

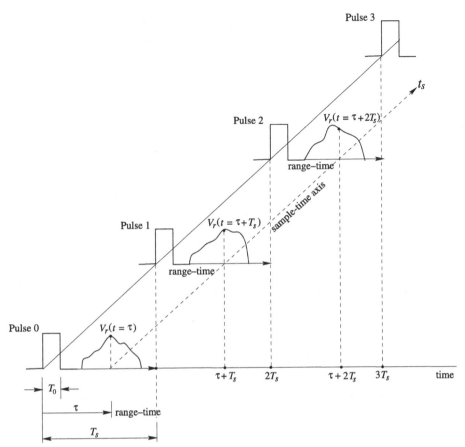

Fig. 5.17. Illustrating the (continuous) range–time axis (τ) and the (discrete) sample–time axis (t_s). The pulse repetition time (PRT) is T_s.

the received voltage due to a periodic pulse train. For a single transmitted pulse the range–time was defined as $\tau = 2r/c$, and the received voltage $V_r(t)$ at $t = \tau$ was due to back scatter from particles located in the resolution volume at range–time τ (or range $= c\tau/2$; see Fig. 5.10). For a periodic pulse train, the received voltage at the same range–time (τ) is given as $V_r(t = \tau)$, $V_r(t = \tau + T_s), \ldots, V_r(t = \tau + mT_s)$ which form a sequence of temporal samples from the same resolution volume. Now define a continuous sample–time axis (t_s) as illustrated in Fig. 5.17. For a given range–time (τ), the voltage samples $V_r(t = \tau + mT_s) = V_r(\tau, t_s = mT_s)$ can be viewed as regularly spaced samples along the continuous sample–time axis (t_s). Thus, it is possible to define an underlying continuous two-dimensional function, $V_r(\tau, t_s)$. The fluctuations of the received voltage along the t_s-axis are determined by the time-varying properties of the particles located in the resolution volume corresponding to a fixed τ.

The sample–time autocorrelation function can be derived from (5.80) by setting $t_1 =$

τ_0 and $t_2 = \tau_0 + mT_s$. Equation (5.80) can be expressed as,

$$R_v(\tau_0, \tau_0 + mT_s) = \langle V_r^*(\tau_0)V_r(\tau_0 + mT_s)\rangle$$

$$= C' \int_0^\infty f_\eta(\tau; mT_s)U_{tr}^*(\tau_0 - \tau)U_{tr}(\tau_0 + mT_s - \tau)\, d\tau \qquad (5.87)$$

where,

$$f_\eta(\tau; mT_s) = \frac{\eta(r; mT_s)}{r^2}\langle e^{-j(4\pi/\lambda)\hat{r}\cdot\vec{v}mT_s}\rangle\bigg|_{r=c\tau_0/2} \qquad (5.88a)$$

$$U_{tr}^*(\tau_0 - \tau)U_{tr}(\tau_0 + mT_s - \tau) = |U_{tr}(\tau_0 - \tau)|^2 \qquad (5.88b)$$

For a rectangular transmitted pulse,

$$|U_{tr}(\tau_0 - \tau)|^2 = 1; \quad \tau_0 - T_0 < \tau < \tau_0 \qquad (5.89a)$$

$$= 0; \quad \text{otherwise} \qquad (5.89b)$$

Assuming uniform f_η over the integration limits $\tau_0 - T_0$ to τ_0,

$$R_v(\tau_0, \tau_0 + mT_s) = C'T_0\frac{\eta(r; mT_s)}{r^2}\left\langle e^{-j(4\pi/\lambda)\hat{r}\cdot\vec{v}mT_s}\right\rangle \qquad (5.90)$$

Recall that $\langle\exp(-j(4\pi/\lambda)\hat{r}\cdot\vec{v}mT_s)\rangle$ is also evaluated at r; i.e. the radial velocity component $\hat{r}\cdot\vec{v}$ is evaluated at r.

When $mT_s = 0$, (5.90) reduces to $R_v(\tau_0, \tau_0)$ which equals $\bar{P}_r(\tau_0)$,

$$R_v(\tau_0, \tau_0) = \bar{P}_r(\tau_0) = (C'T_0)\frac{\eta(r; 0)}{r^2} \qquad (5.91)$$

Note that the above equation is similar to (5.48) with $r \equiv r_0$. The normalized autocorrelation function, or the correlation coefficient at lag mT_s, is expressed as,

$$\rho(\tau_0, \tau_0 + mT_s) = \frac{R_v(\tau_0, \tau_0 + mT_s)}{R_v(\tau_0, \tau_0)} = \left[\frac{\eta(r; mT_s)}{\eta(r; 0)}\right]\langle e^{-j(4\pi/\lambda)\hat{r}\cdot\vec{v}mT_s}\rangle \qquad (5.92a)$$

The correlation coefficient is a product of two separate correlation terms. The first term is controlled by processes such as drop vibration or particle tumbling, whereas the second term is controlled by the particle velocity distribution. For precipitation media, the second term typically dominates the correlation coefficient and it is customary to approximate it as,

$$\rho(\tau_0, \tau_0 + mT_s) = \langle e^{-j(4\pi/\lambda)\hat{r}\cdot\vec{v}mT_s}\rangle \qquad (5.92b)$$

Note that $\tau_0 = 2r/c$ and, henceforth, τ_0 will be dropped for notational convenience. Also, $\hat{r}\cdot\vec{v}$ is the projection of the hydrometeor's velocity vector parallel to the incident beam direction.

From (5.92b), it is clear that $\rho(mT_s)$ depends on the radial velocity distribution of the particles (i.e. the distribution of $\hat{r}\cdot\vec{v} \equiv v$). Let $S(v)$ be the probability density function (pdf) of the radial velocity, which is assumed here to be independent of the scattering amplitude distribution. The characteristic function of v is defined (Papoulis 1965) as,

$$\chi(\omega) = E(e^{j\omega v}) \tag{5.93a}$$

Therefore, $\rho(mT_s)$ in (5.92b) is the characteristic function of v evaluated at $\omega = -(4\pi/\lambda)mT_s$. If $S(v)$ is a Gaussian pdf with mean $\bar{v}$ and standard deviation σ_v, its characteristic function (for example, Papoulis 1965) is,

$$\chi(\omega) = e^{j\omega\bar{v}} e^{-\frac{1}{2}\omega^2\sigma_v^2} \tag{5.93b}$$

The corresponding $\rho(mT_s)$ is,

$$\rho(mT_s) = e^{-j\frac{4\pi}{\lambda}\bar{v}mT_s} \exp\left(-\frac{8\pi^2}{\lambda^2}\sigma_v^2 m^2 T_s^2\right) \tag{5.94}$$

The autocorrelation function of regularly spaced samples of any continuous time signal is the same as the sampled version of the autocorrelation function of the continuous time signal. Therefore, a continuous time equivalent of $\rho(mT_s)$ can be expressed (under adequate sampling, see Section 5.7) as,

$$\rho(t_s) = e^{-j\frac{4\pi}{\lambda}\bar{v}t_s} \exp\left[-\frac{8\pi^2}{\lambda^2}(\sigma_v)^2(t_s)^2\right] \tag{5.95}$$

It is shown later in Section 5.7.2 that the Fourier transform of $\rho(t_s)$ is related to the Doppler spectrum. The coherency time (T_D) of the medium (the subscript "D" here refers to Doppler) is defined as the value of t_s for which $|\rho(t_s)|$ equals e^{-1}. From (5.95),

$$T_D = \frac{\lambda}{2\sqrt{2\pi}\,\sigma_v} \tag{5.96}$$

From (5.94) it is clear that $\arg[\rho(T_s)] = -(4\pi/\lambda)\bar{v}T_s$ and, consequently, the mean radial velocity (under the Gaussian pdf model) can be estimated from $\arg[\rho(T_s)]$. It will be shown later in Section 5.11.1 that $\arg[\rho(T_s)]$ can be used to estimate $\bar{v}$ for any symmetric velocity distribution. The coherency time of the medium (which is dependent on λ and σ_v) must be considered as one parameter which dictates the pulse repetition time (T_s) for coherent phase measurement. This concept will be further elaborated in Section 5.9. Shorter wavelength radars must sample more rapidly (higher pulse repetition frequency (PRF)) than longer wavelength radars. For $\lambda = 10$ cm (the wavelength of the WSR–88D radars) and with a typical value of $\sigma_v = 3$ m s^{-1}, the coherency time is 3.75 ms. The WSR–88D radars, for example, are configured for selectable pulse repetition times from 0.76 to 3 ms (Doviak and Zrnić 1993; see their Table 3.1). At $\lambda = 3$ mm (operating frequency near 100 GHz, the W-band radars used in cloud physics), the T_D is around 0.1 ms and pulse repetition times are around 50 μs.

For the extreme case of Doppler lidars, $\lambda \approx 10$ μm and $T_D \approx 0.3$ μs, which is of the order of the pulsewidth itself (essentially, this precludes estimation of $\bar{v}$ using pulses spaced T_s apart).

The continuous time version of (5.92a) for a fixed range (with $\eta(r; mT_s) \equiv \eta(t_s)$ and $\eta(r; 0) \equiv \eta(0)$) can be expressed as,

$$\rho(t_s) = \frac{\eta(t_s)}{\eta(0)} \langle e^{-j(4\pi/\lambda)\hat{r}\cdot\vec{v}t_s} \rangle \tag{5.97}$$

The $\rho(t_s)$ will have a magnitude of unity at $t_s = 0$, and will decrease as t_s increases. This decrease in $|\rho(t_s)|$ with t_s can be considered as a loss in correlation (or decorrelation), which can occur for a variety of reasons in addition to a velocity distribution. Some important contributions to loss in correlation are listed below:

1. The terminal velocity of hydrometeors varies with size, composition and shape (referred to as particle states). Since a distribution of particle states is expected within the resolution volume, there exists a distribution of $\hat{r}\cdot\vec{v}$ in the term $\langle \exp[-j(4\pi/\lambda)\hat{r}\cdot\vec{v}] \rangle$. Recall that $\hat{r}$ is a unit vector along the incidence direction and the terminal velocity vector is vertically directed ($\vec{v} = -v\hat{z}$) resulting in $\hat{r}\cdot\vec{v} = \sin\theta_e$, where θ_e is the elevation angle. Thus, the loss in correlation is maximal at vertical incidence ($\theta_e = 90°$) and vanishes at $\theta_e = 0°$. Practically, it can be neglected for $\theta_e \leq 10°$.

2. An important contribution to loss in correlation (apart from velocity distribution) is the change in $\hat{r}\cdot\vec{v}$ due to changes in the mean wind speed/direction across the resolution volume (also termed wind shear). Corresponding to the directions parallel and orthogonal to the beam, $(\hat{r}, \hat{\theta}, \hat{\phi})$, the spatial gradient of $\hat{r}\cdot\vec{v}$ can also be resolved along these directions. This topic is dealt with in detail by Nathanson (1969) and Doviak and Zrnić (1993). The loss in correlation can be significant at long ranges and for larger beamwidth antennas ($\theta_1, \phi_1 > 1°$).

Some less important contributions to loss in correlation are:

1. The amplitude term $\eta(t_s)$ contains the scattering matrix element, and non-spherical particles can have time-varying shape and canting angle variations with a resultant loss in correlation. This effect is believed to be small for precipitation particles. Zrnić and Doviak (1989) have shown the effect to be negligible for drop vibration.

2. At long ranges, the angular beam extent can be large ($r_0\theta_1$ or $r_0\phi_1$) and $\hat{r}$, which is the spherical wave direction, changes across the resolution volume. Thus, even for uniform velocity, the radial component changes across the resolution volume leading to loss in correlation.

3. The assumption made so far is that the antenna beam is fixed (with boresight along θ_0, ϕ_0) and that the particles have a velocity distribution. However, there is also a further loss in correlation if the antenna is scanning (or if the antenna is moving, as in an airborne radar system). In the expression for the autocorrelation function

in (5.90), the term C' contains the antenna gain function, which was assumed to be stationary. For a scanning antenna, the boresight direction changes with time (i.e. rotates with angular velocity Ω). For any given particle, the angular weighting due to $f^2(\theta, \phi)$ in (5.78a) changes with time, which results in a loss in correlation. Doviak and Zrnić (1993) (see their Appendix 32) have derived an expression for the loss in correlation assuming a circular Gaussian main lobe, and find that it is proportional to θ_1 (the 3-dB beamwidth). For ground-based radars, this loss in correlation is small for typical one degree beamwidths and normal scan rates (10–$18°\ s^{-1}$).

5.6 Spaced-time, spaced-frequency coherency function

In the previous section, the autocorrelation function derived in (5.97) may be termed the spaced-time autocorrelation function since it involves signals spaced in time. This concept can be extended to develop a spaced-time as well as a spaced-frequency autocorrelation function, which is termed here as the spaced-time, spaced-frequency coherency function of the medium (Ishimaru 1978).

From (5.23d), the received signal $s_r(t)$ from a single particle can be expressed as,

$$s_r(t) = Ae^{-j2\pi(f_0+f)\tau}\, e^{j2\pi(f_0+f)t} \tag{5.98}$$

where the transmitted signal is now a continuous wave sinusoid at the frequency $f_0 + f$. The frequency f_0 can be interpreted as the center frequency (say, within a frequency range of interest) and f as the deviation ($f \ll f_0$). For a large number of particles, the received signal $s_r(t)$ can be written using superposition as,

$$s_r(t) = \left[\sum_k A_k(\tau_k; t)\, e^{-j2\pi(f_0+f)\tau_k}\right] e^{j2\pi(f_0+f)t} \tag{5.99}$$

where $A_k(\tau_k; t)$ is the time-varying scattering amplitude of the kth particle as developed in Section 5.2. The received envelope corresponding to $s_r(t)$ is defined as the frequency response function of the medium (see Section 5.1.3) or $H(f; t)$,

$$H(f; t) = \sum_k A_k(\tau_k; t)e^{-j2\pi(f_0+f)\tau_k} \tag{5.100}$$

Note that $\tau_k = 2r_k/c$. $H(f; t)$ is also the Fourier transform of the impulse response of the medium, $h(\tau; t)$; see Section 5.1.3. However, the transform variables are (f, τ) here. By inspection, $h(\tau; t)$ can be written as,

$$h(\tau; t) = \sum_k A_k(\tau_k; t)\, e^{-j2\pi f_0\tau_k}\delta(\tau - \tau_k) \tag{5.101}$$

since $\delta(\tau - \tau_k) \Longleftrightarrow \exp(-j2\pi f\tau_k)$. It also follows that the received voltage, $V_r(t)$, for an arbitrary transmit waveform, $U_{\text{tr}}(t)$, can be expressed in terms of the impulse response of the medium (see Section 5.1.3),

$$V_r(t) = \int_{-\infty}^{\infty} h(\tau; t) U_{\text{tr}}(t - \tau) \, d\tau \tag{5.102}$$

It is readily seen that using (5.101) in the above equation will result in (5.32), which was arrived at earlier without using the impulse response concept.

For any particular frequency, the function $H(f; t)$ for a given t can be interpreted as a sum of elemental phasors (see Fig. 5.8). As this frequency is shifted by Δf, and if Δf is very small, the set of phasors at f and $f + \Delta f$ will be similarly aligned. As the frequency separation increases, the elemental phasors will be less aligned and if Δf is sufficiently large then $H(f + \Delta f; t)$ will be uncorrelated with $H(f; t)$. The separation frequency Δf at which this correlation decreases to a specified value is called the coherence bandwidth of the medium. The correspondence in phasor alignment in Fig. 5.8 will tend to break down when $\Delta f(\tau_k)_{\max} \approx 1$, which gives a measure of the coherence bandwidth as $c/2r_{\max}$, where $r_{\max}$ is the maximum radial extent of the medium. In general, the coherence bandwidth for back scattering from a layer of a random medium with radial extent $(r_2 - r_1)$ is approximately equal (Ishimaru 1978) to $c/2(r_2 - r_1)$. One implication is that back scatter from a very thin layer of cloud will retain the original pulse shape if the bandwidth of the transmitted waveform is less than the coherence bandwidth of the thin layer. For precipitation measurements using pulsed radar it is not necessary to maintain the shape of the pulse. Instead, narrow pulses (whose bandwidth will usually exceed the coherency bandwidth of the medium) are used to measure the precipitation range profile.

Referring again to $H(f; t)$ in (5.100), consider now the case when $\Delta f = 0$ but time is shifted by Δt. If Δt is sufficiently large, such that Δr_k is a significant fraction of the wavelength, then $H(f; t + \Delta t)$ will be uncorrelated with $H(f; t)$. The correlation structure of $H(f; t)$ can be expressed in terms of the spaced-time, spaced-frequency coherency function defined as,

$$\Gamma(\Delta f; \Delta t) = \langle H^*(f; t) H(f + \Delta f; t + \Delta t) \rangle \tag{5.103}$$

Using $h(\tau; t) \Longleftrightarrow H(f; t)$ results in,

$$\Gamma(\Delta f; \Delta t) = \iint_{-\infty}^{\infty} \langle h^*(\tau_1; t) h(\tau_2; t + \Delta t) \rangle e^{-j2\pi \Delta f \tau_2} e^{j2\pi f(\tau_1 - \tau_2)} \, d\tau_1 \, d\tau_2 \tag{5.104}$$

From (5.101) and using the procedure used to arrive at (5.77a), $\langle h^*(\tau_1; t) h(\tau_2; t + \Delta t) \rangle$ can be written as,

$$\langle h^*(\tau_1; t) h(\tau_2; t + \Delta t) \rangle = R(\tau_1; \Delta t) \delta(\tau_1 - \tau_2) \tag{5.105}$$

where $\delta(\tau_1 - \tau_2)$ is the Dirac delta function. The function $\Gamma(\Delta f; \Delta t)$ can therefore be expressed as,

$$\Gamma(\Delta f; \Delta t) = \int_{-\infty}^{\infty} R(\tau; \Delta t) e^{-j2\pi \Delta f \tau} \, d\tau \tag{5.106a}$$

where,

$$R(\tau; \Delta t) = \langle h^*(\tau; t)h(\tau; t + \Delta t)\rangle \qquad (5.106b)$$

The spaced-frequency coherency function $\Gamma(\Delta f; \Delta t = 0)$ from (5.106a) can be expressed as,

$$\Gamma(\Delta f; \Delta t = 0) = \int_{-\infty}^{\infty} R(\tau; 0)e^{-j2\pi \Delta f \tau}\, d\tau \qquad (5.107a)$$

$$= \int_{-\infty}^{\infty} \langle |A(\tau)|^2\rangle e^{-j2\pi \Delta f \tau}\, d\tau \qquad (5.107b)$$

where Δf is the frequency shift and $\langle |A(\tau)|^2\rangle$ is proportional to the mean scattered power at range–time $\tau = 2r/c$. The above expression is valid for the transmission of two CW signals with carrier frequency shifted by Δf.

For an arbitrary transmitted waveform, $U_{tr}(t)$, it can be shown using (5.101) and (5.102), that the correlation between $V_r(t; f)$ and $V_r(t; f + \Delta f)$ can be written as,

$$\langle V_r^*(t; f)V_r(t; f + \Delta f)\rangle = \int_{-\infty}^{\infty} \langle |A(\tau)|^2\rangle e^{-j2\pi \Delta f \tau}|U_{tr}(t - \tau)|^2\, d\tau \qquad (5.108)$$

Note that for a rectangular pulse of width T_0, $|U_{tr}(t-\tau)|^2$ is non-zero only when $t - T_0 < \tau < t$. In addition, if the mean scattered power is uniform over this range, then it can be taken outside the integral in (5.108). It follows that,

$$\langle |A|^2\rangle \int_{t-T_0}^{t} e^{-j2\pi \Delta f \tau}\, d\tau = \langle |A|^2\rangle \frac{\sin(\pi \Delta f T_0)}{\pi \Delta f T_0} \qquad (5.109a)$$

$$= \langle |A|^2\rangle \operatorname{sinc}(\Delta f T_0) \qquad (5.109b)$$

The above sinc function vanishes when $\Delta f = T_0^{-1}$, or when the frequency separation is the inverse of the pulsewidth. For a typical pulsewidth of 1 μs, the frequency separation is 1 MHz. The received signals (at the same time) from two different transmit pulses with spaced frequency $\Delta f = T_0^{-1}$ are thus uncorrelated (Wallace 1953). This principle has been implemented on the NCAR/ELDORA airborne X-band radar[6] to obtain uncorrelated voltage samples from the same resolution volume at the same time (see Section 5.10 for a discussion of the need for uncorrelated samples to improve the accuracy in the estimation of mean power). This principle can also be implemented by changing the carrier frequency discretely from pulse-to-pulse within a pulse train spaced T_s apart. If T_s is much smaller than the coherency time of the medium, see (5.96), then the voltage samples from the same resolution volume but spaced T_s apart will be nearly uncorrelated. Nevertheless, even if $T_s \approx T_D$ the correlation will be reduced relative to the case if Δf is zero. Several research radars such as the SPANDAR S-band radar located in Wallops Island, Virginia and operated by NASA, as well as the C-band radar operated by Joanneum Research/Technical University of Graz in Austria use frequency shifting from pulse-to-pulse (the total frequency excursion or transmitted bandwidth is fixed). Pulse compression waveforms[7] often use some kind of frequency modulation (e.g. linear FM) within the pulse.

5.7   Sampling the received signal

The received signal at the radar is a complex stochastic signal. Each observation of the received signal is one realization of the underlying complex stochastic process. In most contexts, processing of discrete time signals (or signals sampled in time) is more flexible and often preferable to processing of continuous time signals mostly due to the proliferation of inexpensive and powerful digital signal processors. Sampling occurs naturally in a pulsed radar system. Inferences made after processing the discrete time signal can be converted to those of continuous time signals.

When a train of pulses is transmitted, the received signal observed at range–time τ corresponds to scatterers within a resolution volume located at range $c\tau/2$. It was shown in Section 5.5.2 that for a given range–time τ, the received voltage due to a transmit pulse train spaced T_s apart can be viewed as uniformly spaced discrete time samples (see Fig. 5.18) of an underlying continuous voltage function $V(t_s)$ along the continuous sample–time axis t_s. In the following, background material related to sampling of continuous time signals is presented. For notational simplicity, henceforth in this chapter $V(t_s)$ will be simply referred to as $V(t)$.

Let $V(t)$ be any continuous time signal. Samples of $V(t)$ spaced T_s apart can be written as,

$$V_s(t) = V(t)p(t) \tag{5.110}$$

where $p(t)$ is the impulse-train sampling function defined as,

$$p(t) = \sum_{n=-\infty}^{\infty} \delta(t - nT_s) \tag{5.111}$$

The time T_s is called the sampling period, the frequency $f_s = 1/T_s$ is called the sampling frequency, and $\omega_s = 2\pi f_s$. Figure 5.19 illustrates the sampling process. The signal $V(t)$ is modulated by the sampling function $p(t)$. From the modulation property of Fourier transforms (Proakis and Manolakis 1988), the Fourier transform (commonly referred to as the spectrum) of the sampled signal can be written as,

$$V_s(\omega) = \frac{1}{2\pi} V(\omega) \star P(\omega) \tag{5.112a}$$

where

$$P(\omega) = \frac{2\pi}{T_s} \sum_{k=-\infty}^{\infty} \delta(\omega - k\omega_s) \tag{5.112b}$$

is the Fourier transform of the impulse train. It follows that,

$$V_s(\omega) = \frac{1}{T_s} \sum_{k=-\infty}^{\infty} V(\omega - k\omega_s) \tag{5.112c}$$

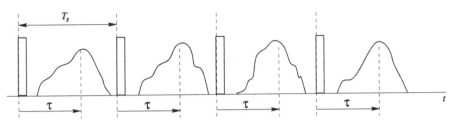

Fig. 5.18. Samples of the complex envelope of the received signal at range time τ. The transmit waveform is a pulse train spaced T_s apart.

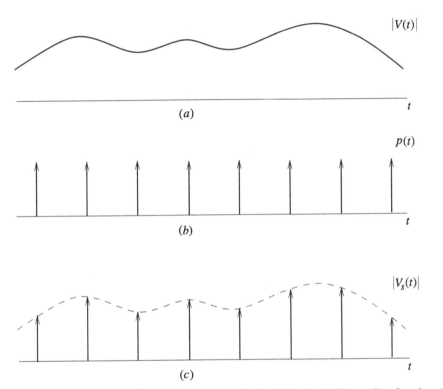

Fig. 5.19. Illustrating sampling of $V(t)$: (a) amplitude of the signal, (b) sampling function, (c) amplitude of the sampled signal.

The Fourier transform of the samples of $V(t)$ is, therefore, a periodic function of frequency consisting of a sum of shifted replicas of $V(\omega)$ scaled by $1/T_s$, as illustrated in Fig. 5.20. If $\omega_s < 2\omega_{max}$, where ω_{max} is the maximum angular frequency component of $V(\omega)$, then there will be a significant overlap in the spectrum of $V_s(t)$ as illustrated in Fig. 5.21. On the contrary, if $\omega_s > 2\omega_{max}$ then $V(t)$ can be recovered exactly from its samples by means of a low pass filter with a cut off frequency higher than ω_{max} and less than $\omega_s - \omega_{max}$, as shown in Fig. 5.22. This restriction on the sampling frequency for full recovery of the continuous time signal is called the sampling theorem, which is strictly valid only for deterministic signals. The minimum sampling frequency is called

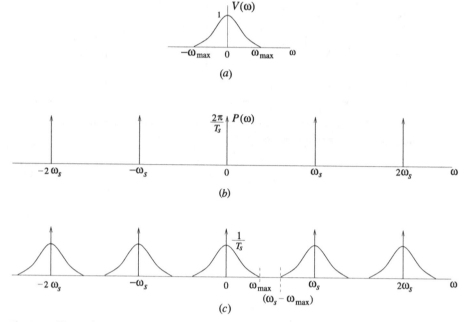

Fig. 5.20. Illustrating sampling in the frequency domain: (a) spectrum of the original signal, (b) spectrum of the sampling function, (c) spectrum of the sampled signal when $\omega_s > 2\omega_{max}$.

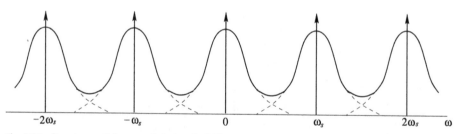

Fig. 5.21. Spectrum of the sampled signal of Fig. 5.20 if $\omega_s < 2\omega_{max}$. This sampling condition creates overlapped spectra.

the Nyquist sampling rate. When the signal is sampled at a rate less than the Nyquist rate, then overlap of the spectrum occurs and is referred to as aliasing.

In the case of a stochastic signal such as the received voltage from precipitation particles, one observation of $V(t)$ is a single realization of an underlying complex stochastic process. The sampling theorem for stochastic signals states that a band limited stochastic signal can be reconstructed from samples taken at the Nyquist rate, and the reconstruction is done in the sense that,

$$E\left[|V(t) - \tilde{V}(t)|^2\right] = 0 \tag{5.113}$$

where $\tilde{V}(t)$ is the reconstructed signal and E stands for statistical expectation.

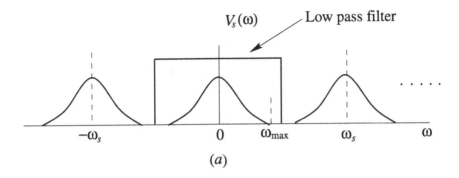

$$(a)$$

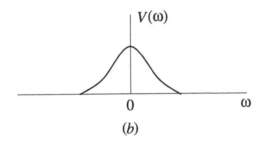

$$(b)$$

Fig. 5.22. (a) Spectrum of the sampled signal. (b) Spectrum at the output of the low pass filter.

5.7.1 Discrete time processing

As mentioned earlier, the proliferation of low cost digital signal processors has made it advantageous to process continuous time signals in discrete time (or from samples). The samples of any continuous time signal $V(t)$ are converted to a discrete time sequence,

$$V[n] = V(nT_s) = V_s(nT_s) \tag{5.114}$$

This conversion from a continuous time signal to a discrete time sequence is illustrated in Fig. 5.23. The Fourier transform of the discrete time sequence $V[n]$ is defined as,

$$V(\Omega) = \sum_{n=-\infty}^{\infty} V[n]e^{-j\Omega n} \tag{5.115}$$

The sampled version of the signal, $V_s(t)$, can be written as,

$$V_s(t) = \sum_{n=-\infty}^{\infty} V[nT_s]\delta(t - nT_s) \tag{5.116}$$

which has the corresponding Fourier transform,

$$V_s(\omega) = \sum_{n=-\infty}^{\infty} V[nT_s]e^{-j\omega nT_s} \tag{5.117}$$

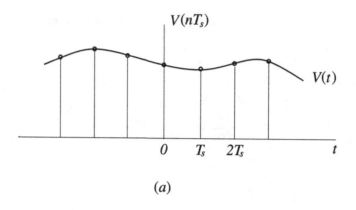

(a)

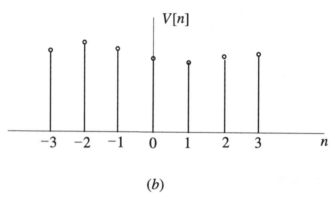

(b)

Fig. 5.23. Conversion of regularly spaced samples of a continuous signal, $V(t)$, to a discrete time sequence, $V[n]$.

Comparing (5.115) and (5.117) it can be seen that $V(\Omega)$ and $V_s(\omega)$ are related by,

$$V(\Omega) = V_s\left(\frac{\Omega}{T_s}\right) \tag{5.118}$$

Figure 5.24 shows the relationship among the spectra of $V(t)$, $V_s(t)$ and $V[n]$ under an adequate sampling rate ($\omega_s > 2\omega_{max}$). It can be seen that $V[\Omega]$ is periodic with a period of 2π. This is a basic property of discrete time Fourier transforms (DTFT). There is a frequency scaling between ω and Ω by a factor $1/T_s$. For example, the sampling frequency ω_s is mapped to 2π. Since $V(\Omega)$ is periodic in 2π the corresponding inverse relation can be written as,

$$V[n] = \frac{1}{2\pi} \int_0^{2\pi} V(\Omega)e^{j\omega n}\, d\Omega \tag{5.119}$$

The Fourier transform pair consisting of (5.115) and (5.119) are the discrete time Fourier transform pair. In practice, signals are observed over a finite time interval. Thus, the sequence $V[n]$ is of finite length N. The N samples of the finite duration

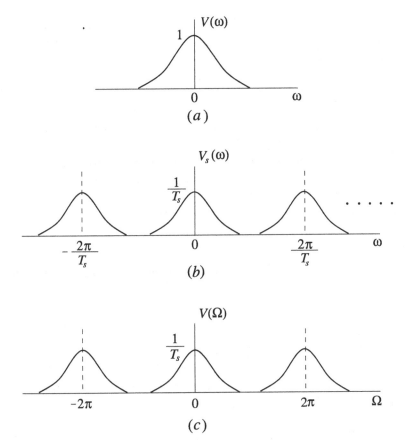

Fig. 5.24. Relationship among the spectra of $V(t)$, $V_s(nT_s)$, and $V[n]$.

signal can be related to the N samples of the spectrum through the discrete Fourier transform (DFT). If the discrete time Fourier transform is sampled at N equispaced intervals over 2π, the discrete Fourier transform pair is written as,

$$V[k] = \sum_{n=0}^{N-1} V[n]e^{-jk(2\pi n/N)} \tag{5.120}$$

and,

$$V[n] = \frac{1}{N} \sum_{k=0}^{N-1} V[k]e^{jk(2\pi n/N)} \tag{5.121}$$

There are two important features of the DFT algorithm described by (5.120). First, it provides a means to compute the spectrum at N points over the spectral domain. In other words, N samples of the spectrum are computed from only N time samples. Second, and perhaps the most important feature, is the availability of very fast algorithms to compute the DFT called the fast Fourier transform (FFT). The FFT algorithms are easily implemented in modern digital signal processors (DSP) enabling full spectral analysis of the received signal.

5.7.2   Power spectral density

As mentioned earlier, the signal received at the radar is a complex stochastic signal. The spectrum of a deterministic signal is given by its Fourier transform, which yields the decomposition of the signal into its frequency components. Equivalently, for stochastic signals the power spectral density provides the decomposition of the signal power into the various frequencies contained in the signal. If $x(t)$ is a wide sense stationary random process (Papoulis 1965) then its power spectral density $S(\omega)$ is defined as,

$$S(\omega) = \lim_{T \to \infty} \frac{1}{2T} E\left\{ |\mathcal{F}[x_T(t)]|^2 \right\}$$

(5.122)

where E stands for expectation and $\mathcal{F}[x_T(t)]$ is the time-limited Fourier transform defined as,

$$\mathcal{F}[x_T(t)] = \int_{-T}^{T} x(t) e^{-j\omega t} \, dt$$

(5.123)

The above two equations indicate how the power spectral density of a stochastic signal is, on "average", related to the Fourier transform. The power spectral density is also related to the autocorrelation function of the signal $x(t)$ via the Fourier transform (Papoulis 1965) as,

$$S(\omega) = \int_{-\infty}^{\infty} R(t) e^{-j\omega t} \, dt$$

(5.124)

where $R(t)$ is the autocorrelation function of the stochastic signal $x(t)$. It can be easily shown that, if $S(\omega)$ is even (or symmetric about $\omega = 0$), then $R(t)$ is a real function.

In Section 1.9, refer to (1.113), it was shown that the Doppler frequency spectrum (or power spectral density) of a single scatterer moving with a constant velocity $\vec{v}$ is given as,

$$S(f) = \sigma_b(-\hat{i}, \hat{i})\delta\left(f + \frac{2\hat{i}\cdot\vec{v}}{\lambda} \right)$$

(5.125)

Thus, there is a one-to-one correspondence between the Doppler frequency (f) and the velocity component parallel to the incident beam direction ($\hat{i}\cdot\vec{v}$, also referred to as the radial velocity). Therefore, the transformation $f = -2\hat{i}\cdot\vec{v}/\lambda = -2v/\lambda$ defines the radial velocity axis for the power spectral density, which can be referred to as the Doppler velocity spectrum.

For example, if a Gaussian-shaped power spectral density as a function of frequency (f) is written as,

$$S(f) = \frac{S_0}{\sigma_f \sqrt{2\pi}} \exp\left[-\frac{(f - \bar{f})^2}{2\sigma_f^2} \right]$$

(5.126)

then the corresponding Doppler velocity spectrum is,

$$S(v) = \frac{S_0}{\sigma_v \sqrt{2\pi}} \exp\left[-\frac{(v - \bar{v})^2}{2\sigma_v^2}\right] \tag{5.127}$$

where $\bar{v} = -\lambda \bar{f}/2$ and $\sigma_v = (\lambda/2)\sigma_f$. Using (5.8) and (5.19), the corresponding autocorrelation function can be derived as,

$$R(\tau) = S_0 \exp\left(\frac{-8\pi^2 \sigma_v^2 \tau^2}{\lambda^2}\right) \exp\left(-j\frac{4\pi \bar{v}\tau}{\lambda}\right) \tag{5.128}$$

Figure 5.25 shows the Doppler velocity spectrum and the magnitude and phase of the autocorrelation function. It can be seen that larger σ_v results in a narrower $R(\tau)$, which follows from the basic property of the Fourier transform. In addition, information about the mean velocity is contained in the phase of $R(\tau)$. In pulsed radar systems, only samples of $V(t)$ are available and the power spectral density is estimated from these samples. It was shown in Section 5.7.1 that samples of $V(t)$ at a fixed sampling rate become the sequence $V[n]$, $n = 0, 1, \ldots, N - 1$. A commonly used estimate of the power spectral density from $V[n]$ is the periodogram defined as,

$$\hat{S}(\Omega) = \frac{1}{N} \left|\sum_{n=0}^{N-1} V(n)e^{-j\Omega n}\right|^2 \tag{5.129}$$

From (5.124), the power spectral density can also be estimated from the autocorrelation function. For a sequence $V[n]$, the autocorrelation function can be estimated as,

$$\hat{R}(l) = \frac{1}{N} \sum_{n=0}^{N-l-1} V(n + l)V^*(n); \quad 0 \leq l < N \tag{5.130a}$$

The above estimate of the autocorrelation function is biased (especially for small values of N) and is asymptotically unbiased for large N ($N \to \infty$). Though the estimate of the autocorrelation function given by (5.130a) is biased, it satisfies the condition necessary to be an autocorrelation function (namely, non-negative definite property; Brockwell and Davis 1991).

The unbiased estimator for the autocorrelation function is given by,

$$\hat{R}(l) = \frac{1}{N - |l|} \sum_{n=0}^{N-l-1} V(n + l)V^*(n); \quad 0 \leq l \leq N \tag{5.130b}$$

However, $\hat{R}(l)$ given by (5.130b) may not satisfy the necessary condition to be an autocorrelation function. Power spectral density can be obtained as the discrete time Fourier transform of the estimated autocorrelation function as,

$$\hat{S}(\Omega) = \sum_{l=-(N-1)}^{N-1} \hat{R}(l)e^{-j\omega l} \tag{5.131}$$

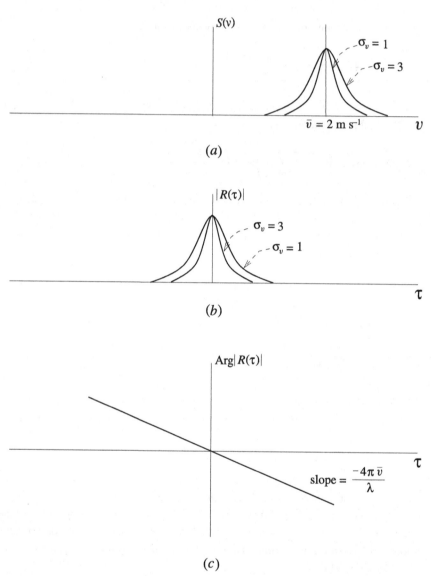

Fig. 5.25. Doppler velocity spectrum (a), and the corresponding autocorrelation function magnitude (b), and phase (c).

If (5.130a) is used in (5.131), then it can be shown with modest algebraic manipulation that (5.129) and (5.131) are identical. The power spectral density, as the name implies, is a density function that describes how the average power is distributed over the various frequency components. Since it is the density of power, it is real and is non-negative, $S(\Omega) \geq 0$. In order to obtain the autocorrelation estimates at all lags the signal has to be observed over infinite time. However, in practice, the autocorrelation function goes to zero as the time lag increases. It can be seen from (5.130a) that for a data record

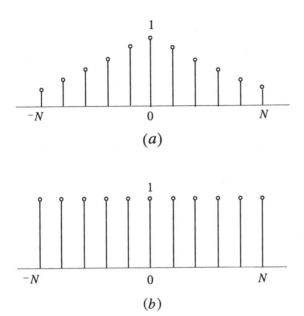

Fig. 5.26. Window functions: (*a*) Bartlett window, and (*b*) rectangular window.

of length N all the estimates of autocorrelation at lags larger than $N - 1$ will be zero. Therefore, the estimate of power spectral density can be written as,

$$\hat{S}(\Omega) = \sum_{l=-\infty}^{\infty} \hat{R}(l)W(l)e^{-j\Omega l} \tag{5.132}$$

where $W(l)$ is a window function.

The expected value of power spectral density can be written as,

$$E[\hat{S}(\Omega)] = \sum_{l=-\infty}^{\infty} W(l)R(l)e^{-j\Omega l} \tag{5.133}$$

The window function $W(l)$ in (5.133) depends on the estimator used for $\hat{R}(l)$. The window function corresponding to the biased estimate of $R(l)$, see (5.130a), is called the Bartlett triangular window given by,

$$W_B(l) = \begin{cases} \dfrac{N - |l|}{N}; & |l| < N \\ 0; & \text{otherwise} \end{cases} \tag{5.134a}$$

The window function $W(l)$ that corresponds to the unbiased estimator of $R(l)$, see (5.130b) is called the rectangular window given by,

$$W_R(l) = \begin{cases} 1; & |l| < N \\ 0; & \text{otherwise} \end{cases} \tag{5.134b}$$

Figure 5.26 shows the Bartlett and rectangular window functions.

Thus, the mean power spectral density is obtained as the Fourier transform of the product of the autocorrelation and window functions. Using the property of the Fourier transform that products in the time domain can be converted to a convolution in the frequency domain,

$$E[\hat{S}(\Omega)] = \frac{1}{2\pi} W(\Omega) \otimes S(\Omega) \tag{5.135}$$

where $W(\Omega)$ is the discrete time Fourier transform of the window function. The symbol $\otimes$ indicates "cyclic convolution". All frequency domain convolutions for discrete time signals are cyclic convolutions because the frequency axis is limited from 0 to 2π. Equation (5.135) shows that the power spectral density estimate is a biased estimator since the expected value of $\hat{S}(\Omega)$ is not $S(\Omega)$.

For the Bartlett (triangular) window,

$$W_B(\Omega) = \frac{1}{N} \frac{\sin^2(N\Omega/2)}{\sin^2(\Omega/2)} \tag{5.136a}$$

and for the rectangular window,

$$W_R(\Omega) = \frac{1}{N} \frac{\sin[(2N-1)\Omega/2]}{\sin(\Omega/2)} \tag{5.136b}$$

It can be seen that even though the estimates of the power spectral density are biased, both estimates are asymptotically unbiased,

$$\lim_{N\to\infty} E\left[\hat{S}(\Omega)\right] = S(\Omega) \tag{5.137}$$

From (5.136a, b) it can be seen that $W_B(\Omega) \geq 0$, whereas $W_R(\Omega)$ can be negative in certain regions. As a result, the expected value of the spectral estimate $\hat{S}(\Omega)$ computed using the unbiased autocorrelation estimate, see (5.130b), can be negative, which is absurd. This is essentially due to the fact that the unbiased autocorrelation function estimate given by (5.130b) does not satisfy the properties of an autocorrelation function.

The variance of the periodogram estimate $\hat{S}(\Omega)$ is given (Brockwell and Davis 1991) by,

$$\text{var}[\hat{S}(\Omega)] \approx [S(\Omega)]^2 \tag{5.138}$$

Typically, FFTs are used to estimate the power spectral density. However, it was shown in (5.121) that the FFT algorithm actually computes the discrete Fourier transform (as opposed to the discrete time Fourier transform). Use of the discrete Fourier transform indicates that samples of $\hat{S}(\Omega)$ are computed at N equally spaced frequencies, namely $\Omega_k = 2\pi k/N, k = 0, \ldots, N-1$. The spectral estimate $\hat{S}(\Omega)$ computed at Ω_k is asymptotically uncorrelated,

$$\lim_{N\to\infty} \text{cov}[\hat{S}(\Omega_{k1}), \hat{S}(\Omega_{k2})] \approx 0 \tag{5.139}$$

where cov represents covariance. Because the standard deviation of $\hat{S}(\Omega)$ is proportional to its mean value, and adjacent estimates of $\hat{S}(\Omega)$ are uncorrelated, it produces widely fluctuating records of power spectral density estimates. There are many techniques available in the literature to improve the basic power spectral density estimates (Brockwell and Davis 1991).

5.8   Noise in radar systems

The concept of noise is used to represent all unwanted signals that enter into a receiver system. There are many potential sources of noise, which may be external or internal to the receiver system. All objects with a physical temperature greater than 0 K generate noise. In a radar system the antenna receives noise contributions from the surroundings. The external noise contribution can come from the following, namely: cosmic background noise, galactic noise, noise due to precipitation, solar noise, Earth noise, noise from nearby structures such as radome and buildings, etc. Internal noise is generated by electrical circuits that are used in the receiver chain. Two types of internal noise are shot noise and thermal noise (Van der Ziel 1970). Noise temperature is a useful concept to characterize the noise received by the system under consideration. The noise power, P_N, due to an external source which is received by a receiver of bandwidth B can be written as,

$$P_N = kT_N B \tag{5.140}$$

where k is Boltzmann's constant $= 1.38 \times 10^{-23}$ J K^{-1}, B is the bandwidth of the receiver in Hz, and T_N is the noise temperature of the source.

Since noise exists at all frequencies, a conventional model for noise is called white noise, where the power spectral density of noise is a constant, $N_0/2$ (see Fig. 5.27). Most models superpose noise on the signal, indicated by the term additive noise. The power spectral density of white noise (delivered to a load whose impedance is matched to the source) is expressed as,

$$S_N(f) = \frac{N_0}{2} = kT_N \tag{5.141}$$

If this noise is input to an ideal bandpass receiver whose bandwidth is B (see Fig. 5.28), then the received noise power from this source is $P_N = kT_N B$.

The radar receiver consists of many devices connected in cascade such as amplifiers, filters and mixers that generate their own noise. Even without any input signal these devices will produce some noise output. Let a two port device (such as an amplifier or filter) be specified with gain G, and bandwidth B. Let a noise source with temperature T_N be connected to its input. The output noise power is then given by,

$$P_N = GkT_N B + N_D \tag{5.142}$$

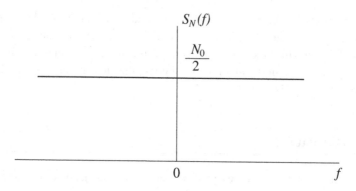

Fig. 5.27. Power spectral density of white noise.

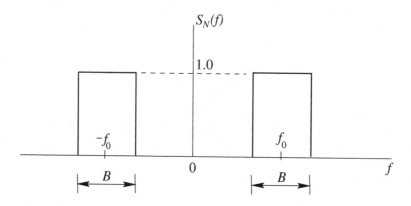

Fig. 5.28. Noise spectrum in an idealized bandpass receiver.

where N_D is the output noise power produced by the device itself. Equation (5.142) can be rewritten as,

$$P_N = GkB\left(T_N + \frac{N_D}{GkB}\right) = GkB(T_N + T_D) \qquad (5.143)$$

The noise produced by the device can be modeled by a noise source of equivalent noise temperature T_D connected to the input (see Fig. 5.29).

The noise figure (F) is a measure used to characterize the internal noise generated by a device or a complete receiver chain. F can be defined in terms of the signal-to-noise ratio at the input and output of a device/system,

$$F = \frac{(\text{SNR})_{\text{in}}}{(\text{SNR})_{\text{out}}} \qquad (5.144)$$

where SNR is the signal-to-noise ratio. It is equivalent to the output noise power divided by the output noise power if the system is noiseless. It is also assumed that the noise sources at the input are at the ambient temperature T_0 (nominal value for T_0 is 290 K).

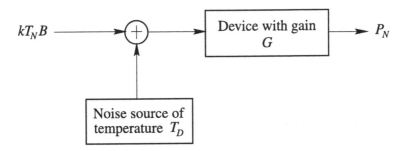

Fig. 5.29. Noise model for a two-port device in terms of noise temperature T_D.

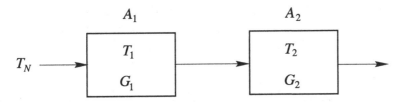

Fig. 5.30. Two-port devices connected in cascade.

Thus F can be expressed as,

$$F = \frac{GkT_0 B + N_D}{GkT_0 B} = \left(1 + \frac{T_D}{T_0}\right)$$ (5.145)

Therefore,

$$T_D = (F - 1)T_0$$ (5.146)

Typically, the radar receiver, as mentioned earlier, consists of devices connected in cascade such as filters, amplifiers and mixers. Consider a system of two devices A_1 and A_2 connected in cascade with noise temperatures T_1, T_2 and gains G_1, G_2, respectively (see Fig. 5.30). The noise power P_{N1} at the output of A_1 is given by (5.143),

$$P_{N1} = G_1 k B(T_N + T_1)$$ (5.147)

This noise is amplified by A_2, and the noise power at the output of A_2 is given by,

$$P_{N2} = G_2 G_1 k B(T_N + T_1) + G_2 k B T_2$$ (5.148a)

$$= G_1 G_2 k B\left(T_N + T_1 + \frac{T_2}{G_1}\right)$$ (5.148b)

The cascade unit has gain $G_1 G_2$ and, therefore, it can be characterized by an equivalent noise temperature,

$$T_e = T_1 + \frac{T_2}{G_1}$$ (5.149)

This result can be generalized to the equivalent noise temperature of n devices connected in cascade as,

$$T_e = T_1 + \frac{T_2}{G_1} + \frac{T_3}{G_1 G_2} + \cdots + \frac{T_n}{G_1 G_2 \cdots G_{n-1}} \qquad (5.150)$$

Similarly, the equivalent noise figure F_e of n devices connected in cascade can be written in terms of the noise figure F_i of device i as,

$$F_e = F_1 + \frac{F_2 - 1}{G_1} + \frac{F_3 - 1}{G_1 G_2} + \cdots + \frac{F_n - 1}{G_1 G_2 \cdots G_{n-1}} \qquad (5.151)$$

It can be seen from (5.150) that if $G_1 \gg 1$, then the devices from A_2 onwards contribute negligibly to the overall noise temperature of the receiver chain. Therefore, it is extremely important to have the first amplifier as a low noise device, typically referred to as a low noise amplifier (LNA), with sufficient gain so as to render the noise contribution from succeeding stages in the receiver negligible.

In addition to active devices with gains, there are also ohmic losses in transmission lines such as waveguide and coaxial cables. For a lossy device with loss L ($L > 1$), the output noise is given by $kT_0 B$, where T_0 is the ambient temperature. With $T_N = T_0$ and gain $= 1/L$, the noise power is given by,

$$kT_0 B = \frac{1}{L} kB(T_0 + T_D) \qquad (5.152)$$

where T_D is the equivalent noise temperature given by,

$$T_D = (L - 1)T_0 \qquad (5.153)$$

Using (5.144), the noise figure F of a lossy device is,

$$F = L \qquad (5.154)$$

In many radar systems a significant segment of transmission line is often used to connect the antenna feed to the low noise amplifier, and then to the rest of the receiver chain as shown in Fig. 5.31. The antenna noise, as measured at the output of the antenna feed is composed of antenna noise received via the complete antenna pattern. The antenna noise temperature is also dependent on the elevation angle. Typical noise temperatures of pencil beam antennas can be of the order of 50 to 100 K (but is widely variable with elevation angle). The transmission line connecting the antenna to the LNA has a loss L, with equivalent noise temperature $T_1 = (L - 1)T_0$. The LNA is specified by its gain (G_2) and noise temperature (T_2). There is wide variation in the noise temperature of LNAs depending on the technology used. Cryogenically cooled parametric amplifiers have noise temperatures as low as 20–30 K, whereas other types of LNAs can have T_N as high as 70–100 K. In Fig. 5.31, the equivalent noise temperature of the LNA and receiver system is $T_e = T_2 + T_3/G_2$. The noise power at the output of the transmission

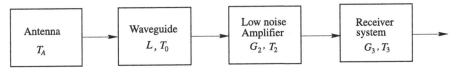

Fig. 5.31. Block diagram showing the gain/loss and noise temperature of antenna, waveguide, low noise amplifier, and receiver.

line is given, from (5.147) by,

$$P_N = \frac{kT_A B}{L} + \frac{(L-1)T_0 k B}{L} \tag{5.155a}$$

$$= kB\left[\frac{T_A}{L} + \frac{(L-1)}{L}T_0\right] \tag{5.155b}$$

Therefore, the noise temperature T'_A measured at the output of the waveguide is

$$T'_A = \frac{T_A}{L} + \frac{L-1}{L}T_0 \tag{5.156}$$

The total noise temperature referred to the input of the LNA is thus,

$$T = \frac{T_A}{L} + \frac{L-1}{L}T_0 + T_2 + \frac{T_3}{G_2} \tag{5.157}$$

The transmission line loss and antenna temperature can be combined with the noise temperature of the receiver chain to specify an equivalent noise temperature referred to the input of the low noise amplifier. This reference point is commonly used to specify the noise temperature of the system.

The power spectral density of noise in radar receivers is often modeled as additive white noise with power spectral density $N_0/2$ (see Fig. 5.27). The autocorrelation function of white noise is given by,

$$R_N(t) = \frac{N_0}{2}\delta(t) \tag{5.158}$$

White noise is only a concept and is not physically realizable since it has infinite power. However, white noise when observed by a finite bandwidth device such as a radar receiver gives rise to band limited white noise. Band limited white noise is characterized by a flat spectrum between $-W$ and W, as shown in Fig. 5.32a. The autocorrelation function of band limited noise is given by $N_0 W \, \mathrm{sinc}(2Wt)$. The autocorrelation function has zero crossings at $t = l/2W; l = \pm 1, \pm 2, \ldots$. If the band limited noise is sampled at the Nyquist rate of $2W$, or sample spacing of $T = 1/2W$, then the autocorrelation function also gets sampled at time interval $T = 1/2W$, see Fig. 5.32b. When the samples are converted to discrete time, then the autocorrelation function is represented by,

$$R_N[n] = 2W\left(\frac{N_0}{2}\right)\delta[n] \tag{5.159}$$

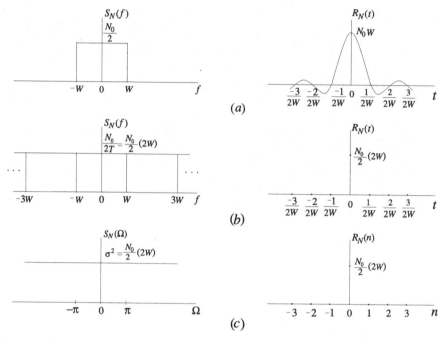

Fig. 5.32. Spectrum and autocorrelation function of band limited white noise. (*a*) Band limited noise spectrum and autocorrelation function. (*b*) Spectrum of band limited noise sampled at the Nyquist sampling rate, and the corresponding autocorrelation function. (*c*) Spectrum and autocorrelation function of the discrete time noise sequence.

where $\delta[n]$ is a discrete time impulse function defined as,

$$\delta[n] = \begin{cases} 1; & n = 0 \\ 0; & n \neq 0 \end{cases} \tag{5.160}$$

Thus, band limited noise with power spectral density $N_0/2$ is converted by sampling to a discrete time noise sequence with noise power equal to $2W N_0/2$, see Fig. 5.32*c*.

5.9   Statistical properties of the received signal

In a pulsed Doppler radar, the received signal can be written as the sum of the individual scattered signals from all the scatterers in the resolution volume. It was shown through (5.32) that each sample of the received signal can be described as a sum of elemental phasors with phases uniformly distributed between $(-\pi, \pi)$. By invoking the central limit theorem (Papoulis 1965), which describes the asymptotic distribution of sums of large numbers of independent random variables, it can be shown that the real and imaginary parts of the received signal are Gaussian distributed with zero mean. The properties of the received signal resemble those of narrowband Gaussian noise encountered in statistical communication theory (Middleton 1960). The received signal

$V(t)$ can be written as,

$$V(t) = I(t) + jQ(t) \tag{5.161}$$

where I and Q are the real and imaginary components (or in-phase and quadrature phase; see Fig. 5.8). The real and imaginary components carry information about the stochastic process and have to satisfy the following properties.

1. The in-phase and quadrature phase components of the received voltage at the same time are uncorrelated,

$$\text{cov}[I(t), Q(t)] = 0 \tag{5.162}$$

where cov indicates covariance.

2. The means of $I(t)$ and $Q(t)$ are zero,

$$E[I(t)] = E[Q(t)] = 0 \tag{5.163}$$

3. The in-phase and quadrature phase components have the same variance,

$$E[I^2(t)] = E[Q^2(t)] = \sigma^2 \tag{5.164}$$

4. The real and imaginary parts have the same autocorrelation function,

$$E[I(t_1)I(t_2)] = E[Q(t_1)Q(t_2)] = \sigma^2 \rho_0(t); \quad t = t_2 - t_1 \tag{5.165}$$

5. Even though the in-phase and quadrature phase components at the same time are uncorrelated, the in-phase and quadrature phase components at two different times are correlated,

$$E[I(t_1)Q(t_2)] = \sigma^2 \alpha_0(t) = -E[I(t_2)Q(t_1)]; \quad t = t_2 - t_1 \tag{5.166}$$

Using the above properties, the covariance matrix of the vector $\mathbf{Z} = [I(t_1) \ Q(t_1)I(t_2)Q(t_2)]^t$ can be written as,

$$E[\mathbf{ZZ}^t] = \sigma^2 \begin{bmatrix} 1 & 0 & \rho_0 & \alpha_0 \\ 0 & 1 & -\alpha_0 & \rho_0 \\ \rho_0 & -\alpha_0 & 1 & 0 \\ \alpha_0 & \rho_0 & 0 & 1 \end{bmatrix} \tag{5.167}$$

Let $\mathbf{V} = [V_1, V_2, \ldots, V_n]^t$ be n time samples of the received signal (the subscript r is dropped here) from a given resolution volume at times $t_1, t_2, \ldots, t_n$, respectively. The complex vector $\mathbf{V}$ can be written in terms of two real Gaussian vectors representing the in-phase and quadrature phase components. The signal sample V_k can be written as,

$$V_k = I_k + jQ_k; \quad k = 1, \ldots, n \tag{5.168a}$$

Similarly, the components of the complex vector $\mathbf{V}$ can be written as,

$$V_k = a_k \exp(j\theta_k); \quad k = 1, \ldots, n \tag{5.168b}$$

where $a_k = |V_k|$. The multivariate density function of $\mathbf{V}$ is given (Middleton 1960) by,

$$f(\mathbf{V}) = \frac{1}{\pi^n \det \mathbf{R}} \exp[-(\mathbf{V}^*)^t \mathbf{R}^{-1} \mathbf{V}] \tag{5.168c}$$

where $\mathbf{R}$ is the covariance matrix of the complex vector $\mathbf{V}$ defined as,

$$\mathbf{R} = E[(\mathbf{V})(\mathbf{V}^*)^t] \tag{5.168d}$$

5.9.1   Probability distribution functions

For a single sample with variance R, the probability density of the complex signal can be written as,

$$f(V) = \frac{1}{2\pi\sigma^2} \exp\left(\frac{-|V|^2}{2\sigma^2}\right) \tag{5.169}$$

It can be seen from (5.164) that $R = 2\sigma^2$. Using transformation of variables, the above equation can be written in terms of amplitude and phase as,

$$f(a, \theta) = \frac{1}{2\pi} \frac{a}{\sigma^2} \exp\left(-\frac{a^2}{2\sigma^2}\right) \tag{5.170a}$$

$$= f_\Theta(\theta) f_A(a) \tag{5.170b}$$

where,

$$f_A(a) = \frac{a}{\sigma^2} \exp\left(-\frac{a^2}{2\sigma^2}\right); \quad a > 0 \tag{5.171a}$$

$$f_\Theta(\theta) = \frac{1}{2\pi}; \quad 0 < \theta \le 2\pi \tag{5.171b}$$

Thus it can be seen that the joint distribution, $f(a, \theta)$, can be written as the product of the (marginal) distributions of the amplitude (called Rayleigh density) and phase (uniform density), indicating that the amplitude and phase are independent.

Similarly, the probability density function of the signal power P can be obtained from transformation of variables, since $P_k = |V_k|^2$. Therefore, the probability density function of signal power is given by,

$$f_P(P) = \frac{1}{2\sigma^2} e^{-P/2\sigma^2}; \quad P > 0 \tag{5.172}$$

The above expression for the density of power shows that it is an exponential distribution. The mean power of the signal is given by $\bar{P} = E(P) = E(|V|^2) = 2\sigma^2$, and $\text{var}(P) = (\bar{P})^2$. Therefore, (5.172) can also be written as,

$$f_P(P) = \frac{1}{\bar{P}} e^{-P/\bar{P}}; \quad P > 0 \tag{5.173}$$

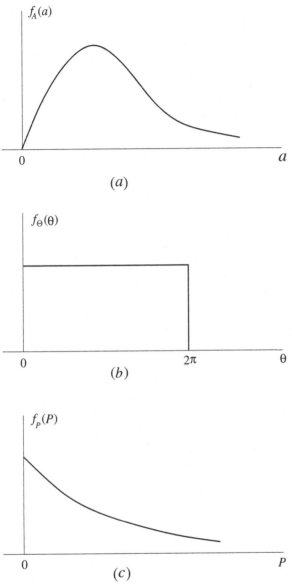

Fig. 5.33. Probability density function of (*a*) signal amplitude, (*b*) signal phase, and (*c*) signal power.

Figure 5.33 shows the probability density function of amplitude, phase and power of the received signal. Thus, in summary: (i) the amplitude of the received signal is distributed as a Rayleigh density, (ii) the phase of the received signal is distributed uniformly between 0 to 2π, (iii) the power of the received signal is distributed as an exponential density, and (iv) the amplitude and phase of the received signal are independent.

The above distributions are used to obtain properties of estimators derived from the received signal. The joint distribution of a sequence of signal amplitudes and phases can be derived in principle, but the expressions are complicated and omitted here. The joint distribution of phases of two adjacent signal samples is important in the understanding of the fluctuations of the Doppler frequency estimate. It will be shown later in Section 5.11 that the mean Doppler frequency can be estimated from the average phase difference between adjacent signal samples.

Let θ_1, θ_2 be the phases of the received signal at times t_1 and t_2, respectively. Then, the joint distribution of phases θ_1, θ_2 can be written (Middleton 1960) as,

$$f(\theta_1, \theta_2) = \frac{1 - \gamma^2(t)}{4\pi^2}(1 - \beta^2)^{-3/2}\left(\beta \sin^{-1} \beta + \frac{\pi\beta}{2} + \sqrt{1 - \beta^2}\right); \quad -1 \leq \beta \leq 1$$

(5.174)

where $\beta = \gamma(t)\cos(\theta_2 - \theta_1 - \theta_0)$, $\theta_0 = \tan^{-1}(\alpha_0/\rho_0)$, and $\gamma(t)$ is the magnitude of the correlation coefficient between V_1 and V_2. Note that the joint distribution of (θ_1, θ_2) is only a function of the magnitude of the correlation coefficient between the samples and $\theta_2 - \theta_1$. Figure 5.34 shows the probability density function of (θ_1, θ_2) with the magnitude of the correlation coefficient between the signal samples as a parameter. The figure demonstrates an important concept, namely that the distribution of phase difference $(\theta_2 - \theta_1)$ between two signal samples is narrow (i.e. small variance) if the correlation between them is high. On the contrary, if the correlation is zero, then the phase difference is uniformly distributed between 0 and 2π, indicating that the phase difference cannot be estimated. This fact demonstrates the need for samples to be correlated in order to estimate the Doppler phase shift. The narrowness of the distribution of $(\theta_2 - \theta_1)$ can be computed from the variance of $(\theta_2 - \theta_1)$, and this variance can be expressed (Middleton 1960) as,

$$\text{var}[(\theta_2 - \theta_1)] = \frac{\pi^2}{6}\{4 - \Omega[\gamma(t)]\} - 2\pi[\sin^{-1}\gamma(t)] + 2[\sin^{-1}\gamma(t)]^2$$

(5.175a)

where,

$$\Omega[\gamma(t)] = \frac{6}{\pi^2}\sum_{n=1}^{\infty}\frac{[\gamma(t)]^{2n}}{n^2}$$

(5.175b)

Figure 5.35 shows the standard deviation of $(\theta_2 - \theta_1)$ as a function of $\gamma(t)$. The standard deviation of $(\theta_2 - \theta_1)$ increases as $\gamma(t)$ decreases. Thus, for accurate estimation of the mean Doppler phase shift, the correlation $\gamma(t)$ between V_1 and V_2 should be high, which imposes restrictions on the sampling rate of the received signal. Similar to the joint distribution of phases θ_1 and θ_2, the joint distribution of amplitudes (a_1, a_2) can be obtained (Middleton 1960) as,

$$f(a_1, a_2) = \frac{a_1 a_2}{\sigma^4(1 - \gamma^2)}e^{-(a_1^2 + a_2^2)/2\sigma^2(1 - \gamma^2)}I_0\left[\frac{\gamma a_1 a_2}{\sigma^2(1 - \gamma^2)}\right]$$

(5.176)

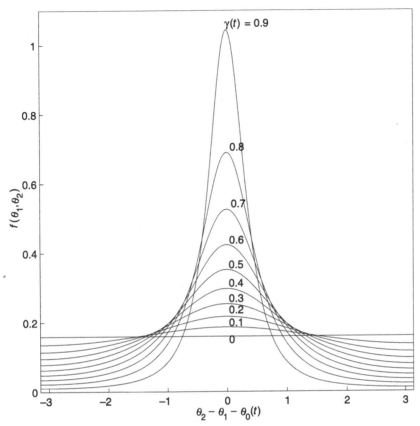

Fig. 5.34. The joint probability density of (θ_1, θ_2) shown as a function of $(\theta_2 - \theta_1 - \theta_0(t))$. Note that γ is the magnitude of the correlation coefficient between the complex signal samples.

where γ is the magnitude of the correlation coefficient between the samples V_1 and V_2, and I_0 is the modified Bessel function of order zero and first kind. The correlation between the power samples can be obtained from (5.176) as, $E(a_1^2 a_2^2) = 4\sigma^2(1 + \gamma^2)$. Since $E(a^2) = 2\sigma^2$, the correlation coefficient between the power samples is γ^2. This result is useful in obtaining the number of samples needed to estimate the mean power of the signal to a prescribed accuracy.

5.9.2 Joint distribution of dual-polarized signals

The joint distribution of dual-polarized signals is determined by the scattering properties of particles at the corresponding polarizations. The received signal at the two polarization states can be written as $V_h^{10}(t)$, $V_v^{10}(t)$, $V_h^{01}(t)$, and $V_v^{01}(t)$ where subscripts h, v indicate horizontal and vertical polarized returns. The superscript 10 indicates that the transmit polarization state is horizontal only, while 01 indicates that the transmit state is vertical only. Thus, the two copolar signals are $V_h^{10}(t)$ and $V_v^{01}(t)$, while $V_h^{01}(t)$ and

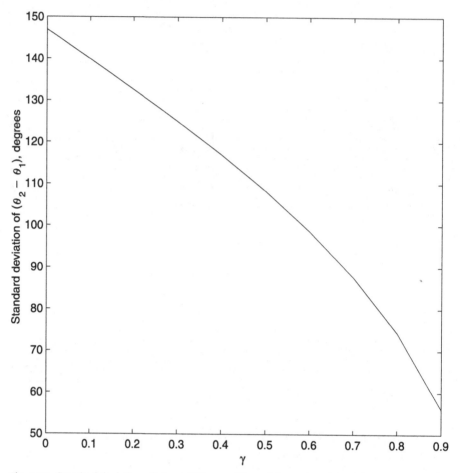

Fig. 5.35. Standard deviation of phase difference $(\theta_2 - \theta_1)$ as a function of the magnitude of the correlation coefficient between the complex signal samples.

$V_v^{10}(t)$ are the two cross-polar signals. These signals can be written in terms of their in-phase and quadrature phase components as,

$$V_h^{10}(t) = I_h^{10}(t) + jQ_h^{10}(t) \qquad (5.177a)$$

$$V_v^{10}(t) = I_v^{10}(t) + jQ_v^{10}(t) \qquad (5.177b)$$

$$V_v^{01}(t) = I_v^{01}(t) + jQ_v^{01}(t) \qquad (5.177c)$$

$$V_h^{01}(t) = I_h^{01}(t) + jQ_h^{01}(t) \qquad (5.177d)$$

From a practical viewpoint it is not possible to observe all four signals at the same time, nevertheless the joint distribution can be specified. Similar to the covariance matrix of the received signal at two different times t_1, t_2 given by (5.167), the covariance matrix of the signals at the same time but at two different polarizations can be constructed as follows. Let **Z** be the vector $[I_1\ Q_1\ I_2\ Q_2]^t$, and let $I_1(t) + jQ_1(t) = V_1(t)$ and

$I_2(t) + jQ_2(t) = V_2(t)$, where $V_1(t)$ and $V_2(t)$ are any two of the four signals listed in (5.177). The covariance matrix of $\mathbf{Z}$ can be written as,

$$E[\mathbf{Z}\mathbf{Z}'] = \begin{bmatrix} \sigma_1^2 & 0 & \sigma_1\sigma_2\rho_{12} & \sigma_1\sigma_2\alpha_{12} \\ 0 & \sigma_1^2 & -\sigma_1\sigma_2\alpha_{12} & \sigma_1\sigma_2\rho_{12} \\ \sigma_1\sigma_2\rho_{12} & -\sigma_1\sigma_2\alpha_{12} & \sigma_2^2 & 0 \\ \sigma_1\sigma_2\alpha_{12} & \sigma_1\sigma_2\rho_{12} & 0 & \sigma_2^2 \end{bmatrix} \tag{5.178}$$

The values of $\sigma_1, \sigma_2, \rho_{12}$, and α_{12} depend on the choice of $V_1(t)$ and $V_2(t)$ from the list in (5.177). If V_1, V_2 are $V_h^{10}(t)$ and $V_v^{01}(t)$, then the covariance matrix terms describes the copolar covariance. Thus,

$$E[(V_h^{10})^* V_v^{01}] = \sqrt{\mathrm{var}(V_h^{10})\,\mathrm{var}(V_v^{01})}\,\rho_{co} \tag{5.179}$$

Substituting for the variance of $V(t)$ in terms of the mean power, (5.179) can be reduced to,

$$E[(V_h^{10})^* V_v^{01}] = \sqrt{P_{co}^h P_{co}^v}\,\rho_{co} \tag{5.180}$$

where ρ_{co} indicates the correlation coefficient between the copolar signals, see also (3.182c). Similarly, the covariance between V_h^{10} and V_v^{10} can be expressed as,

$$E[(V_h^{10})^* V_v^{10}] = \sqrt{P_{co}^h P_{cx}}\,\rho_{cx}^h \tag{5.181}$$

where ρ_{cx}^h indicates the correlation coefficient between copolar and cross-polar signals, see also (4.78).

The correlation coefficients ρ_{co} and ρ_{cx}^h defined in (5.180, 5.181) can be complex, and are the same as those defined in (4.77, 4.78) from physical considerations. Note that for meteorological scatterers the back scattered signal from an ensemble of scatterers at copolar horizontal, $V_h^{10}(t)$, and vertical polarization, $V_v^{01}(t)$, have a high degree of correlation. However, the correlation coefficient between copolar ($V_h^{10}(t)$) and cross-polar ($V_v^{10}(t)$) components is low. Similarly, the correlation coefficient between V_h^{01} and V_v^{01} is low. The ratio of the two copolar received signals at the same time can be separated as magnitude ratio and phase difference as,

$$\frac{V_v^{01}(t)}{V_h^{10}(t)} = \frac{a_v e^{j\theta_v}}{a_h e^{j\theta_h}} \tag{5.182}$$

where a_v and a_h are the amplitudes of the signals. The above equation can be written in terms of the corresponding variances or mean powers as,

$$\frac{a_v e^{j\theta_v}}{a_h e^{j\theta_h}} = \frac{1}{\sqrt{\mathcal{Z}_{dr}}} \frac{a_v/\sqrt{P_{co}^v}}{a_h/\sqrt{P_{co}^h}} e^{j(\theta_v - \theta_h)} \tag{5.183a}$$

$$= \frac{1}{\sqrt{\mathcal{Z}_{dr}}} u e^{j(\theta_v - \theta_h)} \tag{5.183b}$$

where u is the normalized amplitude ratio and $\not z_{dr}$ is the differential reflectivity, e.g. see (3.231c). The probability density function of the normalized amplitude ratio can be expressed as,

$$f(u) = \frac{2u(1+u^2)(1-|\rho_{co}|^2)}{[(1+u^2)^2 - 4|\rho_{co}|^2 u^2]^{3/2}} \tag{5.184}$$

where $|\rho_{co}|^2$ is the correlation between P_{co}^h and P_{co}^v (Bringi et al. 1983). Figure 5.36 shows the probability density function of u with $|\rho_{co}|$ as a parameter. When $|\rho_{co}|^2$ is high then the amplitude ratio has a very narrow distribution. Similarly, the distribution of the differential phase $(\theta_v - \theta_h)$ can be obtained from the joint distribution of the phases (Middleton 1960) as,

$$f(\theta_v, \theta_h) = \frac{(1-|\rho_{co}|^2)}{4\pi^2}(1-\beta^2)^{-3/2}\left(\beta\sin^{-1}\beta + \frac{\pi\beta}{2} + \sqrt{1-\beta^2}\right); \quad -1 \le \beta \le +1 \tag{5.185}$$

where $\beta = |\rho_{co}|\cos(\theta_v - \theta_h - \bar{\theta}_0)$, and $\bar{\theta}_0 = \arg(\rho_{co})$. Note that the joint distribution of θ_v, θ_h is only a function of the difference $(\theta_v - \theta_h)$. Also, note that (5.185) is identical to (5.174) with $|\rho_{co}|^2 \equiv \gamma^2(t)$, $\theta_v \equiv \theta_1$, $\theta_h \equiv \theta_2$. Hence, the pdf of $f(\theta_v, \theta_h)$ is also given by Fig. 5.34, which illustrates an important concept, namely that the distribution of phase difference between the two signals is narrow if the correlation between them is high. If the correlation is zero, then the phase difference is uniformly distributed between $(-\pi, \pi)$. Thus, if the estimate of phase difference between two complex samples is needed, the estimate will be more accurate if the correlation is high. This has important implications for the estimation of differential phase between horizontal and vertical polarized copolar signals, see (4.65). The variance of the distribution indicates the accuracy in the estimate of phase difference, $(\theta_v - \theta_h)$. The variance of $(\theta_v - \theta_h)$ can be expressed (Middleton 1960) as,

$$E[(\theta_v - \theta_h)^2] = \frac{\pi^2}{6}[4 - \Omega(|\rho_{co}|)] - 2\pi[\sin^{-1}(|\rho_{co}|)] + 2[\sin^{-1}(|\rho_{co}|)]^2 \tag{5.186a}$$

where,

$$\Omega(|\rho_{co}|) = \frac{6}{\pi^2}\sum_{n=1}^{\infty}\frac{|\rho_{co}|^{2n}}{n^2} \tag{5.186b}$$

Note that (5.186) is identical to (5.175) with $|\rho_{co}|^2 \equiv \gamma^2(t)$, $\theta_v \equiv \theta_1$, and $\theta_h \equiv \theta_2$ (Kostinski 1994). Hence, the standard deviation of $(\theta_v - \theta_h)$ as a function of $|\rho_{co}|$ is identical to that shown in Fig. 5.35. It can be seen from Fig. 5.35 that the standard deviation of $(\theta_v - \theta_h)$ increases as $|\rho_{co}|$ decreases. This feature is similar to the one encountered in the estimation of Doppler phase shift. In a pulsed radar system that alternates between horizontal and vertical transmitted pulses, the corresponding copolar returns at H and V polarization states are not available at the same time. However, in the

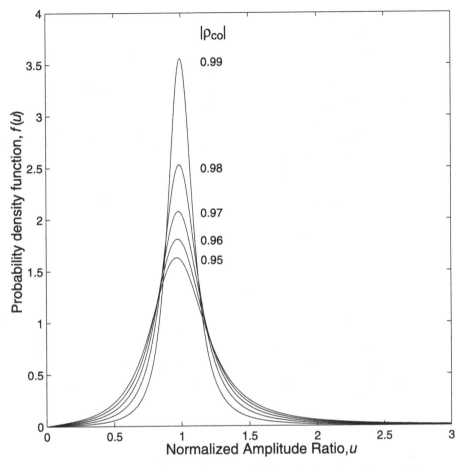

Fig. 5.36. Probability density function of the normalized amplitude ratio, u, for correlation coefficients, $|\rho_{co}|$, from 0.95 to 0.99.

hybrid mode of operation (discussed in Section 4.7), the horizontal and vertical returns are sampled at the same time (under either a circular or 45° linear transmit polarization state). In this mode of operation, it is possible to use the variance formula for the differential phase given in (5.186).

5.10   Estimation of mean power

The mean power of the received signal can be obtained by averaging the instantaneous power samples. The samples of the signal $V(t)$ are obtained as a sequence of voltages $V[n]$, $n = 1, 2, \ldots, N$. The estimate of the mean power $\bar{P}$ can be obtained by averaging the samples of power as,

$$\hat{\bar{P}} = \frac{1}{N} \sum_{n=1}^{N} P_n \tag{5.187}$$

where $P_n = |V[n]|^2$, and $\bar{P}$ is the mean power. The radar reflectivity and the equivalent reflectivity factor (Z_e) are proportional to the mean power of the received signal, see (5.49). As the number of samples used in the estimate of the mean power increases, the estimate will converge to the mean power of the signal. The variance of the mean power estimate can be obtained as,

$$\text{var}[\hat{\bar{P}}] = \frac{1}{N^2} \sum_{i=1}^{N} \sum_{j=1}^{N} \text{cov}(P_i, P_j) \tag{5.188}$$

where $\text{cov}(P_i, P_j)$ denotes the covariance between power samples P_i and P_j. This can be simplified as,

$$\text{var}[\hat{\bar{P}}] = \frac{1}{N} \sum_{l=-(N-1)}^{(N-1)} \left(1 - \frac{|l|}{N}\right) R_p[l] \tag{5.189}$$

where $R_p[l]$ is the autocorrelation function of the power samples at lag l. Based on the statistics of the received signal it was shown in Section 5.9.1 that the correlation coefficient (ρ_p) between any two power samples can be obtained as the magnitude square of the correlation coefficient between the complex signal samples $(|\rho(l)| = \gamma)$, see discussion related to (5.176). Hence,

$$\rho_p[l] = |\rho[l]|^2 \tag{5.190}$$

In addition, the autocovariance at lag l, $R_p[l]$, is defined as,

$$R_p[l] = R_p[0]\rho_p[l] \tag{5.191}$$

where $R_p[0]$ is the autocovariance at lag 0 which is identical to the variance of the power samples, $(\bar{P})^2$, see (5.172). Substituting (5.190) and (5.191) in (5.189), $\text{var}[\hat{\bar{P}}]$ reduces to,

$$\text{var}(\hat{\bar{P}}) = \frac{(\bar{P})^2}{N} \sum_{l=-(N-1)}^{(N-1)} \left(1 - \frac{|l|}{N}\right) \rho_p[l] \tag{5.192}$$

If the signal samples are uncorrelated, then $\rho_p[l] = 0$ when $l \neq 0$, which would yield,

$$\text{var}(\hat{\bar{P}}) = \frac{(\bar{P})^2}{N} \tag{5.193}$$

indicating a factor of N reduction in the variance for N samples. In practice, the samples are correlated, see (5.95). Therefore, the reduction in the variance will be less than the factor N.

If the Doppler spectrum is of Gaussian shape described by parameters $\bar{v}$ and σ_v, see (5.127), then the corresponding autocorrelation of the signal sampled T_s apart, see (5.128), is given by,

$$R[n] = S_0 \exp\left(-\frac{8\pi^2\sigma_v^2 n^2 T_s^2}{\lambda^2}\right) \exp\left(-j\frac{4\pi\bar{v}nT_s}{\lambda}\right); \quad n = 1, 2, \dots, N \qquad (5.194)$$

where the magnitude of the correlation coefficient at lag n is given by,

$$|\rho[n]| = \exp\left(-\frac{8\pi^2\sigma_v^2 n^2 T_s^2}{\lambda^2}\right); \quad n = 1, 2, \dots, N \qquad (5.195)$$

With the above model the standard deviation of $\hat{P}$ can be estimated. Often the received signal power is expressed in the decibel (dB) scale, where,

$$\hat{P}(\text{dB}) = 10\log_{10}(\hat{P}) \qquad (5.196)$$

Perturbation analysis can be used to obtain an approximation for the variance of a non-linear function of a random variable. If $y = f(x)$, where x is a random variable (such as $x \equiv \hat{P}$), then,

$$\text{var}(y) \approx \text{var}(x)\left(\frac{df}{dx}\right)\Bigg|_{x=\bar{x}} \qquad (5.197)$$

The above equation is valid if x has a narrow distribution about its mean. The perturbation approximation is widely used in obtaining the variance of estimators which involve non-linear functions such as $10\log_{10}()$, $\tan^{-1}()$, etc. Applying (5.197) in the context of (5.196) gives,

$$\sigma[\hat{P}(\text{dB})] \approx 10\log_{10}\left[1 + \frac{\sigma(\hat{P})}{\bar{P}}\right] \qquad (5.198)$$

where $\sigma[\hat{P}(\text{dB})]$ is the standard deviation in dB, and $\sigma(\hat{P})$ is the standard deviation of the mean power. The standard deviation of mean power in dB is the same as the standard deviation of the equivalent reflectivity factor (Z_e) in dB (dBZ). Figure 5.37 shows the standard deviation of Z_e (dBZ) at S-band for a Gaussian-shaped spectrum as a function of the number of samples. It can be seen that Z_e (dBZ) can be estimated to an accuracy of around 1 dB for a typical wavelength of 10 cm and PRT of 1 ms. Note that for a fixed N, the standard deviation of Z_e decreases with increasing σ_v. This occurs because adjacent power samples are less correlated when σ_v increases, effectively yielding a greater number of "independent" samples for the estimate.

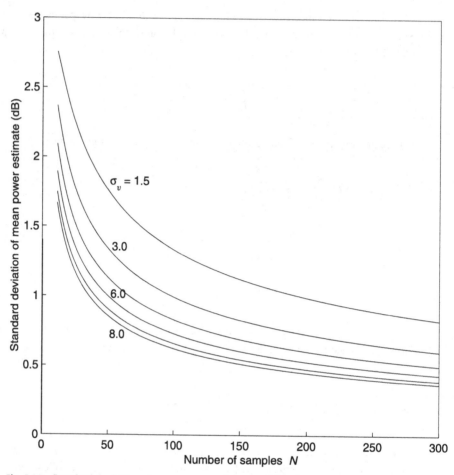

Fig. 5.37. Standard deviation of mean power estimate at S-band ($\lambda = 10$ cm) in dB shown for Gaussian-shaped Doppler spectrum with various σ_v (in m s^{-1}).

5.11   Doppler spectrum (or power spectral density) and estimate of mean velocity

The received signal at time t due to a transmitted pulse is the sum of scattered signals from the ensemble of particles in the resolution volume defined by the three-dimensional weighting function $W(r, \theta, \phi)$; see (5.60a). If the velocity of a single scatterer is $\vec{v}$ and if $\hat{i}$ and $\hat{s}$ are the unit vectors of the incident and scattered field directions, then the Doppler spectrum (or power spectral density) of the single particle is given by $\delta[\omega + k_0(\hat{i} - \hat{s}) \cdot \vec{v}]$ or $\delta[f + (\hat{i} - \hat{s}) \cdot \vec{v}/\lambda]$ (see Section 1.9). In the back scatter direction $\hat{s} = -\hat{i}$, and the Doppler spectrum is a delta function located at $f = -2\hat{i} \cdot \vec{v}/\lambda$, (see Fig. 5.38). A scatterer moving away from the radar will have $\hat{i} \cdot \vec{v} > 0$ and the Doppler spectrum will be at a negative frequency. If $\hat{i} \cdot \vec{v} = 0$, which can happen

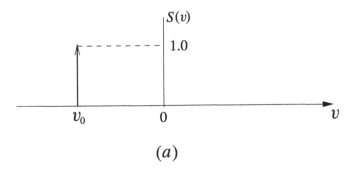

(a)

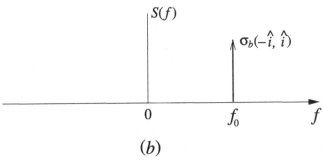

(b)

Fig. 5.38. Doppler velocity (a) and frequency spectrum (b) of a single particle with spectral power $S(f) = \sigma_b(-i, \hat{i})\delta(f - f_0)$. Note that $f_0 = -2v_0/\lambda$.

either when $\hat{i}$ and $\hat{v}$ are orthogonal or when $\vec{v}$ is zero, the Doppler spectrum is at zero frequency.[8] Since the spectral representation is linear, the power spectral density of the signal from an ensemble of independent particles within the resolution volume is a superposition of the power spectral densities due to the individual particles, see (5.7). If all the particles remain in the resolution volume during the observation period and if their velocities remain the same, then there is a stationary Doppler spectrum associated with the resolution volume. Though the individual particles contribute to a line spectrum corresponding to the radial velocity component $(\hat{i} \cdot \vec{v})$, the presence of a large number of particles with different velocities results in a continuum in the frequency domain. The spectral amplitude of a single particle at the frequency $f = -2\hat{i} \cdot \vec{v}/\lambda$ is given by its back scatter cross section, see (1.113).

When the Doppler spectrum is continuous, the spectral power contained in the interval $(f, f + \Delta f)$ is proportional to the sum of back scatter cross sections (or the reflectivity η) of all particles whose Doppler frequency shifts $(f_D = -2\hat{i} \cdot \vec{v}/\lambda)$ lie in the interval $f < f_D \le f + \Delta f$. If $\tilde{S}(v)$ is the actual radial velocity distribution function of the particles in the resolution volume, then the observed Doppler velocity (or frequency) spectrum is weighted by the reflectivity. Therefore, it is commonly referred to as the reflectivity-weighted Doppler velocity (or frequency) spectrum.

Any stochastic (e.g. noise) signal will have an associated power spectral density (loosely referred to as frequency spectrum), but it is not meaningful to associate every spectrum with an equivalent Doppler velocity spectrum. In the case of Doppler radar, the frequencies of the power spectral density can be interpreted as the Doppler frequency shifts which are related to the radial velocities by the Doppler principle ($v = -\lambda f/2$). It follows that the Doppler velocity spectrum ($S(v)$) can be derived from the Doppler frequency spectrum $S(f)$ by a simple transformation of the independent variable according to $S(v) = S(-\lambda f/2)$.

5.11.1 Mean of the Doppler velocity spectrum

In general, the Doppler frequency spectrum is not symmetric (about $f = 0$), which indicates that the autocovariance function is complex; refer to (5.124). If a frequency $\bar{f}$ exists about which the spectrum is symmetric, then that frequency can be defined as the mean frequency of the spectrum. Let $S(f)$ and $R(t)$ be Fourier transform pairs, where $S(f)$ is the Doppler frequency spectrum and $R(t)$ is the autocorrelation function. In that case, using (5.8), $R(t) \exp(-j2\pi \bar{f}t)$ is the Fourier transform pair of $S(f + \bar{f})$. If the spectrum $S(f)$ is symmetric about $\bar{f}$, then $S(f + \bar{f})$ is a symmetric spectrum about the frequency origin $f = 0$ (see Fig. 5.39). For a symmetric spectrum about the origin, the autocovariance function is real, which yields the result that the phase of the autocovariance function is given by $2\pi \bar{f}t$. Therefore, the mean frequency can be derived from the phase of the autocovariance function as,

$$\bar{f} = \frac{\arg[R(t)]}{2\pi t} \tag{5.199}$$

where arg stands for the argument of a complex number. From the mean frequency, the mean velocity can be obtained as,

$$\bar{v} = \frac{-\lambda}{4\pi t} \arg[R(t)] \tag{5.200}$$

If the spectrum is not naturally symmetric about a certain frequency $\bar{f}$, an equivalent $\bar{f}$ can still be defined by equating the argument of $R(t)$ to $2\pi \bar{f}t$. If the spectrum is not naturally symmetric about some $\bar{f}$ then the value of $\bar{f}$ given by (5.199) will depend on the choice of t.

The mean frequency can also be defined as the first moment of the Doppler spectrum as,

$$\bar{f} = \frac{\displaystyle\int_{-\infty}^{\infty} f S(f)\, df}{\displaystyle\int_{-\infty}^{\infty} S(f)\, df} \tag{5.201}$$

Using the properties of the Fourier transform, the nth order moment of the Doppler

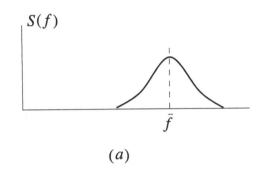

(a)

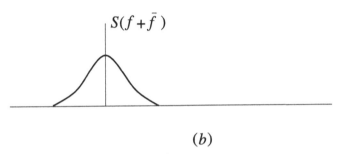

(b)

Fig. 5.39. A Doppler spectrum symmetric about $\bar{f}$ (a) is shifted by $\bar{f}$ so that it is symmetric about the origin (b).

spectrum is given (Papoulis 1965) as,

$$\int f^n S(f)\, df = \frac{1}{(j2\pi)^n} \frac{R^n(0)}{R(0)} \tag{5.202}$$

where $R^n(0)$ is the nth derivative evaluated at lag zero. Therefore, the mean velocity defined from the first moment of the Doppler spectrum can be obtained in terms of the autocorrelation function as,

$$\bar{f} = \frac{1}{j2\pi} \frac{R^1(0)}{R(0)} \tag{5.203}$$

The estimates of $\bar{f}$ (or $\bar{v}$) obtained from (5.199) and (5.203) are the same only if the spectrum is naturally symmetric about a mean frequency. Note that the effect of noise is not considered in (5.203).

5.11.2 Mean velocity estimation from sampled signals

In the case of pulsed Doppler radar, only samples of the received signal are available. Therefore all Fourier transforms convert to discrete time Fourier transforms. The spectral range for discrete time signals is $(-\pi, \pi)$ with the transformation $\omega = \Omega/T_s$, where ω is the frequency axis for the continuous time signals and Ω is the frequency axis for the discrete time signals. The spectrum of the sampled signal is periodic. When

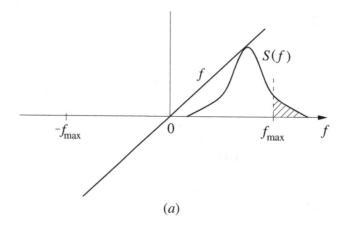

(a)

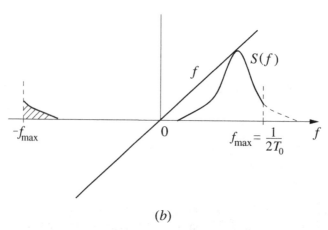

(b)

Fig. 5.40. Effect of aliasing on computation of the mean frequency or mean velocity as a spectral moment. The straight line represents the function "f", which multiplies the true spectrum to determine the first moment: (a) unaliased spectrum, (b) aliased spectrum.

the sampling frequency is not high enough, then aliasing is introduced in the sampled spectrum as shown in Fig. 5.21. Therefore, if the moment-based definition is used, see (5.201) for estimation of mean frequency or mean velocity, the estimate will be biased in the presence of aliasing as shown in Fig. 5.40. The product $f S(f)$ under aliasing will produce biased estimates. Since only signal samples are involved it follows that only samples of the autocorrelation are available. The autocorrelation of the sampled signal at lag n is the same as the autocorrelation of the continuous time signal at time lag nT_s, that is,

$$R[n] = R(nT_s) \tag{5.204a}$$

The autocorrelation function $R[n]$ can also be expressed in terms of the correlation coefficient $\rho[n]$ as,

$$R[n] = R[0]\rho[n] \tag{5.204b}$$

The expression for the discrete time autocovariance can be written in terms of the discrete time power spectral density as,

$$R[n] = \frac{1}{2\pi} \int_{-\pi}^{\pi} S(\Omega)e^{-j\Omega n}\, d\Omega \tag{5.205}$$

Substituting for $R[n]$ from (5.204) and $S(\Omega)$ from (5.118), (5.205) can be rewritten as,

$$R(nT_s) = \int_{-1/2T_s}^{1/2T_s} S(f)e^{-j2\pi f n T_s}\, df \tag{5.206}$$

Since $1/T_s$ is the sampling frequency, it follows from Section 5.7 that the maximum frequency component of the spectrum that can be observed without aliasing is $f_{max} = 1/2T_s$. The transformation $v = -\lambda f/2$ yields the corresponding maximum radial velocity that can be observed without aliasing as $v_{max} = -\lambda f_{max}/2 = -\lambda/4T_s$. Therefore, radial velocities in the range $(-\lambda/4T_s, \lambda/4T_s)$ can be observed without aliasing. The total interval is $\lambda/2T_s$, which is referred to as the unambiguous velocity interval. Since the maximum unambiguous range $r_{max} = cT_s/2$, the product $v_{max}r_{max} = c\lambda/8$, which is fixed by the operating radar wavelength. Thus, increasing r_{max} comes at the expense of decreasing v_{max} and vice versa.[9]

In (5.206), it can be seen that $\exp(j2\pi f n T_s)$ is periodic with $1/T_s$. Therefore, if part of the spectrum is outside the Nyquist frequency and is aliased, it will be multiplied the same way with $S(f)$ as if it was not aliased. Figure 5.41 shows the multiplication of $\exp(j2\pi f T_s)$ or $(\cos 2\pi f T_s + j \sin 2\pi f T_s)$ on the partially aliased spectrum. It can be seen that the mean frequency (or mean velocity) estimate based on the autocorrelation function is not affected by partial aliasing. However, if most of the spectrum is aliased, then the mean frequency or mean velocity estimate will be erroneous. Another important feature of the autocorrelation-based mean velocity estimate is that it is not biased by white noise, which appears as a constant across the spectral domain (see Fig. 5.27); note that when $S(f)$ in (5.206) is replaced by $N_0/2$, then the integral goes to zero. However, it is easily seen that the estimate of mean velocity based on (5.201) will be biased by noise. Special techniques are available to eliminate the noise bias when (5.201) is used (Gossard and Strauch 1983). Even though the autocorrelation-based estimate of $\bar{f}$ (or $\bar{v}$) has the advantages noted above, it is accurate only if the Doppler spectrum is naturally symmetric about some mean velocity $\bar{v}$. The estimate of $\bar{v}$ from the autocorrelation function, see (5.200), is commonly known as the pulse pair estimate.

5.11.3 Pulse pair estimate

Let $V[0], V[1], \ldots, V[N-1]$ be N time samples of the received signal (from a given resolution volume) spaced T_s apart. Then using (5.130a) with $l = 1$, the autocorrelation

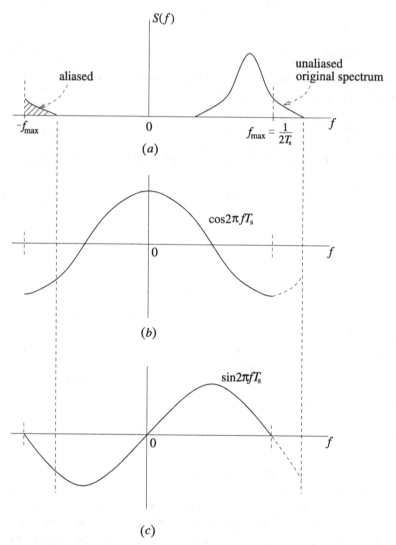

Fig. 5.41. (*a*) Partial aliasing of the Doppler spectrum, (*b*) $\cos 2\pi f T_s$, and (*c*) $\sin 2\pi f T_s$, respectively. Note that the products $S(f) \cos 2\pi f T_s$, $S(f) \sin 2\pi f T_s$ remain the same with or without aliasing.

at lag 1 is estimated as,

$$\hat{R}[1] = \frac{1}{N} \sum_{n=0}^{N-2} V[n+1] V^*[n] \tag{5.207}$$

From (5.200) with $t = T_s$, the mean velocity estimate based on the autocorrelation function at lag 1 is given as,

$$\hat{\bar{v}} = -\frac{\lambda}{4\pi T_s} \arg(\hat{R}[1]) \tag{5.208}$$

Computation of the variance of $\hat{\bar{v}}$ involves computing the variance of the phase of $\hat{R}[1]$. The estimate of $\bar{v}$ from samples of the received signal involves a non-linear function (i.e. argument function) so that analytically exact variance computations are complicated. The perturbation approximation, see (5.197), is typically used under such conditions assuming a narrow distribution of the estimate ($\hat{\bar{v}}$) about its mean. This condition is usually satisfied when a large number of samples are used in the estimate. Appendix 5 describes the procedure for calculating the variance of the phase of $\hat{R}[1]$ based on perturbation analysis. Using the results from Appendix 5, the variance of $\hat{\bar{v}}$ can be expressed as (Zrnić 1977),

$$\text{var}(\hat{\bar{v}}) = \frac{\lambda^2}{32\pi^2 T_s^2 |\rho[1]|^2} \left(\frac{1 - |\rho[1]|^2}{N^2} \right) \sum_{n=-(N-1)}^{(N-1)} |\rho[n]|^2 (N - |n|) \qquad (5.209)$$

where N is the number of samples, T_s is the PRT, and $|\rho[n]|$ is the magnitude of correlation coefficient function defined previously in (5.204b). The presence of additive white noise will decrease $\rho[n]$, as shown below. Let $R_{S+N}[n]$ be defined as the autocorrelation function of signal-plus-noise,

$$R_{S+N}[n] = R[n]; \quad n \neq 0 \qquad (5.210a)$$
$$= R[n] + R_N[n]; \quad n = 0 \qquad (5.210b)$$

where $R_N[n]$ is defined in (5.159). Therefore, the correlation coefficient function at lag n in the presence of additive noise is given by,

$$\rho_{S+N}[n] = \frac{R[n]}{R[0] + R_N[0]} = \rho[n]\left[\frac{1}{1 + (1/\text{SNR})}\right]; \quad n \neq 0 \qquad (5.211a)$$
$$= 1.0; \quad n = 0 \qquad (5.211b)$$

where the signal-to-noise ratio (or SNR) is defined as $R[0]/R_N[0]$. Figure 5.42 shows the standard deviation of $\hat{\bar{v}}$ computed using (5.209) for a Gaussian-shaped spectrum with $|\rho[n]|$ given by (5.195). The abscissa is normalized spectrum width defined as $\sigma_v/(\lambda/2T_s)$. White noise is accounted for by using (5.211). For typical values of $\lambda = 10$ cm, $T_s = 1$ ms, $N = 32$, and $\sigma_v = 2$ m s^{-1}, the mean velocity can be estimated with an accuracy of about 1 m s^{-1}. The presence of noise degrades the velocity estimates as illustrated in Fig. 5.42.

5.11.4 Estimation of σ_v

The power spectral density of meteorological echoes is often approximated by a Gaussian shape, with a mean velocity $\bar{v}$ and standard deviation σ_v. The σ_v provides information on the width of the spectral shape, and is often referred to as the spectrum width. The width of the spectrum controls the magnitude of the autocorrelation function. The parameter σ_v of the Gaussian spectral shape can be estimated from

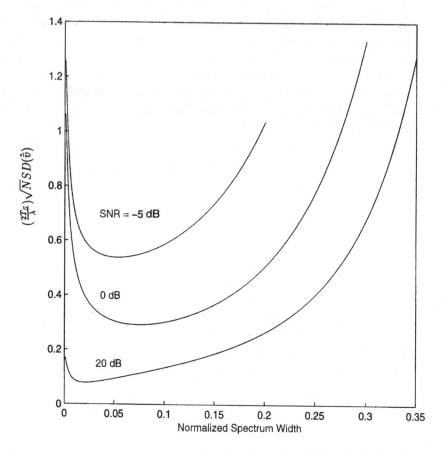

Fig. 5.42. Standard deviation of the estimate of $\bar{v}$ (SD($\hat{\bar{v}}$)) normalized with respect to $(2T_s/\lambda)\sqrt{N}$ as a function of normalized spectrum width, defined as $\sigma_v/(\lambda/2T_s)$.

(5.128) using the magnitudes of the autocorrelation at zero lag (or $R(0)$) and lag one (or $R(1)$) as,

$$\hat{\sigma}_v = \frac{\lambda}{2\pi T_s \sqrt{2}} \left[\ln \left| \frac{R(0)}{R(1)} \right| \right]^{1/2} \tag{5.212}$$

The above expression is valid in the absence of noise. In the presence of noise, $R(0)$ should be replaced by $R_{S+N}(0) - R_N(0)$, see (5.210b). Since only the autocorrelation magnitudes are involved in (5.212), it follows that σ_v can also be estimated by non-Doppler (or incoherent) radars. There are other estimators available to determine σ_v based on $R(1)$ and $R(2)$ (Srivastava et al. 1979).

5.11.5 Estimation of the Doppler spectrum (or power spectral density) from signal samples

The availability of high speed digital signal processors makes it feasible to compute the full Doppler spectrum (as opposed to just mean velocity and spectrum width). Periodograms are most commonly used to estimate the Doppler spectrum. Periodograms are computed from the discrete Fourier transform of the received signal samples, see (5.120). The periodogram estimate of the power spectral density from N signal samples is defined as,

$$\hat{S}[k] = \frac{1}{N} \left| \sum_{n=0}^{N-1} V[n] e^{-j2\pi nk/N} \right|^2 \tag{5.213}$$

where $S[k]$ is the kth frequency sample of the power spectral density corresponding to $\Omega_k = 2\pi k/N$. Note that Ω_k is the frequency of the discrete time sequence corresponding to $\omega_k = (2\pi/T)(k/N)$ (or $f_k = (1/T)(k/N)$). Thus, the periodogram estimate provides N samples of the power spectral density over the Nyquist interval $(1/T_s)$. It was shown in Section 5.7.2 that the spectral components from the periodogram have three properties, namely: (i) they are asymptotically unbiased, (ii) the standard deviation of the spectral estimate is proportional to the mean (and independent of N, the number of samples), and (iii) the spectral estimates are asymptotically uncorrelated. When the number of samples N is large,

$$\lim_{N \to \infty} \hat{S}[k] = \frac{1}{T_s} S(\omega_k) \tag{5.214a}$$

where $\hat{S}[k]$ is the periodogram estimate and $S(\omega_k)$ is the power spectral density (or Doppler spectrum), and,

$$\lim_{N \to \infty} \text{var}[\hat{S}(k)] = \left[\frac{1}{T_s} S(\omega_k) \right]^2 \tag{5.214b}$$

and,

$$\lim_{N \to \infty} \text{cov} \left[\hat{S}[k_1], \hat{S}[k_2] \right] = 0 \tag{5.214c}$$

One of the important inferences from (5.214) is that the variance of the periodogram cannot be reduced by taking a large number of samples, which only serves to increase the velocity or frequency resolution. There are many ways to improve the basic periodogram estimates, which involve some sort of averaging in the spectral domain. The commonly used procedures are described below.

(i) Averaging
If the original sample sequence is N long and is stationary, it can be divided into K sequences each $N/K = L$ long. The K different periodograms from each of these

sequences can be subsequently averaged to reduce the variance of the spectral estimate. The averaged spectral estimate can be obtained as,

$$\hat{S}_A[l] = \frac{1}{K} \sum_{k=1}^{K} \hat{S}_k(l) \tag{5.215}$$

where the lth sample from each of the K spectra are averaged, and S_A is the averaged spectral estimate. Note that the resolution of the spectrum is reduced here because there are only L equispaced samples in the Nyquist interval. This averaging procedure is also called the Bartlett procedure.

(ii) Windowing

A smoothing window can be applied to the estimate of the sample autocorrelation function before computing the Fourier transform to obtain the spectral estimate. Multiplication by a window is equivalent to performing a moving average (or convolution with the Fourier transform of the window function in the frequency domain). This procedure also reduces the variance of the spectral estimates. One advantage of this procedure is that it maintains the spectral resolution, but the spectral estimates at adjacent frequencies will be correlated. The length and shape of the window control the frequency resolution of the resultant Doppler spectrum. The spectral estimate based on the windowed autocorrelation function is defined as,

$$\hat{S}_W[k] = \sum_{n=-L}^{L} W[n]\hat{R}[n]e^{-j2\pi nk/N} \tag{5.216}$$

where $W[n]$ is the window function and $2L$ is the window length, which must be smaller than N. The corresponding frequency domain relationship can be written as,

$$\hat{S}_W[k] = W[k] \otimes \hat{S}[k] \tag{5.217}$$

where $W[k]$ is the DFT of $W[n]$, and $\hat{S}[k]$ is the power spectral density estimate without windowing. The symbol $\otimes$ indicates cyclic convolution, which can be written explicitly as,

$$W[k] \otimes \hat{S}[k] = \frac{1}{N} \sum_{l=1}^{N-1} W[l]\hat{S}[k-l] \tag{5.218}$$

While performing cyclic convolution the functions $W[l]$ and $\hat{S}[k-l]$ are kept periodic with N. There are numerous window functions available in the literature (Aunon and Chandrasekar 1997). Two commonly used windows are the Hamming window ($W_1[n]$) and the Hanning window ($W_2[n]$) given by,

$$W_1[n] = \begin{cases} 0.54 - 0.46\cos(2\pi n/L); & 0 \le n \le L \\ 0; & \text{otherwise} \end{cases} \tag{5.219a}$$

$$W_2[n] = \begin{cases} \frac{1}{2}[1 - \cos(2\pi n/L)]; & 0 \le n \le L \\ 0; & \text{otherwise} \end{cases} \tag{5.219b}$$

These two windows do a very effective job of smoothing the spectra.

The reduction in the variance of the spectral estimates due to windowing is difficult to estimate analytically. However, when the spectral representation of the window is narrow compared to the power spectral density, and if the window length (L) is much smaller than the data record (N), then the variance reduction is approximately given by,

$$\text{variance reduction factor} = \frac{1}{N} \sum_{-(L-1)}^{(L-1)} W^2[n] = \frac{1}{2\pi N} \int_{-\pi}^{\pi} |W(\Omega)|^2 \, d\Omega \qquad (5.220)$$

5.11.6 Maximum likelihood and maximum entropy spectral estimation

In addition to the periodogram-based techniques, advanced statistical techniques can be used for estimating the spectra. These techniques are referred to as modern spectral estimation techniques. Two of the most common modern techniques are described below. Though they appear complicated, the algorithms can be easily implemented in modern radar digital signal processors and they provide more accurate Doppler spectrum estimation.

(i) Maximum likelihood spectral estimation

Spectrum estimation, by definition, is the decomposition of the signal power into its various frequency components. This can be accomplished in hardware by a bank of narrow band filters, each passing only a certain frequency. The implementation of this concept results in the maximum likelihood estimate of the spectrum (Capon 1969), given as,

$$\hat{S}_{\text{ML}}(\Omega) = \frac{1}{e^{*t} R^{-1} e} \qquad (5.221)$$

where e is the generic column vector of exponentials given by $e = [1 \ e^{j\Omega} \ e^{j2\Omega} \dots e^{j(N-1)\Omega}]^t$, and R is the autocorrelation matrix of the data sequence defined as follows. If V is the column vector of signal samples given by $[V(1) \ V(2) \dots V(n)]^t$, then the autocorrelation matrix is defined as,

$$R = E[VV^{*t}] = \begin{bmatrix} R(0) & R(-1) & R(-2) & \dots & R(-N+1) \\ R(1) & R(0) & R(-1) & \dots & R(-N+2) \\ R(2) & R(1) & R(0) & \dots & R(-N+3) \\ \dots & \dots & \dots & \dots & \dots \\ R(N-1) & R(N-2) & R(N-3) & \dots & R(0) \end{bmatrix} \qquad (5.222)$$

The conventional periodogram estimate of the power spectral density can be written using the matrices **e** and **R** as,

$$S(\Omega) = \frac{1}{N} \left| \sum_{n=0}^{N-1} V(n) e^{-j\Omega n} \right|^2 \tag{5.223a}$$

$$= \frac{1}{N} \left[\sum_{n=0}^{N-1} V(n) e^{-j\Omega n} \right] \left[\sum_{n=0}^{N-1} V(n) e^{-j\Omega n} \right]^* \tag{5.223b}$$

Equation (5.223) can be written in a more compact form as,

$$S(\Omega) = \frac{1}{N} \mathbf{e}^{*t} \mathbf{V} \mathbf{V}^{*t} \mathbf{e} \tag{5.224}$$

and its expected value is given by,

$$E[S(\Omega)] = \frac{1}{N} \mathbf{e}^{*t} \mathbf{R} \mathbf{e} \tag{5.225}$$

The estimate of $S(\Omega)$ can be obtained using the estimate of **R** in (5.221) or (5.225). The maximum likelihood estimate can better resolve closely located peaks in the Doppler spectrum as compared to the periodogram-based method.

(ii) Maximum entropy spectral estimation

Whenever the data sequence is N long, the autocorrelation can be computed only for lags less than N. This imposes a rectangular window on the autocorrelation estimate (see 5.130b). This windowing limits the accuracy and resolution of the spectral estimate. To overcome this problem, the autocorrelation function can be extrapolated if done properly.

The maximum entropy spectrum is defined as the one that has maximum entropy, while matching the measured autocorrelation values. Among all the random sequences that have a given autocorrelation function, the Gaussian random sequences have maximum entropy (Shannon and Weaver 1963). The entropy (H) of a Gaussian process with power spectral density $S(\Omega)$ is defined (Burg 1972) as,

$$H = \frac{1}{2\pi} \int_{-\pi}^{\pi} \ln[S(\omega)] \, d\Omega \tag{5.226}$$

The maximum entropy power spectrum is the one that maximizes the entropy given by (5.226) under the condition that the autocorrelation function of $S(\Omega)$ matches with the measured autocorrelation values up to lag N. The maximum entropy spectral estimate is given by,

$$\hat{S}_{ME}(\Omega) = \frac{\sigma_p^2}{\left| 1 - \sum_{k=1}^{p} a_k e^{-j\omega_k} \right|^2} \tag{5.227a}$$

where $a_k, k = 1, 2, \ldots, p$ is obtained as the solution to the matrix equation,

$$\mathbf{a} = \mathbf{r} \mathbf{R}^{-1} \tag{5.227b}$$

where $\mathbf{a} = (a_1, a_2, \ldots, a_p)^t$ is the vector of coefficients, $\mathbf{r} = [R(1), R(2), \ldots, R(p)]^t$ is the vector of autocorrelation values starting from lag 1, and $\mathbf{R}$ is the matrix in (5.222), up to order p and σ_p^2 is the residual error variance.

The maximum likelihood spectral estimate and maximum entropy spectrum are related (Burg 1972) as,

$$\frac{1}{\hat{S}_{\mathrm{ML}}(\Omega)} = \sum_{p=0}^{N-1} \frac{1}{\hat{S}_{\mathrm{ME}}^p(\Omega)} \tag{5.228}$$

where $\hat{S}_{\mathrm{ML}}$ is the maximum likelihood estimate of the spectrum, and $\hat{S}_{\mathrm{ME}}^p$ is the maximum entropy estimate of order p (which matches autocorrelations up to lag p), and N is the length of the data sequence. The relation between $\hat{S}_{\mathrm{ML}}(\Omega)$ and $\hat{S}_{\mathrm{ME}}(\Omega)$ indicates that $\hat{S}_{\mathrm{ML}}$ is some sort of average over $\hat{S}_{\mathrm{ME}}$ of all orders up to $N-1$. This shows that $\hat{S}_{\mathrm{ME}}$ has better resolution than $\hat{S}_{\mathrm{ML}}$.

5.11.7 Spectral processing to estimate $\bar{v}$ and σ_v

The full spectrum of the received signals from a given resolution volume can be computed using one of the many methods discussed in Sections 5.11.5 and 5.11.6. Once that is done, the mean velocity $\bar{v}$ and the parameter σ_v of the spectral shape can be readily estimated as the first and second moments.

Let $S(k)$ be the spectral estimates obtained at $f_1, f_2, \ldots, f_k$ corresponding to velocities $v_1, v_2, \ldots, v_k$. Then the mean velocity can be estimated as,

$$\hat{\bar{v}} = \frac{\displaystyle\sum_{k=0}^{N-1} v_k \hat{S}(k)}{\displaystyle\sum_{k=0}^{N-1} \hat{S}(k)} \tag{5.229a}$$

and the second central moment (σ_v) as,

$$\hat{\sigma}_v^2 = \frac{\displaystyle\sum_{k=0}^{N-1} (v_k - \hat{\bar{v}})^2 \hat{S}(k)}{\displaystyle\sum_{k=0}^{N-1} \hat{S}(k)} \tag{5.229b}$$

To avoid biases due to partial aliasing of the spectrum, the simple expressions given above can be computed about the peak value of the spectral estimate $\hat{S}(k)$, utilizing the fact that the estimates are periodic in N.

5.12 Example of received signal statistics and spectral estimation

Samples of the received signal at a single polarization collected with the CSU–CHILL radar in precipitation are analyzed to illustrate the theory presented in Sections 5.9.1

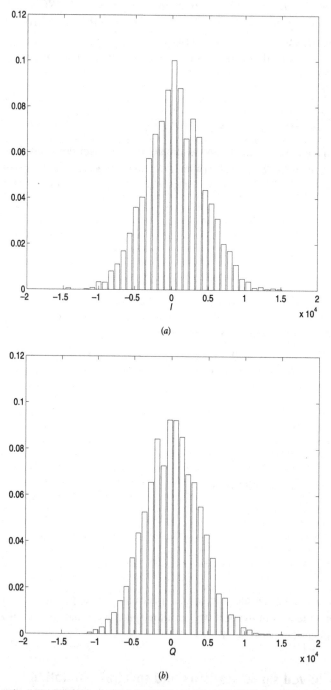

(a)

(b)

Fig. 5.43. Histogram of (a) the in-phase (I), and (b) the quadrature phase (Q) components of the received signal. Note the Gaussian shape with zero mean (see Section 5.9). Data collected in precipitation by the CSU–CHILL radar.

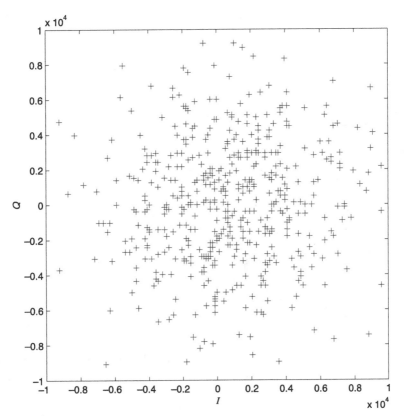

Fig. 5.44. Scatter plot of in-phase versus quadrature phase components. Note the uncorrelated nature of I and Q; see (5.162).

and 5.11. The data samples being presented are based on digitized samples of the in-phase (I) and quadrature phase (Q) components of the signal from one range resolution volume. Figure 5.43a shows the histogram of I, whereas Fig. 5.43b shows the histogram of Q. Both histograms are similar to Gaussian density functions with zero mean as shown in Section 5.9. In addition, the scattergram of I versus Q in Fig. 5.44 is spread equally in all directions with respect to the origin (without any preferential direction) indicating that they are independent as given by (5.162). The histograms in Fig. 5.43a, b look very similar, in concurrence with (5.164). Figure 5.45 shows the histogram of the power samples, ($I^2 + Q^2$). The near linear form of this histogram indicates that the power samples are distributed as exponential, as given in (5.172).

The periodogram estimate of the Doppler spectrum computed from (5.129) is displayed in Fig. 5.46. Note that the fluctuation in the spectral estimates in the decibel scale is similar throughout, indicating that the same graph, when plotted in linear scale, will have fluctuations which increase with the magnitude of $S(v)$ in agreement with (5.138). In addition, the fluctuation of adjacent samples is uncorrelated, as given in (5.139). This periodogram spectral estimate can be smoothed using windowing. The window size determines the degree of smoothing; see (5.216). Figure 5.47a shows a smooth spectral

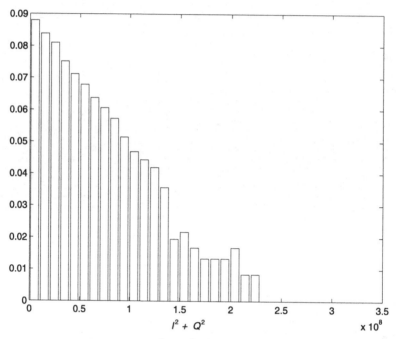

Fig. 5.45. Log of histogram of $(I^2 + Q^2)$. The histogram decreases linearly, indicating that $I^2 + Q^2$ is exponentially distributed.

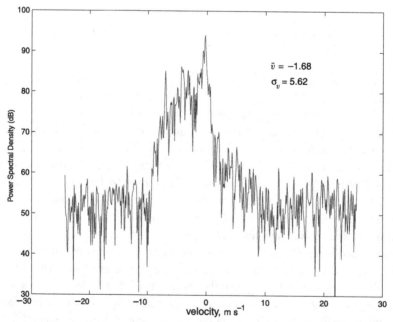

Fig. 5.46. Periodogram estimate of the Doppler velocity spectrum of the received signal, see (5.129). Data collected by the CSU–CHILL radar.

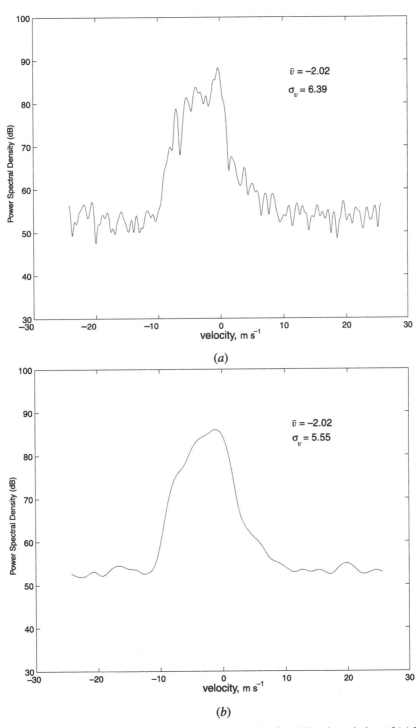

Fig. 5.47. Smoothed estimate of the Doppler spectrum using a Hanning window of (a) length 128 (see Section 5.11.5), and (b) length 32.

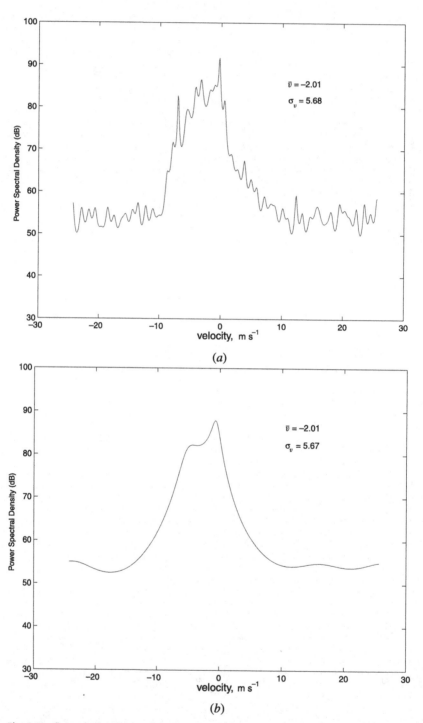

Fig. 5.48. Smoothed estimate of the Doppler spectrum using the maximum entropy method of (*a*) order 50, and (*b*) order 5.

estimate using a Hanning window of length 128; which may be compared to Fig. 5.47b, which uses a Hanning Window of length 32. Similarly, Fig. 5.48a shows the maximum entropy spectral estimate of order 50, whereas Fig. 5.48b shows the maximum entropy spectral estimate of order 5. Each figure also indicates the corresponding $\bar{v}$ and σ_v. Thus, smooth Doppler spectral estimates can be obtained using either windowing, maximum likelihood, or maximum entropy techniques.

Notes

1. The application of Doppler radar to meteorology is extensively covered in Gossard and Strauch (1983), Doviak and Zrnić (1993), or Sauvageot (1992). For a historical review of the early years of Doppler radar in meteorology see Rogers (1990). See, also, reviews provided by Röttger and Larsen (1990) on clear-air applications, by Ray (1990) on convective dynamics, by Marks (1990) on tropical weather systems, and by Browning (1990) on mid-latitude precipitation systems.

2. Section 5.1 provides only a very brief summary. It is assumed that the reader has had prior exposure to basic signal concepts. Background material is available in numerous texts, for example, Haykin and Van Veen (1999).

3. The concept of characteristic lines on a range–time diagram is commonly used to analyze pulses on a transmission line (or time domain reflectometry).

4. Techniques such as systematic and random phase coding are available to overcome the ambiguous range problem (for example, see Zrnić and Mahapatra 1985; Sachidananda and Zrnić 1999).

5. The receiver filter input is actually $V_r(t)$ as expressed in (5.33) and the output signal is $V_o(t)$ as expressed in (5.51). This input–output relation is valid for every scattered pulse. However, for stochastic signals only the mean power loss introduced by the finite bandwidth of the filter is relevant in the radar range equation.

6. The ELDORA radar is described in Hildebrand et al. (1996). The NASA SPANDAR S-band radar utilizes frequency diversity and is also dual-polarized for Z_{dr} measurements (see Goldhirsch et al. 1987). The C-band dual-polarized radar operated by Joanneum Research/Technical University, Graz, also uses frequency diversity. Examples of radar data can be found at,

   http://ias.tu-graz.ac.at/radar/radar.html

7. Pulse compression waveforms have been extensively used in military radars. However, the technique has not been widely used in weather radars. Feasibility studies can be found, for example, in Fetter (1970), Bucci and Urkowitz (1993), or Mudukutore et al. (1998).

8. The spectrum of ground clutter is centered at zero frequency with a narrow width. Filters are routinely used in Doppler weather radars to suppress the effects of ground clutter on the weather signal spectrum, see, for example, Groginsky and Glover (1980), Passarelli and Siggia (1983).

9. Techniques such as staggered pulsing schemes are available to overcome the velocity ambiguity problem (Zrnić and Mahapatra 1985).

6 Dual-polarized radar systems and signal processing algorithms

Dual-polarized radar systems can be configured in different ways depending on the measurement goals and the choice of orthogonal polarization states. From a theoretical perspective, the 3×3 covariance matrix (see Section 3.11) forms a complete set, but only a few research meteorological radars exist at the present time that are configured for this measurement. The circularly polarized radars built at the National Research Council of Canada were essentially configured for coherency matrix measurements (see Section 3.9). In the early 1980s, a number of single-polarized research Doppler radars were upgraded for limited dual-polarization measurements in the linear h/v-basis (for measurement of differential reflectivity and differential propagation phase). Because only copolar signals were involved, the system requirements were much less stringent and significant practical results (e.g. rain rate estimation, hail detection) were obtained fairly quickly (Hall et al. 1980; Bringi et al. 1984; Sachidananda and Zrnić 1986). This chapter discusses a number of dual-polarized radar configurations from a systems perspective. Since antenna performance is critical for achieving high accuracy in the measurement of the "weak" cross-polar signal, both antenna performance characteristics and formulation of radar observables in the presence of system polarization errors are treated. Calibration issues relevant to polarization diversity systems are also discussed. A significant portion of this chapter is devoted to estimation of the elements of the covariance matrix from signal samples under three different pulsing schemes. The accuracy of these covariance matrix estimates is also treated in some detail. This chapter concludes with a section on estimation of specific differential phase (K_{dp}) from range profiles of the differential propagation phase (Φ_{dp}).

6.1 General system aspects

A simplified block diagram of a polarization-agile/dual-receiver radar is shown in Fig. 6.1. Polarization agility refers to the ability to change the transmitted polarization state between any two orthogonal states on a pulse-to-pulse basis. The important microwave hardware components on the transmitting side (Fig. 6.1a) are: (i) the high-power waveguide switch, (ii) the microwave polarizer network, and (iii) the dual-polarized antenna (includes the main reflector, feedhorn, and orthomode transducer, or OMT). The OMT/feedhorn combination has two inputs or ports, with input into one port resulting in a transmitted wave of a certain polarization, and input

294

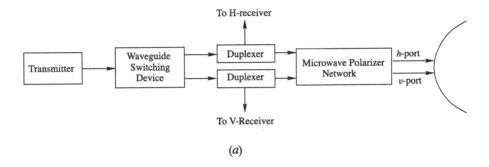

(a)

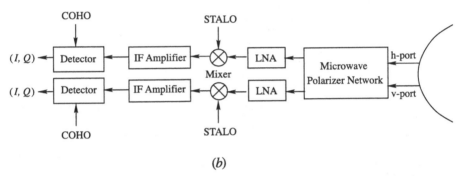

(b)

Fig. 6.1. (a) Simplified block diagram of a polarization-agile radar showing the transmitting side only. When the transmitter is "fired", the duplexer routes the high power pulse to the antenna. It also protects the receiver from the transmitted power. After the transmitted pulse is completed, the backscattered signal at the h-port is routed to the H-receiver, and that at the v-port is routed to the V-receiver. (b) Simplified block diagram of the dual-channel receiver system. The top channel is the H-receiver, while the bottom channel is the V-receiver. STALO, stable local oscillator; COHO, coherent oscillator; LNA, low noise amplifier; IF, intermediate frequency. The duplexers are not shown in this figure.

into the second port resulting in a transmitted wave of the orthogonal polarization. The OMT/feedhorn combination determines the polarization basis which is invariably orthogonal linear $(\hat{h}, \hat{v})$, or orthogonal circular $(\hat{R}, \hat{L})$. Figure 6.2 illustrates the OMT/feedhorn for the linear $(\hat{h}, \hat{v})$ basis. Elliptical polarization, see (3.2c), can be radiated if the input amplitudes, (M_h, M_v), are adjusted to $|M_v| = |M_h| \tan \varepsilon$ and $\arg(M_v) - \arg(M_h) = \delta$. The microwave polarizer network comprises a variable ratio power divider plus a variable phase shifter, which synthesizes the required input wave amplitudes, (M_h, M_v), at the OMT/feedhorn. Rapid switching of the transmitted pulse between orthogonal states is based on a high-power waveguide switch. These waveguide switches can be based on mechanical motor-driven rotary vane switches, or electronically controlled ferrite circulator switches. A typical transmitted waveform is illustrated in Fig. 6.3, where the polarization states are horizontal and vertical.

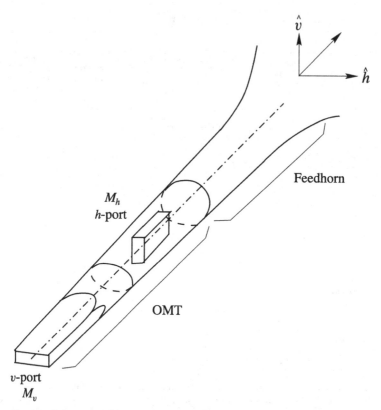

Fig. 6.2. Schematic of feedhorn/orthomode transducer (OMT) for linear polarization. The input wave amplitudes are M_h and M_v. When $M_h = 1$, $M_v = 0$ the radiated polarization is horizontal. When $M_h = 0$, $M_v = 1$, the radiated polarization is vertical. For reception, the component $\hat{h} \cdot \vec{E}_r$ is coupled to the h-port, while $\hat{v} \cdot \vec{E}_r$ is coupled to the v-port where the reflected field $\vec{E}_r$ is expressed in the backscatter system alignment (BSA) convention. Radiation of an elliptically polarized wave is possible by adjusting the wave amplitudes and phases, $\tan \varepsilon = |M_v|/|M_h|$ and $\delta = \arg(M_v) - \arg(M_h)$ at the input ports.

On the receiving side, the dual-channel receivers (Fig. 6.1b) are each essentially identical to the single channel Doppler receiver (e.g. Fig. 1.16b). Also shown in Fig. 6.3 are the voltage samples available at the output of the dual-receivers using the notation developed in Section 5.5.2. One receiver (e.g. the H-receiver) outputs the voltage samples ($V_h^{10}(\tau_0, 0)$, $V_h^{01}(\tau_0, T_s)$, $V_h^{10}(\tau_0, 2T_s)$, $V_h^{01}(\tau_0, 3T_s) \cdots$ etc.), while the second receiver (e.g. the V-receiver) outputs voltage samples ($V_v^{10}(\tau_0, 0)$, $V_v^{01}(\tau_0, T_s)$, $V_v^{10}(\tau_0, 2T_s)$, $V_v^{01}(\tau_0, 3T_s) \cdots$ etc.). The superscript "10" or "01" refers to the transmitted horizontal or vertical polarization state, respectively (($M_h = 1$, $M_v = 0$) or ($M_h = 0$, $M_v = 1$) in (3.68a)). The symbol τ_0 refers to a fixed range–time (or resolution volume), while T_s is the pulse repetition time. It should be clear that the voltage samples at the output of the H-receiver are proportional to S_{hh} and S_{hv} (i.e. $V_h^{10} \propto S_{hh}$, $V_h^{01} \propto S_{hv}$), while at the output of the V-receiver the voltage samples are

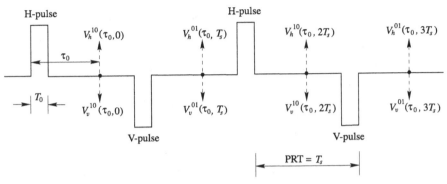

Fig. 6.3. Transmitted waveform showing alternate pulses switched between horizontal and vertical polarization states. Only the rectangular envelopes are shown. The pulsewidth is T_0 and the pulse repetition time (PRT) is T_s. The τ_0 refers to a fixed range time (or range resolution volume). Voltage samples are illustrated at τ_0 and spaced at the PRT. The superscript "10" refers to horizontal transmit, whereas "01" refers to vertical transmit. The subscript "h" or "v" refers to the receive state.

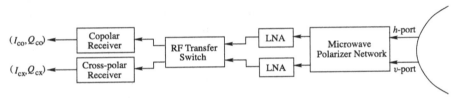

Fig. 6.4. Illustrates RF transfer switch used to route "strong" signals (i.e. V_h^{10} and V_v^{01}) to the copolar receiver, and "weak" signals (i.e. V_h^{01} and V_v^{10}) to the cross-polar receiver. This transfer switch could also be located after the RF mixers (see Fig. 6.1b) and would then be termed the IF transfer switch.

proportional to S_{vh} and S_{vv} ($V_v^{10} \propto S_{vh}$, $V_v^{01} \propto S_{vv}$). Since the off-diagonal elements of the scattering matrix are "weak" relative to the diagonal elements (S_{hh} or S_{vv}), it is preferable from dynamic range considerations and for higher accuracy in Z_{dr}, Φ_{dp}, and $|\rho_{co}|$, to keep the "weak" signals in the one receiver (i.e. called the cross-polar receiver) and the "strong" signals in the other receiver (the copolar receiver). This can be done via additional switching on the receiver side as illustrated in Fig. 6.4. This receiver transfer switch can be located on the RF side (as in the CSU–CHILL radar system), or on the IF side.

The polarization-agile/dual-receiver system as shown in Fig. 6.1 is configured to measure the three real and three complex terms of the covariance matrix in any orthogonal basis, e.g. see (3.180). The mean Doppler velocity and spectral width information are essentially imbedded within the corresponding covariance matrix of the received signal vector[1] (this will be treated in detail in Section 6.4).

Without the high power waveguide switch, the transmitted radiation is of a fixed polarization state as determined by the microwave polarizer network setting (for the moment advanced microwave devices that combine a high power waveguide switch, a polarizer and a receiver switch in one unit are excluded, see Fig. 6.10). With dual-channel receivers, these radar systems are referred to as possessing polarization diversity but not polarization agility, since the polarization basis change (e.g. from linear to circular) can take up to 15 s or so (Hendry et al. 1987). They are capable of measuring the two real and one complex terms of the coherency matrix (see Section 5.4) and form the basis for much of the research done at the National Research Council (NRC) of Canada, and at the Alberta Research Council of Canada.[2] The transmitted polarization state was generally fixed circular (e.g. RHC), and upon reception the measured terms were $\langle |S_{RL}|^2 \rangle$, $\langle |S_{RR}|^2 \rangle$ and $\langle S_{RR} S_{RL}^* \rangle$. However, "slow" switching between RHC and LHC states was usually exercised during data collection.

6.1.1   Polarization-agile/single-receiver system

The simplest dual-polarized system is one configured for polarization agility and reception via a single receiver, as illustrated in Fig. 6.5; this is referred to here as a polarization-agile/single-receiver system (Seliga and Bringi 1976). The microwave network is usually not needed since polarizations other than the natural basis of the horn/OMT (usually horizontal/vertical) are not used. The transmitted waveform is the same as in Fig. 6.3, but the received voltage sequence is $V_h^{10}(\tau_0, 0)$, $V_v^{01}(\tau_0, T_s)$, $V_h^{10}(\tau_0, 2T_s)$, $V_v^{01}(\tau_0, 3T_s) \cdots$ etc. From Fig. 6.5 it is seen that both the transmitted and received signals pass through the waveguide switch, which is often located close to the feedhorn/OMT, or even integrated with the feedhorn as with the Rutherford Appleton Laboratory's (RAL) S-band radar,[3] Chilbolton, UK. Figure 6.6 shows a schematic of the motor-driven rotary vane switch coupled to a turnstile junction and scalar feedhorn. The waveguide switch is a rotary vane divided into six sectors alternating between open section and closed metal section. As the vane rotates, energy is switched alternately through the horizontal and vertical polarization arms marked in the figure. The switch was designed to change the polarization state between horizontal and vertical at a pulse repetition frequency (PRF) of 610 s^{-1}. The isolation of the switch is a measure of its ability to prevent the input signal energy from entering the undesired arm (the arm that is blocked by the closed section of the vane), and is around 30 dB for the RAL switch. Further details of the switch operation are given in Bringi and Hendry (1990).

The motor-driven rotary vane switch does not easily afford the flexibility of different pulse repetition frequencies, or staggering the pulse repetition times. In the early 1980s, an electronically controlled ferrite circulator switch was used as a waveguide switch on the CHILL S-band radar (Seliga and Mueller 1982). Subsequently, a number of single-polarized research Doppler radars[4] in the USA were converted to polarization-agile/single-receiver systems using this type of waveguide switch (also termed a polarization switch). The critical element of these switches is the non-reciprocal latching ferrite phase shifter (e.g. Pozar 1998). Non-reciprocal means that the transmit path phase shift will be different from the receive path phase shift. Figure 6.7a

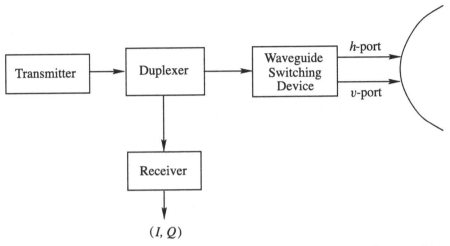

Fig. 6.5. Simple block diagram of a polarization-agile/single-receiver system. The waveguide switch can be either a motor-driven rotating vane, or an electronically controlled ferrite circulator switch.

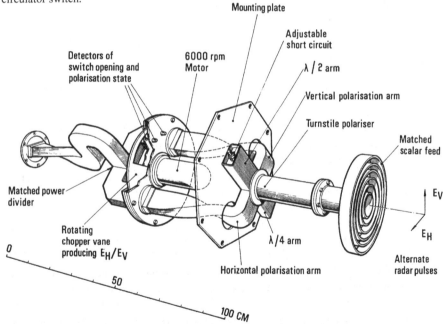

DUAL–POLARISATION RADAR POLARISER SWITCH AND FEED ASSEMBLY

Fig. 6.6. Motor-driven rotary vane switch on the Rutherford Appleton Laboratory's S-band radar located in Chilbolton, UK. Photo courtesy of Dr John Goddard, Rutherford Appleton Laboratory.

gives a schematic of the four-port waveguide switch with phase shifters in the top and bottom waveguide arms set for connecting the transmitter (at port 1) to the output

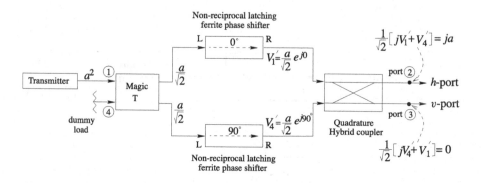

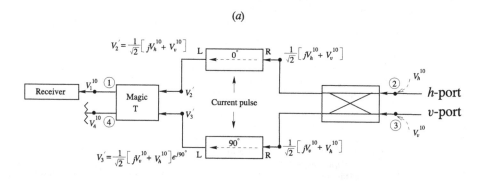

Fig. 6.7. (*a*) Schematic of ferrite circulator switch shown with phase shifters set for the transmitter to be coupled to the output *h*-port corresponding to a horizontally polarized pulse. The arrows in the phase shifter "boxes" reflect the non-reciprocal phase shifts for signal paths from left to right (L → R). For example, the top phase shifter provides a phase shift from L → R of 0°. For the bottom phase shifter, L → R gives a 90° phase shift. The transmitter power is proportional to a^2 (and amplitude is proportional to a). The magic-T splits the power equally and in-phase as shown. The top arm gives a phase shift of 0° while the bottom arm gives a phase shift of 90°. The quadrature hybrid coupler combines the two input wave amplitudes as indicated. The final result, in the ideal case, is perfect coupling from the transmitter (port 1) to the *h*-port of the OMT/feedhorn. In practice, the transmit insertion loss (from port 1 to the *h*-port) is around 0.6 dB, while the transmit isolation between port 1 and the *v*-port is typically around 20 dB. This implies that a small fraction of the transmitted power (typically 1%) is coupled to the undesired *v*-port. (*b*) The switch states are shown (top arm R → L is 0°; bottom arm R → L is 90°) for coupling of the received voltage V_h^{10} at the *h*-port to the receiver at port 1. The magic-T outputs $V_1^{10} = (1/\sqrt{2})(V_2' + V_3')$ and $V_4^{10} = (1/\sqrt{2})(-V_2' + V_3')$ giving $V_1^{10} = jV_h^{10}$ as the desired input to the receiver and $V_4^{10} = -V_v^{10}$ to the dummy load. Because of finite receiver isolation between the *v*-port and port 1, a small fraction (1% or 20-dB isolation) of the undesired power component, $|V_v^{10}|^2$, is coupled into the receiver.

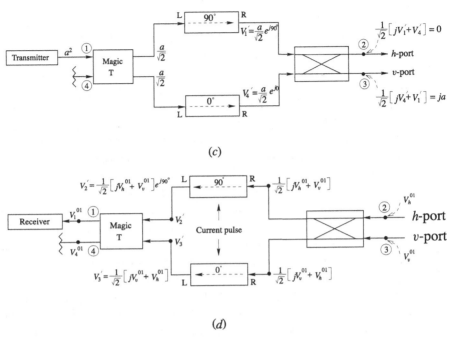

(c)

(d)

Fig. 6.7. (*cont.*) (*c*) The switch states (top arm L → R is 90°; bottom arm L → R is 0°) are shown for coupling of the transmitter to the v-port resulting in radiation of a vertically polarized pulse. Because of finite transmit isolation, a small fraction (typically 1% or 20-dB isolation) of the transmitted power is coupled to the undesired h-port. (*d*) The switch states (top arm R → L is 90°; bottom arm R → L is 0°) are shown for coupling the received voltage V_v^{01} at the v-port to the receiver at port 1. The magic-T outputs $V_1^{01} = \frac{1}{\sqrt{2}}(V_2' + V_3') = -jV_v^{01}$ to the receiver, and $V_4^{01} = \frac{1}{\sqrt{2}}(-V_2' + V_3') = -V_h^{01}$ to the dummy load. Because of finite receiver isolation between the h-port and port 1, a small fraction (typically 1% or 20-dB isolation) of the undesired power component, $|V_h^{01}|^2$, is coupled into the receiver.

Table 6.1. Specification for switchable ferrite circulator (from Seliga and Mueller 1982)

Operating frequency, GHz	2.7–2.8
Typical insertion loss, dB	0.9–1.0
Isolation (forward or reverse), dB	min 20
Peak power, kW	500
Average power, W	500
Switching time, µs	<5
Voltage standing wave ratio (VSWR)	<1.2

port 2 (h-port). As illustrated in the figure, the transmitted power (proportional to a^2; amplitude is proportional to a), is evenly split by the magic-T into two equal in-phase amplitude components. The signals in the top and bottom arms are shifted in phase by 0 and 90°, respectively (for transmission from left to right), and then recombined in the quadrature hybrid junction. The signal paths are traced in the figure and ideal operation results in the transmitted power being coupled to the output h-port (port 2) without any loss, while the output v-port (port 3) has zero power coupled to it. In practice, the transmit insertion loss can be around 0.5 dB, while the transmit isolation is typically around 20–25 dB. The radiated polarization state will, thus, have a small ($\approx$1%) component of the undesired polarization (i.e. vertical polarization) even if the antenna is ideal in terms of its own polarization purity. About 2 μs after the transmitted pulse is radiated, the ferrite phase shifters are electronically configured to their "receive" state using a current pulse that latches the ferrites so that the top arm will give a phase shift of 0° for the receive signal path (from right to left in Fig. 6.7b) in the top arm, and 90° in the bottom arm. This re-configuration is necessary if the received signal in the h-port (port 2) must be coupled to the single receiver at port 1. The receive signal paths are illustrated in the figure assuming that the voltages at the OMT are V_h^{10} (at port 2) and V_v^{10} (at port 3). The signal V_h^{10} at port 2 (h-port) is coupled to port 1, but because of the finite receive isolation, a small fraction (typically 1% for 20-dB isolation) of the undesired signal power, $|V_v^{10}|^2$, at port 3 (v-port) is also coupled to the receiver at port 1. Note that at port 4, which is terminated by a dummy load, the $V_4^{10} = -V_v^{10}$. In principle, a second receiver connected at port 4 would receive the voltage V_v^{10} if the receive isolation were sufficiently high. For relatively low isolation (<20 dB), a small fraction of the undesired power component, $|V_h^{10}|^2$, will cause serious error in the measurement of $|V_v^{10}|^2$ since $|V_v^{10}|^2 \ll |V_h^{10}|^2$.

The phase shifters are now in the correct setting (90° in the top arm and 0° in the bottom arm, see Fig. 6.7c) for the next transmitted pulse to be coupled from the input port 1 to the output port 3 (v-port). After the vertically polarized pulse transmission, the phase shifters are re-configured using a current pulse to latch the ferrites so that the top arm gives 90° phase shift and the bottom arm gives 0° phase shift from right to left signal path (see Fig. 6.7d). The ferrite phase shifters are "switched" just after the transmission of the high power pulse, and thus the switching rate is the pulse repetition frequency. The switching time is typically <15 μs; note that 15 μs corresponds to a minimum range of 2.25 km (see Table 6.1 for a summary of the switch specifications).

The transmit isolation (between ports 1 and 3 in Fig. 6.7a, and between ports 1 and 2 in Fig. 6.7c) is defined as $10\log_{10}(P_1/P_3)$ and $10\log_{10}(P_1/P_2)$, respectively; where $P_1 \equiv$ transmitter power, while P_2 and P_3 are the undesired powers at ports 2 and 3. Typical transmit isolations are around 20 dB. The receive isolation (with reference to Fig. 6.7b) is defined as $10\log_{10}\left(|V_v^{10}|^2/|V_1^{10}|^2\right)$, while with reference to Fig. 6.7d, the receive isolation is defined as $10\log_{10}\left(|V_h^{01}|^2/|V_1^{01}|^2\right)$. In practice, it is difficult to match the transmit and receive isolations to 20 dB each way and fine-tuning is necessary. The ferrite phase shifters are especially sensitive to temperature fluctuations and to achieve stable transmit/receive isolation, the ferrite material must be kept at a constant temperature (Carter et al. 1986).

Since only the copolar voltages, V_1^{10} and V_1^{01}, are received and processed to yield, for example, $Z_{dr} = 10 \log_{10}\left(\langle|V_1^{10}|^2\rangle/\langle|V_1^{01}|^2\rangle\right)$, the error introduced by transmit/receive isolation of 20 dB is, in fact, negligible. For example, consider that a horizontally polarized pulse is transmitted (as in Fig. 6.7a). The wave amplitudes at the OMT/feedhorn can be assumed to be $M_h \approx 1$, $M_v \approx \varepsilon$ (where ε is a small, complex error term), and $10 \log_{10}(1/|\varepsilon|^2)$ equals the transmit isolation defined earlier. Further assume that the antenna is ideal. Upon scattering from a single particle with scattering matrix given as $\mathbf{S}_{BSA}$, the received voltages at the h-port and v-port of the OMT/feedhorn can be written as,

$$V_h^{10} = S_{hh} M_h + S_{hv} M_v \tag{6.1a}$$
$$= S_{hh} + \varepsilon S_{hv} \tag{6.1b}$$
$$V_v^{10} = S_{hv} M_h + S_{vv} M_v \tag{6.1c}$$
$$= S_{hv} + \varepsilon S_{vv} \tag{6.1d}$$

Referring to Fig. 6.7b, the voltage at the receiver (V_1^{10}) can be written as,

$$V_1^{10} = V_h^{10} + \varepsilon V_v^{10} \tag{6.2a}$$
$$= S_{hh} + 2\varepsilon S_{hv} + \varepsilon^2 S_{vv} \tag{6.2b}$$

Note that the receive isolation is also defined as $10 \log_{10}(1/|\varepsilon|^2)$. If a second receiver is connected to port 4, then the cross-polar signal is,

$$V_4^{10} = V_v^{10} + \varepsilon V_h^{10} \tag{6.2c}$$
$$= S_{hv} + \varepsilon S_{vv} + \varepsilon S_{hh} + \varepsilon^2 S_{hv} \tag{6.2d}$$

For simplicity, the transmit and receive complex error terms are assumed to be identical. Similarly, referring to Fig. 6.7c, d it follows that,

$$V_1^{01} = S_{vv} + 2\varepsilon S_{hv} + \varepsilon^2 S_{hh} \tag{6.3}$$

Thus,

$$Z_{dr}^{meas} = 10 \log_{10}\left(\frac{\langle|V_1^{10}|^2\rangle}{\langle|V_1^{01}|^2\rangle}\right) \tag{6.4a}$$

$$= 10 \log_{10} \frac{\{\langle|S_{hh}|^2\rangle + 2\,\mathrm{Re}[(\varepsilon^*)^2\langle S_{hh}S_{vv}^*\rangle] + 4\,\mathrm{Re}[\varepsilon^*\langle S_{hh}S_{hv}^*\rangle]\}}{\{\langle|S_{vv}|^2\rangle + 2\,\mathrm{Re}[(\varepsilon^*)^2\langle S_{vv}S_{hh}^*\rangle] + 4\,\mathrm{Re}[\varepsilon^*\langle S_{vv}S_{hv}^*\rangle]\}} \tag{6.4b}$$

where terms involving $\varepsilon^2 S_{hv}^2$, $\varepsilon^4 S_{vv}^2$, and $\varepsilon^3 S_{hv}S_{vv}$ have been neglected. If the particles satisfy "mirror" reflection symmetry about a mean canting angle of $0°$ as discussed in Section 3.12, then $\langle S_{hh}S_{hv}^*\rangle = \langle S_{vv}S_{hv}^*\rangle = 0$. This is a good approximation for oriented raindrops. Equation (6.4b) reduces to,

$$Z_{dr}^{meas} = 10 \log_{10} \frac{\{\langle|S_{hh}|^2\rangle + 2\,\mathrm{Re}[(\varepsilon^*)^2\langle S_{hh}S_{vv}^*\rangle]\}}{\{\langle|S_{vv}|^2\rangle + 2\,\mathrm{Re}[(\varepsilon^*)^2\langle S_{vv}S_{hh}^*\rangle]\}} \tag{6.5}$$

For a distribution of scatterers, (3.231d) relates the copolar correlation coefficient $|\rho_{co}|$ to $\langle S_{hh}S_{vv}^*\rangle$. Using (3.231d) in (6.5) and using $\delta_{co} \approx 0$ results in,

$$Z_{dr}^{meas} = 10\log_{10}\frac{[\mathcal{Z}_{dr} + 2|\rho_{co}|\sqrt{\mathcal{Z}_{dr}}\,\text{Re}(\varepsilon^*)^2]}{[1 + 2|\rho_{co}|\sqrt{\mathcal{Z}_{dr}}\,\text{Re}(\varepsilon^*)^2]} \qquad (6.6)$$

where $\mathcal{Z}_{dr} = \langle|S_{hh}|^2\rangle/\langle|S_{vv}|^2\rangle$ is the intrinsic differential reflectivity ratio of the particles. For raindrops, typical values of $\mathcal{Z}_{dr}$ and $|\rho_{co}|$ of 2 and 0.985, respectively, can be used. Further, assume that $\text{Re}(\varepsilon^*)^2 \approx |\varepsilon|^2$, which is sufficient to estimate the error in Z_{dr}^{meas}. If the transmit/receive isolation is 20 dB, or $|\varepsilon|^2 = 0.01$, then Z_{dr}^{meas} (dB) is 2.951 (compared to the intrinsic value of 3 dB) and thus the error is negligible. Consideration of errors when the particles satisfy "mirror" reflection symmetry about a non-zero mean canting angle can be evaluated using, for example, (3.226, 3.227) in (6.4b), which would then yield results identical with Metcalf and Ussailis (1984). They found that errors in Z_{dr}^{meas} were less than 0.2 dB for a mean canting angle of 5° and isolation of 20 dB. The error exceeded 0.5 dB only when the isolation fell below 10 dB.

It is easy to show that transmit/receive isolation of the order of 20 dB will cause serious error in the measurement of the linear depolarization ratio. Neglecting $\varepsilon^2 S_{hv}$ in comparison with $\varepsilon(S_{hh} + S_{vv})$ in (6.2d), the cross-polar signal power can be expressed as,

$$\langle|V_4^{10}|^2\rangle = \langle|S_{hv} + \varepsilon(S_{hh} + S_{vv})|^2\rangle \qquad (6.7a)$$
$$= \langle|S_{hv}|^2\rangle + |\varepsilon|^2\langle|S_{hh} + S_{vv}|^2\rangle + 2\,\text{Re}[\varepsilon\langle S_{hv}^*(S_{hh} + S_{vv})\rangle] \qquad (6.7b)$$
$$= \langle|S_{hv}|^2\rangle + |\varepsilon|^2\langle|S_{hh} + S_{vv}|^2\rangle \qquad (6.7c)$$

The last step follows from the assumption of "mirror" reflection symmetry. The measured linear depolarization ratio is,

$$LDR_{vh}^{meas} = 10\log_{10}\left(\frac{\langle|V_4^{10}|^2\rangle}{\langle|V_1^{10}|^2\rangle}\right) = 10\log_{10}\left\{\frac{\langle|S_{hv}|^2\rangle + |\varepsilon|^2\langle|S_{hh} + S_{vv}|^2\rangle}{\langle|S_{hh}|^2\rangle + 2\,\text{Re}[(\varepsilon^*)^2\langle S_{hh}S_{vv}^*\rangle]}\right\} \qquad (6.8a)$$

$$\approx 10\log_{10}\left(\frac{\langle|S_{hv}|^2\rangle}{\langle|S_{hh}|^2\rangle} + |\varepsilon|^2\frac{\langle|S_{hh} + S_{vv}|^2\rangle}{\langle|S_{hh}|^2\rangle}\right)$$
$$\qquad (6.8b)$$

$$\approx 10\log_{10}\left(\frac{\langle|S_{hv}|^2\rangle}{\langle|S_{hh}|^2\rangle} + 4|\varepsilon|^2\right) \qquad (6.8c)$$

The last step involves the further approximation that $S_{hh} \approx S_{vv}$, which is sufficient for the purposes here. If the intrinsic depolarization ratio is, say, -27 dB in rain, then a transmit/receive isolation of 20 dB will give the measured value as -13.7 dB. For the error to be within 1 dB, the isolation must exceed nearly 40 dB. It is very difficult to "tune" the ferrite circulator to such high isolations in both transmit and receive directions. A three-switch configuration (see Fig. 6.9a) has been used to double the transmit isolation compared to a single unit, but the insertion losses are doubled as well.

6.1.2 Polarization diversity systems

Now consider systems defined as possessing polarization diversity (i.e. configured with dual-channel receivers) but not polarization agility. The radiated polarization state is usually fixed at one of the circular states, and the coherency matrix elements in the circular basis are measured. There is no need of a high power waveguide switch in such systems. The early radar systems designed and operated by the National Research Council of Canada, as well as by the Alberta Research Council, were configured as polarization diversity systems. However, these radars were capable of "slowly" switching the radiated polarization state between RHC, LHC, and linear states. More recently, the NOAA/ETL 35-GHz cloud radar is also configured as a dual-receiver system with the radiated polarization state being elliptical. Some examples of the data from this radar were considered in Section 3.8 (see Section 7.3 for more examples).

The basic elements of a polarization diversity system are illustrated in Fig. 6.8a for the circular basis. Figure 6.8b shows the dual-channel receiver configuration. Measurement of the coherency matrix does not require coherent (or coherent on-receive) systems since the Doppler frequency measurement is not needed. The superscript "10" on V_R and V_L in Fig. 6.8b refers to reception when a RHC polarized wave is radiated ($M_R = 1$, $M_L = 0$ in (3.111)). In most types of precipitation, the "weak" signal channel is in receiver 1, and the "strong" signal in receiver 2. Thus to recover the phase between $V_R^{10}(t)$ and $V_L^{10}(t)$ which is implied by the complex cross-correlation, the "strong" signal (at the intermediate frequency) is used as a reference signal for the detector (I/Q demodulator) as opposed to the COHO frequency in Doppler receivers (see Fig. 1.16b and associated discussion of Doppler receiving principles in Section 1.9).

Figure 6.8b shows the essential principles of dual-polarized (non-Doppler) radar receivers and is adapted from McCormick and Hendry (1979b). Conventional superheterodyne receiver principles are used to measure voltages proportional to $|V_R^{10}|$ and $|V_L^{10}|$. Figure 6.8c shows how the phase difference between V_R^{10} and V_L^{10} is obtained. In (b), the antenna port signals are preamplified and mixed with the local oscillator (which may or may not be locked to the transmitter via automatic frequency control circuitry). The output of this stage is at the intermediate frequency (usually 30 or 60 MHz) and is divided into two branches: one of which is the linear IF amplifier, and the other of which is an IF limiter whose function is to provide a constant amplitude reference signal for the phase detector shown in Fig. 6.8c. The IF amplifier outputs are fed into envelope detectors whose output voltages are proportional to $|V_R^{10}|$ and $|V_L^{10}|$, as indicated. The phase difference between V_R^{10} and V_L^{10} is obtained as illustrated in Fig. 6.8c. The IF limited signal from receiver 2 (which is the "strong" signal channel when RHC is radiated) takes the place of the COHO reference in Doppler receivers (see Fig. 1.16b) and thus one branch is shifted by 90° as shown. The IF limited signal from receiver 1 is then compared with the unshifted and 90° phase shifted reference signals. The output of the phase detector gives video voltages proportional to $\cos\theta$ and $\sin\theta$, where θ is the phase difference between V_R^{10} and V_L^{10}. Because of IF limiting, the magnitude of the complex cross-correlation is not obtained at the output of the phase detector, rather it is

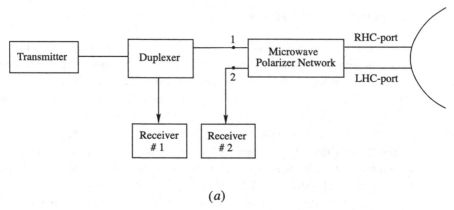

(a)

Fig. 6.8. (a) Simple block diagram of a polarization diversity system (McCormick and Hendry 1979b). The OMT is a turnstile junction (not shown). The polarizer can be adjusted to radiate any fixed elliptical state, though in practice one circular state is usually radiated. Details of the polarizer network are shown in (d). (© 1979 American Geophysical Union.) (b) Conventional superheterodyne receiver principles are used to output video voltages proportional to $|V_R^{10}|$ and $|V_L^{10}|$. The local oscillator may or may not be locked to the transmitter (not shown) via AFC circuitry. From McCormick and Hendry (1979b). (© 1979 American Geophysical Union.) (c) IF limited signals are used in I/Q demodulation to produce output video voltages proportional to $\cos\theta$ and $\sin\theta$, where θ is the phase difference between V_R^{10} and V_L^{10}. (d) Details of the microwave polarizer network which functions as a variable ratio power divider and phase shifter. The transmitted polarization can be set to any elliptical state by adjusting the two rotary phase shifters. The axial ratio of the ellipse is controlled by θ, while ϕ controls its orientation. From Hendry et al. (1987). (© 1987 American Geophysical Union.)

calculated from the cross-correlation between $|V_R^{10}|$ and $|V_L^{10}|$.

The microwave polarizer network (Fig. 6.8d) consists of two high power (continuously variable) rotary phase shifters and two quadrature hybrid junctions. The OMT is a turnstile junction; see Fig. 6.6, which illustrates a turnstile junction as the OMT. The polarizer network in Fig. 6.8d can be described for transmission (McCormick and Hendry 1985) as follows,

$$
\begin{bmatrix} M_R \\ M_L \end{bmatrix} = \begin{bmatrix} m & -n \\ n^* & m^* \end{bmatrix} \begin{bmatrix} M_1 \\ 0 \end{bmatrix}
\tag{6.9}
$$

where $m = \sin\theta e^{j\phi}$ and $n = -j\cos\theta e^{j\phi}$. The ellipticity angle τ, see (3.13), is given by $\tan\tau = (\sin\theta - \cos\theta)/(\sin\theta + \cos\theta)$, and the orientation angle $\psi = (\pi/4 - \phi)$ (see Fig. 3.3b). Thus, there is a one-to-one correspondence between the polarization ellipse parameters and the settings of the rotary phase shifters. The wave amplitudes (M_R, M_L) were used earlier in (3.102) in formulating the radiated electric field. Since the polarizer is reciprocal (no non-reciprocal ferrite phase shifters are used), the voltages at the output of the polarizer, (V_1, V_2), can be related to the received voltages at the RHC/LHC-ports,

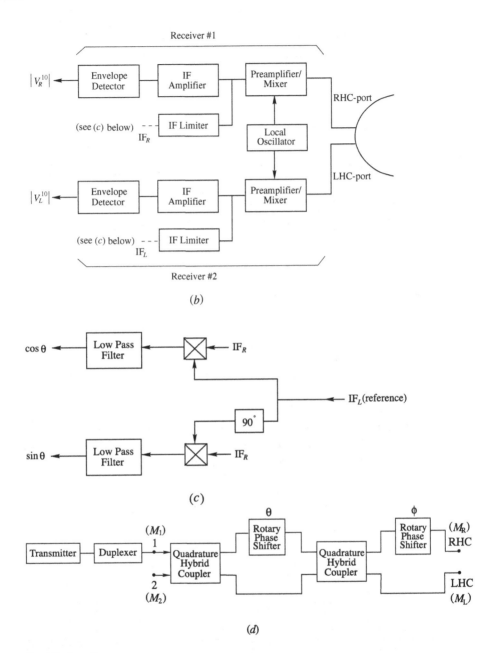

Fig. 6.8. (*cont.*)

(V_R, V_L), by the transpose of the 2×2 matrix defined in (6.9),

$$
\begin{bmatrix} V_1 \\ V_2 \end{bmatrix} = \begin{bmatrix} m & n^* \\ -n & m^* \end{bmatrix} \begin{bmatrix} V_R \\ V_L \end{bmatrix} \tag{6.10}
$$

The relation between (V_R, V_L) and (M_R, M_L) was given in (3.111), which is the radar range equation for a single particle whose scattering matrix is $\mathbf{S}^c_{BSA}$. Using (6.9, 6.10) in (3.111) gives a generalized radar range equation of the form,

$$\begin{bmatrix} V_1 \\ V_2 \end{bmatrix} = \frac{\lambda G}{4\pi r^2}e^{-j2k_0 r}\begin{bmatrix} m & n^* \\ -n & m^* \end{bmatrix}\mathbf{S}^c_{BSA}\begin{bmatrix} m & -n \\ n^* & m^* \end{bmatrix}\begin{bmatrix} M_1 \\ 0 \end{bmatrix} \tag{6.11a}$$

$$= \frac{\lambda G}{4\pi r^2}e^{-j2k_0 r}[\mathbf{S}'_{BSA}]\begin{bmatrix} M_1 \\ 0 \end{bmatrix} \tag{6.11b}$$

which relates the voltages at the receiver inputs to M_1 (which refers to the transmitter exciting port 1 of the microwave polarizer unit, see Fig. 6.8a). The form in (6.11b) is particularly suitable for analyzing slant linear polarizations oriented at any angle (in (6.9) substitute $\theta = 45°$ giving $m = \exp(j\phi)/\sqrt{2}$ and $n = -j\exp(j\phi)/\sqrt{2}$). Thus,

$$[\mathbf{S}'_{BSA}] = \frac{1}{2}\begin{bmatrix} e^{j\phi} & je^{-j\phi} \\ je^{j\phi} & e^{-j\phi} \end{bmatrix}\begin{bmatrix} S_{RR} & S_{RL} \\ S_{LR} & S_{LL} \end{bmatrix}\begin{bmatrix} e^{j\phi} & je^{j\phi} \\ je^{-j\phi} & e^{-j\phi} \end{bmatrix} \tag{6.12}$$

The elements of $\mathbf{S}^c_{BSA}$ were given in (3.93) for an axisymmetric particle canted at angle β. Using (3.93) in (6.12) gives,

$$\mathbf{S}'_{BSA} = \begin{bmatrix} S'_{11} & S'_{12} \\ S'_{21} & S'_{22} \end{bmatrix} = j\frac{(S_{11} + S_{22})}{2}\begin{bmatrix} 1 - v\sin(2\beta - 2\phi) & v\cos(2\beta - 2\phi) \\ v\cos(2\beta - 2\phi) & 1 + v\sin(2\beta - 2\phi) \end{bmatrix} \tag{6.13}$$

where v is defined in (3.97a), and $S_{11}(\psi)$ and $S_{22}(\psi)$ are the principal plane scattering elements defined in (2.94). Recall that ψ is the angle between the incident wave and the direction of the symmetry axis (see Fig. 2.10a). The elements of the ensemble-averaged covariance matrix corresponding to $\mathbf{S}'_{BSA}$ follow from $\mathbf{\Omega} = [S'_{11}\ \sqrt{2}S'_{12}\ S'_{22}]^t$, with $\mathbf{\Sigma}'_{BSA} = \langle\mathbf{\Omega\Omega}^{t*}\rangle$,

$$\langle|S'_{11}|^2\rangle = \langle\eta_c\rangle - 2\,\mathrm{Re}(\langle\eta_c v\rangle)\rho_2\sin(2\beta_0 - \phi)$$
$$+ \frac{\langle\eta_c|v|^2\rangle}{2}[1 - \rho_4\cos(4\beta_0 - 4\phi)] \tag{6.14a}$$

$$\langle S'_{11}S'^*_{12}\rangle = \langle\eta_c v^*\rangle\rho_2\cos(2\beta_0 - 2\phi) - \langle\eta_c|v|^2\rangle\frac{\rho_4}{2}\sin(4\beta_0 - 4\phi) \tag{6.14b}$$

$$\langle|S'_{21}|^2\rangle = \langle|S'_{12}|^2\rangle = \frac{\langle\eta_c|v|^2\rangle}{2}[1 + \rho_4\cos(4\beta_0 - 4\phi)] \tag{6.14c}$$

where $\langle\eta_c\rangle$, $\langle\eta_c v^*\rangle$, and $\langle\eta_c|v|^2\rangle$ are defined in (3.202–3.204). Further, the orientation parameters ρ_2, ρ_4 are defined in (3.211) and it is assumed that the canting angle pdf is symmetric about the mean angle β_0. The rest of the covariance matrix elements can be similarly obtained. From (6.14c), the maximum of $\langle|S'_{12}|^2\rangle$ occurs at $\phi = \beta_0$ (or at a linear polarization state with orientation angle $\psi = \pi/4 - \beta_0$). Recall that a positive β_0 is measured clockwise from the vertical, while a positive ψ is measured counter-clockwise from the horizontal reference direction. Thus, the transmitted linear polarization state which is oriented at 45° relative to the direction of the mean canting angle gives a maximum in the cross-polar power ($\langle|S'_{12}|^2\rangle$, (which is the cross-polar power measured by receiver 2 in Fig. 6.8a). The minimum in $\langle|S'_{12}|^2\rangle$ occurs at $\phi = \beta_0 - \pi/4$, or when $\psi = \pi/2 - \beta_0$. This occurs when the transmitted polarization state is oriented at $90°$ relative to the direction of the mean canting angle. The ratio of the maximum to minimum value of $\langle|S'_{12}|^2\rangle$ is,

$$\frac{\langle|S'_{12}|^2\rangle_{\text{max}}}{\langle|S'_{12}|^2\rangle_{\text{min}}} = \frac{1 + \rho_4}{1 - \rho_4} \qquad (6.15)$$

which was first derived by Hendry et al. (1987). Examples of the variation of $\langle|S'_{12}|^2\rangle$ (or P_{cx}) as a function of the orientation angle ψ of the transmitted linear polarization (or rotating linear polarization) were shown in Fig. 3.16. Such data enable a direct determination of both β_0 and ρ_4.

The complex interchannel correlation coefficient is defined as,

$$\rho_{\text{cx}} = \frac{\langle S'_{11} S'^*_{12}\rangle}{\sqrt{\langle|S'_{11}|^2\rangle\langle|S'_{12}|^2\rangle}} \qquad (6.16)$$

which will be zero when $\phi = \beta_0 - \pi/4$. This condition[5] is identical with the condition for a minimum in the cross-polar power ($P_{\text{cx}} \propto \langle|S'_{12}|^2\rangle$). For hydrometeors possessing an axis of symmetry with symmetric canting angle pdf about the mean β_0, the condition for cross-polar optimal polarization can be taken as $\rho_{\text{cx}} = 0$. Thus, when the transmitted polarization state is linear and oriented at $90°$ relative to the direction of the mean canting angle, the interchannel correlation coefficient, which is proportional to $\langle S'_{11} S'^*_{12}\rangle$, will be zero. For this transmitted polarization state the cross-polar power ($P_{\text{cx}} \propto \langle|S'_{12}|^2\rangle$) will be a minimum (it will be null for a single target; see (3.140)) and it, therefore, defines the cross-polar optimal polarization state. In practice, because of low signal-to-noise ratio and antenna polarization errors, it is difficult to accurately estimate both the cross-polar power and the interchannel correlation coefficient at the cross-polar optimal polarization state. Further experiments are needed to determine whether the measurement of optimal polarization states for distributed media such as precipitation will provide added value in the remote sensing of precipitation type and amount.

6.1.3   Polarization-agile/dual-receiver systems

As shown in Fig. 6.1a, insertion of a waveguide switch ahead of the microwave polarizer network is sufficient to implement polarization agility within a polarization diversity system. The National Research Council of Canada's 10-GHz radar used three latching ferrite circulator switches, as illustrated in Fig. 6.9a. This array provides double the transmit isolation of a single unit, but the insertion loss is doubled as well. Since the received signals do not pass through the waveguide switch, the ferrites can be optimized for maximizing the transmit isolation between the input and the decoupled output port to nearly 60 dB over a narrow bandwidth. Motor-driven rotary vane switches have also

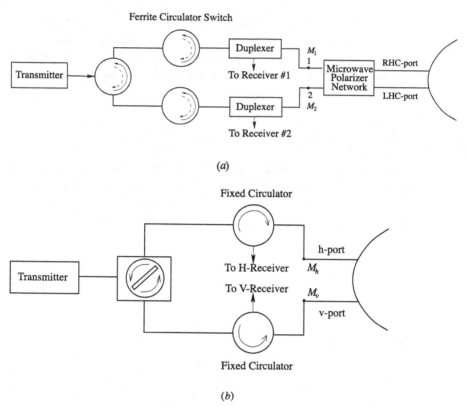

Fig. 6.9. (a) Polarization agility on the NRC 10-GHz radar was based on using an array of three ferrite circulator switches (Hendry et al. 1987), each similar to the one shown in Fig. 6.7. (© 1987 American Geophysical Union.) (b) The NCAR/S-POL radar uses a motor-driven rotating vane waveguide switch (similar to the RAL switch in Fig. 6.6) to achieve polarization agility. Two fixed circulators are used to separate the transmit and receive paths. From Randall et al. (1997).

been used to provide similar high levels of transmit isolation, e.g. the RAL S-band radar and the NCAR/S-POL radar (see Randall et al. 1997). In both these systems, junction circulators are used to separate the transmit and receive paths.[6] Figure 6.9b shows a schematic of the NCAR/S-POL configuration. With reference to (6.11b), the voltages at the dual-receiver outputs can now be expressed as,

$$
\begin{bmatrix} V_1 \\ V_2 \end{bmatrix} = \frac{\lambda G}{4\pi r^2} e^{-j2k_0 r} \begin{bmatrix} S'_{11} & S'_{12} \\ S'_{21} & S'_{22} \end{bmatrix}_{\text{BSA}} \begin{bmatrix} M_1 \\ M_2 \end{bmatrix}
$$

(6.17)

where $[M_1 \ M_2]^t$ can be $[1 \ 0]^t$ or $[0 \ 1]^t$. Again, a transfer switch on the receive side (as in Fig. 6.4) is generally used to route the signals into the copolar or cross-polar receivers.

A different configuration which combines the high power waveguide switch, microwave polarizer network and receiver transfer switch in one unit is illustrated in

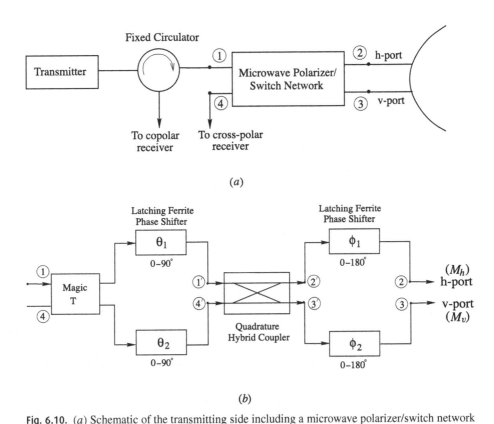

Fig. 6.10. (*a*) Schematic of the transmitting side including a microwave polarizer/switch network that combines the function of high power waveguide switch, polarizer, and receive transfer switch in one unit as used on the German DLR radar system operating at 5.5 GHz (C-band). (*b*) Combined polarizer/switch schematic diagram composed of latching ferrite phase shifters. If the transmitter input power equals a^2, then $V_1' = (a/\sqrt{2})e^{j\theta_1}$, $V_4' = (a/\sqrt{2})e^{j\theta_2}$. At the output ports of the quadrature hybrid coupler,
$V_2' = (1/\sqrt{2})(jV_1' + V_4') = ae^{j[(\theta_1+\theta_2)/2]+j(\pi/4)} \cos[(\theta_1 - \theta_2)/2] + (\pi/4)]$ while
$V_3' = (1/\sqrt{2})(jV_4' + V_1') = ae^{j[(\theta_1+\theta_2)/2]+j(\pi/4)} \sin[(\theta_1 - \theta_2)/2 + (\pi/4)]$. Thus, at ports (2')
and (3') the input power can be divided into any ratio by controlling ($\theta_1 - \theta_2$), i.e. the action is a variable ratio power divider. At port (2), which is the *h*-port, $M_h = (V_2')e^{j\phi_1}$, while at the *v*-port, $M_v = (V_3')e^{j\phi_2}$. Therefore, $|M_v|/|M_h| = \tan[(\theta_1 - \theta_2)/2 + (\pi/4)] = \tan \varepsilon$, whereas $\arg(M_v) - \arg(M_h) = \phi_2 - \phi_1 = \delta$. The radiated electric field has the unit polarization vector given by $\hat{e} = (\cos \varepsilon \, \hat{h} + \sin \varepsilon \, e^{j\delta} \hat{v})$ with complex polarization ratio $\chi = (\tan \varepsilon)e^{j\delta}$, see (3.2c, 3.3). From Schroth et al. (1988).

Fig. 6.10 and is used on the German DLR radar system (Schroth et al. 1988). As illustrated in Fig. 6.10*b* for the transmit path, the radiated polarization ellipse is set by $[(\theta_1 - \theta_2)/2 + \pi/4]$ and ($\phi_2 - \phi_1$), where the complex polarization ratio $\chi = M_v/M_h = (\tan \varepsilon) \exp(j\delta)$, with $\varepsilon = [(\theta_1 - \theta_2)/2 + \pi/4]$ and $\delta = \phi_2 - \phi_1$; see (3.2, 3.3). The two (non-reciprocal) 0–90° phase shifters are capable of providing any phase shift between 0 and 90° in 128 discrete steps. The phase shifters are controlled by digital logic and

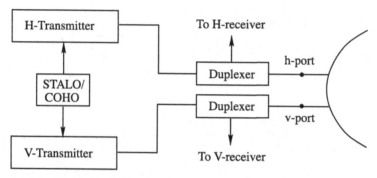

Fig. 6.11. Simple block diagram of the two-transmitter CSU–CHILL radar system. Normally the transmitters are "fired" alternately leading to the transmitted waveform in Fig. 6.3. But they can also be "fired" simultaneously with matched power and controlled phase difference. See Fig. 6.4 for a block diagram of the dual-receiver configuration.

memory with 7-bit command states. The accuracy of the power division at ports $2'$ and $3'$ is provided by the use of a programmable read only memory (PROM) look-up table to command the precise current for each ferrite element. Because of non-reciprocity, two separate look-up tables are provided for transmit and receive paths. The 0–$180°$ phase shifters are also latching ferrites with controlled current. As illustrated in Fig. 6.10b, the transmitter power is coupled to port 1 of a magic-T. During transmission, the ferrites are set to one state optimized for the selected polarization. The switch directs the transmitted power to the h- and v-ports of the OMT/feedhorn in any variable power ratio and phase difference that is desired. The transmit isolation is around 30 dB or better. Immediately after the transmit pulse is completed, the ferrites are switched to the receive state. In this state, the received power at each port of the OMT/feedhorn is phase adjusted and combined in the switch so as to direct the copolar received signal to port 1 and the cross-polar received signal to port 4. Here copolar and cross-polar are defined relative to the transmitted polarization state. The receive isolation is also around 30 dB or better. For the next transmitted pulse whose polarization state is orthogonal to the previous pulse, the ferrites are again set to an optimized state based on a look-up table. Just after transmission, the ferrites are again re-set so that the copolar received power is coupled to port 1 while the cross-polar received power is coupled to port 4. Thus, two switching operations are needed for every transmitted pulse. Also, the received states can be set for the "hybrid" measurements as discussed in Section 4.7 (examples of such data are given in Chandra et al. 1999). Because the transmit/receive isolation of the switch is temperature sensitive, it must be maintained in a controlled environment. In practice, the switch command states may need to be frequently adjusted for optimal performance.

The CSU–CHILL radar system implements polarization agility through two separate (but identical) transmitters, as illustrated schematically in Fig. 6.11, which avoids the necessity for a waveguide switching device. A microwave polarizer network is also not necessary, since in principle, the transmitters can be "fired" simultaneously with controlled phase difference. Normally, the H- and V-transmitters are "fired" alternately, as illustrated in Fig. 6.3. The use of two separate transmitters, while expensive, affords a high degree of operational flexibility and is based on the S-band radar design adopted by the US Air Force at the Rome Laboratory, Rome, New York (Stiefvater et al. 1992). Excellent transmit isolation is possible since the H-transmitter is coupled to the *h*-port of the OMT/feedhorn, while the V-transmitter is coupled to the *v*-port. The transmit isolation is limited by the dual-channel rotating joint and by the OMT, both of which have isolations in excess of 45 dB. The dual-receiver configuration of the CSU–CHILL system is essentially the same as shown in Fig. 6.4. The RF transfer switch in Fig. 6.4 has high isolation (>50 dB). The overall cross-polar performance of the two-transmitter/two-receiver system is thus primarily limited by the polarization errors of the antenna (discussed in Section 6.2).

In the hybrid mode of operation previously discussed in Section 4.7, the two transmitters are "fired" simultaneously with matched power but with arbitrary phase difference. The transmitted polarization state is thus fixed from pulse-to-pulse. On reception the dual-receivers are configured as in Fig. 6.1*b*. The voltages from the H- and V-receivers are used to estimate the elements of the coherency matrix. If the transmitted wave amplitudes at the input to the OMT/feedhorn are $M_h = 1$, $M_v = e^{j\theta}$, then the voltages at the output of the H- and V-receivers can be expressed from (3.68a) as,

$$\begin{bmatrix} V_h^{11} \\ V_v^{11} \end{bmatrix} = \frac{\lambda G}{4\pi r^2} e^{-j2k_0 r} \begin{bmatrix} S_{hh} & S_{hv} \\ S_{vh} & S_{vv} \end{bmatrix}_{BSA} \begin{bmatrix} 1 \\ e^{j\theta} \end{bmatrix} \tag{6.18}$$

The coherency matrix elements are essentially given by $\langle |V_h^{11}|^2 \rangle$, $\langle |V_v^{11}|^2 \rangle$, and $\langle (V_h^{11})^* V_v^{11} \rangle$, as in (4.135). The voltage sequence at the output of the H-receiver is $(V_h^{11}(\tau_0, 0),\ V_h^{11}(\tau_0, T_s),\ V_h^{11}(\tau_0, 2T_s)$, etc.), while at the V-receiver it is $(V_v^{11}(\tau_0, 0),\ V_v^{11}(\tau_0, T_s),\ V_v^{11}(\tau_0, 2T_s)$, etc.). Thus the mean Doppler velocity is estimated from voltage samples at the output of any one of the receivers, which are spaced at the PRT of T_s. The cross-correlation term $\langle (V_h^{11})^* V_v^{11} \rangle$ is also based on simultaneous voltage samples available at the dual-receiver outputs (see also, Section 6.4.5). For examples of data collected with the CSU–CHILL radar in this mode see Brunkow et al. (1997).

The following describes an expanded block diagram of the polarization-agile/dual-receiver CSU–CHILL radar system.[7] While various radar systems are implemented differently, the following example of the CSU–CHILL system presents an idea about the sub-systems of a polarimetric radar. Refer to Section 1.9.2 for a description of a generic pulsed Doppler radar.

A detailed block diagram of the CSU–CHILL transmitter and receiver system is shown in Fig. 6.12. The transmitters share a common STALO. The STALO is mixed with a 50-MHz source derived from the processor clock to produce the transmit frequency. The vertical channel transmit drive signal can be further adjusted by an I/Q vector modulator to permit adjustments of the relative phase and amplitude between the H- and V-transmitted pulses. Either an alternate triggering mode or a

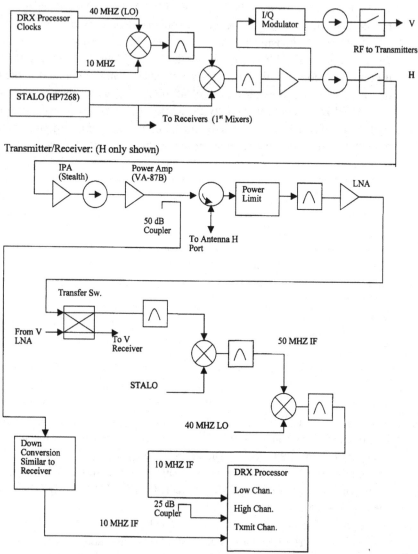

Fig. 6.12. Block diagram of the CSU–CHILL radar system. From Brunkow et al. (2000).

simultaneous triggering mode is used based on pulses generated in the signal processor state machine. When the transmitters are triggered simultaneously, and the appropriate inter-channel phase adjustment is applied, the radiated polarization can be set to slant $\pm45°$ linear or right-hand or left-hand circular polarization states. Separate digital power meters monitor the average RF output level developed by each transmitter. These digitized power readings are inserted into the recorded data stream at 2 s intervals so that accurate reflectivity and differential reflectivity calibrations can be maintained. Once the transmitters are fully stabilized (around 30 minutes), the output power levels generally vary by less than 0.3 dB during the course of a normal data collection period (around 10 hours).

Return signals are passed through a pair of separate, matched receivers. The low noise amplifier (LNA) and first mixers for both receivers are mounted on the same aluminum plate. The temperature of the plate is held at a constant value (20 °C) by a thermostatically controlled Peltier effect cooler. The minimization of temperature variations between the "front end" components of the two receiver channels reduces the tendency for differential drifts to develop in the calibration characteristics of the two receivers.

A solid state transfer switch located after each LNA permits the H- and V-received signals returning from the antenna to be connected to either receiver channel. The transfer switch can be used to route all the copolar signals to one receiver and all the cross-polar signals to the other receiver. In the alternating V- and H-operational mode, this switch is toggled just before each pulse is transmitted. Thus, using a single receiver to develop the copolar radar measurables reduces errors due to calibration differences and drift between the two receivers. However, the stability and linearity of the newly installed digital receivers is believed to be of sufficiently high quality that the transfer switch, while still in the circuit, need not be used. Calibration results indicate less than 0.1-dB difference in gain between the H- and V-receivers over a wide dynamic range.

From the transfer switch, the received signals pass through image rejection filters and are down converted to a 50-MHz IF by mixing with the STALO frequency. A second conversion to 10-MHz IF is performed by mixing with the 40-MHz reference clock from the processor. The IF signal is digitized by 12-bit digitizers running at a rate of 40 MHz. Programmable finite impulse response (FIR) filter chips then apply quadrature detection followed by low-pass filtering to produce in-phase (I) and quadrature (Q) voltage numbers. Each processor performs these functions in a pair of receiver channels separated in sensitivity by 25 dB. When all samples are complete for a given integration period, the set that offers the most sensitivity without excessive saturation is selected for processing. To facilitate this channel selection process, each range gate has a counter to keep track of the number of saturated samples encountered during each integration cycle. The bank of counters is available to the digital signal processors, where the actual decision on which receiver channel to use is made. Directional couplers between the transmitters and the circulators are used to obtain a high quality sample of each transmitter pulse that is down-converted to 10 MHz just as it is done in the receiver. The transmitted pulse is then fed to a third input channel on each processor, where the pulse is digitized at a rate of 40 MHz. The processor uses these samples to calculate the average amplitude and phase of each pulse and, optionally, to use these results to correct the received signals from that pulse for any slight

Table 6.2. CSU–CHILL system characteristics

Antenna	Shape	Parabolic
	Diameter, m	8.5
	Feed type	Scalar horn
	Gain, dB	43[a]
	3-dB beamwidth, deg	1.1
	Maximum sidelobe, dB	−27[b]
	Inter-channel isolation, dB	−45
	ICPR (two-way), dB	−34
Transmitters	Wavelength, cm	11
	Peak power, MW	1
	Final power amplifier	VA-87B/C (Klystron)
	PRT range, ms	800–2500
	Pulse width, μs	0.3–1.0 μs
	Available polarizations	Horizontal
		Vertical
		Slant 45/135°
		Right/left circular
Receivers/	Noise figure, dB	~3.4
digital signal	Noise power, dBm	~−114.0
processing	Dynamic range, dB	~96
	Bandwidth, MHz	10[c]
	Output range resolution, m	45[d]
	Maximum range gates	3000

[a] Includes waveguide loss
[b] In any ϕ plane
[c] Basic bandwidth of the analog section is 10 MHz. Subsequently, the bandwidth is restricted with a programmable filter depending on the transmit pulsewidth. For example, a 1 μs pulse corresponds to a B_6 bandwidth of 1 MHz (3-dB bandwidth of 750 kHz)
[d] This is the minimum output range resolution and is adjustable upward in 15-m intervals

variations in phase and magnitude. The phase stability of this system produces mean phase estimation accuracies of better than 0.1°. A series of digital signal processing chips perform this correction, and produce covariance estimates. The data stream is passed to a workstation for archiving and further distribution. The processors have extensive flexibility and most of the programming can be done using a high level language. The performance characteristics of the CSU–CHILL radar are summarized in Table 6.2.

6.2   Antenna performance characteristics

The distributed nature of precipitation and the fact that the intrinsic reflectivity of precipitation can vary over several orders of magnitude over short distances makes it imperative to use pencil beam antennas of high quality even in single-polarized radar systems. "Smoothing" of the intrinsic reflectivity of precipitation by the main lobe of the antenna is a well-known phenomenon and is essentially a two-dimensional convolution over a spherical surface as indicated by (5.63). Of course, use of narrow beams (3-dB beamwidth $\leq 0.5°$) is preferable and, in practice, is feasible in terms of cost only at the higher frequencies (>5 GHz). Sidelobe-induced errors can occur when strong gradients of the intrinsic reflectivity exist across the main lobe and the close-in sidelobes (typically within $\pm 5°$ from boresight). These errors are generally not obvious in the reflectivity data even with modest close-in sidelobe levels (20-dB down relative to the peak of the main lobe) unless the intrinsic reflectivity gradients are extremely high (as might occur in supercell hailstorms). However, for dual-polarized radars the errors due to poor antenna performance will be much more obvious even in homogeneous precipitation, or in the presence of modest reflectivity gradients. Thus, the performance of the antenna is of prime importance.

The most common antenna type is the parabolic reflector antenna (see Fig. 6.13a) and is the work-horse of operational meteorological radars (e.g. the WSR–88D system; the terminal Doppler weather radar (TDWR) system). Advances in design and manufacturing methodologies in the last two decades have resulted in improved performance characteristics of parabolic reflector antennas. However, blockage by the feedhorn/OMT and the feed support struts generally places a lower bound on the close-in sidelobe levels that can be achieved (typically -25 to -27 dB in the "worst-case" planes). Offset-feed parabolic reflector antennas (see Fig. 6.13b) are clearly superior in terms of sidelobe performance because feed and support strut blockages are eliminated.

Apart from the general requirement of narrow beam and low close-in sidelobe levels necessary for reduction of beam "smoothing" and gradient-induced errors, dual-polarized antennas have additional stringent performance requirements. The radiated electric field when the h-port of a dual-polarized antenna is excited can be expressed (see Section 3.3) as,

$$\vec{E}_h^i = \sqrt{\frac{P_t^h Z_0}{2\pi}} \sqrt{G_h} e^{j\Phi_h} (i_h \hat{h}_i + \varepsilon_h \hat{v}_i) \frac{e^{-jk_0 r}}{r} \tag{6.19a}$$

Similarly, when the v-port is excited, the radiated field is,

$$\vec{E}_v^i = \sqrt{\frac{P_t^v Z_0}{2\pi}} \sqrt{G_v} e^{j\Phi_v} (\varepsilon_v \hat{h}_i + i_v \hat{v}_i) \frac{e^{-jk_0 r}}{r} \tag{6.19b}$$

The convention and notation is the same as in Section 3.3. The symbols Φ_h and Φ_v have been introduced here as phase terms associated with exciting the h- and v-ports,

(a)   (b)

Fig. 6.13. (a) The CSU–CHILL S-band antenna at the manufacturer's test range (Radiation Systems Inc.). (b) The National Research Council of Canada's offset reflector antenna (courtesy of A. Hendry).

respectively. The normalized copolar (power) gain patterns, f_h and f_v, are defined as in (5.38),

$$f_h = G_h/G_{0h} \tag{6.20a}$$
$$f_v = G_v/G_{0v} \tag{6.20b}$$

where G_{0h} and G_{0v} are the peak boresight gains associated with each port. The cross-polar (power) pattern functions are defined as,

$$f_{\varepsilon h} = |\varepsilon_h|^2 f_h \tag{6.21a}$$
$$f_{\varepsilon v} = |\varepsilon_v|^2 f_v \tag{6.21b}$$

Note that the four pattern functions are dependent on $(\theta - \theta_0, \phi - \phi_0)$, where θ_0 and ϕ_0 are the assumed boresight directions. Normally, only the power patterns are available from the antenna manufacturer. The polarization error, $|\varepsilon_{h,v}|^2$, gives the ratio between the radiated cross-polar power to the copolar power at each point in the beam. Hence, it represents the polarization purity of the radiated electric field (from power conservation, $i_{h,v}^2 + |\varepsilon_{h,v}|^2 = 1$ and $|\varepsilon_{h,v}| \ll 1$ for well-designed antennas). McCormick (1981) defines a parameter β_a^2, which accounts for gain and phase inequality between the two ports as,

$$\beta_a^2 = \sqrt{\frac{G_h}{G_v}} e^{j(\Phi_h - \Phi_v)} \tag{6.22a}$$

$$= \sqrt{\frac{G_{0h}}{G_{0v}}} \sqrt{\frac{f_h}{f_v}} e^{j(\Phi_h - \Phi_v)} \tag{6.22b}$$

Note that $\sqrt{G_h} e^{j\Phi_h}$ and $\sqrt{G_v} e^{j\Phi_v}$ in (6.19) can be expressed as,

$$\sqrt{G_h} e^{j\Phi_h} = \beta_a (G_h G_v)^{1/4} e^{j(\Phi_h + \Phi_v)/2} \tag{6.22c}$$

$$\sqrt{G_v} e^{j\Phi_v} = \frac{1}{\beta_a} (G_h G_v)^{1/4} e^{j(\Phi_h + \Phi_v)/2} \tag{6.22d}$$

Antenna patterns are normally measured in the two principal planes of the antenna (these planes contain the boresight direction and the unit vectors $\hat{h}$ or $\hat{v}$ as indicated by the $\beta = 0$ and $90°$ planes in Fig. 6.14). However, it is also important to measure the patterns in other planes (e.g. the $\beta = 45$ and $135°$ planes). In the case of center-feed parabolic reflectors a set of four maxima in f_ε will occur symmetrically placed in the 45 and $135°$ planes (but displaced from the boresight direction). For example, Fig. 6.15a shows the copolar pattern (f_h) and the cross-polar pattern ($f_{\varepsilon h}$) in the $\beta = 135°$ plane for the CSU–CHILL antenna measured at the manufacturer's test range. The peaks in $f_{\varepsilon h}$ occur in this plane (and the $\beta = 45°$ plane) but are displaced away from boresight, the peak values being around -33 to -34 dB. Figure 6.15b shows similar patterns in the $\beta = 90°$ plane. Note now that the cross-polar pattern levels are significantly lower than in the $\beta = 135°$ plane. Close examination of the patterns shows that the close-in copolar sidelobes fall faster with angle away from boresight compared with the $\beta = 135°$ case where the sidelobes do not fall as rapidly. This is because the feed support struts are located in the $\beta = 45$ and $135°$ planes. Hence, both the "worst-case" close-in copolar sidelobe levels and the cross-polar pattern peaks are located in these two planes for the CSU–CHILL radar antenna.

Measurements of $\arg(\beta_a^2)$ are normally not provided by antenna manufacturers. Mudukutore et al. (1995) have given a procedure for measuring the differential phase pattern (i.e. patterns of $\arg(\beta_a^2) = \Phi_h - \Phi_v$) by illuminating the reflector antenna with a continuous wave signal from a standard gain horn located in the far-field. The polarization state of the standard gain horn was set to linear $45°$ to provide equal amplitude/phase excitation at the dual-ports of the reflector antenna. This technique relies on achieving phase coherence between the stable independent signal source used to excite the standard gain horn and the STALO/COHO of the radar over very short time periods (<20 ms). For well-designed antennas the $\arg(\beta_a^2)$ is generally uniform over the central part of the main lobe ($\pm 0.5°$ from boresight for typical $1°$ beam antennas). Large excursions of $\arg(\beta_a^2)$ generally occur near the location of the first sidelobe and beyond.

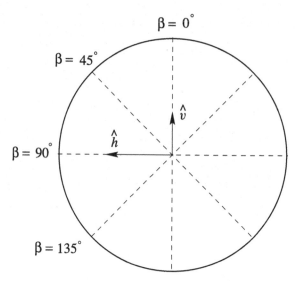

Fig. 6.14. The principal planes ($\beta = 0$ and $90°$) of a prime-focus parabolic reflector antenna are illustrated as viewed from the front of the antenna. The four cross-polar maxima lie in a symmetric configuration in the $\beta = 45$ and $135°$ planes around boresight.

6.2.1   Evaluation of antenna performance assuming spherical particles fill the beam

In weather radar applications, the antenna's dual-polarized performance is usually evaluated by assuming that homogenous spherical scatterers (e.g. very light drizzle) fill the entire beam. The dual-polarized radar range equation was derived in Section 3.4, see (3.67), and is written here (including antenna errors) for a single spherical scatterer,

$$
\begin{bmatrix} V_h \\ V_v \end{bmatrix} = \frac{\lambda}{4\pi r^2} e^{-j2k_0 r} (S) e^{j(\Phi_h + \Phi_v)}
$$

$$
\times \sqrt{G_h G_v} \begin{bmatrix} \beta_a^2(i_h^2 + \varepsilon_h^2) & i_h \varepsilon_v + \varepsilon_h i_v \\ i_h \varepsilon_v + i_v \varepsilon_h & \frac{1}{\beta_a^2}(\varepsilon_v^2 + i_v^2) \end{bmatrix} \begin{bmatrix} M_h \\ M_v \end{bmatrix} \tag{6.23}
$$

where S is the scattering matrix element of a sphere. For a random distribution of spherical scatterers, the Z_{dr} error due to the antenna is expressed as the ratio of $\langle |V_h^{10}|^2 \rangle$ to $\langle |V_v^{01}|^2 \rangle$, where the superscript "10" refers to $M_h = 1$, $M_v = 0$; while "01" refers to $M_h = 0$, $M_v = 1$. Following the derivation in Section 5.4, Z_{dr} can be expressed as,

$$
Z_{dr} = 10 \log_{10} \left(\frac{G_{0h}^2 \displaystyle\iint |i_h^2 + \varepsilon_h^2|^2 f_h^2 \, d\Omega}{G_{0v}^2 \displaystyle\iint |\varepsilon_v^2 + i_v^2|^2 f_v^2 \, d\Omega} \right) \tag{6.24}
$$

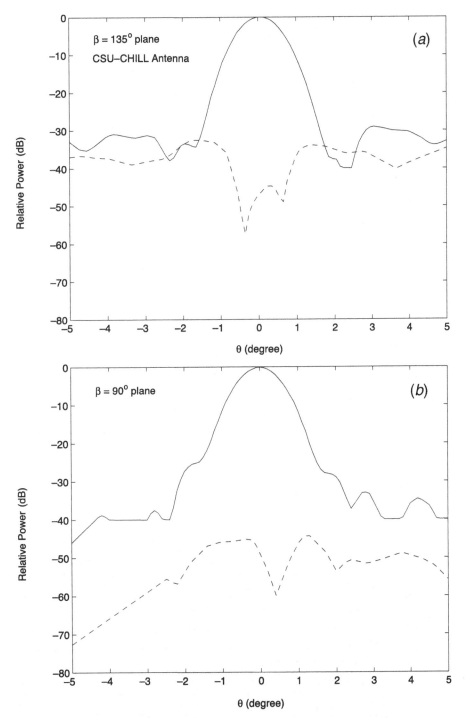

Fig. 6.15. CSU–CHILL antenna patterns in (a) the $\beta = 135°$ plane, and (b) the $\beta = 90°$ plane. Both copolar (solid line) and cross-polar (dashed-line) data are shown. Data were obtained at the manufacturer's test range (Radiation Systems Inc.).

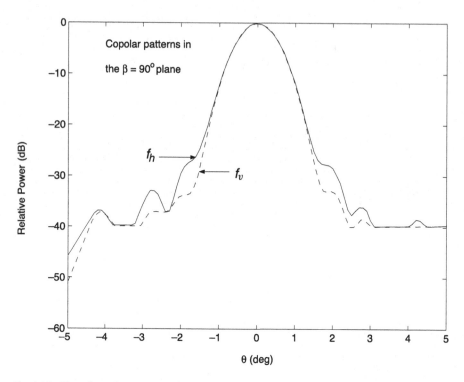

Fig. 6.16. Plot of copolar patterns, f_h and f_v in the $\beta = 90°$ plane to illustrate the pattern match.

To first order in ε, the Z_{dr} error is simply related to the copolar pattern mismatch,

$$Z_{dr} = 10 \log_{10} \left(\frac{G_{0h}^2}{G_{0v}^2} \frac{\int\int f_h^2 \, d\Omega}{\int\int f_v^2 \, d\Omega} \right) \tag{6.25}$$

where $i_h \approx i_v \approx 1$ for well-designed systems. Note that only the copolar patterns are needed to evaluate this error, but patterns must be available in at least four planes ($\beta = 0°, \pm 45°, 90°$; see Fig. 6.14) to obtain a good error estimate. Figure 6.16 shows the pattern match for the CSU–CHILL antenna in the 90° plane. When measurements can be obtained in light rain with the antenna pointed vertically, then a direct estimate of the Z_{dr} error due to the entire radar system (including antenna, polarization switch, receivers) is obtained relatively easily since oblate raindrops at vertical incidence appear "spherical" (see Section 6.3.2 and Fig. 6.22).

To first order in ε, $|\rho_{co}|$ can be expressed as,

$$|\rho_{co}| = \frac{|\langle (V_h^{10})^* V_v^{01} \rangle|}{\sqrt{\langle |V_h^{10}|^2 \rangle \langle |V_v^{01}|^2 \rangle}} \tag{6.26a}$$

$$= \frac{\left| \iint e^{-j2(\Phi_h - \Phi_v)} f_h f_v \, d\Omega \right|}{\sqrt{\iint f_h^2 \, d\Omega \iint f_v^2 \, d\Omega}} \tag{6.26b}$$

Again, the error in $|\rho_{co}|$ (whose ideal value is 1.0 for spherical scatterers) can be evaluated using the copolar power patterns, but the differential phase $(\Phi_h - \Phi_v)$ pattern of the antenna must also be known.

To first order in ε, $\arg\langle (V_h^{10})^*(V_v^{01}) \rangle$ (which is related to $\Psi_{dp} = \Phi_{dp} + \delta_{co}$, see (4.65b)) is written as,

$$\arg\langle (V_h^{10})^* V_v^{01} \rangle = \arg \left(\iint e^{-j2(\Phi_h - \Phi_v)} f_h f_v \, d\Omega \right) \tag{6.27}$$

LDR_{vh} is obtained from (6.23) and can be expressed as,

$$\text{LDR}_{vh} = 10 \log_{10} \left(\frac{\langle |V_v^{10}|^2 \rangle}{\langle |V_h^{10}|^2 \rangle} \right) \tag{6.28a}$$

$$= 10 \log_{10} \left\{ \frac{G_{0h} G_{0v} \iint \left[f_h f_{\varepsilon v} + f_v f_{\varepsilon h} + 2 \operatorname{Re}(\varepsilon_h \varepsilon_v^*) f_h f_v \right] d\Omega}{G_{0h}^2 \iint \left[f_h^2 + f_{\varepsilon h}^2 + 2 f_h^2 \operatorname{Re}(\varepsilon_h^2) \right] d\Omega} \right\} \tag{6.28b}$$

$$\approx 10 \log_{10} \left\{ \frac{G_{0v}}{G_{0h}} \frac{\iint \left[f_h f_{\varepsilon v} + f_v f_{\varepsilon h} + 2 \operatorname{Re}(\varepsilon_h \varepsilon_v^*) f_h f_v \right] d\Omega}{\iint f_h^2 \, d\Omega} \right\} \tag{6.28c}$$

Since the intrinsic LDR_{vh} for spherical scatterers is $-\infty$ dB, it is useful to consider the upper bound of (6.28c), which will occur when $2 \operatorname{Re}(\varepsilon_h \varepsilon_v^*)$ is replaced by $2|\varepsilon_h||\varepsilon_v|$. Hence, LDR_{vh}^{ub} is given by,

$$\text{LDR}_{vh}^{ub} = 10 \log_{10} \left[\frac{G_{0v}}{G_{0h}} \frac{\iint (f_h f_{\varepsilon v} + f_v f_{\varepsilon h} + 2\sqrt{f_h f_v f_{\varepsilon h} f_{\varepsilon v}}) \, d\Omega}{\iint f_h^2 \, d\Omega} \right] \tag{6.29}$$

The upper-bound value can be evaluated if the copolar and cross-polar power patterns are known (Chandrasekar and Keeler 1993). Since the maxima in $f_{\varepsilon h}$ (and $f_{\varepsilon v}$) lie along the $\beta = \pm 45°$ planes for center-fed parabolic reflectors, a further approximation to the upper bound is to integrate the patterns in the worst case plane (say, the $45°$ plane).

The last term to be considered is related to $R_{cx}^h = \langle V_h^{10}\{V_v^{10}\}^* \rangle$ defined in (5.70). This term can be divided (normalized) by $\langle |V_h^{10}|^2 \rangle$, and the corresponding antenna error effects are given as,

$$\frac{\langle V_h^{10}(V_v^{10})^* \rangle}{\langle |V_h^{10}|^2 \rangle} = \frac{\iint G_h G_v (\beta_a^2)(i_h^2 + \varepsilon_h^2)(i_h \varepsilon_v^* + i_v \varepsilon_h^*) \, d\Omega}{\iint G_h^2 |i_h^2 + \varepsilon_h^2|^2 \, d\Omega} \tag{6.30}$$

Neglecting second-order error quantities and using (6.21, 6.22) gives,

$$\frac{\langle V_h^{10}(V_v^{10})^* \rangle}{\langle |V_h^{10}|^2 \rangle} = \sqrt{\frac{G_{0v}}{G_{0h}}} \frac{\iint e^{j(\Phi_h - \Phi_v)} f_h(\sqrt{f_h f_v} \varepsilon_v^* + \sqrt{f_h f_v} \varepsilon_h^*) \, d\Omega}{\iint f_h^2 \, d\Omega} \tag{6.31}$$

Again, an upper bound is obtained by setting $\varepsilon_v^* = |\varepsilon_v|$ and $\varepsilon_h^* = |\varepsilon_h|$ resulting in an expression for the magnitude of the error involving only the copolar and cross-polar pattern functions,

$$\frac{|\langle V_h^{10}(V_v^{10})^* \rangle|^{ub}}{\langle |V_h^{10}|^2 \rangle} = \sqrt{\frac{G_{0v}}{G_{0h}}} \frac{\iint f_h(\sqrt{f_h f_{\varepsilon v}} + \sqrt{f_v f_{\varepsilon h}}) \, d\Omega}{\iint f_h^2 \, d\Omega} \tag{6.32}$$

As mentioned earlier, offset-feed parabolic reflector antennas have been used in research weather radars mainly at frequencies of 5 GHz and higher. Because there are no feed or strut blockages, excellent sidelobe levels (typically better than -30 dB) are achievable in any plane. Also, the sidelobe levels fall off rapidly from boresight in any plane as compared with the center-fed parabolic reflector antenna (see Fig. 6.15a). Figure 6.17 shows the linear copolar and cross-polar patterns of an offset-feed reflector antenna to illustrate the low sidelobe levels. The main disadvantage of the offset geometry is that there is inherent linear cross-polarization in the radiation field (in the plane normal to the offset) even if the feed is ideal. The linear cross-polarization levels can reach a maximum of around -28 to -30 dB. The linear cross-polarization can be compensated for by using a circularly polarized feed but this leads to beam-squinting effects, i.e. the right- and left-hand radiation patterns are shifted in opposite directions, away from the boresight. The magnitude of the shift depends on the geometry but is typically quite small, and the effects on meteorological measurement at close range (<60 km for typical $1°$ beams) are believed to be small. Figure 6.18 illustrates the RHC/LHC pattern mismatch and $|\beta_a|^2$, see (6.22a), for the NRC of Canada's 16.5-GHz radar antenna. In very light drizzle (spherical scatterers) this antenna's circular depolarization ratio limit is estimated to be around -38 to -39 dB. For the CSU–CHILL antenna, the equivalent linear depolarization ratio limit is estimated to be -34 dB.

The linear cross-polarization inherent in offset-feed reflector antennas can be nearly eliminated by the dual-offset reflector design illustrated in Fig. 6.19, which is also referred to as an offset Cassegrain reflector. This design is used on the NOAA/ETL K_a-band (35-GHz) radar. Figure 6.20 shows a picture of the NOAA/ETL antenna. This radar is a polarization diversity system with two receivers (see also, Sections 3.8.1 and 7.3 for examples of data from this radar). The linear depolarization ratio limit in light drizzle is estimated to be around -35 dB.

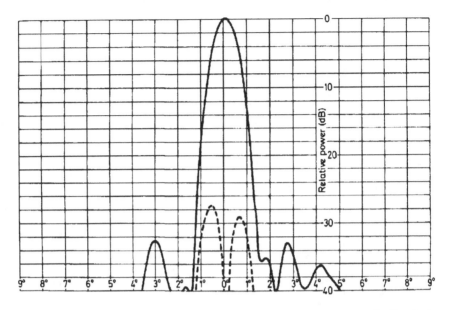

Fig. 6.17. Copolar and cross-polar power patterns for the Italian (IFA) Polar-55C offset-feed reflector antenna (azimuth cut). From Leonardi et al. (1984).

6.2.2   Formulation of radar observables in the presence of system polarization errors

Apart from those imperfections in the polarization purity of the antenna itself considered earlier, polarization errors also arise in the passive microwave circuitry (e.g. if a separate polarizer network, as illustrated in Fig. 6.8d is used), or when a high power waveguide switch is used (e.g. a ferrite circulator switch). Depending on the location of the switch in the system, the errors may occur on transmission only (e.g. see Fig. 6.9a), or on both transmission and reception (e.g. see Fig. 6.10a).

The mathematical formulation of system polarization errors was comprehensively treated by McCormick (1981). Simplifications occur for dual-channel systems operating at circular or linear (horizontal/vertical) bases. This simplification occurs because the signal power in one channel is much smaller than in the other channel in most types of precipitation (typically, these are referred to as the "weak" and "strong" channels, respectively). Errors arise in the radar observables because the radar radiates a small component of the undesired polarization, and upon reception couples a small component of the "strong" signal into the desired "weak" signal channel. Thus, the radar observables that are mostly affected (in the linear h/v-basis) are, for example, LDR_{vh} and R_{cx}^{h}, and to a much lesser extent, the Z_{dr} and ρ_{co} (for definitions of these observables, see Section 3.10). In the circular basis, the observables CDR and (W/W_2) are affected by system polarization errors (for a definition of W/W_2 see Section 4.4). If the beam-averaged system errors are known, and if the precipitation is homogeneous throughout the beam, then the radar observables can be approximately corrected for

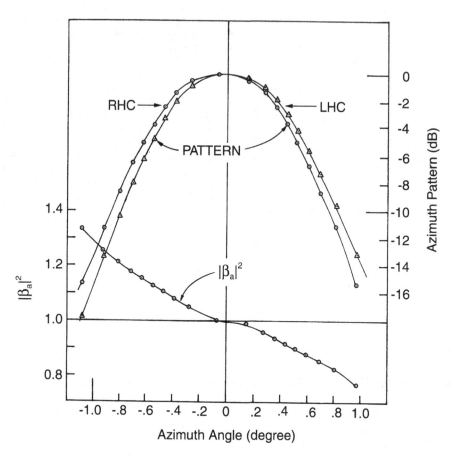

Fig. 6.18. Antenna patterns for the 16.5-GHz radar of the National Research Council of Canada. Note that $|\beta_a|^2$ is the beam inequality between ports. From McCormick and Hendry (1979b). (© 1979 American Geophysical Union.)

system polarization errors. The case of linear (h/v-basis) will be used here to illustrate the formulation.

System polarization errors can be lumped into an error matrix of the form,

$$\mathbf{X} = \begin{bmatrix} i_h & \varepsilon_v \\ \varepsilon_h & i_v \end{bmatrix} \begin{bmatrix} \beta_a & 0 \\ 0 & (\beta_a)^{-1} \end{bmatrix} \tag{6.33}$$

and the radar range equation for a single particle whose scattering matrix is assumed to be $\mathbf{S}_{\mathrm{BSA}}$ can be expressed as,

$$\begin{bmatrix} V_h \\ V_v \end{bmatrix} = \frac{\lambda}{4\pi r^2} e^{-j2k_0 r} \sqrt{G_h G_v} \, e^{j(\Phi_h + \Phi_v)} [\mathbf{X}^t][\mathbf{S}_{\mathrm{BSA}}][\mathbf{X}] \begin{bmatrix} M_1 \\ M_2 \end{bmatrix} \tag{6.34}$$

where the transmitted wave amplitudes are given by $[M_1 \ M_2]^t$, which is either $[1 \ 0]^t$ or $[0 \ 1]^t$ in the usual case. The voltages V_h and V_v are assumed to be at the input to the

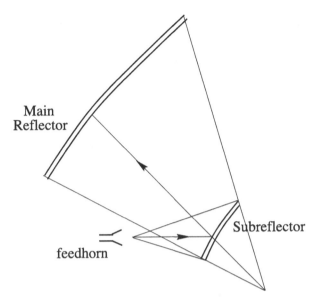

Fig. 6.19. Simple diagram of an offset Cassegrain reflector antenna.

Fig. 6.20. NOAA/ETL dual-polarization, Doppler K_a-band radar's offset Cassegrain antenna. The parabolic dish diameter is 1.05 m. Photo courtesy of Bob Kropfli, NOAA/ETL.

receivers. To first order in ε, the matrix product $\mathbf{X}^t \mathbf{S}_{BSA} \mathbf{X}$ can be expressed as,

$$[\mathbf{X}^t][\mathbf{S}_{BSA}][\mathbf{X}] = \begin{bmatrix} \beta_a^2(S_{hh}i_h^2 + 2S_{hv}i_h\varepsilon_h) & (S_{hv}i_h i_v + S_{hh}i_h\varepsilon_v + S_{vv}i_v\varepsilon_h) \\ (S_{hv}i_h i_v + S_{hh}i_h\varepsilon_v + S_{vv}i_v\varepsilon_h) & (\beta_a)^{-2}(S_{vv}i_v^2 + 2S_{hv}i_v\varepsilon_h) \end{bmatrix}$$

(6.35)

Substituting in (6.34) and letting $[M_1 \ M_2]^t = [1 \ 0]^t$, it follows that,

$$V_h^{10} = \frac{\lambda}{4\pi r^2} e^{-j2k_0 r} G_h e^{j2\Phi_h}(S_{hh}i_h^2 + 2S_{hv}i_h\varepsilon_h)$$

(6.36a)

$$\approx \frac{\lambda}{4\pi r^2} e^{-j2k_0 r} G_h e^{j2\Phi_h} S_{hh}i_h^2$$

(6.36b)

$$V_v^{10} = \frac{\lambda}{4\pi r^2} e^{-j2k_0 r} \sqrt{G_h G_v} e^{j(\Phi_h + \Phi_v)}(S_{hv}i_h i_v + S_{hh}i_h\varepsilon_v + S_{vv}i_v\varepsilon_h)$$

(6.36c)

The measured linear depolarization ratio ($\mathrm{LDR}_{vh}^{\mathrm{meas}}$) is,

$$10^{0.1(\mathrm{LDR}_{vh}^{\mathrm{meas}})} = \frac{\langle |V_v^{10}|^2 \rangle}{\langle |V_h^{10}|^2 \rangle}$$

$$= \frac{\langle G_h G_v \{|S_{hv}|^2 i_h^2 i_v^2 + |S_{hh}i_h\varepsilon_v + S_{vv}i_v\varepsilon_h|^2 + 2\,\mathrm{Re}[i_h i_v S_{hv}^*(S_{hh}i_h\varepsilon_v + S_{vv}i_v\varepsilon_h)]\} \rangle}{\langle G_h^2 |S_{hh}|^2 i_h^2 \rangle}$$

(6.37)

Since well-designed systems will have $i_h \approx i_v \approx 1$, and using (6.20), the above reduces to,

$$10^{0.1(\mathrm{LDR}_{vh}^{\mathrm{meas}})} = \left(\frac{G_{0v}}{G_{0h}}\right) \left\{ \frac{\langle |S_{hv}|^2 f_h f_v \rangle}{\langle |S_{hh}|^2 f_h^2 \rangle} + \frac{\langle |S_{hh}\varepsilon_v + S_{vv}\varepsilon_h|^2 f_h f_v \rangle}{\langle |S_{hh}|^2 f_h^2 \rangle} \right. $$
$$\left. + \frac{2\,\mathrm{Re}[\langle (S_{hv}^* S_{hh}\varepsilon_v + S_{hv}^* S_{vv}\varepsilon_h) f_h f_v \rangle]}{\langle |S_{hh}|^2 f_h^2 \rangle} \right\}$$

(6.38)

Note that (6.28c) derived earlier is a special case of (6.38) if the scatterers are assumed to be spherical and homogeneously distributed within the beam, To simplify (6.38), assume that $\langle S_{hv}^* S_{hh} \rangle \approx \langle S_{hv}^* S_{vv} \rangle$ and $\langle f_h f_v |S_{hh}\varepsilon_v + S_{vv}\varepsilon_h|^2 \rangle \approx \langle |S_{hh}|^2 \rangle \langle f_h f_v |\varepsilon_h + \varepsilon_v|^2 \rangle$. It follows that (6.38) reduces to,

$$10^{0.1(\mathrm{LDR}_{vh}^{\mathrm{meas}})} = \left(\frac{G_{0v}}{G_{0h}}\right) \left\{ \frac{\langle |S_{hv}|^2 \rangle}{\langle |S_{hh}|^2 \rangle} \frac{\iint f_h f_v \, d\Omega}{\iint f_h^2 \, d\Omega} + \frac{\iint f_h f_v |\varepsilon_h + \varepsilon_v|^2 \, d\Omega}{\iint f_h^2 \, d\Omega} \right.$$
$$\left. + 2\,\mathrm{Re}\left[\frac{\langle S_{hv}^* S_{hh} \rangle}{\langle |S_{hh}|^2 \rangle} \frac{\iint (\varepsilon_v + \varepsilon_h) f_h f_v \, d\Omega}{\iint f_h^2 \, d\Omega} \right] \right\}$$

(6.39)

As discussed in Section 3.12, if the precipitation medium satisfies polarization isotropy, or "mirror" reflection symmetry about the vertical direction (mean canting angle is $0°$), then $\langle S_{hv}^* S_{hh} \rangle = 0$ and (6.39) can be used to estimate the true (or intrinsic) linear depolarization ratio of the precipitation as follows,

$$\frac{\langle |S_{hv}|^2 \rangle}{\langle |S_{hh}|^2 \rangle} \approx \left(\frac{G_{0h}}{G_{0v}} \right) 10^{0.1(\mathrm{LDR}_{vh}^{\mathrm{meas}})} - \frac{\iint f_h^2 |\varepsilon_h + \varepsilon_v|^2 \, d\Omega}{\iint f_h^2 \, d\Omega} \tag{6.40}$$

In the above, the copolar pattern is assumed to be rotationally symmetric (i.e. $f_h = f_v$). The "correction" term on the right-hand side of (6.40) is essentially identical to (6.28c) and may be obtained by collecting data separately in very light drizzle conditions (tiny, spherical drops) where $S_{hv} \equiv 0$. It is not clear how accurate (6.40) would be in correcting the measured linear depolarization ratio in other types of precipitation because of the assumptions made. However, (6.40) can be used to determine operationally how low the beam-averaged system polarization errors must be for the measured LDR to be within, say, 1 dB of its true value. If the true value is, for example, -27 dB, then the beam-averaged system polarization error term in (6.40) must satisfy,

$$10 \log_{10} \left(\frac{\iint f_h^2 |\varepsilon_h + \varepsilon_v|^2 \, d\Omega}{\iint f_h^2 \, d\Omega} \right) \leq -33 \ \mathrm{dB} \tag{6.41}$$

or, in other words, the radar should measure LDR in very light rain of around -33 dB. As mentioned earlier, the CSU–CHILL radar normally measures around -34 dB in very light drizzle while the NOAA/ETL radar measures around -35 dB. Note that (6.40) is a generalization of (6.8c), which was derived assuming finite transmit/receive isolation of the ferrite polarization switch but an otherwise ideal antenna.

The observable, $\langle V_h^{10}(V_v^{10})^* \rangle / \langle |V_h^{10}|^2 \rangle$, can be derived from (6.36) and the result is,

$$\frac{\langle V_h^{10}(V_v^{10})^* \rangle}{\langle |V_h^{10}|^2 \rangle} = \sqrt{\frac{G_{0v}}{G_{0h}}} \left[\left(\frac{\langle S_{hh} S_{hv}^* \rangle}{\langle |S_{hh}|^2 \rangle} \right) \frac{\iint e^{j(\Phi_h - \Phi_v)} f_h \sqrt{f_h f_v} \, d\Omega}{\iint f_h^2 \, d\Omega} \right.$$

$$+ \frac{\iint e^{j(\Phi_h - \Phi_v)} f_h \sqrt{f_h f_v} \varepsilon_v^* \, d\Omega}{\iint f_h^2 \, d\Omega}$$

$$\left. + \frac{\rho_{co}}{\sqrt{\mathcal{F}_{dr}}} \frac{\iint e^{j(\Phi_h - \Phi_v)} f_h \sqrt{f_h f_v} \varepsilon_h^* \, d\Omega}{\iint f_h^2 \, d\Omega} \right] \tag{6.42}$$

where ρ_{co} and $\mathfrak{z}_{dr}$ are the true copolar correlation coefficient and differential reflectivity of the precipitation (see Section 3.10 for definitions). It may be assumed that $(\Phi_h - \Phi_v)$ is constant over the main lobe. Rotational symmetry of the copolar patterns may also be assumed to set $f_h = f_v$. With these assumptions, (6.42) simplifies to,

$$\frac{\langle V_h^{10}(V_v^{10})^* \rangle}{\langle |V_h^{10}|^2 \rangle} = \sqrt{\frac{G_{0v}}{G_{0h}}} e^{j(\Phi_h - \Phi_v)} \left(\frac{\langle S_{hh} S_{hv}^* \rangle}{\langle |S_{hh}|^2 \rangle} + \frac{\iint f_h^2 \varepsilon_v^* \, d\Omega}{\iint f_h^2 \, d\Omega} + \frac{\rho_{co}}{\sqrt{\mathfrak{z}_{dr}}} \frac{\iint f_h^2 \varepsilon_h^* \, d\Omega}{\iint f_h^2 \, d\Omega} \right)$$

$$(6.43)$$

In the absence of system cross-polarization errors, it is clear that the magnitude of $\langle V_h^{10}(V_v^{10})^* \rangle$ estimates $|\langle S_{hh} S_{hv}^* \rangle|$. For precipitation satisfying "mirror" reflection symmetry about the vertical (mean canting angle $\approx 0°$) such as rain, $\langle S_{hh} S_{vh}^* \rangle \approx 0$. If measurements in rain indicate a high value of $|\langle V_h^{10}(V_v^{10})^* \rangle|$ (reflected in values of ρ_{cx}^h close to unity, see (4.78)), then it must be due to system cross-polarization errors which tend to "leak" a fraction of the strong signal into the weak signal channel (see Metcalf and Ussailis 1984).

In very light drizzle, where $|\rho_{co}| \to 1$, $\mathfrak{z}_{dr} \to 1$, and $S_{hv} \to 0$, observations will yield an estimate of the beam-averaged first order error,

$$\left[\frac{\langle V_h^{10}(V_v^{10})^* \rangle}{\langle |V_h^{10}|^2 \rangle} \right]_{drizzle} = \sqrt{\frac{G_{0v}}{G_{0h}}} e^{j(\Phi_h - \Phi_v)} \left[\frac{\iint f_h^2 (\varepsilon_h^* + \varepsilon_v^*) \, d\Omega}{\iint f_h^2 \, d\Omega} \right]$$

$$(6.44)$$

which is identical to (6.31) if $f_h = f_v$. However, correction of data using (6.43) collected in other kinds of precipitation is not straightforward since separate estimates of the beam-averaged error terms are required as well as the true values of ρ_{co} and $\mathfrak{z}_{dr}$.

Similar formulations can be derived using (6.35) for the observables used to estimate Z_{dr} and ρ_{co}. For example, neglecting second-order terms in $\varepsilon_{h,v}$,

$$V_v^{01} = \frac{\lambda}{4\pi r^2} e^{-j2k_0 r} G_v e^{j2\Phi_v} (S_{vv} + 2S_{hv}\varepsilon_v)$$

$$(6.45)$$

and,

$$10^{0.1(Z_{dr}^{meas})} = \frac{\langle |V_h^{10}|^2 \rangle}{\langle |V_v^{01}|^2 \rangle}$$

$$\approx \left(\frac{G_{0h}}{G_{0v}} \right)^2 \left[\frac{\langle |S_{hh}|^2 \rangle \iint f_h^2 \, d\Omega + 4 \operatorname{Re}(\langle S_{hh} S_{hv}^* \rangle) \iint f_h^2 \varepsilon_h^* \, d\Omega}{\langle |S_{vv}|^2 \rangle \iint f_v^2 \, d\Omega + 4 \operatorname{Re}(\langle S_{vv} S_{hv}^* \rangle) \iint f_v^2 \varepsilon_v^* \, d\Omega} \right]$$

$$(6.46)$$

Note that (6.46) is a generalization of (6.4b) if only first-order error terms are considered. If the precipitation satisfies "mirror" reflection symmetry about the vertical direction (mean canting angle $= 0°$), or if it satisfies azimuthal symmetry, then there is no first-order cross-polarization error contribution to the measured Z_{dr} (Metcalf and Ussailis 1984). Similarly, the measured copolar correlation coefficient,

$$
\rho_{co}^{meas} = \frac{\langle (V_h^{10})^*(V_v^{01}) \rangle}{\sqrt{\langle |V_h^{10}|^2 \rangle \langle |V_v^{01}|^2 \rangle}} \approx \left[(\rho_{co}) \frac{\iint e^{-j2(\Phi_h - \Phi_v)} f_h f_v \, d\Omega}{\sqrt{\iint f_h^2 \, d\Omega \iint f_v^2 \, d\Omega}} \right.
$$

$$
+ (2\rho_{cx}^h \sqrt{L \not{z}_{dr}}) \frac{\iint e^{-j2(\Phi_h - \Phi_v)} f_h f_v \varepsilon_v \, d\Omega}{\sqrt{\iint f_h^2 \, d\Omega \iint f_v^2 \, d\Omega}}
$$

$$
\left. + (2\rho_{cx}^v \sqrt{L}) \frac{\iint e^{-j2(\Phi_h - \Phi_v)} f_h f_v \varepsilon_h^* \, d\Omega}{\sqrt{\iint f_h^2 \, d\Omega \iint f_v^2 \, d\Omega}} \right] \tag{6.47}
$$

where ρ_{cx}^h and ρ_{cx}^v are intrinsic coefficients defined in (4.78) and L is the intrinsic LDR$_{vh}$ expressed as a ratio. Again, there is no first-order error contribution if precipitation satisfies "mirror" reflection symmetry (mean canting angle $= 0°$), or azimuthal symmetry.

When the spatial distribution of scatterers is inhomogeneous, then gradients of reflectivity across the main lobe and close-in sidelobes induce antenna pattern related errors which bias the polarimetric estimates. Errors in Z_{dr} due to mis-matched copolar patterns (e.g. see Fig. 6.16) together with intrinsic reflectivity gradients across the beam were evaluated by Pointin et al. (1988). As a first approximation, (6.46) may be simplified as,

$$
10^{0.1(Z_{dr}^{meas})} \approx \frac{\iint Z_h(r, \theta, \phi) f_h^2(\theta - \theta_0, \phi - \phi_0) \, d\Omega}{\iint Z_v(r, \theta, \phi) f_v^2(\theta - \theta_0, \phi - \phi_0) \, d\Omega} \tag{6.48}
$$

where $Z_h(r, \theta, \phi)$ and $Z_v(r, \theta, \phi)$ are the intrinsic reflectivities. The form of the above equation implies a convolution over a spherical surface of given radius r between the intrinsic reflectivity and the antenna pattern similar to (5.63) (note that θ_0, ϕ_0 is the boresight direction). Errors can also be more simply evaluated in any one plane, typically, the azimuth or elevation plane (e.g. Herzegh and Carbone 1984).

Errors in LDR due to gradients of cross-polar and copolar reflectivities across the beam can be evaluated by using (6.38) and assuming "mirror" reflection symmetry about a mean canting angle of $0°$, together with $S_{hh}\varepsilon_v \approx S_{vv}\varepsilon_h$, $f_h \approx f_v$, and

$f_h|\varepsilon_h|^2 \approx f_v|\varepsilon_v|^2$ to arrive at,

$$
10^{0.1(\mathrm{LDR}_{vh}^{\mathrm{meas}})} = \frac{\displaystyle\iint Z_{hv}(r,\theta,\phi)f_h^2(\theta-\theta_0,\phi-\phi_0)\,d\Omega}{\displaystyle\iint Z_h(r,\theta,\phi)f_h^2(\theta-\theta_0,\phi-\phi_0)\,d\Omega}
$$
$$
+\;\frac{4\displaystyle\iint Z_h(r,\theta,\phi)f_h(\theta-\theta_0,\phi-\phi_0)f_{\varepsilon h}(\theta-\theta_0,\phi-\phi_0)\,d\Omega}{\displaystyle\iint Z_h(r,\theta,\phi)f_h^2(\theta-\theta_0,\phi-\phi_0)\,d\Omega}
\tag{6.49}
$$

The first term is simply "smoothing" of the intrinsic cross-polar reflectivity, Z_{hv}, by the main lobe of the antenna, while the second term will dominate the error if strong gradients of the copolar reflectivity (Z_h) exist across the product pattern, $f_h f_{\varepsilon h}$. Evaluation of such errors can be found in Hubbert et al. (1998) for model Z_{hv} and Z_h profiles and using CSU–CHILL antenna patterns in the worst-case plane ($\beta = 45$ and $135°$ planes). Similarly, gradient-induced errors in $|\rho_{co}|$ can also be evaluated (e.g. see Hubbert et al. 1998). Experience has shown that data from regions where the cross-beam reflectivity gradients exceed 20–25 dB km^{-1}, especially at ranges exceeding 50 km, must be viewed with caution. Such errors can be reduced by using narrower beam antennas.

Errors in Φ_{dp} can be evaluated by examining $\arg(\rho_{co})$. Using (6.47) with $f_h \approx f_v$, and introducing $\exp(j\Phi_{dp})$ as the intrinsic differential propagation phase, and neglecting the cross-polar error terms results in,

$$
\arg(\rho_{co}^{\mathrm{meas}}) = \Phi_{dp}^{\mathrm{meas}}
$$
$$
= \arg\left[\iint \langle S_{hh}^* S_{vv}\rangle e^{-j2(\Phi_h-\Phi_v)} e^{j2\Phi_{dp}} f_h^2(\theta-\theta_0,\phi-\phi_0)\,d\Omega\right]
\tag{6.50}
$$

Recall that $\Phi_h - \Phi_v$ is the differential phase pattern of the antenna. Equation (6.50) may be approximated by assuming $\Phi_h - \Phi_v = \text{constant}$ and $\langle S_{hh}^* S_{vv}\rangle \approx \langle|S_{hh}|^2\rangle \equiv Z_h$, and integration over a single plane (e.g. azimuthal plane) resulting in,

$$
\Phi_{dp}^{\mathrm{meas}} \approx \arg\left[\int Z_h(r,\theta)e^{j2\Phi_{dp}(r,\theta)} f_h^2(\theta-\theta_0)\,d\theta\right]
\tag{6.51}
$$

Azimuthal gradients in intrinsic Φ_{dp} can cause errors in the resulting estimate of K_{dp} especially at long ranges; for model calculations of the errors involved see Ryzhkov and Zrnić (1998).

6.3　Radar calibration

The weather radar range equation (for a single channel radar) derived in Section 5.3, see (5.57), relates the radar reflectivity (η) to the averaged power at the receiver output

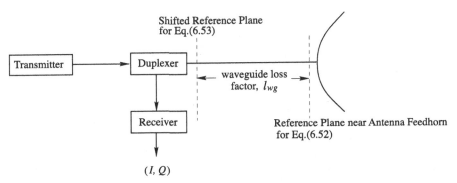

Fig. 6.21. Illustrating the shifted reference plane where receiver calibration is more conveniently performed. The transmitted power is also measured at this shifted reference plane. The peak boresight gain G_0 referred to the antenna port is replaced by G_0/l_{wg} when referred to the shifted reference plane, where l_{wg} is the waveguide loss factor defined to be ≥ 1.0.

$(\bar{P}_o)$ via the radar system constant C,

$$\bar{P}_o = \frac{C\eta}{r_0^2} \tag{6.52a}$$

where

$$C = \left(\frac{cT_0}{2}\right)\left(\frac{G_r}{l_r}\right)\left[\frac{\lambda^2 P_t G_0^2}{(4\pi)^3}\right]\left[\frac{\pi\theta_1\phi_1}{8\ln 2}\right] \tag{6.52b}$$

The above equation implicitly assumes a common reference plane at the antenna feedhorn port for defining the receiver power gain (G_r; from antenna port to output of receiver), the transmitter pulse power (P_t) and the peak antenna gain (G_0). For convenience in calibrating the receiver, the reference plane is usually moved closer to the receiver's input as illustrated in Fig. 6.21. Receiver calibration involves injecting (continuous wave) signals of known amplitude at the reference plane using a highly stable, standard signal generator and recording the digital values (of $I^2 + Q^2$) at the output of the signal processor. Such calibration measures G_r but not l_r. The transmitter pulse power is also measured at this plane, and the antenna gain will then be reduced by the waveguide loss from the antenna port to the new reference plane (i.e. $G_0 \rightarrow G_0/l_{wg}$), where l_{wg} is the one-way waveguide loss factor (≥ 1). If $\bar{P}_{ref}$ is the average received power at the new reference plane, then η can be expressed as,

$$\eta = \frac{(r_0^2)\bar{P}_{ref}}{(cT_0/2)\left[\lambda^2 P_t G_0^2/(4\pi)^3 l_{wg}^2\right](\pi\theta_1\phi_1/8\ln 2)} \tag{6.53}$$

Note that $\bar{P}_o = (G_r/l_r)\bar{P}_{ref}$. Recall that l_r is the finite bandwidth loss factor ($l_r \geq 1.0$ by definition, see (5.56)), which depends on the shape of the receiver filter and the transmitted waveform.[8] For a matched (to a rectangular pulse) filter receiver,

$10 \log_{10}(l_r) = 1.8$ dB (Nathanson and Smith 1972). For a "practical" receiver filter, $10 \log_{10}(l_r) \approx 2.3$ dB (see Fig. 4.8 of Doviak and Zrnić 1993). In practice, the measured power at the reference plane is the sum of the intrinsic signal power plus noise power ($\bar{P}_{S+N} = \bar{P}_{ref} + P_N$). Noise "correction" for η implies replacing $\bar{P}_{ref} \rightarrow \bar{P}_{S+N} - P_N$ in (6.53).

Equation (6.53) can also be written in the form,

$$\eta = C' r_0^2 \bar{P}_{ref} \tag{6.54a}$$

or,

$$10 \log_{10}(\eta) = 10 \log_{10}(C') + 20 \log_{10}(r_0) + 10 \log_{10}(\bar{P}_{ref}) \tag{6.54b}$$

where,

$$C' = \left(\frac{2}{cT_0}\right) \left[\frac{(4\pi)^3 l_{wg}^2}{\lambda^2 P_t G_0^2}\right] \left(\frac{8 \ln 2}{\pi \theta_1 \phi_1}\right) \tag{6.54c}$$

Each term in the above can be separately determined to establish the nominal value of C' (which is also referred to as the radar system constant). The relation between η and the equivalent reflectivity factor (Z_e) is given in (5.49a). It follows that,

$$Z_e = \frac{1}{\pi^5 |K_\omega|^2} \left(\frac{2}{cT_0}\right) \left[\frac{(4\pi)^3 l_{wg}^2}{P_t G_0^2}\right] \left(\frac{8 \ln 2}{\pi \theta_1 \phi_1}\right) \lambda^2 r_0^2 \left(\bar{P}_{S+N} - P_N\right) \tag{6.55}$$

It is customary to define the minimum detectable Z_e at a certain range r_0 when the signal-to-noise ratio at the reference plane is unity, that is, $\bar{P}_{ref}/P_N = 1$. Using this condition and the fact that $P_N = kTB$ at the reference plane, where T is defined in (5.157) and B is the noise bandwidth , (6.55) reduces to,

$$\min(Z_e) = \frac{1}{\pi^5 |K_\omega|^2} \left(\frac{2}{cT_0}\right) \left[\frac{(4\pi)^3 l_{wg}^2}{P_t G_0^2}\right] \left(\frac{8 \ln 2}{\pi \theta_1 \phi_1}\right) \lambda^2 r_0^2 (kTB) \tag{6.56}$$

This equation can be used to compute the minimum detectable Z_e at a certain r_0 if the radar system parameters are known. Typical values of $\min(Z_e)$ range between -20 to -30 dBZ at $r_0 = 10$ km (see, Smith 1986; 1993).

6.3.1 Absolute calibration using a metal sphere

The radar range equation for a point target, see (3.70), located on the antenna's boresight direction and in the far-field can be written as,

$$P_{ref(ms)} = \frac{\lambda^2 P_t G_0^2}{(4\pi)^3 l_{wg}^2} \frac{\sigma_{ms}}{r_{ms}^4} \tag{6.57}$$

where the subscript "ms" stands for metal sphere. The σ_{ms} is the radar cross section of the metal sphere and is given by its geometrical cross section area when its diameter

$\gg \lambda$ (geometrical optics limit value). Note that P_t and the received power, $P_{\text{ref(ms)}}$, are referenced to the same reference plane used in deriving (6.53). As explained earlier, the peak antenna gain $G_0 \to G_0/l_{\text{wg}}$ when referenced to this plane. The received power from the sphere should be the peak value which is attained when the sphere is located at the peak of the three-dimensional weighting function defined in (5.60a). From (6.57), the term G_0^2/l_{wg}^2 can be expressed as,

$$\frac{G_0^2}{l_{\text{wg}}^2} = \frac{(4\pi)^3 r_{\text{ms}}^4 P_{\text{ref(ms)}}}{\lambda^2 P_t \sigma_{\text{ms}}} \tag{6.58a}$$

which when substituted in (6.54c) gives,

$$C' = \left(\frac{1}{P_{\text{ref(ms)}}} \right) \left(\frac{\sigma_{\text{ms}}}{r_{\text{ms}}^4} \right) \left(\frac{8 \ln 2}{\pi \theta_1 \phi_1} \right) \left(\frac{2}{cT_0} \right) \tag{6.58b}$$

Thus the sphere calibration procedure directly establishes the radar system constant C'. The finite bandwidth loss factor, l_r, is determined from (5.56). When (6.58b) is substituted in (6.54a), the reflectivity of precipitation relative to σ_{ms} can be expressed as,

$$\frac{\eta}{\sigma_{\text{ms}}} = \left(\frac{\bar{P}_{\text{ref}}}{P_{\text{ref(ms)}}} \right) \left(\frac{r_0^2}{r_{\text{ms}}^4} \right) \left(\frac{8 \ln 2}{\pi \theta_1 \phi_1} \right) \left(\frac{2}{cT_0} \right) \tag{6.59}$$

which reinforces the fact that the absolute calibration procedure yields the reflectivity of precipitation during operations relative to the radar cross section of the metal sphere. For higher accuracy it is preferable to use the measured antenna patterns to calculate the antenna weighting integral rather than using the Probert–Jones approximation, see (5.47). The final accuracy in reflectivity is generally expected to be within 0.5–1.0 dB accounting for various error sources during both sphere calibration and during operations.

6.3.2 Polarization-agile/single-receiver radar

The only difference between this system and the single-channel system is the waveguide switching device and the dual-polarized antenna. Refer to Fig. 6.5 for a simplified block diagram of the system.

Assume that the metal sphere calibration has been performed with the switch connected to the h-port of the antenna.[9] The additional insertion loss factor ($l_{\text{sw}(h)} \geq 1.0$) introduced by the switch in this state can be merged into the waveguide loss factor, i.e. $l_{\text{wg}} \to l_{\text{wg}} l_{\text{sw}(h)}$.

It was shown earlier, see (6.46), that there is no first-order cross-polarization error contribution to the measured Z_{dr} under conditions of "mirror" reflection symmetry about the vertical, or under azimuthal symmetry. It follows that the copolar reflectivities of the precipitation can be expressed similar to (6.54a),

$$\eta_{hh} = C'_{hh} r_0^2 \bar{P}_{\text{ref}(h)} \tag{6.60a}$$

$$\eta_{vv} = C'_{vv} r_0^2 \bar{P}_{\text{ref}(v)} \tag{6.60b}$$

where the radar constants C'_{hh} and C'_{vv} are given as,

$$C'_{hh} = \left(\frac{2}{cT_0}\right)\left[\frac{(4\pi)^3 l_{wg}^2 l_{sw(h)}^2}{\lambda^2 P_t G_{0h}^2}\right]\frac{1}{\iint f_h^2 \, d\Omega} \tag{6.61a}$$

$$C'_{vv} = \left(\frac{2}{cT_0}\right)\left[\frac{(4\pi)^3 l_{wg}^2 l_{sw(v)}^2}{\lambda^2 P_t G_{0v}^2}\right]\frac{1}{\iint f_v^2 \, d\Omega} \tag{6.61b}$$

Similar to (5.49a), the copolar equivalent reflectivity factors at h and v-polarizations are defined as,

$$Z_{e,hh} = \frac{\lambda^4}{\pi^5 |K_w|^2}\eta_{hh} \equiv Z_h \tag{6.61c}$$

$$Z_{e,vv} = \frac{\lambda^4}{\pi^5 |K_w|^2}\eta_{vv} \equiv Z_v \tag{6.61d}$$

Note that Z_h and Z_v will also refer to the radar-measured copolar equivalent reflectivity factors.

The intrinsic Z_{dr} of the precipitation is,

$$Z_{dr} = 10\log_{10}\left(\frac{\eta_{hh}}{\eta_{vv}}\right) = 10\log_{10}\left(\frac{\bar{P}_{ref(h)}}{\bar{P}_{ref(v)}}\right) + 10\log_{10}\left(\frac{C'_{hh}}{C'_{vv}}\right) \tag{6.62a}$$

$$= 10\log_{10}\left(\frac{\bar{P}_{ref(h)}}{\bar{P}_{ref(v)}}\right) + 10\log_{10}\left(\frac{l_{sw(h)}^2}{l_{sw(v)}^2}\right) + 10\log_{10}\left(\frac{G_{0v}^2}{G_{0h}^2}\right)$$

$$+ 10\log_{10}\left(\frac{\iint f_v^2 \, d\Omega}{\iint f_h^2 \, d\Omega}\right) \tag{6.62b}$$

Ferrite circulator-based waveguide switches are non-reciprocal and as a consequence the transmit (t) and receive (r) insertion loss factors can be different. In this case, replace $l_{sw(h)}^2 \rightarrow l_{sw(h)}^t l_{sw(h)}^r$ while $l_{sw(v)}^2 \rightarrow l_{sw(v)}^t l_{sw(v)}^r$. Noise "correction" of the Z_{dr} estimate in (6.62b) implies replacing $\bar{P}_{ref(h)} \longrightarrow \bar{P}_{S+N(h)} - P_{N(h)}$ and $\bar{P}_{ref(v)} \longrightarrow \bar{P}_{S+N(v)} - P_{N(v)}$, where $P_{N(h)}$ and $P_{N(v)}$ are the noise power estimates corresponding to the h and v states of the switch.

At vertical incidence (see Fig. 6.22a), raindrops are known to satisfy azimuthal symmetry and the intrinsic Z_{dr} should be 0 dB, see (3.194a). Thus, measurements of $10\log_{10}[\bar{P}_{ref(h)}/\bar{P}_{ref(v)}]$ at vertical incidence in rain will establish the overall system "correction" factor inclusive of switch insertion losses, differences in boresight antenna gains associated with the two ports, and pattern mismatch effects. Figure 6.22a shows a schematic of the antenna rotated 360° in azimuth while pointing vertically. Note that the polarization basis also rotates with the antenna and is periodic with

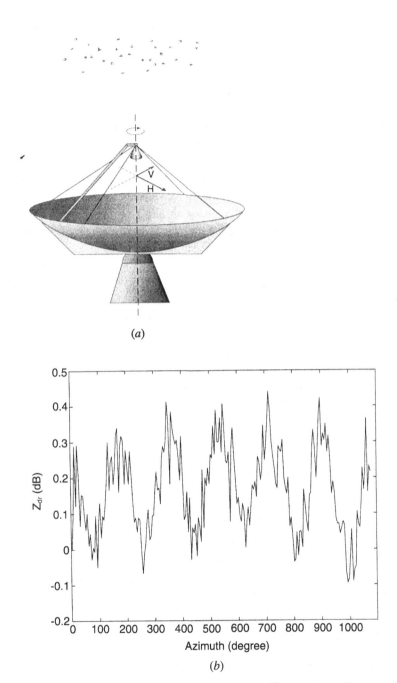

(a)

(b)

Fig. 6.22. (a) Illustrating the use of a vertically pointing beam for calibration of Z_{dr}. The antenna is rotated 360° in azimuth. From Gorgucci et al. (1999b, © 1999 IEEE). (b) Sample radar data acquired for Z_{dr} calibration, as in (a). The system offset is 0.15 dB in this example. From Gorgucci et al. (1999b, © 1999 IEEE).

the azimuth angle, the period being 180°. It follows that the Z_{dr} measurement is also periodic and is independent of the scattering mechanism. Hence, the system "correction" factor can be obtained by averaging the Z_{dr} data over integer multiples of 360°. Figure 6.22*b* shows Z_{dr} measurements in rain as a function of the azimuth angle, from which the overall system Z_{dr} bias was estimated (0.15 dB in this example). Note that data must be collected in the antenna's far-field (range $\geq 2D^2/\lambda$, where D is the antenna diameter). For conventional S-band 1° beam antennas, the range must exceed around 1.5 km. If the antenna diameter is large, or if the antenna cannot be pointed vertically then measurements can be taken in very light drizzle, where the drops are expected to be spherical. An alternative approach is to assume that ice particles in the upper levels of storms approximately satisfy polarization plane isotropy independent of the antenna's elevation angle so that the intrinsic $Z_{dr} \approx$ 0 dB. This method is easier to implement since such radar data are usually collected during operations. Since small ice crystals can be oriented by in-cloud electric fields (see latter part of Section 4.4) the assumption of polarization plane isotropy may not always hold. This method is useful, however, when it is suspected that the ferrite switch's differential insertion loss is unstable during operations (e.g. due to changing temperature). Provided the waveguide switch's differential insertion loss is stable and its isolation exceeds 15–20 dB, a polarization-agile/single-receiver radar is capable of accurate Z_{dr} measurements, typically to within 0.1–0.2 dB.

Polarization-agile/single-receiver radars can also measure the magnitude and phase of the copolar correlation coefficient. Algorithms for estimating $|\rho_{co}|$ and $\arg(\rho_{co})$ will be derived in Section 6.4.3. Receiver noise tends to bias $|\rho_{co}^{meas}|$ to lower values. Liu et al. (1994) have given an expression for $|\rho_{co}^{meas}|$ in terms of the intrinsic $|\rho_{co}|$, the signal-to-noise ratio (or SNR) and the intrinsic Z_{dr}. From simulations they found that to keep the bias in $|\rho_{co}|$ to within ± 0.01, the SNR should typically be higher than 20 dB.

The system differential phase defined below in (6.63b) should be stable during operations while its absolute value is of no significance. While the phase inequality between antenna ports should be stable, the ferrite switch may introduce non-reciprocal phase shifts during transmission and reception. If the sum of transmit and receive phase shifts when the switch state is horizontal is expressed as $\Phi_{sw(h)}^t + \Phi_{sw(h)}^r$, while when the switch state is vertical it is expressed as $\Phi_{sw(v)}^t + \Phi_{sw(v)}^r$, then the phase of the measured copolar correlation coefficient in (6.47) will be modified (again assuming that terms involving $\varepsilon_{h,v}$ are negligible) as,

$$\arg(\rho_{co}^{meas}) = \arg(\rho_{co}) + \arg \iint e^{-j2(\Phi_h - \Phi_v)} f_h f_v \, d\Omega$$

$$- [(\Phi_{sw(h)}^t - \Phi_{sw(v)}^t) + (\Phi_{sw(h)}^r - \Phi_{sw(v)}^r)] \tag{6.63a}$$

$$= \Psi_{dp} - (\Psi_{dp})_{system} \tag{6.63b}$$

where Ψ_{dp} is the intrinsic differential phase and $(\Psi_{dp})_{system}$ is the system differential phase. It follows that stability of the transmit and receive phase differences when the switch state cycles from horizontal to vertical is the important criterion for the stability of the system differential phase. This system differential phase can easily be obtained from the value $\arg(\rho_{\text{co}}^{\text{meas}})$ along the initial portion of rain cells intercepted by the beam. The stability of the system differential phase can also be checked by examination of such data.

6.3.3   Polarization diversity systems

A direct way to calibrate the differential gain and phase between the two receivers of a polarization diversity system is to use an external reference source that illuminates both channels equally. Scanning the sun will provide for equal "noise" power excitation at the two ports since solar radiation is randomly polarized. Alternatively, an external source such as a horn radiating slant 45° polarization to the radar will cause equal in-phase excitation at the two ports. The received powers at the output of the two receivers (see Fig. 6.1b) when scanning the sun can be expressed as,

$$\bar{P}_{h(\text{sun})} \propto \frac{G_{rh}G_{0h} \iint f_h \, d\Omega}{l_{\text{wg}(h)}} \tag{6.64a}$$

$$\bar{P}_{v(\text{sun})} \propto \frac{G_{rv}G_{0v} \iint f_v \, d\Omega}{l_{\text{wg}(v)}} \tag{6.64b}$$

Note that the sun does not "fill" a typical 1° antenna beam; rather the subtended angle is nearly 0.5°. G_{rh} and G_{rv} are the receiver gains, G_{0h} and G_{0v} are the boresight gains associated with the antenna ports and l_{wg} is the waveguide loss factor. The ratio, $\bar{P}_{h(\text{sun})}/\bar{P}_{v(\text{sun})}$, can be used to estimate the true LDR_{vh} of precipitation by "correcting" the measurement as follows,

$$10\log_{10}\left(\frac{\langle|S_{vh}|^2\rangle}{\langle|S_{hh}|^2\rangle}\right) \approx 10\log_{10}\left(\frac{\bar{P}_{\text{cx}}}{\bar{P}_{\text{co}}^h}\right) + 10\log_{10}\left(\frac{\bar{P}_{h(\text{sun})}}{\bar{P}_{v(\text{sun})}}\right) \tag{6.64c}$$

Hourly sun scans with the CSU–CHILL radar have shown that the stability of $10\log_{10}(\bar{P}_{h(\text{sun})}/\bar{P}_{v(\text{sun})})$ is within a few tenths of a dB throughout the day. The approximation in (6.64c) is related to the fact that precipitation fills the entire beam while the sun does not. Also, it is assumed that the polarization error terms (involving $\varepsilon_{h,v}$ in (6.38)) are negligible compared with the intrinsic LDR. Since the cross-polar signals are relatively weak (compared with the copolar signals), it is very important to "correct" the LDR estimate for noise by replacing $\bar{P}_{\text{cx}} \longrightarrow \bar{P}_{S+N(\text{cx})} - P_{N(\text{cx})}$, where $P_{N(\text{cx})}$ is the noise power estimate for the cross-polar receiver.

The calibration of LDR mainly involves the determination of the gain differential between the two receivers. The gain inequality between antenna ports tends to be nearly constant. However, the receivers involve active elements such as amplifiers, and these can suffer from gain drifts of opposite sign. The two receivers must be regularly calibrated by injecting signals of known amplitude from a highly stable source and recording the receiver outputs.

When the radar antenna is illuminated at boresight by slant 45° polarization (e.g. using a standard gain pyramidal horn rotated 45° about its axis), the phase of the correlation between the two received signals at the h-receiver and v-receiver outputs (see Fig. 6.1b) can be expressed as,

$$\arg[(V_h^{\text{slant }45°})(V_v^{\text{slant }45°})^*] = (\Phi_{hs} - \Phi_{vs}) + (\Phi_{0h} - \Phi_{0v}) = (\Phi_h - \Phi_v)_{\text{system}} \quad (6.65)$$

where the differential phase between the h- and v-receivers, $(\Phi_{hs} - \Phi_{vs})$, has been separately written along with the boresight phase inequality between antenna ports $(\Phi_{0h} - \Phi_{0v})$. The sum of these two terms is defined as the differential phase of the system. During precipitation measurements $\arg(R_{\text{cx}}^h)$ can be "corrected" as follows (assuming that the terms involving $\varepsilon_{h,v}$ in (6.42) are negligible compared with the intrinsic value for precipitation),

$$\arg\langle S_{hh} S_{vh}^* \rangle = \arg(R_{\text{cx}}^h) + \arg \left(\iint f_h \sqrt{f_h f_v} e^{j(\Phi_h - \Phi_v)} d\Omega \right)^{-1} - (\Phi_{hs} - \Phi_{vs})$$

$$(6.66a)$$

$$\approx \arg(R_{\text{cx}}^h) - (\Phi_h - \Phi_v)_{\text{system}} \quad (6.66b)$$

The latter step assumes that $\Phi_h - \Phi_v$ is uniform across the main lobe and equal to the value at boresight.

6.3.4 Polarization-agile/dual-receiver systems

Several configurations are possible for such systems as illustrated in Figs. 6.9–6.11. The calibration procedures already discussed in the previous sub-sections are also valid here. The most general configuration is based on two separate transmitters and dual-receivers, as illustrated in Fig. 6.11. The power transmitted in each channel must be continuously monitored and recorded for calibration of Z_{dr}, i.e. a term $10 \log_{10}(P_t^v / P_t^h)$ must be added to the right-hand side of (6.62b). Use of two separate receivers (the H- and V-receivers in Fig. 6.11) implies that the differential gain between receivers, $10 \log_{10}(G_{rv}/G_{rh})$, must be known and added to the right-hand side of (6.62b). It is difficult to match the gain of the two receivers over the entire dynamic range of the weather signals, which can be in the range of 80 dB or more. As mentioned earlier, in some systems an additional RF transfer switch has been used to route the "strong" copolar signals to the same receiver (see Fig. 6.4) to improve the accuracy of the Z_{dr} measurement. With recent improvements in receiver technology (e.g. digital receivers), an additional RF transfer switch may not be necessary.

For calibration of the two LDRs (LDR$_{vh}$ and LDR$_{hv}$), the differential gain between the copolar and cross-polar receivers must be determined for each of the two states of the RF transfer switch. For example, one state (corresponding to "firing" of the H-transmitter) connects the h-port to the copolar receiver and the v-port to the cross-polar receiver. The other state (corresponding to "firing" of the V-transmitter) connects the

h-port to the cross-polar receiver and the v-port to the copolar receiver. The calibration factors for the two LDRs may be determined by pointing at the sun with the RF transfer switch activated. Referring to Fig. 6.4, if the superscript "10" refers to one state of the RF transfer switch and "01" to its second state, the receiver output powers can be expressed as,

$$\bar{P}^{10}_{\text{co(sun)}} \propto \frac{G_{\text{co}} G_{0h} G_{\text{LNA}(h)} \iint f_h \, d\Omega}{l_{\text{wg}(h)}} \tag{6.67a}$$

$$\bar{P}^{10}_{\text{cx(sun)}} \propto \frac{G_{\text{cx}} G_{0v} G_{\text{LNA}(v)} \iint f_v \, d\Omega}{l_{\text{wg}(v)}} \tag{6.67b}$$

$$\bar{P}^{01}_{\text{co(sun)}} \propto \frac{G_{\text{co}} G_{0v} G_{\text{LNA}(v)} \iint f_v \, d\Omega}{l_{\text{wg}(v)}} \tag{6.67c}$$

$$\bar{P}^{01}_{\text{cx(sun)}} \propto \frac{G_{\text{cx}} G_{0h} G_{\text{LNA}(h)} \iint f_h \, d\Omega}{l_{\text{wg}(h)}} \tag{6.67d}$$

where G_{co} and G_{cx} are the gains of the copolar and cross-polar receivers in Fig. 6.4, l_{wg} is the waveguide loss factor assuming that the h- and v-waveguide runs are different, and G_{LNA} is the gain of the two low noise amplifiers which are also assumed to be slightly different. Two calibration factors can be derived from the sun data as,

$$\text{CAL}_{vh} = 10 \log_{10} \left(\frac{\bar{P}^{10}_{\text{co(sun)}}}{\bar{P}^{10}_{\text{cx(sun)}}} \right) \tag{6.68a}$$

$$\text{CAL}_{hv} = 10 \log_{10} \left(\frac{\bar{P}^{01}_{\text{co(sun)}}}{\bar{P}^{01}_{\text{cx(sun)}}} \right) \tag{6.68b}$$

It follows that the intrinsic LDR_{vh} of precipitation can be expressed as,

$$10 \log_{10} \left(\frac{\langle |S_{vh}|^2 \rangle}{\langle |S_{hh}|^2 \rangle} \right) = 10 \log_{10} \left(\frac{\bar{P}_{\text{cx}}}{\bar{P}^h_{\text{co}}} \right)_{10} + \text{CAL}_{vh} \tag{6.69a}$$

Similarly, the intrinsic LDR_{hv} of precipitation can be expressed as,

$$10 \log_{10} \left(\frac{\langle |S_{hv}|^2 \rangle}{\langle |S_{vv}|^2 \rangle} \right) = 10 \log_{10} \left(\frac{\bar{P}_{\text{cx}}}{\bar{P}^v_{\text{co}}} \right)_{01} + \text{CAL}_{hv} \tag{6.69b}$$

It follows that the intrinsic Z_{dr} of precipitation can be obtained from the difference of the intrinsic values of LDR_{hv} and LDR_{vh},

$$Z_{\text{dr}} = 10 \log_{10} \left(\frac{\langle |S_{hh}|^2 \rangle}{\langle |S_{vv}|^2 \rangle} \right) \tag{6.70a}$$

$$= 10 \log_{10} \left(\frac{\langle |S_{hv}|^2 \rangle}{\langle |S_{vv}|^2 \rangle} \right) - 10 \log_{10} \left(\frac{\langle |S_{vh}|^2 \rangle}{\langle |S_{hh}|^2 \rangle} \right) \tag{6.70b}$$

Using (6.69a, b), the intrinsic Z_{dr} of precipitation is,

$$Z_{dr} = 10 \log_{10} \left(\frac{\bar{P}_{cx}}{\bar{P}_{co}^v} \right)_{01} - 10 \log_{10} \left(\frac{\bar{P}_{cx}}{\bar{P}_{co}^h} \right)_{10} + (CAL_{hv} - CAL_{vh}) \qquad (6.71)$$

One advantage of using the above equation for estimating Z_{dr} is that it is independent of the transmitted powers (i.e. the ratio $10 \log_{10}(P_t^v/P_t^h)$). The temporal stability of $(CAL_{hv} - CAL_{vh})$ from hourly sun scans during the day obtained with the CSU–CHILL radar was found to be better than 0.09 dB. As mentioned earlier, noise "correction" can be done by replacing $\bar{P}_{cx} \rightarrow \bar{P}_{S+N(cx)} - P_{N(cx)}$ in (6.69).

The system differential phase, see (6.63), can be determined by illuminating the radar antenna with slant $45°$ linear polarization with the RF transfer switch activated. It follows that two system differential phase values will be obtained. Configurations with a single transmitter (but with a high power waveguide switch to achieve polarization agility) and with dual-receivers, as shown in Fig. 6.9b, can be calibrated in a similar manner.

6.4   Estimation of the covariance matrix

6.4.1   Covariance matrix of the received signal vector

The covariance matrix of particles distributed within a resolution volume was defined in Chapter 3, see (3.183), based on the concept of a target feature vector. In this section, a similar covariance matrix is defined in terms of the vector of received signals. Let $V_{hh}[n]$, $V_{hv}[n]$, $V_{vh}[n]$, and $V_{vv}[n]$, $n = 1, 2, \ldots, N$, be N time samples of the signals V_h^{10}, V_h^{01}, V_v^{10}, and V_v^{01}, respectively, spaced T_s apart. Note that, for a given n, all four samples $(V_{hh}[n], V_{hv}[n], V_{vh}[n],$ and $V_{vv}[n])$ may not always be available. The availability of samples is determined by the pulsing scheme. Figure 6.23 shows a schematic of the copolar and cross-polar received signals.

Let $\mathbf{Z}[n]$ be the vector given by $(V_{hh}[n], V_{vh}[n], V_{hv}[n], V_{vv}[n])^t$. It was shown in Section 5.9 that the mean of $\mathbf{Z}$ is zero. The covariance of $\mathbf{Z}$ can be written as,

$$E[\mathbf{Z}(\mathbf{Z}^*)^t] = \mathbf{K} = E \begin{bmatrix} |V_{hh}|^2 & V_{hh}(V_{vh})^* & V_{hh}(V_{hv})^* & V_{hh}(V_{vv})^* \\ V_{vh}(V_{hh})^* & |V_{vh}|^2 & V_{vh}(V_{hv})^* & V_{vh}(V_{vv})^* \\ V_{hv}(V_{hh})^* & V_{hv}(V_{vh})^* & |V_{hv}|^2 & V_{hv}(V_{vv})^* \\ V_{vv}(V_{hh})^* & V_{vv}(V_{vh})^* & V_{vv}(V_{hv})^* & |V_{vv}|^2 \end{bmatrix} \qquad (6.72)$$

It was shown in Section 5.3 that the received signal sampled at a fixed time delay (τ_0) after the transmit pulse corresponds to a resolution volume centered at a distance $c\tau_0/2$ from the radar. Therefore, the elements of the covariance matrix of the signal vector $\mathbf{Z}$ are directly related to the elements of the covariance matrix characterizing the scatterers in the resolution volume as given by (3.183).

The covariance matrix is conjugate symmetric,

$$\mathbf{K} = (\mathbf{K}^t)^* \qquad (6.73)$$

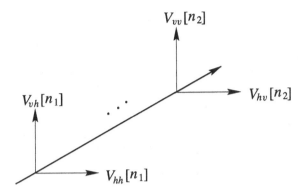

Fig. 6.23. Schematic showing the copolar and cross-polar received signals at sample times corresponding to n_1 and n_2. The samples at n_1 are for a horizontally polarized transmit pulse, whereas the samples at n_2 are for a vertically polarized transmit pulse.

The conjugate symmetry property yields ten independent elements of the matrix $\mathbf{K}$. In addition, by invoking reciprocity it can be seen that the cross-polar signal returns V_{hv} and V_{vh} are identical. This reduces the number of independent elements of the matrix to the list $\langle |V_{hh}|^2 \rangle$, $\langle V_{hh}(V_{vh})^* \rangle$, $\langle V_{hh}(V_{vv})^* \rangle$, $\langle V_{hv}(V_{vv})^* \rangle$, $\langle |V_{hv}|^2 \rangle$, and $\langle |V_{vv}|^2 \rangle$. Thus, the above six quantities estimate the complete covariance matrix of $\mathbf{Z}$. It can be seen that $\langle |V_{hh}|^2 \rangle$, $\langle |V_{vv}|^2 \rangle$, $\langle |V_{hv}|^2 \rangle$ are real, and $\langle V_{hh}(V_{vh})^* \rangle$, $\langle V_{hh}(V_{vv})^* \rangle$, and $\langle V_{hv}(V_{vv})^* \rangle$ are complex. Note that $\langle V_{hh}(V_{vv})^* \rangle$ is the only term that involves two transmission states and its estimation depends critically on the transmit pulse sequence.

6.4.2   Estimation of the covariance matrix from signal samples

The terms of the covariance matrix of the particles distributed within the resolution volume that correspond to the estimation of elements of the covariance matrix of the received signal vector can be written as,

$$P_{co}^h = \langle |V_{hh}|^2 \rangle \Rightarrow \langle |S_{hh}|^2 \rangle \tag{6.74a}$$

$$P_{co}^v = \langle |V_{vv}|^2 \rangle \Rightarrow \langle |S_{vv}|^2 \rangle \tag{6.74b}$$

$$P_{cx} = \langle |V_{hv}|^2 \rangle = \langle |V_{vh}|^2 \rangle \Rightarrow \langle |S_{hv}|^2 \rangle \tag{6.74c}$$

$$R_{cx}^v = \langle V_{vv}(V_{hv})^* \rangle \Rightarrow \langle S_{vv} S_{hv}^* \rangle \tag{6.74d}$$

$$R_{cx}^h = \langle V_{hh}(V_{vh})^* \rangle \Rightarrow \langle S_{hh} S_{vh}^* \rangle \tag{6.74e}$$

$$R_{co} = \langle V_{hh}(V_{vv})^* \rangle \Rightarrow \langle S_{hh} S_{vv}^* \rangle \tag{6.74f}$$

It was shown in Section 4.3, see (4.65), that, in the absence of propagation effects, $\arg[R_{co}^*]$ is the differential phase shift upon scattering, whereas in the presence of propagation effects $\arg[R_{co}^*] = \Phi_{dp} + \delta = \Psi_{dp}$.

The $\mathbf{Z}[n]$, $n = 1, 2, \ldots, N$ is the vector time-series of observations from which the covariance matrix elements listed in (6.74) must be obtained. It will be shown later that these will be estimated from the autocorrelation matrix of $\mathbf{Z}[n]$.

Let $\mathbf{K}[l]$ be the autocorrelation matrix of the signal vector $\mathbf{Z}[n]$,

$$\mathbf{K}[l] = E\{\mathbf{Z}[n + l](\mathbf{Z}^*[n])^t\} \tag{6.75a}$$

$$= E\left\{ \begin{bmatrix} V_{hh}[n + l] \\ V_{vh}[n + l] \\ V_{hv}[n + l] \\ V_{vv}[n + l] \end{bmatrix} \begin{bmatrix} [V_{hh}^*[n] \ V_{vh}^*[n] \ V_{hv}^*[n] \ V_{vv}^*[n]] \end{bmatrix} \right\} \tag{6.75b}$$

$\mathbf{K}[l]$ has 16 terms, which are the auto and cross-correlations of the elements of the signal vector, but from symmetry and reciprocity only six are independent, similar to $\mathbf{K}$ in (6.72). Note that the matrix $\mathbf{K}$ in (6.72) is the same as $\mathbf{K}[0]$. These six independent elements are,

$$E\{V_{hh}[n + l]V_{hh}^*[n]\} = R_{hh,hh}[l] \tag{6.76a}$$

$$E\{V_{hh}[n + l]V_{vh}^*[n]\} = R_{hh,vh}[l] \tag{6.76b}$$

$$E\{V_{hh}[n + l]V_{vv}^*[n]\} = R_{hh,vv}[l] \tag{6.76c}$$

$$E\{V_{vh}[n + l]V_{vh}^*[n]\} = R_{vh,vh}[l] \tag{6.76d}$$

$$E\{V_{vv}[n + l]V_{hv}^*[n]\} = R_{vv,hv}[l] \tag{6.76e}$$

$$E\{V_{vv}[n + l]V_{vv}^*[n]\} = R_{vv,vv}[l] \tag{6.76f}$$

The corresponding correlation coefficient functions will be denoted by $\rho_{hh,hh}[l]$, $\rho_{hh,vh}[l]$, $\rho_{hh,vv}[l]$, $\rho_{vh,vh}[l]$, $\rho_{vv,hv}[l]$, and $\rho_{vv,vv}[l]$. The temporal behavior of the autocorrelation function is determined by the Doppler velocity spectrum and the time-varying behavior of the scattering amplitude, see (5.92a). If the Doppler velocity spectrum dominates this temporal behavior all autocorrelation functions will be similar in shape and the corresponding correlation coefficient functions will be identical. Specifically,

$$\rho_{hh,hh}[l] = \rho_{vv,vv}[l] = \rho_{vh,vh}[l] = \rho[l] \tag{6.77}$$

In addition, the copolar correlation coefficient previously referred to as ρ_{co} is equivalent to $\rho_{vv,hh}[0] = \{\rho_{hh,vv}[0]\}^*$.

In a polarization-agile/dual-receiver system, the transmit polarization state is typically changed according to some fixed pattern that is repeated. The most common transmit pattern is alternating transmission of h and v polarization states. Other possible transmit patterns used are hhv and $hhvv$. When an alternating transmission mode is used, $V_{hh}[n]$ and $V_{vh}[n]$ are only available every other pulse (or when the h polarization state is transmitted). Similarly, $V_{hv}[n]$ and $V_{vv}[n]$ are available only every other pulse (or when the v polarization state is transmitted). Under the hybrid mode of operation, see Section 4.7, $V_{hv}[n]$ and $V_{vh}[n]$ samples are not available. Therefore, for a given pulsing scheme, direct estimates of all the elements of the covariance matrix will not be available.

The pulsing scheme in a dual-polarization radar can be broadly classified into three modes:

1. In the alternating polarization mode the polarization state of the transmit pulse is changed alternately between h and v. This mode does not readily provide two adjacent received signal samples at the same polarization, thus inhibiting the direct estimation of the autocorrelation function at lag T_s (or R[1], see (5.207)).

2. To overcome the above, a block of pulses with a fixed transmit pattern such as $hhvv$, hhv, vvh is repeated: this is denoted as the periodic block pulsing mode.

3. The third pulsing method is the hybrid mode where h and v states are simultaneously transmitted (with equal power, or the slant $45°$ linear state) with simultaneous reception of horizontally and vertically polarized signals.

6.4.3   Estimation of covariance matrix elements under alternate polarization mode

The alternate polarization mode is described by the transmit sequence $\cdots hvhv \cdots$. With no loss of generality, it can be assumed that the first transmit pulse is the h polarization state. Let the total number of received signal samples be $2N$ with N interlaced samples at h or v polarization states. Figure 6.24 shows the sequence of received signals in the alternate polarization mode.

Let $V_{hh}[2n-1]$, $V_{vh}[2n-1]$, $V_{vv}[2n]$, $V_{hv}[2n]$, $n = 1, \ldots, N$ be the samples of the received signal. The estimates of the three mean power elements of the covariance matrix can be written as,

$$\hat{P}_{co}^h = \frac{1}{N}\sum_{n=1}^{N}|V_{hh}[2n-1]|^2 \tag{6.78a}$$

$$\hat{P}_{co}^v = \frac{1}{N}\sum_{n=1}^{N}|V_{vv}[2n]|^2 \tag{6.78b}$$

$$\hat{P}_{cx} = \frac{1}{N}\sum_{n=1}^{N}|V_{vh}[2n-1]|^2 \tag{6.78c}$$

The above power estimates can be used to obtain Z_{dr} and LDR_{vh} as,

$$\hat{Z}_{dr} = 10\log_{10}\left(\frac{\hat{P}_{co}^h}{\hat{P}_{co}^v}\right) \tag{6.78d}$$

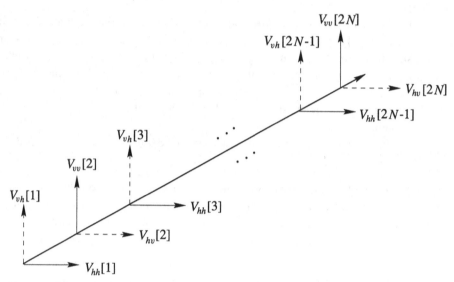

Fig. 6.24. Schematic showing the copolar and cross-polar received signals for the alternating polarization mode with the first transmit pulse at horizontal polarization.

$$\hat{\text{LDR}}_{vh} = 10\log_{10}\left(\frac{\hat{P}_{\text{cx}}}{\hat{P}_{\text{co}}^h}\right) \tag{6.78e}$$

Henceforth, the subscript vh on LDR will be dropped for ease of notation. The two correlations between copolar and cross-polar signals can be estimated as,

$$\hat{R}_{\text{cx}}^h = \frac{1}{N}\sum_{n=1}^{N}V_{hh}[2n-1](V_{vh}[2n-1])^* \tag{6.79a}$$

$$\hat{R}_{\text{cx}}^v = \frac{1}{N}\sum_{n=1}^{N}V_{vv}[2n](V_{hv}[2n])^* \tag{6.79b}$$

As mentioned earlier, direct estimation of the copolar correlation ($R_{hh,vv}[0]$) is not possible in this mode. Nevertheless, it can be computed indirectly as follows. The correlation estimate between two adjacent copolar samples yields the correlation between them at time lag T_s. Therefore, the estimation of $R_{hh,vv}[1]$ can be written as,

$$\hat{R}_{hh,vv}[1] = \frac{1}{N}\sum_{n=1}^{N-1}V_{hh}[2n+1](V_{vv}[2n])^* \tag{6.80a}$$

Similarly, the estimation of $R_{vv,hh}[1]$ can be written as,

$$\hat{R}_{vv,hh}[1] = \frac{1}{N}\sum_{n=1}^{N}V_{vv}[2n](V_{hh}[2n-1])^* \tag{6.80b}$$

The corresponding correlation coefficient estimates can be written as,

$$\hat{\rho}_{hh,vv}[1] = \frac{\hat{R}_{hh,vv}[1]}{\sqrt{\hat{P}_{co}^h \hat{P}_{co}^v}} \tag{6.80c}$$

$$\hat{\rho}_{vv,hh}[1] = \frac{\hat{R}_{vv,hh}[1]}{\sqrt{\hat{P}_{co}^h \hat{P}_{co}^v}} \tag{6.80d}$$

The correlation between signal samples at the same time is dependent only on the corresponding correlation between the horizontal and vertical scattering amplitudes of the scatterers, which in turn is controlled by the size, shape, and orientation distributions. As discussed earlier, the autocorrelation function of signal samples spaced T_s apart is controlled primarily by the Doppler spectrum (see Section 5.11). At low elevation angles, the joint distribution of size, shape, orientation, and velocity(v) can be separated as,

$$f(\text{size, shape, orientation, } v) \approx f_1(\text{size, shape, orientation}) f_2(v) \tag{6.81}$$

where f_1 and f_2 represent the marginal distributions. Under this approximation (see also, Sachidananda and Zrnić 1985) the copolar correlation coefficient at $l = 1$ can be written as,

$$\rho_{hh,vv}[1] = \rho_{hh,vv}[0]\rho_{hh,hh}[1] = \rho_{hh,vv}[0]\rho[1] \tag{6.82a}$$

$$\rho_{vv,hh}[1] = \rho_{vv,hh}[0]\rho_{vv,vv}[1] = \rho_{vv,hh}[0]\rho[1] \tag{6.82b}$$

Note that $\rho_{vv,hh}[0] = (\rho_{hh,vv}[0])^*$. It follows that the phase of $\rho_{hh,vv}[1]$ and $\rho_{vv,hh}[1]$ can be written as,

$$\arg(\rho_{hh,vv}[1]) = \Psi_1 = -\Psi_{dp} + \arg(\rho[1]) \tag{6.83a}$$

$$\arg(\rho_{vv,hh}[1]) = \Psi_2 = \Psi_{dp} + \arg(\rho[1]) \tag{6.83b}$$

where Ψ_{dp} is the phase of $R_{vv,hh}[0]$. It was shown in Section 4.3, see (4.65), that $\Psi_{dp} = \Phi_{dp} + \delta_{co}$, where Φ_{dp} is the differential propagation phase and δ_{co} is differential phase upon scattering. The Ψ_{dp} and $\arg(\rho[1])$ can be estimated from (6.83) (Mueller 1984) as,

$$\hat{\Psi}_{dp} = \frac{1}{2}(\Psi_2 - \Psi_1) \tag{6.84a}$$

$$\arg \hat{\rho}[1] = \arg \hat{R}[1] = \frac{1}{2}(\Psi_2 + \Psi_1) \tag{6.84b}$$

It was shown in Chapter 5, see (5.208), that for symmetric Doppler spectra the mean velocity can be estimated from $\arg \hat{R}[1]$. Therefore, the estimate of mean velocity in the alternating polarization mode is given by,

$$\hat{\bar{v}} = -\frac{\lambda}{4\pi T_s}\frac{1}{2}(\Psi_2 + \Psi_1) \tag{6.85}$$

From (6.82), $\rho_{hh,vv}[0]$ can be estimated if $\rho[1]$ is known. Estimating $\rho[1]$ requires samples of "like" signals spaced T_s apart which are not available in the alternating polarization mode. By "like" signals, it is meant the time-series of only one type of signal such as $V_{hh}[n]$. However, every other copolar received sample is of the same type, indicating that $\rho[2]$ can be estimated directly. If the shape of the Doppler spectrum is Gaussian, then the magnitude of $\rho[n]$ can be written as (refer to (5.128)),

$$|\rho[n]| = \exp\left(\frac{-8\pi^2\sigma_v^2 n^2 T_s^2}{\lambda^2}\right) \tag{6.86}$$

It follows that $|\rho[n_1]|$ can be written in terms of $|\rho[n_2]|$ as,

$$|\rho[n_1]| = |\rho[n_2]|^{(n_1/n_2)^2} \tag{6.87}$$

When $n_2 = 2$ and $n_1 = 1$,

$$|\rho[1]| = |\rho[2]|^{0.25} \tag{6.88}$$

Therefore, under the assumption of Gaussian spectral shape, $|\rho_{vv,hh}[0]|$ can be estimated from $\hat{\rho}_{hh,vv}[1]$ and $\hat{\rho}[2]$, where the estimate of $\hat{\rho}[2]$ is given by,

$$\hat{\rho}[2] = \left[\frac{1}{N}\sum_{n=1}^{N-1} V_{hh}[2n+1](V_{hh}[2n-1])^*\right] / \hat{P}_{co}^h \tag{6.89a}$$

$\hat{\rho}(2)$ can also be estimated from $V_{vv}[n]$ as,

$$\hat{\rho}[2] = \left[\frac{1}{N}\sum_{n=1}^{N-1} V_{vv}[2n+2](V_{vv}[2n])^*\right] / \hat{P}_{co}^v \tag{6.89b}$$

Subsequently, the estimate of $|\rho_{hh,vv}[0]|$ is obtained as,

$$|\hat{\rho}_{hh,vv}[0]| = \frac{|\hat{\rho}_{hh,vv}[1]|}{|\hat{\rho}[2]|^{0.25}} = |\hat{\rho}_{co}| \tag{6.90}$$

Note that the estimate of mean velocity can also be obtained from $\arg \hat{\rho}[2]$ as,

$$\hat{v} = -\frac{\lambda}{4\pi(2T_s)}\arg(\hat{\rho}[2]) \tag{6.91}$$

The algorithms given by (6.84) and (6.91) estimate phases. Any phase estimation is unique only within an interval of 2π (or 360°). Since estimates of Ψ_{dp} and $\arg(R[1])$ in (6.84) involve one-half of the algebraic sum of two phase estimates, they are unique only in an interval of π (or 180°). Nevertheless, some practical constraints can be imposed on the estimates of Ψ_{dp} to resolve this ambiguity. At low frequencies, such as S-band, the angle δ_{co} is very small for raindrops. However, depending on the propagation medium, Φ_{dp} can be very high. For example, propagation through a rain medium can result in Φ_{dp} as high as 180° even at S-band (see Section 4.3). Since Φ_{dp} is a range cumulative quantity it should be a non-decreasing function of range. This constraint can be used to detect ambiguity (or folding) in Ψ_{dp} estimates and can be corrected accordingly. Subsequently, $\arg(R[1])$ (and the mean velocity) can be estimated from (6.83) as (see also, Sachidananda and Zrnić 1988),

$$\hat{\bar{v}} = \frac{-\lambda}{4\pi T_s} \arg\left(\hat{R}_{vv,hh}[1] e^{-j\hat{\Psi}_{dp}} \right) \tag{6.92}$$

This estimate of $\hat{\bar{v}}$ is based on copolar samples spaced T_s apart. If it can be ensured that there is no ambiguity in $\hat{\Psi}_{dp}$, then the maximum unambiguous velocity, v_{max}, is given (see also, Section 5.11.2) by,

$$v_{max} = \pm\frac{\lambda}{4T_s} \tag{6.93}$$

If the mean velocity estimates are obtained from (6.91), where the "like" samples are spaced $2T_s$ apart, the corresponding v_{max} will be half of that given by (6.93), thereby justifying the use of (6.92) for estimating $\bar{v}$.

The above algorithms can be used to estimate Z_{dr}, LDR, Ψ_{dp}, $|\rho_{co}|$, and $\bar{v}$ in the alternating polarization mode of operation. Note, however, that the $|\rho_{co}|$ estimator assumes a Gaussian-shaped Doppler spectrum. If the spectrum is non-Gaussian, then the $|\rho_{co}|$ estimate is biased as shown by Liu et al. (1994).

6.4.4   Estimation of covariance matrix elements under periodic block pulsing mode

The periodic block pulsing mode can be described by a pattern of the transmit pulse sequence that is repeated periodically. Let $p_h(n)$ and $p_v(n)$ be indicator functions that represent the transmit polarization state as,

$$p_h[n] = \begin{cases} 1, & \text{if the transmit polarization state is horizontal} \\ 0, & \text{if the transmit polarization state is vertical} \end{cases} \tag{6.94}$$

$$p_v[n] = \begin{cases} 1, & \text{if the transmit polarization state is vertical} \\ 0, & \text{if the transmit polarization state is horizontal} \end{cases} \tag{6.95}$$

where $n = 1, 2, \ldots, N$. For polarization agile systems,

$$p_v[n] = 1 - p_h[n] \tag{6.96}$$

For an alternating polarization mode, $p_h[n]$ will be of the form $101010\cdots$ and $p_v[n]$ will be of the form $010101\cdots$.

The three mean power estimates can be written as,

$$\hat{P}_{co}^h = \frac{\displaystyle\sum_{n=1}^{N} p_h[n]|V_{hh}[n]|^2}{\displaystyle\sum_{n=1}^{N} p_h[n]} \tag{6.97a}$$

$$\hat{P}^v_{co} = \frac{\sum\limits_{n=1}^{N} p_v[n]|V_{vv}[n]|^2}{\sum\limits_{n=1}^{N} p_v[n]} \tag{6.97b}$$

$$\hat{P}_{cx} = \frac{\sum\limits_{n=1}^{N} p_h[n]|V_{vh}[n]|^2}{\sum\limits_{n=1}^{N} p_h[n]} \tag{6.97c}$$

The two correlation estimates between copolar and cross-polar signals can be obtained as,

$$\hat{R}^h_{cx} = \frac{\sum\limits_{n=1}^{N} p_h[n]V_{hh}[n](V_{vh}[n])^*}{\sum\limits_{n=1}^{N} p_h[n]} \tag{6.98a}$$

$$\hat{R}^v_{cx} = \frac{\sum\limits_{n=1}^{N} p_v[n]V_{vv}[n](V_{hv}[n])^*}{\sum\limits_{n=1}^{N} p_v[n]} \tag{6.98b}$$

As in the alternating polarization mode, the direct estimation of $\hat{R}_{hh,vv}[0]$ is not possible in the periodic block pulsing mode. It can be estimated indirectly using techniques similar to those used in the alternate polarization mode.

In any transmit pattern the polarization state changes between h and v, as well as between v and h, along the sequence. Once these transitions are located, the correlation can be computed. The transition from h to v and v to h can be easily identified from the indicator functions, and the two copolar correlations at one lag can be estimated as follows,

$$\hat{R}_{vv,hh}[1] = \frac{\sum\limits_{n=1}^{N-1} V_{vv}[n+1](V_{hh}[n])^* p_v[n+1]p_h[n]}{\sum\limits_{n=1}^{N} p_v[n+1]p_h[n]} \tag{6.99}$$

The corresponding correlation coefficient is,

$$\hat{\rho}_{vv,hh}[1] = \frac{\hat{R}_{vv,hh}[1]}{(\hat{P}^h_{co}\hat{P}^v_{co})^{1/2}} \tag{6.100}$$

In any periodic block pulsing mode (which is not a strictly alternating mode) such as $hhvv, \ldots, hhhv, \ldots$, or $hvvv, \ldots$, there exists two adjacent "like" signal samples (at the same polarization). Therefore, the autocorrelation function of "like" signals at lag 1 can be estimated as,

$$\hat{R}[1] = \frac{\displaystyle\sum_{n=1}^{N-1} V_{hh}[n+1](V_{hh}[n])^* p_h[n] p_h[n+1]}{\displaystyle\sum_{n=1}^{N-1} p_h[n] p_h[n+1]} \tag{6.101a}$$

or,

$$\hat{R}[1] = \frac{\displaystyle\sum_{n=1}^{N-1} V_{vv}[n+1](V_{vv}[n])^* p_v[n] p_v[n+1]}{\displaystyle\sum_{n=1}^{N-1} p_v[n] p_v[n+1]} \tag{6.101b}$$

The corresponding correlation coefficient is,

$$\hat{\rho}[1] = \frac{\hat{R}_{hh,hh}[1]}{\hat{P}_{\text{co}}^h} = \frac{\hat{R}_{vv,vv}[1]}{\hat{P}_{\text{co}}^v} \tag{6.102}$$

From (6.82), $\hat{\rho}_{vv,hh}[0]$ can be estimated as,

$$\hat{\rho}_{vv,hh}[0] = \frac{\hat{\rho}_{vv,hh}[1]}{\hat{\rho}[1]} \tag{6.103}$$

An estimate of copolar differential phase Ψ_{dp} can be obtained from the phase of $\hat{\rho}_{vv,hh}[0]$ as,

$$\hat{\Psi}_{\text{dp}} = \arg\left(\hat{\rho}_{vv,hh}[0]\right) \tag{6.104}$$

The estimates of radar parameters such as Z_{dr}, LDR, $|\rho_{\text{co}}|$, and $\bar{v}$ can be obtained from (6.97) and (6.99) as,

$$\hat{Z}_{\text{dr}} = 10 \log_{10}\left(\frac{\hat{P}_{\text{co}}^h}{\hat{P}_{\text{co}}^v}\right) \tag{6.105a}$$

$$\hat{\text{LDR}} = 10 \log_{10}\left(\frac{\hat{P}_{\text{cx}}}{\hat{P}_{\text{co}}^h}\right) \tag{6.105b}$$

$$|\hat{\rho}_{\text{co}}| = \frac{|\hat{\rho}_{vv,hh}[1]|}{|\hat{\rho}[1]|} \tag{6.105c}$$

$$\hat{\Psi}_{\text{dp}} = \arg\left(\hat{\rho}_{vv,hh}[0]\right) \tag{6.105d}$$

$$\hat{\bar{v}} = -\frac{\lambda}{4\pi T_s} \arg\left(\hat{R}[1]\right) \tag{6.105e}$$

6.4.5 Estimators of covariance matrix elements in hybrid mode

The hybrid mode can be viewed as simultaneous transmission and reception of both horizontal and vertical polarization states (see Section 4.7). In this mode, the signals that are copolar and cross-polar to the transmitted polarization state are not received. Rather, the elements of the coherency matrix, see (4.130), are measured directly. These elements can be used to estimate the conventional Z_{dr}, $|\rho_{co}|$, and Ψ_{dp}, as discussed in Section 4.7. These estimates will be correct when the propagation and scattering matrices are diagonal (i.e. no cross-coupling). Figure 6.25 illustrates the received signal samples $V_h[n]$ and $V_v[n]$ for the hybrid mode.

The estimates of powers in the h- and v-channels can be written as,

$$\hat{P}_h = \frac{1}{N} \sum_{n=1}^{N} |V_h[n]|^2 \tag{6.106a}$$

$$\hat{P}_v = \frac{1}{N} \sum_{n=1}^{N} |V_v[n]|^2 \tag{6.106b}$$

The covariance between V_h and V_v at lag 0 is,

$$\hat{R}_{v,h}[0] = \frac{1}{N} \sum_{n=1}^{N} V_v[n] V_h^*[n] \tag{6.107}$$

The estimator for Z_{dr} is,

$$\hat{Z}_{dr} = 10 \log_{10} \left(\frac{\hat{P}_h}{\hat{P}_v} \right) \tag{6.108}$$

One major advantage of the hybrid mode is that ρ_{co} can be estimated directly (without invoking (6.81)),

$$\hat{\rho}_{co} = \left[\frac{\hat{R}_{v,h}[0]}{(\hat{P}_h \hat{P}_v)^{1/2}} \right] \tag{6.109}$$

Subsequently, the magnitude and phase of $\hat{\rho}_{co}$ are given by,

$$|\hat{\rho}_{co}| = \frac{|\hat{R}_{v,h}[0]|}{(\hat{P}_h \hat{P}_v)^{1/2}} \tag{6.110a}$$

$$\hat{\Psi}_{dp} = \arg(\hat{R}_{v,h}[0]) \tag{6.110b}$$

The autocorrelation function at lag 1 can be estimated from either $V_h[n]$ or $V_v[n]$ as,

$$\hat{R}[1] = \frac{1}{N} \sum_{n=1}^{N-1} V_h[n+1] V_h^*[n] \tag{6.111}$$

The mean velocity is obtained from $\arg(\hat{R}[1])$,

$$\hat{v} = -\frac{\lambda}{4\pi T_s} \arg(\hat{R}[1]) \tag{6.112}$$

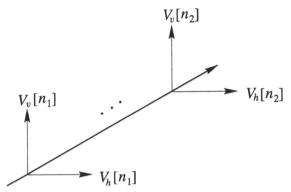

Fig. 6.25. Schematic showing the horizontal and vertical components of the received signal in the hybrid mode. The transmitted polarization state is assumed to be slant 45° linear polarization.

6.5   Variance of the estimates of the covariance matrix elements

It was shown in Section 6.4 that algorithms that estimate the covariance matrix elements depend on the pulsing mode. The variance calculations of the estimates are provided for the three pulsing modes, namely: (i) alternating, (ii) periodic block, and (iii) hybrid. Except for the mean power and covariances, all of the estimates discussed in the previous section involve non-linear functions of mean powers and/or covariances, and are thus biased when a finite number of samples are used in the estimate. Nevertheless, they can be shown to be asymptotically unbiased. The following presents variance computations of the estimators in the alternating and hybrid modes. The variance computation of estimators under the periodic block pulsing mode is provided in Appendix 5.

6.5.1   Alternating mode

(i) Variance of mean power estimates
The variance of power estimates can be obtained similar to that given by (5.192), recognizing that only alternate samples are used,

$$\text{var}(\hat{P}_{\text{co}}^{h}) = \frac{(P_{\text{co}}^{h})^2}{N} \sum_{l=-(N-1)}^{N-1} \left(1 - \frac{|l|}{N}\right) \rho_p(2l) \tag{6.113a}$$

$$\text{var}(\hat{P}_{\text{co}}^{v}) = \frac{(P_{\text{co}}^{v})^2}{N} \sum_{l=-(N-1)}^{N-1} \left(1 - \frac{|l|}{N}\right) \rho_p(2l) \tag{6.113b}$$

$$\text{var}(\hat{P}_{\text{cx}}) = \frac{(P_{\text{cx}})^2}{N} \sum_{l=-(N-1)}^{N-1} \left(1 - \frac{|l|}{N}\right) \rho_p(2l) \tag{6.113c}$$

It was shown in Section 5.10, see (5.190), that the correlation coefficient function of power samples can be obtained as the square of the correlation coefficient function of signal samples. Therefore,

$$P_p[2l] = |\rho[2l]|^2 \tag{6.114}$$

The behavior of the variance of $\hat{P}_{co}^h$, $\hat{P}_{co}^h$, and $\hat{P}_{cx}$ is similar to the mean power estimate described in Section 5.10.

(ii) Variance of $\hat{Z}_{dr}$
The standard deviation of $\hat{Z}_{dr}$ can be derived using the perturbation approximation, see (5.198):

$$SD(\hat{Z}_{dr}) = SD\left[10\log_{10}\left(\frac{\hat{P}_{co}^h}{\hat{P}_{co}^v}\right)\right] \simeq 10\log_{10}\left[1 + \frac{SD(\hat{z}_{dr})}{\hat{z}_{dr}}\right] \tag{6.115}$$

where SD indicates standard deviation and $\hat{z}_{dr}$ is the ratio $\hat{P}_{co}^h/\hat{P}_{co}^v$ (or $Z_{dr} = 10\log_{10}(\hat{z}_{dr})$). The standard deviation of $\hat{P}_{co}^h/\hat{P}_{co}^v$ can be obtained from the perturbation approximation of variance of ratios. If X and Y are two random variables, then the perturbation approximation for their ratio (Papoulis 1965) is,

$$\text{var}\left(\frac{X}{Y}\right) \simeq \left(\frac{\mu_X}{\mu_Y}\right)^2\left[\frac{\text{var}(X)}{\mu_X^2} + \frac{\text{var}(Y)}{\mu_Y^2} - \frac{2\,\text{cov}(X,Y)}{\mu_X\mu_Y}\right] \tag{6.116}$$

where μ_X and μ_Y are the expected values of random variables X and Y, and cov indicates covariance. Application of (6.116) requires the variance of $\hat{P}_{co}^h$, $\hat{P}_{co}^v$, and $\text{cov}(\hat{P}_{co}^h, \hat{P}_{co}^v)$. Variances of $\hat{P}_{co}^h$ and $\hat{P}_{co}^v$ are given in (6.113). The $\text{cov}(\hat{P}_{co}^h, \hat{P}_{co}^v)$ can be obtained as follows,

$$\text{cov}(\hat{P}_{co}^h, \hat{P}_{co}^v) = \frac{1}{N^2}\text{cov}\left(\sum_{n=1}^{N}|V_{hh}[2n-1]|^2, \sum_{n=1}^{N}|V_{vv}[2n]|^2\right) \tag{6.117}$$

The above equation can be reduced to,

$$\text{cov}(\hat{P}_{co}^h, \hat{P}_{co}^v) = \frac{1}{N}\sum_{l=-(N-1)}^{N-1}\left(1 - \frac{|l|}{N}\right)|R_{hh,vv}[2l+1]|^2 \tag{6.118}$$

Substituting (6.118) and (6.113) in (6.116), the variance of $\hat{P}_{co}^h/\hat{P}_{co}^v$ is obtained as,

$$\text{var}\left(\frac{\hat{P}_{co}^h}{\hat{P}_{co}^v}\right) = 2\left(\frac{P_{co}^h}{P_{co}^v}\right)^2\frac{1}{N}\sum_{l=-(N-1)}^{N-1}\left(1 - \frac{|l|}{N}\right)(|\rho[2l]|^2 - |\rho_{hh,vv}[2l+1]|^2) \tag{6.119}$$

From (6.82), $\rho_{hh,vv}[2l+1]$ can be approximated as,

$$\rho_{hh,vv}[2l+1] = \rho_{hh,vv}[0]\rho(2l+1) \tag{6.120}$$

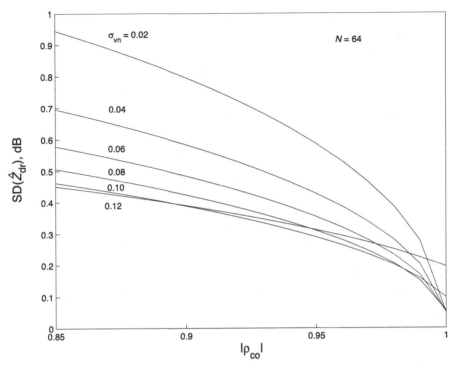

Fig. 6.26. Standard deviation (SD) of $\hat{Z}_{dr}$ versus $|\rho_{co}|$ for the alternating mode for various normalized spectrum widths. Calculations assume Gaussian-shaped Doppler spectrum, with the number of samples $N = 64$. The normalized spectrum width is defined as $\sigma_v/(\lambda/2T_s)$.

Substituting (6.120) in (6.119), it follows that (6.115), which gives the standard deviation of $\hat{Z}_{dr}$, can be expressed as,

$$SD(\hat{Z}_{dr}) = 10\log_{10}\left\{1 + \left[\frac{2}{N}\sum_{l=-(N-1)}^{N-1}\left(1 - \frac{|l|}{N}\right)\right.\right.$$
$$\left.\left. \times \left(|\rho[2l]|^2 - |\rho[2l+1]\rho_{hh,vv}[0]|^2\right)\right]^{1/2}\right\} \qquad (6.121)$$

Figure 6.26 shows the standard deviation of $\hat{Z}_{dr}$ as a function of $|\rho_{hh,vv}[0]| \equiv |\rho_{co}|$ for various normalized widths of a Gaussian-shaped Doppler spectrum, see (5.195). Note the sharp decrease in standard deviation as $|\rho_{co}| \rightarrow 1.0$, which is essentially a manifestation of the statistical behavior of correlated copolar signal samples as illustrated in Fig. 5.36. This important feature makes accurate measurement of Z_{dr} possible (typically to within a few tenths of a dB), which is required for quantitative applications; it also applies to $\hat{\Psi}_{dp}$ and $|\hat{\rho}_{co}|$.

Recall that Z_{dr} is the ratio of copolar sample mean powers at horizontal and vertical polarizations. When $|\rho_{co}| \leq 0.9$, the SD($\hat{Z}_{dr}$) decreases with increasing normalized spectrum width (σ_{vn}) similar to the reduction in standard deviation of reflectivity estimates (see Fig. 5.37). However, when σ_{vn} is small, the adjacent signal samples are more correlated resulting in an interpolatory effect which approximately tends to reconstruct simultaneous samples of h and v polarization with a high degree of cross-correlation. It follows that $\mathrm{SD}(\hat{Z}_{dr})$ will be reduced due to this interpolatory effect. When $|\rho_{co}|$ is close to unity and σ_{vn} is small, this effect dominates, as can be seen in Fig. 6.26. See Sachidananda and Zrnić (1985) for $\mathrm{SD}(\hat{Z}_{dr})$ as a function of N.

The presence of noise can affect Z_{dr} by deteriorating the correlation between samples. The effect of noise can be studied by substituting the autocorrelation and cross-correlation of the signals in the presence of noise, similar to (5.211), as,

$$\tilde{\rho}_{hh,vv}[n] = \rho_{hh,vv}[n] \left[\frac{1}{1+1/\mathrm{SNR}}\right]^{1/2} \left[\frac{1}{1+\tilde{z}_{dr}}\right]^{1/2} \tag{6.122}$$

where $\tilde{\rho}_{hh,vv}[n]$ is the correlation coefficient function in the presence of noise and SNR is the signal-to-noise ratio.

(iii) Variance of $\hat{\Psi}_{dp}$

Since Ψ_{dp} is estimated from the difference between $\hat{\Psi}_2$ and $\hat{\Psi}_1$, see (6.84a), its variance can be expressed (Doviak and Zrnić 1993) as,

$$\mathrm{var}(\hat{\Psi}_{dp}) = \frac{1}{4}[\mathrm{var}(\Psi_1) + \mathrm{var}(\Psi_2) - 2\,\mathrm{cov}(\Psi_1, \Psi_2)] \tag{6.123}$$

Variances of Ψ_1 and Ψ_2 can be obtained in a manner similar to the method used for obtaining the variance of Doppler phase shift estimates when a staggered PRT pulsing scheme is used (Zrnić 1977; see also, Appendix 5 of this text). The covariance of Ψ_1 and Ψ_2 is also similar to the covariance of Doppler shift estimates from correlated pulse pairs in a staggered PRT scheme. In Appendix 5, the variances of Ψ_1 and Ψ_2 are derived as,

$$\mathrm{var}(\Psi_1) = \mathrm{var}(\Psi_2) = \frac{1 - |\rho[1]|^2 |\rho_{co}|^2}{2N^2 |\rho[1]|^2 |\rho_{co}|^2} \sum_{n=-(N-1)}^{N-1} (N - |n|)|\rho[2n]|^2 \tag{6.124}$$

Similarly, the covariance of Ψ_1 and Ψ_2 is derived as,

$$\mathrm{cov}(\Psi_1, \Psi_2) = \frac{|\rho_{co}|^2 - |\rho[1]|^2}{2N^2 |\rho[1]|^2 |\rho_{co}|^2} \sum_{n=-(N-1)}^{N-1} (N - |n|)|\rho[2n + 1]|^2 \tag{6.125}$$

Substituting (6.124) and (6.125) into (6.123) the variance of Ψ_{dp} can be calculated. Figure 6.27 shows the standard deviation of Ψ_{dp} as a function of $|\rho_{co}|$ for various normalized spectrum widths (σ_{vn}) of a Gaussian-shaped spectrum. It can be seen that Ψ_{dp} can be estimated to an accuracy of a few degrees if $|\rho_{co}|$ exceeds 0.95. The behavior of $\mathrm{SD}(\hat{\Psi}_{dp})$ in Fig. 6.27 is very similar to $\mathrm{SD}(\hat{Z}_{dr})$ in Fig. 6.26 for the same reasons described therein. The sharp decrease in $\mathrm{SD}(\hat{\Psi}_{dp})$ as $|\rho_{co}| \rightarrow 1.0$ is essentially a manifestation of the statistical behavior of the phase differential between copolar signal samples, as illustrated in Fig. 5.34. See Sachidananda and Zrnić (1986) for standard deviation of $\hat{\Psi}_{dp}$ as a function N.

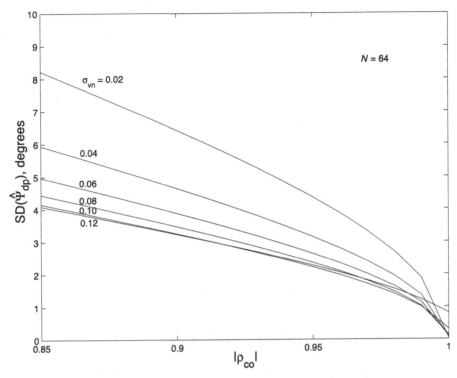

Fig. 6.27. Standard deviation (SD) of $\hat{\Psi}_{dp}$ versus $|\rho_{co}|$ for alternating mode for various normalized spectrum widths. Calculations assume Gaussian-shaped Doppler spectrum, with number of samples $N = 64$.

(iv) Variance of $\hat{v}$

The variance of $\hat{v}$ can be computed following a procedure similar to that used for deriving the variance of $\hat{\Psi}_{dp}$. The variance of the mean velocity estimate given by (6.85) can be written as,

$$\text{var}[\hat{v}] = \frac{\lambda^2}{16\pi^2 T_s^2} \frac{1}{4} [\text{var}(\Psi_1) + \text{var}(\Psi_2) + 2\,\text{cov}(\Psi_1, \Psi_2)] \tag{6.126a}$$

$$= \frac{\lambda^2}{16\pi^2 T_s^2} \frac{1}{2} [\text{var}(\Psi_1) + \text{cov}(\Psi_1, \Psi_2)] \tag{6.126b}$$

Since $|\rho_{co}|$ in most types of precipitation exceeds 0.95, the correlation between Ψ_1 and Ψ_2 is close to unity, and as a consequence the $\text{var}(\hat{v})$ in the alternating polarization mode is nearly the same as the non-alternating (or single polarization) mode at the same PRT. Refer to Fig. 5.42 for the standard deviation curves.

(v) Variance of $|\hat{\rho}_{co}|$

The variance of $|\hat{\rho}_{co}|$, with $|\hat{\rho}_{co}|$ given by (6.90), can be written using perturbation analysis as,

$$\frac{\text{var}(|\hat{\rho}_{co}|)}{|\rho_{co}|^2} = \frac{\text{var}(|\hat{\rho}_{hh,vv}[1]|)}{|\rho_{hh,vv}[1]|^2} + \frac{1}{16}\frac{\text{var}(|\hat{\rho}[2]|)}{|\rho[2]|^2} - \frac{1}{2}\frac{\text{cov}(|\hat{\rho}_{hh,vv}[1]|, |\hat{\rho}[2]|)}{|\rho_{hh,vv}[1]||\rho[2]|} \quad (6.127)$$

Each of the terms on the right-hand side of (6.127) can be expanded using results from Appendix 5 (see also, Liu et al. 1994),

$$\frac{\text{var}(|\hat{\rho}_{hh,vv}[1]|)}{|\rho_{hh,vv}[1]|^2} = \frac{1}{2}\text{Re}\left[E\left(\frac{\hat{R}_{hh,vv}[1]}{R_{hh,vv}[1]}\right)^2 + E\left|\frac{\hat{R}_{hh,vv}[1]}{R_{hh,vv}[1]}\right|^2\right]$$

$$- \text{Re}\left[\text{cov}\left(\frac{\hat{R}_{hh,hh}[0]}{R_{hh,hh}[0]}, \frac{\hat{R}_{hh,vv}[1]}{R_{hh,vv}[1]}\right)\right] - \text{Re}\left[\text{cov}\left(\frac{\hat{R}_{vv,vv}[0]}{R_{vv,vv}[0]}, \frac{\hat{R}_{hh,vv}[1]}{R_{hh,vv}[1]}\right)\right]$$

$$+ \frac{1}{4}\left[\frac{\text{var}(R_{hh,hh}[0])}{(R_{hh,hh}[0])^2} + \frac{\text{var}(\hat{R}_{vv,vv}[0])}{(R_{vv,vv}[0])^2} + \frac{2\,\text{cov}(\hat{R}_{hh,hh}[0], \hat{R}_{vv,vv}[0])}{R_{hh,hh}[0]R_{vv,vv}[0]}\right] \quad (6.128a)$$

$$\frac{\text{var}(|\hat{\rho}[2]|)}{|\rho[2]|^2} = \frac{1}{2}\text{Re}\left[E\left(\frac{\hat{R}_{hh,hh}[2]}{R_{hh,hh}[2]}\right)^2 + E\left|\frac{\hat{R}_{hh,hh}[2]}{R_{hh,hh}[2]}\right|^2\right]$$

$$- 2\,\text{Re}\left[\text{cov}\left(\frac{\hat{R}_{hh,hh}[0]}{R_{hh,hh}[0]}, \frac{\hat{R}_{hh,hh}[2]}{R_{hh,hh}[2]}\right)\right] + \text{var}\left[\frac{\hat{R}_{hh,hh}[0]}{R_{hh,hh}[0]}\right] \quad (6.128b)$$

$$\frac{\text{cov}(|\hat{\rho}_{hh,vv}[1]|, |\hat{\rho}[2]|)}{|\rho_{hh,vv}[1]||\rho[2]|} = \frac{1}{2}\left[\frac{\text{var}(\hat{R}_{hh,hh}[0])}{R_{hh,hh}[0]} + \frac{\text{cov}(\hat{R}_{hh,hh}[0], \hat{R}_{vv,vv}[0])}{R_{hh,hh}[0]R_{vv,vv}[0]}\right]$$

$$+ \frac{1}{2}\text{Re}\left\{\frac{\text{cov}(\hat{R}_{hh,vv}[1], \hat{R}_{hh,hh}[2])}{R_{hh,vv}[1]R_{hh,hh}[2]} + \frac{\text{cov}[\hat{R}_{hh,vv}[1], (R_{hh,hh}[2])^*]}{R_{hh,vv}[1](R_{hh,hh}[2])^*}\right\}$$

$$- \frac{1}{2}\text{Re}\left[\frac{\text{cov}(\hat{R}_{hh,hh}[0], \hat{R}_{hh,hh}[2])}{R_{hh,hh}[0]R_{hh,hh}[2]} + \frac{\text{cov}(\hat{R}_{vv,vv}[0], \hat{R}_{hh,hh}[2])}{R_{vv,vv}[0]R_{hh,hh}[2]}\right]$$

$$- \text{Re}\left[\frac{\text{cov}(\hat{R}_{hh,hh}[0], \hat{R}_{hh,vv}[1])}{R_{hh,hh}[0]R_{hh,vv}[1]}\right] \quad (6.128c)$$

All of the terms in the above equations involve variance and covariance of covariance estimates, which are derived in Appendix 5. Using these in (6.128) and substituting in (6.127), results in the standard deviation of $|\hat{\rho}_{co}|$. Figure 6.28 shows plots of the SD($|\hat{\rho}_{co}|$) versus N, and $|\rho_{co}|$. For typical precipitation types ($|\rho_{co}| > 0.95$), the standard error is of the order of a few percent. Figure 6.28b shows that the standard deviation decreases nearly linearly with increase in $|\rho_{co}|$. The typical measurement accuracy of $|\hat{\rho}_{co}|$ of a few percent is sufficient for applications such as hail detection and hydrometeor classification. However, note that substantial deviation from the assumed

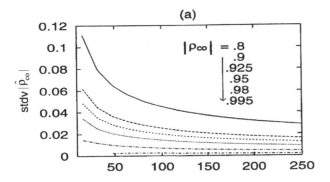

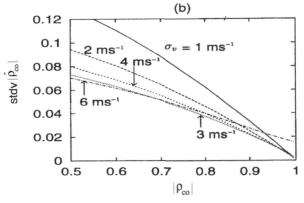

Fig. 6.28. Standard deviation of $|\hat{\rho}_{co}|$ for the alternating mode versus: (*a*) number of sample pairs N with $\sigma_{vn} = 0.06$, (*b*) $|\rho_{co}|$ for $N = 128$. Calculations assume a Gaussian-shaped Doppler spectrum. From Liu et al. (1994).

Gaussian-shaped spectra (e.g. double-peaked spectra, or very broad spectra) can result in significant bias in the estimator. See Liu et al. (1994) for a more comprehensive analysis of the bias and variance of $|\hat{\rho}_{co}|$.

(vi) Variance of LD̂R

The standard deviation of LD̂R can be written similar to that of $\hat{Z}_{dr}$ as,

$$SD(\text{LD̂R}) = 10\log_{10}\left[1 + \sqrt{\text{var}\left(\frac{\hat{P}_{cx}}{\hat{P}_{co}^h}\right)}\left(\frac{P_{cx}}{P_{co}^h}\right)^{-1}\right] \tag{6.129}$$

Using (6.116),

$$\text{var}\left(\frac{\hat{P}_{cx}}{\hat{P}_{co}^h}\right) \simeq \left(\frac{P_{cx}}{P_{co}^h}\right)^2\left[\frac{\text{var}(\hat{P}_{cx})}{(P_{cx})^2} + \frac{\text{var}(\hat{P}_{co}^h)}{(P_{co}^h)^2} - \frac{2\,\text{cov}(\hat{P}_{cx}, \hat{P}_{co}^h)}{P_{cx}P_{co}^h}\right] \tag{6.130}$$

where $\hat{P}_{co}^h$ and $\hat{P}_{cx}$ are given by (6.78a) and (6.78c), and their variances are given in (6.113a) and (6.113c), respectively. The covariance term can be written as,

$$\text{cov}(\hat{P}_{cx}, \hat{P}_{co}^h) = \frac{P_{cx}P_{co}^h}{N} \sum_{l=-(N-1)}^{N-1} \left(1 - \frac{|l|}{N}\right) |\rho_{hh,vh}[2l]|^2 \tag{6.131}$$

For most types of precipitation, $|\rho_{hh,vh}[2l]|^2 \approx 0$ (see Section 3.12). In this case,

$$\text{SD}(\hat{\text{LDR}}) \simeq 10 \log_{10} \left\{ 1 + \left[\frac{2}{N} \sum_{l=-(N-1)}^{N-1} \left(1 - \frac{|l|}{N}\right) |\rho(2l)|^2 \right]^{1/2} \right\} \tag{6.132a}$$

$$= \sqrt{2}\,\text{SD}\left[10 \log_{10}(\hat{P}_{co}^h) \right] \tag{6.132b}$$

(vii) Variance of $\hat{R}_{cx}^h$, $|\rho_{cx}^h|$, and $\arg \hat{R}_{cx}^h$

The variance of $\hat{R}_{cx}^h$ can be readily obtained from Appendix 5 as,

$$\text{var}[\hat{R}_{cx}^h] = \frac{1}{N^2} \sum_{n=-(N-1)}^{N-1} (N - |l|) R_{hh,hh}[l] (R_{vh,vh}[l])^* \tag{6.133}$$

The quantities of practical interest are the magnitude and phase of the cross-polar correlation coefficient corresponding to $\hat{R}_{cx}^h$.

The variance of the magnitude of the correlation coefficient ρ_{cx}^h can be written as,

$$\text{var}(|\hat{\rho}_{cx}^h|) = \text{var}\left[\frac{|\hat{R}_{cx}^h|}{(\hat{P}_{co}^h \hat{P}_{cx})^{1/2}} \right] \tag{6.134}$$

Using perturbation analysis of the magnitude of a complex function (as discussed in Appendix 5), the variance of $|\hat{\rho}_{cx}^h|$ can be expressed as

$$\text{var}[|\hat{\rho}_{cx}^h|] = \frac{1}{2} \text{Re}\left[E\left(\frac{\hat{R}_{cx}^h}{R_{cx}^h}\right)^2 + E\left|\frac{\hat{R}_{cx}^h}{R_{cx}^h}\right|^2 \right] - \text{Re}\left[\text{cov}\left(\frac{\hat{P}_{co}^h}{P_{co}^h}, \frac{\hat{R}_{cx}^h}{R_{cx}^h}\right) \right]$$
$$- \text{Re}\left[\text{cov}\left(\frac{\hat{P}_{cx}}{P_{cx}}, \frac{\hat{R}_{cx}^h}{R_{cx}^h}\right) \right] + \frac{1}{4}\left[\frac{\text{var}[\hat{P}_{co}^h]}{(P_{co}^h)^2} + \frac{\text{var}[\hat{P}_{cx}]}{(P_{cx})^2} + \frac{2\,\text{cov}(\hat{P}_{co}^h, \hat{P}_{cx})}{P_{co}^h P_{cx}} \right] \tag{6.135}$$

It can be seen from (6.135) that all terms involve variance and covariance of covariance estimates. Using results given in Appendix 5, the variance in the above estimates can be calculated from the autocorrelation function of the copolar and cross-polar signals. The variance of $|\hat{\rho}_{cx}^h|$ can be simplified as,

$$\text{var}[|\hat{\rho}_{cx}^h|] = \frac{(1 - |\rho_{cx}^h|^2)^2}{2N^2|\rho_{cx}^h|^2} \sum_{n=-(N-1)}^{(N-1)} (N - |n|)|\rho[n]|^2 \tag{6.136}$$

The variance of $\arg \hat{R}_{cx}^h$ can be expressed (based on results of Appendix 5) as,

$$\text{var}[\arg(\hat{R}_{cx}^h)] = \frac{1}{2} E \left\{ \left[\text{Re} \left(\left| \frac{\hat{R}_{cx}^h}{R_{cx}^h} \right|^2 - \left(\frac{\hat{R}_{cx}^h}{R_{cx}^h} \right) \right) \right]^2 \right\} \tag{6.137}$$

Again this can be simplified to,

$$\text{var}[\arg(\hat{R}_{cx}^h)] = \frac{1 - |\rho_{cx}^h|^2}{|\rho_{cx}^h|^2} \frac{1}{2N^2} \sum_{n=-(N-1)}^{(N-1)} (N - |n|)|\rho[n]|^2 \tag{6.138}$$

It will be shown later, in Section 6.5.3, that the variance formula derived here will also be applicable for the hybrid mode, and results will be shown later. Most types of precipitation result in fairly small values of $|\rho_{cx}^h|$ because of azimuthal symmetry, or "mirror" reflection symmetry about a mean canting angle of $0°$. This results in large variance of $|\hat{\rho}_{cx}^h|$ and $\arg \hat{R}_{cx}^h$. Therefore, $\arg \hat{R}_{cx}^h$ is generally unusable for estimating the differential propagation phase, see also (4.78). The variance formulas derived here are more practically useful if the polarization basis is other than horizontal/vertical (for example, RHC/LHC or slant $45°/135°$).

6.5.2   Hybrid mode

(i) Variance of mean power and Z_{dr} estimates
The variance of $\hat{P}_h$ and $\hat{P}_v$ is similar to (6.113),

$$\text{var}(\hat{P}_h) = \frac{(P_h)^2}{N} \sum_{l=-(N-1)}^{(N-1)} \left(1 - \frac{|l|}{N}\right) \rho_p[l] \tag{6.139a}$$

$$\text{var}(\hat{P}_v) = \frac{(P_v)^2}{N} \sum_{l=-(N-1)}^{(N-1)} \left(1 - \frac{|l|}{N}\right) \rho_p[l] \tag{6.139b}$$

Variance of $\hat{z}_{dr}$ can be obtained from the above and (6.116) as,

$$\text{var}(\hat{z}_{dr}) = \frac{2}{N}(z_{dr})^2 \sum_{l=-(N-1)}^{(N-1)} \left(1 - \frac{|l|}{N}\right) \left(|\rho[l]|^2 - |\rho_{hh,vv}[l]|^2\right) \tag{6.140}$$

Using the approximation for $\rho_{hh,vv}[l]$ given in (6.120), the above can be further simplified to,

$$\text{var}(\hat{z}_{dr}) = \frac{2}{N}(z_{dr})^2(1 - |\rho_{co}|^2) \sum_{l=-(N-1)}^{(N-1)} \left(1 - \frac{|l|}{N}\right) |\rho(l)|^2 \tag{6.141}$$

The standard deviation of $\hat{Z}_{dr}$ can be obtained from the above and (6.115) as,

$$\text{SD}(\hat{Z}_{dr}) = 10 \log_{10} \left\{ 1 + \left[\frac{2}{N}(1 - |\rho_{co}|^2) \sum_{l=-(N-1)}^{N-1} \left(1 - \frac{|l|}{N}\right) |\rho(l)|^2 \right]^{1/2} \right\} \tag{6.142}$$

Figure 6.29a shows the standard deviation of $\hat{Z}_{dr}$ (dB) in the hybrid mode as a function of $|\rho_{co}|$, and may be compared with Fig. 6.26 for the alternating mode (see also, Gingras et al. 1997). The SD($\hat{Z}_{dr}$) behaves similarly in the two modes. However, unlike the alternating mode, in the hybrid mode SD($\hat{Z}_{dr}$) decreases with increasing σ_{vn} for all values of $|\rho_{co}|$. Also, the SD($\hat{Z}_{dr}$) $\rightarrow$ 0 as $|\rho_{co}| \rightarrow$ 1 independent of the value of σ_{vn}, which is a direct manifestation of the statistical behavior previously noted in reference to Fig. 5.36. Among the parameters such as N, σ_{vn}, and $|\rho_{co}|$, the SD($\hat{Z}_{dr}$) is most influenced by $|\rho_{co}|$. This is generally also true for SD($\hat{\Psi}_{dp}$) and SD($|\hat{\rho}_{co}|$).

Figure 6.29b shows SD($\hat{Z}_{dr}$) versus N in the hybrid mode for various values of normalized spectrum widths assuming a Gaussian-shaped spectrum (note that all variance computations are based on a Gaussian-shaped spectrum). Similar to the alternating mode, the Z_{dr} can be measured to an accuracy of a few tenths of a dB assuming that the propagation and scattering matrices are diagonal in the h/v-basis (see Section 4.7). Figure 6.29c compares SD($\hat{Z}_{dr}$) versus σ_{vn} for the two modes; note that the dwell times are matched (i.e. 64 samples in alternating mode gives the same dwell time as 128 samples in the hybrid mode). Figure 6.29c shows that for $\sigma_{vn} > 0.1$, the SD($\hat{Z}_{dr}$) in the hybrid mode is smaller than in the alternating mode. Figure 6.29d shows that for a fixed $|\rho_{co}|$, the difference between the two modes is significant when σ_{vn} is large and N is small.

(ii) Variance in the estimates of Ψ_{dp} and $|\rho_{co}|$

The variance of $\hat{\Psi}_{dp}$ is the same as the variance of the phase of $\hat{R}_{cx}^{h}$ and $\hat{R}_{cx}^{v}$, see (6.137). Therefore, following (6.138), the variance of $\hat{\Psi}_{dp}$ can be written as,

$$\text{var}[\hat{\Psi}_{dp}] = \left(\frac{1 - |\rho_{co}|^2}{|\rho_{co}|^2}\right) \frac{1}{2N^2} \sum_{n=-(N-1)}^{(N-1)} (N - |n|)|\rho[n]|^2 \qquad (6.143)$$

The above equation is an extension of the exact result given in (5.186), which is the SD($\hat{\Psi}_{dp}$) for $N = 1$ (see Gingras et al. 1997). Figure 6.30a shows the standard deviation of $\hat{\Psi}_{dp}$ as a function of $|\rho_{co}|$ and the accuracy is 2–5° for $|\rho_{co}| > 0.95$. Again, the standard deviation $\rightarrow$ 0 as $|\rho_{co}| \rightarrow$ 1 independent of σ_{vn}, which is a direct manifestation of the statistical behavior previously noted in reference to Fig. 5.34. Figure 6.30b shows SD($\hat{\Psi}_{dp}$) versus N for various values of normalized spectrum widths (σ_{vn}). Figure 6.30c compares SD($\hat{\Psi}_{dp}$) versus σ_{vn} between the hybrid and alternating modes. For large σ_{vn}, there is a significant advantage afforded by the hybrid mode. For small σ_{vn}, both modes give the same standard deviation. Similar to the results of Fig. 6.29d, Fig. 6.30d shows that for a fixed $|\rho_{co}|$, the difference between the two modes is significant when σ_{vn} is large and N is small. Holt et al. (1999) demonstrated this result experimentally using CSU–CHILL radar data. Since $\sigma_{vn} = \sigma_v(2T_s/\lambda)$, the hybrid mode will be more advantageous as the radar operating frequency increases (assuming the PRT $= T_s$ is fixed).

The variance computation of $|\hat{\rho}_{co}|$ in the hybrid mode is the same as the variance computation of $|\rho_{cx}^{h}|$ or $|\rho_{cx}^{v}|$, see (6.136). Therefore, following (6.136), the variance of

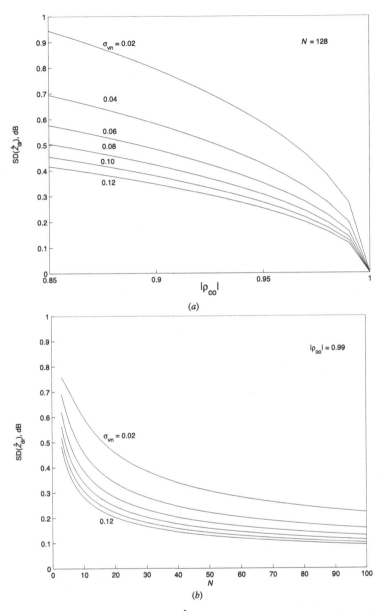

Fig. 6.29. (*a*) Standard deviation of $\hat{Z}_{dr}$ versus $|\rho_{co}|$ for hybrid mode for various normalized spectrum widths. Calculations assume Gaussian-shaped Doppler spectrum, with number of samples $N = 128$. (*b*) Standard deviation of $\hat{Z}_{dr}$ versus N for hybrid mode for various normalized spectrum widths. Calculations assume Gaussian-shaped Doppler spectrum, with $|\rho_{co}| = 0.99$. (*c*) Standard deviation of $\hat{Z}_{dr}$ versus σ_{vn} for hybrid mode (———) and alternating mode (– – –), shown for matched dwell times (with $N = 64$) for various $|\rho_{co}|$. Calculations assume Gaussian-shaped Doppler spectrum. (*d*) Standard deviation of $\hat{Z}_{dr}$ versus N for hybrid mode (———) and alternating mode (– – –), shown for $\sigma_{vn} = 0.02$ and 0.12. Calculations assume Gaussian-shaped Doppler spectrum, with $|\rho_{co}| = 0.99$. Note the higher values of SD($\hat{Z}_{dr}$) for alternate mode when σ_{vn} is 0.12.

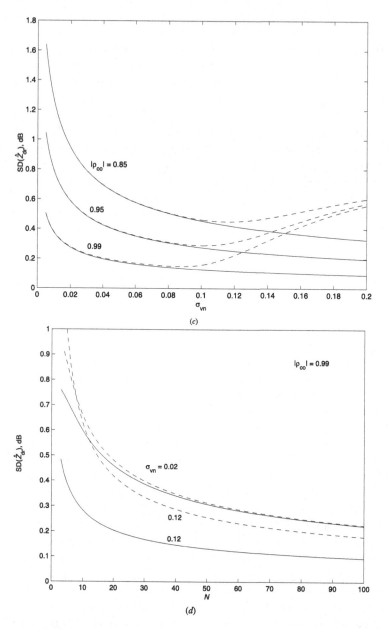

Fig. 6.29. (*cont.*)

$|\hat{\rho}_{co}|$ can be written as,

$$\text{var}[|\hat{\rho}_{co}|] = \frac{(1 - |\rho_{co}|^2)^2}{2N^2|\rho_{co}|^2} \sum_{n=-(N-1)}^{(N-1)} (N - |n|)|\rho[n]|^2 \tag{6.144}$$

Figure 6.31*a* shows the SD($|\hat{\rho}_{co}|$) as a function of N for various values of $|\rho_{co}|$, which

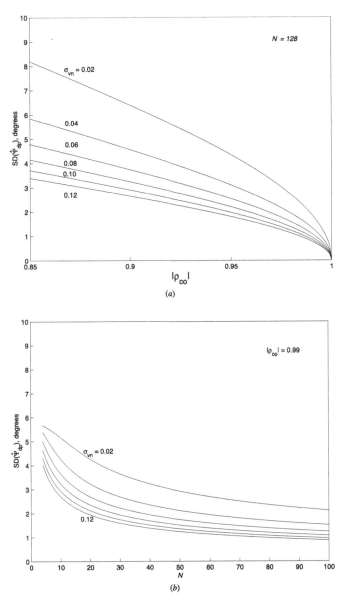

Fig. 6.30. (*a*) Standard deviation of $\hat{\Psi}_{dp}$ versus $|\rho_{co}|$ for hybrid mode for various normalized spectrum widths. Calculations assume Gaussian-shaped Doppler spectrum, with number of samples $N = 128$. (*b*) Standard deviation of $\hat{\Psi}_{dp}$ versus N for hybrid mode for various normalized spectrum widths. Calculations assume Gaussian-shaped Doppler spectrum, with $|\rho_{co}| = 0.99$. (*c*) Standard deviation of $\hat{\Psi}_{dp}$ versus σ_{vn} for hybrid mode (———) and alternating mode (– – –), shown for matched dwell times (with $N = 128$) for various $|\rho_{co}|$. Calculations assume Gaussian-shaped Doppler spectrum. (*d*) Standard deviation of $\hat{\Psi}_{dp}$ versus N for hybrid mode (———) and alternating mode (– – –), shown for $\sigma_{vn} = 0.02$ and 0.12. Calculations assume Gaussian-shaped Doppler spectrum, with $|\rho_{co}| = 0.99$. Note the higher values of SD($\hat{\Psi}_{dp}$) for alternate mode when σ_{vn} is large.

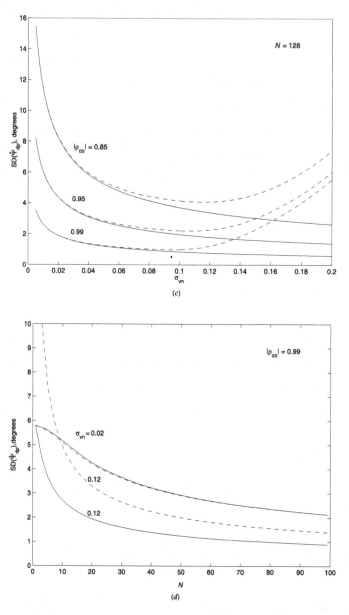

Fig. 6.30. (*cont.*)

may be compared with Fig. 6.28b. Note that the accuracy is around a few percent for $|\rho_{co}| > 0.95$. Figure 6.31b shows SD($|\hat{\rho}_{co}|$) versus $|\rho_{co}|$ and can be compared with Fig. 6.28b. Note the steady decrease of SD($|\hat{\rho}_{co}|$) in the hybrid mode versus increase in $|\rho_{co}|$ with SD($|\hat{\rho}_{co}|$) $\to 0$ as $|\rho_{co}| \to 1$, independent of the value of σ_{vn}. This is because the estimator given by (6.109) is a direct estimate of ρ_{co} by definition.

Some general comments regarding the hybrid and alternating modes are appropriate

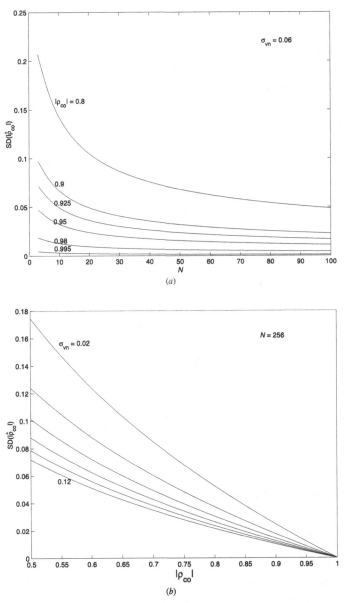

Fig. 6.31. Standard deviation of (a) $|\hat{\rho}_{co}|$ versus N for hybrid mode for various values of $|\rho_{co}|$. Calculations assume Gaussian-shaped Doppler spectrum with $\sigma_{vn} = 0.06$. (b) $|\hat{\rho}_{co}|$ versus $|\rho_{co}|$ for hybrid mode for various normalized spectrum widths. Calculations assume Gaussian-shaped Doppler spectrum with number of samples $N = 256$.

here. In the hybrid mode, the horizontal and vertical components are sampled simultaneously. Therefore, in the absence of cross-coupling, estimates of Z_{dr}, Ψ_{dp}, and $|\rho_{co}|$ are inherently more accurate than in any non-simultaneous sampling mode. However, this advantage is significant only when the Doppler spectrum widths are large and $|\rho_{co}|$ is close to unity. When the Doppler spectrum width is small the temporal samples are highly correlated. This temporal correlation results in an interpolatory effect between samples when the alternating mode is used, which effectively reconstructs the "missing" simultaneous samples yielding fairly accurate estimates of Z_{dr}, Ψ_{dp}, and $|\rho_{co}|$. However, this advantage is lost for large Doppler spectrum widths. In practice, the hybrid mode will be more advantageous when short dwell times need to be used.

The dwell time for conventional Doppler radars is primarily determined by the desired accuracy in the reflectivity estimate, which is typically ± 1 dB. This accuracy is controlled by the number of samples used in the estimate, the PRT and Doppler spectrum width (assuming a fixed wavelength of operation). Typically, the dwell times are of the order of tens of milliseconds at S-band. In practice, the antenna scan rate is governed by the dwell time and by the desired azimuth (or elevation) resolution ($\Delta\theta$),

$$\text{Antenna scan rate} = \frac{\Delta\theta}{\text{Dwell time}} = \frac{\Delta\theta}{N(\text{PRT})} \qquad (6.145)$$

In the case of dual-polarized radars, the dwell time is primarily determined by the desired accuracy in Z_{dr} (a few tenths of a dB) and Ψ_{dp} (a few degrees). In addition to the factors listed above, both the pulsing mode and $|\rho_{co}|$ are involved in determining the accuracy in Z_{dr} and Ψ_{dp}. Typically, the dwell times are of the order of 100 ms for the alternating mode, which implies slower antenna scan rates compared to a conventional Doppler radar. Techniques are available to reduce the dwell time for polarimetric radars such as frequency diversity, or use of pulse compression waveform combined with range averaging (Krehbiel and Brook 1979; Sachidananda and Zrnić 1985; Mudukutore et al. 1998; Brunkow et al. 2000).

6.6   Estimation of specific differential phase (K_{dp})

Specific differential phase can be estimated as the range derivative of Φ_{dp}. Also, recall that $\Psi_{dp} = \Phi_{dp} + \delta_{co}$, see (4.65). Therefore, when δ_{co} is small or relatively constant over range compared with Φ_{dp}, then the profile of Ψ_{dp} can be used to estimate K_{dp}. Let $\Psi_{dp}(r)$ and $\Phi_{dp}(r)$ be the values of Ψ_{dp} and Φ_{dp} at range r, respectively. Then the finite difference estimate of K_{dp} can be written as,

$$\hat{K}_{dp} = \frac{\Phi_{dp}(r_2) - \Phi_{dp}(r_1)}{2(r_2 - r_1)} \simeq \frac{\hat{\Psi}_{dp}(r_2) - \hat{\Psi}_{dp}(r_1)}{2(r_2 - r_1)} \qquad (6.146)$$

The estimate of K_{dp} obtained from (6.146) is valid between the ranges r_1, r_2. Note that K_{dp} is the differential propagation phase per unit distance of the medium and (6.146) is consistent with this fundamental definition given in (4.8). Care must be exercised since several articles (e.g. Smyth et al. 1999) do not follow this fundamental definition, i.e. they define K_{dp} as twice the differential propagation phase per unit distance. The Φ_{dp} is a range cumulative quantity, which implies that range profiles of Φ_{dp} are smooth. The fluctuations of Ψ_{dp} with range are mostly due to fluctuations in the estimate of Ψ_{dp}, as well as to range variation of δ_{co}. These statistical fluctuations depend on the Doppler spectrum and $|\rho_{co}|$. In rainfall, $|\rho_{co}|$ is fairly high (tends to unity, see Section 7.1.3) and for a sufficient number of samples the standard deviation of Ψ_{dp} is of the order of few degrees (see Fig. 6.27).

The standard deviation of $\hat{K}_{dp}$ given by (6.146) can be written as,

$$SD[\hat{K}_{dp}] = \frac{SD[\hat{\Psi}_{dp}]}{\sqrt{2}\Delta r} \tag{6.147}$$

where Δr is the range interval $(r_2 - r_1)$. If Δr is of the order of 2 km and $SD(\Psi_{dp})$ is $2°$ then $SD(K_{dp})$ is of the order of $0.707°$ km^{-1} (or $1/\sqrt{2}$). However, for applications such as rain rate estimation, K_{dp} needs to be estimated more accurately.

The K_{dp} profile can be estimated from the Ψ_{dp} profile by more accurate methods than the simple finite difference procedure given by (6.146). Two general approaches are used to improve the accuracy of the K_{dp} estimate, namely: (i) smoothing the range profile of Ψ_{dp}, and (ii) regression-based estimates.

6.6.1   Smoothing the range profile of Ψ_{dp}

Techniques such as range averaging and weighted smoothing can be described in terms of filtering. In filter terminology, fluctuations of adjacent range samples of Ψ_{dp} can be interpreted as a "high frequency" component of the Ψ_{dp} range profile. Therefore, a low pass filter would eliminate these high frequency fluctuations and retain the mean trend. The exact filter to be used depends on the range sample interval and the degree of smoothing desired (Hubbert et al. 1993; Hubbert and Bringi 1995).

Figure 6.32 shows the magnitude response function of two infinite impulse response (IIR) filters as a function of inverse frequency (Proakis and Manolakis 1988). The response function "A" strongly attenuates spatial scales less than 375 m by 10 dB or more. This filter is useful in smoothing the high frequency fluctuations that occur along adjacent resolution volumes. The filter curve marked "B" strongly attenuates spatial scales less than 1.5 km by 10 dB or more, and is useful in preserving the mean trend of Ψ_{dp} from which K_{dp} may be estimated using (6.146). For example, Fig. 6.33a shows range profiles of "raw" Ψ_{dp} and the smoothing obtained using filter A, while Fig. 6.33b shows the result of applying filter B. The accuracy of deriving K_{dp} by a direct application of filter B to the "raw" Ψ_{dp} range profile (assuming $\delta_{co} \approx 0°$) and using (6.146) is estimated to be 0.3–0.4 ° km^{-1} (see Fig. 4.8b for an example of derived K_{dp}). The above smoothing approach can be directly used only if δ_{co} is negligible.

An iterative filtering technique of estimating non-zero δ_{co} superposed on a monotonic increasing Φ_{dp} range profile has been described by Hubbert and Bringi (1995). A finite impulse response (FIR) filter, shown in Fig. 6.32b, is used since it does not introduce any range delay (as opposed to IIR filters, which introduce delays that must be corrected). Estimates of K_{dp} will be biased unless range profiles of Ψ_{dp} are first

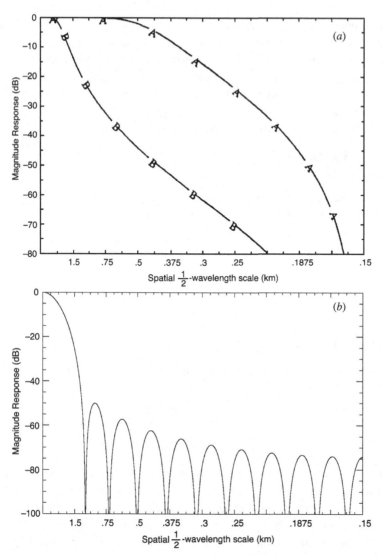

Fig. 6.32. (*a*) Magnitude of the filter transfer functions that are used to filter range profiles of Ψ_{dp}. The resolution volumes are assumed to be spaced 150-m apart. These filters are infinite impulse response (IIR) third-order Butterworth filters. Filter "A" is a "light" filter designed to filter out the rapid gate-to-gate fluctuations, while "B" is a "heavy" filter designed to preserve the mean increasing trend of Ψ_{dp} with range. (*b*) as in (*a*), except the magnitude response of a twentieth-order finite impulse response filter (FIR) is shown. (*a*) From Hubbert et al. (1993), and (*b*) from Hubbert and Bringi (1995).

"corrected" for effects due to non-zero δ_{co}. Figure 6.34*a* shows the "raw" Ψ_{dp} range profile (dashed curve) and the same profile after filtering. A large "bump" in Ψ_{dp} in the range interval 48.0–50.5 km can be noted, which is likely due to δ_{co}. This large phase perturbation will cause erroneous estimates of K_{dp} (including negative values)

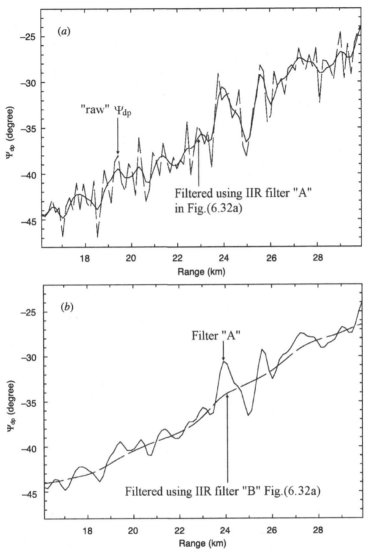

Fig. 6.33. (*a*) Example of filtering "raw" Ψ_{dp} data using the filter marked "A" in Fig. 6.32*a*. Radar data from the C-band DLR radar located near Munich, Germany. (*b*) As in (*a*), except the filter marked "B" in Fig. 6.32*a* is used. The input to this filter is the filtered curve in (*a*). From Hubbert et al. (1993).

if used as it is. Figure 6.34*b* shows the corrected range profile using the iterative filter algorithm described by Hubbert and Bringi (1995). The "bump" in Fig. 6.34*a* has been removed, and a better estimate of K_{dp} can be made from the filtered (solid line) in Fig. 6.34*b*. Note how only the large phase excursions of the raw Ψ_{dp} profile have been eliminated while the remainder of the profile (from 52–57 km) has been left unchanged, thus preserving the more subtle mean variations of the original "raw" profile

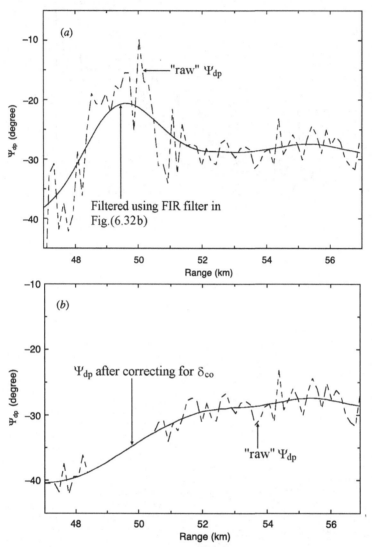

Fig. 6.34. (*a*) Example range profile of "raw" Ψ_{dp} and filtered version using the FIR filter in Fig. 6.32*b*. (*b*) Result after applying an iterative filter algorithm to correct the Ψ_{dp} for back scatter differential phase (δ_{co}) in the range interval 48–50.5 km. The corrected solid line is Φ_{dp}, from which K_{dp} can be estimated. (*c*) Estimate of δ_{co} obtained by differencing the "raw" Ψ_{dp} in (*a*) and the solid curve in (*b*), i.e. $\delta_{co} = \Psi_{dp} - \Phi_{dp}$. (*d*) Example of the iterative filtering technique showing results after 1, 3, 5, 7, 9, 11, and 13 iterations (curves A–G). The "raw" Ψ_{dp} is the dashed curve in (*a*). (*e*) As in (*d*) except profiles are filtered using the FIR filter in Fig. 6.32*b*. From Hubbert and Bringi (1995).

in that area. The δ_{co} is estimated by simply differencing the filtered data (solid curve) in Fig. 6.34*b* from the original raw Ψ_{dp} (dashed curve) in Fig. 6.34*a*. The resulting profile

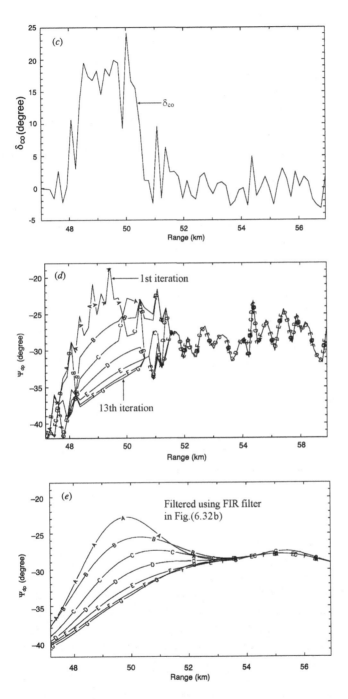

Fig. 6.34. (*cont.*)

of δ_{co} is shown in Fig. 6.34c. The large $20°$ phase excursions around 49.0–50.5 km are well above the measurement fluctuations estimated to have a standard error of 3–4°.

Figure 6.34*d, e* shows how the iterative algorithm would work on this example range profile.

6.6.2   Regression-based estimates of K_{dp}

When δ_{co} is negligible, an alternate approach to estimating K_{dp} involves applying a linear or non-linear regression fit to the range profile of Ψ_{dp}. The range profile, $\Psi_{dp}(r)$, can be approximated by a polynomial of the form,

$$\Psi_{dp}(r) \simeq \beta_0 + \beta_1 r + \beta_2 r^2 + \cdots + \beta_n r^n \tag{6.148}$$

where the coefficients $\beta_0, \beta_1, \ldots, \beta_n$ can be estimated from a least square fit to the profile of $\Psi_{dp}(r)$. Let $r_1, r_2, \ldots, r_n$ be the ranges at which Ψ_{dp} is observed. The observations can be written in a matrix form as,

$$\Psi_{dp} \simeq X\beta \tag{6.149a}$$

where,

$$X = \begin{bmatrix} 1 & r_1 & r_1^2 & \cdots & r_1^n \\ 1 & r_2 & r_2^2 & \cdots & r_2^n \\ \vdots & \vdots & \vdots & \cdots & \vdots \\ 1 & r_N & r_N^2 & \cdots & r_N^n \end{bmatrix} \tag{6.149b}$$

$$\beta = [\beta_0\ \beta_1\ \cdots\ \beta_n]^t \tag{6.149c}$$

$$\Psi_{dp} = [\Psi_{dp}(r_1)\ \Psi_{dp}(r_2)\ldots\Psi_{dp}(r_N)]^t \tag{6.149d}$$

In the above, n is the degree of the polynomial, N is the number of range samples, and β is the vector of polynomial coefficients. β can be estimated from regression analysis as,

$$\hat{\beta} = [X^t X]^{-1} X^t \Psi_{dp} \tag{6.149e}$$

Subsequently, K_{dp} can be estimated as,

$$\hat{K}_{dp} = \beta_1 + 2\beta_2 r + 3\beta_3 r^2 + \cdots + n\beta_n r^{n-1} \tag{6.150}$$

In practice, it is better to limit the order of the polynomial and apply the regression over small range segments instead of one polynomial fit to the entire range. For short range segments, K_{dp} can be estimated from a linear approximation of $\Psi_{dp}(r)$ as β_1.

Let $\Psi_{dp}(r_i), i = 1, \ldots, N$ be the N range samples of $\Psi_{dp}(r_i)$ at the corresponding range location r_i. Then the estimate of K_{dp} based on β_1 only can be written as,

$$\hat{K}_{dp} = \frac{\sum\limits_{i=1}^{N} [\Psi_{dp}(r_i) - \bar{\Psi}_{dp}](r_i - \bar{r})}{2\sum\limits_{i=1}^{N}(r_i - \bar{r})^2} \tag{6.151}$$

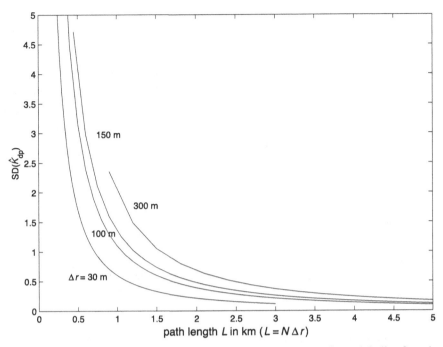

Fig. 6.35. Standard deviation of K_{dp} estimate obtained as the slope of a straight line fit to the Ψ_{dp} profile, shown as a function of path length. The various curves show the effect of different range sample spacings for a fixed path length.

where,

$$\bar{\Psi}_{dp} = \frac{1}{N} \sum_{i=1}^{N} \Psi_{dp}(r_i), \qquad (6.152a)$$

$$\bar{r} = \frac{1}{N} \sum_{i=1}^{N} r_i \qquad (6.152b)$$

and,

$$\sum_{i=1}^{N} (r_i - \bar{r})^2 = \left[\frac{N(N-1)(N+1)}{12}\right](\Delta r)^2 \qquad (6.152c)$$

If the samples are uniformly spaced in range by Δr, then,

$$\hat{K}_{dp} = \frac{\displaystyle\sum_{i=1}^{N}[\Psi_{dp}(r_i) - \bar{\Psi}_{dp}][i - (N+1)/2]\Delta r}{\frac{1}{6}N(N-1)(N+1)(\Delta r)^2} \qquad (6.153)$$

Subsequently, the standard deviation of $\hat{K}_{dp}$ can be derived from (6.153) (Gorgucci et al.

1999a),

$$SD(\hat{K}_{dp}) = \frac{SD(\hat{\Psi}_{dp})}{\sqrt{[N(N-1)(N+1)/3]}} \frac{1}{\Delta r} \tag{6.154}$$

Equation (6.154) can be compared with (6.147). Whenever $N > 2$, then the straight line fit to the $\Psi_{dp}(r)$ profile yields more accurate estimates of K_{dp} with increasing N. Equation (6.154) can also be written in terms of the path length L over which K_{dp} is estimated (see also, Balakrishnan and Zrnić 1990a) as,

$$SD(\hat{K}_{dp}) = \frac{SD(\hat{\Psi}_{dp})}{L} \sqrt{\frac{3}{[N - (1/N)]}} \tag{6.155}$$

Figure 6.35 shows the plot of $SD(\hat{K}_{dp})$ as a function of L ($L = N\Delta r$) for several values of Δr. Note that K_{dp} can be estimated to an accuracy of $0.5°$ km^{-1} over a 2-km path with Δr of 150 m. For a fixed L, the $SD(\hat{K}_{dp})$ can be reduced by decreasing Δr, provided the samples can be made nearly independent, e.g. by transmitting a narrower pulse.

Notes

1. The Doppler spectral shape differences between the copolar and cross-polar components of the received signal are not considered here. These have been measured at linear polarization and at vertical incidence by Battan and Theiss (1970). Theoretical formulation of the Doppler spectra of the right-hand and left-hand circular components of the received signal can be found in Warner and Rogers (1977) and Metcalf and Echard (1978); for examples of such data see Metcalf (1986). In the case of linear (h/v) basis, the mean Doppler velocities corresponding to signal returns (hh, vv, and hv) have been analyzed by Wilson et al. (1997).

2. The 16.5- and 10-GHz circularly polarized radars designed and developed by the National Research Council (NRC) of Canada by McCormick, Hendry and coworkers were primarily intended for studying propagation phenomena for communication applications. The NRC also designed the S-band circularly polarized radar for the Alberta Research Council, which was primarily intended for hail detection. The system specifications of these radars are given in Bringi and Hendry (1990). These radars are no longer in service.

3. On a historical note, the first "fast switched" Z_{dr} data were obtained in 1978 with the RAL S-band radar located in the UK and these initial data were reported by Bringi et al. (1978). However, the first "slow-switched" Z_{dr} data were acquired earlier with the CHILL radar during the summer of 1977, and these initial data were reported by Seliga et al. (1979).

4. The second radar in the USA to be converted to a polarization-agile/single-receiver radar was the NCAR/CP-2 radar in 1983. Both the CHILL and CP-2 radars were originally designed for dual-wavelength operation for hail detection. However, their antennas were designed for dual-polarization and this made it relatively easy to modify these radars for Z_{dr} measurement capability. The third radar in the USA to be converted to a polarization-agile/single-receiver system was the S-band research radar operated by the National Severe Storms Laboratory, Cimarron, Oklahoma (Carter et al. 1986).

5. Tragl (1990) has derived the exact conditions for the optimal polarizations of a random, reciprocal medium characterized by an averaged covariance matrix.

6. The C-POL radar operated by the Bureau of Meteorology Research Center in Australia uses a similar configuration except that a ferrite polarization switch is used (Keenan et al. 1998). Examples of data from this radar are provided in Section 7.4.

7. Further details are given in Brunkow et al. (2000). See, also, the CSU–CHILL web site at,

http://www.chill.colostate.edu

8. Refer to the above web site for details on calculating l_r for the CSU–CHILL radar.

9. Other techniques are available for determining system gain, for example, using the sun as a reference source (see Pratte and Ferraro 1989). The self-consistency of Z_h, Z_{dr}, and K_{dp} in rain can be used to "fine tune" the radar constant. This methodology was proposed by Aydin et al. (1983) for dual-wavelength radar and has been adapted for dual-polarization (see, for example, Gorgucci et al. 1999b).

7 The polarimetric basis for characterizing precipitation

The conventional single-polarized Doppler radar uses the measurement of radar reflectivity, radial velocity, and storm structure to infer some aspects of hydrometeor types and amounts. With the advent of dual-polarized radar techniques it is generally possible to achieve significantly higher accuracies in the estimation of hydrometeor types, and in some cases of hydrometeor amounts. A description of these techniques and their rationale, within the scattering and propagation theory presented in Chapters 3 and 4, is the main subject of this chapter. For certain important classes of hydrometeors, the information that can be obtained from dual-polarized radar is so dramatic that it is now considered to be an indispensable tool for the study of the formation and evolution of precipitation.

The determination of hydrometeor types and amounts can be formulated in terms of the elements of the covariance matrix. The covariance matrix defined in (3.183) is averaged over the particle size distribution $N(D)$ (here D is the characteristic size of a hydrometeor) and over the joint probability density function of particle states (typically inclusive of shape, orientation, and density). The radar elevation angle is an independent variable and the radar operating frequency is fixed. Thus, if $N(D)$ and the particle state distributions are assumed, then the covariance matrix (and radar observables derived from the matrix, such as differential reflectivity and linear depolarization ratio) can be computed as a function of elevation angle and operating frequency. Such calculations have been used extensively in the literature to compare with measurements, or to simply set bounds on the inference of hydrometeor types and amounts. Similarly, the transmission matrix defined in (4.42), representing the propagation of a dual-polarized radar beam and derived parameters, such as absolute attenuation, differential attenuation, and differential phase, can be computed for comparison with measurements. Such calculations are indeed important but they depend on a priori assumptions on $N(D)$ and particle state distributions, which are often *ad hoc* and may only approximate the state of actual precipitation within the cloud.

Cloud models (either explicit "bin" resolved microphysical models or "bulk" microphysical models) can be used to predict hydrometeor concentrations and amounts. Particle state distributions (principally shape and orientation) are assumed and the covariance and transmission matrices are calculated, from which related radar and propagation parameters are obtained. The use of a cloud model to diagnose the radar observables makes it possible to test hypotheses regarding the microphysical origin of certain polarimetric radar signatures. Sophisticated cloud models have been similarly coupled to radiative transfer models to diagnose brightness temperature at cloud tops at various frequencies for application to retrieval algorithms for airborne or satellite multifrequency radiometers (e.g. the Tropical Rainfall Measurement Mission (TRMM) satellite sensors).

As mentioned earlier, the goal of this chapter is to set a theoretical framework for the estimation of hydrometeor types and amounts. This framework is built on a mix of techniques such as scattering simulations, cloud microphysics, *in-situ* observations of hydrometeors and empirically based insights. It is beyond the scope of this chapter to discuss, in any depth, the formation of different hydrometeor types and their evolution, other than in a rudimentary manner.[1] While it is recognized that the environment under which the precipitation forms and evolves is an important constraint it will not be considered in this chapter.[2] When necessary the updraft/downdraft structure within storm cells will be invoked to explain precipitation evolution inferred from polarimetric radar measurements.

7.1 Rain

The most important characteristic polarization signatures of rain at low radar elevation angles are the differential reflectivity ($Z_{dr} = 10 \log_{10}(\mathcal{z}_{dr})$) and the specific differential phase (K_{dp}, $^\circ$ km^{-1}). These characteristic signatures are a consequence of the approximately oblate spheroidal shapes of raindrops coupled with a nearly vertical orientation of their symmetry axes forming an anisotropic medium. The microphysical origin of these signatures is discussed in relation to the drop size distribution. The application of Z_{dr} and K_{dp} to the measurement of rainfall rate will be considered in Chapter 8. At higher frequencies ($\geq$5 GHz), the differential attenuation (A_{dp}) can cause significant reduction of the Z_{dr} signal, which must be corrected. Of course, absolute attenuation causes the measured power at horizontal polarization to decrease with increasing propagation distance into the rain storm, and, thus, the measured reflectivity must also be corrected. Techniques for correction of reflectivity and Z_{dr} will be covered in Section 7.4.

7.1.1 The equilibrium shape and orientation of raindrops: implications for Z_{dr} and K_{dp}

Assuming that the air flow is steady, the equilibrium shape of a raindrop is determined by a balance of forces on the interface involving hydrostatic, surface tension, and aerodynamic forces. A simple hydrostatic model derived by Green (1975) assumes this shape to be a priori oblate spheroidal, and force balance between hydrostatic and surface tension alone applied along the equator of the spheroid yields an analytical expression for the axis ratio (defined here as b/a, with b being the semi-minor axis length and a the semi-major axis length; see Fig. 2.7*b*) as a function of the volume-equivalent spherical diameter $D = 2(b)^{1/3}(a)^{2/3}$,

$$D = 2\left(\frac{\sigma}{g\rho_w}\right)^{1/2} \frac{[(b/a)^{-2} - 2(b/a)^{-1/3} + 1]^{1/2}}{(b/a)^{1/6}} \qquad (7.1)$$

where σ is the surface tension (72.25 erg cm^{-2}), g is the acceleration due to gravity (981 cm s^{-2}) and ρ_w is the water density (1 g cm^{-3}). A more accurate description of equilibrium shapes has been derived by Beard and Chuang (1987), using a numerical model, who included aerodynamic effects; their shapes are shown in Fig. 7.1 for D from 1 to 6 mm. While the larger drops have slightly flattened bases, it is sufficient for radar applications to calculate an effective axis ratio as the ratio between the maximum vertical and horizontal chords. Figure 7.2 shows the effective axis ratio versus D for the Beard–Chuang numerical model and also includes a number of experimentally derived axis ratios as well as the earlier perturbation model results of Pruppacher and Pitter (1971). A useful linear fit to the wind-tunnel data of Pruppacher and Beard (1970) is,

$$\frac{b}{a} = 1.03 - 0.062(D); \quad 1 \le D \le 9 \text{ mm} \qquad (7.2)$$

A polynomial fit to the numerical model of Beard and Chuang (1987) is,

$$\frac{b}{a} = 1.0048 + 5.7 \times 10^{-4}(D) - 2.628 \times 10^{-2}(D^2) + 3.682 \times 10^{-3}(D^3)$$
$$- 1.677 \times 10^{-4}(D^4); \quad 0 \le D \le 7 \text{ mm} \qquad (7.3)$$

The above two equations are compared in Fig. 7.3 and are seen to be in good agreement for $D \ge 4$ mm. For drops with $D \le 4$ mm, the linear fit in (7.2) gives slightly more oblate axis ratios as compared with the Beard–Chuang model (this point will be discussed further in Section 7.1.2).

Large raindrops have been observed with airborne particle imaging sensors, with the largest ones detected being around 8 mm in Hawaiian rainbands (Beard et al. 1986) and off the Florida coast (Baumgardner and Colpitt 1995). In calculations involving Z_{dr} and K_{dp} the maximum drop diameters used in the literature have ranged between 6 and 10 mm, with 8 mm being commonly used.

In the absence of wind shear and turbulence, raindrops are expected to fall with their symmetry axes vertical. In the presence of uniform wind shear, the mean canting angle could be non-zero (Brussaard 1976). The rotating linear polarization data in heavy rain in Fig. 3.16a clearly shows that the mean canting angle is close to 0°. The large swing in the cross-polar power implies a high degree of orientation of drops with standard deviation of the canting angle estimated to be 6° assuming a Gaussian model. On theoretical grounds, Beard and Jameson (1983) showed that the Gaussian canting angle model is valid with a mean $\approx$0° and standard deviation $\le$5°. In most calculations involving Z_{dr} and K_{dp} it is customary to assume that drops are in near perfect alignment, having a mean canting angle of 0°, which is accurate for narrow distributions (Holt 1984). However, this model is not sufficient for calculations involving the linear

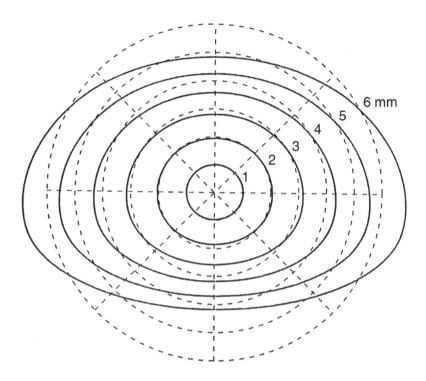

Fig. 7.1. Equilibrium drop shapes for drop diameters of 1–6 mm. From Beard and Chuang (1987).

depolarization ratio, and a finite standard deviation, typically between 5 and 10°, is sufficient to explain measurements in most rain types, as will be shown later. The early circularly polarized observations of Hendry et al. (1976) also demonstrated the high degree of orientation of raindrops about a mean canting angle near 0°.

Assume a perfectly aligned, oblate raindrop model and let the radar elevation angle be near 0° (horizontal incidence as illustrated in Fig. 2.8b). From (2.54), the differential reflectivity (in dB) of a single drop of size D and axis ratio $r = b/a$ is,

$$Z_{dr} = 10\log_{10}\frac{|S_{hh}|^2}{|S_{vv}|^2} \tag{7.4a}$$

or

$$\mathfrak{z}_{dr} = 10^{0.1(Z_{dr})} = \frac{|S_{hh}(r, D)|^2}{|S_{vv}(r, D)|^2} \tag{7.4b}$$

where the functional dependencies of the scattering amplitude on the size and axis ratio are explicitly shown. For a collection of identical drops of n per unit volume, $\mathfrak{z}_{dr}$ is

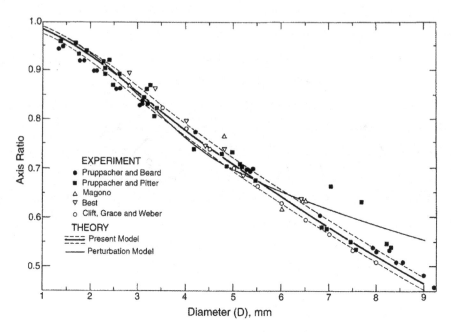

Fig. 7.2. Experimental and model results for axis ratio (b/a) versus D. The dashed lines are the upper and lower bounds for the Beard–Chuang model. From Beard and Chuang (1987).

given (see (3.168)) as,

$$\mathcal{Z}_{dr} = \frac{n|S_{hh}(r, D)|^2}{n|S_{vv}(r, D)|^2} \tag{7.5a}$$

and in the Rayleigh–Gans limit, the scattering amplitudes can be expressed as,

$$S_{hh}(r, D) = \frac{k_0^2}{4\pi} \frac{V(\varepsilon_r - 1)}{\left[1 + \frac{1}{2}(1 - \lambda_z)(\varepsilon_r - 1)\right]} \tag{7.5b}$$

$$S_{vv}(r, D) = \frac{k_0^2}{4\pi} \frac{V(\varepsilon_r - 1)}{[1 + \lambda_z(\varepsilon_r - 1)]} \tag{7.5c}$$

where V is the drop volume and the depolarizing factor λ_z defined in (A1.22b) is repeated here,

$$\lambda_z = \frac{1 + f^2}{f^2}\left(1 - \frac{1}{f}\tan^{-1}f\right); \quad f^2 = \frac{1}{r^2} - 1 \tag{7.5d}$$

Note that λ_z is only a function of the axis ratio r. This separation of the drop volume and the axis ratio function as a product in the expression for the scattering amplitudes is a direct consequence of Rayleigh–Gans scattering. Note that λ_z is implicitly a function of drop diameter via the equilibrium shape relation in (7.1–7.3).

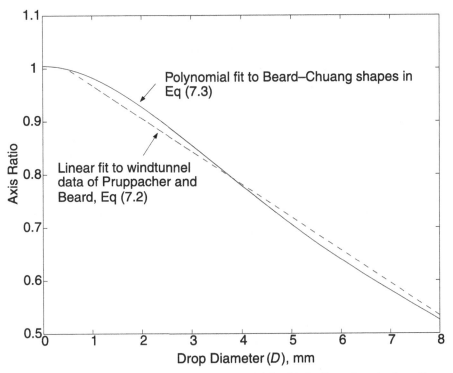

Fig. 7.3. Comparison of the Beard–Chuang polynomial fit with the linear fit to the data of Pruppacher and Beard (1970).

The specific differential phase of a collection of identical drops with size D and axis ratio r is given in (4.8),

$$K_{dp} = \frac{2\pi n}{k_0} \, \mathrm{Re}[\hat{h} \cdot \vec{f}(r, D) - \hat{v} \cdot \vec{f}(r, D)] \tag{7.6}$$

where n is the number of drops per unit volume and $\vec{f}$ is the forward scattering amplitude. Note that Re stands for the real part of a complex number. For Rayleigh–Gans scattering, (7.6) reduces to (see also (4.10)),

$$K_{dp} = \frac{n k_0}{2} V \, \mathrm{Re}\left[\frac{(\varepsilon_r - 1)}{1 + \frac{1}{2}(1 - \lambda_z)(\varepsilon_r - 1)} - \frac{(\varepsilon_r - 1)}{1 + \lambda_z(\varepsilon_r - 1)} \right] \tag{7.7}$$

and again the drop volume and axis ratio functional dependence are separable. It should be clear that for a collection of identical drops (or a monodisperse distribution) the differential reflectivity ($\mathfrak{z}_{dr}$) being a ratio quantity is independent of n, and also the drop volume V cancels out, see (7.5b, c). In contrast, K_{dp} is a product of both n and the drop volume. As a consequence, for a collection of identical drops whose axis ratio is close to 1 (e.g. $D \le 1.5$ mm), the differential reflectivity $\mathfrak{z}_{dr} \to 1$, whereas K_{dp} can be quite large (depending on n) even if the difference term in square brackets in (7.7) becomes small.

For a given drop size distribution $N(D)$, it follows from (3.161) and (4.39) that,

$$n|S_{hh}(r, D)|^2 \rightarrow \int N(D)|S_{hh}(r, D)|^2 \, dD \qquad (7.8a)$$

$$n|S_{vv}(r, D)|^2 \rightarrow \int N(D)|S_{vv}(r, D)|^2 \, dD \qquad (7.8b)$$

$$n \, \mathrm{Re}[\hat{h}\cdot\vec{f}(r, D) - \hat{v}\cdot\vec{f}(r, D)] \rightarrow \int N(D) \, \mathrm{Re}[\hat{h}\cdot\vec{f}(r, D) - \vec{v}\cdot\vec{f}(r, D)] \, dD$$

$$\qquad (7.8c)$$

Substituting Rayleigh limit expressions for the scattering amplitudes and writing the drop volume as $(\pi/6)D^3$ results in,

$$\mathcal{Z}_{dr} = \frac{\displaystyle\int |S_{hh}(r, D)|^2 N(D) \, dD}{\displaystyle\int |S_{vv}(r, D)|^2 N(D) \, dD} \qquad (7.9a)$$

$$= \frac{\displaystyle\int D^6 N(D) \left| (\varepsilon_r - 1)/[1 + \tfrac{1}{2}(1 - \lambda_z)(\varepsilon_r - 1)] \right|^2 dD}{\displaystyle\int D^6 N(D) |(\varepsilon_r - 1)/[1 + \lambda_z(\varepsilon_r - 1)]|^2 \, dD} \qquad (7.9b)$$

$$K_{dp} = \frac{2\pi}{k_0} \, \mathrm{Re} \int N(D) \left[\hat{h}\cdot\vec{f}(r, D) - \vec{v}\cdot\vec{f}(r, D) \right] dD \qquad (7.10a)$$

$$= \frac{\pi k_0}{12} \, \mathrm{Re} \left\{ \int D^3 N(D) \left[\frac{\varepsilon_r - 1}{1 + \tfrac{1}{2}(1 - \lambda_z)(\varepsilon_r - 1)} - \frac{\varepsilon_r - 1}{1 + \lambda_z(\varepsilon_r - 1)} \right] dD \right\}$$

$$\qquad (7.10b)$$

Note the weighting of the drop size distribution by D^6 in the two integrals in (7.9) and by D^3 in the integral involving K_{dp}. It follows that if λ_z is constant (e.g. all drops have the same axis ratio), the differential reflectivity ($\mathcal{Z}_{dr}$) is independent of the actual form of $N(D)$ and is dependent only on the axis ratio. In contrast, K_{dp} is proportional to the product of the third moment of $N(D)$ and a function of the axis ratio only. If the drops have equilibrium shapes with axis ratio a function of D (e.g. (7.2)), then λ_z is an implicit function of D in the integrals above, and both $\mathcal{Z}_{dr}$ and K_{dp} become functions of $N(D)$ though to different extents.

A commonly assumed form for $N(D)$ is the exponential, $N(D) = N_0 \exp(-\Lambda D)$, and it is obvious that if this particular form is used in (7.9), then the N_0 parameter cancels out in the ratio while it does not in (7.10). Also, if a particular equilibrium axis ratio versus D relation is assumed (e.g. (7.2)) then $\mathcal{Z}_{dr}$ is a function of Λ (the slope of the exponential dsd) only, whereas K_{dp} can be expressed as a product of N_0 with a function of Λ. This dependence of $\mathcal{Z}_{dr}$ on Λ, and of K_{dp}/N_0 on Λ was first shown by Seliga and Bringi (1976; 1978). The exponential form for $N(D)$ is generally applicable if a large number of drop size distributions are averaged in space or time. An example is shown

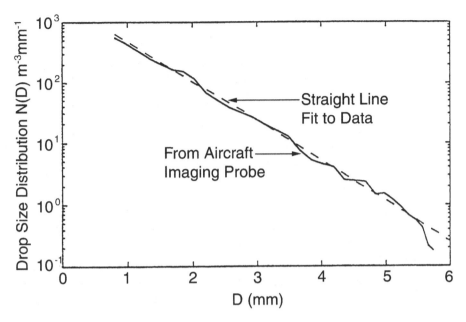

Fig. 7.4. Averaged drop size distribution from aircraft imaging probe in Florida rain cells. The straight line fit is $N(D) = 2155 \exp(-1.505D)$. The sample volume is 62 140 liters. From Bringi et al. (1998, © 1998 IEEE).

in Fig. 7.4 based on aircraft observations in Florida storm cells and the straight line fit gives $N_0 = 2155$ m^{-3} mm^{-1} and $\Lambda = 1.505$ mm^{-1} in this case (the classic Marshall and Palmer (1948) dsd has N_0 fixed at 8000 m^{-3} mm^{-1}).

The slope of an exponential dsd can be related to the median volume diameter (D_0) of the distribution. The D_0 is defined such that drops less than D_0 contribute to half the total rainwater content (W), or,

$$\frac{\pi}{6}\rho_w \int_0^{D_0} D^3 N(D)\,dD = \frac{1}{2}\frac{\pi}{6}\rho_w \int_0^{\infty} D^3 N(D)\,dD = \frac{1}{2}(W) \qquad (7.11)$$

For an exponential dsd it turns out that $\Lambda D_0 = 3.67$. Hence, Z_{dr} can be expressed as a function of the median volume diameter (D_0), and it was this relation that was expressed in graphical form by Seliga and Bringi (1976). This microphysical link between a radar observable and a characteristic diameter of the raindrop size distribution is indeed an important one and has been extensively studied. Careful intercomparisons between radar measurement of Z_{dr} and D_0 derived from surface drop size meters (or disdrometers) have shown that D_0 can be estimated from Z_{dr} with normalized standard errors of typically 10–15% (Goddard and Cherry 1984; Aydin et al. 1987; see also, Fig. 7.6 later).

Because the exponential dsd form is not representative of "instantaneous" drop size distributions, the more general gamma form[3] is used by Ulbrich (1983) to study natural variations of the dsd,

$$N(D) = N_0 D^\mu \exp(-\Lambda D); \quad \mu > -1 \tag{7.12}$$

where $\Lambda D_0 = 3.67 + \mu$. Note that this form differs from the classical gamma form defined later in (7.27). When substituted in (7.9), it is clear that ζ_{dr} becomes a function of Λ and μ. The parameter μ controls the shape of the dsd. The mass-weighted mean diameter (D_m) of the dsd is defined as,

$$D_m = \frac{\int D^4 N(D)\, dD}{\int D^3 N(D)\, dD} \tag{7.13}$$

For $N(D)$ of the form in (7.12), it is easy to show that $\Lambda D_m = 4 + \mu$; thus D_m is close to D_0. It is useful to relate the Rayleigh–Gans limit Z_{dr} to D_m (or D_0) for the family of gamma distributions. Since D_m is dependent on both Λ and μ, a mean fit can be arrived at relating ζ_{dr} and D_m assuming $-1 < \mu \le 5$, $0.5 \le D_0 \le 2.5$ mm and N_0 chosen to be consistent with thunderstorm rainfall rates. In addition, the Beard–Chuang equilibrium shape relation (7.3) is used to arrive at the following power-law fits,

$$D_m = 1.619(Z_{dr})^{0.485}; \quad \text{mm} \tag{7.14a}$$
$$D_0 = 1.529(Z_{dr})^{0.467}; \quad \text{mm} \tag{7.14b}$$

Note that Z_{dr} is in decibels in the above fits and that these fits are valid at S-band (frequency near 3 GHz). A careful intercomparison of radar measured Z_{dr} with D_m calculated from drop size distributions measured by an airborne particle imaging sensor in thunderstorm rain cells is shown in Fig. 7.5 along with (7.14a). Details of this intercomparison can be found in Bringi et al. (1998). Note the excellent agreement between radar measurements and simulations. The bias between D_m-radar (via radar measured Z_{dr} and the power-law fit) and the D_m from the measured dsds is only 0.1 mm, while the normalized standard error (nse) is 11.1%. This nse is consistent with theoretical evaluation of the standard error in the estimate of D_m from simulations which include error due to both dsd fluctuations as well as radar measurement error in Z_{dr}. Figure 7.6 shows the expected nse versus D_m based on simulations, together with the experimentally derived nse of 11.1% at the average measured D_m of 2.5 mm (note that Fig. 7.4 is the average dsd from all the penetrations). The single mean data point is in good agreement with the simulation results (the measurement error in Z_{dr} in the simulations is assumed to be ±0.2 dB), and is a remarkable result considering the various other error sources that can affect radar measurements, and the large differences in resolution volumes sampled by the radar and the airborne sensor. The data of Fig. 7.6 also show that the accuracy of radar Z_{dr} measurements should be very high (between 0.1 and 0.2 dB) for application to retrieval of D_m (and also, for rain rate estimation, see Section 6.5.1 for the variance of Z_{dr} estimates, and Section 6.3.2 under radar calibration). The airborne/radar results shown here are consistent with earlier intercomparisons using surface disdrometers.

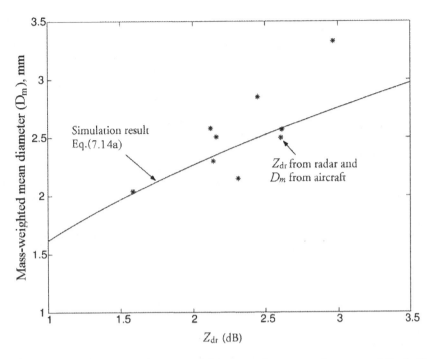

Fig. 7.5. Comparison of D_m from aircraft imaging probe versus radar measured Z_{dr} in Florida rain cells. The Z_{dr} data are from the NCAR/CP-2 radar during the convective and precipitation/electrification experiment (CaPE). The aircraft is the University of Wyoming King Air.

Consider now (7.10), which defines the specific differential phase, K_{dp}. Confining to Rayleigh–Gans scattering, observe that the integral can be substantially simplified by recognizing that,

$$\mathrm{Re}\left[\frac{\varepsilon_r - 1}{1 + \frac{1}{2}(1 - \lambda_z)(\varepsilon_r - 1)} - \frac{\varepsilon_r - 1}{1 + \lambda_z(\varepsilon_r - 1)} \right] \approx C(1 - r) \tag{7.15}$$

where C is approximately constant (varying between 3.3 and 4.2 with r from 1 to 0.5). This range for C is valid for ε_r of water at microwave frequencies in the range 3–30 GHz. Substituting in (7.10) results in,

$$K_{dp} = \frac{\pi k_0 C}{12} \int D^3 (1 - r) N(D) \, dD \tag{7.16a}$$

$$= \left(\frac{\pi}{\lambda}\right) \frac{C}{\rho_w} \int \frac{\pi}{6} \rho_w D^3 (1 - r) N(D) \, dD \tag{7.16b}$$

$$= \left(\frac{\pi}{\lambda}\right) C \left(\frac{W}{\rho_w}\right) \left[1 - \frac{\int r D^3 N(D) dD}{\int D^3 N(D) \, dD} \right] \tag{7.16c}$$

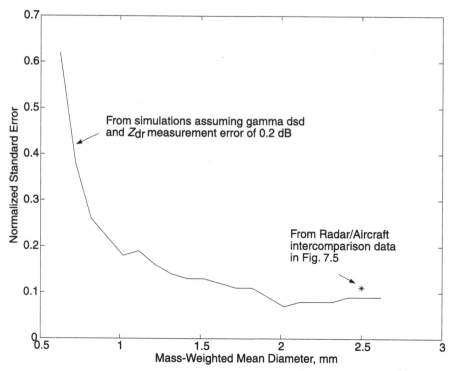

Fig. 7.6. Comparison of the theoretically expected normalized standard error in D_m with the single mean data point from radar/aircraft intercomparisons shown in Fig. 7.5.

where the rainwater content (W) is defined in (7.11) and ρ_w is the water density. Note that λ, above, is the wavelength and should not be confused with the depolarizing factor λ_z. The ratio of integrals in (7.16c) can be defined as a mass-weighted mean axis ratio $(\bar{r}_m)$. In terms of conventional units for W in g m^{-3}, $\rho_w = 1$ g cm^{-3} and λ in meters, the K_{dp} in units of $^\circ$ km^{-1} can be written as,

$$K_{dp} = \left(\frac{180}{\lambda}\right) 10^{-3} C W (1 - \bar{r}_m); \quad ^\circ \text{ km}^{-1} \tag{7.17}$$

where $C \approx 3.75$ is both dimensionless and independent of wavelength. The above equation is an important result which links the specific differential phase with the rainwater content multiplied by the deviation of the mass-weighted mean axis ratio from unity (Jameson 1985). As alluded to earlier, even if $1 - \bar{r}_m$ equals 0.05, for example, due to a monodisperse distribution of approximately 1.5-mm drops, and if the rainwater content is sufficiently high, then K_{dp} can rise to measurable values especially at shorter wavelengths. This is a manifestation of coherent addition of waves in the forward scattered direction (see also, Section 1.10). If the equilibrium axis ratio versus D relation of (7.2) is used in (7.16c), then the K_{dp} is nearly linearly related to the product of W and the mass-weighted mean diameter (D_m; see (7.13)) independent of

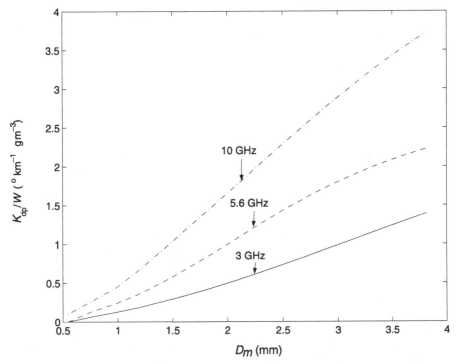

Fig. 7.7. Calculations of K_{dp}/W versus D_m for exponential dsds using the T-matrix method (see Appendix 3). The Beard–Chuang equilibrium shapes were used with maximum drop diameter of 8 mm. The canting angle distribution is Gaussian with zero mean and standard deviation of $7°$.

the form of $N(D)$,

$$K_{dp} \approx \left(\frac{180}{\lambda}\right) 10^{-3} C W (0.062 D_m); \quad °\,km^{-1} \tag{7.18}$$

The units of D_m are in mm in the above equation, while W is in g m^{-3} and λ in meters. While (7.17) was derived here in the Rayleigh–Gans limit, the general form is valid up to at least 13 GHz. Figure 7.7 shows calculations of (K_{dp}/W) versus D_m at 3, 5.6, and 10 GHz for an exponential dsd using the Beard and Chuang shapes with maximum drop diameters of 8 mm. The T-matrix method (see Appendix 3) is used, since the Rayleigh–Gans limit no longer applies at 5.6 GHz and higher frequencies for the larger drops (≥ 5 mm). The linearity implied by (7.18) is quite remarkable, especially at 10 GHz. Also, the theoretical slope of K_{dp}/W versus D_m implied by (7.18) is in relatively good agreement with the result of calculations in Fig. 7.7.

It is also possible to link the differential reflectivity $\mathcal{Z}_{dr}$ in the Rayleigh–Gans limit to a reflectivity-factor weighted mean axis ratio by observing that (7.9) may be simplified using the approximation,

$$\frac{|1 + \lambda_z(\varepsilon_r - 1)|^2}{|1 + \frac{1}{2}(1 - \lambda_z)(\varepsilon_r - 1)|^2} (r^{7/3}) \approx 1 \tag{7.19}$$

where the left-hand side varies between 0.99 and 0.94 for r from 1 to 0.5. It follows that the inverse of $\mathcal{z}_{dr}$ may be expressed as,

$$(\mathcal{z}_{dr})^{-1} = \frac{\int r^{7/3} D^6 N(D) \left| (\varepsilon_r - 1)/[1 + \frac{1}{2}(1 - \lambda_z)(\varepsilon_r - 1)] \right|^2 dD}{\int D^6 N(D) \left| (\varepsilon_r - 1)/[1 + \frac{1}{2}(1 - \lambda_z)(\varepsilon_r - 1)] \right|^2 dD} \tag{7.20a}$$

$$\approx \frac{\int r^{7/3} D^6 N(D)\, dD}{\int D^6 N(D)\, dD} \tag{7.20b}$$

$$= \overline{r_z^{7/3}} \tag{7.20c}$$

where $\overline{r_z^{7/3}}$ may be defined as the reflectivity-factor weighted mean of $r^{7/3}$. The reflectivity factor Z was defined in (3.166) for spherical drops. Numerical integration shows that,

$$Z = \int D^6 N(D)\, dD \approx \int D^6 N(D) \left| \frac{\varepsilon_r - 1}{1 + \frac{1}{2}(1 - \lambda_z)(\varepsilon_r - 1)} \right|^2 dD \tag{7.21}$$

where the right-hand side integral, in essence, defines the reflectivity factor for oblate drops at horizontal polarization. Equation (7.20c) is useful in illustrating the microphysical link between $(\mathcal{z}_{dr})^{-1}$ and the reflectivity-factor weighted mean axis ratio, $\bar{r}_z$, defined as,

$$\bar{r}_z = \frac{\int r D^6 N(D)\, dD}{\int D^6 N(D)\, dD} \tag{7.22}$$

It follows that for narrow distributions of axis ratio, $\bar{r}_z \approx (\mathcal{z}_{dr})^{-3/7} = [10^{-0.1(Z_{dr})}]^{3/7}$, as derived by Jameson (1983), where $Z_{dr} = 10 \log_{10}(\mathcal{z}_{dr})$. This relation is useful in estimating the sensitivity of Z_{dr} relative to different drop shape versus size relations; in particular, when drops are oscillating in non-steady flow. It is also useful in deducing that the Z_{dr} due a mixture of oblate raindrops and spherical hailstones will tend to 0 dB since $\bar{r}_z$ for the mixture will tend to be dominated by the D^6-weighting of the larger hailstones (whose axis ratio is close to 1) (for example, this may be visualized quite readily by referring to Fig. 2.8b). In contrast, it is easy to see from (7.16b) that spherical hailstones do not contribute to K_{dp} at all and, therefore, K_{dp} is only sensitive to the oriented oblate raindrops (see also the discussion related to (4.52)).

7.1.2   Raindrop oscillations: implications for Z_{dr} and K_{dp}

The actual shapes of raindrops in unsteady flow are expected to differ from the equilibrium shapes assumed in the previous section. Based on field studies of Tokay and Beard (1996) there is clear evidence that raindrops from 1 to 4 mm oscillate, but the cause of oscillation and the resultant distribution of oscillation amplitudes, especially for the larger drops, is still uncertain. Laboratory studies of the axis ratios of artificially generated water drops (Kubesh and Beard 1993), which have fallen a sufficient distance for viscous damping of the initial oscillation due to the drop generator to be completed, show that raindrops tend to oscillate in the steady state in two preferred modes of the fundamental harmonic: (a) the axisymmetric mode oscillations between oblate–prolate shapes, which produce two-sided scatter in the axis ratio about the equilibrium value; and (b) (asymmetric) transverse mode oscillations, which produce mostly one-sided scatter about the equilibrium value. Three-dimensional drop shapes for these two modes are depicted in Beard (1984) and are discussed in Beard and Kubesh (1991). These modes are composed of spherical harmonic perturbations of the surface, the particular modes being $n = 2, m = 0$ for the axisymmetric oblate–prolate variation; and $n = 2, m = 1$ for the (asymmetric) transverse mode (spherical harmonics are defined in Appendix 2 in another context). Beard and Kubesh (1991) show that the time-varying axis ratio for these two modes can be expressed as,

$$\left[\frac{b(t)}{a(t)}\right]_{n=2,m=0} = \frac{a_0 + A \sin \omega t}{a_0 - 0.5A \sin \omega t} \tag{7.23a}$$

$$\left[\frac{b(t)}{a(t)}\right]_{n=2,m=1} = \frac{a_0 + A'|\sin \omega t|}{a_0 + 0.5A'|\sin \omega t|} \tag{7.23b}$$

where a_0 is the undistorted drop radius; A, A' are oscillation amplitudes; and ω is the oscillation frequency (varying from a few hundred Hz for $D \leq 2$ mm to a few tens of Hz for $D \geq 5$ mm; Pruppacher and Klett 1997). The axis ratio in (7.23a) varies nearly symmetrically about unity with ωt, whereas in (7.30b) the axis ratio exceeds unity for all ωt. In practice, these oscillations should occur about the equilibrium axis ratio, and thus the axisymmetric mode should exhibit two-sided scatter about the equilibrium axis ratio, while the transverse mode should exhibit one-sided scatter. Note, however, that if the oscillation amplitudes are fixed, the expected probability distribution of axis ratios should have a maximum near the extremes. For the axisymmetric mode, the distribution should be "saddle-shaped" about the mean, while for the transverse mode the distribution should be shaped like a half "saddle" because of the one-sided scatter.

Figure 7.8 shows laboratory results of axis ratios for $D = 2.5, 2.9, 3.6$, and 4.0 mm from Andsager et al. (1999) with axisymmetric mode oscillations inferred for $D = 2.5$ and 3.6 mm (two-sided scatter about the equilibrium axis ratio). Transverse mode oscillations are inferred for $D = 2.9$ and 4.0 mm, because the scatter is mostly (but not entirely) one-sided above the shaded region. Note that the axis ratio distributions are nearly symmetric about their mean and are not "saddle-shaped" as expected for harmonic oscillation. A simple explanation for the observed distributions is that they were obtained from drops oscillating with a range of amplitudes; with most drops experiencing only small-amplitude oscillations, while a relatively low fraction of drops were oscillating with amplitudes near the maximum. The axis ratio distribution of

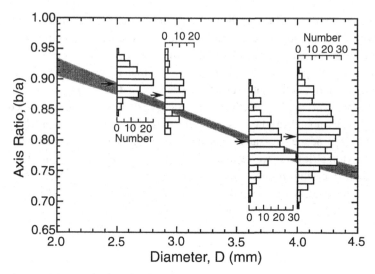

Fig. 7.8. Measured axis ratios for $D = 2.5, 2.9, 3.6$, and 4 mm. Histograms of axis ratio are shown, with the bold arrow indicating the mean value. The shaded region covers the axis ratio range in Fig. 7.2. From Andsager et al. (1999).

natural raindrops supports this viewpoint, as shown in Fig. 7.9, which is taken from Bringi et al. (1998). Over 3500 drop images with $D \geq 2$ mm from an airborne particle imaging sensor were carefully analyzed using techniques developed by Chandrasekar et al. (1988) to estimate the axis ratio (these same data were used in Figs. 7.5 and 7.6, described earlier). Figure 7.9 shows a histogram of $(b/a)/\langle b/a \rangle$ for all drops with $D \geq 2$ mm. The measured axis ratio was divided by the mean axis ratio (represented by angle brackets) in the corresponding size interval to obtain the histogram. These data (whose average $N(D)$ is shown in Fig. 7.4) were obtained around a few hundred meters above the surface in Florida rainshafts with rain rates between 10 and 60 mm h^{-1}. The mode of the distribution is clearly near unity, and the shape is remarkably similar to that obtained by Chandrasekar et al. (1988) from summer rainshowers in the High Plains of the USA in conditions of low to moderate rain rates (1–15 mm h^{-1}).

Andsager et al. (1999) have combined their laboratory axis ratio results with the earlier small-to-moderate diameter drop results of Beard et al. (1991) and Kubesh and Beard (1993), as illustrated in Fig. 7.10. This figure also includes the natural raindrop shape results of Chandrasekar et al. (1988) for $D \leq 4$ mm. There is a systematic upward shift in the mean axis ratio relative to the equilibrium shape (shaded) in the interval $1 \leq D \leq 4$ mm based on these composite results and this shift can be explained by transverse mode oscillations. Andsager et al. (1999) have provided a fit to all the laboratory data as well as to the Chandrasekar et al. (1988) data (total of 27 data points) to arrive at a mean axis ratio versus D relation,

$$\frac{b}{a} = 1.012 - 0.014\,45D - 0.010\,28(D^2); \quad 1 \leq D \leq 4.0 \text{ mm} \tag{7.24}$$

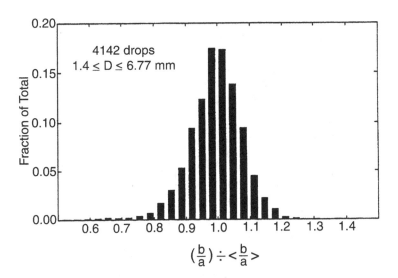

Fig. 7.9. Histogram of $(b/a)/\langle b/a \rangle$, where $\langle b/a \rangle$ is the mean value. The measured axis ratio is divided by the mean value in the corresponding diameter interval. Axis ratios were deduced from drop images recorded by an aircraft imaging probe. From Bringi et al. (1998, © 1998 IEEE).

where the standard derivation is 0.01. For $D < 1$ or $D > 4$ mm they suggest the use of the Beard–Chuang equilibrium shapes, see (7.3). The laboratory results actually show both axisymmetric and transverse oscillation modes in the steady state depending on drop size: for example, axisymmetric for 1.5, 2.5, and 3.6 mm; transverse for 1.1–1.3, 2, 2.5, 2.9, and 4 mm. The two different modes for the one common size of 2.5 mm (transverse mode obtained by Kubesh and Beard (1993)) and axisymmetric mode by Andsager et al. (1999)) was reconciled by appeal to different initial conditions since different drop generators were used. This was generalized by Andsager et al. (1999) to the hypothesis that oscillation modes of raindrops are sensitive to initial conditions (e.g. to off-center collisions) and that the predominant oscillation mode is expected to be transverse rather than axisymmetric. However, this hypothesis has not yet been validated by field data.

Since the differential reflectivity ($\mathfrak{z}_{dr}$) is a sensitive measure of the reflectivity-weighted mean axis ratio, see (7.22), $\mathfrak{z}_{dr} = (\bar{r}_z)^{-7/3}$, it follows that even a slight upward shift in the mean axis ratio for a given D will reduce $\mathfrak{z}_{dr}$ by measurable amounts (recall that a carefully calibrated radar can measure $10 \log_{10}(\mathfrak{z}_{dr})$ to an accuracy of ± 0.1 dB). Similarly, K_{dp} being proportional to $(1 - \bar{r}_m)$ will also be reduced by this small upward shift in mean axis ratios, see (7.17). Such a small upward shift was first introduced by Goddard et al. (1982) and by Goddard and Cherry (1984) to reconcile a small difference (a few tenths of a dB) in the radar-measured $\mathfrak{z}_{dr}$ versus $\mathfrak{z}_{dr}$ computed from disdrometer measurements of $N(D)$ and assuming the equilibrium axis ratio relation in (7.2). Their empirical axis ratio adjustment is also depicted in Fig. 7.10, which is remarkably in the

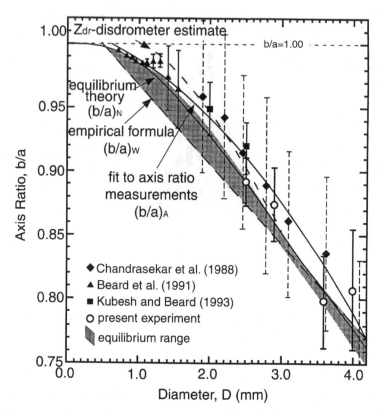

Fig. 7.10. Raindrop axis ratios versus drop diameter from various sources. Mean and standard deviation are shown for the aircraft imaging probe data (diamonds) of Chandrasekar et al. (1988), the laboratory measurements of (triangles) Beard et al. (1991), (squares) Kubesh and Beard (1993), and (open circles) Andsager et al. (1999). Also shown are the Beard–Chuang (1987) model result $(b/a)_N$, the linear fit to the data of Pruppacher and Beard (1970) $(b/a)_W$, and the polynomial fit to the data of Andsager et al. $(b/a)_A$. The dashed line marked as "Z_{dr}-disdrometer estimate" is based on the empirical adjustment of Goddard and Cherry (1984). The shaded region is the same as in Fig. 7.8. From Andsager et al. (1999).

same direction as the laboratory data and the data of Chandrasekar et al. (1988). These radar/disdrometer intercomparison data were obtained in very light stratiform rain and the measured Z_{dr} was generally less than 1–1.5 dB.

The only axis ratio results for larger drops ($D \geq 4$ mm) are based on the airborne drop image measurements of Chandrasekar et al. (1988) and Bringi et al. (1998) shown in Fig. 7.11. Drops larger than 4 mm were found to have mean axis ratios that were very close to the lower bound result of Beard and Chuang (1987) and to the empirical fit recommended by Clift et al. (1978) (see Fig. 7.2). It follows that the intercomparison between ζ_{dr} and D_m previously shown in Fig. 7.5 is consistent with equilibrium shapes, and recourse to transverse mode oscillations is not necessary to explain these data.

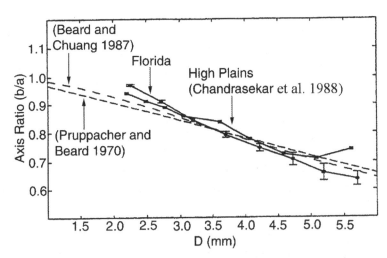

Fig. 7.11. Mean raindrop axis ratio versus D based on aircraft imaging probe data from Florida rain cells and from the high plains, USA. For the Florida data, both mean and 95% confidence interval are shown. For the high plains data only the mean is shown. From Bringi et al. (1998, © 1998 IEEE).

Because of the presence of large drops of equilibrium shape in these higher rain rate events (10–60 mm h^{-1}), the upward shift in Fig. 7.10 for $D \leq 3$ mm is unlikely to have caused the reflectivity-weighted mean axis ratio ($\bar{r}_z$) to have increased significantly enough to cause a major bias in the resulting intercomparison between D_m and $\bar{r}_{dr}$ in Fig. 7.5. Recall that the experimental bias (between D_m and D_m-derived from $\bar{r}_{dr}$ via (7.14a)) was only 0.1 mm. Any systematic upward shift in axis ratio over the entire range of observed D would have caused the bias to increase quite dramatically (of the order of 0.5 mm or more) and would be difficult to explain.

   The mechanism of drop oscillation in small drops (1–1.5 mm) is related to a resonance between the frequency at which eddies are shed in the drop's wake and the natural oscillation frequency of the drop (Gunn 1949), also referred to as resonance maintained intrinsically by vortex shedding (Beard et al. 1989). For larger drops (>2 mm) a large mismatch exists between the eddy shedding frequency and the natural oscillation frequency (Pruppacher and Klett 1997). Although aerodynamic feedback mechanisms have been proposed by Kubesh and Beard (1993) for the excitation of steady state oscillations in large drops, there appears to be no direct evidence at this time linking vortex shedding and oscillations. However, Tokay and Beard (1996) have evaluated extrinsic causes of oscillations such as, (i) collisions with tiny drops, and (ii) wind shear and turbulence. The details of their calculations are complicated, but a concise summary is given in Pruppacher and Klett (1997). In essence, their field data show that a significant number of oscillating drops were observed via photographic measurements at night in Illinois rain showers even though their theoretical estimate of the number of oscillating drops that would be detected in their camera sample volume due to collisions was an order of magnitude less. Thus, it appears that collisional forcing could not have been a major cause of drop oscillations under their experimental conditions (light to moderate rain rate). Tokay and Beard (1996) also concluded, based on theoretical analysis, that axis ratio changes produced by turbulence and wind shear would be negligible, and thus could not constitute an extrinsic forcing mechanism capable of sustaining drop oscillations against viscous damping. In very heavy rainfall with substantial concentrations of drizzle-sized drops (around 0.2–0.5 mm), the theoretical calculations of Beard and Johnson (1984), based on a balance between the excess kinetic energy resulting from drop collisions and viscous damping of the oscillation energy, showed that collisional-forcing can be a dominant mechanism for sustaining large drop oscillations. Such large drop oscillations were not evident in the axis ratio data from Florida (see Fig. 7.11) where the rain rates, though not intense, included several cases of 40–60 mm h^{-1}. It is possible that sub-cloud evaporation may have depleted the drizzle-sized drops thereby reducing the collision rate.

Illingworth and Caylor (1989) used radar measurements of reflectivity and differential reflectivity in heavy rain to conclude that the Beard–Chuang equilibrium shapes, see (7.3), provided the best fit to their radar data. Their result is shown in Fig. 7.12, along with a theoretical curve based on the gamma $N(D)$ together with the Beard–Chuang equilibrium axis ratios. Note that they calculated an average Z_{dr} for each 2-dB increment in Z_h between 30 and 60 dBZ (rain rate in the range 4–160 mm h^{-1}). The agreement between their averaged radar data and theory is quite remarkable when the Beard–Chuang equilibrium shapes, with a maximum drop diameter of 10 mm, are used. Any impact of sustained large drop oscillations due to collisional-forcing or maintained by aerodynamic feedback mechanisms would in fact cause the predicted Z_{dr} to be significantly less than the measured value for a given Z_h, especially for $Z_h > 50$ dBZ. In fact, these data were used by Illingworth and Caylor (1989) to show that the perturbation model (see Fig. 7.2) of Pruppacher and Pitter (1971) would cause the predicted Z_{dr} to decrease by 0.5 dB or more (for $Z_h > 55$ dBZ) relative to the averaged radar data. The use of an upward shift in axis ratios due to asymmetric oscillations would generate an even larger bias with respect to their data.

As mentioned earlier, K_{dp} is also sensitive to the mean axis ratio versus D relation since $K_{dp} \propto (1 - \bar{r}_m)$. For example, a simple linear relation of the form $b/a = 1 - \gamma D$, where the slope $\gamma = 0.062$ for equilibrium shapes (from (7.2)) whereas γ would be smaller to account for asymmetric oscillations, will serve to illustrate the sensitivity. From (7.17) and using the definition of $\bar{r}_m$ from (7.16c), $WD_m = C'(1/\gamma)K_{dp}$, where C' is a constant. Thus, any change in the slope of the mean axis ratio curve will directly affect estimates of WD_m from K_{dp} measurements. It will be shown in Chapter 8 that the still-air rain rate (R) can, on average, be linearly related to K_{dp}. This linear relation can be expressed as $R = C''(1/\gamma)K_{dp}$ and, hence, the coefficient C''/γ will depend on the shape of the mean axis ratio versus D relation. Several recent comparisons between radar-derived rain rates using K_{dp}, with rain gages, found a systematic underestimate in the radar-derived rain rates when the Pruppacher and Beard equilibrium shapes were used (Petersen et al. 1999; May et al. 1999). The May et al. (1999) comparisons used

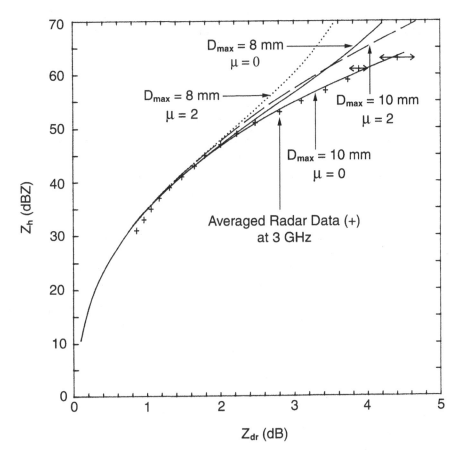

Fig. 7.12. Reflectivity at horizontal polarization (Z_h) versus Z_{dr}. The averaged radar data are from the S-band (near 3 GHz) Chilbolton radar, UK. The simulated curves are based on gamma drop size distributions, see (7.12), with the indicated assumption on the maximum drop diameter in the distribution. The Beard–Chuang axis ratio model is used with all drops being perfectly oriented with the mean canting angle of $0°$. From Illingworth and Caylor (1989).

K_{dp} data at 5.6 GHz in heavier convective rain events in the tropics (during the Maritime Continent Thunderstorm Experiment). The Petersen et al. (1999) study used K_{dp} at 3 GHz in a flash-flood-producing "tropical" storm event in Fort Collins, Colorado. One possible explanation for the underestimate is an upward shift in axis ratio due to transverse oscillations in tropical storms (in effect, increasing the coefficient, C''/γ, in the linear R–K_{dp} relation). A radar-based estimate of γ has been developed by Gorgucci et al. (2000) which is in agreement with this hypothesis.

Within the approximation of a linear axis ratio versus D relation of the form $r = 1 - \gamma D$, and for the gamma form of the drop size distribution it is easy to show that $W =$

$(C'/\gamma)K_{dp}/D_m$ from (7.18) is independent of γ if D_m is estimated from differential reflectivity (ζ_{dr}). Note first from (7.22) that the reflectivity-weighted mean axis ratio, $\bar{r}_z = 1 - \gamma D_z$, where D_z is the reflectivity-weighted mean diameter,

$$D_z = \frac{\int D^7 N(D)\, dD}{\int D^6 N(D)\, dD} \tag{7.25}$$

It follows from $\bar{r}_z \approx (\zeta_{dr})^{-3/7}$ that $D_z = (1/\gamma)(1 - \zeta_{dr}^{-3/7})$. Next, note that the mass-weighted and reflectivity-weighted mean diameters are simply related for a gamma size distribution. The gamma size distribution is written now in classical form,

$$N(D) = n_c f_D(D) \tag{7.26}$$

where n_c is the number concentration and $f_D(D)$ is the gamma pdf,

$$f_D(D) = \frac{\Lambda^\nu}{\Gamma(\nu)} D^{\nu-1} e^{-\Lambda D}; \quad \nu > 0 \tag{7.27}$$

The integral of $f_D(D)$ from 0 to ∞ is 1. The shape of the dsd is controlled by ν, with $\nu = 1$ corresponding to the exponential. The following formula will be useful,

$$\int_0^\infty D^p f_D(D)\, dD = \frac{\Gamma(\nu + p)}{\Gamma(\nu)} \frac{1}{\Lambda^p} \tag{7.28a}$$

from which,

$$D_m = \frac{\Gamma(\nu + 4)}{\Gamma(\nu + 3)} \frac{\Gamma(\nu + 6)}{\Gamma(\nu + 7)} D_z \tag{7.28b}$$

$$= f(\nu) D_z \tag{7.28c}$$

where for large ν, $f(\nu)$ varies slowly with ν, the shape of the gamma distribution. Substituting (7.28c) in $W = (C'/\gamma)K_{dp}/D_m$ gives,

$$W = \frac{C'}{\gamma} \frac{K_{dp}}{f(\nu) D_z} \tag{7.29a}$$

$$= \frac{C' K_{dp}}{\gamma f(\nu) \frac{1}{\gamma}(1 - \zeta_{dr}^{-3/7})} \tag{7.29b}$$

$$= \frac{C' K_{dp}}{f(\nu)(1 - \zeta_{dr}^{-3/7})} \tag{7.29c}$$

and it is clear that γ (the slope of the mean axis ratio versus D curve) cancels out, yielding an estimate for the rainwater content which ideally would be independent of small upward shifts of the axis ratio from equilibrium shapes. It is also reasonable to expect that rain rate algorithms expressed in terms of both K_{dp} and ζ_{dr} (to be discussed in Chapter 8) should be less sensitive than algorithms involving K_{dp} only. In essence, transverse oscillations would tend to increase the axis ratio towards sphericity over more or less the entire range of D, which will cause K_{dp} to decrease (relative to its value for equilibrium shapes) for a given $N(D)$. However, there is a compensating effect, since the term $(1 - \bar{z}_{dr}^{-3/7})$ also reduces keeping the ratio in (7.29c) approximately the same.

So far it has been implicitly assumed that only the mean axis ratio versus D relation of oscillating drops is important as far as $\bar{z}_{dr}$ and K_{dp} are concerned. If the axis ratio is treated as a particle state variable then (7.8) can be generalized, see also (3.161), as follows,

$$n|S_{hh}(r, D)|^2 \rightarrow \int N(D) \int_r |S_{hh}(r, D)|^2 p(r) \, dr \, dD \tag{7.30a}$$

$$n|S_{vv}(r, D)|^2 \rightarrow \int N(D) \int_r |S_{vv}(r, D)|^2 p(r) \, dr \, dD \tag{7.30b}$$

$$n \, \text{Re}[\hat{h}\cdot\vec{f}(r, D) - \hat{v}\cdot\vec{f}(r, D)] \rightarrow \int N(D) \int_r \text{Re}[\hat{h}\cdot\vec{f}(r, D) - \vec{v}\cdot\hat{f}(r, D)]p(r) \, dr \, dD \tag{7.30c}$$

where $p(r)$ is the pdf of the axis ratio. The Rayleigh limit approximations in (7.15) and (7.19) may be expressed more generally as,

$$\frac{1}{D^3} \text{Re}[\hat{h}\cdot\vec{f}(r, D) - \hat{v}\cdot\vec{f}(r, D)] \approx C(1 - r) \tag{7.31a}$$

$$\frac{|S_{hh}(r, D)|^2}{|S_{vv}(r, D)|^2}(r^{7/3}) \approx 1 \tag{7.31b}$$

Thus,

$$K_{dp} = \frac{2\pi C}{k_0} \int D^3 N(D) \int_r (1 - r)p(r) \, dr \, dD$$

$$= \frac{2\pi C}{k_0} \int D^3 N(D)[1 - E(r)] \, dD \tag{7.32a}$$

$$(\bar{z}_{dr})^{-1} = \frac{\int N(D) \int_r r^{7/3}|S_{hh}(r, D)|^2 p(r) \, dr \, dD}{\int N(D) \int_r |S_{hh}(r, D)|^2 p(r) \, dr \, dD} \tag{7.32b}$$

$$\approx \frac{\int D^6 N(D)E(r^{7/3}) \, dD}{\int D^6 N(D) \, dD} \tag{7.32c}$$

where E stands for expectation.

The previously defined mass-weighted mean axis ratio, see (7.16c), is now generalized as,

$$\bar{r}_m = \frac{\int D^3 E(r) N(D) \, dD}{\int D^3 N(D) \, dD} \tag{7.33}$$

so that K_{dp} depends on only the mean axis ratio versus D function and is independent of the variance of the axis ratio distribution. Similarly, the reflectivity-factor weighted mean of $r^{7/3}$ (previously defined in (7.20)) is now generalized as,

$$\overline{r_z^{7/3}} = \frac{\int D^6 E(r^{7/3}) N(D) \, dD}{\int D^6 N(D) \, dD} \tag{7.34}$$

Also, the reflectivity-factor weighted axis ratio ($\bar{r}_z$) previously defined in (7.22) is now generalized as,

$$\bar{r}_z = \frac{\int D^6 E(r) N(D) \, dD}{\int D^6 N(D) \, dD} \tag{7.35}$$

Jameson (1983) showed that, in practice, $\overline{r_z^{7/3}}$ would be close to $(\bar{r}_z)^{7/3}$, and thus it follows that $(\bar{z}_{dr})^{-1} = (\bar{r}_z)^{7/3}$ will not depend on the variance of the axis ratio distribution, at least for narrow distributions.

The issue of raindrop oscillations and their impact on the quantitative interpretation of K_{dp} and Z_{dr} is not entirely resolved, especially for drops ≥ 4 mm. There is now little question that drops up to 8 mm are found in convective rain cells. Field evidence is conclusive that drops between 1 and 4 mm do oscillate, and the fit in this size range proposed by Andsager et al. (1999) will account for the upward shift in the mean axis ratio (relative to equilibrium shapes) due to oscillations. For drops larger than around 4 mm, it appears that the Beard–Chuang equilibrium axis ratios should be used for now unless there is strong evidence of collisionally-forced oscillations as may occur in very high rain rate conditions with a large concentration of drizzle-sized drops (0.2–0.5 mm). In practice, algorithms for rain rate and rainwater content based on combining K_{dp} and Z_{dr} should be relatively insensitive to the precise form of the axis ratio versus D function, in addition to being insensitive to the actual form of $N(D)$.

7.1.3   Other polarimetric observables

While the most important quantitative polarimetric observables[4] for measuring rain characteristics are Z_{dr} and K_{dp}, the other linear polarization observables such as the linear depolarization ratio (LDR), and the copolar correlation coefficient magnitude and phase ($|\rho_{co}|$, δ_{co}), may at times, provide valuable qualitative information especially when rain is mixed with frozen precipitation (refer to Section 3.10 for a definition of these observables). Deviations from values normally expected in pure rain are often used to identify such mixed phase precipitation. Hence, it is useful to examine the range of values that may be expected in rain for LDR$_{vh}$, $|\rho_{co}|$, and δ_{co}, which, in turn, will depend on frequency (i.e. whether the Rayleigh–Gans limit applies or not).

It should be clear that perfectly oriented drops (also referred to as a constant canting model, the angle being normally assumed to equal 0°), or drops undergoing axisymmetric oscillations, cannot generate a cross-polar return for LDR$_{vh}$ (recall that the subscript "vh" stands for horizontal transmit and vertical receive). On the other hand, asymmetric oscillations as well as a canting angle pdf (usually assumed as Gaussian with $\bar{\beta} = 0$, σ_β) will generate cross-polar returns and finite LDR$_{vh}$. Enhanced LDR caused by multiple scattering effects at 35 GHz has been investigated by Ito et al. (1995). As in the previous sub-sections low radar elevation angles are assumed.

In the Rayleigh–Gans limit and using (2.89), (2.91), both with $\psi = 90°$, and (3.169), it follows that L (defined here for convenience via LDR$_{vh} = 10 \log_{10} L$) can be expressed as,

$$L = 10^{0.1(\text{LDR}_{vh})} \tag{7.36a}$$

$$= \frac{\langle n|S_{vh}|^2\rangle}{\langle n|S_{hh}|^2\rangle} \tag{7.36b}$$

$$= \frac{\langle n|\alpha_{z_b} - \alpha|^2 \sin^2\beta \cos^2\beta\rangle}{\langle n|\alpha \cos^2\beta + \alpha_{z_b} \sin^2\beta|^2\rangle} \tag{7.36c}$$

$$= \frac{\langle nV^2 \left|(\varepsilon_r - 1)/[1 + \lambda_z(\varepsilon_r - 1)] - (\varepsilon_r - 1)/[1 + \frac{1}{2}(1 - \lambda_z)(\varepsilon_r - 1)]\right|^2 \sin^2\beta \cos^2\beta\rangle}{\langle nV^2 \left|(\varepsilon_r - 1)\cos^2\beta/[1 + \frac{1}{2}(1 - \lambda_z)(\varepsilon_r - 1)] + (\varepsilon_r - 1)\sin^2\beta/[1 + \lambda_z(\varepsilon_r - 1)]\right|^2\rangle} \tag{7.36d}$$

The angle brackets denote ensemble averaging over $N(D)$, the canting angle distribution and the axis ratio distribution, see (3.161). The denominator of L is insensitive to averaging over the canting angle and axis ratio distributions, and can be approximated by the reflectivity factor Z, see (7.21). The numerator of L is proportional to the cross-polar reflectivity (or cross-polar scattered power) and it is sensitive to both canting angle and axis ratio distributions. To illustrate this sensitivity, first note that, similar to (7.15),

$$\left|\frac{\varepsilon_r - 1}{1 + \lambda_z(\varepsilon_r - 1)} - \frac{\varepsilon_r - 1}{1 + \frac{1}{2}(1 - \lambda_z)(\varepsilon_r - 1)}\right| \approx C(1 - r) \tag{7.37}$$

where C is nearly constant (varies between 3.3 and 4.2 from $r = 1$ to 0.5). Writing the drop volume as $V = (\pi/6)D^3$ and using (7.37), the numerator of L in (7.36b) can be expressed as,

$$\langle n|S_{vh}|^2\rangle = K \int D^6 N(D) \int (1 - r)^2 p(r)\, dr$$

$$\times \left[\int (1 - \cos 4\beta) p(\beta) \, d\beta \, dD \right] \tag{7.38a}$$

$$= KZ(1 - \rho_4)(1 - 2\bar{r}_z + \overline{r_z^2}) \tag{7.38b}$$

$$= KZ(1 - \rho_4)[(1 - \bar{r}_z)^2 + \text{var}(r_z)] \tag{7.38c}$$

where K is a constant, ρ_4 is an orientation parameter previously defined in (3.201b), $\bar{r}_z$ is defined in (7.35) and $\overline{r_z^2}$ can be obtained from (7.35) by substituting $E(r^2)$ for $E(r)$. For narrow Gaussian canting angle pdf, the orientation parameter $\rho_4 = \exp(-8\sigma_\beta^2)$. If the mean axis ratio versus D relation is expressed as $\bar{r}_z \approx 1 - \gamma D_z$ as before, then,

$$L = K'[1 - \exp(-8\sigma_\beta^2)][\gamma^2 D_z^2 + \text{var}(r_z)] \tag{7.39}$$

where K' is a constant. For small σ_β^2 as is expected in rain, $L \propto \sigma_\beta^2$, or in logarithmic scale LDR_{vh} will increase linearly as $20 \log(\sigma_\beta)$ for small σ_β. For example, as σ_β increases from 5 to $10°$, LDR_{vh} will increase by 6 dB. Recall from (6.132b) that the measurement accuracy in LDR_{vh} is expected to be typically ± 1.5 dB, and thus the 6 dB increase should be easily measured. L is also sensitive to the product $\gamma^2 D_z^2$. Recall from (7.28c) that $D_z = D_m / f(\nu)$, where ν is the shape parameter of the gamma dsd, and that for large ν, $f(\nu)$ varies slowly with ν. Thus, for the case of no oscillations, L is proportional to the square of the mass-weighted mean diameter, D_m^2. For example, as D_m increases from 1 to 2 mm, LDR_{vh} increases by 6 dB. The sensitivity to γ, the slope of the mean axis ratio versus D curve is not as much. For example, as γ decreases from its equilibrium shape ($\gamma = 0.062$) to 0.044, a value representative of the upward shift due to transverse oscillations, the LDR_{vh} will decrease by 3 dB. It has been assumed so far that $\text{var}(r_z)$ is negligible in estimating the sensitivity of L to D_m and γ. From Fig. 7.10, the range of $\text{var}(r_z)$ appears to be 0.0025–0.005 and its contribution relative to $\gamma^2 D_z^2$ may be comparable depending on the value of D_z. For example, for an exponential distribution for which $f(\nu) = 0.57$ and assuming $D_m = 1.0$ mm with $\gamma = 0.062$, the product $\gamma^2 D_z^2 = 0.011$. In contrast to $\mathcal{Z}_{dr}$ and K_{dp}, the linear depolarization ratio will depend on both the mean and variance of the reflectivity-factor weighted axis ratio.

The expression derived earlier in (3.234) gives a general relation between L, $\mathcal{Z}_{dr}$, ρ_4, and $|\rho_{co}|$ for Rayleigh–Gans scatterers. In rainfall, $|\rho_{co}| \to 1$ for equilibrium shaped drops, and for narrow Gaussian distributions (3.234) reduces to,

$$L \approx \frac{1}{4} \left[1 - \exp(-8\sigma_\beta^2) \right] [1 - (\mathcal{Z}_{dr})^{-1/2}]^2 \tag{7.40a}$$

$$= \frac{1}{4} \left[1 - \exp(-8\sigma_\beta^2) \right] [1 - (\bar{r}_z)^{7/6}]^2 \tag{7.40b}$$

$$\approx \frac{1}{4} [1 - \exp(-8\sigma_\beta^2)](\gamma^2 D_z^2) \tag{7.40c}$$

which is essentially the same as (7.39) for no oscillations (note also that $K' = 1/4$ in (7.39) by comparison with (7.40c)). The lower bound for L will occur in light

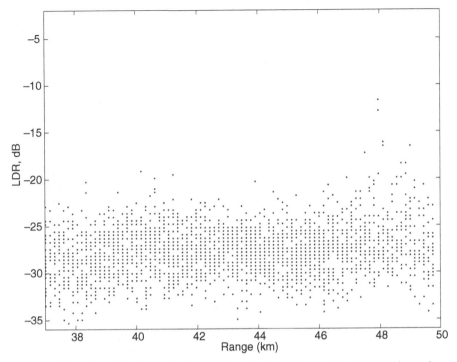

Fig. 7.13. Linear depolarization ratio data (LDR$_{vh}$; horizontal transmit/vertical receive) using the CSU–CHILL radar operating at 2.7 GHz. These data are from an unusual "tropical" rain event in northern Colorado. Data are from resolution volumes along the beam as the antenna scanned past the rain cell at low elevation angle.

rain rate conditions and will be limited by cross-polarization errors in the system (see Section 6.2) with high performance systems capable of a lower bound in the range of −35 dB. In heavy rain rates the maximum expected values for L may be estimated from (7.40c) assuming $\sigma_\beta \approx 10°$, $\gamma = 0.062$ for equilibrium shapes, and $D_z = 7/\Lambda$ for an exponential dsd. In heavy rain rate conditions, the slope (Λ) of the dsd for large drops tends to a constant value of around 2 mm^{-1} (or 20 cm^{-1}; Hu and Srivastava 1995), characteristic of equilibrium drop size distributions. If oscillations are neglected, (7.40c) yields a LDR$_{vh} = 10\log_{10}(L)$ of −26 dB which is expected to be an upper-bound value in heavy rainfall conditions at long wavelengths (S-band frequencies). If oscillations are assumed, then from (7.39) and using a high value for var(r_z) ≈ 0.01 will yield a value of −25 dB. Figure 7.13 shows measurements of LDR$_{vh}$ made with the CSU–CHILL radar, which operates at 2.7 GHz. These data are plotted as a function of range (or range profiles) from a number of "beams" as the antenna scans across the intense part of a rainshaft. The lower-bound value for LDR$_{vh}$ is seen to be around −33 to −34 dB.

Consider now the magnitude of the copolar correlation coefficient ($|\rho_{co}|$) for oblate drops (in the Rayleigh limit) which are perfectly aligned with zero canting angle but can be oscillating. From (3.231d) and setting $\delta_{co} = 0°$, and using (7.31b) results in,

$$|\rho_{co}| = \frac{|\langle n S_{hh} S_{vv} \rangle|}{(\langle n|S_{hh}|^2 \rangle \langle n|S_{vv}|^2 \rangle)^{1/2}} \tag{7.41a}$$

$$= \frac{\sqrt{\bar{z}_{dr}}}{Z_h} \int N(D) \int |S_{hh}|^2 r^{7/6} p(r) \, dr \, dD \tag{7.41b}$$

$$\approx \frac{\overline{r_z^{7/6}}}{(\overline{r_z^{7/3}})^{1/2}} \tag{7.41c}$$

where $\overline{r_z^{7/3}}$ is defined in (7.34) and $\overline{r_z^{7/6}}$ follows similarly. For narrow distributions in axis ratio an approximate expression for $|\rho_{co}|$ in rainfall may be obtained from (7.41c) as,

$$|\rho_{co}| \approx 1 - \frac{1}{2} \frac{\text{var}(r_z)}{(\bar{r}_z)^2} \tag{7.42}$$

For example, if $\text{var}(r_z) \approx 0.01$ and $\bar{r}_z = 0.8$, then $|\rho_{co}| \approx 0.992$.

Even if drops are not oscillating but are distributed in size, the equilibrium axis ratio versus D relation will cause $|\rho_{co}|$ to deviate from unity. From $r_z = 1 - \gamma D_z$ as used earlier, it follows that $\text{var}(r_z) = \gamma^2 \text{var}(D_z)$. Hence, for equilibrium-shaped drops, (7.42) reduces to,

$$|\rho_{co}| \approx 1 - \frac{1}{2} \frac{\gamma^2 \text{var}(D_z)}{(1 - \gamma D_z)^2} \tag{7.43}$$

For gamma dsd, (7.28c) can be used to relate D_m and D_z. Thus, in terms of the mass-weighted mean diameter (D_m), (7.43) may be expressed as,

$$|\rho_{co}| \approx 1 - \frac{1}{2} \frac{\gamma^2 [f(v)]^{-2} \text{var}(D_m)}{\{1 - \gamma[D_m/f(v)]\}^2} \tag{7.44}$$

where $f(v)$ is defined via (7.28b, c) and for large v is a slowly varying function of v, the shape parameter of the gamma dsd. Recall that γ is the slope of the mean axis ratio versus D relation ($\gamma = 0.062$ from (7.2)). For gamma dsd, $\text{var}(D_m)$ is,

$$\text{var}(D_m) = \frac{D_m^2}{(3 + v)} \tag{7.45}$$

Thus, (7.44) can be expressed as,

$$|\rho_{co}| \approx 1 - \frac{1}{2} \frac{\gamma^2 [f(v)]^{-2} (D_m^2)}{\{1 - \gamma[D_m/f(v)]\}^2 (v + 3)} \tag{7.46}$$

At high rain rates, D_m tends to a stable value around 2 mm (the equilibrium dsd referred to earlier). If the dsd shape is exponential ($v = 1$; $f(v) = 4/7$), and if $\gamma = 0.062$, then (7.46) yields $|\rho_{co}| \approx 0.99$ in the Rayleigh–Gans limit as a lower bound at high rain

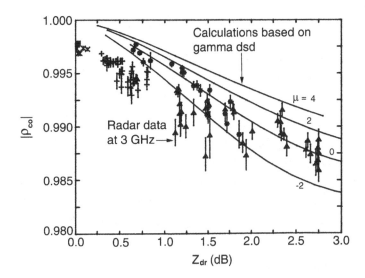

Fig. 7.14. The magnitude of the copolar correlation coefficient ($|\rho_{co}|$) versus Z_{dr} in rain using data from the S-band Chilbolton radar, UK. Both mean and standard error are shown. The gamma dsd is defined in (7.12). From Illingworth and Caylor (1991).

rates if the drops are not oscillating. If ν increases to 3 then for the same D_m and γ, $|\rho_{co}| \approx 0.995$.

The expression for $|\rho_{co}|$ can be generalized by combining the variance contribution due to drop oscillations together with the contribution from the size distribution assuming that they are uncorrelated,

$$|\rho_{co}| \approx 1 - \frac{1}{2} \frac{\left[\!\left[\mathrm{var}(r_z) + \gamma^2 D_m^2 / \{[f(\nu)]^2 (\nu + 3)\} \right]\!\right]}{\{1 - \gamma D_m / [f(\nu)]\}^2} \tag{7.47}$$

The axis ratio data of Chandrasekar et al. (1988) partly shown in Fig. 7.10 suggests that var(r_z) is nearly independent of drop diameter (for $D \geq 2$ mm) with a value around 0.005. For an exponential size distribution, the variance contribution from the size distribution equals 0.005 when $D_m = 1.3$ mm. For larger D_m values the size distribution effect dominates the drop oscillation component, and the deviation of $|\rho_{co}|$ from unity, or $1 - |\rho_{co}|$, is very sensitive to both D_m and ν. It must be emphasized that (7.47) is only an approximate equation since it is based on simplifying assumptions, see (7.41, 7.42). The main aim here is to show the functional dependence of $|\rho_{co}|$ on the size distribution and variance of axis ratios.

Figure 7.14 shows measurements of $|\rho_{co}|$ versus Z_{dr} at 3 GHz using the Chilbolton radar, UK. These data were acquired in rain using long dwell times (15 seconds at each 75 m range resolution volume). Rather than using the alternate polarization mode estimator in (6.90), the copolar power at the two polarization states was recorded for each transmitted pulse and at each range resolution volume (the "time-series" mode of data collection). A frequency domain interpolation scheme was used to calculate

$|\hat{\rho}_{hh,vv}[0]|$ rather than using (6.90); see also, Liu et al. (1994) for a description of the interpolation method. Figure 7.14 indicates that at 3 GHz (where raindrops are in the Rayleigh scattering limit), the lower bound for $|\rho_{co}|$ in rainfall is around 0.985.

Several factors limit the accuracy with which $|\rho_{co}|$ can be measured. These are: (i) signal fluctuations (see Figs. 6.28, 6.31); (ii) low signal-to-noise ratio, see (6.122); (iii) non-Gaussian spectra (see Liu et al. 1994); and (iv) antenna performance, see (6.26). Typically, the accuracy of $|\rho_{co}|$ using the alternate polarization mode, see (6.90), is around ± 0.01, which is not sufficient for quantitative application (e.g. to estimate ν). However, large amplitude drop oscillations with var(r_z) exceeding 0.02, for example, should be detectable.

At higher frequencies, Mie scattering effects from large drops will tend to be the dominant cause for $|\rho_{co}|$ to deviate from unity. Recall from (3.231d) that $|\rho_{co}|$ depends on $|\langle n S_{hh}^* S_{vv} \rangle|$, and assuming no drop oscillation or canting angle distribution, may be expressed as,

$$|\rho_{co}| = \frac{\sqrt{3_{dr}}}{Z_h} \left| \int N(D) S_{hh}^*(r, D) S_{vv}(r, D) \, dD \right| \tag{7.48a}$$

$$= \frac{\sqrt{3_{dr}}}{Z_h} \left| \int N(D) |S_{hh}(r, D) S_{vv}(r, D)| e^{j\{\delta_{vv}(r,D) - \delta_{hh}(r,D)\}} \, dD \right| \tag{7.48b}$$

where $\delta_{vv} - \delta_{hh} = \arg(S_{hh}^* S_{vv})$ is the differential scattering phase shift for a single drop of axis ratio r and diameter D, see also (4.65). Note that the functional dependence of $\delta_{vv} - \delta_{hh}$ on D is critical, as otherwise it can be brought outside the integral and, then, would not impact $|\rho_{co}|$. Figure 7.15a shows calculations of $|\rho_{co}|$ versus Z_{dr} at 5.6 GHz, while Fig. 7.15b shows δ_{co} versus Z_{dr}, for exponential size distributions, see also Scarchilli et al. (1993). Recall that $\delta_{co} = \arg\langle n S_{hh}^* S_{vv} \rangle$, is defined as,[5]

$$\delta_{co} = \arg \left[\int N(D) |S_{hh}(r, D) S_{vv}(r, D)| e^{j[\delta_{vv}(r,D) - \delta_{hh}(r,D)]} \, dD \right] \tag{7.49}$$

It is possible to estimate δ_{co} by examining range profiles of Ψ_{dp} (recall from (4.65) that $\Psi_{dp} = \Phi_{dp} + \delta_{co}$), where Φ_{dp} is the differential propagation phase which must either stay constant or increase monotonically with range. Any perturbation in Ψ_{dp} in excess of statistical fluctuations, which are superposed on the monotonically increasing component of Φ_{dp}, may be attributed to δ_{co} (see Section 6.6).

7.1.4   Raindrop size distributions and simulated radar parameters

The raindrop size distribution (or simply dsd) plays an important role in determining the radar parameters considered so far. The reflectivity factor (for spherical drops) is the sixth moment of the dsd, see (7.21). The differential reflectivity is related to the reflectivity-factor weighted mean diameter, see (7.25). The specific differential phase is proportional to the product of the rainwater content (third moment of the

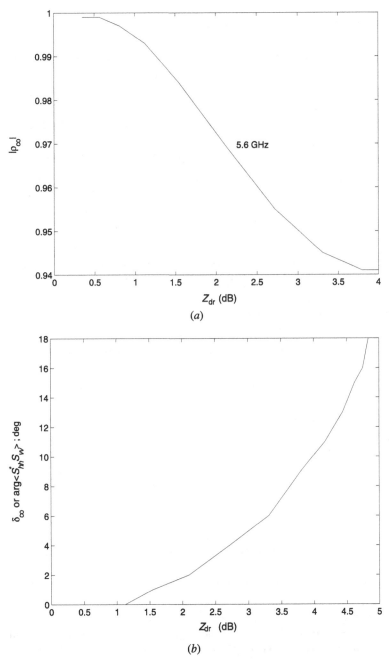

Fig. 7.15. Calculations at 5.6 GHz (C-band) using exponential drop size distributions and the T-matrix method: (a) $|\rho_{co}|$ versus Z_{dr}, and (b) δ_{co} versus Z_{dr}. Beard–Chuang equilibrium shapes are used with Gaussian canting angle distribution with the mean $= 0°$ and standard deviation of $7°$.

dsd) and the mass-weighted mean diameter, see (7.18). The linear depolarization ratio is proportional to the square of the reflectivity-factor weighted mean diameter (D_z), see (7.40c). The magnitude of the copolar correlation coefficient is related to both D_z and its variance, see (7.43). These radar parameters (and K_{dp} to a lesser extent) are strongly influenced by the concentration of large drops (or on the tail of the dsd), which is difficult to measure with drop size measuring devices such as disdrometers or airborne imaging probes due to their poor sampling volumes and the (typically) very low concentration of large drops in natural precipitation. There is a large body of experimental dsd data using such instruments and it is well-known that the dsd fluctuates considerably at small space and time scales (e.g. Ulbrich 1983; Ulbrich and Atlas 1998). From a radar perspective, it is the accurate parameterization of the tail of the dsd that is of importance. In contrast, from a microphysical perspective the drizzle and moderately sized drop concentrations control processes such as drop growth via coalescence, drop breakup, and evaporation (e.g. see Hu and Srivastava 1995). It is these processes that determine the shape of the dsd over the entire range of drop diameters (from 0.1 to 6–8 mm), and simple analytical forms such as the exponential (e.g. Marshall and Palmer 1948) or the gamma (Ulbrich 1983) are often excellent representations of the dsd, see (7.12).

The radar parameters, including K_{dp}, can essentially be expressed in terms of the rainwater content (W); the mass-weighted mean diameter (D_m), see (7.13); and the standard deviation of D_m (σ_m; also referred to as the width of the dsd). The ratio, σ_m/D_m, can be used as a measure of dsd shape, and for the gamma dsd (defined in (7.12)),

$$\frac{\sigma_m}{D_m} = \left[\frac{\int (D - D_m)^2 D^{3+\mu} e^{-\Lambda D} \, dD}{D_m^2 \int D^{3+\mu} e^{-\Lambda D} \, dD} \right]^{1/2} \tag{7.50a}$$

$$= \frac{1}{(4+\mu)^{1/2}} \tag{7.50b}$$

As mentioned earlier in relation to (7.12, 7.13), the mass-weighted mean diameter (D_m) is closely related to the median volume diameter (D_0) defined by (7.11). For a gamma dsd,

$$\Lambda D_0 = 3.67 + \mu \tag{7.51a}$$
$$\Lambda D_m = 4 + \mu \tag{7.51b}$$

where it is assumed that the size integration goes from 0 to ∞. Henceforth, this will be assumed unless otherwise stated.

Consider first the exponential dsd expressed as,

$$N(D) = N_0 \exp(-\Lambda D) \tag{7.52a}$$

$$= N_0 \exp\left(-3.67 \frac{D}{D_0}\right); \quad mm^{-1} \, m^{-3} \tag{7.52b}$$

This dsd is most often plotted on semi-logarithmic axes, and N_0 is termed the intercept parameter, while Λ is the slope parameter of the resulting straight line. Typically, N_0 is expressed in mm^{-1} m^{-3}, while Λ is in mm^{-1} (D_0 in mm). However, it is also common to express N_0 in cm^{-4} and Λ in cm^{-1}. It is well-known that when measured size distributions are substantially averaged in space and/or time, the exponential shape is a good approximation to the averaged dsd shape (e.g. Fig. 7.4). The form in (7.52b) suggests the concept of a scaling diameter (D/D_0) for the dsd. The concept of scaling $N(D)$ was introduced by Sekhon and Srivastava (1971) in order to compare the shapes of different distributions with widely varying rainwater contents (W). The rainwater content (g m^{-3}) is defined as,

$$W = 10^{-3}\frac{\pi}{6}\rho_\omega \int_0^\infty D^3 N_0 \exp(-\Lambda D)\,dD \tag{7.53a}$$

$$= 10^{-3}\frac{\pi\rho_w N_0 \Gamma(4)}{6\Lambda^4} \tag{7.53b}$$

where ρ_ω is 1 g cm^{-3}, D is in mm and N_0 in mm^{-1} m^{-3}. It follows that,

$$N_0 = 10^3 \frac{\Lambda^4 W}{\pi\rho_\omega} \tag{7.54a}$$

$$= 10^3 \frac{(3.67)^4}{\pi\rho_\omega}\left(\frac{W}{D_0^4}\right) \tag{7.54b}$$

Hence, $N(D)$ (in mm^{-1} m^{-3}) can be re-expressed as,

$$N(D) = 10^3 \frac{(3.67)^4}{\pi\rho_\omega}\left(\frac{W}{D_0^4}\right)\exp\left(-3.67\frac{D}{D_0}\right) \tag{7.55}$$

The normalized (or scaled) $N(D)$ is expressed as,

$$N_{\text{norm}}(D) = \left(\frac{\rho_\omega D_0^4}{10^3 W}\right) N(D) \tag{7.56a}$$

$$= \frac{(3.67)^4}{\pi}\exp\left(-3.67\frac{D}{D_0}\right) \tag{7.56b}$$

It follows that a semi-logarithmic plot of $N_{\text{norm}}(D)$ versus D/D_0 will be a straight line with slope of -3.67 and intercept of $(3.67)^4/\pi$ independent of the rainwater content. Thus, the shapes of measured distributions with widely differing rainwater contents may be compared and plotted on the same graph (e.g. see Willis 1984).

Similarly, for a gamma dsd (in mm^{-1} m^{-3}) expressed as,

$$N(D) = N_0 D^\mu \exp(-\Lambda D) \tag{7.57a}$$

$$= N_0 D^\mu \exp\left[-(3.67 + \mu)\frac{D}{D_0}\right] \tag{7.57b}$$

the water content (in g m^{-3}) is derived as,

$$W = 10^{-3} \frac{\pi}{6} \frac{\rho_\omega N_0 \Gamma(\mu+4)}{\Lambda^{\mu+4}} \tag{7.58a}$$

$$= 10^{-3} \frac{\pi}{6} \frac{\rho_\omega N_0 \Gamma(\mu+4)}{(3.67+\mu)^{\mu+4}} D_0^{\mu+4} \tag{7.58b}$$

and the normalized[6] $N(D)$ is expressed (Willis 1984) as,

$$N_{\text{norm}}(D) = \left(\frac{\rho_\omega D_0^4}{10^3 W} \right) N(D)$$

$$= \frac{6}{\pi} \frac{(3.67+\mu)^{\mu+4}}{\Gamma(\mu+4)} \left(\frac{D}{D_0} \right)^\mu \exp\left[-(3.67+\mu)\frac{D}{D_0} \right] \tag{7.59}$$

From (7.59), $N(D)$ can be re-expressed as,

$$N(D) = \frac{6W 10^3}{\pi \rho_\omega D_0^4} \frac{(3.67+\mu)^{\mu+4}}{\Gamma(\mu+4)} \left(\frac{D}{D_0} \right)^\mu \exp\left[-(3.67+\mu)\frac{D}{D_0} \right] \tag{7.60a}$$

$$= \frac{6W 10^3}{\pi \rho_\omega D_0^4} \frac{(3.67)^4}{(3.67)^4} \frac{(3.67+\mu)^{\mu+4}}{\Gamma(\mu+4)} \left(\frac{D}{D_0} \right)^\mu \exp\left[-(3.67+\mu)\frac{D}{D_0} \right] \tag{7.60b}$$

$$= N_w \frac{6}{(3.67)^4} \frac{(3.67+\mu)^{\mu+4}}{\Gamma(\mu+4)} \left(\frac{D}{D_0} \right)^\mu \exp\left[-(3.67+\mu)\frac{D}{D_0} \right] \tag{7.60c}$$

where N_w (in mm^{-1} m^{-3}) is an "intercept" parameter defined as,

$$N_w = \frac{(3.67)^4}{\pi \rho_\omega} \left(\frac{10^3 W}{D_0^4} \right) \tag{7.61}$$

Comparing N_w with N_0 in (7.54b), it is clear that N_w is the same as the intercept parameter N_0 of an equivalent exponential dsd which has the same W and D_0 as the gamma dsd. In compact form, (7.60c) may be expressed as,

$$N(D) = N_w f(\mu) \left(\frac{D}{D_0} \right)^\mu \exp\left[-(3.67+\mu)\frac{D}{D_0} \right] \tag{7.62a}$$

with

$$f(\mu) = \frac{6}{(3.67)^4} \frac{(3.67+\mu)^{\mu+4}}{\Gamma(\mu+4)} \tag{7.62b}$$

Note that $f(0) = 1$. Note also that N_0 in (7.57) can be expressed as,

$$N_0 = N_w f(\mu) D_0^{-\mu}; \quad \text{mm}^{-1-\mu}\text{ m}^{-3} \tag{7.63}$$

One disadvantage of N_0 in (7.57) is that its unit will depend on μ, whereas N_w is in conventional units of mm^{-1} m^{-3}. This is because the normalized diameter, D/D_0, is used in (7.62a), whereas this is not the case in (7.57).

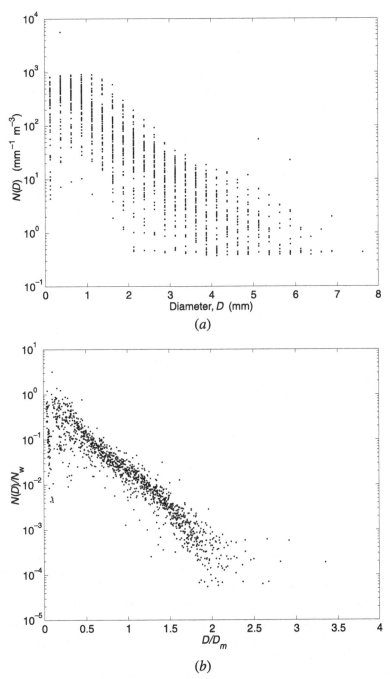

Fig. 7.16. (*a*) $N(D)$ versus D for 70 2-minute averaged drop size distributions measured by a 2D-video disdrometer in convective rain cells in Colorado; (*b*) corresponding plot of $N(D)/N_w$ versus D/D_m.

For a measured dsd then, the water content (W) and the mass-weighted mean diameter (D_m), or D_0, can be calculated. Then using (7.61) the N_w is estimated. Figure 7.16a shows $N(D)$ versus D measured by a 2D-video disdrometer, whereas Fig. 7.16b shows $N(D)/N_w$ versus D/D_m. The normalization reduces the scatter in the data and is useful in comparing the shapes of distributions with widely different rain rates. Estimates of μ for the measured dsd can be determined by minimizing the absolute deviation between $N(D)/N_w$ and the normalized gamma form in (7.62a). A large number of measured distributions from Darwin, Australia, were analyzed in this manner and a scatter plot of N_w versus D_0 is shown in Fig. 7.17a. Each data point represents a 3-minute averaged dsd measured by a JW disdrometer[7] (Joss and Waldvogel 1967). While there is considerable scatter, there is a tendency for N_w to decrease with increasing D_0. For reference the Marshall–Palmer N_0-value is 8000 mm^{-1} m^{-3}. For a given D_0, the N_w values span less than three orders of magnitude. Histograms of $\log_{10}(N_w)$ and D_0 are shown in Fig. 7.17b, c, with the modal value of N_w being approximately 10 000 mm^{-1} m^{-3}, which is close to the Marshall-Palmer value of N_0, while the modal value of D_0 is around 1.25 mm. These data are characterized by both convective and stratiform rainfall types though the statistics is dominated by stratiform rain. The histogram of μ-values is shown in Fig. 7.18. The modal value appears to be near unity. There is a long tail of high μ-values (>8) which is an artifact of the estimation method when the number of drop size categories in which drops are found is small (this typically occurs when the rain rate is very small; see Ulbrich and Atlas 1998). Also, some of the high μ-values are caused by undercounting of small drops by the disdrometer at high rain rates.

Once the dsd is parameterized in the form given by (7.62a) it is straightforward to compute radar parameters such as Z_h, Z_{dr}, K_{dp}, LDR, $|\rho_{co}|$, and δ_{co}, as well as the specific attenuation at horizontal polarization (A_h), see (1.133d), and the specific differential attenuation (A_{dp}), see (4.7b). Note that Z_h here is the equivalent reflectivity factor at h-polarization, see (6.61c). The axis ratio versus D relation is based on Andsager et al. (1999) given in (7.24) for D up to 4 mm. Beyond 4 mm, the equilibrium Beard–Chuang shapes in (7.3) are used. The canting angle distribution is assumed to be Gaussian in shape with mean zero and standard deviation of 10°. The radar elevation angle is assumed to be zero and the frequency is chosen as 5.5 GHz (the C-band range). The dielectric constant of water is obtained from Ray (1972) at a temperature of 20 °C. Because Mie scattering effects are important at C-band especially for large drops ($D \geq 5$ mm), the numerical T-matrix scattering method (see Appendix 3) is used. Since numerical integration is used, a finite maximum diameter (D_{max}) must be chosen. It is sufficient to use $D_{max} = 2.5D_m$, which is based on a histogram of D_{max}/D_m for the Darwin data set (not shown here). The modal value was found to be around 1.75 with negligible occurrence of values exceeding 2.5. From the measured dsd, the D_{max} was obtained as the diameter corresponding to the largest drop size category in which drops were found.

The still-air rain rate is defined as,

$$R = \frac{\pi}{6} \int v(D) D^3 N(D) \, dD \tag{7.64}$$

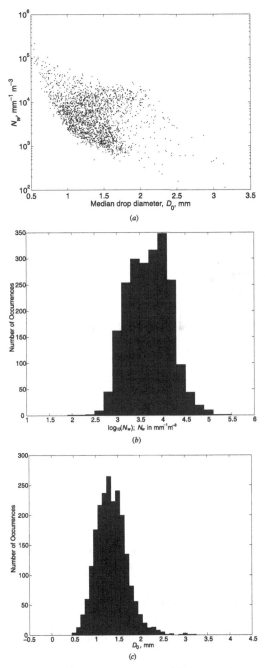

Fig. 7.17. (*a*) Scatter plot of N_w versus D_0 based on 3-minute averaged dsds using a Joss disdrometer from Darwin, Australia. These data represent nearly an entire season of rainfall. Disdrometer data were provided by BMRC, Australia. (*b*) Histogram of $10\log_{10}(N_w)$. (*c*) Histogram of D_0.

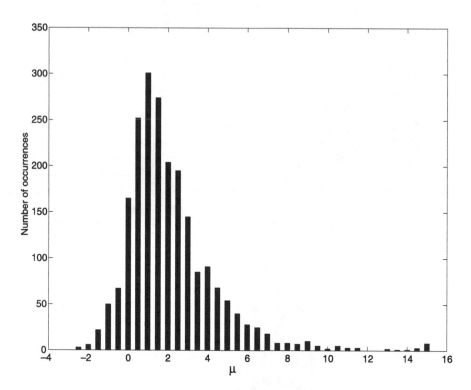

Fig. 7.18. Histogram of μ using the same data set as in Fig. 7.17.

where $v(D)$ is the drop terminal velocity which (at sea-level; Gunn and Kinzer 1949) can be expressed as (Atlas and Ulbrich 1977) a power law,

$$v(D) = 3.78D^{0.67}; \quad \mathrm{m\,s}^{-1} \tag{7.65a}$$

A different form is given by Atlas et al. (1973),

$$v(D) = 9.65 - 10.3\exp(-0.6D); \quad \mathrm{m\,s}^{-1} \tag{7.65b}$$

In (7.65), D is in units of mm, and the terminal velocity is at sea-level. The conventional unit for rain rate is mm h^{-1} in which case (7.64) becomes,

$$R = 0.6\pi \times 10^{-3} \int v(D)D^3 N(D)\,dD; \quad \mathrm{mm\,h}^{-1} \tag{7.66a}$$

with $v(D)$ in m s^{-1}, D in mm and $N(D)\,dD$ the number of drops m^{-3} in the interval D to $D + dD$.

The altitude factor for adjusting the terminal velocity from sea-level depends on air density and drop diameter (Beard 1985). The rain rate expression is modified as follows:

$$R = 0.6\pi \times 10^{-3} \int \left(\frac{\rho_0}{\rho}\right)^{m(D)} v(D)D^3 N(D)\,dD \tag{7.66b}$$

where $m(D) = 0.375 + 0.025D$, where D is in mm; ρ_0 is the air density at sea-level; while ρ is the value at the specified altitude. Beard (1985) showed that a mean $\bar{m}$ could be defined as,

$$R = 0.6\pi \times 10^{-3}\left(\frac{\rho_0}{\rho}\right)^{\bar{m}} \int v(D)D^3 N(D)\,dD \tag{7.66c}$$

where $\bar{m}$ varies from 0.41–0.43.

As mentioned earlier, the measured distributions were obtained with a JW disdrometer in Darwin, Australia. For each 3-minute averaged dsd, the N_w, D_m and μ were calculated (see Figs. 7.17, 7.18) and D_{max} was set at $2.5D_m$ for each of the fitted dsd. Figure 7.19 shows a scatterplot of Z_h (in $mm^6\ m^{-3}$) versus R (in mm h^{-1}) where the data have been stratified into two ranges of D_0, i.e. 1.25–1.50 and 1.75–2.0 mm. It is clear that straight line fits to the data in each of the two D_0 ranges will give nearly the same slope, but with different intercepts. Thus, if a power law of the form $Z_h = aR^b$ is proposed (which yields a straight line on a log–log plot), then b will be nearly the same whereas a will differ for the fits to the data in the two D_0 ranges (this behavior was noted by Ulbrich and Atlas 1998). Figure 7.19 clearly shows that for a given rain rate, the reflectivity will increase as D_0 increases, and therefore an independent estimate of D_0 (or D_m) is necessary to avoid errors in the retrieval of R using Z_h–R power-law relations. This additional information can be provided by Z_{dr} as shown in the scatterplot of Z_{dr} versus D_m in Fig. 7.20. The scatter can be explained by the variability in the μ-values, but a mean fit of the form in (7.14) can be used (recall that (7.14) is valid at 3-GHz frequency, while the simulations here are performed at 5.5 GHz). Note also that Z_{dr} is independent of N_w, since it is the ratio of reflectivities at h- and v-polarizations. It will be shown in Chapter 8 that rain rate algorithms based on both Z_h and Z_{dr} will be substantially more accurate than Z_h–R power laws, especially for larger D_m values.

Following Ulbrich and Atlas (1998), it is easy to show analytically the behavior of Z_h and R depicted in Fig. 7.19. Assuming spherical drops, the reflectivity factor (Z in (7.21)) can be expressed as,

$$Z = N_w f(\mu)D_0^{-\mu} \int_0^\infty D^{6+\mu} \exp\left[-(3.67 + \mu)\frac{D}{D_0}\right]dD \tag{7.67a}$$

$$= N_w f(\mu)D_0^{-\mu}\Gamma(7 + \mu)\frac{D_0^{7+\mu}}{(3.67 + \mu)^{7+\mu}} \tag{7.67b}$$

$$= N_w f(\mu)\Gamma(7 + \mu)\frac{D_0^7}{(3.67 + \mu)^{7+\mu}}; \quad mm^6\ m^{-3} \tag{7.67c}$$

Similarly, using (7.65a), the rain rate can be expressed as,

$$R = (0.6 \times 10^{-3}\pi)(3.78)N_w f(\mu)D_0^{-\mu} \int_0^\infty D^{3.67+\mu} \exp\left[-(3.67 + \mu)\frac{D}{D_0}\right]dD \tag{7.68a}$$

$$= (0.6 \times 10^{-3}\pi)(3.78)N_w f(\mu)\Gamma(4.67 + \mu)\frac{D_0^{4.67}}{(3.67 + \mu)^{4.67+\mu}}; \quad mm\ h^{-1} \tag{7.68b}$$

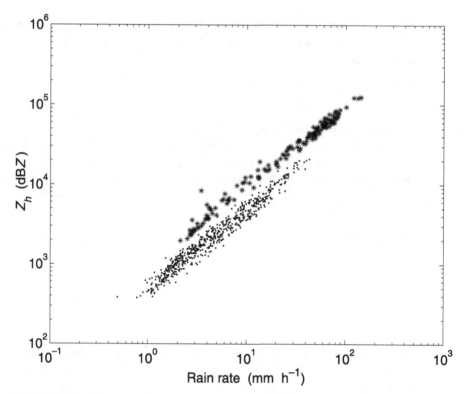

Fig. 7.19. Reflectivity (at horizontal polarization and 5.5 GHz) versus rain rate using normalized gamma fits to 3-minute averaged dsds from Darwin, Australia. Data are stratified into two different intervals of D_0: (∗) represents the range $1.75 < D_0 < 2$ mm, (•) represents the range $1.25 < D_0 < 1.5$ mm.

It follows that,

$$\frac{Z}{R} = \frac{1}{(0.6 \times 10^{-3}\pi)(3.78)} \frac{\Gamma(7+\mu)}{\Gamma(4.67+\mu)} \left(\frac{D_0}{3.67+\mu}\right)^{2.33} \tag{7.69a}$$

$$= F(\mu)D_0^{2.33} \tag{7.69b}$$

where (7.51) has also been used. Ulbrich and Atlas (1998) show that $F(\mu)$ varies very slowly with μ and, thus, when data are stratified by D_0, the log–log plot of Z versus R will be a straight line whose slope is unity and whose coefficient a (in $Z = aR^b$) will vary as $D_0^{2.33}$ in agreement with the results of Fig. 7.19.

Alternatively, D_0 can be eliminated between (7.67c) and (7.68b). To simplify, Z and R are expressed as,

$$\frac{Z}{N_w} = F_z(\mu)D_0^7 \tag{7.70a}$$

$$\frac{R}{N_w} = F_R(\mu)D_0^{4.67} \tag{7.70b}$$

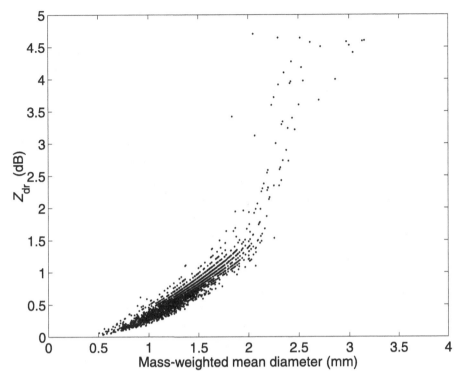

Fig. 7.20. Z_{dr} (at 5.6 GHz) versus D_m. The axis ratios used are based on the Andsager et al. (1999) model, see (7.24), for $1 \le D \le 4$ mm, and the Beard–Chuang model, see (7.3), for $D < 1$ and $D > 4$ mm. Gaussian canting angle distribution is assumed with mean $= 0°$ and standard deviation $= 10°$.

where

$$F_z(\mu) = \frac{f(\mu)\Gamma(7+\mu)}{(3.67+\mu)^{7+\mu}} \tag{7.70c}$$

$$F_R(\mu) = (0.6 \times 10^{-3}\pi)(3.78)f(\mu)\frac{\Gamma(4.67+\mu)}{(3.67+\mu)^{4.67+\mu}} \tag{7.70d}$$

Recall that $f(\mu)$ is given in (7.62b). From (7.70a, b) it follows that,

$$\frac{Z}{N_w} = \frac{F_z(\mu)}{[F_R(\mu)]^{7/4.67}}\left(\frac{R}{N_w}\right)^{7/4.67} \tag{7.71}$$

Thus, when Z and R are normalized by N_w, the resulting straight line in the log–log plot will have a constant slope (independent of μ) with very little μ-dependence of the "intercept", since it can be shown that $F_z(\mu)/[F_R(\mu)]^{1.5}$ will vary slowly with μ. Figure 7.21 shows the log–log plot of Z_h/N_w versus R/N_w for the Darwin data set. The data points lie on nearly the same straight line whose slope (based on a least squares fit) was found to be 1.52, which is very close to the theoretical value of 1.5 in (7.71). Thus, when Z and R are normalized by N_w, the "Z–R" relation may be written (Testud et al.

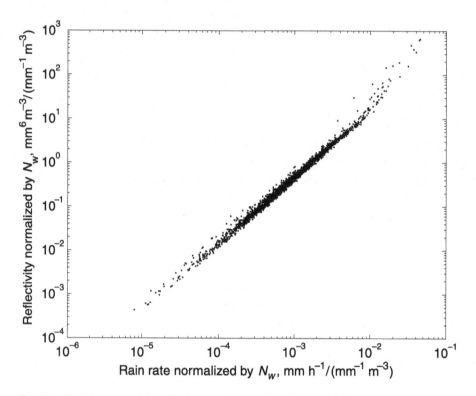

Fig. 7.21. Z_h/N_w versus R/N_w for the same data set shown in Fig. 7.19.

2000) as,

$$\frac{Z}{N_w} = a\left(\frac{R}{N_w}\right)^b \tag{7.72a}$$

or,

$$Z = a(N_w)^{1-b} R^b \tag{7.72b}$$

For a Marshall–Palmer dsd ($\mu = 0$) where $N_w = 8000$ mm^{-1} m^{-3}, the multiplicative factor from (7.71) is $aN_w^{1-b} \approx 237$ while $b = 7/4.67 \approx 1.5$. Thus, the variability in the multiplicative factor in Z–R relations can be traced to variations in N_w. For comparison, the Z–R relation used for the WSR–88D radar system is $Z = 300R^{1.4}$. It follows that the exponent b in (7.72a) may be accurately determined from a log–log plot of Z/N_w versus R/N_w. This result may be generalized to any two moments of the dsd. For example, if

$$P_n = N_w f(\mu) D_0^{-\mu} \int_0^\infty D^{n+\mu} \exp\left[-(3.67 + \mu)\frac{D}{D_0}\right] dD \tag{7.73a}$$

$$P_m = N_w f(\mu) D_0^{-\mu} \int_0^\infty D^{m+\mu} \exp\left[-(3.67 + \mu)\frac{D}{D_0}\right] dD \tag{7.73b}$$

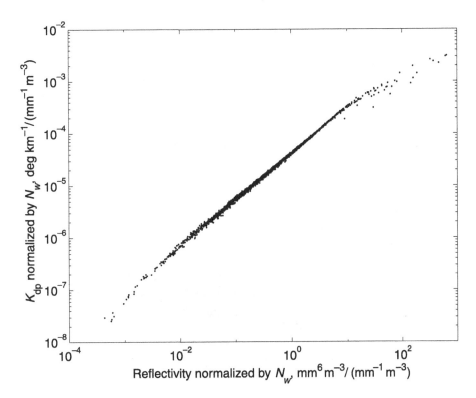

Fig. 7.22. K_{dp}/N_w versus Z_h/N_w for the same data set shown in Fig. 7.20.

then,

$$\frac{p_n}{N_w} = \frac{F_n(\mu)}{[F_m(\mu)]^{(n+1)/(m+1)}} \left(\frac{p_m}{N_w}\right)^{(n+1)/(m+1)} \tag{7.73c}$$

and the exponent (i.e. the slope in the log–log plot of p_n versus p_m) is μ-independent (Testud et al. 2000). It follows that the relation between p_n and p_m will be nearly independent of N_w if $(n + 1/m + 1)$ is close to unity.

Finally, several other scatterplots are shown, e.g. K_{dp}/N_w versus Z_h/N_w in Fig. 7.22, Z_{dr} versus Z_h in Fig. 7.23, and Z_{dr} versus K_{dp} in Fig. 7.24. From Fig. 7.24 it is apparent that data points with $Z_{dr} > 3$ dB correspond to $K_{dp} < 3°$ km^{-1}, whereas Z_{dr} is generally between 1 and 1.5 dB for $K_{dp} > 4°$ km^{-1}. If the average Z_{dr} is calculated for small increments of K_{dp}, for $K_{dp} > 3°$ km^{-1}, it is apparent that the average Z_{dr} tends to a steady value around 1–1.5 dB (or average D_0 between 1.5–2.0 mm, see Fig. 7.20). From (7.18), this means that for this data set K_{dp} is linearly related to water content since the average D_0 (or D_m) is nearly constant at around 2 mm (for high $K_{dp} > 3°$ km^{-1}).

Parameterization of $N(D)$ in the gamma form appears to be suitable for stratiform and convective rain events particularly when there is substantial depth between the melting level and the surface, e.g. in the tropics where the 0 °C level is near 4.5–5.0-km altitude. Such depth is conducive to the formation of the equilibrium distribution, which

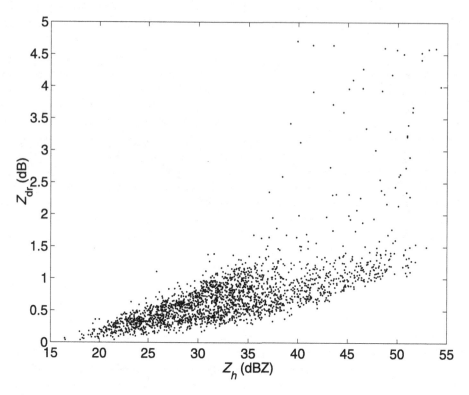

Fig. 7.23. Z_{dr} versus Z_h for the same data set shown in Fig. 7.20.

results when there is a balance between coalescence growth and collisional breakup of drops (Hu and Srivastava 1995). The shape of the equilibrium distribution tends to fall within the gamma shape with positive μ when $D > 1$ mm. As the melting level lowers in altitude (e.g. in the mid-latitudes) and the ice phase is more involved in rain production (i.e. through melting of graupel and small hail), the $N(D)$ shape tends to deviate from the gamma shape. Other factors such as strong downdrafts and sorting of drops by horizontal winds can also lead to non-gamma shapes. For example, the average $N(D)$ measured with a two-dimensional video distrometer[8] from a summer-time convective shower in Colorado is shown in Fig. 7.25, and it is clear that the relatively high concentration of drops with $D > 4$ mm makes it difficult to fit the dsd with positive μ-values of a gamma distribution.

Sempere-Torres et al. (1994) have proposed a general formulation for $N(D)$ based on a scaling law that provides for an alternate method of normalizing $N(D)$ without a priori assuming the gamma shape. A brief description of their formulation is provided here. First a reference (or scaling) integral variable is chosen, e.g. the water content W (it can also be rain rate or any moment of the dsd). The drop size distribution function

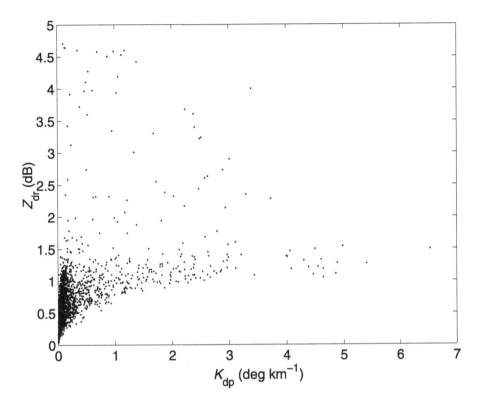

Fig. 7.24. Z_{dr} versus K_{dp} for the same data set shown in Fig. 7.20.

is expressed as,

$$N(D; W) = W^{\alpha} g\left(\frac{D}{W^{\beta}}\right) \tag{7.74a}$$

$$= W^{\alpha} g(x) \tag{7.74b}$$

where α and β are constants, and $g(x)$ is a general distribution function whose form is independent of W. Note that the reference variable is chosen as W here, only for illustrative purposes. The diameter is scaled as $x = D/W^{\beta}$. Self-consistency relations are evident since W is defined as,

$$W = \frac{\pi}{6} \rho_{\omega} \int_0^{\infty} D^3 N(D; W) \, dD \tag{7.75a}$$

$$= \frac{\pi}{6} \rho_{\omega} \int_0^{\infty} D^3 W^{\alpha} g\left(\frac{D}{W^{\beta}}\right) dD \tag{7.75b}$$

$$= \frac{\pi}{6} \rho_{\omega} W^{4\beta+\alpha} \int_0^{\infty} x^3 g(x) \, dx \tag{7.75c}$$

Self-consistency demands that,

$$4\beta + \alpha = 1 \tag{7.76a}$$

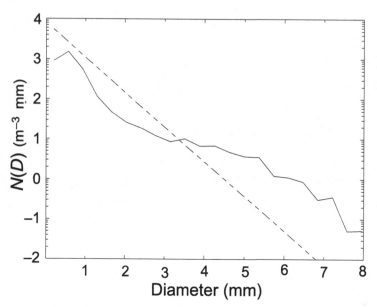

Fig. 7.25. Drop size distribution data in a typical summer rain shower in northern Colorado using a 2D video disdrometer. Data have been averaged over 11 minutes, the average rain rate being 30 mm h^{-1} (maximum 1-minute rain rate of 65 mm h^{-1}). Dashed line is an exponential Marshall–Palmer fit assuming $N_0 = 8000$ mm^{-1} m^{-3} and average rain rate of 30 mm h^{-1}. From Hubbert et al. (1999).

$$\frac{\pi}{6}\rho_w \int_0^\infty x^3 g(x)\,dx = 1 \tag{7.76b}$$

which implies that α and β are not independent, and that the third moment of $g(x)$ is constrained.

Now consider any moment of the dsd,

$$p_n = \int_0^\infty D^n N(D; W)\,dD \tag{7.77a}$$

$$= \int_0^\infty D^n W^\alpha g\left(\frac{D}{W^\beta}\right)\,dD \tag{7.77b}$$

$$= W^{\alpha+\beta(n+1)}\int_0^\infty x^n g(x)\,dx \tag{7.77c}$$

$$= a_n W^{b_n} \tag{7.77d}$$

where,

$$b_n = \alpha + \beta(n + 1) \tag{7.78a}$$

$$a_n = \int_0^\infty x^n g(x)\,dx \tag{7.78b}$$

Thus, any moment of the dsd is expressible as a power law involving the reference variable W. If the reference variable were, for example, rain rate (R), then the reflectivity factor (for spherical drops) being the sixth moment would be expressible as a power law, $Z = aR^b$, provided the terminal velocity relation of the form in (7.65a) is used.

One method of determining α and β from the measured size distributions is to compute different moments such as p_0–p_6 from the data and find the corresponding b_n based on a straight line fit to $\log p_n = \log a_n + b_n \log W$. Once b_0–b_6 are found, then a simple linear regression between b_n and $(n + 1)$ will yield α and β. A model for $g(x)$ can then be proposed, which need not fall in the gamma family. Examples may be found in Sempere-Torres et al. (1999).

The equilibrium distribution referred to earlier can be expressed as,

$$N(D; W) = Wg(D) \tag{7.79}$$

It follows that $\alpha = 1$, $\beta = 0$ in (7.74a). From (7.77, 7.78) it follows that any moment of $N(D; W)$, e.g. p_n, is linearly related to W since $b_n = 1$. It also follows that the mass-weighted mean diameter $D_m = p_4/p_3$ is constant for such distributions. Recalling the discussion relating to Fig. 7.24 where the average Z_{dr} (or D_m) tends to a steady value when K_{dp} exceeds around $3°$ km^{-1}, it may be deduced that radar detection of equilibrium distributions is possible (see also, List 1988). This also follows from (7.18) since K_{dp} will be linearly related to W if D_m tends to a constant value.

7.1.5   Vertical structure of Z_{dr} and K_{dp}

Measurements of Z_{dr} in the early 1980s in the UK showed that vertical columns of positive Z_{dr} could be observed within convective storm cells that often extended several kilometers above the environmental $0\,°C$ level (Hall et al. 1984; Illingworth et al. 1987). These columns were interpreted as raindrops being lofted by the updraft into the colder regions of the cloud where they freeze and lose their orientational stability leading to a rapid decrease in Z_{dr}. Subsequent field data[9] involving instrumented aircraft in northern Alabama and in Florida have largely confirmed this interpretation. A good example of a positive Z_{dr} column seen in a vertical section through a growing storm cell in Florida is shown in Fig. 7.26a, where the column is nearly centered within the peak updrafts within the cell. The 1–2-dB contour of Z_{dr} reaches an altitude of 6.5 km (around $-10\,°C$) with peak Z_{dr} in this vertical plane reaching 3.6 dB at 1.7-km altitude. The peak reflectivity reached 45 dBZ at 2-km altitude. This cell was penetrated by two King Air[10] aircraft (NKA: NCAR King Air; WKA: University of Wyoming King Air) as shown at altitudes of 5.5 and 4.0-km, respectively. Along these tracks the reflectivity peaked at 37 dBZ with peaks in Z_{dr} of 2.7 dB. The updraft magnitudes within the central portion of the NKA penetration were around 12 m s^{-1}, while they were 6–9 m s^{-1} along

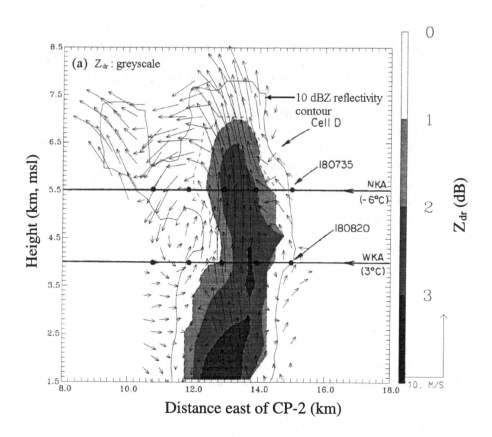

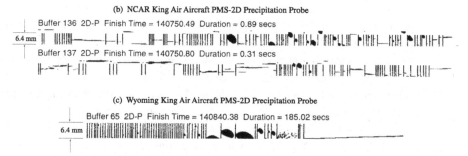

Fig. 7.26. (*a*) Vertical section of radar data (from the NCAR CP-2 radar) in a vigorous growing cell in Florida. Greyscales depict Z_{dr} values (from 0 to 3 dB) with darker shades representing larger values. An outer 10-dBZ reflectivity contour is also shown. Arrows depict triple-Doppler derived wind vectors in the vertical plane. Two aircraft penetration tracks are shown (NKA: NCAR King Air; WKA: University of Wyoming King Air) with solid dots spaced at 10-second increments. (*b*) Drop images from an imaging probe on the NKA. (*c*) Drop images from an imaging probe on the WKA. From Bringi et al. (1997).

the WKA penetration. Particle images recorded by imaging probes[11] on both aircraft are shown in Fig. 7.26b,c. Smooth elliptical images indicate liquid drops along both penetration segments (maximum D ranging from 4–8 mm). These observations show that warm rain processes involving collision/coalescence of drops were dominant within the positive Z_{dr} column. Size sorting due to growth of some drops as they fall through the updraft (i.e. those drops whose terminal velocity exceeds the updraft speed and, thus, are exposed to high liquid water contents) are largely responsible for the observed vertical structure of Z_h and Z_{dr} within the column. Raindrops can grow to large sizes in such columns (up to 8 mm), but are found in very low concentrations.

The average Z_{dr} versus Z_h within the volume of such columns undergoes a distinct time evolution from the early growth phase to the mature phase as shown in Fig. 7.27 for a storm cell in northern Alabama. In the early phase (curve marked 1534) the average Z_{dr} is large (3–4 dB) with Z_h in the range 30–40 dBZ. As time increases, the average Z_{dr} (for a given Z_h) decreases for the first 15 minutes or so (from 1534 to 1548) after which the average Z_{dr} tends to fall on the dashed line, which corresponds to the Marshall–Palmer dsd. This evolution is consistent with weakening updrafts and the beginning of downdrafts in the cell. The dsd evolves from a low concentration of large drops to an exponential shaped Marshall–Palmer dsd (Caylor and Illingworth 1987). The exponential shape results from increasing drop collisions, which lead to drop breakup and production of smaller drops. The difference in the vertical structure of Z_{dr} between the vigorous growth phase and the mature (steady) phase of a storm cell is more vividly shown in Fig. 7.28 which depicts the contoured frequency by altitude diagram (or CFAD) of Z_{dr} during: (i) the vigorous growth phase, and (ii) the mature phase of a Florida storm cell. Note the larger Z_{dr} values occurring within the volume of the cell in Fig. 7.28a, especially below 5-km altitude, whereas during the steady phase shown in Fig. 7.28b the Z_{dr} values are largely confined within a narrow range (<2 dB). Near the surface the modal Z_{dr} value in the mature phase is around 1.25 dB with most values between 1 and 1.5 dB, indicating that an equilibrium drop size distribution was probably reached. From a practical viewpoint, it should be clear that different Z–R relations would be valid for the vigorous growth phase versus the mature phase, i.e. the coefficient a in $Z = aR^b$ will be significantly less for the mature phase (corresponding to smaller Z_{dr} or D_m), relative to the growth phase (see also Fig. 7.19 or (7.69)). Such systematic variation in Z–R relations was clearly documented by Atlas et al. (1999).

The vertical structure of Z_{dr} together with K_{dp} in convective storm cells (see Fig. 7.29) has important applications to the study of precipitation physics, since both D_m and the rainwater content (W) can be retrieved. This example is from a storm cell that produced a flash-flood in Fort Collins, Colorado. It should be noted that D_m and W are not necessarily (spatially) correlated within the storm cell, the peak in W generally occurring at higher altitudes compared with the peak in D_m. Generally, drop sorting with larger drops falling against the updraft will cause D_m to peak at lower levels of the cell while the main water mass tends to accumulate at higher altitudes near the particle

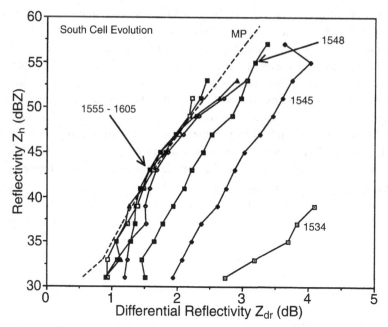

Fig. 7.27. The time evolution (from 1534 to 1605) of average Z_{dr} versus Z_h within the rain volume of a storm cell in northern Alabama. Radar data were obtained using the NCAR CP-2 radar. From Bringi et al. (1991).

balance level (where the terminal velocity corresponding to D_m at that level is roughly equal to the updraft speed).

7.2   Convective precipitation

The dominant polarization signatures in convective storms generally occur at and below the melting level where ice hydrometeors melt to form rain. In the case of hailstorms it is possible to detect the hailshaft below the melting level in contrast to the surrounding rain areas. As described in the previous section, positive Z_{dr} columns whose tops can extend to several kilometers above the 0 °C level appear to play an important microphysical role in the development of convective storms by providing a source of frozen drop embryos which can later grow into large hail. The mixed phase region (typically extending from 0 °C to around −10°C) consisting of supercooled raindrops, partially freezing raindrops, and other forms of wet ice hydrometeors can be detected by enhanced values of LDR, which often form a "cap" on the positive Z_{dr} column. As described in the previous section, enhanced regions of positive K_{dp} and the related large water content values near 0 °C are indicative of water loading at these levels, which, in turn, contributes to the initiation of the downdraft within the storm. Ice crystals in the upper levels of convective storms are often vertically aligned by strong electric fields

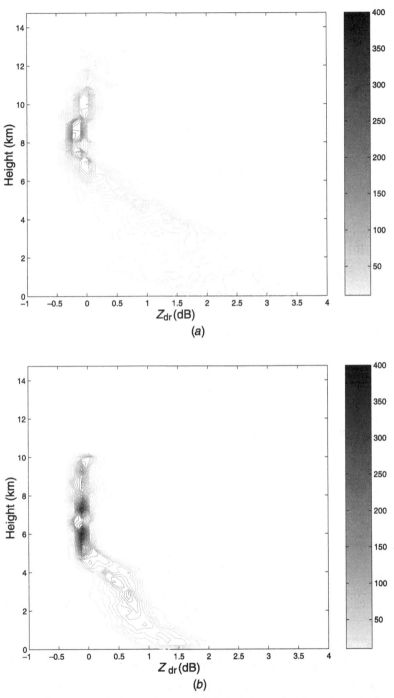

Fig. 7.28. Contoured frequency by altitude (CFAD) diagram of Z_{dr} using radar data from the NCAR CP-2 radar in a Florida storm cell: (*a*) vigorous growth phase, and (*b*) mature phase.

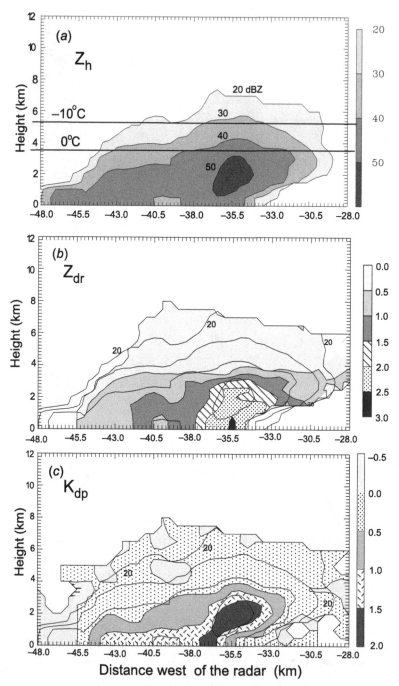

Fig. 7.29. Vertical section of radar data using the CSU–CHILL radar from a "tropical" type storm that produced a flash-flood in Fort Collins, Colorado. (*a*) Reflectivity (Z_h) at horizontal polarization, (*b*) Z_{dr} (in dB), and (*c*) K_{dp} (in ° km^{-1}).

leading to characteristic signatures of negative K_{dp} prior to a lightning discharge (for example, see Fig. 4.16a).

The qualitative application of polarimetric signatures towards further understanding of these physical processes in convective storms has been more prevalent than quantitative applications involving determination of hydrometeor amounts, especially in the ice phase. The reason is mainly due to the difficulty in retrieving the ice particle size distribution (or even parameters of the distribution such as total concentration or the mass-weighted mean diameter) and particle density from the polarimetric measurements. Further complications arise because a mixture of hydrometeor types is often found within the radar resolution volume with each component possibly having different size/shape/orientation/density distributions. For convenience, ordinary convective storms will be treated first, followed by hailstorms. The main ice hydrometeor types considered will be graupel and hail (including wet or melting graupel/hail) as well as frozen raindrops.

7.2.1   Ordinary convective storms

The term ordinary convective storm refers here to single or multiple cell storms that do not produce hail. An individual cell is generally short-lived (e.g. 20–30 minutes) and typically follows the evolution of air-mass type storms described by Byers and Braham (1949). The term cell as used here refers to a compact structure of reflectivity as observed in horizontal sections of radar data, which is generally coincident with updrafts during the vigorous growth phase. The predominant ice hydrometeor type in such ordinary storms is observed to be graupel particles. The origin of such particles is often a large ice crystal that accretes supercooled cloud droplets as it falls. These cloud droplets freeze upon impact with the crystal and this deposit is called "rime". As this riming growth of ice crystals continues it leads to the formation of graupel particles, which are typically conical in shape. A stable fall mode is with the apex pointed vertically upward, although they are more likely to oscillate and tumble while falling. As these particles melt, water percolates into the interior and their fall mode becomes more stable; in the limit of complete melting the raindrops are oblate and stably oriented. This is shown in Fig. 7.30 via particle images of (a) melting conical graupel, and (b) oblate raindrops from an aircraft imaging instrument, the penetration being at 15 °C through a Colorado storm cell. The resultant increase in Z_{dr} as graupel particles melt to form oblate raindrops has a clear polarimetric signature which was observed in the early 1980s by Hall et al. (1984) in the UK. For example, this signature can be noted in Fig. 7.28 starting at a height of 5 km (which is near the 0 °C level) below which the modal value of Z_{dr} increases with decreasing height.

The melting of graupel particles is essentially a thermodynamic process that has been modeled in detail by Rasmussen and Heymsfield (1987) and depends on the initial particle size and density as well as on the sounding profile, especially the relative humidity below cloud base. There is also a dependence on the vertical air speed i.e. whether the particles are in a downdraft, for example. Typically, dry graupel particles

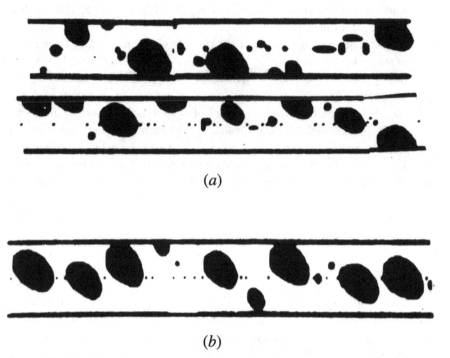

(a)

(b)

Fig. 7.30. Particle images from an airborne imaging probe in a convective storm in Fort Collins, Colorado: (a) melting graupel, (b) raindrops. Vertical distance between bars is 6.4 mm. From Bringi et al. (1986a).

are of medium density ($\sim$0.5 g cm^{-3}), although their density is likely to vary with size. An example of model calculations depicting the change in Z_{dr} and LDR_{vh} due to melting of conical graupel into oblate drops is shown in Fig. 7.31.

Figure 7.32 shows constant altitude plan position indicator (PPI) data of reflectivity in a mature convective cell in Colorado, which was penetrated by the SDSM&T T-28 aircraft[12] at an altitude of 4 km (above ground level). The T-28 was equipped with a high volume particle spectrometer (HVPS) which showed predominantly dry graupel images up to 15 mm in maximum dimension (a typical image and the 30-s averaged size distribution are also shown in Fig. 7.32b and c, respectively). In this particular experiment, the T-28 aircraft penetrations were closely coordinated with CSU–CHILL radar scans resulting in excellent spatial and temporal resolution for a period of 40 minutes during which the HVPS showed predominantly graupel images.

It was shown in Section 2.3 (see Fig. 2.8c) that LDR is very sensitive to particle density. One method of estimating the average density is by comparing the radar measurement of equivalent reflectivity factor (Z_e; see (5.49a)) at each grid point closest to the aircraft track with that computed from the HVPS-measured particle size

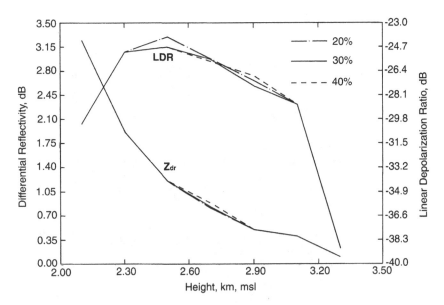

Fig. 7.31. Calculations at 3 GHz depicting the change in Z_{dr} and LDR_{vh} due to melting of conical graupel into oblate raindrops. From Vivekanandan et al. (1990).

distribution data. The $Z_e(\text{dsd})$ is computed from,

$$Z_e(\text{dsd}) = \frac{\lambda^4}{\pi^5 |K_w|^2} \eta \tag{7.80a}$$

$$= \frac{\lambda^4}{\pi^5 |K_w|^2} \int N(D)\sigma_b(D)\,dD \tag{7.80b}$$

where $|K_w|^2$ is the dielectric factor for water and the particles are assumed to be spherical for the purposes of this comparison. Note that σ_b is the back scatter cross section of the graupel particles and $N(D)$ is the measured size distribution (dsd refers here to the particle size distribution). Since σ_b depends on the particles' dielectric constant which in turn is related to particle density, it follows that $Z_e(\text{dsd})$ can be iterated with density until an optimal match with the measured Z_e along the aircraft track is achieved. If the particles fall in the Rayleigh scattering limit, then $\sigma_b(D)$ in (7.80b) can be replaced by (1.51b) resulting in,

$$Z_e(\text{dsd}) = \frac{|K_p|^2}{|K_w|^2} \int D^6 N(D)\,dD; \quad \text{mm}^6\,\text{m}^{-3} \tag{7.81}$$

where $|K_p|^2$ and $|K_w|^2$ are the dielectric factors of the particle and water, respectively. As discussed in Section 1.6, the dielectric constant of dry graupel particles can be computed using the Maxwell-Garnet mixing formula for a two-phase mixture of ice and air. Using (1.69) it is easy to show that $|K_p|^2 \approx \rho_p^2 |K_{ice}|^2$, where $|K_{ice}|^2$ is the dielectric

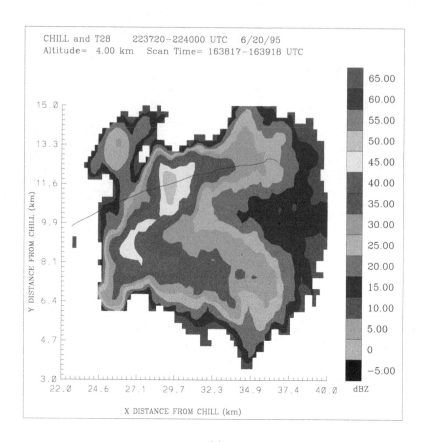

CHILL and T28   223720-224000 UTC   6/20/95
Altitude= 4.00 km   Scan Time= 163817-163918 UTC

(a)

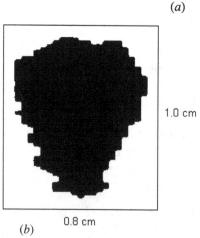

(b)

Fig. 7.32. (*a*) Constant altitude PPI (or CAPPI) data of reflectivity (Z_h) using the CSU–CHILL radar in a convective storm in northern Colorado. The CAPPI is at 4-km above ground level. The T-28 aircraft penetration track is shown. (*b*) Typical graupel image from the high volume particle spectrometer (HVPS) on the T-28. (*c*) Particle size distribution averaged over 30 seconds from HVPS data. $N(D)$ is in $mm^{-1}\ m^{-3}$.

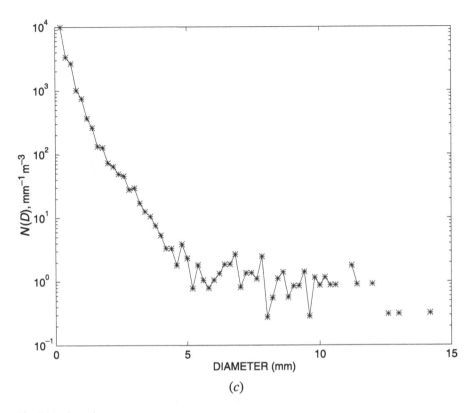

Fig. 7.32. (*cont.*)

factor for solid ice and ρ_p is the particle density (see also, Smith 1984). Hence,

$$Z_e(\text{dsd}) = \rho_p^2 \frac{|K_{\text{ice}}|^2}{|K_w|^2} \int D^6 N(D)\, dD; \quad \text{mm}^6\, \text{m}^{-3} \tag{7.82}$$

and the dependence on particle density is more transparent in the Rayleigh scattering limit. At 3-GHz frequency and at a temperature of $0\,^\circ$C it is easy to show that $|K_w|^2 \approx 0.93$ while $|K_{\text{ice}}|^2 \approx 0.17$.

Figure 7.33 shows a plot of Z_h (equivalent reflectivity factor) measured by radar compared with that calculated from the HVPS-measured dsd for all penetration segments after an optimal density match of 0.56 g cm^{-3} was determined. Figure 7.34 compares the radar measured LDR$_{vh}$ with computed values assuming a conical shape for the graupel particles (estimated from images, a typical image being shown in Fig. 7.32b), and assuming an isotropic orientation distribution. The agreement in LDR$_{vh}$ is quite reasonable considering the assumptions involved. The measured LDR$_{vh}$ for dry graupel with average density of around 0.5 g cm^{-3} lies in the range -25 to -27 dB. The average values of measured Z_{dr} and K_{dp} were close to zero (not shown here). Thus, it appears that dry graupel may be modeled with a covariance matrix assuming polarization plane symmetry as in (3.193), the only polarimetric variable of

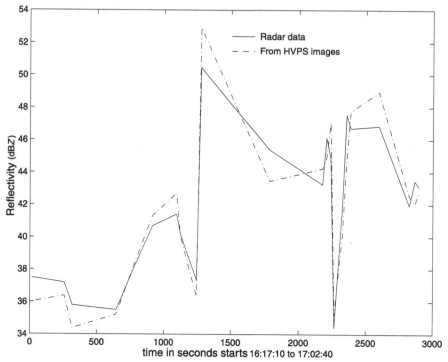

Fig. 7.33. Comparison of measured radar reflectivity (Z_h) with equivalent reflectivity calculated from HVPS-measured spectra, using (7.82), after an optimal density match of $\rho_p = 0.56$ g cm^{-3}. The horizontal axis is time along the T-28 penetration tracks. These data include a number of cell penetrations.

significance then being LDR and, of course, the equivalent reflectivity factor (or Z_h). From a hydrometeor classification viewpoint using 3-GHz radar data, the range of Z_h and LDR for dry graupel with average density of 0.5 g cm^{-3} in the upper parts of convective cells would be, respectively, 30–50 dBZ and -25 to -27 dB. If the average densities were lower, then the LDR values would be further reduced (e.g. see Fig. 2.8c). The wide range for Z_h should account for variability in both $N(D)$ as well as particle density.

From a cloud microphysical application, the ice–water content (IWC) is an important parameter, which is defined as,

$$\text{IWC} = (10^{-3})\frac{\pi}{6}\rho_p \int D^3 N(D)\,dD; \quad \text{g m}^{-3} \tag{7.83}$$

where D is in mm, $N(D)$ in mm^{-1} m^{-3}, and ρ_p in g cm^{-3}. If the graupel particle size distribution is modeled as a gamma function of the form in (7.62), and using (7.73c), it is clear that a power law between Z_e and IWC can be derived as,

$$\frac{\text{IWC}}{N_p} = \frac{\pi}{6}(10^{-3})\rho_p^{-1/7}\frac{F_3(\mu)}{[F_Z(\mu)]^{4/7}}\left(\frac{|K_w|^2}{|K_{\text{ice}}|^2}\frac{Z_e}{N_p}\right)^{4/7} \tag{7.84a}$$

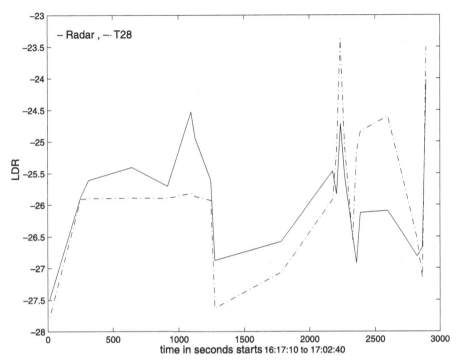

Fig. 7.34. Comparison of radar measured LDR_{vh} with equivalent reflectivity calculated from HVPS-measured spectra, using (7.82), after an optimal density match of $\rho_p = 0.56\text{ g cm}^{-3}$. Calculations assume conical graupel shape and isotropic orientation distribution. The T-matrix method is used at a frequency of 3 GHz.

where,

$$F_3(\mu) = \frac{f(\mu)\Gamma(4+\mu)}{(3.67+\mu)^{4+\mu}} \qquad (7.84b)$$

and $F_Z(\mu)$ and $f(\mu)$ were previously defined in (7.70c) and (7.62b), respectively. Note that N_p is similar to N_w defined in (7.61) except that W and ρ_w are to be replaced by IWC and ρ_p, respectively. Note that the dependence on particle density is very weak as expected, since IWC is related to the equivalent reflectivity factor. The IWC–Z_e relation can be expressed in the form, IWC $= a(Z_e)^b$ where $b = 0.57$ for the gamma dsd model whereas the multiplicative coefficient a is usually empirically determined. For an exponential size spectrum ($\mu = 0$) with $N_p = 8000\text{ mm}^{-1}\text{ m}^{-3}$, and with particle density $= 0.5\text{ g cm}^{-3}$, the coefficient $a \approx 9.7 \times 10^{-3}$. The variability in a may be traced to variations in N_p, and to a much lesser extent to μ or to particle density. Typically, the coefficient a can range from 1 to 10×10^{-3}. It may also be estimated from a simple fit to radar-measured Z_e along the aircraft flight track and the corresponding IWC deduced from the measured particle size distributions. Hence, radar-based estimates of IWC may then be obtained from measured Z_e in those regions of the storm cell where aircraft data are not available assuming, of course, that the multiplicative coefficient is the same in these other regions.

During the vigorous growth phase of convective cells, strong updrafts loft raindrops to cold temperatures where they freeze. These frozen drops often experience rapid growth by accreting cloud water. Partially freezing drops and other forms of wet ice particles cause enhanced LDR signatures around the -5 to $-10\,^{\circ}$C level, a region that "caps" the tops of positive Z_{dr} columns. Figure 7.35a shows a vertical section through a vigorous convective cell in Florida which was penetrated by the NOAA P-3 aircraft as indicated. The positive Z_{dr} column (shown in greyscale) is clearly evident with values between 1 and 2 dB extending up to 6.5 km altitude (or $-11\,^{\circ}$C). The peak reflectivity is 53 dBZ at 5.2-km altitude, while the peak Z_{dr} in this vertical cut plane is 3.7 dB at around 4-km altitude. The wind vectors show a broad updraft with a width of 4 km and peak speed of around 13 m s^{-1} within the plane of the aircraft penetration. Figure 7.35b shows both Z_{dr} (in contours) and LDR (in greyscale), the peak LDR of -17 dB occurring a few hundred meters below the P-3 track and within the updraft core. The P-3 aircraft track penetrates through enhanced LDR in the range -18 to -21 dB. It should be emphasized that these LDR data were acquired at a frequency of 10 GHz (X-band radar), but reflectivity and Z_{dr} were acquired at 3 GHz (S-band radar). Hydrometeor images from an imaging probe on the aircraft showed a mix of liquid drops (elliptically shaped images) and frozen drops (near circular images), see Fig. 7.35c.

The freezing of raindrops has been modeled by Smith et al. (1999) under realistic in-cloud conditions. They found that the freezing process of typical drops could take several minutes (when initialized at $-5\,^{\circ}$C) and that the particles could ascend 1 to 2 km in the updraft during this period. These model results and other *in situ* hydrometeor data in Florida storm cells described by Smith et al. (1999) support the interpretation that enhanced LDR in the mixed phase region capping positive Z_{dr} columns is likely to be due to partially freezing drops. The signature of the positive Z_{dr} column capped by enhanced LDR is a time-dependent phenomenon reflecting the microphysical and kinematic evolution within the convective cell. For example, Fig. 7.36 shows the time evolution of the vertical structure of Z_{dr} (in contours) and LDR (in greyscale) of a convective cell (the same situation shown in Fig. 7.35) over a period of about 15 minutes. At 1808 (the early growth phase) the enhanced LDR region is just beginning, with peak LDR of -20.6 dB near the top of a well-developed positive Z_{dr} column. At 1810 (the vigorous phase) the enhanced LDR signature has expanded and by 1821 it has disappeared along with the positive Z_{dr} column structure. This last vertical section of Z_{dr} is reflective of the mature (steady) phase of the cell (see also, Fig. 7.28b) which is dominated by graupel particles above the 0$\,^{\circ}$C level; these particles melt to form raindrops whose Z_{dr} values lie in a narrow range between 1 and 2 dB. The LDR values are typically less than -25 dB during this phase.

The mixed phase region can be approximately modeled as a mixture of oriented oblate raindrops (with canting angle $\approx 0\,^{\circ}$) and partially frozen/wet ice particles with an isotropic orientation distribution. In such a model, the total equivalent copolar

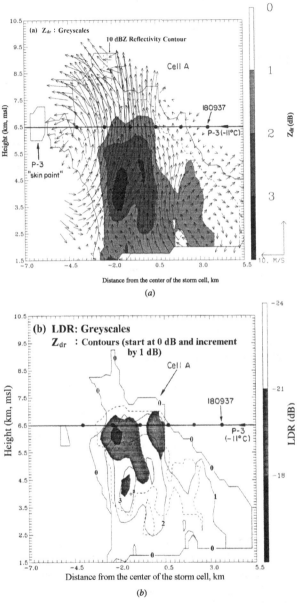

Fig. 7.35. (*a*) Vertical section of radar data (from the NCAR CP-2 radar) through the core of a vigorous convective cell in Florida. Greyscale depicts Z_{dr}. The outer 10-dBZ reflectivity contour is also shown. Arrows depict triple-Doppler derived wind vectors in the vertical plane. The NOAA P-3 aircraft track is shown with solid dots spaced at 10-second increments. (*b*) Vertical section of LDR in greyscales with darker shades representing higher depolarization levels (note that LDR measurements are at a frequency of 10 GHz). Contours depict Z_{dr}, with contours starting at 0 dB and incrementing by 1 dB. (*c*) Hydrometeor images from a cloud imaging probe on the P-3. Bringi et al. (1997).

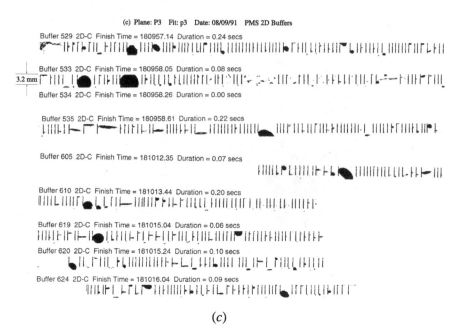

(c)

Fig. 7.35. (cont.)

reflectivity factors, see (6.61c, d), may be written as,

$$Z_h = Z_h^{\text{rain}} + Z^{\text{ice}}; \quad \text{mm}^6\,\text{m}^{-3} \tag{7.85a}$$
$$Z_v = Z_v^{\text{rain}} + Z^{\text{ice}}; \quad \text{mm}^6\,\text{m}^{-3} \tag{7.85b}$$

where because of the isotropic orientation assumption for ice particles there is no polarization dependence of Z^{ice}. The difference reflectivity factor Z_{dp} (in dB) is defined as,

$$Z_{\text{dp}} = 10\log_{10}(Z_h - Z_v) \tag{7.86a}$$
$$= 10\log_{10}(Z_h^{\text{rain}} - Z_v^{\text{rain}}) \tag{7.86b}$$

It follows that Z_{dp} is only sensitive to the oriented oblate raindrops in the mixture, similar to K_{dp} (see discussion related to (4.54)). Furthermore, the statistical correlation coefficient (due to measurement fluctuations) between radar-measured Z_h and Z_{dp} is very high and tends to 1. In addition, in rain-only regions, Z_{dp} is also physically highly correlated with Z_h^{rain} (the correlation coefficient tending to 1) even though the underlying drop size distribution, $N(D)$, can vary over a very wide range encompassing, for example, rain rates from 1 to 300 mm h^{-1}. This combination of very high statistical correlation (from a radar measurement fluctuation viewpoint) combined with very high physical correlation (from the dsd variability viewpoint) between Z_{dp} and Z_h^{rain}, gives rise to a very tight clustering of corresponding radar data about a straight line, which may be referred to as a "rain line", expressible as $Z_{\text{dp}} = a(10\log_{10} Z_h^{\text{rain}}) + b$. An example of radar data from rain is shown in Fig. 7.37.

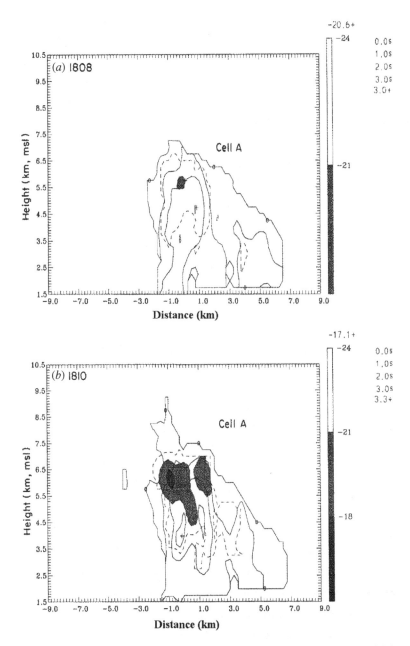

Fig. 7.36. Time evolution of vertical sections of radar data (from the NCAR CP-2 radar) through the core of a storm cell in Florida. LDR is shown in greyscales with darker shades representing higher depolarization levels. Contours depict Z_{dr}, with contours starting at 0 dB and incrementing by 1 dB. This cell is the same as shown in Fig. 7.35. For time periods: (*a*) 1808, (*b*) 1810, (*c*) 1814, and (*d*) 1821.

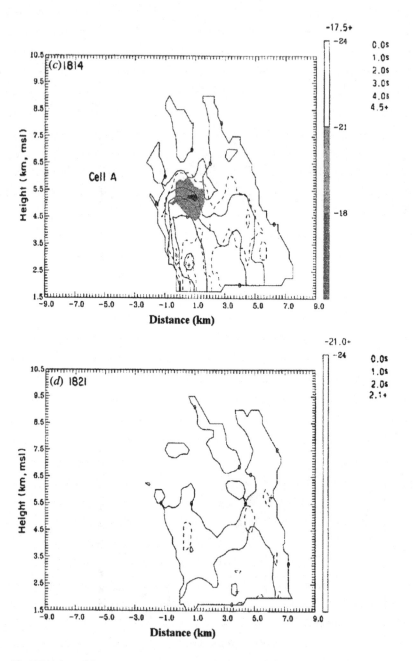

Fig. 7.36. (*cont.*)

When the radar resolution volume is composed of rain mixed with ice, as might occur in the mixed-phase region of a storm cell, the measured total Z_h will exceed Z_h^{rain}, whereas Z_{dp} will be approximately the same since it is insensitive to ice. The resulting

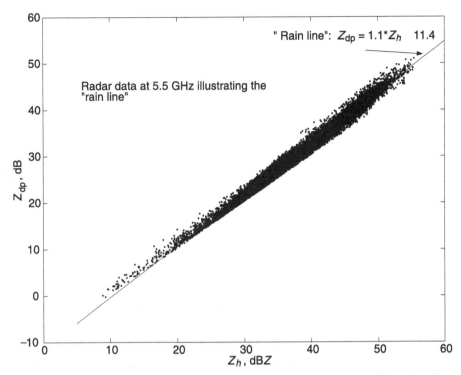

Fig. 7.37. Radar data at C-band (frequency near 5.5 GHz) illustrating the "rain line". Data from the BMRC C-POL radar located in Darwin, Australia. Solid line is a least-squares fit to the data. C-POL data were provided by the BMRC, Melbourne, Australia.

pair of data points (Z_{dp}, Z_h) from a resolution volume in the mixed-phase region will lie to the right of the rain line as illustrated in Fig. 7.38 where ΔZ (in dB) is the horizontal "deviation" from the rain line. It follows that,

$$10 \log_{10} Z_h = 10 \log_{10}(Z_h^{\text{rain}}) + \Delta Z \tag{7.87a}$$

$$= 10 \log_{10}(Z_h - Z^{\text{ice}}) + \Delta Z; \quad \text{dB}Z \tag{7.87b}$$

or,

$$\frac{Z^{\text{ice}}}{Z_h} = f = 1 - 10^{-0.1(\Delta Z)}; \quad \Delta Z \geq 0 \text{ dB} \tag{7.87c}$$

where f is the ratio of the equivalent ice reflectivity factor to the total equivalent reflectivity factor and may be referred to as the "ice fraction" ($0 \leq f \leq 1.0$). For example, when $\Delta Z = 3$ dB, the ice fraction is 50%, whereas for $\Delta Z = 10$ dB, $f = 90\%$. This means that the rain line must be very accurate, i.e. the slope a and intercept b in $Z_{dp} = a(\log_{10} Z_h^{\text{rain}}) + b$ should be estimated from the radar data in rain-only regions (at low altitudes within the storm cell). This will ensure that the deviation from the rain line in the rain-only regions is "calibrated" and will on average be 0 dB (the

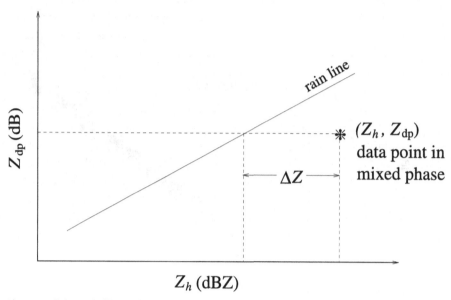

Fig. 7.38. Schematic illustrating the concept of "deviation" from the rain line. The deviation is ΔZ (in dB); see also (7.87).

standard error is estimated to be <1.0 dB). This in turn will ensure that ΔZ in the mixed-phase region will be accurate relative to the calibrated rain line established from data in rain-only regions.

Figure 7.39*a, b* shows vertical profiles of the average f (the ice fraction) as a function of time for a Florida convective storm cell (the same cell shown previously in Figs. 7.35 and 7.36) over a period from 1803 (early phase) to 1811 (vigorous growth phase) to 1820 (mature phase). This time sequence may be compared with the time sequence in Fig. 7.36. Such profiles reflect the height zone where the transition from rain to frozen hydrometeors occurs within the storm cell depending on their evolution.

Once the ice fraction is estimated, the ice–water content may be computed from power laws of the form in (7.84) with $Z_e \rightarrow fZ_e$, i.e. $IWC = a(fZ_e)^b$ where Z_e is the total equivalent reflectivity factor measured by radar (Z_e is the same as Z_h in (7.85a)). As explained earlier, the exponent b (for gamma dsd) is given as $4/7 = 0.57$, while the coefficient a may be established either empirically or through coordinated radar/aircraft data as explained earlier in relation to (7.84). Tong et al. (1998) modified the empirical relation provided by Sikdar et al. (1974) to estimate the vertical profile of total IWC (in kg) within each layer of a storm cell (the same cell discussed earlier in relation to Figs. 7.35, 7.36) and their results are shown in Fig. 7.39*c, d*. As the storm evolves from the early phase (1803) to the vigorous growth phase (1811–1813) to the mature phase (1821), the vertical profile of layer-total IWC in the cell also evolves in response to the microphysical and kinematic evolution. There is very little ice mass in the early phase (prior to 1805), followed by rapid ice mass build-up in the height layer from 5 to 6 km from 1809 to 1815. After 1816, the ice-mass peaks reduce during the mature

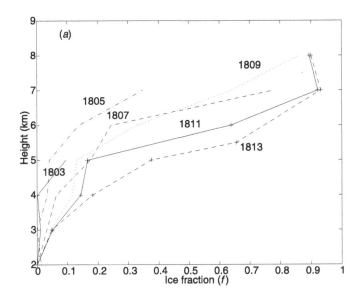

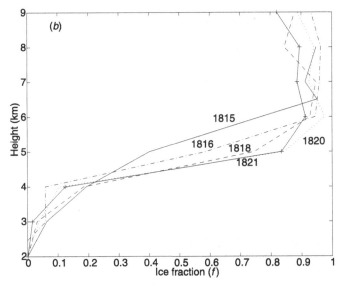

Fig. 7.39. Vertical profiles of the average ice fraction (f) as a function of time for the storm cell shown previously in Figs. 7.35 and 7.36. Time from: (a) 1803 to 1813, (b) 1815 to 1821. (c), (d) show vertical profiles of layer-total IWC as a function of time in the same storm cell. From Tong et al. (1998).

and dissipating phase (1818–1823). Such vertical profiles can also be used to estimate the time evolution of the total (vertically integrated) ice mass in the storm cell from which the time derivative of total ice mass in the cell can be derived. This estimate of

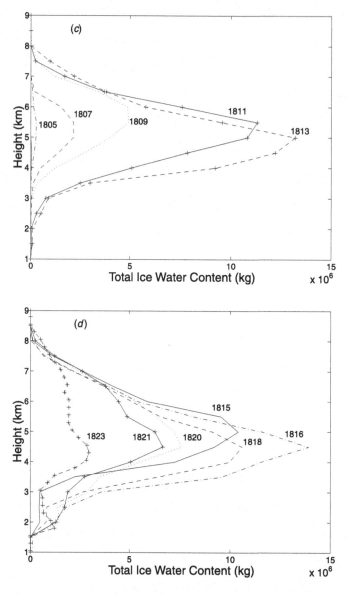

Fig. 7.39. (*cont.*)

d(IWC)$/dt$ is important in ice-budget studies of convective storm cells.

7.2.2   Hailstorms

Distinct polarimetric radar signatures have been observed in hailstorms since at least the early 1970s (based largely on circular polarized radar), and from the early 1980s based on linear polarization radar. The Alberta Research Council's S-band circularly polarized radar was designed and built for hail detection and was used for many years for hail research (e.g. Barge 1974; Al-Jumily et al. 1991). A detailed study of hailstorms was conducted under the auspices of the US National Hail Research Experiment (NHRE) in northeastern Colorado from 1972 to 1976 (Knight et al. 1982). Two dual-wavelength radars operating at wavelengths of 10 and 3 cm were deployed for this experiment (the NCAR/CP-2 and the University of Chicago–Illinois State Water Survey's CHILL radar). The essential idea is that hailstones with a diameter of around 10 mm fall in the Rayleigh scattering regime at the 10-cm wavelength ($\pi D/\lambda \approx 0.3$), whereas they fall in the Mie scattering regime at the 3-cm wavelength ($\pi D/\lambda \approx 1.0$). The ratio of the equivalent reflectivity factors at these two wavelengths ($Z_{e,10}/Z_{e,3}$) can be used to detect large hail reliably, provided matched antenna beams are used for the two radars, and provided the 3-cm wavelength returned signal power can be corrected for attenuation due to rain/wet ice along the propagation path. Practical difficulties related to mismatched antenna beams (and related errors caused by reflectivity gradients across the beam), as well as inaccurate correction of propagation-induced attenuation effects frequently resulted in ambiguous hail signatures (Rinehart and Tuttle 1984). Since this book is focused on polarimetric radar, dual-wavelength radar techniques will not be further discussed. For descriptions of combined polarimetric/dual-wavelength radar observations of hailstorms refer to Bringi et al. (1986b) and Tuttle et al. (1989).

Hailstones are characterized by wide variability in their size and shape distributions. The exponential size distribution, see (7.52a), appears to be a good approximation when large samples of hailstones are considered (Ulbrich and Atlas 1982). Cheng and English (1983) used the exponential shape and arrived at the relation $N_0 = 115\Lambda^{3.63}$ (mm^{-1} m^{-3}) from their data analysis but it is very unlikely that such a relationship is universally valid. The gamma model is likely to be a better approximation for analysis of time-resolved hail measurements. Figure 7.40 shows an example of hail size spectrum (from a 5-minute period) measured during a severe hailstorm in northern Colorado with an estimated slope of 0.3 mm^{-1} between 10 and 35 mm (Hubbert et al. 1998). A gamma shape with positive μ would be a better fit for diameters ≤ 10 mm. Mie scattering effects as well as lack of a well-defined relationship between hail shape and size have made it difficult, so far, to arrive at formal methods for estimating the parameters of the hail size distribution using polarimetric measurements.

Knight (1986) was able to document the average non-sphericity of hailstones from a large database acquired in Oklahoma and northeast Colorado. The short and large dimensions were measured from photographs of thin sections of hailstones, and a "shape factor" for approximately spheroidal stones was defined as the ratio of the shortest to longest dimensions. For stones with near-conical shapes, the long dimension was defined along the cone axis, while the shortest dimension was taken to be the maximum dimension perpendicular to the cone axis. Figure 7.41 shows the results from Knight's (1986) analysis. There is a general decrease in the shape factor with increase in the longest dimension up to around 25–30 mm. Beyond this, the shape factor is nearly constant (at 0.75) for the Oklahoma data while the Colorado data show a local maximum (to 0.85) near 35–40 mm with a decreasing trend thereafter. The most

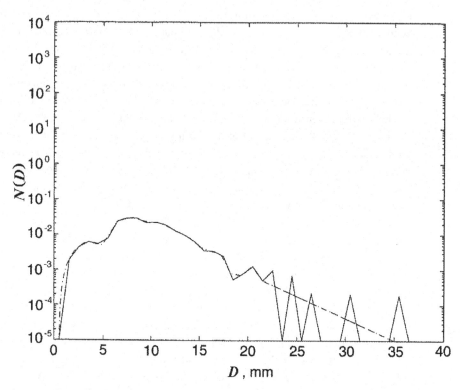

Fig. 7.40. Example of a measured hailstone size distribution over a 5-minute period in a severe storm in northern Colorado. $N(D)$ is in units of $mm^{-1}\ m^{-3}$. Radar data from this storm are shown in Figs. 7.56 and 7.58. The 5-minute collection interval is marked in 7.58c, from 600 to 900 s in that figure. From Hubbert et al. (1998).

commonly observed shapes for large hailstones (longest dimension $\geq$20 mm) appears to be oblate spheroidal, in which case the shape factor in Fig. 7.41 may be interpreted as the axis ratio (b/a in Fig. 2.7b). Wang et al. (1987) found (for a data set from one hailfall event) that a conical shape (see Fig. 7.42) was most representative, with a ratio of the short dimension to the long dimension (as defined by Knight 1986) of $\approx$0.85 (modal value), the longest dimension in their sample being less than about 20 mm. Knight also mentions that Colorado hailstones were nearly conical until they attained a size of 20–30 mm and, from Fig. 7.41, the shape factor is consistent with the Wang et al. (1987) analysis. Thus, it appears that conical hailstones have their long dimension (base to apex) larger than the short dimension (perpendicular to the cone's axis). Zrnić et al. (1993) argue that the equilibrium orientation of such conical stones should be with the apex pointed upward or downward. For a low elevation angle radar beam as illustrated in Fig. 2.7a and with the cone apex oriented along the Z-axis, the resulting monodisperse differential reflectivity (Z_{dr}, dB) would tend to be negative. A more realistic orientation

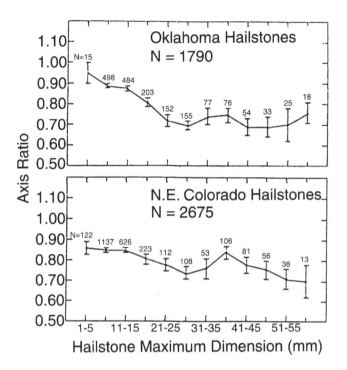

Fig. 7.41. Axis ratios of hailstones versus their longest dimensions. The number of hailstones in each size interval and the 95% confidence interval are marked: (top) for Oklahoma hailstones, $N = 1790$; (bottom) for northeast Colorado hailstones, $N = 2675$. From Knight (1986).

distribution would be the Fisher distribution with $\bar{\theta}_b = \bar{\phi}_b = 0$ (see (2.75) and Fig. 2.6*a*, which defines θ_b, ϕ_b).

The falling behavior of large oblate hailstones was analyzed by Knight and Knight (1970) based on a detailed examination and interpretation of the internal structure, since the symmetry of growth layers can provide important information on possible fall modes. Hailstones grow in size from the embryo stage (typically a frozen drop or a conical graupel particle), primarily by accretion of a unidirectional flux of supercooled droplets. Since the basic spheroidal symmetry is not changed by growth it implies that the fall mode is regular (not random) and that symmetrically equivalent points on the surface are equally exposed to this unidirectional flux (symmetrically equivalent points lie on any two circles about the minor axis which are equidistant from the equatorial plane, Kry and List 1974). Growth layers are often thickest along the equator and thinnest at the poles (see Fig. 7.43), which implies that the equator must have been directed more nearly downward compared to the poles. Based on a theoretical analysis, Kry and List (1974) conclude that:

> a freely falling spheroid spinning about its minor axis and with its total angular momentum horizontal can have an angular motion called symmetric gyration, in

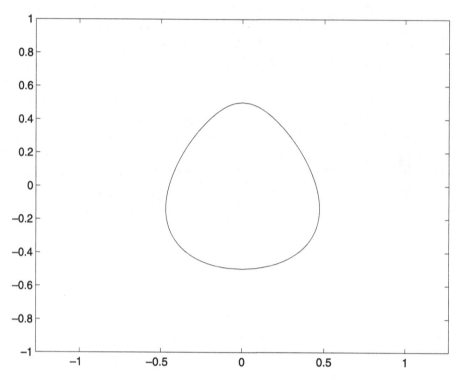

Fig. 7.42. Model shape of conical hailstones. From Wang et al. (1987).

which symmetrical equivalent points on the surface are equally exposed to the flow. The minor axes trace out a cone which is symmetric about the horizontal plane through the center of the spheroid at a rate which is of the same order as the spin rate but is not necessarily an integral multiple of it.

Figure 7.44 illustrates schematically the implied conical trace of the minor axis. It turns out that the Fisher orientation distribution in (2.73) with $\bar{\theta}_b = \bar{\phi}_b = (\pi/2)$ (recall that θ_b and ϕ_b are defined in Fig. 2.6a) can be used to compute polarimetric radar parameters due to such a fall mode. In the case of a radar beam at low elevation angle, and for distributions of such falling spheroids, it would be more realistic to assume the axial distribution defined in (2.81). Such a distribution will yield negative Z_{dr} (in dB) at 3 GHz (S-band) for an equivalent diameter of the oblates up to around 40 mm or so (negative Z_{dr} observations in large hail will be discussed later).

The dielectric constant of hailstones will vary with composition ranging from solid ice to "spongy" ice which can be modeled as a two-phase mixture of water imbedded in an ice matrix (see the Maxwell-Garnet formula in Section 1.6). "Wet" hailstones can be modeled either as possessing the dielectric constant of water, which is the simplest approximation, or as a two-layer body with an inner solid ice core surrounded by an oblate coat of water.

Experiments conducted by Rasmussen et al. (1984) have shown that melting ice

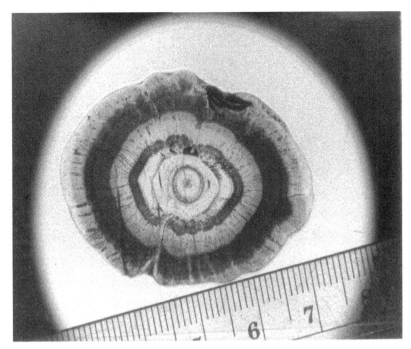

Fig. 7.43. Photomicrograph through a large hailstone showing the various growth layers. This stone was collected during the severe hailstorm event shown in Figs. 7.56 and 7.58.

spheres shed part of their melt water if their initial diameter is greater than 9 mm (at sea-level). There is a maximum amount of water that can be retained on the surface of a given ice core mass (in the form of a water torus around the equator) before shedding begins. This maximum amount is termed the "equilibrium" amount; if the actual amount of water exceeds this equilibrium amount then shedding occurs in the form of tiny drops (0.5–1.5 mm in diameter). Figure 7.45 shows a schematic diagram of the stages of melting experienced by high- and low-density ice particles. Figure 7.46 shows an example of an HVPS image of a melting hailstone obtained during a T-28 aircraft penetration of a hailshaft in northeastern Colorado near the melting level; the image shows what appears to be a water torus around an ice core.

As the melting of small hail (i.e. with diameters $\leq$10–12 mm) proceeds, the melt water can form an oblate shell around the diminishing inner ice core which greatly increases its orientational stability leading to what appears to the radar as "giant" raindrops. Larger melting hailstones, as mentioned earlier, shed off their excess water until the equilibrium amount is attained, following which the retained melt water will tend to stabilize their orientation. Figure 7.47 shows schematically the melting of initial 6- and 10-mm ice spheres using the Rasmussen and Heymsfield (1987) melting model and a typical environmental sounding profile from the southeastern USA. Figure 7.48 shows how the differential reflectivity (Z_{dr}) of single particles would increase with decreasing height below the 0 °C level (4.5-km altitude in this case). This figure

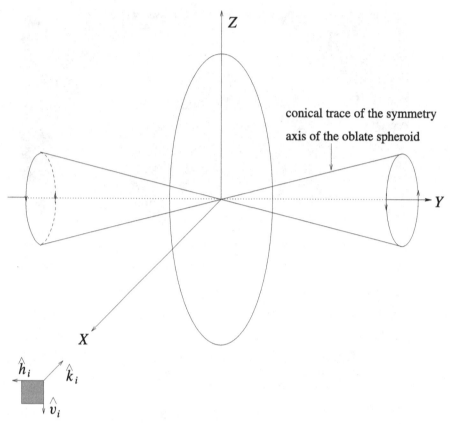

Fig. 7.44. Illustrating the symmetric gyrating fall mode behavior of large hailstones proposed by Kry and List (1974).

compares Z_{dr} for identical 6- and 10-mm sized spherical ice particles (model hailstones) with corresponding low density (0.4 g cm^{-3}) conical-shaped graupel particles. The high-density hailstones have to fall further from the initialization altitude (4.5 km) before they melt completely, relative to the lower-density graupel particles. Note the rather large values of Z_{dr} (4 dB) for the 6-mm hail particle upon complete melting after a fall of 2.5 km from the 0 °C level compared with 2.5 dB for the graupel particle (after a corresponding fall of 2 km from the 0 °C level). These calculations were performed for a frequency of 3 GHz.

At a frequency of 5.5 GHz (C-band) resonance effects due to Mie scattering are manifested for melting hail particles in the diameter range 6–8 mm causing Z_{dr} to increase substantially relative to a fully melted particle of the same axis ratio. For example, Fig. 7.49 shows Z_{dr} versus height for a 7-mm spherical hail particle initialized at 4.3 km (environmental 0 °C level). Also shown is the diameter of the inner ice core from the Rasmussen–Heymsfield melting model. Note the large increase in Z_{dr} to 8 dB near 1.5 km where the inner ice core diameter is 4.6 mm and the overall axis ratio is 0.58. This ice core is assumed to be completely contained within an oblate water

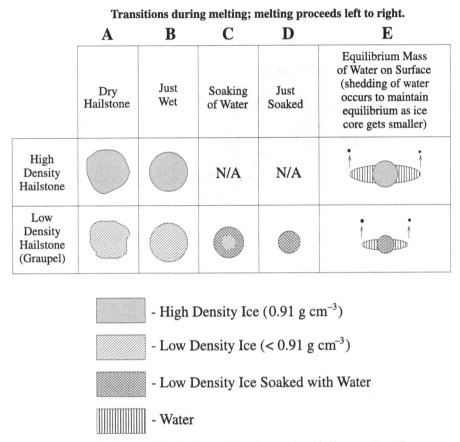

Fig. 7.45. Illustrating the transition during melting of graupel and hailstones. From Rasmussen and Heymsfield (1987).

shell as depicted in Fig. 7.47 (third from left). Calculations show that this strong Z_{dr} enhancement at C-band disappears for diameters less than 6 mm or greater than around 9 mm. Large values of differential reflectivity (6–8 dB) have been observed at C-band in convective precipitation shafts (see, for example, Meischner et al. 1991b).

For a given size distribution of hailstones initialized at the 0 °C level and allowed to melt according to the Rasmussen–Heymsfield model, the vertical profile of Z_{dr} will strongly depend on the maximum initial hailstone size, the sounding profile, and the magnitude of the downdraft within the precipitation shaft. If the initial maximum hail diameter is less than around 12 mm or so, and if the sounding profile is representative of summer-time conditions in southeastern USA (i.e. warm cloud base temperatures, high relative humidity below cloud base), then most of the hailstones will melt by the time they reach the surface, and the vertical profile of Z_{dr} would be similar to the model calculations shown in Fig. 7.48. As the initial maximum hail diameter increases, those melting hailstones exceeding about 12 mm in diameter will tend to shed off their melt water and appear spherical in shape causing the resultant differential reflectivity

1.8 cm

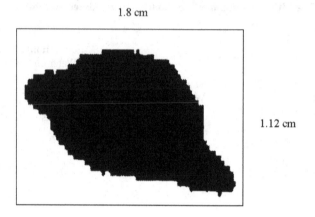

1.12 cm

Fig. 7.46. Example of a melting hailstone image as recorded by the high volume particle spectrometer (HVPS). The HVPS was mounted on the T-28 aircraft. Further examples are shown in Fig. 7.62*b*.

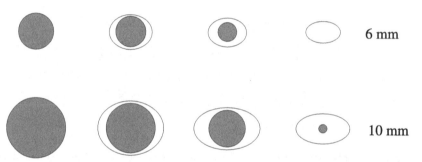

6 mm

10 mm

Fig. 7.47. Illustrating the melting of initial 6- and 10-mm spherical hailstones using the Rasmussen–Heymsfield melting model. The assumed environmental sounding is typical of summer-time conditions in southeastern USA.

to be biased towards 0 dB (recall that Z_{dr} is a measure of the reflectivity-weighted mean axis ratio, and that the axis ratio of the larger hailstones would tend to be near unity). This bias towards lower Z_{dr} values with increasing initial hail diameter would be more pronounced for sounding profiles representative of the USA High Plains region where the cloud base temperatures are colder and where the relative humidity below cloud base is lower than in the southeastern USA. Downdrafts within the hailshaft will further reduce the time available for melting causing the resultant Z_{dr} to be biased towards lower values (i.e. tending to 0 dB). Figure 7.50 shows a vertical section through a convective storm near Huntsville, Alabama, that produced a damaging downdraft (microburst) at the surface. A narrow vertical region of reduced Z_{dr} values (≤ 1 dB) is clearly evident at the range of 15–16 km (marked by a solid bold arrow at the bottom of the figure) extending from the surface to 4 km. Wakimoto and Bringi (1988) have referred to this signature as a "Z_{dr}-hole". Note that the Z_{dr}-hole is contained within

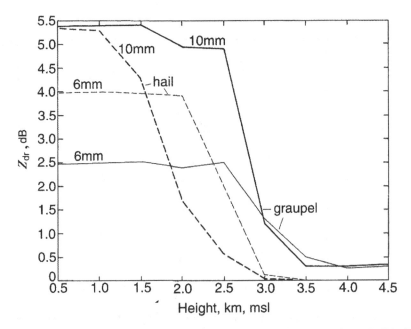

Fig. 7.48. The differential reflectivity of single melting particles (Z_{dr}, dB) at 3-GHz frequency as they fall in still air and melt according to the Rasmussen–Heymsfield model. Particles are initialized at the 0 °C level (4.5 km height above ground level in this case). Initial graupel density is 0.4 g cm^{-3}, while it is 0.92 g cm^{-3} for hailstones. Environmental sounding is assumed to be typical of summer-time conditions in the southeastern US. From Vivekanandan et al. (1990).

reflectivity values exceeding 50 dBZ (shaded region). Surrounding the Z_{dr}-hole, the Z_{dr} values increase quickly and are more representative of values associated with large raindrops (2–4 dB) formed from melting of smaller hail/graupel.

Hailshafts below the melting level can easily be detected via the Z_{dr}-hole signature, which has been well-documented in the literature (e.g. Bringi et al. 1986b). Aydin et al. (1986) have introduced a more quantitative boundary drawn on the two-dimensional plane defined by Z_h along the vertical axis and Z_{dr} along the horizontal axis, as illustrated in Fig. 7.51. This figure shows simulated values of (Z_h, Z_{dr})-pairs calculated from surface disdrometer measurements of the drop size distribution obtained in central Illinois and near Boulder, Colorado. The straight line segments tend to form an upper bound to the rain data, i.e. for a given Z_{dr} the corresponding Z_h in rain must be below the straight line segment boundary. If Z_h values exceed the boundary then it is deduced that they are due to ice hydrometeors. Values near the boundary may be interpreted as rain mixed with ice hydrometeors. This idea is related to the deviation from the rain line illustrated in Fig. 7.38, which estimates the fraction of ice in mixed-phase precipitation. Aydin et al. (1986) define H_{dr} as,

$$H_{dr} = Z_h - f(Z_{dr}); \quad dB \tag{7.88}$$

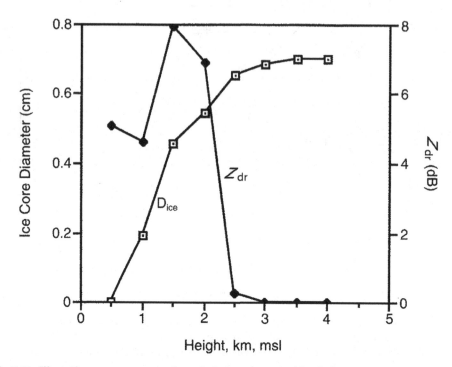

Fig. 7.49. Illustrating resonance scattering calculations due to melting hailstones at 5.5-GHz frequency (C-band). Spherical 7-mm diameter ice spheres are initialized at the 0 °C level (4.5 km height above ground level in this case) and allowed to melt according to the Rasmussen–Heymsfield melting model in still air. D_{ice} refers to the inner ice core diameter (see also, Fig. 7.47), while Z_{dr} is the differential reflectivity of a single particle. Resonant scattering causes enhanced Z_{dr} between 1.5- and 2.0-km height. From Vivekanandan et al. (1990).

where Z_h is the measured reflectivity (in dBZ) while $f(Z_{dr})$ defines the straight line boundary segment (at 3-GHz frequency),

$$f(Z_{dr}) = \begin{cases} 27; & Z_{dr} \leq 0 \text{ dB} \\ aZ_{dr} + 27; & 0 < Z_{dr} \leq b \\ 60; & Z_{dr} > b \end{cases} \tag{7.89}$$

The three different boundaries in Fig. 7.51a are based on the equilibrium raindrop axis model, as well as on two different drop oscillation models. The equilibrium axis ratio assumption gives $a = 16.5$, $b = 2$ dB. The two oscillation models give (a, b) of $(19, 1.74)$ and $(19.5, 1.67)$, respectively. In practice, use of the boundary based on the equilibrium model should be sufficient. Figure 7.51b,c shows radar data along a single beam (or a range profile) at low elevation angle , through the core of a hailstorm to illustrate the magnitude of H_{dr} as well as its sharp gradient near the rain/hail boundary. From theory, it appears unlikely that the magnitude of H_{dr} itself can be used to estimate the damage potential of hailstorms since (to a first approximation) spherical

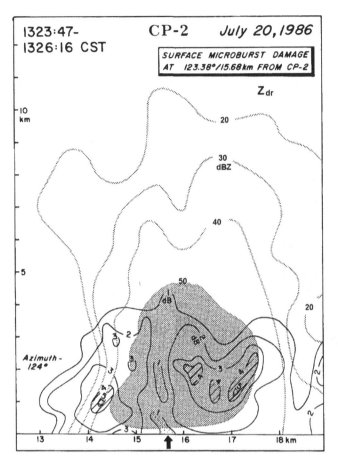

Fig. 7.50. Vertical section of reflectivity and Z_{dr} (from the NCAR CP-2 radar) in a microburst-producing storm cell. The arrow at the bottom marks the "Z_{dr}-hole" signature. From Wakimoto and Bringi (1988).

hail will yield $H_{dr} = Z_h - 27$, and then H_{dr} depends only on Z_h (which in turn will depend on both N_0 and Λ of the size distribution for an exponential shape assumption). Nevertheless, an exploratory study by Brandes and Vivekanandan (1998) found that H_{dr} was correlated with maximum hail size (correlation coefficient of 0.63) in a study of two Colorado hailstorms with coordinated hail chase vehicles and volunteer observers. Their data are shown in Fig. 7.52 as a plot of H_{dr} from radar versus maximum hail diameter recorded by observers. Even though there is uncertainty due to comparing point ground observations with radar data, H_{dr} is clearly correlated with the maximum hail diameter. The correlation coefficient with the equivalent reflectivity factor (Z_h) was much smaller (0.33) and hence H_{dr} is likely to be a better indicator of hail size than reflectivity alone. However, if hailstones fall in a mode which yields positive Z_{dr} in the same range as observed for oblate raindrops, then the H_{dr} method for hail detection will

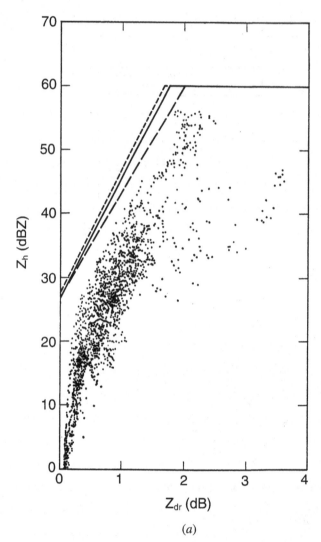

(a)

Fig. 7.51. (a) Calculations of Z_h versus Z_{dr} at 3-GHz frequency based on measured drop size distributions. Straight line segments on the figure form an upper-bound to the rain data, i.e. for a given Z_{dr} in rain, the corresponding Z_h must be below the straight line segments. (b) Radar data (Z_h and Z_{dr}) along a single beam (from the NCAR CP-2 radar) through a severe convective storm in Boulder, Colorado. (c) H_{dr} calculated from (7.88, 7.89) versus range along the same beam. Large values of H_{dr} indicate hail regions. From Aydin et al. (1986).

not work (Smyth et al. 1999). The utility of H_{dr}, in practice, is related to its simplicity in detecting hailshafts at altitudes below the melting level; at higher altitudes within the storm it quickly loses its informative content and becomes strictly proportional to Z_h only, since most forms of frozen hydrometeors typically yield average $Z_{dr} \approx 0$ dB.

Once the Z_{dr}-hole signature has been detected, additional measurements of LDR, $|\rho_{co}|$, and K_{dp} within the hailshaft may provide important indications of hail severity

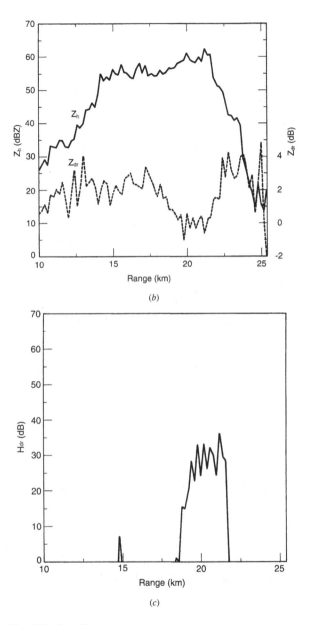

Fig. 7.51. (*cont.*)

and if the hailshaft is accompanied by heavy rainfall (Balakrishnan and Zrnić 1990a,b). The covariance matrix in the linear polarization basis of such a two-component mixture of hydrometeor types may be modeled as a superposition of the individual covariance matrices of oblate rain (satisfying "mirror" reflection symmetry) and hail (to a first approximation satisfying polarization plane isotropy; see Section 3.12). The covariance

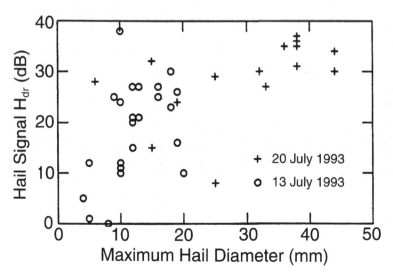

Fig. 7.52. Radar-based values of H_{dr} correlated with surface observations of maximum hail diameters in two hailstorms in northern Colorado. Radar observations were made at 3-GHz frequency with the NCAR CP-2 radar. From Brandes and Vivekanandan (1998).

matrix for hail (subscript a below) follows from (3.193) and may be expressed as,

$$
\left(\sum_{\mathrm{BSA}}\right)_{\mathrm{hail}}
= \begin{bmatrix}
\langle n_a|S_{hh}|_a^2\rangle & 0 & \{\langle n_a|S_{hh}|_a^2\rangle - 2\langle n_a|S_{hv}|_a^2\rangle\} \\
0 & 2\langle n_a|S_{hv}|_a^2\rangle & 0 \\
\{\langle n_a|S_{hh}|_a^2\rangle - 2\langle n_a|S_{hv}|_a^2\rangle\} & 0 & \langle n_a|S_{hh}|_a^2\rangle
\end{bmatrix}
$$
(7.90)

The equivalent reflectivity factor of the hail component, $(Z_e)_{\mathrm{hail}}$, may be expressed in terms of $\langle n_a|S_{hh}|_a^2\rangle$, see (5.49a), while the differential reflectivity ($\mathfrak{z}_{\mathrm{dr}}$ expressed as a ratio), linear depolarization ratio (L), and copolar correlation coefficient magnitude ($|\rho_{\mathrm{co}}|$) of the hail component are,

$$(\mathfrak{z}_{\mathrm{dr}})_{\mathrm{hail}} = 1 \tag{7.91a}$$

$$(L)_{\mathrm{hail}} = \frac{\langle n_a|S_{hv}|_a^2\rangle}{\langle n_a|S_{hh}|_a^2\rangle} \tag{7.91b}$$

$$|\rho_{\mathrm{co}}|_{\mathrm{hail}} = 1 - 2(L)_{\mathrm{hail}} \tag{7.91c}$$

The covariance matrix for the rain component (subscript b below) assuming Rayleigh scattering and "mirror" reflection symmetry about a mean canting angle of $0°$ can be expressed, see (4.107), as,

$$\left(\sum_{\text{BSA}}\right)_{\text{rain}} = \begin{bmatrix} \langle n_b|S_{hh}|_b^2\rangle & 0 & \langle n_b|S_{hh}|_b\,|S_{vv}|_b\rangle \\ 0 & 2\langle n_b|S_{hv}|_b^2\rangle & 0 \\ \langle n_b|S_{hh}|_b\,|S_{vv}|_b\rangle & 0 & \langle n_b|S_{vv}|_b^2\rangle \end{bmatrix} \tag{7.92}$$

The equivalent reflectivity factor of the rain component, $(Z_h)_{\text{rain}}$, may be expressed in terms of $\langle n_b|S_{hh}|_b^2\rangle$, while the differential reflectivity $(\mathcal{Z}_{\text{dr}})$, linear depolarization ratio (L), and copolar correlation coefficient magnitude $(|\rho_{\text{co}}|)$ of the rain component are,

$$(\mathcal{Z}_{\text{dr}})_{\text{rain}} = \frac{\langle n_b|S_{hh}|_b^2\rangle}{\langle n_b|S_{vv}|_b^2\rangle} \tag{7.93a}$$

$$(L)_{\text{rain}} = \frac{\langle n_b|S_{hv}|_b^2\rangle}{\langle n_b|S_{hh}|_b^2\rangle} \tag{7.93b}$$

$$|\rho_{\text{co}}|_{\text{rain}} = \frac{\langle n_b|S_{hh}|_b\,|S_{vv}|_b\rangle}{\sqrt{\langle n_b|S_{hh}|_b^2\rangle\,\langle n_b|S_{vv}|_b^2\rangle}} \tag{7.93c}$$

The covariance matrix of the mixture of hail and rain is the sum of (7.90) and (7.92) and with modest algebraic manipulation can be expressed as,

$$\left(\sum_{\text{BSA}}\right)_{\text{mix}} = (Z_e)_{\text{mix}} \begin{bmatrix} 1 & 0 & |\rho_{\text{co}}|_{\text{mix}} \\ 0 & 2(L)_{\text{mix}} & 0 \\ |\rho_{\text{co}}|_{\text{mix}} & 0 & 1/(\mathcal{Z}_{\text{dr}})_{\text{mix}} \end{bmatrix} \tag{7.94}$$

where,

$$(Z_e)_{\text{mix}} = (Z_e)_{\text{hail}} + (Z_h)_{\text{rain}} \tag{7.95a}$$

$$(\mathcal{Z}_{\text{dr}})_{\text{mix}} = \frac{1 + [(Z_h)_{\text{rain}}/(Z_e)_{\text{hail}}]}{1 + [1/(\mathcal{Z}_{\text{dr}})_{\text{rain}}][(Z_h)_{\text{rain}}/(Z_e)_{\text{hail}}]} \tag{7.95b}$$

$$(L)_{\text{mix}} = \frac{(L)_{\text{hail}} + (L)_{\text{rain}}[(Z_h)_{\text{rain}}/(Z_e)_{\text{hail}}]}{1 + [(Z_h)_{\text{rain}}/(Z_e)_{\text{hail}}]} \tag{7.95c}$$

$$|\rho_{\text{co}}|_{\text{mix}} = \frac{|\rho_{\text{co}}|_{\text{hail}} + |\rho_{\text{co}}|_{\text{rain}}[(Z_h)_{\text{rain}}/(Z_e)_{\text{hail}}](\mathcal{Z}_{\text{dr}})_{\text{rain}}^{-1/2}}{\{1 + [(Z_h)_{\text{rain}}/(Z_e)_{\text{hail}}]\}\,(\mathcal{Z}_{\text{dr}})_{\text{mix}}^{-1/2}} \tag{7.95d}$$

Note that $(Z_h)_{\text{rain}}/(Z_e)_{\text{hail}}$ may also be expressed in terms of the ice fraction, f, defined in (7.87c),

$$\frac{(Z_h)_{\text{rain}}}{(Z_e)_{\text{hail}}} = \frac{1-f}{f} \tag{7.96a}$$

$$= \frac{1}{10^{0.1(\Delta Z)} - 1} \tag{7.96b}$$

It is illustrative to express (7.95) in terms of the ice fraction f, as,

$$(\mathcal{Z}_{\text{dr}})_{\text{mix}} = \frac{(\mathcal{Z}_{\text{dr}})_{\text{rain}}}{1 + f[(\mathcal{Z}_{\text{dr}})_{\text{rain}} - 1]} \tag{7.97a}$$

$$(L)_{\text{mix}} = f(L)_{\text{hail}} + (1 - f)(L)_{\text{rain}} \tag{7.97b}$$

$$|\rho_{\text{co}}|_{\text{mix}} = \frac{f|\rho_{\text{co}}|_{\text{hail}} + (1 - f)|\rho_{\text{co}}|_{\text{rain}}(\mathcal{Z}_{\text{dr}})_{\text{rain}}^{-1/2}}{(\mathcal{Z}_{\text{dr}})_{\text{mix}}^{-1/2}} \tag{7.97c}$$

It is interesting to note that in this rain/hail mixture model, only the linear depolarization ratio satisfies a simple mixing formula. If the ice fraction can be determined from measurement of ΔZ (which is the horizontal deviation from the rain line, see Fig. 7.38), then $(\mathcal{Z}_{\text{dr}})_{\text{rain}}$ can be derived from (7.97a) since $(\mathcal{Z}_{\text{dr}})_{\text{mix}}$ is measured. However, it is not possible to separately estimate $(L)_{\text{hail}}$, $(L)_{\text{rain}}$ and $|\rho_{\text{co}}|_{\text{rain}}$ since only two additional equations are available (7.97b, c).

As the hailshaft or Z_{dr}-hole is approached from the surrounding rain regions, it is clear that $(\mathcal{Z}_{\text{dr}})_{\text{mix}}$ will reduce and tend to unity as the ice fraction tends to unity. The linear depolarization ratio will increase as the hailshaft is approached since $(L)_{\text{hail}}$ will generally exceed $(L)_{\text{rain}}$. Recall that the linear depolarization ratio in rain is unlikely to exceed -26 dB or so (see Fig. 7.13) at 3 GHz (S-band). The $|\rho_{\text{co}}|_{\text{mix}}$ will also decrease as the hailshaft is approached from the surrounding rain region. Again, recall that $|\rho_{\text{co}}|_{\text{rain}}$ is generally larger than 0.985 (see discussion related to (7.47)), whereas $|\rho_{\text{co}}|_{\text{hail}} = 1 - 2(L)_{\text{hail}}$. Figure 7.53a shows how $|\rho_{\text{co}}|_{\text{mix}}$ varies with the ice fraction f with $(L)_{\text{hail}}$ as a parameter, and for fixed $|\rho_{\text{co}}|_{\text{rain}} = 0.99$ and $(\mathcal{Z}_{\text{dr}})_{\text{rain}} = 2.0$. For large $(L)_{\text{hail}}$ there is a steady decrease in $|\rho_{\text{co}}|_{\text{mix}}$ as f increases from 0 (rain only case) to 1 (hail only case), the value at $f = 1$ being governed by the linear depolarization ratio of hail (for $(L)_{\text{hail}} = -15$ dB, $|\rho_{\text{co}}|_{\text{mix}}$ when $f = 1$ is 0.937, see (7.91c)). For smaller values of $(L)_{\text{hail}}$, there is a minimum in $|\rho_{\text{co}}|_{\text{mix}}$ which occurs as f varies from 0 to 1.

Observations at S-band show that lower values of $|\rho_{\text{co}}|$ are possible in hail, e.g. as low as 0.90. Within the context of the mixture model proposed here, the $|\rho_{\text{co}}|_{\text{mix}}$ can attain a value which is lower than $|\rho_{\text{co}}|_{\text{hail}}$ or $|\rho_{\text{co}}|_{\text{rain}}$ if $(\delta_{\text{co}})_{\text{rain}}$ is non-zero. This can occur at S-band (and is more likely at C-band) if partially melting hailstones (≤ 12–15 mm) retain an oblate shape and are oriented, like "giant" raindrops. Mie scattering from such giant raindrops will result in non-zero δ_{co}, see (7.49). Equation (7.97c) may be easily modified to account for non-zero δ_{co},

$$|\rho_{\text{co}}|_{\text{mix}} = \frac{\left| f|\rho_{\text{co}}|_{\text{hail}} + (1 - f)|\rho_{\text{co}}|_{\text{rain}}e^{-j\delta_{\text{co}}}(\mathcal{Z}_{\text{dr}})_{\text{rain}}^{-1/2} \right|}{(\mathcal{Z}_{\text{dr}})_{\text{mix}}^{-1/2}} = \tag{7.98a}$$

$$\frac{\left[f^2|\rho_{\text{co}}|_{\text{hail}}^2 + 2f(1-f)|\rho_{\text{co}}|_{\text{hail}}|\rho_{\text{co}}|_{\text{rain}}(\mathcal{Z}_{\text{dr}})_{\text{rain}}^{-1/2}\cos\delta_{\text{co}} + (1-f)^2|\rho_{\text{co}}|_{\text{rain}}^2(\mathcal{Z}_{\text{dr}})_{\text{rain}}^{-1} \right]^{1/2}}{(\mathcal{Z}_{\text{dr}})_{\text{mix}}^{-1/2}} \tag{7.98b}$$

Figure 7.53b shows sample plots of $|\rho_{\text{co}}|_{\text{mix}}$ versus f assuming various values of δ_{co} to illustrate how a non-zero scattering differential phase caused by Mie scattering can lower $|\rho_{\text{co}}|_{\text{mix}}$. In general, the $|\rho_{\text{co}}|_{\text{mix}}$ of a rain–ice mixture will decrease significantly when the reflectivity-weighted contributions from the two components are nearly equal,

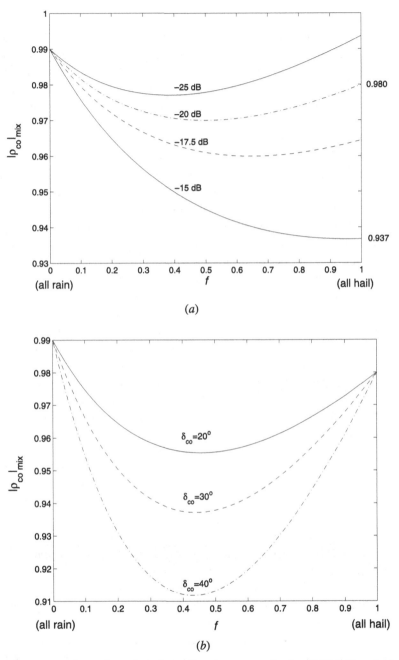

Fig. 7.53. (*a*) Illustrating calculations of $|\rho_{co}|_{mix}$ versus f using (7.97c). For rain, it is assumed that $|\rho_{co}|_{rain} = 0.99$ and $(\xi_{dr})_{rain} = 2.0$ ($Z_{dr} = 3$ dB). (*b*) Illustrating application of (7.98). For hail, the $|\rho_{co}|_{hail}$ is assumed to be 0.98 (consistent with LDR = -20 dB). Rain assumptions are as in (*a*).

or when $f \approx 0.5$. However, the minimum value attained will depend upon having significant δ_{co} for one of the components.

There have been numerous observations of negative Z_{dr} (lower bound at S-band being ≈ -1.5 dB) occurring with high reflectivity within the cores of hailshafts (e.g. Bringi et al. 1984; Aydin et al. 1986). As discussed earlier, the orientation model depicted in Fig. 7.44 can result in negative Z_{dr} at low elevation angles (depending on size, axis ratio, and dielectric constant), and this means that the covariance matrix in (7.90), and in (7.91), is not strictly valid. Figure 7.54 shows scattering calculations at S-band from dry, wet, and spongy oblate hailstones whose orientation model is simpler than that shown in Fig. 7.44, i.e. $\bar{\theta}_b = 90°$ while the pdf of ϕ_b is uniform in $(0, 2\pi)$; recall that (θ_b, ϕ_b) specify the orientation of the symmetry axis of the spheroid as illustrated in Fig. 2.6a. For axis ratios of 0.8 (which is typical based on Fig. 7.41), the differential reflectivity is around 0 to -1 dB, and is only weakly dependent on hail diameters up to 40 mm. Beyond 40 mm, strong resonances cause Z_{dr} to vary rapidly with size. This weak dependence on diameter (up to 40 mm) suggests that negative Z_{dr} by itself may not be useful in estimating the damage potential caused by hail. However, Husson and Pointin (1989) analyzed (Z_h, Z_{dr}) radar data in coordination with a hailpad network and found that estimates of D_{max} (the maximum hail diameter recorded by the hailpad) could be improved if the Z_{dr} value (at the corresponding resolution volume nearest the hailpad where peak reflectivity occurred) was used as compared with using the peak reflectivity alone (correlation coefficient improved from 0.68 to 0.86 which is significant). Figure 7.55 shows examples of the time evolution of Z_h and Z_{dr} over three hailpads (marked E5, E55, and F4) which experienced light, strong, and very strong hailfall intensity, respectively. At the location of the peak Z_h in each case, the corresponding Z_{dr} values were, respectively, 1.0, 0.6, and -0.6 dB. While negative Z_{dr} values in hailstorms have been frequently observed, Smyth et al. (1999) have reported positive Z_{dr} values (≈ 3 dB) coincident with surface observations of oblate hail (disc-shaped stones) with axis ratio around 0.6 and longest dimension around 32 mm. They invoke a fall mode, where the major axis is describing a cone about the horizontal (rather than the minor axis depicted in Fig. 7.44). Such a fall mode will give rise to positive Z_{dr} values and is consistent with the fall mode postulated by Browning and Beimers (1967).

Considerable additional information for estimating large hail within a hailshaft is provided by K_{dp}, LDR, and $|\rho_{co}|$. This is well-illustrated in Figs. 7.56 and 7.57. In Fig. 7.56, a vertical section of reflectivity is shown through the core of a severe hailstorm which produced large hail (the size distribution was shown earlier in Fig. 7.40). Along the cascade (marked X), the profiles of Z_{dr} and $|\rho_{co}|$ are shown in Fig. 7.57a, while (b) shows K_{dp} and LDR$_{vh}$. Note the large increase in LDR$_{vh}$ below the 0 °C level (from -22 to -15 dB near the surface), whereas Z_{dr} becomes negative (from around zero to -1 dB) and $|\rho_{co}|$ decreases from 0.96 to 0.88. The core of this hailstorm passed over an instrumented van, and visual observations of hail size (as well as rain accumulation with a gage) were available for a period of nearly 50 minutes (Hubbert et al. 1998). Radar observations over the van location are shown in Fig. 7.58 as a

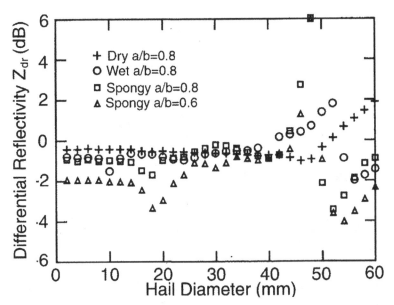

Fig. 7.54. Calculations of Z_{dr} (at 3-GHz frequency) versus hail diameter. The hailstones are assumed to be oblate in shape (with a fixed axis ratio of 0.8) with orientation similar to Fig. 7.44, except there is no conical gyration (i.e. $\bar{\theta}_b = 90°$, ϕ_b is uniform in 0 to 2π). Spongy hail refers to dielectric constant with 40% water fraction. From Balakrishnan and Zrnić (1990b).

function of time, with visual observations of precipitation type marked in (c). From 1803–1808, hailstones were manually collected using a roof-mounted net on the van before the severity of hailfall pulled the netting from its frame (the size distribution in Fig. 7.40 was obtained from analysis of these collected stones). Visual observations of golfball-sized hail started around 1809 (i.e. 1000 s in panel Fig. 7.58(c)) and ended at around 1827 (i.e. at 2000 s). This golfball-sized hail is associated with negative Z_{dr}, low $|\rho_{co}|$ ($\leq$0.93), and enhanced LDR$_{vh}$ (≥ -18 dB). The polarimetric radar data prior to, during, and after the large hailfall period are in good agreement with scattering models. From 0 to 1000 s, the LDR$_{vh}$ increases gradually from -27 to -20 dB, in agreement with visual observations of increasing hail size from pea to marble to golfball. The K_{dp} (which can be related to rain rate, see Chapter 8) increases from 0.75 to 2° km^{-1}, with Z_{dr} in the range 1.3–2.5 dB. The $|\rho_{co}|$ decreases from 0.96 to 0.93 in agreement with the predictions of (7.97c). After the large hailfall period, Z_{dr} shows an increasing trend to more positive values while LDR$_{vh}$ decreases to -27 dB, and $|\rho_{co}|$ increases to 0.97–0.98, all characteristic of the transition from rain mixed with hail to rain only, see (7.97).

Figure 7.58d shows rain accumulation using a K_{dp}-based algorithm (see Section 8.1.2) and a Z_h-based algorithm ($Z_h = 300R^{1.4}$). The rain accumulation using the K_{dp}-based estimator is in excellent agreement with gage data in this particular event (total accumulation of 40 mm), while the Z_h–R relation overestimates by a factor of

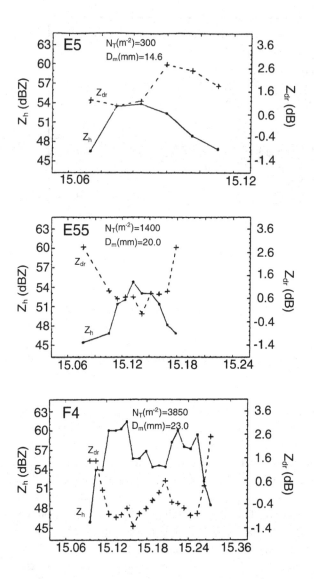

Fig. 7.55. Time variation of Z_h and Z_{dr} above three hailpads as measured by the S-band ANATOL radar in France. N_T refers to the number of hailstones per m² and D_{max} is the maximum hail size as recorded by the three hailpads (E5, E55, and F4). From Husson and Pointin (1989).

two (the maximum Z_h is limited to 55 dBZ; rain rate algorithms will be discussed in more detail in Chapter 8).

The radar identification of hail growth at mid-to-upper levels of hailstorms is of considerable importance as opposed to detection of the hailshaft near the surface.

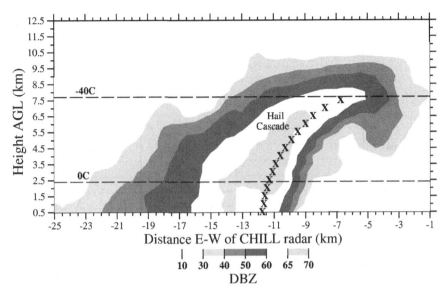

Fig. 7.56. Vertical section of Z_h (measured by the CSU–CHILL radar) through the core of a severe hailstorm in northern Colorado. The Xs mark the hail cascade. From Carey and Rutledge (1998). (© 1998 American Geophysical Union.)

Circular polarization radar measurements at S-band of CDR and ORTT, see (3.214), have been shown to be useful in detection of hail aloft, in particular, ORTT values are low ranging from 10–40%, while CDR ≥ -20 dB. Low values of ORTT in hail have been interpreted as implying a high degree of randomness in the orientation distribution, which may be associated with frozen or partly frozen hydrometeors (Hendry et al. 1976). However, from (3.214d) the measured ORTT is a product of ρ_2 (an orientation parameter defined in (3.201a)) and a "shape" factor $[|\bar{v}|/(\overline{v^2})^{1/2}]$. Hence, ORTT can also decrease if there is a large variance in the shape of hailstones and this is likely to occur in natural hailfall. Al-Jumily et al. (1991) showed that, after correction for differential propagation phase in rain, the parameters ORTT and Z_{dr} (derived from circular polarization data, see (3.228a)), were the most useful in identification of hail, together with the reflectivity η_c, see (3.202a).

The linear depolarization ratio (LDR) is also useful in detection of hail growth aloft, especially if the hail is in "wet" growth. Wet growth occurs on a hailstone when all the accreted supercooled water does not immediately freeze but accumulates on or inside the stone. Partially freezing raindrops can also cause an increase in LDR. Figure 2.8c showed the large sensitivity of LDR to the dielectric constant. At S-band, there is an increase in LDR of typically 3–5 dB as hail becomes wet. Kennedy et al. (in press) suggest an S-band LDR_{vh} threshold of -25 dB at mid-to-upper levels of a thunderstorm (around 0 to $-20\,°C$ levels) for the detection of growing hail. Enhancement of LDR above this threshold by 3–6 dB is likely due to an increase in the dielectric constant and is representative of spongy hail (water embedded in an ice matrix). Kennedy et al.

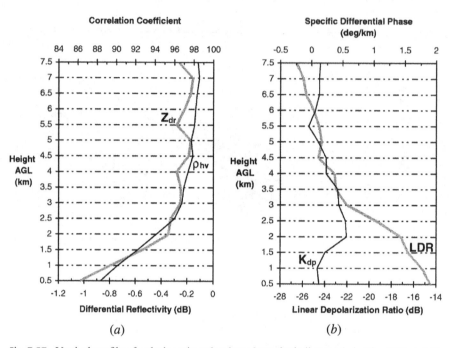

Fig. 7.57. Vertical profile of polarimetric radar data along the hail cascade in Fig. 7.56: (*a*) Z_{dr} and $|\rho_{co}|$, (*b*) K_{dp} and LDR$_{vh}$. From Carey and Rutledge (1998). (©️ 1998 American Geophysical Union.)

(in press) found that this LDR enhancement aloft occurred around 10 minutes before a hailshaft could be detected at the surface. A vertical cross section along the main updraft axis of a supercell storm in eastern Colorado is shown in Fig. 7.59 to illustrate enhanced LDR (shown as solid dots where LDR > -25 dB) within the storm core. A positive Z_{dr} column is noted, which extends 1 km above the 0 °C level. A cap of enhanced LDR is noted on the Z_{dr} column top and above where rapid accretional growth of hail is occurring. The Z_{dr}-hole signature is clearly identified near the surface of the hailshaft (Z_{dr} near 0 dB, reflectivity >60 dBZ; at a horizontal distance of 3 km). As one progresses upward in altitude from the Z_{dr}-hole, this region is surrounded by LDR > -25 dB, which extends up to 8 km or so.

At C-band (frequencies near 5.0 GHz), Mie scattering by spongy hail causes LDR to increase significantly, as shown by vertical sections of radar data through the core of a hailstorm in Fig. 7.60: (*a*) shows contours of Z_h with Z_{dr} in greyscale; while (*b*) shows contours of Z_h, with LDR in greyscale. The positive Z_{dr} column is clearly evident below the reflectivity overhang. A pocket of enhanced LDR (-18 to -14 dB) is noted at mid-levels (near a height of 5.5 km) while the -22 to -18 dB contour extends from 7.5 km to the surface. Conway and Zrnić (1993) demonstrated the use of trajectory analysis (in a Colorado supercell storm) to show that positive Z_{dr} columns had an important role to play in embryo production and hail growth, and their analysis is consistent with the data shown in Figs. 7.59 and 7.60. Hubbert et al. (1998) plotted

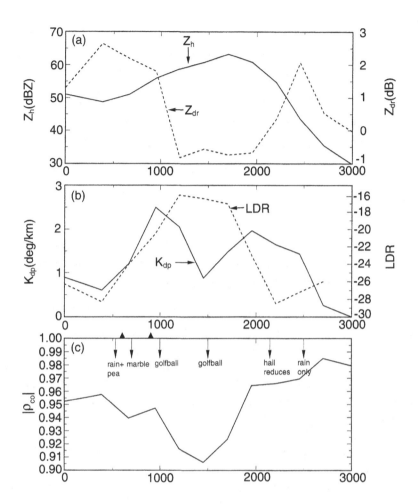

Fig. 7.58. Time series of polarimetric radar data (*a–c*) over the chase van location (see Fig. 7.61 for van location). Data from the CSU–CHILL radar at 1° elevation angle. Visual observations (made by van personnel) of hydrometeor type are marked in (*c*). The two solid triangles mark the time interval during which hailstones were collected (size distributions were shown earlier in Fig. 7.40). (*d*) Rain accumulation measured with a Young capacitance gage on the van (the gap in the data was caused by a power supply problem to the personal computer). $R(K_{dp})$ and $R(Z_h)$ refer to rain accumulation using K_{dp}–R and Z_h–R algorithms (discussed in Chapter 8). From Hubbert et al. (1998).

swaths of enhanced LDR (≥ -20 dB) near the surface, and positive Z_{dr} (≥ 1 dB) at 3.5-km altitude, for an hour-long period of a supercell hailstorm shown earlier in Fig. 7.56. These swaths were created by selecting, at each grid point, the maximum value from all the analyzed volumes of data during the hour-long period, and are shown in Fig. 7.61*a*. The LDR swath locates the hailswath near the surface. The positive Z_{dr} column swath is

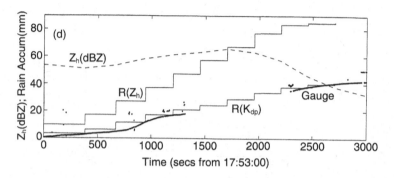

Fig. 7.58. (*cont.*)

located southeast of the hailswath and its location across the main inflow region places the drop embryos in a favorable position for the production of large hail, in general agreement with the trajectory analysis of Miller et al. (1988; 1990) and Conway and Zrnić (1993).

Figure 7.61*b* shows a swath of the LDR "cap" region (≥ -25 dB) at 4.5-km altitude, and a swath of K_{dp} ($\geq 0.5°$ km^{-1}) at 2.5 km (0 °C level). The swath of the LDR cap is contained within the positive Z_{dr} column swath and it marks a mixed-phase region composed of supercooled rain, partially freezing drops, and wet graupel. The positive K_{dp} swath is located approximately between the surface hailswath and the positive Z_{dr} column swath. Since the surface hailswath is generally coincident with the storm's main downdraft, while the positive Z_{dr} column is located near the low-to-mid level updraft, the K_{dp} swath appears in the updraft/downdraft interface region. It is hypothesized that some of the positive K_{dp} values at this level (near 0 °C level) may be due to 0.5–1.5-mm drops shed by wet hail (see also, Fig. 7.45). Even though these drops are nearly spherical (axis ratio >0.95 or so), the fact that copious numbers of drops are shed will result in a measurable K_{dp} (refer, also, to the discussion related to (7.7) and (7.17)). If this hypothesis is true, then these shed drops are in a favorable region to be carried aloft, freeze, and become embryos for large hail (Rasmussen and Heymsfield 1987; Miller et al. 1988; Loney et al. 1999).

As a final example, data from coordinated CSU–CHILL radar and T-28 aircraft penetrations near the melting level (around 0 °C) of a Colorado hailshaft are shown in Fig. 7.62: (*a*) shows a constant altitude PPI section of reflectivity at an altitude of 2.5 km (above ground level) together with the T-28 penetration track (the "skin paint" of the aircraft itself may be noted at $X = 6.4$ km, $Y = 35$ km); (*b*) shows samples of HVPS images of raindrops mixed with hailstones (within the high reflectivity part of the penetration) with very smooth edges of the larger particles indicative of the onset of melting; (*c*) shows the size distribution (from a 1-minute interval) with maximum detected hail diameters up to 15 mm. During this particular penetration (third of a sequence of four successive penetrations of the same region) the peak Z_h was nearly 60 dBZ, with corresponding Z_{dr}, LDR, and K_{dp} of 0 dB, -22.5 dB, and 0.5° km^{-1}, respectively. Figure 7.63*a* shows the radar-measured Z_h (along four penetration segments, but shown

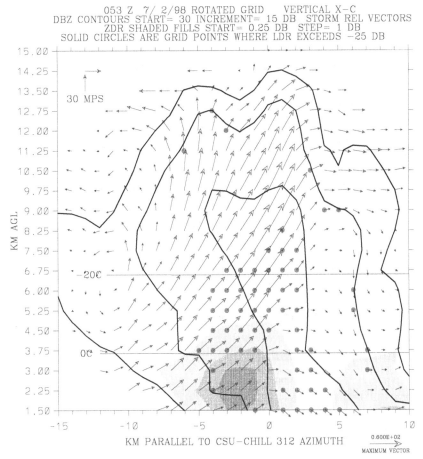

Fig. 7.59. Vertical section of radar data (using the CSU–CHILL radar) through the core of a severe hailstorm in northern Colorado. Wind vectors in this plane are based on dual-Doppler synthesis using the CSU–CHILL and Pawnee radar. Z_{dr} is shown in greyscales, with darker shades corresponding to more positive values. Solid dots represent grid points where LDR_{vh} exceeds −25 dB. Figure prepared by Patrick Kennedy of the CSU–CHILL staff.

along a single time axis) compared with calculations of the equivalent reflectivity factor from the measured size distributions along corresponding segments with an optimal adjustment of particle density ($\rho_p \approx 0.93$ g cm^{-3}); this optimal adjustment procedure was described earlier in relation to (7.80). Figure 7.63b shows comparisons of the radar-measured LDR_{vh} against computed LDR using the measured size distributions (details are available in Abou-El-Magd et al. 2000). Comparing (Fig. 7.63a, b) it may be noted that when $Z_h \geq 50$ dBZ, the corresponding LDR typically ranges between −22 and −24 dB (which may be compared with lower LDR values of around −26 dB due to dry, lower density graupel, see Fig. 7.34). These data generally support the LDR threshold of −25 dB used by Kennedy et al. (in press) to differentiate between spongy/wet hail aloft from lower-density graupel (see also, Fig. 7.59).

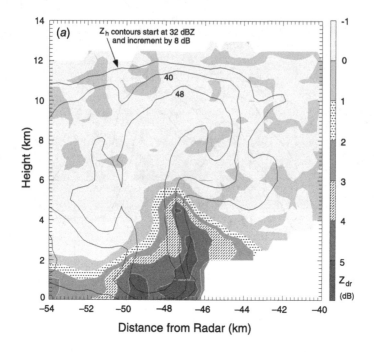

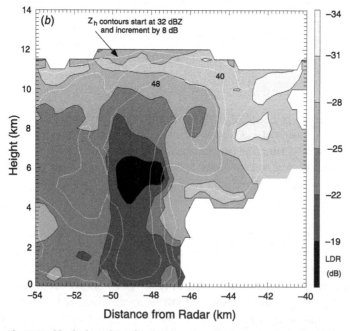

Fig. 7.60. Vertical section of radar data through the core of a hailstorm near Munich, Germany. Data taken with the C-band DLR polarimetric radar. Contours of: (*a*) Z_h, with Z_{dr} in greyscale; (*b*) Z_h, with LDR in greyscale. From Hubbert et al. (1995).

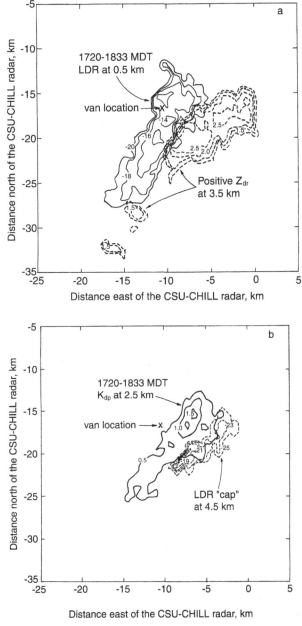

Fig. 7.61. Swaths of radar signatures (from the CSU–CHILL radar) for a severe hailstorm case in northern Colorado (see also, Figs. 7.56 and 7.58). Swaths are constructed based on the maximum value of a radar observable at each grid point from all the analyzed volumes. (*a*) Swaths of LDR > −20 dB at the 0.5-km height (above ground level). It represents the surface hailswath. Dashed lines are swaths of Z_{dr} > 1 dB at the 3.5-km altitude level representing the positive Z_{dr} column locations. (*b*) Swath of K_{dp} > 0.5° km^{-1} at 2.5-km altitude, possibly representing drops shed by wet hail. Dashed lines show LDR > −25 dB at the 4.5-km altitude, representing the mixed-phase interface region. From Hubbert et al. (1998).

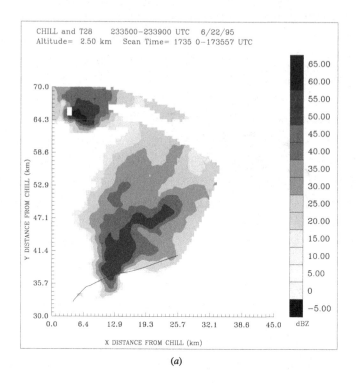

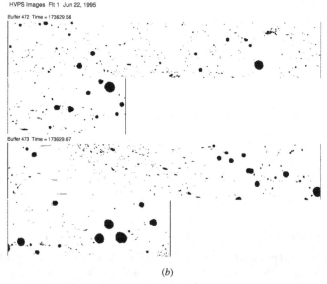

Fig. 7.62. (a) Constant altitude PPI (CAPPI) of reflectivity at 2.5-km altitude (above ground level) together with the T-28 aircraft penetration track. (b) HVPS images of hailstones within the high reflectivity portion of the penetration. Vertical dotted lines on the extreme right-hand side of the image buffers represent 4.5 cm. (c) Size distribution calculated from HVPS images for a 1-minute interval along the penetration track. $N(D)$ is in m^{-3} mm^{-1}.

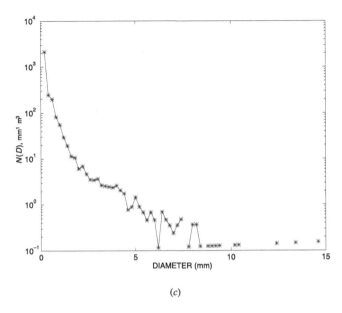

(c)

Fig. 7.62. (*cont.*)

7.3   Stratiform precipitation

From a radar perspective, the vertical profile of Z_e in stratiform precipitation is generally representative of rain with a "bright-band" (BB), which is a frequent occurrence in the mid-latitudes (e.g. Fabry and Zawadzki 1995). The Z_e is nearly constant from the surface up to the melting level, where Z_e increases in the BB. Above the BB, Z_e generally decreases with height (see Fig. 7.64). A characteristic kinematic feature of stratiform precipitation observed with vertical pointing Doppler radars is the rapid increase in particle fall speed in the BB as the low density snow particles melt to form raindrops (see Fig. 7.65, which also shows the BB). The mesoscale vertical air motion in stratiform precipitation is generally very weak, with draft magnitudes on the order of tens of centimeters per second (Houze 1993).

At heights well above the BB (i.e. at cold temperatures $\leq -5\,^{\circ}$C), ice crystals can grow by a process called deposition (water vapor diffusion to ice particles). While there are a large number of crystal types that can form depending upon humidity and temperature (e.g. Pruppacher and Klett 1997), in the Rayleigh scattering approximation they can be modeled as "plate-like" or "column-like" scatterers. Plate-like shapes are usually characterized by two dimensions (a "thickness" h, and equivalent circular "diameter" L with $h/L \ll 1$). In the case of column-like crystals, L would be the maximum dimension along the axis of the column and h the equivalent circular "diameter" of the cross section of the column with $L/h \gg 1$. Matrosov et al. (1996) have summarized a number of experimental observations by various investigators relating h and L by a power law of the form $h = \alpha L^{\beta}$. A further simplification can

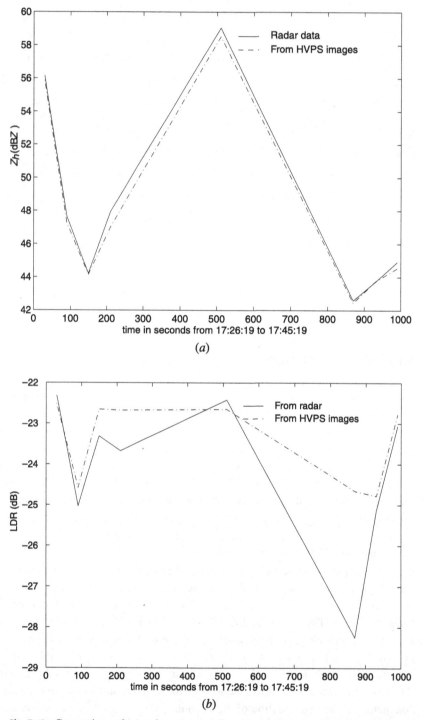

Fig. 7.63. Comparison of (a) radar measured Z_h with equivalent reflectivity calculated from HVPS-based size distributions for four successive T-28 penetrations, (b) comparison of LDR.

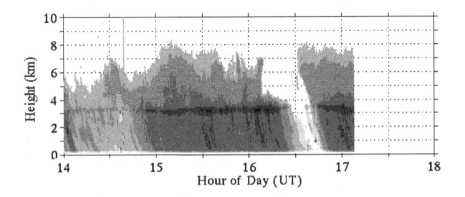

Fig. 7.64. Time–height cross section of equivalent reflectivity measured by a vertically pointing 915-MHz Doppler radar (or Profiler) showing the bright-band signature. Reflectivity contours are shaded, the highest shade being 10 dBZ, with increments of 10 dB. From Gage et al. (1999).

be made by assuming that plate-like shapes can be approximated as oblate spheroidal (axis ratio $b/a \equiv h/L \ll 1$, see Fig. 2.7b), whereas column-like shapes are prolate spheroidal (axis ratio $b/a \equiv h/L \gg 1$, see Fig. 2.7a). The fall behavior of pristine plate-like crystals may be modeled as a narrow Gaussian pdf in θ_b (mean $\bar{\theta}_b = 0°$; small σ) with ϕ_b being uniform in $(0, 2\pi)$ (the angles θ_b and ϕ_b refer to the orientation of the symmetry axis of the spheroid, see Fig. 2.6a). For column-like crystals, the fall behavior may be approximated as shown in Fig. 2.7c, with narrow Gaussian pdf in θ_b (mean $\bar{\theta}_b = 90°$; small σ) and ϕ_b uniform in $(0, 2\pi)$. At low radar elevation angles, one characteristic signature of pristine ice crystals is an increase in the differential propagation phase (Φ_{dp}) with increasing range (or positive K_{dp}). This signature was described in Section 4.4, and was shown to be derived from circular polarization radar data (see Fig. 4.14). Of course, a Φ_{dp} increase has also been observed at conventional linear horizontal/vertical polarization basis at frequencies ranging from 3 to 10 GHz. A good example is shown in Fig. 7.66 from an experiment conducted in the Arctic region with an X-band dual-polarized radar (Hudak et al. 1999). The antenna elevation angle was fixed at 3.5° and the height was derived from the slant range (slant range $\approx$65 km for a height of 4 km). The three panels show Z_h, Z_{dr}, and Φ_{dp}, and it can be noted that while Z_h and Z_{dr} are nearly uniform, Φ_{dp} increases by 40° over a slant range of 65 km (or average $K_{dp} \approx 0.3°$ km^{-1}). The temperature decreased from -1 °C at the surface to -12 °C at 4-km height. These data are representative of light snow, which fell for 18 hours.

Ryzhkov et al. (1998) have reported on the average relation between Z_{dr} and K_{dp} at 3-GHz frequency in "cold" snow versus "warm" snow events in Oklahoma (see Fig. 7.67). They noted that "cold" snow events (surface temperature < -5 °C) were dominated by pristine and moderately aggregated crystals, while "warm" events (surface temperature ≈ 0 °C) contained heavily aggregated snow with much higher

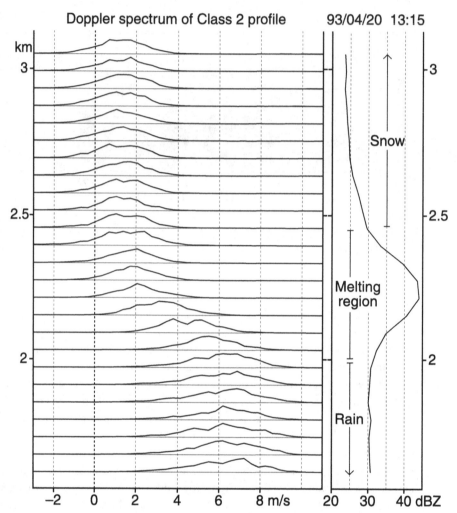

Fig. 7.65. Doppler spectrum at vertical incidence using a 915-MHz profiler. The horizontal axis is Doppler velocity (positive values imply downward vertical velocity). The vertical profile of equivalent reflectivity is shown on the right panel and shows the bright-band. Snow particles fall relatively slowly ($\sim$1.5 m s^{-1}) and melting gives rise to the bright-band. The velocity increase is mostly confined to the bottom of the bright-band. From Fabry et al. (1994).

equivalent reflectivity factors. If it is assumed that the heavily aggregated snow satisfies polarization plane isotropy (with $Z_{dr} \rightarrow 0$ dB, see Section 3.12.2); then it follows that in a mixture with oriented crystals, the Z_{dr} will be biased towards the Z_{dr} of the larger snow aggregates, whereas K_{dp} will only be sensitive to the oriented pristine crystals. This situation is exactly similar to a mixture of spherical hail and oblate raindrops; see (7.95b). Thus, in Fig. 7.67 the "cold" events have slightly larger Z_{dr} (on average by 0.5–0.7 dB) than the "warm" events for the same K_{dp}. It also follows that the scatterplot of Z_h versus Z_{dr} from regions of pristine ice crystals will be substantially different from

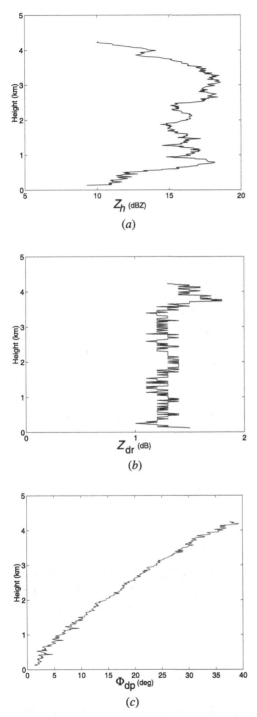

Fig. 7.66. Vertical profile of polarimetric radar data at 10 GHz in snow from the Arctic region using the Canadian IPIX radar: (a) reflectivity (Z_h), (b) Z_{dr}, and (c) Φ_{dp}. Height profiles were reconstructed from low elevation angle scans. From Hudak et al. (1999).

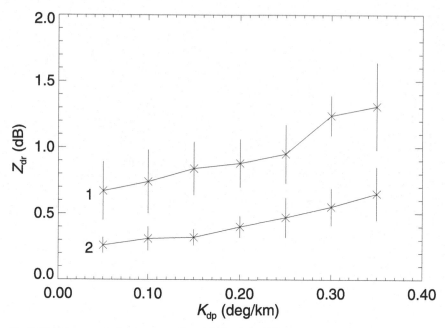

Fig. 7.67. Average relation between Z_{dr} and K_{dp} in "cold" (1) and "warm" (2) snow events in Oklahoma using the NSSL/Cimarron S-band radar. From Ryzhkov et al. (1998).

regions of heavily aggregated snow particles, as shown in Fig. 7.68. These radar data were at C-band and were coordinated with instrumented aircraft equipped with imaging probes from which particle types were classified as indicated in Fig. 7.68b, d. In regions known to be dominated by dendritic crystals (which are approximately "plate-like"), the Z_{dr} is positive (0.5–4 dB) and Z_h is relatively low (−5–10 dBZ). In regions dominated by larger aggregates of snow crystals and rimed crystals, $Z_{dr} \approx 0$ dB while Z_h is larger (10–20 dBZ). Early observations of this signature at S-band were reported by Bader et al. (1987). Qualitatively, such polarimetric data may be used to study aggregation processes in winter precipitation events.

Quantitative application of K_{dp} data has been proposed for estimating the ice water content of pristine ice crystals (Vivekanandan et al. 1994; Ryzhkov et al. 1998). Similar to (7.15), and assuming that plate-like crystals are oblate spheroids,

$$\mathrm{Re}\left[\frac{\varepsilon_r - 1}{1 + \frac{1}{2}(1 - \lambda_z)(\varepsilon_r - 1)} - \frac{\varepsilon_r - 1}{1 + \lambda_z(\varepsilon_r - 1)}\right] \approx C\rho_p^2(1 - r) \qquad (7.99)$$

where C is nearly constant (varying from 1.35 to 1.85 over a range of r from 0.01 to 0.2), ρ_p is the crystal density, and r is the axis ratio ($r = b/a = h/L \ll 1$). If D is the equal-volume spherical diameter of the oblate crystal, $D = 2(h)^{1/3}(L)^{2/3}$, and if $N(D)$

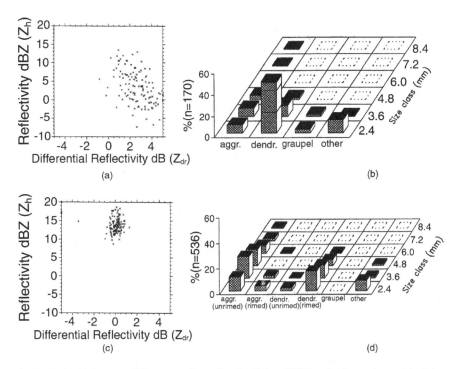

Fig. 7.68. (a, b) Scatter of Z_h versus Z_{dr} using the C-band DLR radar located near Munich, Germany. (c, d) Results of hydrometeor classification using data from a particle imaging probe on the NCAR King Air aircraft. (a) and (b) represent data from a region of predominantly dendritic ice crystals, while (c) and (d) represent data from predominantly heavily aggregated snow particles. From Vivekanandan et al. (1994, © 1994 IEEE).

is the size distribution, then substituting (7.99) in (7.10b) gives,

$$K_{dp} \approx \frac{\pi k_0 C}{12} \int \rho_p^2 D^3 (1 - r) N(D) \, dD \tag{7.100a}$$

$$= \frac{\pi}{\lambda} C \int \rho_p^2 \frac{\pi}{6} D^3 (1 - r) N(D) \, dD \tag{7.100b}$$

As a first approximation, assume that crystal density and axis ratio are independent of D in which case,

$$K_{dp} = 10^{-3} \left(\frac{180°}{\lambda} \right) C \rho_p (\text{IWC})(1 - r); \quad ° \, \text{km}^{-1} \tag{7.101}$$

where IWC is the ice water content in g m^{-3}, λ is the wavelength in m, ρ_p is the particle density in g cm^{-3}, and $C \approx 1.6$ (g cm^{-3})$^{-2}$ for Rayleigh scattering. If ρ_p is chosen to be 0.107 g cm^{-3}, then assuming $r \to 0$, (7.101) reduces to the simple relation IWC $\approx 3.2 K_{dp}$ at S-band ($\lambda = 0.10$ m) applied by Ryzhkov et al. (1998) to compare radar-derived IWC against *in situ* measurements by aircraft. For a fixed ρ_p, the linear relation IWC $= \alpha K_{dp}$ was found (based on numerical scattering calculations) to be an excellent approximation at 95 and 220 GHz for hexagonal ice plates and stellar crystals (Aydin and Tang 1997). In practice, the quantitative application of (7.101) is limited by lack of knowledge of how ρ_p changes with crystal dimension (L) as aggregation becomes important, together with uncertainty in identification of crystal type as well as changes in fall mode behavior as crystals start to rime and aggregate.

Similar to the "rain line" described in Section 7.2, see (7.87) and Fig. 7.38, Z_h–Z_{dp} relations for plate- and column-like crystals yield "ice crystal lines" about which radar data (Z_h versus Z_{dp}) are tightly clustered, as illustrated in Fig. 7.69a (Meischner et al. 1991a; Aydin and Walsh 1998). The solid and dashed lines in Fig. 7.69a are based on numerical scattering calculations at 95 GHz for stellar and hexagonal column shapes, respectively, assuming gamma size distributions of the form in (7.57a) with sizes in the range 0.03–2 mm, $0 \leq \mu \leq 2$, and IWC from 0.001 to 1 g m^{-3}. The solid and dashed lines are mean fits to the numerical scattering-based values of Z_h and Z_{dp}. The data points in Fig. 7.69a are from radar measurements at 95 GHz (University of Wyoming's airborne 95-GHz dual-polarized radar), which are seen to be tightly clustered near the theoretical solid line for stellar crystals. Sample particle images from probes on the aircraft (see Fig. 7.69b) show predominant occurrence of pristine stellar crystals. It follows that as aggregation proceeds, a "deviation" from the "ice crystal line" may be defined, similar to (7.87), which can be used to determine the "aggregate fraction" in a mixture of pristine ice crystals and aggregated snow particles (the correspondence with (7.87) may be established by recognizing that the term "ice crystal" corresponds to "rain" while the term "aggregates" corresponds to "ice").

When precipitation is horizontally homogeneous, such as in stratiform events, polarization measurements taken as a function of radar elevation angle afford another basis for distinguishing between plate-like, column-like, and spherical (drizzle) scatterers (Vivekanandan et al. 1994; see also Section 3.8.1). Figure 7.70 illustrates the viewing geometry and expected fall modes for hexagonal plates and columns. Recall that the orientation pdf of plates can be modeled as a narrow Gaussian in θ_b (mean $\bar{\theta}_b = 0°$; small σ), whereas for columns mean $\bar{\theta}_b = 90°$ and σ is small); with ϕ_b being uniform in $(0, 2\pi)$ in both cases. Reinking et al. (1999) have used the 45°/135° linear polarization basis to measure the slant LDR and to distinguish between supercooled drizzle, plate-like crystals, and column-like crystals by using the elevation angle dependence of slant LDR. Slant LDR may be defined using the definition in (3.134), which is repeated here for convenience,

$$\sigma_{rt} = 4\pi |\hat{e}_r \cdot (\mathbf{S}_{BSA})\hat{e}_t|^2 \qquad (7.102)$$

where $\hat{e}_t$ is the unit polarization vector of the transmitted wave, see (3.136), and $\hat{e}_r$ corresponds to the back scattered wave. The copolar radar cross section is defined by (7.102) with $\hat{e}_r = \hat{e}_t$. Using (3.136), the 45° slant linear polarization unit vector is given

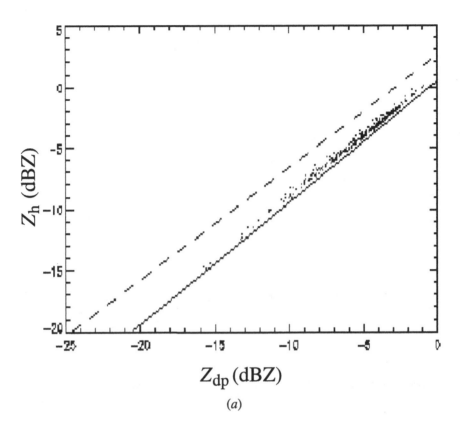

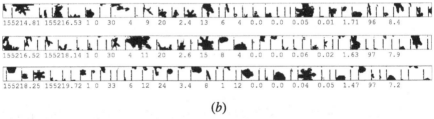

(b)

Fig. 7.69. (a) Graph illustrating the "ice crystal" line, using Z_h versus Z_{dp}. The solid line is a mean fit to stellar crystals at 95 GHz using gamma size distributions. The dashed line is a similar fit to hexagonal column crystals. The data points are from a 95-GHz radar on board the University of Wyoming King Air aircraft. (b) Sample images from a particle imaging probe on the aircraft showing a predominance of stellar crystals. From Aydin and Walsh (1998).

by substituting $\psi_t = 45°$, $\tau_t = 0°$ (see also, Fig. 3.3b),

$$\hat{e}_t = \hat{e}_r = \frac{1}{\sqrt{2}} \begin{bmatrix} 1 \\ 1 \end{bmatrix} \tag{7.103}$$

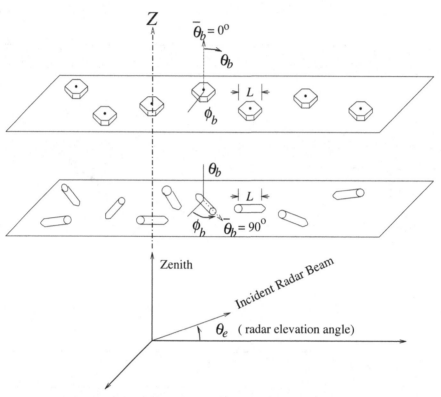

Fig. 7.70. Illustrating the approximate fall modes of plate- and column-like crystals. Elevation angle dependence of polarimetric radar data can be used to distinguish between these crystal types.

and (7.102) reduces to,

$$\sigma_{co} = \pi \left| \begin{bmatrix} 1 & 1 \end{bmatrix} \begin{bmatrix} S_{hh} & S_{hv} \\ S_{vh} & S_{vv} \end{bmatrix}_{BSA} \begin{bmatrix} 1 \\ 1 \end{bmatrix} \right|^2 \tag{7.104a}$$

$$= \pi \left| S_{hh} + 2S_{hv} + S_{vv} \right|^2 \tag{7.104b}$$

The cross-polar radar cross section is defined by (7.102) with $\hat{e}_r$ being orthogonal to $\hat{e}_t$. From (3.136), the unit vector orthogonal to $\hat{e}_t$ is obtained by replacing $\psi_t \to \psi_t + \pi/2$ and $\tau_t \to -\tau_t$. It follows that $\hat{e}_r$, which is orthogonal to $\hat{e}_t$, is given by,

$$\hat{e}_r = \frac{1}{\sqrt{2}} \begin{bmatrix} -1 \\ 1 \end{bmatrix} \tag{7.105}$$

and (7.102) reduces to,

$$\sigma_{cx} = \pi \left| \begin{bmatrix} -1 & 1 \end{bmatrix} \begin{bmatrix} S_{hh} & S_{hv} \\ S_{vh} & S_{vv} \end{bmatrix}_{BSA} \begin{bmatrix} 1 \\ 1 \end{bmatrix} \right|^2 \tag{7.106a}$$

$$= \pi \left| S_{vv} - S_{hh} \right|^2 \tag{7.106b}$$

The slant LDR is then defined as,

$$\text{SLDR} = 10\log_{10}\left(\frac{\sigma_{\text{cx}}}{\sigma_{\text{co}}}\right); \quad \text{dB} \tag{7.107}$$

To determine the variation of SLDR with radar elevation angle for plates and columns, the back scatter matrix in the Rayleigh limit for plates and needles will be used. First, consider plates whose polarizability elements are given in Table A1.1 (see Appendix 1). From (2.53) and using (2.13), together with $\theta_b = 0°$ (all plates oriented and of the same size) results in,

$$(S_{hh})_{\text{BSA}} = \frac{k_0^2}{4\pi\varepsilon_0}(\alpha) = \frac{k_0^2}{4\pi\varepsilon_0}V\varepsilon_0(\varepsilon_r - 1) \tag{7.108a}$$

$$(S_{vh})_{\text{BSA}} = (S_{hv})_{\text{BSA}} = 0 \tag{7.108b}$$

$$(S_{vv})_{\text{BSA}} = \frac{k_0^2}{4\pi\varepsilon_0}\left[\alpha + (\alpha_{z_b} - \alpha)\sin^2\theta_i\right] \tag{7.108c}$$

$$= \frac{k_0^2}{4\pi\varepsilon_0}V\varepsilon_0(\varepsilon_r - 1)\left[1 + \left(\frac{1}{\varepsilon_r} - 1\right)\cos^2\theta_e\right] \tag{7.108d}$$

Note that the angle of incidence, θ_i, and radar elevation angle, θ_e, are related by $\theta_e = 90° - \theta_i$. Substituting (7.108) in (7.104) and (7.106) results in,

$$\text{SLDR(plates)} = 10\log_{10}\left[\frac{\left|-1 + \sin^2\theta_e + (\cos^2\theta_e/\varepsilon_r)\right|^2}{\left|1 + \sin^2\theta_e + (\cos^2\theta_e/\varepsilon_r)\right|^2}\right]; \quad \text{dB} \tag{7.109}$$

Figure 7.71a shows calculations of SLDR for plates versus θ_e assuming $\varepsilon_r = 1.75$ and that the system LDR limit is -28 dB (see Section 6.2). Thus, the measurement of SLDR versus θ_e should give a characteristic signature from plate-like scatterers and this is indeed the case as shown in Fig. 7.71b. These data are representative of dendritic crystals (approximately plate-like) and were collected by the NOAA/ETL 35-GHz cloud radar. Note that the radar's transmitted polarization state differed slightly from slant 45° (i.e. τ_t in (3.136) was not exactly 0° but had slight ellipticity) and this places a lower bound on the measured SLDR of -28 to -29 dB. The basic elevation angle dependence of SLDR in (7.109) is sufficient to explain the radar measurements (see also, Fig. 3.12; except there the measurement of elliptical depolarization ratio was shown).

The polarizability elements of needles are also given in Table A1.1. From (2.53), the scattering matrix elements assuming $\theta_b = 90°$ are given by,

$$(S_{hh})_{\text{BSA}} = K\left[\frac{2}{\varepsilon_r + 1} + \left(\frac{\varepsilon_r - 1}{\varepsilon_r + 1}\right)\sin^2\phi_b\right] \tag{7.110a}$$

$$(S_{hv})_{\text{BSA}} = (S_{vh})_{\text{BSA}} = K\left\{\left[\frac{\varepsilon_r - 1}{2(\varepsilon_r + 1)}\right]\sin\theta_e\sin2\phi_b\right\} \tag{7.110b}$$

$$(S_{vv})_{\text{BSA}} = K\left[\frac{2}{\varepsilon_r + 1} + \left(\frac{\varepsilon_r - 1}{\varepsilon_r + 1}\right)\sin^2\theta_e\cos^2\phi_b\right] \tag{7.110c}$$

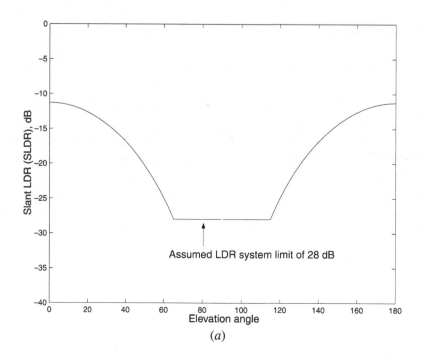

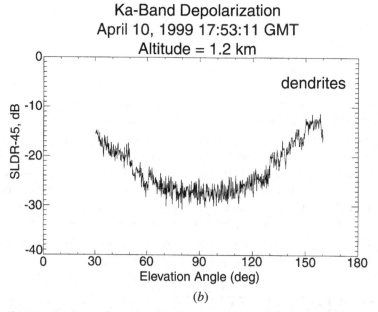

Fig. 7.71. (*a*) Calculations (for plate-like crystals) of slant LDR versus elevation angle in the Rayleigh–Gans limit using eq. (7.109) with $\varepsilon_r = 1.75$. The system LDR limit is assumed to be -28 dB to facilitate comparison with radar measurements. (*b*) Radar measurements of slant LDR in a region of plate-like crystals. The transmitted polarization state was not exactly slant $45°$, but had slight ellipticity. See also, Reinking et al. (1999).

where $K = (k_0^2/4\pi\varepsilon_0)V\varepsilon_0(\varepsilon_r - 1)$. Substituting in (7.104, 7.106) and using ensemble averaging to account for the uniform pdf of ϕ_b in $(0, 2\pi)$ yields,

$$\sigma_{cx} = \pi K^2 \left(\frac{\varepsilon_r - 1}{\varepsilon_r + 1}\right)^2 \left[\frac{3}{8}(\sin^4\theta_e + 1) - \frac{\sin^2\theta_e}{4}\right] \tag{7.111a}$$

$$\sigma_{co} = \pi K^2 \left\{ 4\left(\frac{2}{\varepsilon_r} + 1\right)^2 + \left(\frac{\varepsilon_r - 1}{\varepsilon_r + 1}\right)^2 \left[\frac{3}{8}(\sin^4\theta_e + 1) + \sin^2\theta_e\right] \right.$$
$$\left. + \frac{8(\varepsilon_r - 1)}{(\varepsilon_r + 1)^2} \left[\frac{1}{2}(1 + \sin^2\theta_e)\right] \right\} \tag{7.111b}$$

The plot of SLDR versus θ_e assuming $\varepsilon_r = 1.5$ is shown in Fig. 7.72a. Note the very weak dependence on θ_e; in fact SLDR is nearly uniform around -24 dB. Figure 7.72b shows measurements of SLDR versus θ_e through a layer of columns which are in good agreement with the calculations shown in Fig. 7.72a. Also shown in Fig. 7.73 is radar data through drizzle (spherical drops) where SLDR reduces to -28 to -29 dB (or the system limit). If the transmitted polarization state was perfectly slant linear at 45°, then SLDR for spherical drops would tend to reach the system limit for this radar of around -35 dB (see Section 6.2.1) independent of θ_e. This technique of examining the variation of SLDR as a function of θ_e in horizontally homogeneous (but vertically stratified) precipitation can be used to detect supercooled drizzle and to distinguish between plate-like and column-like crystals. Remote detection of freezing drizzle is important for application to aircraft icing, and radar polarimetry offers a practical solution to this aviation hazard.

Polarimetric radar signatures of the bright-band have been extensively studied starting with cross-polarization observations of the melting layer by Browne and Robinson (1952). At vertical incidence, the bright-band can be clearly identified by enhanced linear depolarization ratio and the corresponding reduction in $|\rho_{co}|$; for example, see Fig. 3.15. At vertical incidence, the bright-band appears to be azimuthally symmetric with $|\rho_{co}| = 1$–$2 \times 10^{0.1(LDR)}$. One characteristic of the bright-band is the enhancement of equivalent reflectivity factor (Z_e) in the bright-band relative to the uniform rain below. This enhancement (ΔZ_e, dB), which can range from 3–15 dB (for typical radar frequencies and approximate Rayleigh scattering) is reflective of aggregation and melting processes within the bright-band. Generally, ΔZ_e is largest when very large, low density snow flakes melt to form small drops, which is also associated with a sharp transition in reflectivity-weighted fall speed (typically, from 1 m s^{-1} to around 5 m s^{-1}). As the large snow flakes melt, and combined with their irregular shapes, the LDR increases rapidly reaching a peak within the bright-band (at a slightly lower height compared with the location of the peak in Z_e). There is a sharp decrease in LDR due to the nearly spherical drops below the bright-band. The copolar correlation coefficient, $|\rho_{co}|$, also shows a sharp transition in the bright-band, but at low elevation angles the relation $|\rho_{co}| = 1$–$2 \times 10^{0.1(LDR)}$ is generally not satisfied (as it is for vertical incidence, see, for example, Fig. 3.15). The minimum in $|\rho_{co}|$ typically occurs at a slightly lower

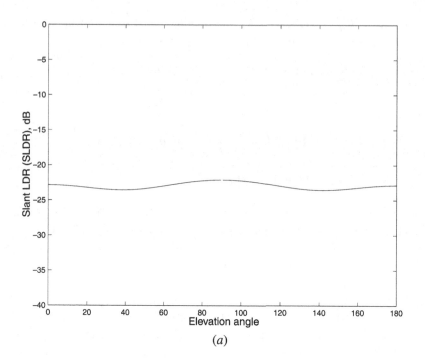

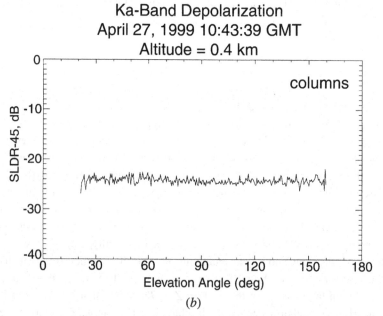

Fig. 7.72. (*a*) Calculations (for needle-like crystals) of slant LDR versus elevation angle in eq. (7.111), assuming $\varepsilon_r = 1.5$. (*b*) Radar measurements of slant LDR in a region of column-like crystals. See Reinking et al. (1999).

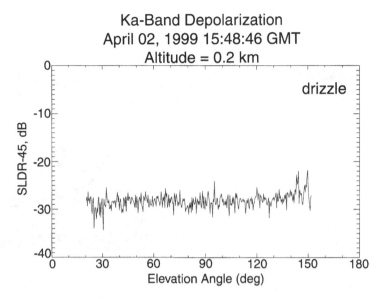

Fig. 7.73. Slant $45°$ radar data in a drizzle region. The system limit is around -28 to -29 dB because the transmitted polarization state had slight ellipticity. See Reinking et al. (1999).

height relative to the height of the LDR peak. Since a mixture of particle types is present in the bright-band, e.g. wet, irregular snow, and drizzle-sized drops, the $|\rho_{co}|_{mix}$ formula in (7.98) can be adapted for this application by replacing "hail" with wet, irregular snow. The $|\rho_{co}|_{mix}$ will attain a minimum when the reflectivity-weighted contributions from the two components of the mixture are nearly matched. Figure 7.74 shows typical vertical structures of Z_h, Z_{dr}, LDR_{vh}, and $|\rho_{co}|$ at S-band (3-GHz) frequency in stratiform precipitation with the bright-band peak in Z_h located at 2.5 km (above ground in Colorado). Figure 7.75 shows the histogram of $|\rho_{co}|$ from drizzle (below the bright-band) and from within the bright-band showing clear separation between the two regions.

The characteristics of the bright-band change with the microphysical evolution, e.g. aggregation processes, breakup processes, degree of riming, etc., which are not directly related in any simple way to the polarimetric signatures. Thus, the quantitative application of radar polarimetry to deducing the microphysics of the bright-band has been rather limited. Qualitatively, it appears that the enhanced LDR signatures together with large positive ΔZ_e offer the potential of distinguishing between the classic bright-band versus the case where weak imbedded convection may enhance riming processes of snow crystals eventually leading to formation of graupel. In the latter case, the LDR values and ΔZ_e are not as high as in the classic bright-band (where $\Delta Z_e \sim 10$ dB and LDR ~ -15 dB). Figure 7.76 shows histograms of ΔZ_e stratified by LDR values in the bright-band based on 11 days of radar data (from the UK) at 3-GHz frequency. It is clear that the most frequent occurrence of ΔZ_e in the range 10–12 dB occurs when the peak LDR in the bright-band is around -15 dB. As the peak

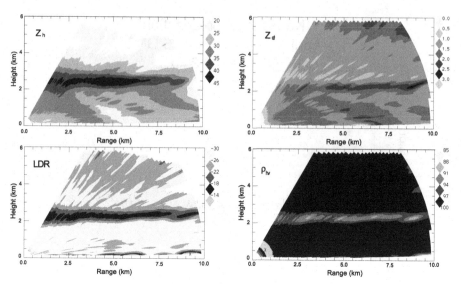

Fig. 7.74. Range–height indicator (RHI) scan data using the CSU–CHILL radar depicting Z_h, Z_{dr}, LDR_{vh}, and $|\rho_{co}|$ in a stratiform event with some imbedded convection. From Beaver and Bringi (1997, © 1997 IEEE).

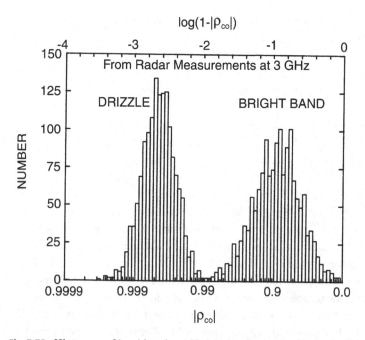

Fig. 7.75. Histogram of $|\rho_{co}|$ in rain and bright-band using the Rutherford Appleton Laboratory's S-band radar, Chilbolton, UK. From Illingworth and Caylor (1991).

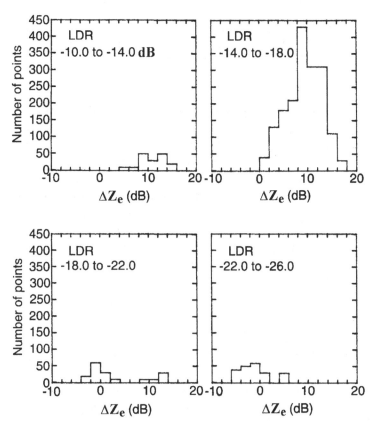

Fig. 7.76. Histograms of the enhancement of reflectivity in the bright-band (ΔZ_e, dB) relative to the rain below as a function of LDR in the bright-band. Data from the S-band RAL/Chilbolton radar, UK. From Caylor et al. (1989).

LDR decreases to -20 dB, the frequency of occurrence of $\Delta Z_e > 3$ dB reduces rather dramatically.

The attenuation of radio waves due to the bright-band has been extensively investigated by the wave propagation research community because of the need to model satellite–Earth propagation links.[13] Since the total attenuation depends partly on the microphysics of the particles in the bright-band, which is quite complicated (e.g. Fabry and Zawadzki 1995), the attenuation models depend on environmental parameters as well as snow size distribution data above the bright-band, which are not generally known. Another important consequence of the bright-band is related to overestimation of surface rain rate as the radar beam penetrates the bright-band at long ranges. Depending on the magnitude of ΔZ_e, the rain rate using Z–R relations can be significantly overestimated. A related application is the choice of coefficients in Z–R relations depending upon the microphysics dominating the bright-band, e.g. distinguishing between large, low density snow melting to form rain versus rimed particles melting to form rain (Waldvogel et al. 1995). The degree of riming and ΔZ_e appears to be negatively correlated, and the peak LDR value in the bright-band may offer a method to remotely detect this feature. Because $|\rho_{co}|$ and LDR are related in the bright-band, the $|\rho_{co}|$ signature could also be used to detect this feature.

7.4   The estimation of attenuation and differential attenuation in rain using Φ_{dp}

The absorption and scattering of electromagnetic waves due to precipitation has been studied since the early 1940s nearly coincidental with the beginning of radar (Ryde 1946; see also, Atlas and Ulbrich 1990). The extinction cross section of particles determines the power loss suffered by the incident wave due to absorption and scattering, see (1.57). For frequencies corresponding to Rayleigh scattering, the absorption cross section of a spherical drop is proportional to its volume, and it dominates the extinction cross section, see (1.59) and Section 2.5. The specific attenuation for such drops using (1.133d) and (2.134c) is expressed as,

$$A = 4.343 \times 10^3 \int \sigma_{ext}(D)N(D)\,dD \qquad (7.112a)$$

$$= 4.343 \times 10^{-6}\left(\frac{18\pi}{\lambda}\right)\frac{\varepsilon_r''}{|\varepsilon_r + 2|^2}\int \frac{\pi}{6}D^3 N(D)\,dD; \quad \text{dB km}^{-1} \qquad (7.112b)$$

where λ is in m, $\sigma_{ext}(D)$ is in m^2 and $N(D)\,dD$ is in m^{-3} in (7.112a) whereas D is in mm in (7.112b). In terms of water content W, see (7.53), (7.112b) can be written as,

$$A = 4.343 \times 10^{-3}\left(\frac{18\pi}{\lambda}\right)\frac{\varepsilon_r''}{|\varepsilon_r + 2|^2}\frac{W}{\rho_w}; \quad \text{dB km}^{-1} \qquad (7.113)$$

where λ is in m, W is in units of g m^{-3} and ρ_w is the water density (1 g cm^{-3}). The relative permittivity of water is $\varepsilon_r = \varepsilon_r' - j\varepsilon_r''$. The Rayleigh limit expression for specific attenuation in (7.113) is directly proportional to the total liquid water content, which is generally the sum of the cloud water content and the rainwater content. The imaginary part (ε_r'') of the relative permittivity of water is a function of wavelength and water temperature (T) and its functional form is $\varepsilon_r''(\lambda, T)$ (Ray 1972). The absorption spectrum of water is such that in the microwave frequency band, ε_r'' is inversely proportional to temperature and, hence, for a given W and operating wavelength, say, S-band ($\lambda \approx 0.1$ m), the specific attenuation can vary by about a factor of 1.5–2.0 for temperatures in the range 0–30 °C (e.g. see Jameson 1992).

At frequencies greater than 3 GHz but less than around 15 GHz, a better approximation for σ_{ext} up to order $(D/\lambda)^5$ can be obtained using the Mie expansion coefficients a_1^s, b_1^s, and a_2^s, see (2.119, 2.130),

$$\sigma_{ext} = \frac{2\pi}{k_0^2}\sum_{n=1}^{\infty}(2n+1)\,\text{Re}(a_n^s + b_n^s) \qquad (7.114a)$$

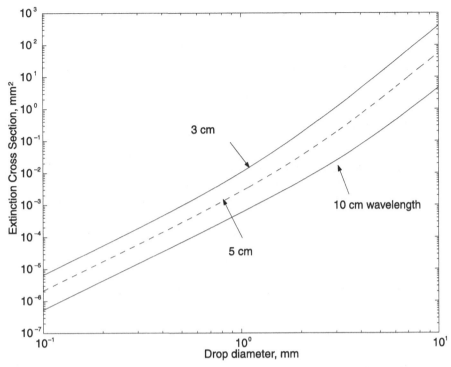

Fig. 7.77. Extinction cross section of spherical drops versus drop diameter using the low frequency expansion in (7.114c).

$$\approx \frac{2\pi}{k_0^2}[3\operatorname{Re}(a_1^s + b_1^s) + 5\operatorname{Re}(a_2^s)] \qquad (7.114b)$$

$$= \frac{6\pi}{\lambda}\left(\frac{\pi}{6}D^3\right)\operatorname{Re}\left\{i\left(\frac{\varepsilon_r-1}{\varepsilon_r+2}\right)\left[1+\left(\frac{\pi D}{\lambda}\right)^2\left(T+U+\frac{5}{3}W\right)\right]\right\} \qquad (7.114c)$$

where T, U, and W are defined in (2.131). Note that "Re" stands for real part of a complex number and $i = \sqrt{-1}$. Figure 7.77 shows σ_{ext} (in mm^2) versus drop diameter for wavelengths of 10, 5, and 3 cm. A power-law fit of the form $\sigma_{ext} = C_\lambda D^n$ for $0.1 \leq D \leq 8$ mm can be used as a first approximation with $n = 3.5$, 3.9, and 4.1 at 10-, 5- and 3-cm wavelengths, respectively. However, for large drops with $5 \leq D \leq 10$ mm, the corresponding n-values are 4.6, 4.8, and 4.9, respectively. To simplify the discussion and to develop the form of the relation between attenuation and the differential propagation phase, let $n \approx 4$. It follows that (7.112a) can be expressed as,

$$A = (4.343 \times 10^3)\int \sigma_{ext}(D)N(D)\,dD; \quad \text{dB km}^{-1} \qquad (7.115a)$$

$$\approx 4.343 \times 10^3 C_\lambda \int D^4 N(D)\,dD \qquad (7.115b)$$

Because of the temperature dependence of $\varepsilon_r''(\lambda, T)$ in (7.114c), both C_λ and the exponent n are temperature dependent (e.g. see Atlas and Ulbrich 1977).

To develop attenuation-correction procedures based on the differential propagation phase, (7.18) is repeated below,

$$K_{dp} \approx \left(\frac{180°}{\lambda}\right) 10^{-3} C W (0.062 D_m); \quad ° \, km^{-1} \tag{7.116}$$

Since WD_m is essentially proportional to the fourth moment of the dsd, see (7.13) for the definition of D_m, it follows from (7.115b) that specific attenuation ($A \approx A_h$) is nearly linearly related to K_{dp} for frequencies from 5 GHz to at least 19 GHz, and is of the form $A_h = \alpha K_{dp}$. The coefficient α depends on water temperature and frequency; however, the temperature dependence is weak for frequencies ≥ 10 GHz (Jameson 1992).

7.4.1 Simple attenuation correction of reflectivity and Z_{dr}

The procedure for correcting the measured equivalent reflectivity factor is as follows. From (4.102a), which is repeated here,

$$\langle n|S_{hh}'|^2 \rangle = \langle n|S_{hh}|^2 \rangle e^{-4k_{im}^h r} \tag{7.117}$$

where the prime refers to the propagation-modified covariance matrix element assuming a homogeneous path of length r, see (1.129b, 4.68b) for definition of k_{im}^h. Since the equivalent reflectivity factor (Z_h) is proportional to $\langle n|S_{hh}|^2 \rangle$, the attenuated reflectivity factor (Z_h') at range r can be expressed as,

$$10 \log_{10}[Z_h'(r)] = 10 \log_{10}[Z_h(r)] - 40 k_{im}^h r (\log_{10} e) \tag{7.118a}$$

$$= 10 \log_{10}[Z_h(r)] - 2 k_{im}^h r (8.686) \tag{7.118b}$$

where k_{im}^h is in units of m^{-1} and r is in m. If the path length is r (in km) and A_h (in dB km^{-1}) as defined in (4.76), then (7.118b) reduces to,

$$10 \log_{10}[Z_h'(r)] = 10 \log_{10}[Z_h(r)] - 2 A_h r \tag{7.119}$$

For an inhomogeneous path (i.e. A_h varies along the path), (7.119) may be modified as,

$$10 \log_{10}[Z_h'(r)] = 10 \log_{10}[Z_h(r)] - 2 \int_0^r A_h(s) \, ds \tag{7.120a}$$

$$= 10 \log_{10}[Z_h(r)] - 2\alpha \int_0^r K_{dp}(s) \, ds \tag{7.120b}$$

$$= 10 \log_{10}[Z_h(r)] - \alpha[\Phi_{dp}(r) - \Phi_{dp}(0)] \tag{7.120c}$$

where the linear relation $A_h = \alpha K_{dp}(r)$ has been used along the propagation path. Thus, the corrected Z_h at a range r is obtained as,

$$10 \log_{10}[Z_h(r)] = 10 \log_{10}[Z_h'(r)] + \alpha[\Phi_{dp}(r) - \Phi_{dp}(0)] \tag{7.121}$$

and the procedure is not dependent on the functional form of $K_{dp}(r)$ at each point along the path. Note that in the Rayleigh limit, use of (7.113) and (7.116) will yield $A_h = \alpha K_{dp}^{0.8}$, and at S-band, scattering calculations based on gamma dsd simulations show a $K_{dp}^{0.85}$ dependence (Jameson 1992; Testud et al. 2000). However, at S-band frequencies the attenuation is generally small except in unusual situations and a linear $A_h = \alpha K_{dp}$ fit to gamma dsds is sufficient for correction purposes (Bringi et al. 1990; Ryzhkov and Zrnić 1995a). It is also evident from (7.115b) and (7.116), that the coefficient α will be inversely proportional to the slope of the axis ratio versus diameter relation (earlier referred to as γ, which is 0.062 for the Pruppacher–Beard formula in (7.2); see also, (7.29a)). This implies that a reduction in γ (due to transverse oscillations, for example) will cause α to increase relative to equilibrium axis ratios. It is also evident that K_{dp} is insensitive to cloud water and to drizzle-sized drops less than 0.5 mm, while the specific attenuation is proportional to the water content contributed by these two components, see (7.113), and to the water content contributed by the larger raindrops (or attenuation is proportional to total water content from all components). Because of these uncertainties coupled with the primary dependence of specific attenuation on temperature, a direct application of (7.121) using a constant α-value based on scattering simulations can only lead to an approximate correction. A range of α-values have been reported in the literature especially at C-band, and a summary is shown as a "band" in the two-dimensional (A_h, K_{dp}) space, see Fig. 7.78. Table 7.1 shows temperature-averaged (0–30 °C) fits to $A_h = \alpha K_{dp}^b$ based on scattering simulations using gamma dsds over a wide frequency range.

The cumulative effect of differential attenuation (A_{dp}), see (4.7), on Z_{dr} is especially evident at C-band and higher frequencies (see Fig. 4.9). The correction for Z_{dr} is based on (4.104b) for a homogeneous propagation path, which is repeated here,

$$Z'_{dr}(r) = Z_{dr}(r) - 2A_{dp}r \tag{7.122a}$$

which can be generalized for an inhomogeneous path as,

$$Z'_{dr}(r) = Z_{dr}(r) - 2\int_0^r A_{dp}(s)\,ds \tag{7.122b}$$

Holt (1988) and Bringi et al. (1990) proposed a linear relation of the form $A_{dp} = \beta K_{dp}$ for Z_{dr}-correction. Scattering simulations based on gamma dsds show that the linearity is a good approximation over a wide frequency range (from 2.8 to 19 GHz; Jameson 1992) but the coefficient β is temperature-dependent at the lower frequencies varying by a factor of around 2 (for 0–30 °C) at 2.8 and 5.5 GHz, with only weak dependence on temperature at higher frequencies. Figure 7.78 illustrates the "band" of β reported in the literature at C-band (5.5 GHz), while Table 7.1 gives the temperature-corrected (0–30 °C) fits to $A_{dp} = \beta K_{dp}^b$ based on scattering simulations using gamma dsds over a wide frequency. Similar to (7.121), the corrected Z_{dr} at range r is obtained from (7.122b) as,

$$Z_{dr}(r) = Z'_{dr}(r) + \beta[\Phi_{dp}(r) - \Phi_{dp}(0)] \tag{7.123}$$

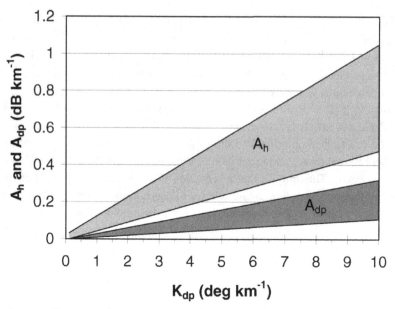

Fig. 7.78. Summary of A_h and A_{dp} versus K_{dp} at C-band (frequency around 5.5 GHz). The shaded "bands" are based on numerous relations at C-band reported in the literature as summarized by Carey et al. (2000).

Table 7.1. Temperature-averaged (0–30 °C) power-law fits of the form $A_h = \alpha K_{dp}^b$ and $A_{dp} = \beta K_{dp}^b$

Frequency (GHz)	α-value	b-value
For α-values[a]		
2.8	0.017	0.84
5.5	0.073	0.99
9.3	0.233	1.02
13.8	0.44	0.97
19.35	0.72	0.99
For β-values		
2.8	0.003	1.05
5.5	0.013	1.23
9.3	0.033	1.15
13.8	0.054	1.13
19.35	0.088	1.17

[a] Data from Jameson (1992).

The use of a constant β-value in the above is expected to only lead to an approximate correction, similar to the use of a constant α-value in (7.121). However, the Φ_{dp}-based methods lead to numerically stable correction algorithms as compared with variants of the Hitschfeld and Bordan (1954) scheme proposed by Aydin et al. (1989) and by Gorgucci et al. (1996) at C-band. Figure 7.79 shows an example of correction of Z_h and Z_{dr} at C-band for a single range profile through the core of a convective rain cell.[14]

Ryzhkov and Zrnić (1995a) have proposed using scatterplots of measured Z'_h and Z'_{dr} versus Φ_{dp} from a large number of beams of data and to fit any decreasing trend by straight lines whose slopes are equated to α and β, respectively. In this empirical method, the difference between $10\log_{10}[Z'_h(r)]$ and $10\log_{10}[Z_h(r)]$ is defined as $\Delta Z_h(r)$,

$$\Delta Z_h(r) = 10\log_{10}[Z'_h(r)] - 10\log_{10}[Z_h(r)]; \quad dB \qquad (7.124a)$$
$$= -\alpha\Phi_{dp}(r) \qquad (7.124b)$$

where the system $\Phi_{dp}(0)$ is set to $0°$ with no loss of generality. Similarly, for Z_{dr},

$$\Delta Z_{dr}(r) = Z'_{dr}(r) - Z_{dr}(r); \quad dB \qquad (7.125a)$$
$$= -\beta\Phi_{dp}(r) \qquad (7.125b)$$

If it is assumed that the rain medium along the propagation path (from 0 to r) is uniform (or the intrinsic values of Z_h, Z_{dr}, and K_{dp} are constant with r), then estimates of α and β can be determined from the measurements by fitting a straight line to pairs of data (Z'_h, Φ_{dp}) and (Z'_{dr}, Φ_{dp}) obtained from each resolution volume along the path. Ryzhkov and Zrnić (1995a) suggest that data points be chosen so that the corresponding K_{dp} lie in a narrow interval ($1–2°$ km^{-1} at S-band). However, there is no unique way to ensure from the measurements that the intrinsic Z_h or Z_{dr} is not changing with range. Carey et al. (2000) have argued that such intrinsic changes of Z_h and Z_{dr} with range only contribute to the scatter about the straight line fit and do not bias the estimates of α and β, provided the data points (Z'_h, Φ_{dp}) and (Z'_{dr}, Φ_{dp}) are carefully selected. While this empirical method is difficult to implement and can be strongly influenced by outliers in the data, it does tend to estimate the average values of α and β, which are in the range predicted by scattering simulations using gamma dsds (see Table 7.1).

Ryzhkov and Zrnić (1995a) estimated α and β at S-band from polarimetric data collected in an unusually intense rain event (a squall line) and found that their (α, β) values were nearly double the theoretical values established from simulations (see Table 7.1), and attributed it to "giant" raindrops ($D > 8$ mm) with ice cores (for example, see Fig. 7.49 and related discussion). However, Smyth and Illingworth (1998) argued that the large (α, β) values could be explained by invoking exponential dsds with very large D_0-values (up to around 7 mm). Referring to Fig. 7.77 and the discussion related to (7.114, 7.115), note that σ_{ext} varies as D^5 for drops with diameters between 5 and 10 mm, rather than D^4 as assumed in (7.115b). Hence, for an exponential distribution A is proportional to D_0^6, whereas K_{dp}, see (7.116), is still proportional to WD_m or to D_0^5. It follows that A is proportional to $D_0 K_{dp}$, and the coefficient α (in $A_h = \alpha K_{dp}$) will

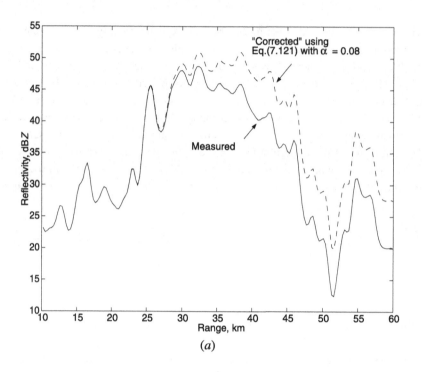

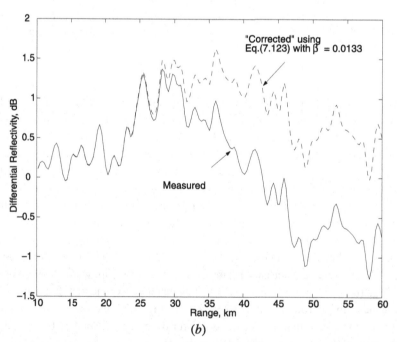

Fig. 7.79. Example of correction of radar data at C-band (frequency around 5.5 GHz) for a single range profile through a convective rain cell. Data from the BMRC C-POL radar: (*a*) Z_h-correction using eq. (7.121) with $\alpha = 0.08$, (*b*) Z_{dr}-correction using eq. (7.123) with $\beta = 0.0133$.

increase with D_0. This dependence of α on D_0 will only occur for dsds with large $D_{0}s$ (≥ 2.5 mm). Similarly, β (in $A_{dp} = \beta K_{dp}$) will increase as D_0 increases beyond 2.5 mm. Figure 7.80 shows scattering simulation results at C-band for A_h versus K_{dp}, and A_{dp} versus K_{dp}, stratified by D_0. If large D_0-values occur along the propagation path, then the attenuation-correction procedure is more complicated since α and β are no longer constant but depend on D_0, which itself is unknown. Carey et al. (2000) argue that such "big drop" zones along the propagation path can be detected at C-band, and they empirically increase α and β in these sections of the path to account for the enhanced attenuation and differential attenuation.

7.4.2   Attenuation correction using constraints

(i) Z_{dr} constraint

At C-band (frequency around 5.5 GHz) and higher frequencies, the cumulative effect of differential attenuation caused by propagation through an intense rain cell will cause Z_{dr} to become negative at range locations beyond the cell (for example, see Fig. 4.9) where typically the reflectivity and rain rates have reduced considerably relative to the intense cell. If the range location at which the measured Z'_{dr} becomes negative is r_m, then Smyth and Illingworth (1998) proposed that $\Delta Z_{dr}(r_m)$ from (7.125) can be expressed as,

$$\Delta Z_{dr}(r_m) = Z'_{dr}(r_m) - Z_{dr}(r_m) \tag{7.126a}$$

$$\approx Z'_{dr}(r_m) \tag{7.126b}$$

$$= -\beta[\Phi_{dp}(r_m) - \Phi_{dp}(0)] \tag{7.126c}$$

where the intrinsic value of $Z_{dr}(r_m)$ is assumed to be 0 dB. An estimate of β can then be obtained as,

$$\hat{\beta} = \frac{|Z'_{dr}(r_m)|}{[\Phi_{dp}(r_m) - \Phi_{dp}(0)]} \tag{7.127}$$

The specific differential attenuation, $A_{dp}(r)$, at each range location along the propagation path can be estimated as,

$$\hat{A}_{dp}(r) = \hat{\beta} K_{dp}(r) \tag{7.128a}$$

$$= \frac{|Z'_{dr}(r_m)|}{[\Phi_{dp}(r_m) - \Phi_{dp}(0)]} K_{dp}(r) \tag{7.128b}$$

The measured $Z'_{dr}(r)$ can be corrected using (7.122b),

$$\hat{Z}_{dr}(r) = Z'_{dr}(r) + 2 \int_0^r \hat{A}_{dp}(s) \, ds \tag{7.129a}$$

$$= Z'_{dr}(r) + \frac{2|Z'_{dr}(r_m)|}{[\Phi_{dp}(r_m) - \Phi_{dp}(0)]} \int_0^r K_{dp}(s) \, ds \tag{7.129b}$$

$$= Z'_{dr}(r) + \frac{|Z'_{dr}(r_m)|}{[\Phi_{dp}(r_m) - \Phi_{dp}(0)]} [\Phi_{dp}(r) - \Phi_{dp}(0)] \tag{7.129c}$$

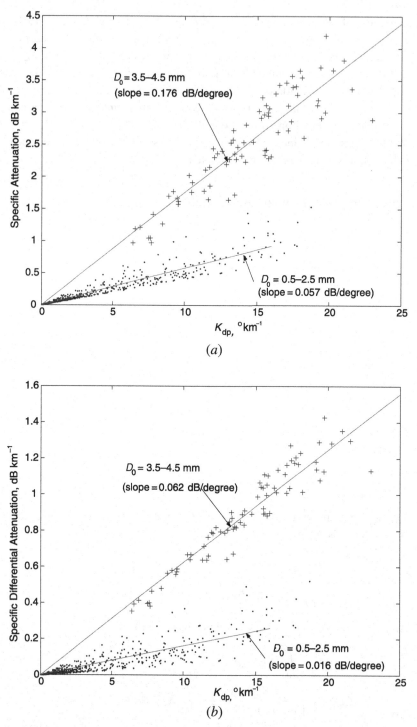

Fig. 7.80. Gamma dsd simulation of (a) A_h versus K_{dp} and (b) A_{dp} versus K_{dp} at C-band stratified into two D_0 ranges to illustrate the "big drop" effect.

In essence, this method of correcting the measured Z'_{dr} uses the constraint that the intrinsic Z_{dr} on the far side of an intense rain cell should tend to 0 dB, representative of light drizzle conditions. If this constraint cannot be established, then $\hat{\beta}$ must be estimated as,

$$\hat{\beta} = \frac{|Z'_{dr}(r_m) - Z_{dr}(r_m)|}{[\Phi_{dp}(r_m) - \Phi_{dp}(0)]} = \frac{|\Delta Z_{dr}(r_m)|}{[\Phi_{dp}(r_m) - \Phi_{dp}(0)]} \tag{7.130}$$

where the intrinsic value of $Z_{dr}(r_m)$ must be established by other physical constraints. Note that $\Delta Z_{dr}(r_m) = Z'_{dr}(r_m) - Z_{dr}(r_m)$ must be negative (in decibel units) in (7.130) because oblate raindrops can only cause positive values of A_{dp} (or $A_h > A_v$). One advantage of this scheme is that it is based on radar measurements of $Z'_{dr}(r)$ and $\Phi_{dp}(r)$ at consecutive range locations along the beam, and allows β to change from beam to beam. An example of Z_{dr}-correction at C-band is given in Fig. 7.81.

To correct for reflectivity at C-band, Smyth and Illingworth (1998) suggest using the linear relation $A_h = (\alpha/\beta)A_{dp} = \kappa A_{dp}$, where $\kappa = \alpha/\beta$ is expected to be relatively insensitive to temperature variations (around 20% variation from 0 to 30 °C). From Table 7.1, the ratio $\alpha/\beta = 5.6$ at 5.5 GHz (and equals 5.66 at 2.8 GHz). Thus, once $\hat{A}_{dp}(r)$ is estimated, see (7.128), at each range location along the beam, the specific attenuation can be estimated as,

$$\hat{A}_h(r) = \kappa \hat{A}_{dp}(r) \tag{7.131a}$$

$$= \kappa \frac{|Z'_{dr}(r_m)|}{[\Phi_{dp}(r_m) - \Phi_{dp}(0)]} K_{dp}(r) \tag{7.131b}$$

where $\kappa = 5.6$ (at 5.5 GHz). From (7.120a), the corrected reflectivity estimate is obtained as,

$$10\log_{10}[\hat{Z}_h(r)] = 10\log_{10}[Z'_h(r)] + 2\int_0^r \hat{A}_h(s)\,ds \tag{7.132a}$$

$$= 10\log_{10}[Z'_h(r)] + \kappa \frac{|Z'_{dr}(r_m)|}{[\Phi_{dp}(r_m) - \Phi_{dp}(0)]}[\Phi_{dp}(r) - \Phi_{dp}(0)] \tag{7.132b}$$

where κ, defined by $A_h = \kappa A_{dp}$, may be computed from scattering simulations based on gamma dsds.

(ii) Φ_{dp} constraint

A different approach to attenuation correction (but only for correction of the measured reflectivity) has been proposed by Testud et al. (2000) primarily for application to C-band radar using a Φ_{dp} constraint. This method (ZPHI algorithm) is based on "rain profiling" algorithms developed for spaceborne radar which operate at frequencies greater than 10 GHz (e.g. the Tropical Rainfall Measurement Mission, or TRMM radar; Kozu et al. 1991). In the case of the TRMM radar, the ocean surface acts as a reference whose radar cross section is stable and known to within a few decibels. By comparing

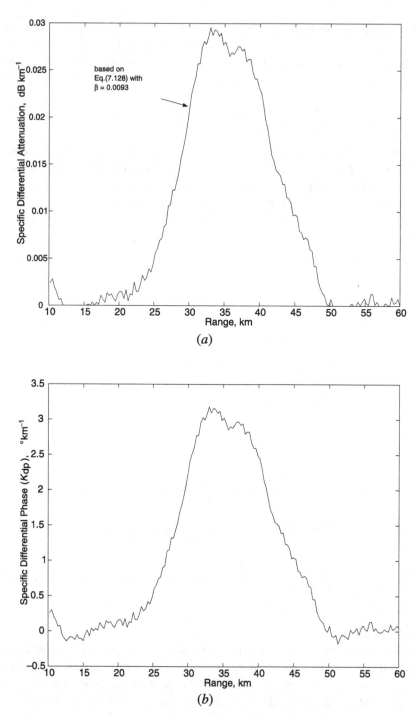

Fig. 7.81. (*a*) Example illustrating estimation of A_{dp} using (7.128). The β-value for this beam of data is 0.0093. Radar data is from the BMRC C-POL radar (same data as used in Fig. 7.79). (*b*) The range profile of K_{dp} used in (7.128a). (*c*) Measured and corrected Z_{dr} range profiles using (7.129c). Compare with Fig. 7.79*b*.

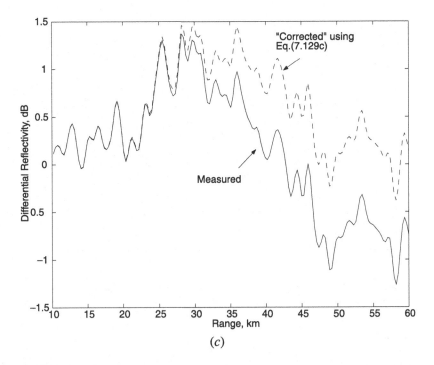

(c)

Fig. 7.81. (*cont.*)

the back scattered signal from the surface in the presence of rain relative to rain-free areas, the path-integrated attenuation is estimated. In the case of ground radars with polarimetric capability the value of $\Phi_{dp}(r_m)$ at locations beyond the attenuating rain cell can be used as a constraint.

The ZPHI and similar rain-profiling methods are based on a technique originally proposed by Hitschfeld and Bordan (1954). The specific attenuation A_h (in dB km^{-1}) is related to Z_h (in mm^6 m^{-3}) by means of a power law,

$$A_h(r) = a[Z_h(r)]^b \tag{7.133}$$

For gamma dsds, scattering simulations show that b is nearly constant for a given frequency whereas a depends on temperature as well as the parameter N_w (see Section 7.1.4 and (7.61)). The ZPHI method of attenuation correction does not involve a, but it does assume that b is constant. Next, a linear relation between A_h (in dB km^{-1}) and K_{dp} (in $^\circ$ km^{-1}) is assumed with known α-coefficient,

$$A_h(r) = \alpha K_{dp}(r) \tag{7.134}$$

As discussed earlier, this linear relation is an excellent approximation at C- and X-bands, though α depends on temperature (see Fig. 7.78), and on the assumed drop axis ratio versus diameter relation.

Following Marzoug and Amayenc (1994), and starting from (7.117), the measured reflectivity at range r is expressed as (for ease of notation the subscript h will be dropped

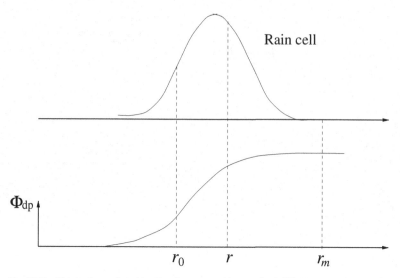

Fig. 7.82. Illustration related to the Φ_{dp} constraint method. The ranges r_0, r, and r_m are indicated relative to the rain cell and Φ_{dp} range profile.

for now),

$$Z'(r) = Z(r)e^{-4k_{im}r} \tag{7.135a}$$

$$= Z(r)e^{-4\frac{Ar}{8.686}} \tag{7.135b}$$

$$= Z(r)e^{-0.46Ar} \tag{7.135c}$$

Note that A is the specific attenuation, which is related to k_{im} by $A = 8.686k_{im}$ (here the units of A are dB m^{-1} and k_{im} is in m^{-1}). However, in (7.135c), it is obvious that A can be in units of dB km^{-1} with r in km. For an inhomogeneous path,

$$Z'(r) = Z(r)e^{-0.46 \int_0^r A(s)\,ds} \tag{7.136}$$

If r_i and r_j are two range locations along the propagation path, then a function $f(r_i; r_j)$ is introduced, which is defined as,

$$f(r_i; r_j) = e^{-0.46b \int_{r_0}^r A(s)\,ds} \tag{7.137a}$$

If r_i is a fixed reference range r_0, and r_j is the variable range r (see Fig. 7.82), then,

$$f(r_0; r) = e^{-0.46b \int_{r_0}^r A(s)ds} \tag{7.137b}$$

$$f(0; r) = e^{-0.46b \int_0^r A(s)ds} \tag{7.137c}$$

$$f(r_0; r) = f(0; r)/f(0; r_0) \tag{7.137d}$$

$$Z'(r) = Z(r)[f(0; r)]^{1/b} \tag{7.137e}$$

The differential equation satisfied by $f(r_0; r)$ may be easily derived as,

$$\frac{df(r_0; r)}{dr} = (-0.46b)f(r_0; r)\left[\frac{d}{dr}\int_{r_0}^r A(s)\,ds\right] \tag{7.138a}$$

$$= -0.46bf(r_0; r)A(r) \tag{7.138b}$$

Using (7.133) and (7.137d, e),

$$\frac{df(r_0; r)}{dr} = -0.46ba[Z(r)]^b \frac{f(0; r)}{f(0; r_0)} \tag{7.139a}$$

$$= -0.46ba \frac{[Z'(r)]^b}{f(0; r_0)} \tag{7.139b}$$

Integrating from r_0 to r results in,

$$\int_{r_0}^r \frac{df(r_0; r)}{dr} dr = \frac{-0.46ab}{f(0; r_0)} \int_{r_0}^r [Z'(r)]^b \, dr \tag{7.140a}$$

or,

$$f(r_0; r) - f(r_0; r_0) = \frac{-0.46ab}{f(0; r_0)} \int_{r_0}^r [Z'(r)]^b \, dr \tag{7.140b}$$

or,

$$f(r_0; r) = 1 - \frac{a}{f(0; r_0)} I(r_0; r) \tag{7.140c}$$

where $I(r_0; r)$ is defined as,

$$I(r_0; r) = 0.46b \int_{r_0}^r [Z'(r)]^b \, dr \tag{7.140d}$$

The intent is to find a solution for the specific attenuation $A(r)$ in terms of the specific attenuation at the reference range, $A(r_0)$, and the measured reflectivities $Z'(r_0)$ and $Z'(r)$. Hence, starting from (7.133) and using (7.137d, e) and (7.140c),

$$A(r) = a[Z(r)]^b \tag{7.141a}$$

$$= \frac{a[Z'(r)]^b}{f(0; r)} \tag{7.141b}$$

$$= \frac{a[Z'(r)]^b}{f(r_0; r)f(0; r_0)} \tag{7.141c}$$

$$= \frac{a[Z'(r)]^b}{f(0; r_0) - aI(r_0; r)} \tag{7.141d}$$

The function $f(0; r_0)$ is now expressed in terms of $A(r_0)$ and $Z'(r_0)$ as follows,

$$Z'(r_0) = Z(r_0)[f(0; r_0)]^{1/b} \tag{7.142a}$$

or,

$$f(0; r_0) = \left[\frac{Z'(r_0)}{Z(r_0)}\right]^b = \frac{a}{A(r_0)}[Z'(r_0)]^b \tag{7.142b}$$

Substituting the above expression for $f(0; r_0)$ in (7.141d) gives the desired result,

$$A(r) = A(r_0) \left\{ \frac{[Z'(r)]^b}{[Z'(r_0)]^b - A(r_0)I(r_0; r)} \right\} \tag{7.143}$$

Note that the coefficient a cancels out, and the specific attenuation $A(r)$ at each range location along the propagation path is expressed in terms of the measured reflectivity $Z'(r)$ and the values of $A(r_0)$ and $Z'(r_0)$ at the reference range r_0. A direct application of (7.143) requires $A(r_0) \neq 0$. If $A(r_0) \to 0$, then (7.143) is of the form $0/0$ and must be evaluated appropriately.

The second part of the ZPHI algorithm is to constrain the reference value of $A(r_0)$ based on the measured differential propagation phase. Figure 7.82 illustrates an idealized propagation path through a rain cell and defines r_0, the variable range r, and the range r_m which refers to the range location beyond the rain cell where $\Phi_{dp}(r_m)$ reaches its final value upon which the constraint is based. Starting from (7.134) and integrating from r_0 to r_m results in,

$$\int_{r_0}^{r_m} A(s)\, ds = \alpha \int_{r_0}^{r_m} K_{dp}(s)\, ds \tag{7.144a}$$

$$= \frac{\alpha}{2} \int_{r_0}^{r_m} 2K_{dp}(s)\, ds \tag{7.144b}$$

$$= \frac{\alpha}{2}[\Phi_{dp}(r_m) - \Phi_{dp}(r_0)] \tag{7.144c}$$

$$= \frac{\alpha}{2} \Delta\Phi_{dp}(r_0; r_m) \tag{7.144d}$$

The left-hand side of the above equation can be integrated using the differential equation satisfied by $f(r_0; r)$ given in (7.138b),

$$\int_{r_0}^{r_m} A(s)\, ds = \frac{-1}{0.46b} \int_{r_0}^{r_m} \frac{d}{dr}[\ln f(r_0; r)]\, dr \tag{7.145a}$$

$$= \frac{-1}{0.46b}[\ln f(r_0; r_m) - \ln(r_0; r_0)] \tag{7.145b}$$

$$= \frac{-1}{0.46b}[\ln f(r_0; r_m)] \tag{7.145c}$$

Note that $f(r_0; r_0) = 1$. Hence, (7.144d) may be expressed as,

$$\ln f(r_0; r_m) = \frac{-0.46b\alpha}{2} \Delta\Phi_{dp}(r_0; r_m) \tag{7.146a}$$

or,

$$f(r_0; r_m) = e^{-0.23b\alpha\,\Delta\Phi_{dp}(r_0; r_m)} \tag{7.146b}$$

Using the expression for $f(r_0; r_m)$ from (7.140c) in the above results in,

$$1 - \frac{a}{f(0; r_0)} I(r_0; r_m) = e^{-0.23b\alpha\,\Delta\Phi_{dp}(r_0; r_m)} \tag{7.146c}$$

Using (7.142b) for $f(0; r_0)$ in the above gives,

$$1 - \frac{A(r_0)}{[Z'(r_0)]^b} I(r_0; r_m) = e^{-0.23b\alpha\,\Delta\Phi_{\mathrm{dp}}(r_0;r_m)} \tag{7.147a}$$

or,

$$A(r_0) = \frac{[Z'(r_0)]^b}{I(r_0; r_m)} \left(1 - e^{-0.23b\alpha\,\Delta\Phi_{\mathrm{dp}}(r_0;r_m)}\right) \tag{7.147b}$$

$$= \frac{[Z'(r_0)]^b}{I(r_0; r_m)} \left(1 - 10^{-0.1(b\alpha)\,\Delta\Phi_{\mathrm{dp}}(r_0;r_m)}\right) \tag{7.147c}$$

Thus, the reference value of $A(r_0)$ is constrained by the change in differential propagation phase from r_0 to r_m. Substituting (7.147c) in (7.143) gives after some modest algebraic manipulation the result,

$$A_h(r) = \frac{[Z_h'(r)]^b \left(1 - 10^{-0.1(b\alpha)\,\Delta\Phi_{\mathrm{dp}}(r_0;r_m)}\right)}{I(r_0; r_m) - \left(1 - 10^{-0.1(b\alpha)\,\Delta\Phi_{\mathrm{dp}}(r_0;r_m)}\right) I(r_0; r)} \tag{7.148}$$

Note that the subscript h (for horizontal polarization) is re-introduced here. The above expression gives the specific attenuation, $A_h(r)$ in dB km^{-1}, at each range location along the propagation path (between r_0 and r_m) in terms of the measured reflectivity $Z_h'(r)$ in mm^6 m^{-3} and $\Delta\Phi_{\mathrm{dp}} = \Phi_{\mathrm{dp}}(r_m) - \Phi_{\mathrm{dp}}(r_0)$ in degrees, with the range r in km. Recall that $I(r_0; r)$ is defined in (7.140d). The exponent b and coefficient α are defined in (7.133, 7.134), respectively; at C-band (frequency near 5.5 GHz), b is very close to 0.8, whereas the temperature-averaged α is given in Table 7.1. Note also that α can vary across the shaded "band" in Fig. 7.78, though here it is assumed to be constant. Equation (7.148) is given in a slightly different form as compared with Testud et al. (2000). Their form may be obtained by recognizing that,

$$I(r_0; r) = I(r_0; r_m) - I(r; r_m) \tag{7.149}$$

Substituting (7.149) in (7.148) and with modest algebraic manipulation results in,

$$A_h(r) = \frac{[Z_h'(r)]^b \left(10^{0.1(b\alpha)\,\Delta\Phi_{\mathrm{dp}}(r_0;r_m)} - 1\right)}{I(r_0; r_m) + \left(10^{0.1(b\alpha)\,\Delta\Phi_{\mathrm{dp}}(r_0;r_m)} - 1\right) I(r; r_m)} \tag{7.150}$$

Once $A_h(r)$ is calculated at each range location from r_0 to r_m (see Fig. 7.82), the reflectivity can be corrected using (7.120a),

$$10\log_{10}[Z_h(r)] = 10\log_{10}[Z_h'(r)] + 2\int_{r_0}^{r} A_h(s)\,ds \tag{7.151}$$

yielding an attenuation-correction algorithm that is constrained by the change in differential propagation phase across the range interval r_0 to r_m. This algorithm is sensitive to the dependence of the α-coefficient on drop temperature and to the form of the mean axis ratio versus diameter relation (see Section 7.1.2). Also, the assumption of a power-law relation between A_h and Z_h means that if certain range resolution volumes along the propagation path contain hail or other mixed phase hydrometeors, then the A_h in those resolution volumes will be biased. This bias will not be present if A_h is derived as in (7.131b) since K_{dp} is relatively insensitive to hail in a rain/hail mixture. Since the probability of occurrence of hail and mixed phase hydrometeors within the main precipitation shaft of convective storms in the mid-latitudes is quite high, (7.150) must be used with caution by segmenting the hail/mixed-phase range locations using polarimetric-based detection schemes discussed in Section 7.2.2. Figure 7.83 illustrates the retrieval of $A_h(r)$ at C-band using (7.150) along with $Z'_h(r)$ and the corrected $Z_h(r)$. This example uses the same initial radar data previously used in Figs. 7.79 and 7.81.

(iii) Combined Φ_{dp}–Z_{dr} constraint

One potential limitation of the ZPHI method, i.e. the assumption of a constant α in the relation $A_h = \alpha K_{dp}$, may be easily overcome by iterating α over a range corresponding to the shaded "band" in Fig. 7.78. At each α-value, the $\Phi_{dp}^c(r; \alpha)$ is computed from,

$$\Phi_{dp}^c(r; \alpha) = 2 \int_{r_0}^{r} \frac{A_h(s; \alpha)}{\alpha} \, ds; \quad \alpha_{min} \le \alpha \le \alpha_{max} \tag{7.152}$$

where $A_h(s; \alpha)$ is obtained from (7.150) for each value of α in the range $(\alpha_{min}, \alpha_{max})$. These "constructed" Φ_{dp}^c range profiles are obtained by varying r from r_0 to r_m in (7.152). The optimal value of α may be selected by minimizing the difference between the constructed Φ_{dp}^c range profiles and a filtered version of the measured Φ_{dp} (see Section 6.6.1) over the range (r_0, r_m),

$$\text{Error} = \sum_{j=1}^{N} |\Phi_{dp}^{filt}(r_j) - \Phi_{dp}^c(r_j; \alpha)| \tag{7.153}$$

where $r_1 \equiv r_0$ and $r_N \equiv r_m$. As discussed in Section 6.6.1 it is possible to adaptively filter the "raw" differential phase data so that back scatter phase shifts (δ_{co}; due, for example, to melting hail) can be corrected, at the same time yielding a smoothed Φ_{dp} range profile that can be used in minimizing the error in (7.153). Figure 7.84a illustrates the constructed $\Phi_{dp}^c(r; \alpha)$ profiles for several values of α, together with the adaptively filtered Φ_{dp}; while Fig. 7.84b shows the error versus α, indicating that an optimal α-value exists (α_{opt}). Figure 7.84c shows the optimal constructed profile, $\Phi_{dp}^c(r; \alpha_{opt})$, compared with the "raw" measured Φ_{dp}, and excellent agreement may be noted. Once the optimal specific attenuation range profile, $A_h(r; \alpha_{opt})$ is calculated, then the measured reflectivity can be corrected using (7.151). As a test of this method, Fig. 7.85 shows K_{dp} versus Z_h (corrected) from a large number of range resolution volumes along with scattering simulations based on measured drop size distributions (see Section 7.1.4). The excellent agreement between the dsd simulations and radar data is an indirect validation of this algorithm. Such an agreement is not possible without proper attenuation-correction of Z_h.

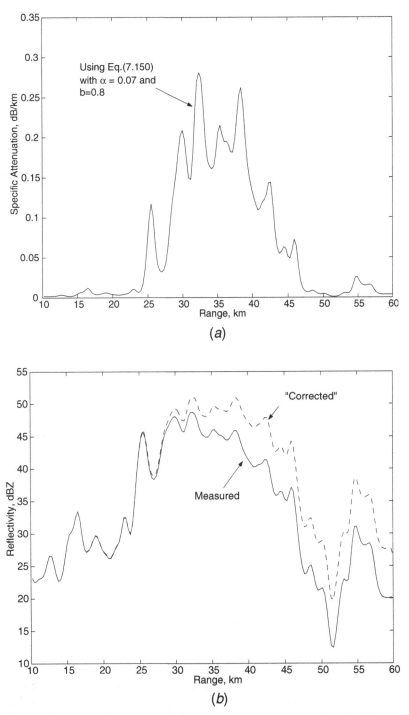

Fig. 7.83. (*a*) Example illustrating the retrieval of A_h using the ZPHI method. Measurements are from the BMRC C-POL radar. (*b*) Correction of Z_h using eq. (7.151). Compare with Fig. 7.79*a*.

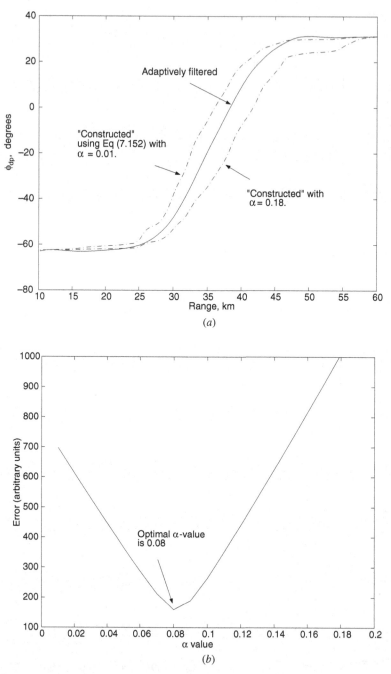

Fig. 7.84. (*a*) Example of "constructed" Ψ_{dp} range profiles using eq. (7.152). The adaptively filtered Φ_{dp} is based on filtering the "raw" Φ_{dp} range profile (see Section 6.6). (*b*) The error in eq. (7.153) versus α. (*c*) The optimal constructed profile of Φ_{dp} compared with the "raw" measured Φ_{dp}. Data from the BMRC C-POL radar.

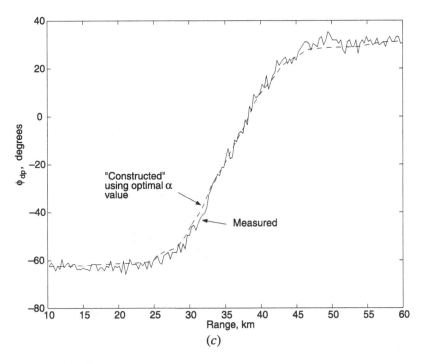

Fig. 7.84. (*cont.*)

A similar procedure for correcting the measured Z'_{dr} using a Z_{dr} constraint at a far range r_m but without assuming a constant β-value for the beam can be described as follows. In contrast to the assumption $Z_{dr}(r_m) \approx 0$ dB used in (7.126), the Z_{dr} constraint is now based on the value of $Z_h(r_m)$, which is available since the reflectivity correction is already done with an optimal α-value. On average, there is a relation between Z_{dr} and Z_h in rain (e.g. see Fig. 7.51a) that can be used to estimate an average value of $Z_{dr}(r_m)$ from $Z_h(r_m)$. This average relationship (which will depend on frequency) can be based on scattering simulations using measured drop size distributions or assuming a gamma model. At C-band, the following equations are based on measured drop size distributions (see Section 7.1.4),

$$\bar{Z}_{dr}(r_m) = \begin{cases} 0; & Z_h(r_m) \leq 20 \text{ dB}Z \\ 0.048Z_h(r_m) - 0.774; & 20 < Z_h(r_m) \leq 45 \text{ dB}Z \end{cases} \qquad (7.154)$$

The overbar on Z_{dr} refers to an average value, while $Z_h(r_m)$ is in dBZ in the above. Once the $Z_{dr}(r_m)$ constraint is established, the initial estimate of β for the beam can be based on (7.130), which is appropriate if $Z'_{dr}(r_m)$ is negative due to large differential propagation phase from r_0 to r_m. If the differential propagation phase is not sufficiently large (or below a threshold value) then it is appropriate to start with a β-value corresponding to the lower-bound in the "band" shown in Fig. 7.78; at C-band this lower bound value is around 0.01 dB per degree. The optimal β-value for a particular beam is found by starting with a first guess for $A_{dp}(r; \beta)$ at each range

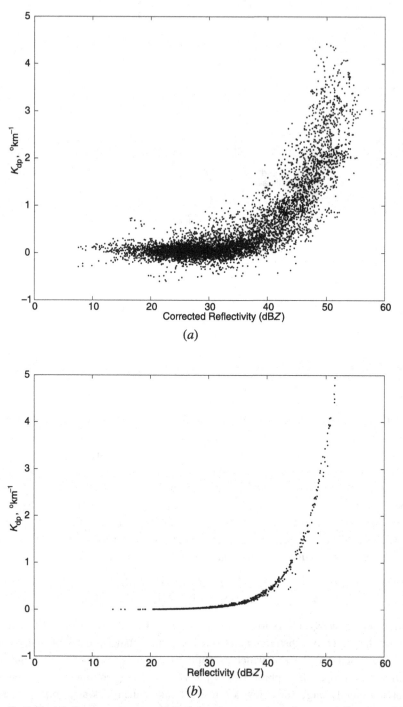

Fig. 7.85. (a) K_{dp} versus corrected Z_h from a large number of range profiles through a convective rain cell. Data from the BMRC C-POL radar. Z_h-correction is based on the technique illustrated in Fig. 7.84. (b) Scattering simulations of K_{dp} versus Z_h at C-band based on measured drop size distributions. Compare with (a).

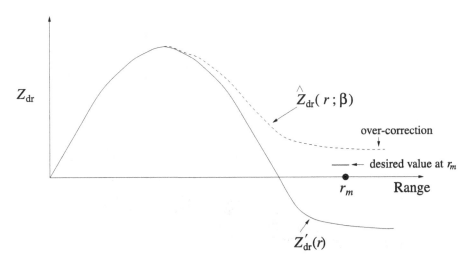

Fig. 7.86. Illustration relative to estimating the optimal value for β using eq. (7.156).

location from r_0 to r_m using (7.131a) expressed as,

$$A_{dp}(r; \beta) = \frac{\beta}{\alpha_{opt}} A_h(r; \alpha_{opt}) \qquad (7.155)$$

Using this first guess of A_{dp}, the first estimate of $\hat{Z}_{dr}(r_m; \beta)$ is calculated using (7.129a),

$$\hat{Z}_{dr}(r_m; \beta) = Z'_{dr}(r_m) + \frac{2\beta}{\alpha_{opt}} \int_{r0}^{r_m} A_h(s; \alpha_{opt}) \, ds \qquad (7.156)$$

This first estimate of $\hat{Z}_{dr}(r_m; \beta)$ is compared against the constraint value $\bar{Z}_{dr}(r_m)$ determined earlier, see (7.154). If $\hat{Z}_{dr}(r_m; \beta)$ is larger than the constraint value, then the cumulative differential attenuation was over-predicted and the next value of β in (7.155) is adjusted to a lower value and the steps (7.155, 7.156) are repeated until an optimal β is found that results in $\hat{Z}_{dr}(r_m; \beta) - \bar{Z}_{dr}(r_m)$ being less than a pre-selected tolerance (e.g. 0.2 dB). If the cumulative differential attenuation is under-predicted $(\hat{Z}_{dr}(r_m; \beta) < \bar{Z}_{dr}(r_m))$, then the β-value may be successively increased in (7.155) until an optimal β is found. Figure 7.86 schematically illustrates this adjustment procedure for the case where the initial β leads to "over-correction". Once the optimal β is estimated, then the final corrected Z_{dr} at each range location is obtained from,

$$\hat{Z}_{dr}(r; \beta_{opt}) = Z'_{dr}(r) + 2\frac{\beta_{opt}}{\alpha_{opt}} \int_{r0}^{r} A_h(s; \alpha_{opt}) \, ds \qquad (7.157)$$

As a validation of the combined Φ_{dp}–Z_{dr} constraint method, Fig. 7.87 shows a plot of (corrected) Z_{dr} versus (corrected) Z_h from radar along with scattering simulations based on measured dsds at C-band (using the same data set as in Fig. 7.85). Once again, note the excellent agreement between dsd simulations and corrected radar data, which

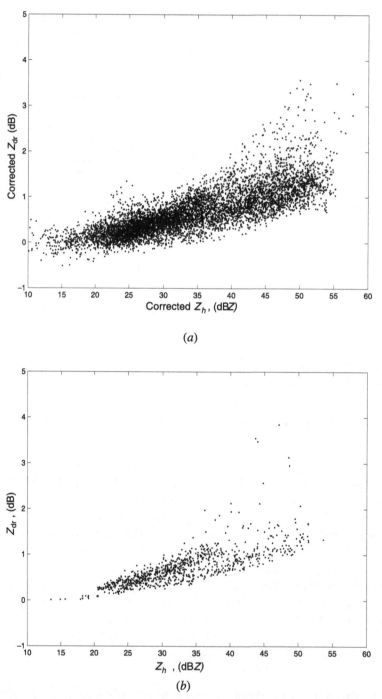

Fig. 7.87. (*a*) Corrected-Z_{dr} versus corrected-Z_h using C-POL radar data from a large number of range profiles through a convective rain cell. Z_{dr} is corrected based on eq. (7.157). (*b*) Scattering simulations of Z_{dr} versus Z_h at C-band based on measured drop size distributions. Compare with (*a*).

is not possible without good attenuation-correction schemes. Finally, the optimally established A_{dp} is plotted against A_h and compared against scattering simulations (see Fig. 7.88). The "linear" appearance of the data is due to (7.155), and the spread in slope reflects the change from beam to beam, which lies within the "bands" implied in Fig. 7.78.

Attenuation correction is an important step at frequencies corresponding to C-band (around 5.5 GHz) and higher, especially in convective storms. It may also be necessary at S-band (near 3 GHz) when the propagation path includes multiple rain cells of high intensity. Rain rate algorithms that use Z_h only, or Z_h and Z_{dr}, which will be discussed in Chapter 8, assume that attenuation-correction has been performed on the data. It is also obvious that K_{dp} is unaffected by attenuation effects, which is an important advantage in estimating rain rate using K_{dp}. The estimation of $A_h(r; \alpha_{opt})$ at each range resolution volume using the Φ_{dp}-constraint procedure allows corresponding rain rate (or rainwater content) estimates at each range resolution volume (via A_h–R relations), which avoids "smoothing" effects inherent in the estimate of K_{dp} (see Section 6.6).

7.5   Hydrometeor classification

It was shown in previous sections that polarimetric radar measurements are sensitive to the types, shapes, sizes, as well as the fall behavior of hydrometeors. Therefore, considerable information about the hydrometeors is contained in the covariance matrix measurements, which can be used to infer some microphysical properties. One such inference is identification of hydrometeor types. The physical basis for such classification is discussed in Sections 7.1–7.3. However, such inference from covariance matrix measurements does not have a unique solution, because the signature set from different types of hydrometeors is not mutually exclusive. Nevertheless, polarimetric radar measurements have been used in arriving at inferences on "bulk" hydrometeor types within a limited context as shown by some examples reported in the literature given below.

In Section 7.2, H_{dr}, see (7.88), was used to identify rain and hail regions whereas Z_{dp}, see (7.86), was used to quantitatively estimate the ice fraction in a rain/ice mixture. Bringi et al. (1984) utilized reflectivity and Z_{dr} measurements to detect hail, and showed that differentiation between regions of hail and rainfall is possible in convective storms. Hall et al. (1984) used the vertical structure of Z_{dr}, and reflectivity measurements to identify the transition from ice to rain.

Bringi et al. (1986a,b) studied the profiles of Z_{dr}, linear depolarization ratio (LDR) and reflectivity at horizontal polarization (Z_h) through the core of convective storms and found that these three measurements were useful in identifying graupel and hail regions. They also showed that the vertical structure of Z_{dr} (below the melting level), LDR (above the melting level), and the use of dual-wavelength techniques can provide information on hailshaft structure and vertical extent. A fairly detailed study of the vertical profiles of Z_h, Z_{dr}, and dual-wavelength measurements in terms of the size,

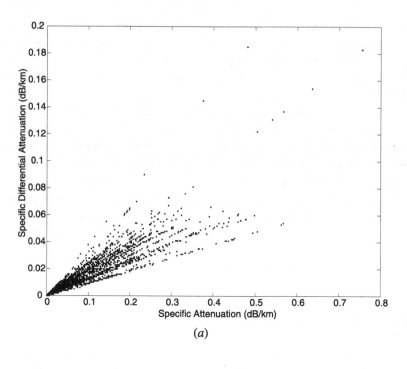

(a)

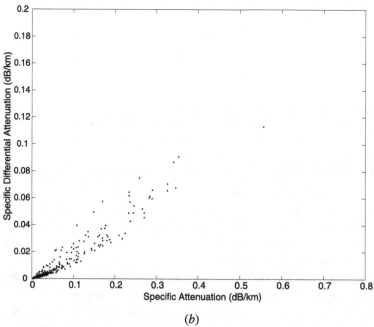

(b)

Fig. 7.88. (a) A_{dp} versus A_h obtained as a result of the combined Φ_{dp}–Z_{dr} constraint method. Radar data from the BMRC C-POL radar. (b) Scattering simulations of A_{dp} versus A_h at C-band based on measured drop size distributions. Compare with (a).

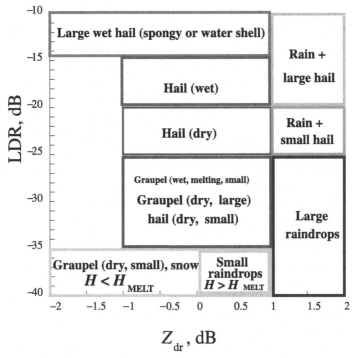

Fig. 7.89. Hydrometeor classification decision boundary in LDR, Z_{dr} plane. The boundaries correspond to the decision scheme used by Höller et al. (1994) and are valid at C-band (frequency near 5.5 GHz).

shape, and fall behavior of hailstones was presented by Aydin et al. (1990) for hail identification.

The physical basis for the signature of $|\rho_{co}|$ in hail as well as rain/hail mixtures is given in Section 7.2, see (7.91c, 7.97c), whereas the corresponding signatures of K_{dp} and $|\rho_{co}|$ in rain were given in Section 7.1. In addition, the physical basis for the K_{dp} signature in ice crystals is provided in Section 7.3. Zrnić et al. (1993) examined K_{dp}, the back scatter differential phase (δ_{co}), $|\rho_{co}|$, and Z_{dr} data in a severe hailstorm and demonstrated that these measurements could be used to detect hail, as well as to identify mixed-phase hydrometeors.

7.5.1   Empirical algorithms for classifying hydrometeor types

Building on the results of hydrometeor classification in the existing literature, Straka and Zrnić (1993), Höller et al. (1994), and Zeng et al. (in press) described techniques to identify different hydrometeors. All of the above used the decision tree method in which predefined boundaries were used to construct the classification methodology. Höller et al. (1994) used empirical Boolean decision logic to classify hydrometeor types, based on Z_{dr} and LDR and the height of the melting level in a hailstorm. The decision scheme used by Höller et al. is shown in Fig. 7.89. Straka and Zrnić (1993) presented a Boolean decision logic based on five polarimetric radar observables, namely Z_h, Z_{dr}, $|\rho_{co}|$, K_{dp}, and LDR. The decision space for this scheme is shown in Fig. 7.90. A cursory glance at Fig. 7.90 demonstrates that such detailed classification between numerous hydrometeor types using Boolean logic may not be accurate because the covariance matrices from different hydrometeor types are not mutually exclusive. In addition, the covariance matrix for the same set of hydrometeors could be different for different frequencies, and the covariance matrix can be altered due to propagation through the precipitation medium. Zeng et al. (in press) developed a simpler scheme with fewer hydrometeor categories. They also introduced an "ambiguous" category to account for some ambiguity in the decision method. In addition, they presented two different classification schemes valid for heights above and below the 0 °C level. Figure 7.91 shows their classification scheme at S-band.

7.5.2   Fuzzy logic classification

There are a number of methods that can be potentially used for hydrometeor identification, such as Boolean decision tree, classical statistical decision theory, neural networks, or fuzzy logic. As stated earlier, the covariance matrix is not mutually exclusive for different hydrometeor types. It follows that the decision tree method, which is essentially based on threshold boundaries and Boolean logic, may not be adequate for the hydrometeor classification problem. In addition, the Boolean logic decision scheme does not provide allowance for measurement errors. Statistical decision theory is another potential technique that can be considered for the hydrometeor classification problem. However, statistical models are difficult to construct. For example, the statistical model for rain identification can be expressed as,

$$P(C = \text{Rain}/Z_h, Z_{dr}, K_{dp}, \text{LDR}, |\rho_{co}|)$$

$$= \frac{f(Z_h, Z_{dr}, K_{dp}, \text{LDR}, |\rho_{co}|/C = \text{Rain}) \times p(C = \text{Rain})}{f(Z_h, Z_{dr}, K_{dp}, \text{LDR}, |\rho_{co}|)} \tag{7.158}$$

where C is the hydrometeor type, $P(C = \text{Rain})$ is the prior probability of rain, $f(Z_h, Z_{dr}, K_{dp}, \text{LDR}, |\rho_{co}|)$ is the joint probability density of the five radar polarimetric parameters, and $f(Z_h, Z_{dr}, K_{dp}, \text{LDR}, |\rho_{co}|/C = \text{Rain})$ is the joint probability density function of the five polarimetric radar parameters under the condition of rain. In practice, it is very difficult to obtain the prior probability and the probability density functions.

Use of fuzzy logic schemes for hydrometeor classification has many inherent advantages over other methods. Fuzzy logic uses simple rules to describe the system of interest rather than analytical equations, thus it is easy to implement for hydrometeor classification. The fuzzy logic system has been developed such that it can reach distinct

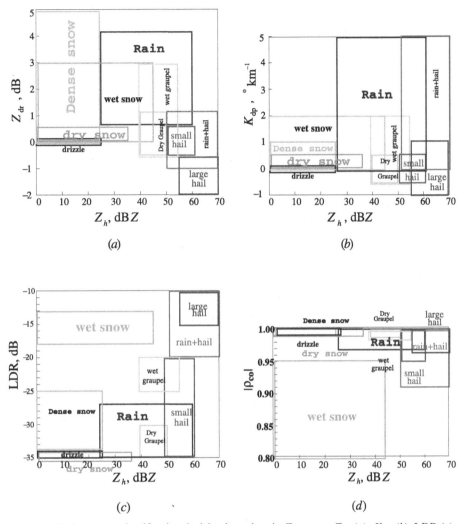

Fig. 7.90. Hydrometeor classification decision boundary in Z_h versus Z_{dr} (a), K_{dp} (b), LDR (c), and $|\rho_{co}|$ (d) space, respectively. The boundaries correspond to the decision scheme used by Straka and Zrnić (1993), and are valid at S-band (frequency near 3 GHz).

decisions based on overlapping or "noise contaminated" data. Therefore, the fuzzy logic scheme is a well-suited technique for the problem of hydrometeor classification.

(i) Configuration of a general fuzzy logic system

The fuzzy logic system consists of four parts, namely fuzzification, rule inference, aggregation, and defuzzification (Zadeh 1983). The block diagram of a general fuzzy logic system is shown in Fig. 7.92 where, $x_1, x_2, \ldots, x_n$ stand for n "crisp" (or distinct) inputs, and y is the crisp output. The functions of the various blocks in the fuzzy logic system are as follows.

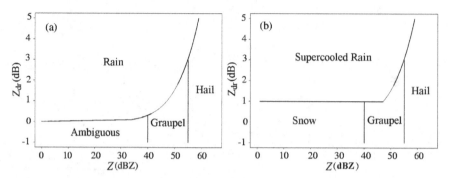

Fig. 7.91. Hydrometeor classification decision boundary in Z_h, Z_{dr} space: (*a*) below and (*b*) above the 0 ° level. The boundaries correspond to the decision scheme used by Zeng et al. (in press), and are valid at S-band.

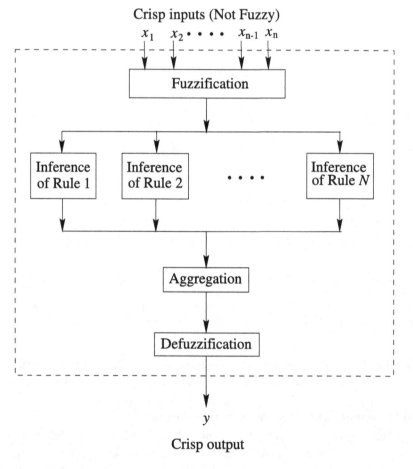

Fig. 7.92. Block diagram of a general fuzzy logic system.

The function of the "fuzzification" block is to convert the crisp inputs (or precise measurements) to fuzzy sets with a corresponding membership degree. A specific crisp input can belong to different fuzzy sets but with different membership degrees (or truth values). The most important component in fuzzification is the *membership function*, which is used to describe the relationship of the crisp input and the fuzzy sets in the input domain. The definition of membership function is: $\mu_A(x)$, i.e. the membership function of fuzzy set A (for a fuzzy variable x), whose value is the degree to which x is a member of set A.

In a fuzzy logic system, rules are used to linguistically describe the complex relationship between the input and output fuzzy variables in the form of IF–THEN statements. Typically, the rule is composed of several "antecedents" in the IF statement and one or several "consequents" in the THEN statement. The process of deducing the "strength" of these consequents from the "strength" of the antecedents is called rule inference. The most commonly used inference methods are correlation minimum, correlation product, and min–max (Heske and Heske 1996). Several rules (instead of one single rule) can be used to describe a fuzzy logic system. This set of rules is called a "rule base". Complete knowledge about a fuzzy model is contained in its rule base and the membership functions. Inference methods can be used to derive the "strength" of each rule, following which the aggregation method can be used to determine an overall fuzzy region. Two commonly used aggregation methods are additive aggregation and max aggregation.

The output of the aggregation process is a fuzzy set, but in many applications (such as hydrometeor classification) it is necessary to find a crisp value that best represents the fuzzy output set, and this process is called defuzzification.

(ii) Architecture of a fuzzy hydrometeor classifier

To implement hydrometeor classification using fuzzy logic (or fuzzy hydrometeor classifier, referred to as FHC), the four general blocks (fuzzification, rule inference, aggregation, and defuzzification) need to be specified. The block diagram of the FHC is shown in Fig. 7.93, where Z_h, Z_{dr}, K_{dp}, LDR, $|\rho_{co}|$, and altitude of the observation (H), are the six inputs, and the hydrometeor class (C) is the output. The FHC will infer the hydrometeor type C based on the rule base from the six inputs. Table 7.2 shows a possible list of ten classes which can be used in the inference of hydrometeor types in convective storms. Referring to Figs. 7.89–7.91, note that different sets of hydrometeor classes have been used depending on the application context and availability of radar observations. The FHC is general enough that it can be developed for any set of hydrometeor class depending upon the application. However, it is better to limit the number of hydrometeor types to be compatible with storm type and geographical location. In the following, details of the classification procedure using, for example, the ten classes listed in Table 7.2, are provided. With no loss of generality, this procedure can be applied to any classification set based on any set of input measurements. For example, an additional important input could be the temperature profile with altitude.

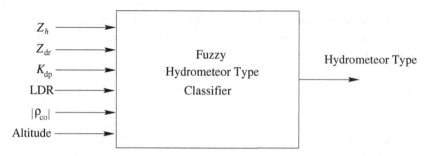

Fig. 7.93. Fuzzy hydrometeor classifier.

Table 7.2. Output of the fuzzy classifier system

Hydrometeor type	Classifier output C
Drizzle	1
Rain	2
Low density snow	3
High density snow	4
Wet snow	5
Dry graupel	6
Wet graupel	7
Small hail	8
Large hail	9
Rain + hail	10

(iii) Classification procedure

A detailed block diagram of the fuzzy hydrometeor classifier is shown in Fig. 7.94 which depicts the four general blocks referred to earlier. First, the five radar measurements and altitude are fuzzified by using membership functions (MBFs). Corresponding to the ten hydrometeor classes, there are ten MBFs for each of the input variables in the system. After fuzzification, the IF–THEN rule inference is carried out based on the rule base for the classification system. Rule aggregation is applied to achieve the total effect of all the rules. The last step is defuzzification, which converts the aggregation result to a single hydrometeor type.

The purpose of fuzzification is to convert the precise input measurements to fuzzy sets with corresponding membership degree. The specification of membership functions is critical to the classification performance. Ten fuzzy sets corresponding to the ten hydrometeor types are specified for each of the six input variables for convective storms. For snowstorms, a reduced number of classes, e.g. five fuzzy sets corresponding to the hydrometeor types (drizzle, rain, dry snow, ice crystal, and wet snow) can be specified for each of the six input variables. Each fuzzy set is represented by a membership function, denoted as MBF_{i_j}, where the index i corresponds to the six inputs, and the

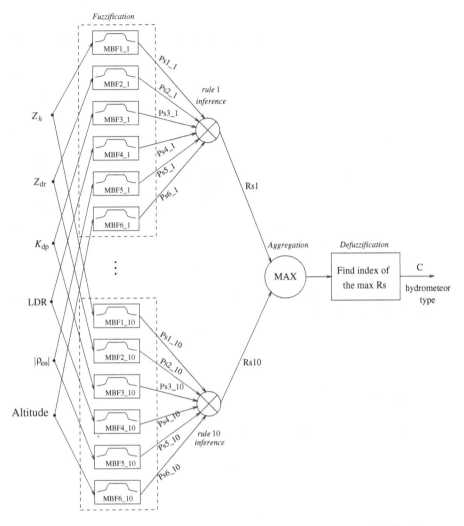

Fig. 7.94. Detailed block diagram of the fuzzy logic system for hydrometeor classification.

index j corresponds to the fuzzy sets. The index j takes values 1–10 for convective storms and 1 to 5 for snowstorms.

Several functional forms can provide adequate representation of membership functions, such as triangular, trapezoidal (e.g. Vivekanandan et al. 1999), Gaussian, or beta functions. Liu and Chandrasekar (2000) used beta functions to describe membership functions for the following reasons. In this application, most membership functions have a wide flat region in which the maximum value is 1. For example, the reflectivity of rain can vary over a wide range (say, 25–60 dBZ). Therefore, when Z_h is the input, the rain MBF should be flat in this range and taper off outside this range (see rain MBF in Fig. 7.95b). The beta function has the required characteristics to represent this form. In addition, beta functions have a long tail, which improves the robustness of the FHC.

The derivative of the beta function is continuous and this feature is useful for automatic adjustment of its parameters based on feedback from *in situ* verification data. The beta membership function is given by,

$$\text{beta}(x, m, a, b) = \frac{1}{1 + \{[(x - m)/a]^2\}^b} \tag{7.159}$$

There are three parameters that define the shape of a beta function, namely, the center m, the width a, and the slope b (see Fig. 7.95a).

Ten membership functions of Z_h are shown in Fig. 7.95b, which represent the ten fuzzy sets of Z_h. For each curve, the horizontal axis is the value of Z_h, and the vertical axis stands for the membership degree of Z_h corresponding to that fuzzy set. In a fuzzy logic system, fuzzy sets instead of precise measurements are used to represent the input variables. For example, a Z_h of 40 dBZ belongs to the drizzle set with membership degree 0; to the rain set with membership degree 1; ..., to the dry graupel set with membership degree 0.8; and to the rain/hail mixture with membership degree 0.

Multi-dimensional membership functions can be used to represent combinational fuzzy sets. For example, if Z_{dr} is physically independent of Z_h, then one-dimensional membership functions of Z_h and Z_{dr} are adequate to represent the fuzzy sets. Let $f_{\text{rain\_}zh}$ be the one-dimensional membership function for a rain set with respect to the input variable Z_h. The $f_{\text{rain\_}zh}$ corresponding to the beta function in (7.159) can be written as,

$$f_{\text{rain\_}zh}(Z_h) = \frac{1}{1 + \{[(Z_h - 42.5)/18.2]^2\}^{18.3}} \tag{7.160}$$

Similarly, $f_{\text{rain\_}zdr}$, the one-dimensional membership function for a rain set with respect to Z_{dr}, can be written as,

$$f_{\text{rain\_}zdr}(Z_{dr}) = \frac{1}{1 + \{[(Z_{dr} - 2.25)/1.8]^2\}^{16.2}} \tag{7.161}$$

If Z_h and Z_{dr} are treated as independent variables for rain, then in a two-dimensional Z_h–Z_{dr} space, the membership function for rain can be expressed as the product of $f_{\text{rain\_}zh}$ and $f_{\text{rain\_}zdr}$,

$$f_{\text{rain\_}zh,zdr}(Z_h, Z_{dr}) = f_{\text{rain\_}zh}(Z_h) \times f_{\text{rain\_}zdr}(Z_{dr}) \tag{7.162}$$

However, Z_h and Z_{dr} are not independent in rainfall since Z_{dr} is proportional to the mass-weighted mean diameter (D_m; see (7.14)), whereas Z_h is proportional to the product of rainwater content and D_m^3, see (7.69b). For example, Fig. 7.23 shows that the maximum excursion of Z_{dr} changes with Z_h. Figure 7.96a shows contours of the two-dimensional membership function given in (7.162) along with observed Z_h–Z_{dr} scatter plot at S-band. Based on Fig. 7.96a, it is preferable to modify the two-dimensional membership function as illustrated in Fig. 7.96b. For non-rain classes, the Z_{dr} membership functions can be considered to be independent (see Fig. 7.97). Other membership functions corresponding to LDR, K_{dp}, and $|\rho_{co}|$ are given in Liu

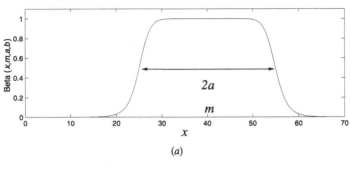

(a)

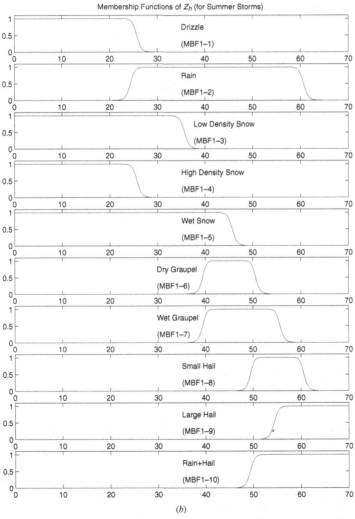

(b)

Fig. 7.95. (*a*) Beta membership function. (*b*) Illustration of the fuzzification of Z_h to its ten fuzzy sets.

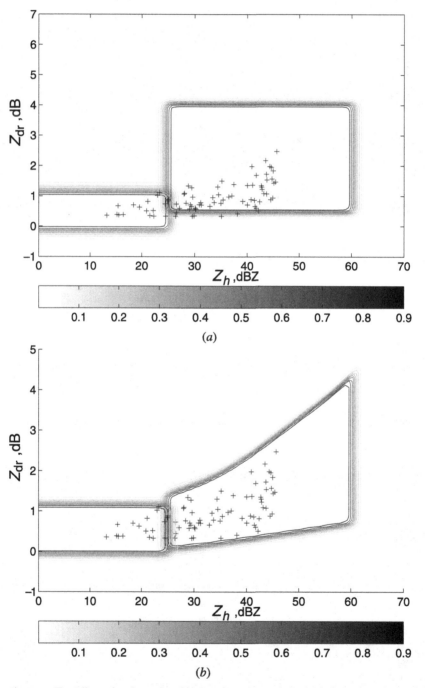

Fig. 7.96. Two-dimensional membership function of (*a*) rain and drizzle treating Z_h and Z_{dr} as independent variables (also shown is the radar-observed scatter plot of Z_h versus Z_{dr}), (*b*) rain modified to account for the correlation between Z_h and Z_{dr} in rain. The gray scale indicates the contour level (nine contours 0.1–0.9).

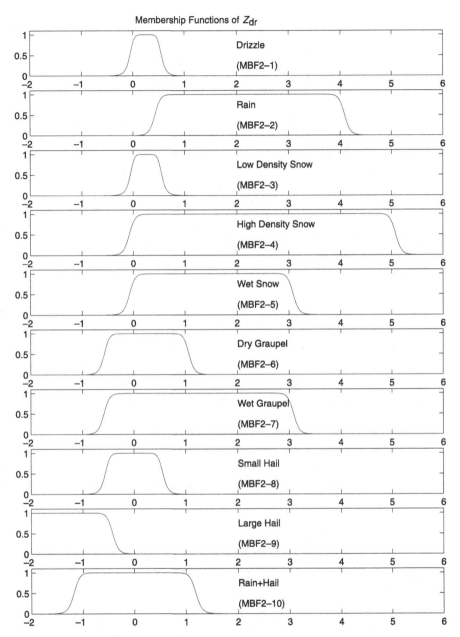

Fig. 7.97. Illustrating the fuzzification of Z_{dr} to its ten fuzzy sets at S-band (frequency near 3 GHz).

and Chandrasekar (2000). As noted earlier, the trapezoidal form for the membership functions was used by Vivekanandan et al. (1999). Figure 7.98 shows the trapezoidal boundaries for their hydrometeor classes in the Z_h–Z_{dr} and K_{dp}–Z_h planes.

Prior knowledge about the hydrometeor classification problem is incorporated in the FHC in the form of IF–THEN rules and membership functions. The IF–THEN rules can be written as follows:

IF (Z_h IS MBF1_j AND Z_{dr} IS MBF2_j AND K_{dp} IS MBF3_j AND LDR IS MBF4_j AND $|\rho_{co}|$ IS MBF5_j AND Height IS MBF6_j) THEN Hydrometeor Class is j

The strength for the six antecedent propositions could be obtained from the fuzzification block as PS_{i_j}, where the index i represents the five measurements and the altitude, and index j represents the classes. For example, PS_{1_2} is the strength of reflectivity for rain. Here, product intersection operation is used to get the strength of the IF–SIDE (of the IF–THEN rule), and the correlation product inference method is used to get the rule strength (Heske and Heske 1996). The truth value of the IF–SIDE is used to scale the consequent fuzzy set. In this case, the strength of the rule is equal to the strength of the IF–SIDE because the output is a single hydrometeor class. Therefore, the strength of rule j (RS_j) can be obtained as the product of the strength of the individual propositions as,

$$RS_j = \prod_{i=1}^{6} PS_{i\_j} \tag{7.163}$$

The "maximum aggregation" method is used to get the net fuzzy result from the individual rule inference results. The maximum aggregation procedure takes only the consequent with the highest truth value. Therefore, the aggregation result is the maximum rule strength of the various RS_js defined in (7.163).

(iv) Training the fuzzy hydrometeor type classifier

The performance of the FHC depends critically on the shape of the membership functions. A combination of empirical and theoretical knowledge of current state-of-the-art should be used to construct the membership functions. Manually adjusting the functions is tedious. It would be useful to develop a system that learns from data and can adjust the membership functions automatically. This can be achieved by the neuro-fuzzy hydrometeor type classifier developed by Liu and Chandrasekar (2000). The combination of neural network and fuzzy logic is called a hybrid neuro-fuzzy system. Refer to Section 8.5 for a discussion of neural networks as applied to rain rate estimation.

In a neuro-fuzzy system, the fuzzy logic part can be modeled as a multilayer feedforward neural network, with five layers, namely, the input layer (consisting of input variables), the IF layer (fuzzification layer), the THEN layer (RULE inference layer), the aggregation/defuzzification layer, and the output layer. Under this model, neural network learning algorithms can be used to learn the parameters of the system. One implementation of the neuro-fuzzy hydrometeor classifier (NFHC) is shown in Fig. 7.99. If the blocks shown by dotted lines are ignored, then it is the same as the configuration of the FHC. The solid lines form the feedforward path, and the dotted lines form the backpropagation path. The mis-classification error is back-propagated

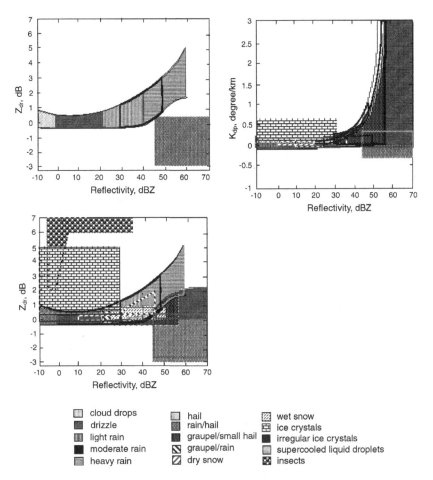

Fig. 7.98. Hydrometeor classification decision boundary in the Z_h–Z_{dr} and Z_h–K_{dp} planes. The boundary marked by a dark line indicates rain. The boundaries for 15 different classes are shown (Vivekanandan et al. 1999) and are valid at S-band (frequency near 3 GHz).

to the IF-layer to adjust the parameters of the membership functions. This is a very effective and efficient procedure to build a good FHC over time, provided sufficient *in situ* validation data are available.

(v) Examples of FHC use

Examples presented in this section use radar data from CSU–CHILL in two summer convective storms and a winter storm with associated *in situ* observations. The three storm events are: a severe hailstorm on June 7, 1995; a convective storm on June 22, 1995; and a snow event on February 18, 1997.

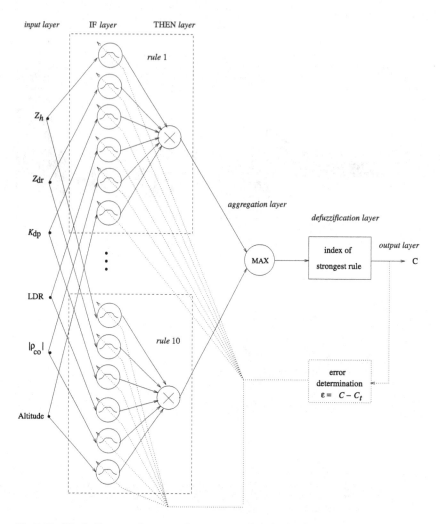

Fig. 7.99. Block diagram of a neuro-fuzzy system for a hydrometeor classifier.

The fuzzy hydrometeor classifier was applied to the hailstorm case described in Section 7.2.2 (see Fig. 7.56). Figure 7.100 shows a vertical section of reflectivity (Z_h), and Fig. 7.101 shows the classification result. Hail/rain mixture, wet graupel, and rain are found on the ground as well as at low altitudes, small hail up to mid-levels and snow at high altitudes. This classification is in general agreement with supercell storm structure (e.g. Conway and Zrnić 1993; Miller et al. 1990). The FHC output at the location of an instrumented van (see also Fig. 7.58) was in good agreement with surface observations. The fuzzy hydrometeor classifier was also applied to another hailstorm case described in Section 7.2.2 (see Fig. 7.62). *In situ* verification data were provided by hail images from T-28 aircraft penetrations as discussed in Section 7.2.2. Figure 7.102 shows the hydrometeor classification results in a horizontal plane at the T-28 penetration

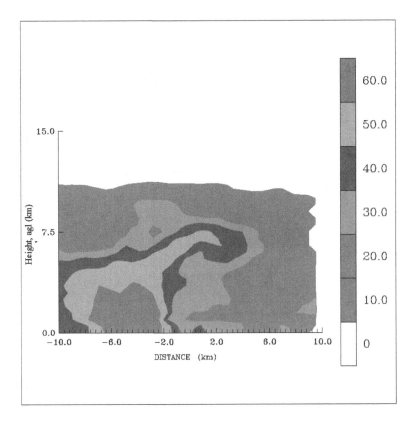

Fig. 7.100. Vertical section of reflectivity measurements through a hailstorm observed by CSU–CHILL radar (see also, Section 7.2.2).

altitude. The solid line on Fig. 7.102 is the flight track of the T-28 aircraft. Figure 7.103 compares the hydrometeor classification results with *in situ* observations from the T-28 imaging probe, and good agreement is noted.

The last example shows hydrometeor classification in a snowstorm. Figure 7.104 shows a vertical section of reflectivity (Z_h) from CSU–CHILL radar measurements in a snowstorm observed on February 18, 1997. The hydrometeor classification shown in Fig. 7.105 indicates rain and wet snow below 1 km, and dry snow and oriented ice crystals above 1 km. A transition from rain to snow may be noted at a distance of 45 km, which was validated by ground observations. Further details of the above classification examples can be found in Liu and Chandrasekar (2000).

Polarimetric techniques have now reached a level of maturity leading to confidence in development of algorithms for hydrometeor classification in real-time. Such advances also present new challenges for algorithm validation using *in situ* observations, which at the present time are insufficient. However, when more observations become available hydrometeor classification techniques can be improved and made more robust.

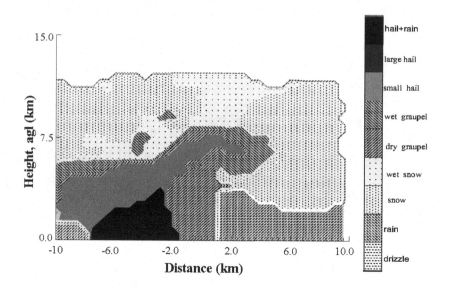

Fig. 7.101. Vertical section of hydrometeor classification of the June 7, 1995 hailstorm (see also Section 7.2.2).

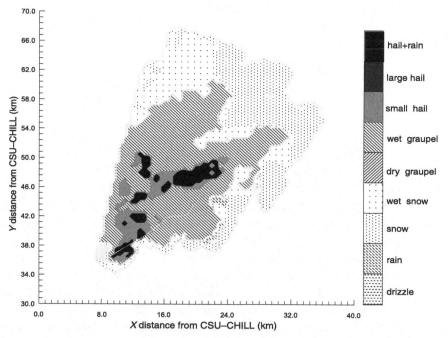

Fig. 7.102. Horizontal section of hydrometeor classification at the aircraft altitude for the June 22, 1995 hailstorm observed by the CSU–CHILL radar (see also, Fig. 7.62).

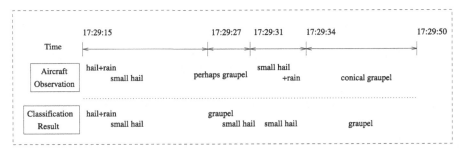

Fig. 7.103. Comparison of the hydrometeor classification results with *in situ* observations along the aircraft track for the June 22, 1995 hailstorm. The time series corresponds to the aircraft track shown in Fig. 7.102.

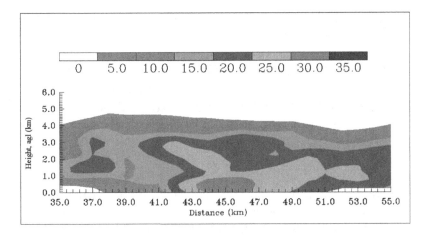

Fig. 7.104. Vertical section of reflectivity through a snowstorm observed by CSU–CHILL radar on February 18, 1997.

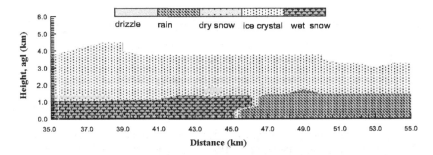

Fig. 7.105. Vertical section of hydrometeor classification for the snowstorm of February 18, 1997. The hydrometeor classes are: drizzle; dry snow; oriented ice crystals; wet snow; dry graupel; wet graupel. Note the transition from wet snow to rain at 45 km.

Notes

1. A comprehensive reference on cloud and precipitation physics is given in Pruppacher and Klett (1997). Also see the review by Jameson and Johnson (1990) on cloud microphysics and radar.

2. A comprehensive reference on cloud dynamics is provided in Cotton and Anthes (1989). See, also, the review by Ray (1990) on convective dynamics.

3. Other forms, e.g. the log-normal, can also be used to describe the drop size distribution; see, for example, Feingold and Levin (1986).

4. While the predominant emphasis in this chapter is on the linear polarization basis, the circular bases can also be used to derive Z_{dr} and K_{dp} as discussed in Chapter 4.

5. This suggested definition of δ_{co} under the $\exp(j\omega t)$ time convention is consistent with $\Psi_{dp} = \Phi_{dp} + \delta_{co}$. Thus, positive values of δ_{co} will cause a local increase in the range profile of Ψ_{dp}, with Φ_{dp} being the monotonic increasing component due to differential propagation phase in rain (see Section 6.6.1).

6. Several conference papers have appeared recently that deal with application of the normalized gamma dsd, e.g. Illingworth and Blackman (1999) or Dou et al. (1999).

7. The Joss–Waldvogel disdrometer has been evaluated, for example, by Sheppard (1990).

8. The two-dimensional video disdrometer is a relatively new instrument manufactured by Joanneum Research, Graz, Austria. A description of this instrument and sample data can be found at,

   http://www.distrometer.at

9. The Cooperative Huntsville Meteorological Experiment (COHMEX) was conducted in northern Alabama during the summer of 1986 (Dodge et al. 1986). The Convective and Precipitation/Electrification Experiment (CaPE) was conducted in central Florida during the summer of 1991. Polarimetric data were acquired with the NCAR/CP-2 radar in both experiments. The NCAR/CP-2 radar is no longer in service having been replaced by the S-POL radar.

10. A description of the University of Wyoming King Air aircraft can be found at,

   http://www-das.uwyo.edu/atsc/facilities

   The NCAR King Air aircraft is no longer in service.

11. These imaging probes are the two-dimensional PMS precipitation probes (Knollenberg 1981). See also, images shown in Figs. 7.30 and 7.35.

12. The South Dakota School of Mines and Technology (SDSM&T) T-28 aircraft is described at,

   http://www.hpcnet.org/sdsmt/scholarly

   Click on "Institute of Atmospheric Sciences" and then proceed to the T-28 home page. The high volume particle spectrometer (HVPS) is manufactured by SPECinc, Boulder, Colorado, see Lawson et al. (1993) and the SPECinc web site at,

   http://www.specinc.com

13. See Chapter V of the *OPEX Reference Handbook on Radar*, which can be downloaded from,

    http://www.estec.esa.nl/xewww/cost255/opex.htm

14. The C-POL radar is described by Keenan et al. (1998). This radar is normally located in Darwin, Australia. The sample data used in this section to illustrate different attenuation-correction procedures were provided by the Bureau of Meteorology Research Center (BMRC), Melbourne, Australia.

8 Radar rainfall estimation

The detection and measurement of precipitation has been pursued since the beginnings of radar, and the early history has been summarized by Atlas and Ulbrich (1990). Joss and Waldvogel (1990) have reviewed the use of reflectivity at a single polarization to estimate rainfall. Dual-wavelength radars have been used to measure attenuation from which rainfall is estimated (Eccles and Mueller 1973). More recently, attenuation-based methods have been evaluated both theoretically and experimentally for airborne and spaceborne radars (for example, see Meneghini and Kozu 1990). This chapter focuses on dual-polarization radar methods to estimate rainfall (Seliga and Bringi 1976; 1978) based on the microphysical properties of rain discussed in Chapter 7.

Rainfall measurement techniques[1] can be broadly classified as: (i) physically based, and (ii) statistical/engineering based. Physically based rainfall algorithms (as defined here) rely on physical models of the rain medium without any feedback from rain gage observations, whereas statistical/engineering solutions are derived using such feedback either directly or indirectly. Both techniques are considered in this chapter.

The main advantage of using radar for precipitation estimation is that radars can obtain measurements over large areas (about $10\,000$ km^2) with fairly high temporal and spatial resolution. Just substituting a gage for each radar spatial sample (150 m resolution in range and one-degree resolution in azimuth) would require more than one-quarter of a million gages over a 150-km radius. These measurements are sent to a central location at the speed of light by "natural" networks. In addition, radars can provide fairly rapid updates of the three-dimensional structure of precipitation (5–6 minutes for the WSR–88D volume scan). Because of these advantages, radar measurements of precipitation have enjoyed wide-spread operational usage and will remain so in the future.

8.1 Physically based parametric rain rate estimation algorithms

The distribution of raindrop sizes and shapes forms the building blocks for deriving physically based rain rate algorithms. The raindrop size distribution describes the probability density/distribution function of raindrops. In practice, the normalized histogram of raindrop sizes converges to the probability density function. A gamma distribution model (or a similar model such as log-normal distribution) can adequately describe many of the natural variations in the shape of raindrop size distributions. The gamma raindrop size distribution (dsd) can be expressed as,

$$N(D) = n_c f_D(D); \quad m^{-3} \, mm^{-1} \tag{8.1}$$

where $N(D)$ is the number of raindrops per unit volume per unit size interval (D to $D + \Delta D$), n_c is the concentration, and $f_D(D)$ is the probability density function. When $f_D(D)$ is of gamma form, it is given by,

$$f_D(D) = \frac{\Lambda^{\mu+1}}{\Gamma(\mu+1)} e^{-\Lambda D} D^\mu; \quad \mu > -1 \tag{8.2}$$

where Λ and μ are parameters of the gamma pdf (i.e. the same as (7.27) except $\mu \equiv \nu - 1$). Any other gamma form, such as the form introduced by Ulbrich (1983) or the form in (7.62a), can be derived from this fundamental notion of drop size distribution. It should be noted that any function used to describe $N(D)$ when integrated over the dsd, must yield the concentration, to qualify as a dsd function. This property is a direct result of the fact that any probability density function should integrate to unity. When $\mu = 0$, the gamma form reduces to the exponential form as $N(D) = n_c \Lambda e^{-\Lambda D}$. The relation between D_0, μ, and Λ is given in (7.51a). An alternate form for $f_D(D)$ in terms of D_0 can be written as,

$$f_D(D) = \frac{(3.67 + \mu)^{\mu+1}}{\Gamma(\mu+1)D_0} \left(\frac{D}{D_0} \right)^\mu e^{[-(3.7+\mu)\frac{D}{D_0}]} \tag{8.3}$$

It can be seen that $f_D(D)$ has units of size^{-1}.

It was shown in Chapter 7 that based on the size distribution and shape of raindrops the various polarimetric radar observables can be calculated. The most commonly used polarimetric measurements for rainfall estimation are the reflectivity factor (say at horizontal polarization), the differential reflectivity, and the specific differential phase. For rain, Z_{dr} and K_{dp} can be written as integrals of back scatter and forward scatter amplitudes over the dsd. It follows from (7.9) that,

$$Z_{dr} = 10 \log_{10} \left[\frac{\int \sigma_{hh}(D)N(D) \, dD}{\int \sigma_{vv}(D)N(D) \, dD} \right]; \quad dB \tag{8.4}$$

where $\sigma_{hh} = 4\pi |S_{hh}|^2$ and $\sigma_{vv} = 4\pi |S_{vv}|^2$ are the radar cross sections at horizontal and vertical polarization, respectively. Similarly, it follows from (7.10) that,

$$K_{dp} = \frac{180\lambda}{\pi} \int \text{Re}\left[f_h(D) - f_v(D) \right] N(D) \, dD; \quad ^\circ \, km^{-1} \tag{8.5}$$

where $f_h = \hat{h} \cdot \vec{f}$, $f_v = \hat{v} \cdot \vec{f}$ are the forward scattering amplitudes at horizontal and vertical polarization, respectively. Rain rate is the volume flux of water per unit area. The still-air rain rate defined in (7.66a) is repeated here,

$$R = 0.6\pi \times 10^{-3} \int v(D) D^3 N(D) \, dD; \quad mm \, h^{-1} \tag{8.6}$$

where $v(D)$ is the terminal fall velocity of raindrops, which depends on parameters such as air density, size, and shape of raindrops. However, for algorithmic purposes the terminal fall velocity can be approximated by a power-law expression of the form,

$$v(D) = \alpha D^{\beta}; \quad \text{m s}^{-1} \tag{8.7}$$

where D is in mm and typical values of α used in the literature vary from 3.6 to 4.2, and β from 0.6 to 0.67 for rain (Sekhon and Srivastava 1971; Atlas and Ulbrich 1977). The most commonly used sea-level values for α and β are 3.78 and 0.67, respectively, for rain rate estimates (Atlas and Ulbrich 1977). In this section, the radar-based rain rate algorithms are assumed to apply at sea-level. However, at higher altitudes the rain rate can be adjusted according to (7.66c). With this model, several forms of rainfall algorithms can be developed depending on the radar measurements used in the estimation. The simplest rainfall algorithm is the Z–R relation, where R is estimated from Z,

$$R(Z) \simeq cZ^{a}; \quad \text{mm h}^{-1} \tag{8.8a}$$

where Z is in units of $\text{mm}^6 \text{ m}^{-3}$. The physical basis of the power-law form of Z–R algorithms was explained in Chapter 7, where it was shown that for a gamma family of dsds the exponent should be nearly 0.67 (or 4.67/7). However, the constant c could vary widely depending on the variability of N_w, see (7.61). The range of c and a reported in the literature is very large (Battan 1973). However, two Z–R algorithms worth mentioning are, namely, the one based on the Marshall–Palmer dsd given by,

$$R_{\text{MP}}(Z) = 0.0365Z^{0.625} \tag{8.8b}$$

and the Z–R relation used by WSR–88D radars,[2]

$$R_{\text{WSR}}(Z) = 0.017Z^{0.714} \tag{8.8c}$$

To avoid hail contamination, the maximum Z is usually truncated at 55 dBZ when using these algorithms. The wide variability of Z–R algorithms has led to the development of alternate methods for rainfall estimation. Based on the physical principles described in Section 7.1, the three radar measurements Z_h, Z_{dr}, and K_{dp} can be used in various combinations to estimate rain rate. Radar rainfall algorithms can be broadly classified into four categories, namely, $R(Z)$, $R(Z, Z_{\text{dr}})$, $R(K_{\text{dp}})$, and $R(K_{\text{dp}}, Z_{\text{dr}})$. Each of the above-mentioned algorithms has advantages and disadvantages. First, $R(Z)$ is the most commonly used algorithm because most radars operate at one polarization. Reflectivity-based algorithms require accurate knowledge of the radar constant (see Section 6.3) and are prone to errors in absolute calibration. Differential reflectivity (Z_{dr}) is a relative power measurement, and it can be measured accurately without being affected by absolute calibration errors (see Section 6.3.2). However, since it is a relative power measurement any rainfall algorithm that uses Z_{dr} needs to be used in conjunction with Z_h or K_{dp}. Algorithms to estimate rainfall from K_{dp} have several advantages, which originate from K_{dp} being derived from phase measurements, e.g. it is unaffected by absolute calibration error, attenuation, and unbiased if rain is mixed with spherical hail (see Section 4.2). However, from Section 6.6 it is clear that K_{dp} is relatively noisy at low rain rates. Also, K_{dp} is derived as the range derivative of Φ_{dp} and, in practice, is estimated over a finite path. Therefore, there is a tradeoff between the accuracy and range resolution of K_{dp}-based rainfall estimates. The advantages and disadvantages of each polarimetric measurement translate into the error structure of algorithms involving Z_h, Z_{dr}, and K_{dp}. The following sections present detailed evaluation of various rainfall algorithms.

Whenever reflectivity is measured, it is done at a specific polarization such as horizontal, vertical, or circular. Until polarization measurements became prevalent, the polarization state of Z measurements was mostly ignored (Zrnić and Balakrishnan 1990). However, Z can vary by several decibels depending on the polarization state. Note that all rain rate algorithms developed in this section use Z_h even though it is sometimes referred to as Z.

8.1.1 $R(Z_h, Z_{dr})$ **algorithm**

It was shown in Chapter 7 that for a given raindrop size distribution the reflectivity factor can be derived as the sixth moment, see (7.21). For a gamma dsd, Z/R is expressed, see (7.69b) as,

$$\frac{Z}{R} = F(\mu) D_m^{2.33} \tag{8.9}$$

From (7.14a), note that D_m can be related to Z_{dr} as a power law of the form $D_m = a(Z_{dr})^b$, where a and b will depend on the frequency band. It follows that the rain rate can be expressed in the form,

$$R = \frac{c}{F(\mu)} Z Z_{dr}^{-2.33b} \tag{8.10}$$

where c is a constant. This simplistic approach can be further refined by pursuing an estimator of the form,

$$R = c Z_h^a Z_{dr}^b \tag{8.11}$$

Though the above form yields a good estimator from theory, measurement errors in Z_{dr} can cause large errors in R especially when mean Z_{dr} tends to 0 dB. A more robust estimator can be constructed using z_{dr} ($z_{dr} = 10^{0.1(Z_{dr})}$) of the form,

$$R = c\, Z_h^{a_1}\, z_{dr}^{b_1}; \quad \mathrm{mm\ h}^{-1} \tag{8.12a}$$

or,

$$R = c_1\, Z_h^{a_1}\, 10^{0.1 b_1 Z_{dr}}; \quad \mathrm{mm\ h}^{-1} \tag{8.12b}$$

Table 8.1. Coefficients of $R(Z_h, Z_{dr})$ rainfall algorithm at S-, C-, and X-bands

Frequency (GHz)/band	c_1	a_1	b_1
3, S-band	6.7×10^{-3}	0.93	-3.43
5.45, C-band	5.8×10^{-3}	0.91	-2.09
10, X-band	3.9×10^{-3}	1.07	-5.97

where Z_h is in units of mm^6 m^{-3} and Z_{dr} is in dB (Gorgucci et al. 1994). The coefficients c_1, a_1, and b_1 can be determined by performing a non-linear regression analysis if a table of R, Z_h, and Z_{dr} is available.

Simulations are now used to construct such a table from which the coefficients can be estimated. Two methods have generally been used in this construction. One is based on measured drop size distributions using surface instruments or aircraft imaging probes (e.g. see Section 7.1.4). The other is based on generating different dsds by varying the parameters of an assumed theoretical distribution. Since the gamma model appears to adequately describe natural variations in the dsd, different gamma dsds are simulated by independently varying the parameters N_w, D_0, and μ over the following ranges,

$$10^3 \leq N_w \leq 10^5; \quad mm^{-1} \, m^{-3}$$

$$0.5 \leq D_0 \leq 2.5; \quad mm$$

$$-1 \leq \mu \leq 5$$

with the constraint of maximum $R \leq 300$ mm h^{-1}. Note that these ranges of N_w, D_0, and μ can be modified to simulate any set of dsds of interest. While D_0 and μ are uniformly varied over their respective ranges, log N_w is varied uniformly over its range. This is done to avoid biasing the dsd population towards large N_w values. This range falls within the range of parameters (N_0, D_0, μ) suggested by Ulbrich (1983); refer to (7.63), which relates N_0 to N_w. The radar variables Z_h, Z_{dr}, and K_{dp} are calculated using the T-matrix method (see Appendix 3) assuming the following:

- Orientation distribution:[3]
  $p(\theta_b)$ is Gaussian with $\bar{\theta}_b = 0$, $\sigma = 7°$
  $p(\phi_b)$ is uniform in $(0, 2\pi)$

- Raindrop axis ratio:

  Equilibrium raindrop shape model,[4] see (7.3); (Beard and Chuang 1987)

The dielectric constant of water at a temperature of $20\,°C$ and frequencies of 3 (S-band), 5.5 (C-band), and 10 (X-band) GHz are used (Ray 1972). From the generated table of R, Z_h, and Z_{dr}, non-linear regression is performed to obtain the coefficients c_1, a_1, and b_1 given in Table 8.1. Note that it is preferable to perform non-linear regression instead of taking the logarithm of (8.12) and doing linear regression. Doing linear regression after logarithmic transformation will alter the coefficient values because the rain rate weighting will be different (see also, Campos and Zawadzki 1999).

8.1.2 $R(K_{dp})$ **algorithm**

It was shown in Chapter 7 that K_{dp} in rain is proportional to the product of rainwater content and the mass-weighted mean diameter, see (7.18). This also indicates that K_{dp} is related to the fourth moment of the dsd. Subsequently, in (7.73c) it was shown that a power-law expression relating any moment of the dsd and rainfall rate can be derived after normalizing with N_w. Therefore, a general $R(K_{dp})$ estimator can be written in the form (see also Sachidananda and Zrnić 1986),

$$R(K_{dp}) = c\,K_{dp}^b; \quad \text{mm h}^{-1} \tag{8.13}$$

where K_{dp} is in units of $^\circ$ km^{-1}. The parameters c and b can be evaluated in a manner similar to that described in the previous section. For a given water content (W) and mass-weighted mean diameter (D_m), it was shown in (7.18) that K_{dp} is inversely proportional to wavelength in the Rayleigh limit. Therefore, using a frequency-scaling argument, the K_{dp}-based rain rate estimate can be derived of the form,

$$R(K_{dp}) = 129 \left(\frac{K_{dp}}{f}\right)^{b_2}; \quad \text{mm h}^{-1} \tag{8.14}$$

where f is in GHz. The frequency-scaling argument is generally valid up to 13 GHz (see discussion related to (7.18)). At 3-GHz frequency, (8.14) reduces to $R(K_{dp}) = 50.7(K_{dp})^{0.85}$ for the Beard–Chuang equilibrium shape model, whereas the corresponding equation using the Pruppacher–Beard equilibrium shape model, see (7.2), is $R(K_{dp}) = 40.5(K_{dp})^{0.85}$.

8.1.3 $R(K_{dp}, Z_{dr})$ **algorithm**

In order to develop the form of this algorithm, (7.18) is repeated here for convenience,

$$K_{dp} = \left(\frac{180}{\lambda}\right) 10^{-3} C W (0.062 D_m); \quad ^\circ \text{km}^{-1} \tag{8.15}$$

From (7.73c) it is known that for a gamma dsd model, W can be related to rain rate as,

$$\frac{W}{N_w} = \frac{F_3(\mu)}{[F_{3.67}(\mu)]^{4/4.67}} \left(\frac{R}{N_w}\right)^{4/4.67} \tag{8.16}$$

In addition, D_m is proportional to D_0, which is directly related to Z_{dr} in dB, see (7.14a). Substitution of (8.16) in (8.15) and expressing D_m in terms of Z_{dr} yields a $R(K_{dp}, Z_{dr})$ relation of the form,

$$R(K_{dp}, Z_{dr}) = c\,K_{dp}^a\,Z_{dr}^b; \quad \text{mm h}^{-1} \tag{8.17}$$

This algorithm has large errors when Z_{dr} is close to 0 dB (because b is negative). A

Table 8.2. Coefficients of $R(K_{dp}, Z_{dr})$ rainfall algorithm at S-, C-, and X-bands

Frequency (GHz)/band	c_3	a_3	b_3
3, S-band	90.8	0.93	−1.69
5.45, C-band	37.9	0.89	−0.72
10, X-band	28.6	0.95	−1.37

more desirable form using differential reflectivity in linear scale (z_{dr}) can be written (Gorgucci and Scarchilli 1997) as,

$$R(K_{dp}, Z_{dr}) = c_3 \, K_{dp}^{a_3} \, z_{dr}^{b_3}; \quad \text{mm h}^{-1} \tag{8.18a}$$

or,

$$R(K_{dp}, Z_{dr}) = c_3 \, K_{dp}^{a_3} \, 10^{0.1 b_3 Z_{dr}}; \quad \text{mm h}^{-1} \tag{8.18b}$$

where Z_{dr} is in dB in (8.18b). Once again regression analysis similar to that used in Section 8.1.1 is used to compute the coefficients c_3, a_3, and b_3 given in Table 8.2.

8.1.4   Example of polarimetric radar–raingage comparison

Numerous examples of comparing polarimetric radar estimates of rainfall with gage measurements are available in the literature. Some of them (not an exhaustive list) include Seliga et al. (1981), Gorgucci et al. (1995; 1996), Aydin et al. (1987; 1995), Ryzhkov and Zrnić (1995b; 1996), Brandes et al. (1997). The following shows an example of comparisons between physically based parametric radar rainfall estimates with ground raingage measurements during a flash flood event in Fort Collins, Colorado on July 28, 1997 (see also, Fig. 7.29, which shows vertical sections of Z_{dr} and K_{dp} through this storm cell). The $R(K_{dp})$ estimate is based on (8.14), while $R(Z_h, Z_{dr})$ and $R(K_{dp}, Z_{dr})$ are based on Tables 8.1 and 8.2, all at S-band.

The Fort Collins flash flood event has been described in detail by Petersen et al. (1999). Maximum accumulation of rainfall in the western part of the city was around 250 mm in a 6-hour period. Figure 8.1 compares the three polarimetric estimators, $R(K_{dp})$, $R(K_{dp}, Z_{dr})$, and $R(Z_h, Z_{dr})$ against gage data (this particular gage was located very near the region of peak accumulation). Since K_{dp} and Z_{dr} are generally noisy when $Z_h < 30$ dBZ, the Z–R relation in (8.8c) is used whenever this occurs. Also, the standard deviation of the K_{dp} estimator is around 0.3–0.4° km^{-1} (see Section 6.6); so when $K_{dp} < 0.25°$ km^{-1}, the $R(K_{dp})$ and $R(K_{dp}, Z_{dr})$ algorithms are not used, rather they are replaced by the Z–R relation. At the gage location, the Z_h was nearly always >30 dBZ and $K_{dp} > 0.25°$ km^{-1} during the times shown, so this replacement is not a relevant factor. Figure 8.1 shows that the three algorithms are

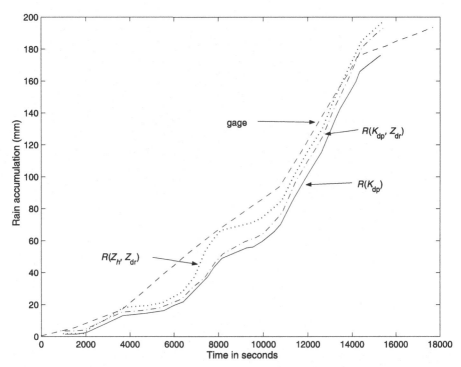

Fig. 8.1. Rainfall accumulation over the Fort Collins City Water Department gage using three different algorithms, namely, $R(K_{dp})$, $R(K_{dp}, Z_{dr})$, and $R(Z_h, Z_{dr})$. The data were collected by the CSU–CHILL radar during the Fort Collins flash flood event. The starting time is 18:00 hours (local time).

internally consistent with the $R(K_{dp})$, giving a slight underestimate relative to the gage at this location.

Another comparison of these three algorithms together with the $R(Z_h)$ algorithm is shown in Fig. 8.2a, b which illustrates rain accumulation maps over a 20 × 20 km area centered on the intersection of Taft and Drake streets in Fort Collins (near the location of peak rainfall accumulation). While the spatial patterns from all four algorithms are similar, the peak accumulation value using $R(Z_h)$ is substantially lower than the other algorithms, and was lower than the gage-measured peak accumulation (260 mm) by around 50–60%. In order of increasing accuracy relative to the gage peak, the peak accumulation from $R(K_{dp})$, $R(K_{dp}, Z_{dr})$, and $R(Z_h, Z_{dr})$ was 215, 240, and 265 mm, respectively. Petersen et al. (1999) point out that the $R(Z_h)$ underestimate was internally consistent among the two WSR–88D radars (located at Cheyenne, Wyoming, and Denver, Colorado) as well as with the CSU–CHILL radar. In this particular event, the $R(Z_h, Z_{dr})$ algorithm performed the best, both in terms of peak accumulation as well as in the spatial pattern of rainfall around the peak. It is likely that smoothing of the differential propagation phase decreased the peak K_{dp} values in these compact intense rain cells (~4–5 km in extent; see also, Gorgucci et al. 1999a).

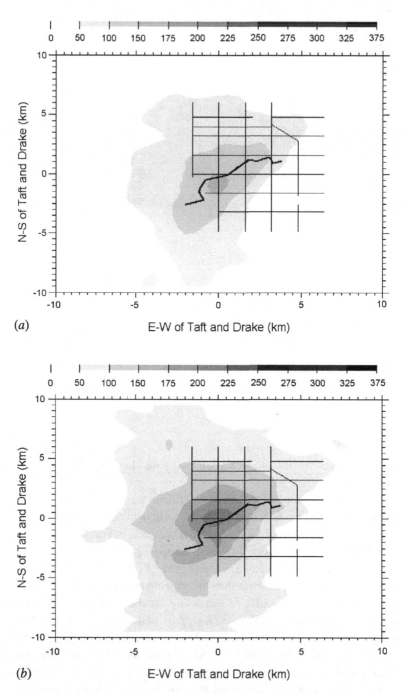

(a) E-W of Taft and Drake (km)

(b) E-W of Taft and Drake (km)

Fig. 8.2. Storm total rainfall in mm from 17:30 to 22:15 hours MDT. The lines on the picture indicate the street map of the city of Fort Collins. The dark line shows the Spring Creek, which flooded and caused the flash flood. (a) $R_{\text{WSR}}(Z)$ estimate, see (8.8c), (b) $R(K_{\text{dp}})$ estimate, (c) $R(K_{\text{dp}}, Z_{\text{dr}})$ estimate, and (d) $R(Z_h, Z_{\text{dr}})$ estimate; (courtesy of Dr Walt Petersen, Colorado State University).

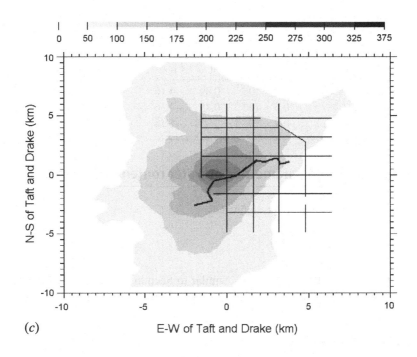

(c)

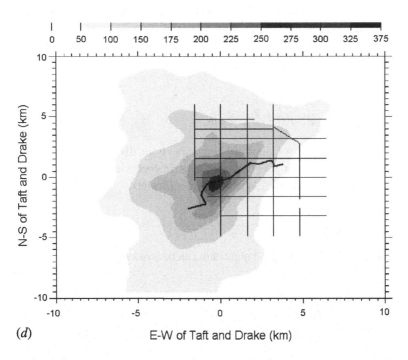

(d)

Fig. 8.2. (*cont.*)

Table 8.3. Coefficients of $W(Z, Z_{dr})$ rainfall algorithm at S-, C-, and X-bands

Frequency (GHz)/band	c_1	a_1	b_1
3, S-band	0.7×10^{-3}	0.89	−4.16
5.45, C-band	0.6×10^{-3}	0.85	−2.36
10, X-band	0.9×10^{-3}	0.95	−6.18

8.2   Physically based parametric rainwater content algorithms

The rainwater content (W) is defined as,

$$W = \rho_w \frac{\pi}{6} \int D^3 N(D)\, dD \tag{8.19}$$

where ρ_w is the density of water. Similar to Section 8.1, W can be parameterized in terms of (Z_h, Z_{dr}), K_{dp}, as well as (K_{dp}, Z_{dr}). These three algorithms are given below.

8.2.1 $W(Z_h, Z_{dr})$ algorithm

Using principles similar to those used in Section 8.1.1 it can be argued that this algorithm can be written in the form,

$$W(Z_h, Z_{dr}) = c_1 Z_h^{a_1} \mathzr_{dr}^{b_1}; \quad \text{g m}^{-3} \tag{8.20a}$$

or

$$W(Z_h, Z_{dr}) = c_1 Z_h^{a_1} 10^{0.1 b_1 Z_{dr}}; \quad \text{g m}^{-3} \tag{8.20b}$$

where Z_h is in units of $\text{mm}^6\ \text{m}^{-3}$ and Z_{dr} is in dB. For notational simplicity and similarity with Section 8.1, the same generic coefficients are used for the rainwater content algorithms. Table 8.3 gives these coefficients.

8.2.2 $W(K_{dp})$ algorithm

Based on (7.73c) the $W(K_{dp})$ algorithm can be written in the form,

$$W(K_{dp}) = c K_{dp}^b; \quad \text{g m}^{-3} \tag{8.21a}$$

$$= c_2 \left(\frac{K_{dp}}{f} \right)^{b_2}; \quad \text{g m}^{-3} \tag{8.21b}$$

where f is in GHz and K_{dp} in $^{\circ}\ \text{km}^{-1}$. The coefficient and exponent, c_2, b_2, are 3.565 and 0.77, respectively.

Table 8.4. Coefficients of $W(K_{dp}, Z_{dr})$ algorithm at S-, C-, and X-bands

Frequency (GHz)/band	c_3	a_3	b_3
3, S-band	6.05	0.88	−2.52
5.45, C-band	2.32	0.83	−1.11
10, X-band	2.13	0.91	−2.19

8.2.3 $W(K_{dp}, Z_{dr})$ algorithm

This algorithm can be written in the form (Jameson and Caylor 1994),

$$W(K_{dp}, Z_{dr}) = c_3 K_{dp}^{a_3} z_{dr}^{b_3}; \quad \text{g m}^{-3} \tag{8.22a}$$

or,

$$W(K_{dp}, Z_{dr}) = c_3 K_{dp}^{a_3} 10^{0.1 b_3 Z_{dr}}; \quad \text{g m}^{-3} \tag{8.22b}$$

where Z_{dr} is in dB and K_{dp} in $^\circ$ km^{-1}. These coefficients are given in Table 8.4.

8.3 Error structure and practical issues related to rain rate algorithms using Z_h, Z_{dr}, and K_{dp}

The four types of parametric rain rate algorithms introduced in Section 8.1 use different combinations of the three measurements (Z_h, Z_{dr}, and K_{dp}). As explained in Section 8.1, each measurement comes with its own advantages and disadvantages which are attributable to both rain microphysical characteristics and measurement errors. These features contribute to the resulting error structure of rainfall algorithms.

The following general analysis can be used to evaluate unbiased estimators of rain rate. Let $\hat{R}$ be the generic estimate from one of the four algorithms described in Section 8.1. The fluctuation of the error in $\hat{R}$ about the true rain rate R can be written as (Chandrasekar et al. 1993),

$$\hat{R} = R + \varepsilon_p + \varepsilon_m = R + \varepsilon_T \tag{8.23}$$

where ε_p is due to the error due to the parametric form of $\hat{R}$, and ε_m is due to the error in radar measurements. This simple model assumes that ε_p and ε_m are zero mean, uncorrelated random variables to ensure zero bias in $\hat{R}$. Bias errors in the radar measurement of Z_h and Z_{dr} will result in a biased estimator $\hat{R}$; this is separately modeled later in this section. For the unbiased estimator $\hat{R}$,

$$E(\hat{R}) = R \tag{8.24a}$$
$$\text{var}(\hat{R} - R) = \sigma^2(\varepsilon_p) + \sigma^2(\varepsilon_m) \tag{8.24b}$$

The variance term $\sigma^2(\varepsilon_m)$ can be minimized by spatial averaging of $\hat{R}$, under the assumption that the dsd is nearly homogeneous. In general, $Z–R$ algorithms have larger $\sigma^2(\varepsilon_p)$ compared with polarimetric algorithms. In addition, $\sigma^2(\varepsilon_p)$ generally changes as a function of R. The following analysis[5] evaluates the impact of each measurement, namely, Z_h, Z_{dr}, and K_{dp} on errors in $\hat{R}$.

8.3.1   Errors in Z_h

Errors in Z_h can be of two types, namely, random measurement errors and systematic bias errors. Random measurement errors can be reduced by spatial averaging whereas systematic biases will remain even after such averaging. The $Z–R$ algorithm in (8.8a) is of the form,

$$R(Z) = c\, Z^a \tag{8.25}$$

Assuming this is an unbiased $Z–R$ algorithm, $\sigma^2(\varepsilon_m)$ can be expressed as,

$$\frac{\sigma(\varepsilon_m)}{R} = a\left(\frac{\sigma_z}{Z}\right) \tag{8.26}$$

where σ_z is the standard deviation in the measurement of Z. Using (5.198), if the standard error in the measurement of reflectivity is about 0.8 dB, then $(\sigma(\varepsilon_m)/R)$ is about 15%. If it is assumed that the rain is uniform and an N point spatial average is done to estimate Z, then $\sigma(\varepsilon_m)$ will decrease by a factor of $\sqrt{N}$.

The $\sigma(\varepsilon_p)$ can be estimated as the variance of the difference between $\hat{R}$ and R in the absence of measurement errors for widely varying dsds (or other factors). Figure 8.3 shows an estimate of $\sigma(\varepsilon_p)$ based on simulations (at S-band) that were used in Section 8.1. Note that $\sigma(\varepsilon_p)$ is of the order 40% when rain rate is about 50 mm h^{-1}. The mean $\sigma(\varepsilon_p)$ averaged over all dsds considered in the simulations is about 45%. It should be noted that the $\sigma(\varepsilon_p)$ estimate shown here is for the best $Z–R$ algorithm. By "best" it refers to the algorithm that yields the minimum mean squared error between $\hat{R}$ and R in the absence of measurement error. Figure 8.3 also shows $\sigma(\varepsilon_m)$ and the net error, $\sigma(\varepsilon_T)$, for the best $Z–R$ algorithm as a function of R. The error of the best $Z–R$ algorithm is around 45% when R is 50 mm h^{-1}. Even though $Z–R$ algorithms have large $\sigma^2(\varepsilon_p)$, if instantaneous short time/spatial scale rain rates are not needed and if all the observations include sufficient samples such that $\langle\varepsilon_p\rangle = 0$ for any R, then the rainfall accumulation can be estimated fairly well. This argument essentially justifies using a single but unbiased $Z–R$ algorithm over large spatial and temporal scales.

The radar range equation given in Section 5.3, see (5.57b), defines the radar constant C. Accurate knowledge of the radar constant is critically important for unbiased estimation of Z. Any error in the radar constant directly translates into a bias in the estimate of Z, which subsequently results in bias in the estimate of R. Let ΔZ (dB) be the bias in the estimate of Z. The corresponding error in R can be calculated as,

$$R(Z) = c\, Z^a\, 10^{0.1a\Delta Z} \tag{8.27a}$$

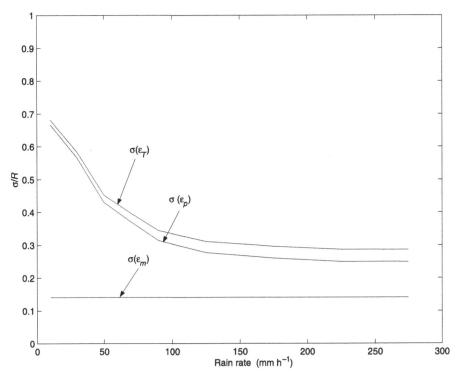

Fig. 8.3. Normalized standard deviation of $R(Z) = c\,Z^a$ as a function of R. $\sigma(\varepsilon_m)$ is the standard deviation due to measurement error, whereas $\sigma(\varepsilon_p)$ is the standard deviation due to parameterization. $\sigma(\varepsilon_T)$ is the total error due to ε_p and ε_m.

or, equivalently, the relative bias error is given by,

$$\frac{\Delta R}{R} = \frac{c\,Z^a\,10^{0.1a\Delta Z} - c\,Z^a}{c\,Z^a} \tag{8.27b}$$

$$= \left(10^{0.1a\Delta Z} - 1\right) \tag{8.27c}$$

For example, a ± 1-dB error in the estimated radar constant will result in $+18$ to -15% error in the rain rate estimate. A ± 2-dB error will result in $+39$ to -28% error in the rain rate estimate. Figure 8.4 shows the percentage error in the WSR–88D algorithm, see (8.8c), as a function of bias error in Z. Unlike the measurement error, the bias error cannot be eliminated by spatial or temporal averaging. Figure 8.4 demonstrates the importance of accurate absolute radar calibration. For example, an underestimate in $\hat{R}$ due to an absolute calibration error in Z could easily underestimate the threat of flood events.

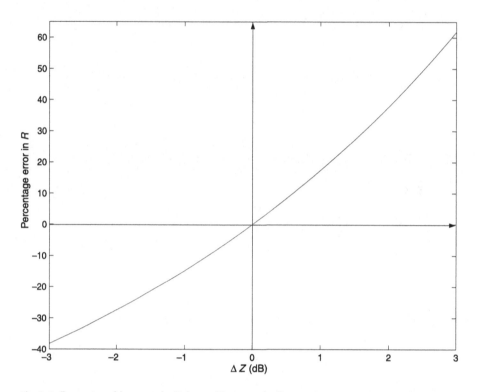

Fig. 8.4. Percentage bias error in R due to bias error in Z.

8.3.2 Error in estimate of K_{dp}

K_{dp} is a unique measurement in terms of the error structure of the rain rate estimate. First, K_{dp} is estimated from range profiles of Φ_{dp}, which is a phase measurement, see (6.146). Being a phase measurement it is immune to any error in the absolute calibration, which is by far the single most important advantage. In addition, it is unaffected by attenuation caused by precipitation along the propagation path. K_{dp} is estimated as the range derivative of Φ_{dp}. Thus, properties inherent to the numerical estimate of derivatives affect K_{dp}-based rain rate estimates. It was shown in Section 6.6 that K_{dp} can be estimated to an accuracy of around 0.3–$0.4°$ km^{-1} depending on the path length over which K_{dp} is estimated. Figure 6.35 indicates that the measurement error in K_{dp} increases rapidly as the path over which it is estimated decreases below 2 km. This results in a tradeoff between the accuracy and range resolution of K_{dp}. The $R(K_{dp})$ estimator is of the form (8.13),

$$R(K_{dp}) = c \, K_{dp}^b \qquad (8.28)$$

Figure 8.5 shows $\sigma(\varepsilon_p)$ estimated from simulations at S-band. Note that $\sigma(\varepsilon_p)$ is around 25% when rain rate is about 50 mm h^{-1}. The mean $\sigma(\varepsilon_p)$ averaged over all rain rates

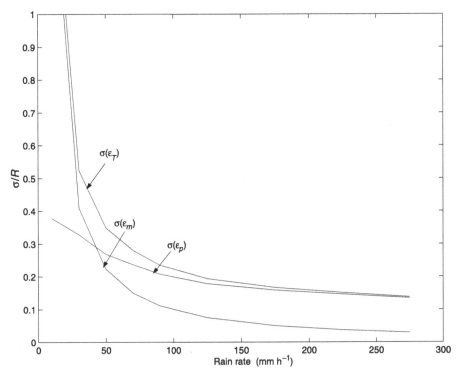

Fig. 8.5. Normalized standard deviation of $R(K_{dp}) = c\,K_{dp}^b$ as a function of R for S-band. $\sigma(\varepsilon_m)$ is the standard deviation due to measurement error, whereas $\sigma(\varepsilon_p)$ is the standard deviation due to parameterization. $\sigma(\varepsilon_T)$ is the total error due to ε_p and ε_m.

is about 27%. The $\sigma(\varepsilon_m)$ can be expressed as,

$$\frac{\sigma(\varepsilon_m)}{R} = b\frac{\sigma(K_{dp})}{K_{dp}} \tag{8.29a}$$

where $\sigma(K_{dp})$ is the standard deviation in the estimate of K_{dp} given by (6.155). Substituting (8.28) and (6.155) into (8.29a) yields,

$$\frac{\sigma(\varepsilon_m)}{R} = b\frac{\sigma(\Phi_{dp})}{L}\sqrt{\frac{3}{[N - (1/N)]}}\left(\frac{c}{R}\right)^{1/b} \tag{8.29b}$$

Several important conclusions can be drawn from (8.29b). For a given rain rate, the accuracy of $R(K_{dp})$ can be improved by increasing L, or increasing N (or decreasing Δr for a fixed L; note that $L = N\Delta r$). This is a direct consequence of properties of the estimate of K_{dp} discussed in Section 6.6. If $\sigma(\varepsilon_m)/R$ is to be kept small, say 20%, then the path length required for different range sample intervals (Δr) can be evaluated from (8.29b) and is shown in Fig. 8.6 for S-band (frequency near 3 GHz). It is clear that at low rain rates, long paths are required for accurate estimation of rain rate using K_{dp}. This error structure is unique to K_{dp} (Gorgucci et al. 1999a). Figure 8.5 shows $\sigma(\varepsilon_m)$

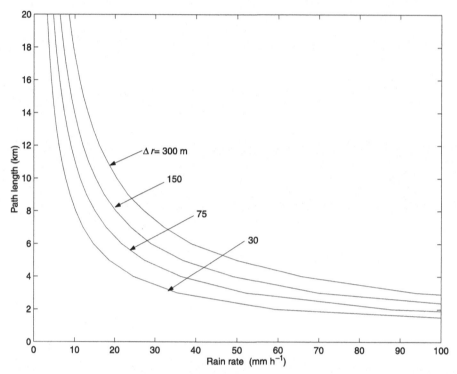

Fig. 8.6. Minimum path length (in km) required at S-band to estimate $R(K_{dp})$. The normalized measurement standard deviation ($\sigma(\varepsilon_m)$) of $R(K_{dp})$ is restricted to be less than 20% for different range sampling (Δr). Note that a smaller Δr also indicates a corresponding smaller pulse duration ($\Delta r = cT_0/2$).

when Δr and L are chosen to be 150 m and 3 km, respectively. Also shown is $\sigma(\varepsilon_T)$ for $R(K_{dp})$ as a function of R. It should be noted here that for a proper comparison of $\sigma(\varepsilon_m)$ of $R(K_{dp})$ with other algorithms, the $\sigma(\varepsilon_m)$ of the other algorithms should be reduced by $\sqrt{N}$, where $N = (L/\Delta r)$.

It was shown in Section 4.2 that when rain is mixed with spherical hail then the K_{dp} observed corresponds to the rain part of the mixture. Therefore, K_{dp} can be used to estimate the rain rate in rain/hail mixtures (e.g. refer to Fig. 7.58; see also, Aydin et al. 1995). Other advantages of K_{dp} have been discussed by Zrnić and Ryzhkov (1996).

8.3.3   Errors in Z_{dr}

Similar to errors in Z_h, errors in Z_{dr} can be of two types, namely, random measurement errors and bias errors. Any error in Z_{dr} will propagate into rainfall estimates that use Z_{dr}. Unlike absolute calibration, the calibration of Z_{dr} is fairly straightforward because it is a relative power measurement (see Section 6.3.2). The random measurement error in Z_{dr} is of the order of a few tenths of a dB (see Fig. 6.29). Since Z_{dr} is a relative power measurement, it is always used with either Z_h or K_{dp} in rain rate algorithms. The rain

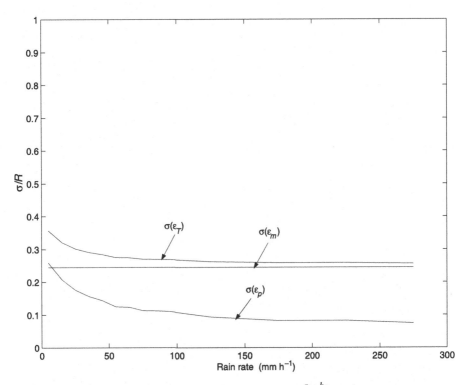

Fig. 8.7. Normalized standard deviation of $R(Z_h, Z_{dr}) = c_1 Z_h^{a_1} \xi_{dr}^{b_1}$ as a function of R for S-band. $\sigma(\varepsilon_m)$ is the standard deviation due to measurement error, whereas $\sigma(\varepsilon_p)$ is the standard deviation due to parameterization. $\sigma(\varepsilon_T)$ is the total error due to ε_p and ε_m.

rate algorithm that uses Z_{dr} and Z_h is of the form (8.12a),

$$R(Z_h, Z_{dr}) = c_1 Z_h^{a_1} \mathcal{Z}_{dr}^{b_1} \tag{8.30}$$

Figure 8.7 shows $\sigma(\varepsilon_p)$ as a function of R from simulations at S-band (frequency near 3 GHz). When the rain rate is about 50 mm h^{-1}, $\sigma(\varepsilon_p)$ is around 15%. The mean $\sigma(\varepsilon_p)$ averaged over all rain rates is also about 15%. Using perturbation analysis the variance of ε_m is expressed as,

$$\frac{\sigma^2(\varepsilon_m)}{R^2} = a_1^2 \left[\frac{\sigma^2(Z_h)}{Z_h^2} \right] + b_1^2 \left[\frac{\sigma^2(\mathcal{Z}_{dr})}{\mathcal{Z}_{dr}^2} \right] \tag{8.31}$$

Using (5.198) and (8.31), when the standard error in the measurement of reflectivity is 0.8 dB and error in the measurement of Z_{dr} is 0.2 dB then $\sigma(\varepsilon_m)/R$ is about 24%. With an error in Z_h of 0.8 dB, $\sigma(\varepsilon_m)$ of $R(Z_h, Z_{dr})$ is about 9% more than $\sigma(\varepsilon_m)$ of $R(Z_h)$, where the increase in the standard deviation is due to the measurement error of Z_{dr}. The usage of Z_{dr} in addition to Z_h can be justified if $\sigma(\varepsilon_p)$ of $R(Z_h, Z_{dr})$ is significantly lower than $\sigma(\varepsilon_p)$ of $R(Z_h)$, which is indeed the case. Figure 8.7 shows $\sigma(\varepsilon_p)$ as well

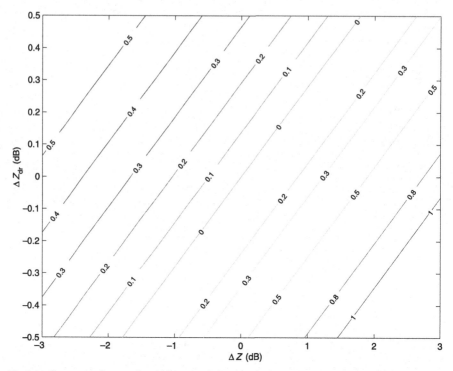

Fig. 8.8. Contours of normalized bias error in ($\Delta R/R$), due to biases in Z_h (ΔZ, dB) and Z_{dr} (ΔZ_{dr}, dB). The line marked 0 indicates zero bias error.

as $\sigma(\varepsilon_T)$ of $R(Z_h, Z_{dr})$ at S-band as a function of R. In the presence of measurement errors, $R(Z_h, Z_{dr})$ has an accuracy of about 30% at 50 mm h^{-1}.

In addition to ε_p and ε_m, bias errors can also influence the accuracy of $R(Z_h, Z_{dr})$. Accurate knowledge of the radar constant and Z_{dr} calibration bias are important in the estimation of Z_h and Z_{dr} (see Section 6.3). Let ΔZ and ΔZ_{dr} be the bias (in decibels) in the estimate of Z_h and Z_{dr}, respectively. The corresponding biased estimate, $R(Z_h, Z_{dr})$, can be calculated from,

$$R(Z_h, Z_{dr}) = c_1 Z_h^{a_1} z_{dr}^{b_1} 10^{0.1 a_1 \Delta Z} 10^{0.1 b_1 \Delta Z_{dr}} \tag{8.32a}$$

or, equivalently, the relative bias error is given by,

$$\frac{\Delta R}{R} = \left(10^{0.1 a_1 \Delta Z} 10^{0.1 b_1 \Delta Z_{dr}} - 1\right) \tag{8.32b}$$

It was shown in Table 8.1 that a_1 and b_1 are of opposite sign. For example, a $\Delta Z = 1$ dB and a $\Delta Z_{dr} = 0.2$ dB will result in +6% bias in R, which is much less than a bias in Z_h alone would have caused. Same-sign bias in Z_h and Z_{dr} tends to cancel each other, whereas opposite-sign bias in Z_h and Z_{dr} is additive. Figure 8.8 shows contour plots of percentage bias in R due to biases in Z_h and Z_{dr} at S-band. The contour lines increase the most when ΔZ is positive and ΔZ_{dr} is negative, or vice versa. In order to limit the

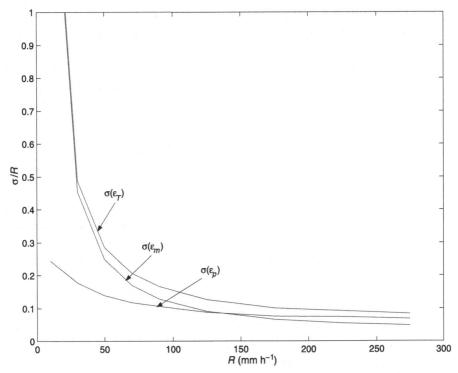

Fig. 8.9. Normalized standard deviation of $R(K_{dp}, Z_{dr})$ as a function of R for S-band. $\sigma(\varepsilon_m)$ is the standard deviation due to measurement error, whereas $\sigma(\varepsilon_p)$ is the standard deviation due to parameterization. $\sigma(\varepsilon_T)$ is the total error due to ε_p and ε_m.

bias error to that of random measurement error (say about 25%), the Z_{dr} bias should be within -0.3 dB to $+0.4$ dB in the absence of any bias in Z_h.

The rain rate algorithm that uses Z_{dr} and K_{dp} is of the form (8.18),

$$R(K_{dp}, Z_{dr}) = c_3 \, K_{dp}^{a_3} \, \mathfrak{z}_{dr}^{b_3} \tag{8.33}$$

Figure 8.9 shows $\sigma(\varepsilon_p)$ as a function of R from simulations at S-band. When the rain rate is about 50 mm h^{-1}, $\sigma(\varepsilon_p)$ of $R(K_{dp}, Z_{dr})$ is of the order of 15%. The mean $\sigma(\varepsilon_p)$ averaged over all rain rates is also about 15%. The evaluation of measurement error in $R(K_{dp}, Z_{dr})$ is not straightforward because K_{dp} is estimated over a path (say, $L = N\Delta r$), whereas N samples of Z_{dr} are available over the same path. To account for the errors properly, the error of the path-averaged Z_{dr} should be considered. Perturbation analysis yields the variance of ε_m of $R(K_{dp}, Z_{dr})$ as,

$$\frac{\text{var}(\varepsilon_m)}{R^2} = a_3^2 \frac{\text{var}(K_{dp})}{K_{dp}^2} + b_3^2 \frac{\text{var}(\mathfrak{z}_{dr})}{\mathfrak{z}_{dr}^2} \tag{8.34}$$

Based on (6.155) it can be inferred that $\text{var}(K_{dp})$ critically depends on the path over which it is estimated. If Δr and L are specified, then $\text{var}(K_{dp})$ can be estimated from

(6.155). In addition, a path of length L has $N = (L/\Delta r)$ estimates of $\bar{\mathcal{J}}_{dr}$. The variance of $\mathcal{J}_{dr}$ is also reduced by a factor of $\sqrt{N}$. Substituting (6.119) and (6.155) into (8.34), $var(\varepsilon_m)/R^2$ can be evaluated. Figure 8.9 shows the normalized standard deviation of ε_m at S-band as a function of rain rate assuming $L = 3$ km, $\Delta r = 0.15$ km, and standard error in Z_{dr} of 0.2 dB. Note that $R(K_{dp}, Z_{dr})$ can be estimated to an accuracy better than $R(K_{dp})$ for all rain rates. At $R = 50$ mm h^{-1}, $\sigma(\varepsilon_T)/R$ for $R(K_{dp}, Z_{dr})$ is about 30%.

8.4   Statistical procedures for rainfall estimation

The physically based rain rate algorithms presented in Sections 8.1 and 8.2 give "instantaneous" rain rate estimates at the location of the resolution volume, which is always at a finite height above the ground. Ground instruments such as raingages collect rain over a finite time interval to estimate rain rate. In addition, they can be considered as point measurements on the ground. In general, the rainfall field evolves as it descends to the ground. Based on these and other factors, it follows that there will generally be differences between radar rain rate estimates made aloft and gage measurements at the ground[6] (Zawadzki 1984). One school of thought considers radar rainfall estimation as a statistical/engineering estimation problem. This section briefly describes three such techniques, namely, (i) area–time integral method, (ii) probability matched rain rate algorithm, and (iii) neural-network rain rate algorithms.

8.4.1   Area–time integral method

Let $R(x, y)$ be the spatial rain rate function. The volume flux (F) of rain can be obtained as,

$$F = \iint R(x, y) \, dx \, dy \tag{8.35}$$

If only rain rates above a threshold R_T are included in the above integral then,

$$F(R_T) = \iint R(x, y) u(R - R_T) \, dx \, dy \tag{8.36}$$

where $u(R - R_T)$ is a unit step function defined as,

$$u(R - R_T) = 1; \quad R > R_T$$
$$= 0; \quad \text{otherwise} \tag{8.37}$$

The average rain rate over an area can be computed as,

$$\bar{R} = \frac{\iint R(x, y) \, dx \, dy}{\iint u(R) \, dx \, dy} = \frac{F(0)}{A(0)} \tag{8.38a}$$

where $A(0)$ is the rain area. Similarly, the average rain rate above a threshold R_T is,

$$\bar{R}(R_T) = \frac{\iint R(x, y)u(R - R_T)\,dx\,dy}{\iint u(R - R_T)\,dx\,dy} = \frac{F(R_T)}{A(R_T)} \tag{8.38b}$$

Note that $\iint u(R - R_T)\,dx\,dy$ computes the area when the rain rate $R > R_T$. The volume flux of rain $F(0)$ can be written as the sum of volume fluxes when $R > R_T$ as well as when $R \le R_T$. Therefore, from (8.35) and (8.36),

$$F = F(R_T) + \iint R(x, y)[1 - u(R - R_T)]\,dx\,dy \tag{8.39a}$$

The rain volume fraction, $f(R_T)$, contributed by rain rates above the threshold R_T can be defined as,

$$f(R_T) = \frac{F(R_T)}{F(0)} \tag{8.39b}$$

The average rain rate from (8.38) and (8.39) can be written as,

$$\bar{R} = \frac{F(0)}{A(0)} = \frac{F(R_T)}{f(R_T)A(0)} = \left[\frac{\bar{R}(R_T)}{f(R_T)}\right]\frac{A(R_T)}{A(0)} \tag{8.40}$$

The terms $\bar{R}(R_T)$ and $f(R_T)$, correspond to a specific rain rate field. Invoking the law of large numbers, if the sample histogram of R converges to the natural probability density function of R, then,

$$\bar{R}(R_T) \text{ converges to } \int_{R_T}^{\infty} R\,g(R)\,dR = E(R; R > R_T) \tag{8.41a}$$

where $g(R)$ is the pdf of R and $E(R; R > R_T)$ is the expected value of R under the condition $R > R_T$. Similarly,

$$f(R_T) \text{ converges to } \frac{\int_{R_T}^{\infty} g_R(R)\,dR}{\int_{0}^{\infty} g_R(R)\,dR} = p(R > R_T) \tag{8.41b}$$

where $p(R > R_T)$ is the probability $R > R_T$. For a given probability density function of R, the ratio $E(R; R > R_T)/p(R > R_T)$ is uniquely defined. Equation (8.40) can be rewritten as,

$$E(R) = \frac{E(R; R > R_T)}{p(R > R_T)}\left[\frac{A(R_T)}{A(0)}\right] \tag{8.42}$$

where $A(R_T)/A(0)$ is the fractional area occupied by $R > R_T$. The above equation provides an estimator for mean rain rate, $E(R)$. Use of this estimator involves computing statistics over rain rate fields. This estimation can be easily done using radar because of its ability to collect data over large areas, provided there is a good algorithm to estimate R from radar measurements. The volume of rain can be integrated over time to yield rain volume as,

$$\int F(0)\, dt = \int\!\!\int\!\!\int R(x, y)\, dx\, dy\, dt \tag{8.43a}$$

$$= \int \bar{R} A(0)\, dt \tag{8.43b}$$

Using (8.40), the above reduces to,

$$\int F(0)\, dt = \int \left[\frac{\bar{R}(R_T)}{f(R_T)} \right] A(R_T)\, dt \tag{8.44}$$

The first term inside the brackets in the integral of (8.44) can be taken outside the integral if it is stationary in time. Therefore,

$$\int F(0)\, dt = \frac{E(R; R > R_T)}{p(R > R_T)} \int A(R_T)\, dt \tag{8.45a}$$

$$= C(R_T)[\text{ATI}] \tag{8.45b}$$

where ATI stands for area–time integral (Doneaud et al. 1984; Atlas et al. 1990). There are two practical issues regarding the application of (8.45). First, the coefficient, $C(R_T)$, needs to be estimated. Second, the threshold area, $A(R_T)$, corresponding to a specific R_T is needed. Radar measurements can be used for the computation of ATI, where a specific value of R_T is mapped to a corresponding threshold in equivalent reflectivity factor Z. Variations of this method are described by Rosenfeld et al. (1990). Nevertheless, the basic problem of independent and representative estimation of $C(R_T)$ is critical to the application of ATI techniques. An important consideration for the success of (8.45) is that the pdf of components contributing to the integral should converge to a steady value, and $C(R_T)$ should remain stationary. Raghavan and Chandrasekar (1994) have presented a technique to independently evaluate $C(R_T)$ using polarimetric radar measurements. They estimate $C(R_T)$ as,

$$C(R_T) = \frac{\displaystyle\int F(0)\, dt}{\text{ATI}} \tag{8.46}$$

where ATI could be estimated using a reflectivity threshold, whereas $F(0)$ could be estimated using the differential propagation phase Φ_{dp}. In principle, a linear relation between rain rate and K_{dp} is required ($R = c\, K_{\text{dp}}$) even though the optimum non-linear fit at 3 GHz is of the form $R = c\, K_{\text{dp}}^{a}$, with $a = 0.85$. Nevertheless, a linear fit can be derived at the expense of slightly increased $\sigma(\varepsilon_p)$, see (8.23). From (8.35),

$$F(0) = \int\!\!\int R(x, y)\, dx\, dy \tag{8.47a}$$

$$= \int\!\!\int c K_{\text{dp}}(x, y)\, dx\, dy \tag{8.47b}$$

Converting to polar coordinates,

$$F(0) = c \iint K_{dp}(r, \theta) r \, dr \, d\theta \tag{8.48}$$

Integrating by parts and using $\Phi_{dp}(r, \theta) = \int 2K_{dp}(r, \theta) \, dr$ results in,

$$F(0) = \frac{c}{2} \int \left\{ [r_2 \Phi_{dp}(r_2, \theta) - r_1 \Phi_{dp}(r_1, \theta)] - \int_{r_1}^{r_2} \Phi_{dp}(r, \theta) \, dr \right\} d\theta \tag{8.49}$$

For low elevation angles, the polar angle θ is approximately the azimuthal pointing angle of the radar. If the rain cell is contained between ranges (r_1, r_2) and between azimuthal angles (θ_1, θ_2), and if the range samples are Δr apart and the azimuthal samples are $\Delta\theta$ apart, then the above equation can be numerically approximated as,

$$F(0) = \frac{c}{2} \sum_j \left\{ [r_2 \Phi_{dp}(r_2, \theta) - r_1 \Phi_{dp}(r_1, \theta)] - \sum_i \Phi_{dp}(r, \theta) \Delta r \right\} \Delta\theta \tag{8.50}$$

Note that the above equation involves only Φ_{dp}, which is directly measured by a polarimetric radar. Substituting $F(0)$ in (8.50) into (8.46), $C(R_T)$ can be estimated since ATI is easily computed from measurements of Z. Thus, measurements of Z and Φ_{dp} provide a mechanism to estimate $C(R_T)$ without using an a priori Z–R relation.

Note that (8.50) can be used to directly compute $\int F(0) \, dt$ over any area of interest, such as a watershed, without the necessity of calculating rain rate at each resolution volume. This is a significant application of Φ_{dp} data, which goes together with the basic advantages of Φ_{dp} measurements (see Section 8.2). Also, note that the $F(0)$ estimator in (8.50) is different from the one proposed by Ryzhkov et al. (2000) which is,

$$F(0) = \frac{40.6}{2} \sum_j \Delta\theta \frac{(r_1 + r_2)}{2} [2(r_2 - r_1)]^{1-0.866} [\Phi_{dp}(r_2, \theta_j) - \Phi_{dp}(r_1, \theta_j)]^{0.866} \tag{8.51}$$

where the relation $R = 40.6(K_{dp})^{0.866}$ is used. The above formula is easily derived by assuming that K_{dp} is constant in the range interval r_1 to r_2.

8.4.2   Probability matching method

The probability matching method (PMM) is a statistical procedure to derive the average relationship between radar measurements and ground rain rates. This method essentially matches the cumulative distribution function of R and Z, yielding a Z–R relation (Calheiros and Zawadzki 1987). The principle behind the probability matching method is fairly straightforward. Let $F_Z(z)$ be the cumulative probability distribution function (CDF) of reflectivity. Let Z and R be related by the function $R = H(Z)$, where H is the functional transformation relating R and Z. Then, the cumulative probability

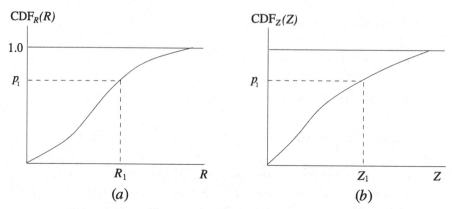

Fig. 8.10. The probability distribution functions of (*a*) R and (*b*) Z (dB). A probability value p_1 corresponds to the pair R_1 and Z_1, which describe $R = H(Z)$.

distribution function of rain rate, $F_R(R)$, can be constructed as follows. Let $H(Z)$ be a monotone increasing function of Z, then,

$$\mathrm{CDF}_R(R) = p(\text{Rain rate} \le R) = p[H(Z) \le R]$$
$$= p[Z \le H^{-1}(R)] = \mathrm{CDF}_Z[H^{-1}(R)] \qquad (8.52a)$$

Therefore,

$$\mathrm{CDF}_R(R) = \mathrm{CDF}_Z[H^{-1}(R)] \qquad (8.52b)$$

The above result can be stated as follows: if there is a unique and monotone increasing function $H(Z)$ that yields R, then $F_R(R)$ and $F_Z(z)$ are related according to (8.52b). If the experimental CDFs of Z and R can be determined from observations of the population of interest (e.g. seasonal, regional, etc.), then from (8.52b), an $R–H(Z)$ relation can be derived. Figure 8.10 illustrates the procedure of constructing $R = H(Z)$ from data. Corresponding to a probability value p_1, the values of R_1 and Z_1 can be obtained from $\mathrm{CDF}_R(R)$ and $\mathrm{CDF}_Z(z)$, then the pair R_1, Z_1 forms the relation,

$$R_1 = H(Z)|_{\text{at } Z=Z_1} \qquad (8.53)$$

The PMM technique for obtaining a $Z–R$ relation has its advantages and disadvantages. First, the procedure assumes that there exists a monotonically increasing unique $H(Z)$. Second, the construction of the experimental CDFs, $F_R(R)$ and $F_Z(Z)$, from data is critical and depends on the sensitivity of the instruments (radar and gage) involved. For example, above a specific gage location, different radars may have different minimum detectable reflectivities. The major advantage of PMM is that as long as the same radar and gages are used for development of the $R–H(Z)$ relation and its evaluation, then all the instrumentation and location-dependent problems will be replicated and, therefore, the evaluation results will be self-consistent. For example, if the radar reflectivity calibration is off by 2 dB, the construction of the $R–H(Z)$ relation will be adjusted accordingly. For application of this PMM technique, only the CDFs of Z and R have to be known and these can be estimated independently from radar and gage observations. However, if the CDFs are constructed using coordinated radar and gage measurements from the same space–time "window" then it is called Window PMM (or WPMM; Rosenfeld et al. 1994).

The probability matching method can be extended to polarimetric radar measurements. However, slight modifications have to be done to the procedure. Rain rate is not a two-dimensional monotonically increasing function with increasing (Z, Z_{dr}) or (K_{dp}, Z_{dr}). Parametric forms can be assumed, such as those given in Section 8.2, and the corresponding coefficients can be derived using the probability matching method. This technique is a compromise between the non-parametric PMM and a parametric rain rate algorithm derived from physical principles.

Probability statements of the type made in (8.52) can be made for (Z, Z_{dr}) or (K_{dp}, Z_{dr}) as,

$$F_R(R) = P[\text{Rain rate} \leq R] \tag{8.54a}$$

$$= P[H(Z, Z_{dr}) \leq R] \tag{8.54b}$$

$$= P[Z \leq z, Z_{dr} \geq z_{dr}] \tag{8.54c}$$

The above CDF can be computed from the joint density function of Z and Z_{dr}. In practice, the two-dimensional CDF of Z and Z_{dr} (or K_{dp} and Z_{dr}) cannot be inverted easily from data. Therefore, it is better to assume a parametric form for $R(Z, Z_{dr})$ or $R(K_{dp}, Z_{dr})$ and to estimate these coefficients such that the difference between the gage-based CDF of R and the radar-based CDF of R is minimized. Gorgucci et al. (1995) applied this procedure to polarimetric radar and gage data to derive a probability-matched rain rate algorithm at S-band. Figure 8.11 compares the two CDFs, one of which is based on $R(Z, Z_{dr})$ and the other based on rain rate data from 20 tipping bucket gages. The polarimetric PMM (or PPMM) based algorithm derived by Gorgucci et al. (1995) at S-band is,

$$R(Z_h, Z_{dr}) = 0.01 \, Z_h^{0.914} (z_{dr})^{-3.77} \tag{8.55}$$

They also showed that this $R(Z_h, Z_{dr})$ algorithm matched the CDF of R better than any $Z–R$ power-law relation for their data set. The above rain rate algorithm was independently validated by Bringi et al. (1998) by comparing against rain rate derived from aircraft-based dsd measurements in Florida rainshafts. This comparison is shown in Fig. 8.12, and resulted in negligible bias and a normalized standard error (nse) of 31.8%. This nse is consistent with the normalized standard deviation of $R(Z_h, Z_{dr})$ shown in Fig. 8.7, which gives $\sigma(\varepsilon_T)/R$ of 30% at $R \approx 40$ mm h^{-1}.

8.5 Neural-network-based radar estimation of rainfall

The previous sections described two completely different approaches to rain rate estimation, namely, (i) physically based parametric algorithms, and (ii) algorithms

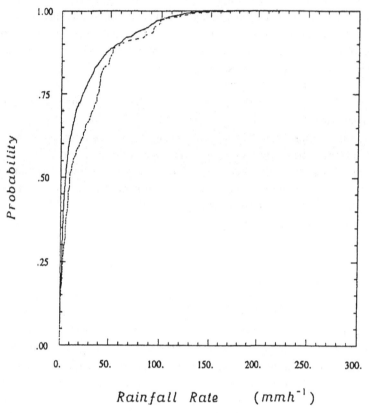

Fig. 8.11. Experimental cumulative distribution function (CDF) of rain rate. The solid line shows CDF obtained from gage data, whereas the dotted line shows the CDF obtained from radar data using probability-matched $R(Z_h, Z_{dr})$. From Gorgucci et al. (1995).

based on statistical methods. In this section an engineering approach is described based on neural networks.

Rainfall on the ground is generally dependent on the four-dimensional structure of precipitation aloft (i.e. three-dimensional and time). In principle, one can obtain a functional relation between rain rate on the ground and the four-dimensional radar observations aloft. However, it is difficult to express this functional relationship in a useful form. Neural networks provide a mechanism to solve this complex functional approximation problem. Using ground measurements of rain rate as the desired output, the neural network builds the functional approximation between radar observations aloft and ground rain rate observations. The development of this functional relationship is based on a "training" data set. When the network is trained sufficiently, then it is ready for application. The theoretical basis of rainfall estimation from neural networks is the universal approximation theorem, which states that a multilayer perceptron neural network is capable of performing any non-linear input–output mapping (Funahashi 1989). The neural-network-based algorithms are also fairly robust and error tolerant.

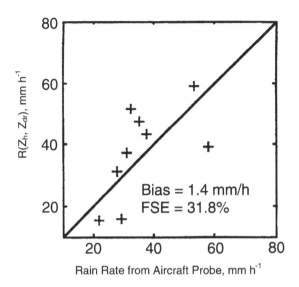

Fig. 8.12. Scatter plot of $R(Z_h, Z_{dr})$ from radar using (8.55) versus rain rate from airborne imaging probe. From Bringi et al. (1998, © 1998 IEEE).

There are two important aspects of using neural networks for radar rain rate estimation, namely, (i) training the network, and (ii) generalization capability of the network. Figure 8.13 shows a block diagram of the general principle of the neural network used here. In the training process, radar data are applied to the network as input, and the corresponding raingage data are used as the desired output (Fig. 8.13a). The network is modified based on the backward error propagation according to a learning algorithm, and this process is repeated until the error between the network output and the desired output meets the prescribed accuracy. When this training process is complete, the network is ready for application. Rain rate estimates can be obtained if radar data are applied to the network at this stage (Fig. 8.13b).

Two types of neural networks that have been used for radar rain rate estimates are: (i) the back-propagation neural network (BPN), and (ii) the radial basis function (RBF) neural network (Xiao and Chandrasekar 1997). Figure 8.14 shows the structure of a three-layer back-propagation network. One of the disadvantages of BPN is that the training process is computationally demanding and the process is tedious. However, once trained, the BPN has very good generalization capability and it will work fairly well for new data that are not part of the original training data set. Xiao and Chandrasekar (1997) successfully developed a BPN for radar rain rate estimation.

Since neural networks learn from training data, it would be ideal if the network could be gradually modified to account for seasonal changes in rainfall type. One way to address this problem is by retraining the network from the beginning as soon as new data is made available. However, the training data set will grow very quickly and

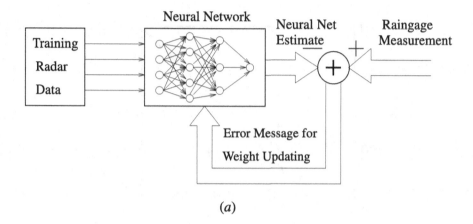

(a)

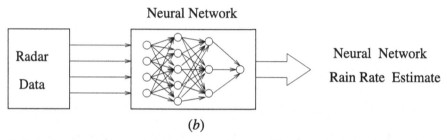

(b)

Fig. 8.13. Development of neural network for rain rate estimation: (a) training the neural network for rain rate estimation, (b) application of the neural network for rain rate estimation.

this is not a practical solution. It would be useful to have a neural network that could be adjusted slightly to incorporate infusion of new data. The structure of the BPN is not conducive to such a process. An alternate type of neural network that allows this adjustment is the radial basis function (RBF) network (Liu et al. in press).

A RBF network has three layers, namely, (i) the input layer, (ii) the hidden layer consisting of the radial basis functions, and (iii) the output layer that forms the linear combination of the hidden layer output. The structure of a RBF network for rain rate estimation is shown in Fig. 8.15. A commonly used radial basis function is the Gaussian function expressed as,

$$h_j(\vec{x}) = \exp\left[-\sum_{i=1}^{n}\left(\frac{x_i - c_{ij}}{r_{ij}}\right)^2\right] \qquad (8.56)$$

The output $f(\vec{x})$ for an input vector $\vec{x}$ is given by,

$$f(\vec{x}) = \sum_{j=1}^{m} w_j h_j(\vec{x}) \qquad (8.57)$$

where $\vec{x} = [x_1, x_2, \ldots, x_n]^t$ is the input vector (e.g. radar data), $\vec{r}_j = [r_{1j}, r_{2j}, \ldots, r_{nj}]^t$

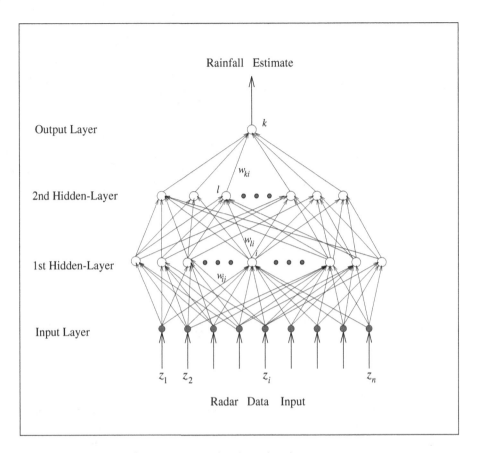

Rainfall   Estimate

Output Layer

2nd Hidden-Layer

1st Hidden-Layer

Input Layer

z_1 z_2 z_i z_n

Radar   Data   Input

Fig. 8.14. Back-propagation network showing the various layers.

is the size vector of neuron j, m is the number of neurons in the hidden layer, and w_j is
the weight from neuron j to the rain rate output.

8.5.1   Input data to neural network

A convenient way to input data to the network can be radar measurements (e.g. Z_h, Z_{dr},
or K_{dp}) interpolated to a regularly spaced Cartesian grid. Figure 8.16 shows a schematic
of input data for the neural network. As an example, radar measurements over a square
grid with a spacing of Δd at various heights $h_1, \ldots, h_n$ can be used as input to the
neural network. Xiao and Chandrasekar (1997) used data over a $3 \times 3\text{-km}^2$ square grid
with Δd of 1 km at four heights below the melting level. The size of a neural network
grows with the size of the input vector. The network can be simplified by using full grid
data at the lowest altitude and using only the mean vertical profile, so that information
about horizontal and vertical variability of the radar measurements is provided to the
network.

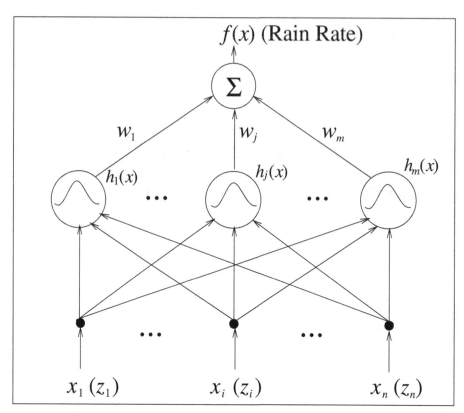

Fig. 8.15. Radial basis function (RBF) neural network. The input is $x_1, x_2, \ldots, x_n$. w_1, $w_2, \ldots, w_m$ are the weights. $h_1, h_2, \ldots, h_m$ are the radial basis function, see (8.56).

8.5.2   Training the network

For training/developing a RBF network the following three parameters need to be determined: (i) the center vector of all the neurons in the hidden layer $\vec{c}_j$, (ii) the size vector (or width) of the neurons in the hidden layer $\vec{r}_j$, and (iii) the weights from the hidden layer to the output ($w_1, w_2, \ldots, w_n$). Once these parameters are determined, the network development is complete.

The most common learning algorithm for a RBF network is the self-organization least square technique which is used to determine the center vectors for the hidden neurons (Haykin 1999). Once the parameters in the hidden layer are determined, the weight vector from the hidden layer to the output layer can be obtained by the linear least square method. This combination learning algorithm is efficient. However, there is no guarantee that the trained RBF network has good generalization capability. Details of this algorithm can be found in Liu et al. (in press).

The size vectors $\vec{r}_j$ must be determined in conjunction with the center vectors $\vec{c}_j$. The generalization capability of the RBF neural network is sensitive to the size vector. If the size vector is small, the network will function very well with the training data set

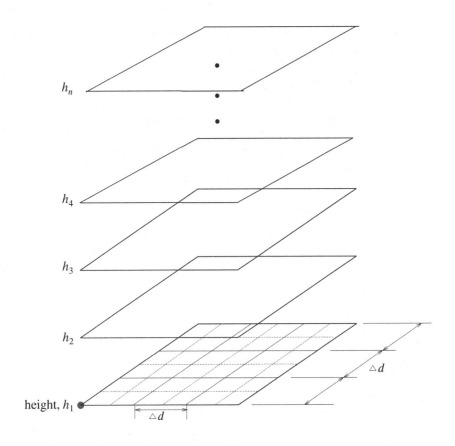

Fig. 8.16. Schematic of data on a regular Cartesian grid, which are input to the neural network.

but will have poor generalization capability. On the other hand, if it is too large, then the network will be over-generalized. Therefore, an appropriate size vector should be determined by several trials.

A subset ($i = 1, p$) of the training input ($\vec{x}$) to the network is chosen as the center of the RBFs. The method starts with an empty subset and adds one basis function at a time. The sum of squared error (S) is used to determine convergence according to the least square algorithm,

$$S = \sum_{i=1}^{p} \left[\hat{y}_i - f(\vec{x}_i) \right]^2 \tag{8.58}$$

where $\hat{y}$ is the target output. Lowest prediction error is the convergence criterion that is

used to determine if any additional radial basis functions are needed or not. The network
has the lowest prediction error when the optimum subset of radial basis functions is
chosen. Standard measures can be used to compute prediction error, e.g. final prediction
error (FPE). When this measure stops decreasing, then no more radial basis functions
should be added to the hidden layer.

If the centers and sizes of the radial basis functions are fixed, then the determination
of weights, w_j, is straightforward. The w_js are determined by minimizing the sum of
the squared error S given by,

$$S = \sum_{i=1}^{p} \left[\hat{y}_i - \sum_{j=1}^{m} w_j h_j(\vec{x}) \right]^2 \tag{8.59}$$

The optimum w_j is given by the generalized inverse equation,

$$\hat{\mathbf{w}} = [\hat{w}_1 \ \hat{w}_2 \ \cdots \ \hat{w}_m]^t = [\mathbf{H}^t\mathbf{H}]^{-1}\mathbf{H}^t\hat{\mathbf{Y}} \tag{8.60}$$

where $\mathbf{H}$ is the matrix of basis functions given by,

$$\mathbf{H} = \begin{bmatrix} h_1(\vec{x}_1) & h_2(\vec{x}_1) & \ldots & h_m(\vec{x}_1) \\ h_1(\vec{x}_2) & h_2(\vec{x}_2) & \ldots & h_m(\vec{x}_2) \\ \vdots & \vdots & \ldots & \vdots \\ h_1(\vec{x}_p) & h_2(\vec{x}_p) & \ldots & h_m(\vec{x}_p) \end{bmatrix} \tag{8.61}$$

and $\hat{\mathbf{Y}}$ is the output vector.

An RBF network is trained with a training data set. As time passes by more data
becomes available. One way to incorporate the new information from the additional
data into the network is by simply combining the new data with the old training data set
to form a new larger training data set, and then retrain the network. The most important
part in the retraining process is searching for the optimum center set from the new
training data set, and this process is computationally tedious. Another disadvantage is
that a simple retraining process will not give higher priority to the latest data and, thus,
cannot make sure that the new network can trace any changes in the mean relationship
between radar input and gage output. Based on these reasons, it is better to use an
adaptive RBF neural network for rain rate estimation.

Liu et al. (in press) implemented the adaptive neural network algorithm for rain rate
estimation for data collected with the WSR–88D radar located in Melbourne, Florida.
Twenty days of radar and gage data during the month of August were used to set up
the RBF network. Subsequently, the network was used in the adaptive mode, where the
network was updated by the end of the day, to be ready for application the following
day. Figure 8.17 shows a schematic diagram of the adaptive RBF network. Figure 8.18
shows the composite hourly rain rate for ten days obtained from the adaptive neural
network over a network of gages (with a maximum range of 200 km from the radar).
The same figure also shows the estimates from rain gages. It can be seen that the
adaptive neural network estimates rain rate very well. The bottom panel in Fig. 8.18
shows the corresponding estimates from $R_{\text{WSR}}(Z)$ for comparison. The normalized

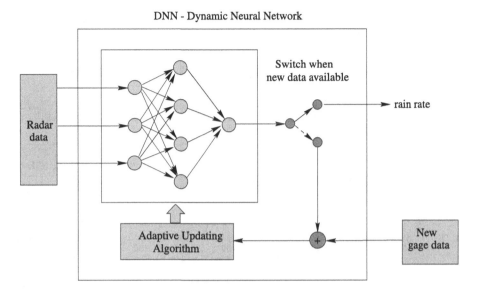

Fig. 8.17. Structure of the adaptive neural network for rain rate estimation.

standard error in the hourly estimate of rainfall from the adaptive RBF is 27% whereas from $R_{WSR}(Z)$ it is 50%. They also compared the daily rainfall accumulation over the gages and the results are shown in Fig. 8.19, which shows again that the adaptive neural network estimated daily accumulation very well to an accuracy of 11%. They extended their analysis to another month in September, where the adaptive RBF network estimated mean hourly rain rate and daily rainfall accumulation to an accuracy of 34% and 17%, respectively. The corresponding results for $R_{WSR}(Z)$ were 49% and 44%, respectively. Thus, the adaptive neural network technique in this study estimated rainfall fairly accurately. The proliferation of cheaper and faster computers could make this technique a practical alternative for radar rainfall estimation.

8.6   Some general comments on radar rainfall estimation

Rainfall estimation by radar can be fundamentally viewed from two different aspects, namely, physically based inferences and engineering solutions. The physically based approach attempts to solve the inverse electromagnetic problem of obtaining (resolution volume averaged) rain rate from back scatter and forward scatter measurements such as Z_h, Z_{dr}, and K_{dp}, together with an underlying rain model. Engineering solutions on the other hand obtain the best possible rain rate or rain accumulation using radar data with some feedback from gages. Though not stated in this form, this fundamental distinction was recognized by Zawadzki (1984). The fundamental difference between physically based algorithms and engineering solutions as defined here is the absence or presence

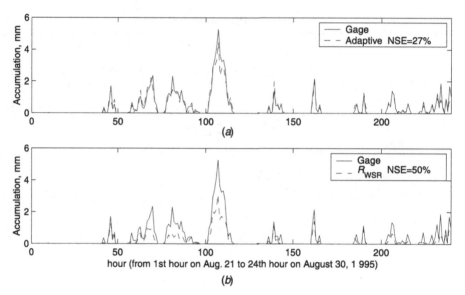

Fig. 8.18. Hourly rain rate comparison over a network of gages: (*a*) compares gage data with adaptive neural-network-based estimate, (*b*) compares gage data with $R_{\mathrm{WSR}}(Z)$ from (8.8c). From Liu et al. (in press).

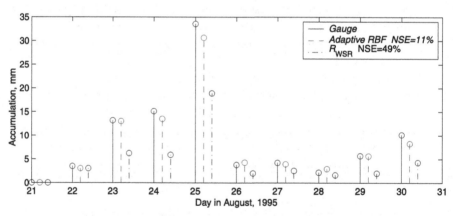

Fig. 8.19. Mean daily accumulations over the gage locations computed by the adaptive neural network compared against raingage data for ten days. Estimates using $R_{\mathrm{WSR}}(Z)$, see (8.8c) are also shown. From Liu et al. (in press).

of feedback from gages. Historically, it appears that these two methods have been used interchangeably without any clear distinction. Use of a $Z-R$ relation or polarimetric estimates of the form (8.12), (8.14), or (8.18) implies physically based algorithms.

Physically based approaches have many advantages. First, polarimetric techniques can easily distinguish rain from other types of frozen hydrometeors such as hail. While such a distinction is important in the study of precipitation physics, it is also an important step prior to application of rainfall algorithms (including those based on statistical methods). In addition, K_{dp} can be used to estimate the rain rate even when the resolution volume is composed of rain mixed with frozen hydrometeors. Physically based algorithms go beyond just rainfall estimation at the ground. They provide the vertical structure of rain rate (or water content) and are useful in the study of the physics of precipitation processes as described in Chapter 7.

Engineering solutions have focused primarily on accurate estimation of rainfall on the ground. Based on some form of feedback from gages, such solutions may involve simple tuning of algorithm coefficients, or deriving a non-parametric $Z-R$ relation (e.g. the PMM), or more sophisticated adaptive neural network training. For long-term averages over large areas, engineering solutions tend to work well because the feedback mechanism ensures that rainfall estimates are unbiased, and that fluctuations about the mean are suppressed in the averaging process. However, if individual storm total rainfall, or if short-term rainfall estimation is important such as in flash flood or extreme rainfall events, then the physically based algorithms are likely to be the best choice. This was demonstrated[7] in the Fort Collins flash flood event, which is one of the few well-documented cases with polarimetric radar measurements.

Notes

1. Considerable literature exists on rainfall measurement techniques which fall into these two categories (see, for example, Wilson and Brandes 1979; Collier 1989; or Smith 1990 and references cited therein). It is beyond the scope of this book to adequately reference this extensive literature.

2. The WSR–88D rainfall algorithm is described by Fulton et al. (1998). For a complete discussion of WSR–88D algorithms see WSR–88D Algorithms, OSF Applications Branch, NOAA at,

   http://www.osf.noaa.gov/app/appl.htm

3. The standard deviation of $7°$ used here is based on the discussion in Section 3.14.4 but is also believed to be applicable to mid-latitude rainfall. A higher value ($10-15°$) could probably be used for tropical rainfall.

4. The Beard–Chuang equilibrium shape model appears to be suitable for mid-latitude rainfall, whereas the Andsager et al. drop oscillation fit, see (7.24), for $1 \leq D \leq 4$ mm appears to be suitable for tropical rainfall. However, care should be exercised when choosing σ (standard deviation of canting angle) and the mean axis ratio versus D relation to ensure consistency.

5. It is assumed that other errors in Z_h, Z_{dr}, and K_{dp} such as due to ground clutter, beam blockage, or gradients of rain rate across the antenna beam are negligible (see Section 6.2.2).

6. Extensive literature is available on this topic, for example, see Zawadzki (1975), Austin (1987), Aydin et al. (1987), Bolen et al. (1997; 1998), Anagnostou et al. (1999).

7. For another flash flood example see Brandes et al. (1997).

Appendix 1
Review of electrostatics

One of the simplest charge configurations in electrostatics is the elementary electric dipole consisting of two point charges of equal magnitude and opposite sign separated by a small distance. Such dipoles are useful to characterize a polarized atom or molecule in which the statistical rest position of the entire negative charge group is slightly separated from the statistical rest position of the entire positive charge group. These two charge groups, which are held together by strong interatomic forces, are electrically neutral (net charge equals zero), but can get distorted and oriented by external forces. This model is useful to characterize simple dielectric materials and is often referred to as the bound-charge model.

Using Coulomb's law for point charges, and for distances large compared with the charge separation distance d (see Fig. A1.1), the electrostatic potential (Ψ) of an elementary dipole is given as,

$$\Psi = \frac{1}{4\pi\varepsilon_0} \frac{p\cos\theta}{r^2} \tag{A1.1}$$

where the dipole moment $p = qd$. In terms of the vector dipole moment, ($\vec{p} = q\vec{d}$), where $\vec{d}$ is a vector connecting the negative and positive charges,

$$\Psi(\vec{r}) = \frac{1}{4\pi\varepsilon_0} \frac{\vec{p}\cdot\vec{r}}{r^3} \tag{A1.2}$$

If the dipole $\vec{p}$ is located at $\vec{r}'$, then a simple translation of origin (see Fig. A1.2) gives the more general expression,

$$\Psi(\vec{r}) = \frac{1}{4\pi\varepsilon_0} \frac{\vec{p}\cdot(\vec{r}-\vec{r}')}{|\vec{r}-\vec{r}'|^3} \tag{A1.3}$$

The position vectors $\vec{r} \equiv (x, y, z)$ and $\vec{r}' \equiv (x', y', z')$ are used to locate the observation point and source point, respectively, while $R = |\vec{r} - \vec{r}'|$ is the distance between these points. The gradient operators, $\nabla \equiv [\hat{x}(\partial/\partial x) + \hat{y}(\partial/\partial y) + \hat{z}(\partial/\partial z)]$ and $\nabla' \equiv [\hat{x}(\partial/\partial x') + \hat{y}(\partial/\partial y') + \hat{z}(\partial/\partial z')]$, and Laplacian operators, $\nabla^2 \equiv [(\partial^2/\partial x^2) + (\partial^2/\partial y^2) + (\partial^2/\partial z^2)]$ and $\nabla'^2 \equiv [(\partial^2/\partial x'^2) + (\partial^2/\partial y'^2) + (\partial^2/\partial z'^2)]$ are used in electrostatics and, more generally, in potential theory. Equation (A1.3) can be written as $\Psi(\vec{r}) = (1/4\pi\varepsilon_0)\vec{p}(\vec{r}')\cdot\nabla'(1/R)$, which is a general expression for the potential due to a point dipole located at $\vec{r}'$.

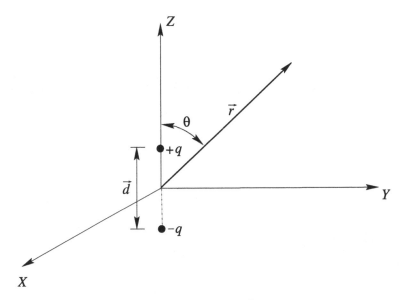

Fig. A1.1. Elementary dipole with moment $\vec{p} = q\vec{d}$. Note that $r \gg d$.

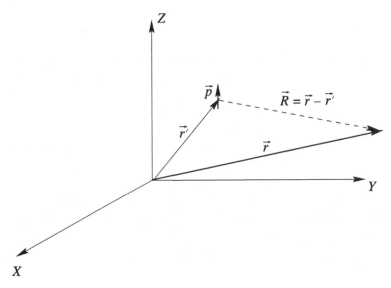

Fig. A1.2. Point dipole moment located at $\vec{r}'$. Note that $R = |\vec{r} - \vec{r}'|$.

The macroscopic description of a dielectric material involves the definition of a vector volume density function $(\vec{P})$ called the volume density of polarization defined

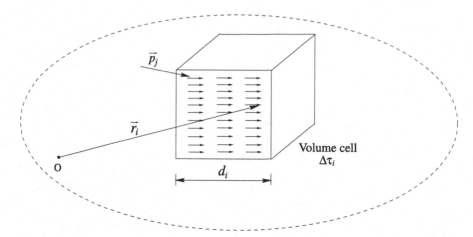

Fig. A1.3. Volume cell $\Delta\tau_i$ drawn within dielectric material with $\Delta\tau_i = d_i^3$. Elementary dipole moment of polarized atom is $\vec{p}_j$.

(King and Prasad 1986) as,

$$\vec{P}(\vec{r}_i) = \frac{\displaystyle\sum_{j=1}^{N} \vec{p}_j}{\Delta\tau_i} \tag{A1.4}$$

where $\Delta\tau_i$ is an elementary volume cell, and $\vec{p}_j$ is the elementary dipole moment of the jth polarized atom (or molecule) inside $\Delta\tau_i$. Note that the volume cell (see Fig. A1.3) must be large enough to contain a statistically large number of polarized atoms ($\Delta\tau_i \approx d_i^3$; $d_i \gg$ "mean free path") but also small enough to characterize the changes in $\vec{P}$ on scales of L ($d_i \ll L$; L is a "laboratory" scale). The cell walls of $\Delta\tau_i$ are defined so that only complete elementary dipoles are contained within it, i.e. the volume $\Delta\tau_i$ is electrically neutral. A continuous function $\vec{P}(\vec{r})$ may be obtained by interpolation of the discrete values of $\vec{P}(\vec{r}_i)$, which are defined at the center of identical volume cells within the dielectric material. Thus, (A1.3) can be generalized by superposition to express the potential due to a continuous volume density of polarization ($\vec{P}$) as,

$$\Psi(\vec{r}) = \frac{1}{4\pi\varepsilon_0} \int_\tau d\vec{p} \cdot \nabla'\left(\frac{1}{R}\right) = \frac{1}{4\pi\varepsilon_0} \int_\tau \vec{P}(\vec{r}')d\tau' \cdot \nabla'\left(\frac{1}{R}\right) \tag{A1.5}$$

where τ is the volume of the material (see Fig. A1.4) and $d\vec{p} = \vec{P} \, d\tau'$. It is assumed that external electrical forces maintain $\vec{P}$ but these are not included in (A1.5), or $\vec{P}$ is due to a permanently polarized material (external forces are not required to maintain $\vec{P}$). Note that $\vec{P}$ is weighted by the gradient of $1/R$ in the integrand. It is possible to mathematically convert this integral into two integrals with $1/R$, or inverse distance, weighting using the vector identity $\nabla' \cdot (\alpha\vec{A}) = \alpha\nabla' \cdot \vec{A} + \nabla'\alpha \cdot \vec{A}$ and the divergence

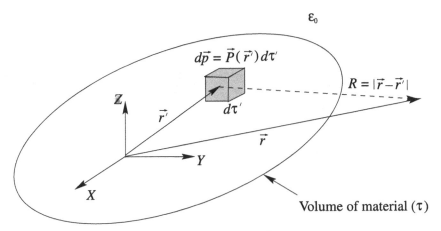

Fig. A1.4. Continuous volume density of polarization $\vec{P}(\vec{r}')$ is defined within the material volume τ.

theorem. Let $\alpha = 1/R = 1/|\vec{r} - \vec{r}'|$ and $\vec{A}(\vec{r}') \equiv \vec{P}(\vec{r}')$. Thus,

$$\vec{P}(\vec{r}') \cdot \nabla' \left(\frac{1}{R} \right) = \nabla' \cdot \left[\frac{\vec{P}(\vec{r}')}{R} \right] - \frac{1}{R} \nabla' \cdot \vec{P}(\vec{r}') \tag{A1.6}$$

Substituting in (A1.5) and using the divergence theorem,

$$\int_\tau \nabla' \cdot \vec{A}(\vec{r}') \, d\tau' = \oint_S \vec{A}(\vec{r}') \cdot \hat{n} \, dS' \tag{A1.7}$$

results in,

$$\Psi(\vec{r}) = \frac{1}{4\pi \varepsilon_0} \int_\tau \frac{-\nabla' \cdot \vec{P}(\vec{r}') \, d\tau'}{R} + \frac{1}{4\pi \varepsilon_0} \oint_S \frac{\hat{n} \cdot \vec{P}(\vec{r}')}{R} \, dS' \tag{A1.8}$$

Since the potential due to a point charge q located at $\vec{r}'$ is simply given by $q/4\pi \varepsilon_0 |\vec{r} - \vec{r}'| = q/4\pi \varepsilon_0 R$, we can identify $-\nabla' \cdot \vec{P}(\vec{r}')$ as a volume density of "bound charge", or ρ_b, and $\hat{n} \cdot \vec{P}(\vec{r}')$ as the surface density of "bound charge", or η_b. Thus, in summary, the electrical description of a polarized dielectric in terms of $\vec{P}$ defined only within volume cells that are electric neutral can be made mathematically equivalent to a distribution of "fictitious" volume density of bound charge $\rho_b = -\nabla \cdot \vec{P}$ in the interior, and a fictitious volume density of bound charge $\eta_b = \hat{n} \cdot \vec{P}$ on the interface between the dielectric and empty space (see Fig. A1.5). To an observer outside the dielectric these two representations are equivalent.

In terms of a continuous density function $\vec{P}$, the electrostatic differential equations for the electric field ($\vec{E}$) within a material can be written as,

$$\begin{aligned} \nabla \times \vec{E} &= 0 \\ \varepsilon_0 \nabla \cdot \vec{E} &= -\nabla \cdot \vec{P} \end{aligned} \tag{A1.9}$$

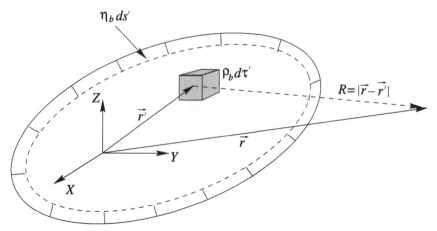

Fig. A1.5. Illustrating volume density of bound charge, ρ_b, within interior, and surface density of bound charge, η_b, on the interface.

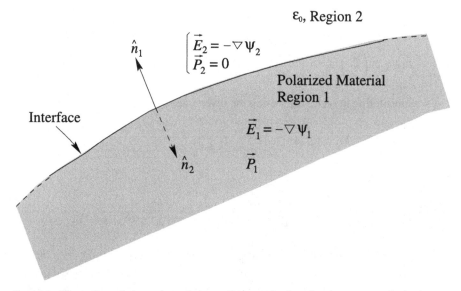

Fig. A1.6. Illustration relative to boundary condition on the interface between a polarized material and empty space, see (A1.10).

from which $\vec{E} = -\nabla\Psi$ and $\varepsilon_0\nabla^2\Psi = \nabla\cdot\vec{P}$. In empty space where $\vec{P} = 0$, the Laplace equation $\nabla^2\Psi = 0$ is valid. The boundary conditions on the interface between the dielectric and empty space are given (see Fig. A1.6) as,

$$\hat{n}_1 \times \vec{E}_1 + \hat{n}_2 \times \vec{E}_2 = 0, \quad \text{or} \quad \Psi_1 = \Psi_2 \tag{A1.10a}$$

$$\varepsilon_0(\hat{n}_1\cdot\vec{E}_1 + \hat{n}_2\cdot\vec{E}_2) = -\hat{n}_1\cdot\vec{P}_1 \tag{A1.10b}$$

Note the convention regarding the unit normals $\hat{n}_{1,2}$. Several comments are appropriate here. It is assumed that within the interior of the dielectric there is no volume density of free charge (ρ_f) nor is there any surface density of free charge (η_f) on the interface. The term free charge refers to the free-charge model (as opposed to the bound-charge model) where the elementary particles possess an abundance of valence electrons, e.g. conducting materials. It is also assumed that there is no surface dipole density on the interface. Since $\rho_b = -\nabla \cdot \vec{P}$ within the dielectric interior, the Poisson equation takes on its more familiar form $\varepsilon_0 \nabla^2 \Psi = -\rho_b$. The boundary condition (A1.10b) can be written in compact notation as $\varepsilon_0 \hat{n} \cdot \vec{E} = -\eta_b$ since on the interface $\eta_b = \hat{n} \cdot \vec{P}$. Thus, the normal component of the electric field is discontinuous across the interface by an amount equal to η_b / ε_0. The boundary condition (A1.10a) can be written in compact notation as $\hat{n} \times \vec{E} = 0$, which implies that the tangential electric field is continuous across the interface. Since $\vec{E} = -\nabla \Psi$ this implies that the potential is continuous across the interface.

So far we have not considered the external field needed to polarize the dielectric particle. Let this external or applied field be $\vec{E}_a$, which is assumed to be known and defined in empty space. It is also termed the incident field ($\vec{E}_i$) and, correspondingly, the incident potential is Ψ_i where $\vec{E}_i = -\nabla \Psi_i$. An initially unpolarized dielectric particle is brought in the presence of this incident field. The response of the dielectric is that it gets polarized with an unknown volume density of polarization $\vec{P}$. Mathematically, this is equivalent to the particle interior being charged with $\rho_b = -\nabla \cdot \vec{P}$ and the interface being charged with $\eta_b = \hat{n} \cdot \vec{P}$. In any case, the initially unperturbed incident field $\vec{E}_i$ (and potential Ψ_i), is now perturbed by the presence of the polarized particle; and, by superposition, the total electric field (and potential Ψ) at all points (either inside or external to the dielectric particle) can be written as $\vec{E}_T = \vec{E}_i + \vec{E}_p$, where $\vec{E}_T$ is the total field and $\vec{E}_p$ is the perturbation field. Similarly, $\Psi_T = \Psi_i + \Psi_p$. Inside the particle, $\varepsilon_0 \nabla \cdot \vec{E}_p = -\nabla \cdot \vec{P}$, while outside the particle $\varepsilon_0 \nabla \cdot \vec{E}_p = 0$. Also, inside the particle $\varepsilon_0 \nabla \cdot \vec{E}_i = 0$ since the sources for the incident field are external to the particle (see Fig. A1.7a). Thus, inside the particle, the total electric field satisfies $\varepsilon_0 \nabla \cdot \vec{E}_T = -\nabla \cdot \vec{P}$; this total field inside will be termed $\vec{E}_T^{\text{in}} = -\nabla \Psi_T^{\text{in}}$. In the case of a linear dielectric material, $\vec{P}$ is linearly related to $\vec{E}_T^{\text{in}}$ by,

$$\vec{P} = \varepsilon_0 \chi_e \vec{E}_T^{\text{in}} \tag{A1.11}$$

where χ_e is the electric susceptibility of the material. The field $\vec{E}_T^{\text{in}}$ is the macroscopic total electric field inside the particle, but it is not the field that polarizes an individual atom or molecule (this is called the local field and is discussed in Section 1.6 under the Clausius–Mosotti equation). Thus, inside the particle, $\varepsilon_0 \nabla \cdot \vec{E}_T = -\nabla \cdot \vec{P}$ becomes $\nabla \cdot (\varepsilon_0 \vec{E}_T + \vec{P}) = 0$, or,

$$\nabla \cdot [\varepsilon_0 (1 + \chi_e) \vec{E}_T^{\text{in}}] = 0 \tag{A1.12a}$$
$$\varepsilon_0 (1 + \chi_e) \nabla^2 \Psi_T^{\text{in}} = 0 \tag{A1.12b}$$

The term $1 + \chi_e = \varepsilon_r$ is the relative permittivity or dielectric constant of the material.

External to the particle the total electric field is written as $\vec{E}_T^{\text{out}} = \vec{E}_i + \vec{E}_p$, or, $\Psi_T^{\text{out}} = \Psi_i + \Psi_p$. From $\varepsilon_0 \nabla \cdot \vec{E}_p = 0$, we get $\nabla^2 \Psi_p = 0$ external to the particle. Thus,

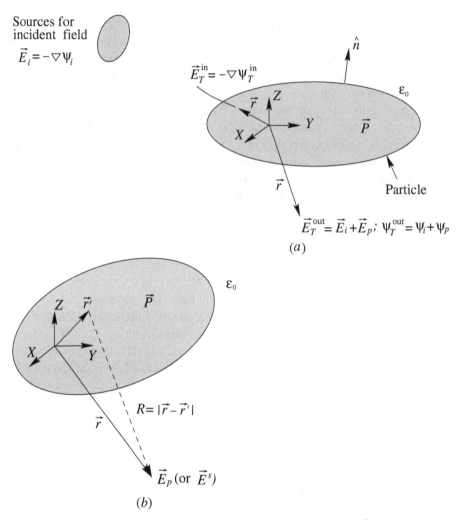

Fig. A1.7. (*a*) Dielectric particle in the presence of an incident electric field $\vec{E}_i$ becomes polarized and causes a perturbation field $\vec{E}_p = -\nabla\psi_p$. (*b*) Illustration showing that the "source" for the perturbation field, $\vec{E}_p$, is the volume density of polarization, $\vec{P}$, within the particle.

the perturbation potential satisfies the Laplace equation in the external region. The expression for Ψ_p is the same as that given in (A1.5); thus, the total potential outside is given as,

$$\Psi_T^{\text{out}}(\vec{r}) = \Psi_i(\vec{r}) + \frac{1}{4\pi\varepsilon_0} \int_\tau \vec{P}(\vec{r}')\cdot\nabla'\left(\frac{1}{R}\right) d\tau' \tag{A1.13a}$$

$$= \Psi_i(\vec{r}) + \frac{1}{4\pi}(\varepsilon_r - 1) \int_\tau \vec{E}_T^{\text{in}}(\vec{r}')\cdot\nabla'\left(\frac{1}{R}\right) d\tau' \tag{A1.13b}$$

and,

$$\vec{E}_p = -\nabla \Psi_p = -\frac{1}{4\pi\varepsilon_0} \nabla \int_\tau \vec{P}(\vec{r}') \cdot \nabla' \left(\frac{1}{R}\right) d\tau' \tag{A1.14a}$$

$$= \frac{(\varepsilon_r - 1)}{4\pi} \nabla \int_\tau \vec{E}_T^{in}(\vec{r}') \cdot \nabla' \left(\frac{1}{R}\right) d\tau' \tag{A1.14b}$$

where we have used (A1.11) valid for a linear material. Note that $\vec{r}$ is external to the particle, while $\vec{r}'$ is the variable of the integration within the particle. Thus, (A1.14) is an integral representation for the perturbation field outside the particle in terms of the unknown total field (or equivalently $\vec{P}$) inside the particle. The "sources" for the perturbation field $\vec{E}_p$ can be thought of as the unknown volume density of polarization $\vec{P}$ (see Fig. A1.7b). The $\vec{P}$ itself can be thought of as being maintained by external sources ($\vec{E}_i$).

The formal boundary value problem (see Fig. A1.7a) can be stated as follows. Inside the particle,

$$\nabla^2 \Psi_T^{in} = 0 \tag{A1.15a}$$

Outside the particle,

$$\nabla^2 \Psi_p = 0 \tag{A1.15b}$$

At the interface, the total potential is continuous,

$$\Psi_p + \Psi_i = \Psi_T^{in} \tag{A1.16a}$$

and the normal component of the total electric field satisfies,

$$\varepsilon_0(\hat{n}_1 \cdot \vec{E}_T^{in} + \hat{n}_2 \cdot \vec{E}_T^{out}) = -\hat{n}_1 \cdot \vec{P}_1 = -\hat{n}_1 \cdot \varepsilon_0 \chi_e \vec{E}_T^{in} \tag{A1.16b}$$

Thus,

$$\varepsilon_0 \varepsilon_r \hat{n}_1 \cdot \vec{E}_T^{in} + \varepsilon_0 \hat{n}_2 \cdot \vec{E}_T^{out} = 0 \tag{A1.17}$$

In terms of potentials,

$$\varepsilon_0 \varepsilon_r \hat{n}_1 \cdot \nabla \Psi_T^{in} + \varepsilon_0 \hat{n}_2 \cdot \nabla (\Psi_i + \Psi_p) = 0 \tag{A1.18}$$

If $\hat{n}_1 = \hat{n} = -\hat{n}_2$, where $\hat{n}$ points outward from the particle boundary to the external region (see Fig. A1.6) then,

$$\varepsilon_0 \varepsilon_r \frac{\partial \Psi_T^{in}}{\partial n} = \varepsilon_0 \frac{\partial}{\partial n} (\Psi_i + \Psi_p) \tag{A1.19}$$

The separation of variables technique can be used to solve the boundary value problem stated above for simple geometries such as spheres and spheroids. In spherical coordinates, separation of variables for the potential $\Psi(r, \theta)$ in the case of no ϕ-dependence (azimuthal symmetry) can be written as $\Psi(r, \theta) = R(r)F(\theta)$. A few simple eigenfunctions are $(1/r)$, $\cos\theta/r^2$, and $r\cos\theta$ corresponding, respectively, to the Coulomb potential for a unit point charge at the origin, the point dipole at the origin, and the potential corresponding to a uniform z-directed electric field (since $z = r\cos\theta$ and $\nabla(r\cos\theta) = \nabla(z) = -\hat{z}$). With these simple eigenfunctions, three related problems can be solved: (a) a permanently polarized sphere with $\vec{P} = P_0\hat{z}$ which introduces the concept of the depolarizing electric field within the sphere, (b) the complementary problem of the electric field inside a spherical cavity in an uniformly polarized medium of infinite extent that clarifies the concept of the "local" electric field used in deriving the Clausius–Mosotti equation (and dielectric mixing formula) alluded to earlier, and (c) the perturbation electric field by a dielectric sphere placed in an uniform incident electric field that will later be related to Rayleigh scattering.

A1.1   Permanently polarized sphere

Consider a permanently polarized sphere of radius a with $\vec{P} = \hat{z}P_0$, P_0 being a constant, as shown in Fig. A1.8a. External to (or outside) the sphere the electric field $\vec{E}_0$ satisfies (from A1.9), $\nabla \times \vec{E}_0 = 0$, $\varepsilon_0 \nabla \cdot \vec{E}_0 = 0$, or $\nabla^2 \Psi_0 = 0$, where $\vec{E}_0 = -\nabla \Psi_0$. Inside the sphere the electric field $\vec{E}_{in}$ satisfies $\nabla \times \vec{E}_{in} = 0$, $\varepsilon_0 \nabla \cdot \vec{E}_{in} = -\nabla \cdot \vec{P} = -\nabla \cdot (\hat{z}P_0) = 0$, so that $\nabla^2 \Psi_{in} = 0$, where $\vec{E}_{in} = -\nabla \Psi_{in}$. On the boundary of the sphere at $r = a$, the boundary conditions (from (A1.10)) are,

$$\Psi_0 = \Psi_{in} \tag{A1.20a}$$

$$\varepsilon_0[\hat{r}\cdot\vec{E}_{in} + (-\hat{r})\cdot\vec{E}_0] = -\hat{r}\cdot P_0\hat{z} = -P_0\cos\theta \tag{A1.20b}$$

Note that $\rho_b = 0$ in the interior and $\eta_b = P_0\cos\theta$ on the interface, see Fig. A1.8b. In terms of simple eigenfunctions the potential outside the sphere is of dipole form and can be written as $\Psi_0(r, \theta) = A_0\cos\theta/r^2$, where A_0 is an unknown constant. Because $\vec{P}$ inside is uniform, the potential inside should correspond to an uniform electric field, or, $\Psi_{in} = B_0 r\cos\theta$ with B_0 an unknown constant. It is easily seen that since $z = r\cos\theta$ and $\vec{E}_{in} = -\nabla\Psi_{in}$, the field should be constant and z-directed ($\vec{E}_{in} = -B_0\hat{z}$). The unknown constants A_0 and B_0 are determined from (A1.20). A few simple steps yield $A_0 = (P_0/3\varepsilon_0)a^3$ and $B_0 = P_0/3\varepsilon_0$. Thus,

$$\psi_0(r, \theta) = \frac{P_0 a^3}{3\varepsilon_0}\frac{\cos\theta}{r^2} = \frac{4\pi a^3}{3}\frac{P_0\cos\theta}{4\pi\varepsilon_0 r^2} \tag{A1.21a}$$

$$\Psi_{in}(r, \theta) = \frac{P_0}{3\varepsilon_0}r\cos\theta, \quad \vec{E}_{in} = -\frac{P_0}{3\varepsilon_0}\hat{z} \tag{A1.21b}$$

External to the sphere, the potential appears to correspond to a $\hat{z}$-directed point dipole at the origin of moment equal to $P_0 V$, where V is the sphere volume. Inside the sphere, the electric field is uniform and directed opposite to $\vec{P}$. This is the so-called "depolarizing" field and the factor $(1/3)$ is called the depolarizing factor (λ) for a sphere.

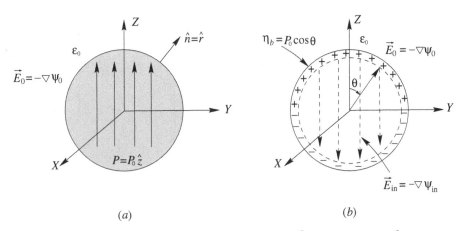

Fig. A1.8. (*a*) Permanently polarized sphere of radius a with $\vec{P} = P_0\hat{z}$ inside and $\vec{E}_0 = -\nabla\psi_0$ outside. (*b*) Surface density of charge η_b on interface with $\eta_b = \hat{n}\cdot\vec{P} = P_0\cos\theta$.

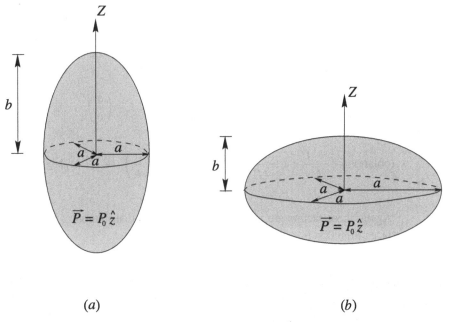

Fig. A1.9. Illustrating permanently polarized spheroids with $\vec{P} = P_0\hat{z}$ inside: (*a*) prolate spheroid with symmetry axis along Z-axis, (*b*) oblate spheroid with symmetry axis along Z-axis.

The origin of this field is related to the surface density of charge $\eta_b = P_0\cos\theta$ sketched in Fig. A1.8*b*, where the top and bottom hemispheres are positively and negatively charged, respectively. This situation is analogous to charges of opposite sign on the two plates of a parallel plate capacitor giving rise to uniform electric field between the plates.

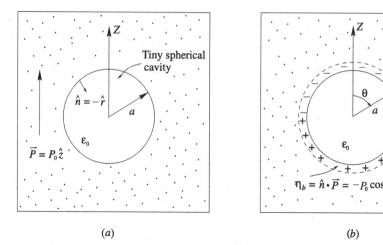

Fig. A1.10. (*a*) Tiny spherical "cavity" cut in a permanently polarized block of material (shaded area) with $\vec{P} = P_0\hat{z}$. (*b*) Surface density of charge (η_b) on cavity/material interface.

The corresponding depolarizing factor (λ_z) for a uniformly polarized spheroid with $\vec{P} = P_0\hat{z}$, where $\hat{z}$ is along the spheroid symmetry axis (see Fig. A1.9), can be obtained in a similar manner using eigenfunctions of the Laplace equation in prolate or oblate spheroidal coordinate systems (Van Bladel 1985),

$$\lambda_z(\text{prolate}) = \frac{1-e^2}{e^2}\left(-1 + \frac{1}{2e}\ln\frac{1+e}{1-e}\right); \quad e^2 = 1 - \left(\frac{a}{b}\right)^2, \quad 0 < \frac{a}{b} \le 1$$

(A1.22a)

$$\lambda_z(\text{oblate}) = \frac{1+f^2}{f^2}\left(1 - \frac{1}{f}\tan^{-1}f\right); \quad f^2 = \left(\frac{a}{b}\right)^2 - 1, \quad a/b \ge 1$$

(A1.22b)

Again, these factors are intimately related to the distribution of η_b on the spheroid surfaces from $0 \le \theta < \pi/2$ and $\pi/2 < \theta \le \pi$.

A1.2   Spherical cavity

The complementary situation to Section A1.1 above is the case of an infinite medium which is permanently polarized with $\vec{P} = P_0\hat{z}$ (see Fig. A1.10a) in which a spherical cavity of radius a is cut out. In the absence of the cavity, the volume density of bound charge $\rho_b = -\nabla\cdot\vec{P} = 0$. Since the medium is infinite there are no surfaces where $\eta_b = \hat{n}\cdot\vec{P}$ can exist. Thus, in the absence of the cavity the potential and electric field are zero in this medium. Once the cavity is cut, η_b is generated on the interface and equals $\hat{n}\cdot P_0\hat{z}$, where $\hat{n} = -\hat{r}$. Note that the unit normal $\hat{n}$ is always outward drawn with respect to the material (Fig. A1.6). Thus, η_b on the cavity surface equals $-P_0\cos\theta$, and the top and bottom hemispheres are, respectively, negatively and positively charged. External

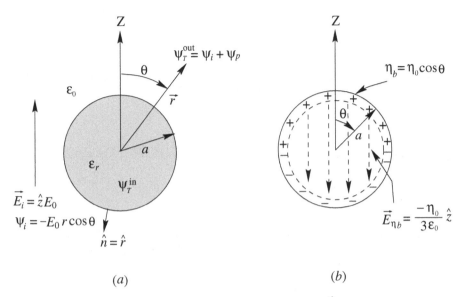

Fig. A1.11. (*a*) Dielectric sphere placed in an uniform electric field $\vec{E}_i = E_0\hat{z}$. The perturbation potential is ψ_p external to the sphere. (*b*) Illustrates the surface density of charge, η_b, on the dielectric interface, and the component electric field due to η_b inside the sphere.

to the cavity ($r > a$) we assume that the permanent polarization is unaltered; thus, $\rho_b = -\nabla \cdot \vec{P}$ continues to be equal to zero. From our solution in Section A1.1, we can conclude that the electric field inside the cavity must be constant and equal to $(P_0/3\varepsilon_0)\hat{z}$, while outside the cavity the electric field is equivalent to a point dipole of moment $\vec{p} = -P_0 V \hat{z}$. Thus, the electric field within the cavity has jumped to $(P_0/3\varepsilon_0)\hat{z}$ from its initially zero value (no cavity case) and its direction is parallel to $\vec{P}$. This example is complementary to the case of the permanently polarized sphere. Similarly, the electric field within spheroidal cavities with z-directed symmetry axes can be obtained and is related to λ_z as given by (A1.22).

A1.3 Dielectric sphere in a uniform incident field

Consider the situation shown in Fig. A1.11a, with the incident field being $\vec{E}_i = E_0\hat{z}$ and $\Psi_i = -E_0 z = -E_0 r \cos\theta$. The reference for the potential is the plane $z = 0$ or the XY-plane. The sphere is characterized by relative permittivity ε_r, in contrast to Section A1.1 where it was permanently polarized. The boundary value problem is stated in (A1.15a, b) with boundary condition (A1.16a, A1.19) on $r = a$. Because the incident potential is an eigenfunction of the form $r \cos\theta$, the perturbation potential is of a dipole form $\Psi_p = A_0 \cos\theta/r^2$, while the total potential inside is of the form $\Psi_T^{in} = B_0 r \cos\theta$. Thus, the total electric field inside is also uniform, $\vec{E}_T^{in} = -B_0\hat{z}$. Boundary condition

(A1.16a) on $r = a$ gives,

$$\frac{A_0 \cos\theta}{a^2} - E_0 a \cos\theta = B_0 a \cos\theta \tag{A1.23a}$$

and (A1.19) (note that $\partial/\partial n \equiv \partial/\partial r$ here) gives,

$$\varepsilon_0 \varepsilon_r (B_0 \cos\theta) = \varepsilon_0 \left(-E_0 \cos\theta - \frac{2}{a^3} A_0 \cos\theta \right) \tag{A1.23b}$$

Again, a few simple steps yield,

$$A_0 = \left(\frac{\varepsilon_r - 1}{\varepsilon_r + 2} \right) E_0 a^3; \quad B_0 = \frac{-3 E_0}{\varepsilon_r + 2} \tag{A1.24}$$

Thus, the total electric field inside is $\vec{E}_T^{in} = -B_0 \hat{z} = \hat{z} 3 E_0/(\varepsilon_r + 2)$, which is less than $\vec{E}_i$ as $\varepsilon_r \geq 1$. The magnitude of the ratio of $\vec{E}_T^{in}$ to $\vec{E}_i$, defined as Λ, is a constant and can be written as,

$$\Lambda = \frac{1}{[(\varepsilon_r/3) + (2/3)]} = \frac{1}{(\varepsilon_r - 1)\lambda + 1} \tag{A1.25a}$$

where $\lambda = 1/3$ is the depolarizing factor for the sphere arrived at in Section A1.1. Similarly, for prolate and oblate spheroids with symmetry axes along the z-direction, Λ_z can be written as,

$$\Lambda_z = \frac{1}{(\varepsilon_r - 1)\lambda_z + 1} \tag{A1.25b}$$

with λ_z defined in (A1.22a) for prolates and in (A1.22b) for oblates.

Returning to the case of the dielectric sphere, the reduction of the total electric field inside relative to $\vec{E}_i$ must be caused by an oppositely directed component electric field $(\vec{E}_{\eta_b})$ which is due to the surface density of bound charge, $\eta_b = \eta_0 \cos\theta$, on the interface as indicated in Fig. A1.11b. The constant η_0 is derived from $\eta_b = \hat{r} \cdot \vec{P}$ on $r = a$, where $\vec{P} = \varepsilon_0(\varepsilon_r - 1)\vec{E}_T^{in}$, which gives $\eta_0 = 3\varepsilon_0 E_0(\varepsilon_r - 1)/(\varepsilon_r + 2)$.

External to the sphere, the perturbation potential Ψ_p is given as,

$$\Psi_p = \frac{A_0 \cos\theta}{r^2} = \left(\frac{\varepsilon_r - 1}{\varepsilon_r + 2} \right) E_0 a^3 \frac{\cos\theta}{r^2} \tag{A1.26a}$$

$$= 4\pi \varepsilon_0 E_0 a^3 \left(\frac{\varepsilon_r - 1}{\varepsilon_r + 2} \right) \frac{\cos\theta}{4\pi \varepsilon_0 r^2} \tag{A1.26b}$$

$$= 3\varepsilon_0 E_0 V \left(\frac{\varepsilon_r - 1}{\varepsilon_r + 2} \right) \frac{\cos\theta}{4\pi \varepsilon_0 r^2} \tag{A1.26c}$$

Thus, the dielectric sphere can be replaced by a point dipole at the origin of moment $\vec{p} = \hat{z} 3\varepsilon_0 E_0 V(\varepsilon_r - 1/\varepsilon_r + 2)$, where V is the sphere volume. Since $\vec{E}_i = \hat{z} E_0$, the polarizability, α, of the sphere is defined by $\vec{p} = \alpha \vec{E}_i$ where,

$$\alpha = 3\varepsilon_0 V \left(\frac{\varepsilon_r - 1}{\varepsilon_r + 2} \right) \tag{A1.27a}$$

Table A1.1. λ and Λ factors for needles, plates, and spheres (DeWolf et al. 1990)

	λ_x	λ_y	λ_z	Λ_x	Λ_y	Λ_z
Needles	1/2	1/2	0	$2/(\varepsilon_r + 1)$	$2/(\varepsilon_r + 1)$	1
Plates	0	0	1	1	1	$1/\varepsilon_r$
Spheres	1/3	1/3	1/3	$3/(\varepsilon_r + 2)$	$3/(\varepsilon_r + 2)$	$3/(\varepsilon_r + 2)$

$$= V\varepsilon_0(\varepsilon_r - 1)\frac{1}{1 + (1/3)(\varepsilon_r - 1)} \tag{A1.27b}$$

$$= V\varepsilon_0(\varepsilon_r - 1)\frac{1}{1 + \lambda(\varepsilon_r - 1)} = V\varepsilon_0(\varepsilon_r - 1)\Lambda \tag{A1.27c}$$

For prolates or oblates with volume V and with symmetry axis along the z-direction (refer to Fig. A1.9) and $\vec{E}_i = E_0\hat{z}$, the polarizability, α_z, is written as,

$$\alpha_z = V\varepsilon_0(\varepsilon_r - 1)\frac{1}{1 + \lambda_z(\varepsilon_r - 1)} = V\varepsilon_0(\varepsilon_r - 1)\Lambda_z \tag{A1.28}$$

When $\vec{E}_i = E_0\hat{x}$ or $E_0\hat{y}$, the depolarizing factors satisfy $\lambda_x = \lambda_y = (1 - \lambda_z)/2$. The corresponding polarizabilities α_x or α_y are given as,

$$\alpha_x = \alpha_y = V\varepsilon_0(\varepsilon_r - 1)\frac{1}{1 + \lambda_{x,y}(\varepsilon_r - 1)} = V\varepsilon_0(\varepsilon_r - 1)\Lambda_{x,y} \tag{A1.29}$$

Thus, the polarizability matrix for prolates/oblates oriented as in Fig. A1.9 relates the components of the dipole moment to the components of the incident field by,

$$\begin{bmatrix} p_x \\ p_y \\ p_z \end{bmatrix} = V\varepsilon_0(\varepsilon_r - 1)\begin{bmatrix} \Lambda_x & 0 & 0 \\ 0 & \Lambda_y & 0 \\ 0 & 0 & \Lambda_z \end{bmatrix}\begin{bmatrix} E_{0x} \\ E_{0y} \\ E_{0z} \end{bmatrix} \tag{A1.30}$$

where $\vec{E}_i \equiv (E_{0x}, E_{0y}, E_{0z})$. In compact notation, $\vec{p} = V\varepsilon_0(\varepsilon_r - 1)\bar{\bar{\Lambda}}\cdot\vec{E}_i$, where $\bar{\bar{\alpha}} = V\varepsilon_0(\varepsilon_r - 1)\bar{\bar{\Lambda}}$ is the polarizability tensor. From DeWolf et al. (1990), Table A1.1 summarizes the λ and Λ factors for needles $(a/b \to 0)$, plates $(a/b \to \infty)$, and spheres $(a/b = 1)$.

When the time variations are slow (e.g. sufficiently low frequencies in the periodic case), the laws of electrostatics given in (A1.9, A1.10) can be extended to the electroquasistatic (EQS) approximation as follows:

$$\nabla \times \vec{E}(\vec{r}, t) = 0; \quad \varepsilon_0\nabla\cdot\vec{E}(\vec{r}, t) = -\nabla\cdot\vec{P}(\vec{r}, t) \tag{A1.31a}$$

$$\vec{E}(\vec{r}, t) = -\nabla\Psi(\vec{r}, t); \quad \varepsilon_0\nabla^2\Psi(\vec{r}, t) = \nabla\cdot\vec{P}(\vec{r}, t) \tag{A1.31b}$$

The boundary conditions given in (A1.10) continue to hold. The conductivity (σ) of the material under consideration is assumed to be zero hence no Ohmic currents $(\vec{J} = \sigma\vec{E})$

flow. It is useful, however, to introduce the concept of polarization current, $\vec{J}_p$, which, similar to the Ohmic current density vector $\vec{J}$, satisfies the equation of continuity or charge conservation law (King and Prasad 1986) as,

$$\nabla \cdot \vec{J}_p + \frac{\partial \rho_b}{\partial t} = 0 \tag{A1.32}$$

Since $\rho_b = -\nabla \cdot \vec{P}$, it follows that $\vec{J}_p = \partial \vec{P}/\partial t$.

The boundary value problem of a dielectric sphere in an incident field given by $\vec{E}^i = \hat{z}E_0(t)$ can be solved in exactly the same manner as before except that the constants A_0, B_0 are now $A_0(t)$ and $B_0(t)$. Thus, the perturbation potential is $\Psi_p(\vec{r}, t) = A_0(t) \cos\theta/r^2$ and the total potential inside the sphere is $\Psi_T^{\text{in}}(\vec{r}, t) = B_0(t)r \cos\theta$. Applying the boundary condition as in (A1.23), the solutions for $A_0(t)$ and $B_0(t)$ are,

$$A_0(t) = \left(\frac{\varepsilon_r - 1}{\varepsilon_r + 2}\right) E_0(t) a^3; \quad B_0(t) = \frac{-3E_0(t)}{\varepsilon_r + 2} \tag{A1.33}$$

The total electric field inside the dielectric is $\vec{E}_T^{\text{in}} = \hat{z}3E_0(t)/\varepsilon_r + 2$ and the polarization current $\vec{J}_p = \partial \vec{P}/\partial t = \varepsilon_0(\varepsilon_r - 1)\partial \vec{E}_T^{\text{in}}/\partial t = [\hat{z}3V\varepsilon_0(\varepsilon_r - 1)/(\varepsilon_r + 2)]\partial E_0/\partial t$. The dipole moment of the sphere is $\vec{p}(t) = \hat{z}3\varepsilon_0 E_0(t)V(\varepsilon_r - 1)/(\varepsilon_r + 2)$.

If $E_0(t)$ is periodic in time of the form $E_0 \cos(\omega t + \theta)$, then the dipole moment of the sphere is in time-phase with the incident field while the polarization current is 90° out of time-phase. It is conventional to express the incident field in sinusoidal steady state as $\vec{E}^i(t) = \hat{z}\, \text{Re}\, E_0 \exp(j\theta)\exp(j\omega t)$. Suppressing the Re (real part of a complex number) and $\exp(j\omega t)$, the incident field phasor-vector (or complex vector) is expressed as $\vec{E}^{ic} = \hat{z}\tilde{E}_0$, where $\tilde{E}_0 = E_0\exp(j\theta)$ and the superscript c means a complex vector. In Section 1.3, this incident field is related to the incident field of a plane wave propagating along the $+y$-axis and linearly polarized along the $\hat{z}$-direction (see Fig. A1.11). The coefficients $A_0(t)$ and $B_0(t)$ in (A1.33) also are phasors $\tilde{A}_0$ and $\tilde{B}_0$. The dipole moment is a phasor-vector $\vec{p}^c = [3\varepsilon_0 V(\varepsilon_r - 1)/(\varepsilon_r + 2)]\vec{E}^{ic}$ and the polarization current $\vec{J}_p^c = [3\varepsilon_0(\varepsilon_r - 1)/(\varepsilon_r + 2)]Vjw\vec{E}^{ic}$. The superscript c is generally dropped if the sinusoidal steady state context is clear. Thus, in the EQS approximation the direction of the dipole moment will reverse periodically and in time-phase with the direction of the incident electric field because ε_r is assumed to be a real constant. The spatial variation of the potentials and fields are exactly the same as in electrostatics and are essentially decoupled from the time variations. The EQS approximation is valid when the sphere diameter $\ll 2\pi c/\omega$, where c is the velocity of light. In real materials the dipole moment $\vec{p}(t)$ will not be in time-phase with $\vec{E}^i(t)$, rather there will be phase lag which can be expressed in the time-domain as $\vec{p}(t)$ being proportional to $\hat{z}E_0(\cos \omega t + \theta - \theta_p)$, where θ_p is the phase lag angle and is itself a function of angular frequency ω. Such a description can be modeled by treating the relative permittivity as complex ($\varepsilon_r^c = \varepsilon_r' - j\varepsilon_r''$) where the real and imaginary parts are functions of ω. The $(-j)$ factor multiplying ε_r'' is to ensure a time-phase lag with respect to $\vec{E}^i$ in the $\exp(j\omega t)$ time convention adopted here. If $\exp(-j\omega t)$ time convention is adopted then $\varepsilon_r^c = \varepsilon_r' + j\varepsilon_r''$.

Appendix 2
Review of vector spherical harmonics and multipole expansion of the electromagnetic field

The normalized spherical harmonics are defined (Jackson 1975) as,

$$Y_n^m(\theta, \phi) = \gamma_{mn}^{1/2} P_n^m(\cos\theta) e^{jm\phi} \qquad (A2.1)$$

where $P_n^m(z)$ with $z = \cos\theta$ are the associated Legendre functions and $\gamma_{mn}^{1/2}$ is a normalization factor. It is convenient to consider the real and imaginary parts of $Y_n^m(\theta, \phi)$ separately rather than to use complex harmonics. Hence, the "even" and "odd" normalized spherical harmonics are defined as,

$$Y_{(\text{even})n}^m(\theta, \phi) = \gamma_{mn}^{1/2} P_n^m(\cos\theta) \cos m\phi \qquad (A2.2a)$$

$$Y_{(\text{odd})n}^m(\theta, \phi) = \gamma_{mn}^{1/2} P_n^m(\cos\theta) \sin m\phi \qquad (A2.2b)$$

A third subscript is added to Y_{eon}^m, where $e \equiv$ "even" and $o \equiv$ "odd", and for compactness is written as $Y_{\sigma n}^m(\theta, \phi)$ with σ taking on "even" or "odd" values. The indices $n = 0, 1, 2, \ldots, \infty$, while $m = 0, 1, 2, \ldots, n$. Any function $\vec{f}(\theta, \phi)$ on a spherical surface can be expanded using the spherical harmonic basis,

$$f(\theta, \phi) = \sum_{\sigma=e,o} \sum_{n=0}^{\infty} \sum_{m=0}^{n} a_{\sigma mn} Y_{\sigma n}^m(\theta, \phi) \qquad (A2.3)$$

where $a_{\sigma mn}$ are the expansion coefficients. The normalization factor is,

$$\gamma_{mn}^{1/2} = \sqrt{\frac{(2n+1)(n-m)!\varepsilon_m}{4\pi(n+m)!}} \qquad (A2.4)$$

with $\varepsilon_m = 2$ for $m > 0$ and $\varepsilon_0 = 1$. The spherical harmonics are also used to construct a general solution to the Laplace equation $\nabla^2 \psi(r, \theta, \phi) = 0$,

$$\psi(r, \theta, \phi) = \sum_{\sigma} \sum_{n=0}^{\infty} \sum_{m=0}^{n} \left[a_{\sigma mn} r^n + b_{\sigma mn} r^{-(n+1)} \right] Y_{\sigma n}(\theta, \phi) \qquad (A2.5a)$$

$$= \sum_{\sigma mn} \left[a_{\sigma mn} r^n + b_{\sigma mn} r^{-(n+1)} \right] Y_{\sigma n}^m(\theta, \phi) \qquad (A2.5b)$$

where the triple summation is denoted as a single sum for compactness. The $\sigma \equiv e, n = 1, m = 0$ terms in (A2.5a) are of the familiar dipole form $r \cos \theta$ and $r^{-2} \cos \theta$ encountered in Appendix 1. In a similar fashion, a general solution to the scalar Helmholtz equation, $(\nabla^2 + k^2)\psi(r, \theta, \phi) = 0$ can be constructed as,

$$\psi(r, \theta, \phi) = \sum_{\sigma nm} \left[a_{\sigma mn} j_n(\rho) + b_{\sigma mn} h_n^{(2)}(\rho) \right] Y_{\sigma n}^m(\theta, \phi) \tag{A2.6}$$

where $\rho = kr$, and j_n and $h_n^{(2)}$ are the spherical Bessel and (second kind) Hankel functions which play a similar role in the solution as the terms r^n and $r^{-(n+1)}$ play in (A2.5). The Hankel function of second kind is chosen since it represents outgoing spherical waves, as $\rho \to \infty$ in the $\exp(i\omega t)$ time convention. The notation $i = \sqrt{-1}$ is used here rather than the usual j so as not to confuse with the notation $j_n(\rho)$ for the spherical Bessel function. As $\rho \to \infty$,

$$\lim_{\rho \to \infty} h_n^{(2)}(\rho) = (i)^{n+1} \frac{e^{-i\rho}}{\rho} \tag{A2.7}$$

Refer to Appendix 4 of Van Bladel (1985) for an excellent summary of Legendre and Bessel functions.

Finally, consider the general solution to the vector Helmholtz equation as in (1.8) valid for the $(\vec{E}, \vec{B})$ fields within a homogeneous dielectric region with wave number k,

$$\nabla \times \nabla \times \vec{E} - k^2 \vec{E} = 0 \tag{A2.8a}$$

$$\nabla \times \nabla \times \vec{B} - k^2 \vec{B} = 0 \tag{A2.8b}$$

The general solution is given as an expansion in terms of the vector spherical wave functions or simply multipoles, which are defined as,

$$\vec{M}_{\sigma mn}(r, \theta, \phi) = \nabla \left[z_n(\rho) Y_{\sigma n}^m(\theta, \phi) \right] \times \vec{r} \tag{A2.9a}$$

$$\vec{N}_{\sigma mn}(r, \theta, \phi) = \frac{1}{r} \nabla \times \vec{M}_{\sigma mn} \tag{A2.9b}$$

where $z_n(\rho)$ represents spherical Bessel or (2nd kind) Hankel functions. The choice of which radial function to use will be discussed later. It is noted that $j_n(\rho)$ is finite at the origin ($\rho = 0$), whereas $h_n^{(2)} \equiv j_n - i n_n$ "blows" up at the origin. The notation Rg $\vec{M}_{\sigma mn}$, Rg $\vec{N}_{\sigma mn}$ is used when $z_n = j_n(\rho)$ is used in (A2.9a) where Rg stands for "regular" at the origin. The vector spherical wave functions $\vec{M}$ and $\vec{N}$ satisfy the vector Helmholtz equation. They can also be expressed by separating the radial ($\rho = kr$) and angular spherical (θ, ϕ) parts as,

$$\vec{M}_{\sigma mn}(r, \theta, \phi) = z_n(\rho)\vec{m}_{\sigma mn}(\theta, \phi) \tag{A2.10a}$$

where,

$$\vec{m}_{emn}(\theta, \phi) = \gamma_{mn}^{1/2} \left[-m \frac{P_n^m(\cos \theta)}{\sin \theta} \sin m\phi \hat{\theta} - \frac{d}{d\theta} P_n^m(\cos \theta) \cos m\phi \hat{\phi} \right] \tag{A2.10b}$$

$$\vec{m}_{omn}(\theta, \phi) = \gamma_{mn}^{1/2} \left[\frac{m P_n^m(\cos \theta)}{\sin \theta} \cos m\phi \hat{\theta} - \frac{d}{d\theta} P_n^m(\cos \theta) \sin m\phi \hat{\phi} \right] \tag{A2.10c}$$

After tedious algebra, the vector wave function $\vec{N}_{\sigma mn}(\theta, \phi)$ can also be separated into radial and angular parts and expressed compactly as,

$$\vec{N}_{\sigma mn}(\theta, \phi) = \frac{z_n(\rho)}{\rho} \vec{l}_{\sigma mn}(\theta, \phi) + \frac{[\rho z_n(\rho)]'}{\rho} \hat{r} \times \vec{m}_{\sigma mn}(\theta, \phi) \tag{A2.11}$$

where the notation $[\rho z_n(\rho)]'$ is a compact form for $d[\rho z_n(\rho)]/d\rho$, and $\vec{l}_{\sigma mn}(\theta, \phi)$ is defined as,

$$\vec{l}_{\sigma mn}(\theta, \phi) = n(n+1) Y_{\sigma n}^m(\theta, \phi) \hat{r} \tag{A2.12}$$

The behavior of $\vec{M}$ and $\vec{N}$ as $\rho \to \infty$ is based on the choice of the second kind Hankel function $h_n^{(2)}(\rho)$ for $z_n(\rho)$. Thus,

$$\lim_{\rho \to \infty} \vec{M}_{\sigma mn}(\rho, \theta, \phi) = \lim_{\rho \to \infty} h_n^{(2)}(\rho) \vec{m}_{\sigma mn}(\theta, \phi)$$

$$= (i)^{n+1} \frac{e^{-i\rho}}{\rho} \vec{m}_{\sigma mn}(\theta, \phi) \tag{A2.13a}$$

$$\lim_{\rho \to \infty} \vec{N}_{\sigma mn}(\rho, \theta, \phi) = \lim_{\rho \to \infty} \frac{h_n^{(2)}(\rho)}{\rho} \vec{l}_{\sigma mn}(\theta, \phi) + \lim_{\rho \to \infty} \frac{[\rho h_n^{(2)}(\rho)]'}{\rho} \hat{r} \times \vec{m}_{\sigma mn}(\theta, \phi)$$

$$= i^n \frac{e^{-i\rho}}{\rho} \hat{r} \times \vec{m}_{\sigma mn}(\theta, \phi) \tag{A2.13b}$$

It follows that as $\rho \to \infty$, $\vec{N}_{\sigma mn} = -i(\hat{r} \times \vec{M}_{\sigma mn})$, which is similar to the relation between $\vec{E}^s$ and $\vec{H}^s$ in the far-field given in (1.27).

A firmer grasp of multipole radiation can be obtained by considering far-field radiation by an electric dipole $\vec{p}$ located at the origin for which the exact solution is given in (1.32) with $\rho = k_0 r$,

$$\vec{E}^s = \frac{k_0^2}{4\pi}(\omega Z_0) \left[\vec{p} - \hat{r}(\hat{r} \cdot \vec{p})\right] \frac{e^{-i\rho}}{\rho} \tag{A2.14a}$$

$$\vec{H}^s = \frac{1}{Z_0}(\hat{r} \times \vec{E}^s) = \frac{\omega k_0^2}{4\pi}(\hat{r} \times \vec{p}) \frac{e^{-i\rho}}{\rho} \tag{A2.14b}$$

When $\vec{p} = p_z \hat{z}$ it is easily verified that,

$$\vec{E}^s = \frac{k_0^2 \omega Z_0}{4\pi} \frac{e^{-i\rho}}{\rho} p_z \left(-\hat{\theta} \sin\theta\right) \tag{A2.15}$$

For $\vec{p} = p_x \hat{x}$ it is again verified that,

$$\vec{E}^s = \frac{k_0^2 \omega Z_0}{4\pi} \frac{e^{-i\rho}}{\rho} p_x \left(\hat{\theta} \cos\theta \cos\phi - \hat{\phi} \sin\phi\right) \tag{A2.16}$$

Finally, for $\vec{p} = p_y \hat{y}$,

$$\vec{E}^s = \frac{k_0^2 \omega Z_0}{4\pi} \frac{e^{-i\rho}}{\rho} p_y \left(\hat{\theta} \cos\theta \sin\phi + \hat{\phi} \cos\phi\right) \tag{A2.17}$$

By comparing the angular part in parentheses in (A2.15–A2.17) to the angular part of $\vec{m}_{\sigma mn}(\theta, \phi)$, and $\hat{r} \times \hat{m}_{\sigma mn}(\theta, \phi)$ and recognizing that $\cos\theta = P_1^0(\cos\theta)$ and $\sin\theta = P_1^1(\cos\theta) = -d[P_1^0(\cos\theta)]/d\theta$, it is easily verified that,

$$-\hat{\theta}\sin\theta = \sqrt{\frac{4\pi}{3}}\hat{r} \times \vec{m}_{e01}(\theta, \phi) \tag{A2.18}$$

Thus, the multipole expansion of $\vec{E}^s$ and $\vec{H}^s$ for a $\hat{z}$-directed electric dipole at the origin is,

$$\vec{E}^s = \sqrt{\frac{4\pi}{3}}\frac{k_0^2 \omega Z_0}{4\pi} p_z \frac{e^{-i\rho}}{\rho}\hat{r} \times \vec{m}_{e01}(\theta, \phi) \tag{A2.19a}$$

$$= K_1 \lim_{\rho\to\infty} \vec{N}_{e01}(\rho, \theta, \phi) \tag{A2.19b}$$

where K_1 is a constant. Similarly, for an $\hat{x}$-directed electric dipole,

$$\vec{E}^s = K_2 \lim_{\rho\to\infty} \vec{N}_{e11}(\rho, \theta, \phi) \tag{A2.20}$$

Finally, for a y-directed dipole,

$$\vec{E}^s = K_3 \lim_{\rho\to\infty} \vec{N}_{o11}(\rho, \theta, \phi) \tag{A2.21}$$

The approach here is intuitive and clearly shows that the vector wave functions with $n = 1; m = 0, 1$, i.e. $\vec{N}_{e01}, \vec{N}_{e11}, \vec{N}_{o11}$ are related to the $\vec{E}^s$ caused by electric dipole radiation. The functions $\vec{M}_{e01}, \vec{M}_{e11}, \vec{M}_{o11}$ are related to $\vec{E}^s$ due to magnetic dipole radiation. Higher order vector wave functions ($n = 2; m = 0, 1, 2$) are related to quadropole radiation. Thus, any general far-field $\vec{E}^s$ in free space with vector scattering amplitude $\vec{f}(\theta, \phi)$ can be expressed as a superposition of multipoles (with $\rho = k_0 r$),

$$\vec{E}^s = K\frac{e^{-i\rho}}{\rho}\vec{f}(\theta, \phi) = \sum_{\sigma mn}\left[f_{\sigma mn}(\lim_{\rho\to\infty}\vec{M}_{\sigma mn}) + g_{\sigma mn}(\lim_{\rho\to\infty}\vec{N}_{\sigma mn})\right] \tag{A2.22a}$$

$$= \frac{e^{-i\rho}}{\rho}\sum_{\sigma mn}f_{\sigma mn}\left[i^{(n+1)}\vec{m}_{\sigma mn}(\theta, \phi)\right] + g_{\sigma mn}\left[i^{(n+1)}\hat{r} \times \vec{m}_{\sigma mn}(\theta, \phi)\right] \tag{A2.22b}$$

The expansion coefficients $(f_{\sigma mn}, g_{\sigma mn})$ are obtained by using the orthogonality of $\vec{m}_{\sigma mn}(\theta, \phi)$ and $\hat{r} \times \vec{m}_{\sigma mn}(\theta, \phi)$ over a spherical surface, which are given as,

$$f_{\sigma mn} = \frac{1}{i^{n+1}n(n+1)}\int_0^\pi\int_0^{2\pi}K\vec{f}(\theta, \phi)\cdot\vec{m}_{\sigma mn}(\theta, \phi)\,d\Omega \tag{A2.23a}$$

$$g_{\sigma mn} = \frac{1}{i^n n(n+1)}\int_0^\pi\int_0^{2\pi}K\vec{f}(\theta, \phi)\cdot\hat{r} \times \vec{m}_{\sigma mn}(\theta, \phi)\,d\Omega \tag{A2.23b}$$

where $d\Omega = \sin\theta\,d\theta\,d\phi$, $\vec{m}_{\sigma mn}$ is given in (A2.10), and $\hat{r} \times \vec{m}_{\sigma mn}$ is also to be obtained from (A2.10). Once the expansion coefficients are known using the far-field vector amplitude, the electric field (in the "near-field" region) external to the smallest spherical surface encapsulating the radiation source is obtained as,

$$\vec{E}^s = \sum_{\sigma mn} \left[f_{\sigma mn} \vec{M}_{\sigma mn}(\rho, \theta, \phi) + g_{\sigma mn} \vec{N}_{\sigma mn}(\rho, \theta, \phi) \right] \tag{A2.24}$$

In reverse sense, measurement of the near-field $\vec{E}^s$ over a spherical surface surrounding the source can be used to establish the far-field amplitude, and is an established technique for antennas. In the case of scattering by a dielectric particle, the "source" for the scattered field $\vec{E}^s$ is the volume density of polarization $\vec{P}$ maintained within the particle. Thus, external to the smallest spherical surface encapsulating the particle, the scattered electric field can be expanded as in (A2.24). This is illustrated in Fig. A2.1a. Note that $z_n(\rho) = h_n^{(2)}(\rho)$ is the proper radial function to be used in (A2.24) since ρ is allowed to tend to ∞.

The general multipole expansion of the $\vec{E}$ field that satisfies the vector Helmholtz equation in (A2.8) must include both the spherical Bessel function as well as the second kind Hankel functions. The multipole expansion for $\vec{E}$ is written as,

$$\vec{E}(\rho, \theta, \phi) = \sum_{\sigma mn} \left(c_{\sigma mn} \operatorname{Rg} \vec{M}_{\sigma mn} + d_{\sigma mn} \operatorname{Rg} \vec{N}_{\sigma mn} + f_{\sigma mn} \vec{M}_{\sigma mn} + g_{\sigma mn} \vec{N}_{\sigma mn} \right) \tag{A2.25}$$

where $\rho = kr$ and k can be complex. In any domain including the origin, the coefficients $(f_{\sigma mn}, g_{\sigma mn})$ are zero, while in any domain not including the origin but extending to ∞, the coefficients $(c_{\sigma mn}, d_{\sigma mn})$ are zero. In any other bounded domain excluding the origin, all four coefficients are in general non-zero. For example, the total electric field inside the dielectric particle under plane wave excitation illustrated in Fig. A2.1b can be expanded as,

$$\vec{E}_T^{\text{in}}(\rho = kr, \theta, \phi) = \sum_{\sigma mn} \left(c_{\sigma mn} \operatorname{Rg} \vec{M}_{\sigma mn} + d_{\sigma mn} \operatorname{Rg} \vec{N}_{\sigma mn} \right) \tag{A2.26}$$

The general incident plane wave field, $\vec{E}^i = E_0 \hat{e}_i \exp(-jk_0 \hat{i} \cdot \vec{r}_i)$, is by definition the field that exists in the absence of the particle. The sources of the incident field are assumed to be at infinite distance from the particle. Thus, the general incident plane wave can be expanded in multipoles,

$$\vec{E}^i(\rho = k_0 r, \theta, \phi) = \sum_{\sigma mn} \left(a_{\sigma mn} \operatorname{Rg} \vec{M}_{\sigma mn} + b_{\sigma mn} \operatorname{Rg} \vec{N}_{\sigma mn} \right) \tag{A2.27}$$

where the expansion coefficients are provided in Barber and Hill (1990). A particularly simple expansion for a plane wave incident field of the form $\vec{E}^i = \hat{x} E_0 \exp(-ik_0 z)$ corresponding to vertical incidence ($\theta_i = \phi_i = 0$; $\hat{x} = \hat{v}_i$, see Fig. 2.4) is,

$$\hat{x} E_0 e^{-ik_0 z} = 2\pi E_0 \sum_{n=1}^{\infty} (-i)^n \gamma_{1n}^{1/2} \left[\operatorname{Rg} \vec{M}_{o1n}(k_0 r, \theta, \phi) + i \operatorname{Rg} \vec{N}_{e1n}(k_0 r, \theta, \phi) \right] \tag{A2.28a}$$

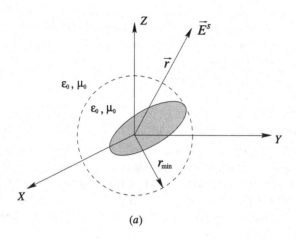

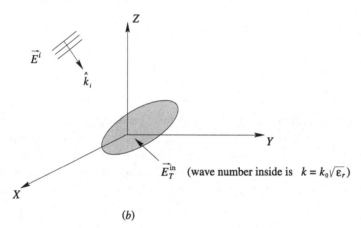

Fig. A2.1. (*a*) Illustrating region of validity ($r > r_{\min}$) for expansion of $\vec{E}^s$ as given in (A2.24), (*b*) total electric field inside the particle $\vec{E}_T^{\text{in}}$ is expanded as in (A2.26) while the incident field geometry is relevant to the T-matrix.

If $\vec{E}^i = \hat{y} E_0 e^{-ik_0 z}$ (with $\hat{y} = \hat{h}_i$),

$$\hat{y} E_0 e^{-ik_0 z} = 2\pi E_0 \sum_{n=1}^{\infty} (-i)^n \gamma_{1n}^{1/2} \left[-\operatorname{Rg} \vec{M}_{e1n}(k_0 r, \theta, \phi) + i \operatorname{Rg} \vec{N}_{o1n}(k_0 r, \theta, \phi) \right]$$

$$(\text{A2.28b})$$

Note that only the mode $m = 1$ is involved when propagation is along the Z-axis. The expansion coefficients for a general plane wave are more complex and involve modes from $m = 0, 1, \ldots, n$. They can be found, for example, in Barber and Hill (1990) or Tsang et al. (1985).

Appendix 3
T-matrix method

The transition (or T-) matrix method is a numerical scattering solution first developed by Waterman (1971), and is also referred to as the extended boundary condition method (EBCM). Barber and Yeh (1975) give a clear exposition of the method for dielectric particles, and the computational method for rotationally symmetric particles is fully described in the book by Barber and Hill (1990). Only a brief overview of the method is provided here since it has been used extensively in the past for radar meteorological applications (e.g. Seliga and Bringi 1978).

Consider a rotationally symmetric dielectric particle with its axis of symmetry oriented along the $+Z$-axis, and illuminated by an incident plane wave, $\vec{E}^i = \hat{x} E_0 e^{-ik_0 z}$, as illustrated in Fig. A3.1. The incident plane wave is expanded into multipoles as in (A2.28a),

$$\hat{x} E_0 e^{-ik_0 z} = \sum_{n=1}^{\infty} \left[a_{o1n} \operatorname{Rg} \vec{M}_{o1n}(k_0 r, \theta, \phi) + b_{e1n} \operatorname{Rg} \vec{N}_{e1n}(k_0 r, \theta, \phi) \right] \tag{A3.1}$$

where $a_{o1n} = (-i)^n \gamma_{1n}^{1/2} 2\pi E_0$ and $b_{e1n} = i a_{o1n}$. The total electric field inside the particle (interior to an inscribing sphere) is expanded into multipoles with $k = k_0 \sqrt{\varepsilon_r}$ as in (2.102),

$$\vec{E}_T^{in} = \sum_{n'=1}^{\infty} \left[c_{o1n'} \operatorname{Rg} \vec{M}_{o1n'}(kr, \theta, \phi) + d_{e1n'} \operatorname{Rg} \vec{N}_{e1n'}(kr, \theta, \phi) \right] \tag{A3.2}$$

Finally, the scattered electric field (exterior to the spherical surface circumscribing the particle) is expanded as,

$$\vec{E}^s = \sum_{n=1}^{\infty} \left[f_{o1n} \vec{M}_{o1n}(k_0 r, \theta, \phi) + g_{e1n} \vec{N}_{e1n}(k_0 r, \theta, \phi) \right] \tag{A3.3}$$

So far, the multipole expansions are exactly the same as for the Mie solution, except the regions of validity for $\vec{E}_T^{in}$ and $\vec{E}^s$ are as shown in Fig. A3.1. However, since the particle surface is not now a coordinate surface of the form $r = a$, the boundary conditions in (2.103a, b) cannot be applied to determine the unknown expansion coefficients. The so-called extended boundary condition is used to express the unknown expansion coefficients of the total electric field inside the particle (actually, interior to the inscribing sphere) to the known expansion coefficients of the incident field.

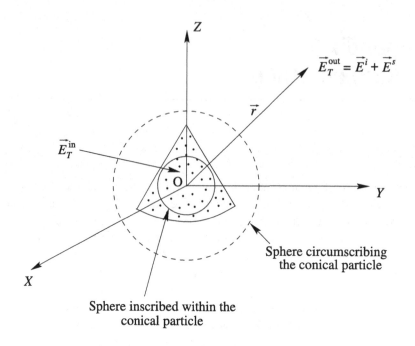

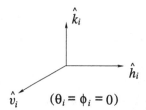

$(\theta_i = \phi_i = 0)$

Fig. A3.1. Geometry relevant to the T-matrix. The extended boundary condition is satisfied within the inscribed sphere. The scattered field expansion is valid exterior to the circumscribed sphere.

Following the notation of Barber and Hill (1990) and referring to Barber and Yeh (1975) for details,

$$
\begin{bmatrix} K_{o1n,o1n'} + \sqrt{\varepsilon_r}J_{o1n,o1n'} & L_{o1n,e1n'} + \sqrt{\varepsilon_r}I_{o1n,e1n'} \\ I_{e1n,o1n'} + \sqrt{\varepsilon_r}L_{e1n,o1n'} & J_{e1n,e1n'} + \sqrt{\varepsilon_r}K_{e1n,e1n'} \end{bmatrix} \begin{bmatrix} c_{o1n'} \\ d_{e1n'} \end{bmatrix}
$$
$$
= \begin{bmatrix} in(n+1)a_{o1n} \\ in(n+1)b_{e1n} \end{bmatrix} \tag{A3.4}
$$

where I, J, K, and L are two-dimensional integrals involving the vector spherical harmonics, which must be numerically calculated over the particle surface. For example, a general expression for K is,

$$
K_{\sigma mn, \sigma' m' n'} = k_0^2 \oint_s \hat{n} \cdot \left[\vec{N}_{\sigma mn}(k_0 r, \theta, \phi) \times \mathrm{Rg}\, \vec{M}_{\sigma' m' n'}(kr, \theta, \phi) \right] dS \tag{A3.5}
$$

where $\hat{n}$ is a unit normal to the particle surface, $r(\theta)$ describes the particle shape and is independent of ϕ for rotationally symmetric particles, and $k = k_0\sqrt{\varepsilon_r}$. Complete expressions for I, J, K, and L are given in Barber and Hill (1990). Note that in (A3.5) the choices $\sigma, \sigma' \equiv$ odd and $m, m' = 1$, lead to $K_{o1n,o1n'}$ in (A3.4).

Next, the scattered field expansion coefficients (exterior to the circumscribing sphere) are expressed in terms of the expansion coefficients of the total interior field as,

$$
\begin{bmatrix} -in(n+1)f_{o1n} \\ -in(n+1)g_{e1n} \end{bmatrix} = \begin{bmatrix} K'_{o1n,o1n'} + \sqrt{\varepsilon_r}J'_{o1n,o1n'} & L'_{o1n,e1n'} + \sqrt{\varepsilon_r}I'_{o1n,e1n'} \\ I'_{e1n,o1n'} + \sqrt{\varepsilon_r}L_{e1n,o1n'} & J'_{e1n,e1n'} + \sqrt{\varepsilon_r}K'_{e1n,e1n'} \end{bmatrix}
$$

$$
\times \begin{bmatrix} c_{o1n'} \\ d_{e1n'} \end{bmatrix} \tag{A3.6}
$$

where the $K'_{\sigma mn,\sigma'm'n'}$ matrix is defined as,

$$
K'_{\sigma mn,\sigma'm'n'} = k_0^2 \oint_S \hat{n} \cdot \left[\text{Rg}\,\vec{N}_{\sigma mn}(k_0 r, \theta, \phi) \times \text{Rg}\,\vec{M}_{\sigma'm'n'}(kr, \theta, \phi)\right] dS \tag{A3.7}
$$

which is obtained from $K_{\sigma mn,\sigma'm'n'}$ defined in (A3.5) by replacing $\vec{N}_{\sigma mn}$ by $\text{Rg}\,\vec{N}_{\sigma mn}$. Again, refer to Barber and Hill (1990) for full details on the form of the I', J', K', and L' terms. Defining the 2×2 matrix in (A3.4) as $[\mathbf{B}]$ and the 2×2 matrix in (A3.6) as $[\mathbf{A}]$, the scattered field expansion coefficients can be directly related to the incident field expansion coefficients as,

$$
\begin{bmatrix} f_{o1n} \\ g_{e1n} \end{bmatrix} = -[\mathbf{B}][\mathbf{A}]^{-1} \begin{bmatrix} (-i)^n \gamma_{1n}^{1/2} 2\pi E_0 \\ -(-i)^{n+1}\gamma_{1n}^{1/2} 2\pi E_0 \end{bmatrix} \tag{A3.8}
$$

The matrix $[\mathbf{T}] = [\mathbf{B}][\mathbf{A}]^{-1}$ is defined as the transition or T-matrix. In the form given in (A3.8), it is valid for an incident plane wave of the form $\vec{E}^i = \hat{x}E_0 \exp(-jk_0 z)$ with particle symmetry axis along OZ (see Fig. A3.1).

The far-field scattered wave can be expressed similar to (2.106),

$$
\vec{E}^s = \frac{e^{-ik_0 r}}{k_0 r} \sum_{n=1}^{\infty} \left[i^{n+1}f_{o1n}\vec{m}_{o1n}(\theta_s, \phi_s) + i^n g_{e1n}\hat{r} \times \vec{m}_{e1n}(\theta_s, \phi_s)\right] \tag{A3.9a}
$$

$$
= \frac{e^{-ik_0 r}}{r}\frac{1}{k_0} \sum_{n=1}^{\infty} \gamma_{1n}^{1/2} \left\{ \hat{\theta}_s \left[i^{n+1}f_{o1n}\frac{P_n^1(\cos\theta_s)}{\sin\theta_s} + i^n g_{e1n}\frac{dP_n^1(\cos\theta_s)}{d\theta_s}\right] \right.
$$

$$
\left. \times \cos m\phi_s - \hat{\phi}_s \left[i^{n+1}f_{o1n}\frac{dP_n^1(\cos\theta_s)}{d\theta_s} + i^n g_{e1n}\frac{P_n^1(\cos\theta_s)}{d\theta_s}\right]\sin m\phi_s \right\} \tag{A3.9b}
$$

Note the close correspondence between (A3.9b) and (2.106) for the Mie solution. In the case of the Mie solution, the scattered field expansion coefficients $(\alpha_{o1n}, \beta_{e1n})$ were given in closed form in (2.104a, b), while in the case of the T-matrix method, the corresponding coefficients (f_{o1n}, g_{e1n}) must be obtained numerically from (A3.8).

Note that n and n' both vary from 1 to n_{max}, where the first estimate of n_{max} is given by (2.127) with $k_0 a$ being replaced by $k_0 r_{eff}$, where r_{eff} is the volume-equivalent spherical radius of the rotationally symmetric particle (Barber and Hill 1990).

The method of handling the situation where the particle symmetry axis is not aligned along the OZ-axis is also given in Barber and Hill (1990) by using a "body" reference frame $X_b Y_b Z_b$ as was illustrated in Fig. 2.6b. Similar to the procedure for the oriented spheroid in the Rayleigh limit discussed in Section 2.3, the T-matrix is first computed in the "body" coordinate system. Since the incident wave is assumed to still propagate along the OZ-axis, as in Fig. A3.1, in the "body" reference frame the incident wave can no longer be expanded as in (A3.1) but the general expansion in (A2.27) must be used. The form of the T-matrix is still given by (A3.8), as it relates the scattered field expansion coefficients ($f_{\sigma mn}$, $g_{\sigma mn}$) to the incident plane wave expansion coefficients ($a_{\sigma mn}$, $b_{\sigma mn}$). This general form may be written as,

$$
\begin{bmatrix} f_{\sigma mn} \\ g_{\sigma mn} \end{bmatrix} = [\mathbf{T}] \begin{bmatrix} a_{\sigma mn} \\ b_{\sigma mn} \end{bmatrix}
\tag{A3.10}
$$

and the details are referred to in Barber and Hill (1990). Once the T-matrix is computed in the "body" reference frame, the scattered field in the "body" frame is re-expressed in the XYZ frame.

In radar meteorological applications (frequencies 3–10 GHz), the T-matrix numerical method has been primarily used to calculate scattering from oblate raindrops, oblate wet hail, and conical graupel. The convergence of the numerical scheme is poor for oblate axis ratios less than around 0.2, depending on the size parameter and relative permittivity. The T-matrix method has also been extended to two-layer particles and applied to model scattering by melting hail and freezing of raindrops at typical radar frequencies between 3 and 10 GHz. Scattering calculations for hydrometeors of specified axis ratio, canting angle distribution, and dielectric constant can be performed at the Website:

www.engr.colostate.edu/ece/radar_education

Appendix 4
Solution for the transmission matrix

The differential equation governing the coherent propagation of an electromagnetic wave along the positive Z-axis was given in (4.40), which is repeated here,

$$\frac{d}{dz} \begin{bmatrix} E_h(z) \\ E_v(z) \end{bmatrix} = \begin{bmatrix} -jk_0 + P_{hh} & P_{hv} \\ P_{vh} & -jk_0 + P_{vv} \end{bmatrix} \begin{bmatrix} E_h(z) \\ E_v(z) \end{bmatrix} \qquad (A4.1)$$

where **P** was defined in (4.39). Let

$$\vec{E} = \begin{bmatrix} E_h(z) \\ E_v(z) \end{bmatrix}$$

and **M** be defined such that (A4.1) can be written as,

$$\frac{d\vec{E}}{dz} = [\mathbf{M}]\vec{E} \qquad (A4.2a)$$

$$= \begin{bmatrix} M_{11} & M_{12} \\ M_{21} & M_{22} \end{bmatrix} \qquad (A4.2b)$$

A general solution to (A4.2a) can be written (e.g. Kreyszig 1988) as,

$$\vec{E} = \vec{X}e^{\lambda z} \qquad (A4.3)$$

where,

$$\vec{X} = \begin{bmatrix} X_h \\ X_v \end{bmatrix}$$

which when substituted in (A4.2a) gives,

$$\lambda \vec{X}e^{\lambda z} = [\mathbf{M}]\vec{X}e^{\lambda z} \qquad (A4.4a)$$

or,

$$\lambda \vec{X} = [\mathbf{M}]\vec{X} \qquad (A4.4b)$$

which is the equation for determining the eigenvalues of the matrix **M**. The corresponding solutions for $\vec{X}$ are called the eigenvectors of **M**. The eigenvalues are determined by,

$$\det(\mathbf{M} - \lambda \mathbf{I}) = 0 \qquad (A4.5)$$

where $\mathbf{I}$ is the 2×2 unit matrix. Thus,

$$\det[\mathbf{M} - \lambda\mathbf{I}] = \begin{vmatrix} M_{11} - \lambda & M_{12} \\ M_{21} & M_{22} - \lambda \end{vmatrix} = 0 \tag{A4.6a}$$

or,

$$(M_{11} - \lambda)(M_{22} - \lambda) - M_{12}M_{21} = 0 \tag{A4.6b}$$

The eigenvalues are solutions to the simple quadratic,

$$\lambda^2 - \lambda(M_{11} + M_{22}) + M_{11}M_{22} - M_{12}M_{21} = 0 \tag{A4.7a}$$

with,

$$\lambda_1 = \frac{(M_{11} + M_{22})}{2} + \sqrt{\frac{(M_{11} - M_{22})^2}{4} + M_{12}M_{21}} \tag{A4.7b}$$

$$\lambda_2 = \frac{(M_{11} + M_{22})}{2} - \sqrt{\frac{(M_{11} - M_{22})^2}{4} + M_{12}M_{21}} \tag{A4.7c}$$

Noting that,

$$\frac{M_{11} + M_{22}}{4} = -jk_0 + \frac{P_{hh} + P_{vv}}{2} \tag{A4.8a}$$

and,

$$\sqrt{\frac{(M_{11} - M_{22})^2}{4} + M_{12}M_{21}} = \frac{1}{2}\sqrt{(P_{hh} - P_{vv})^2 + 4P_{hv}P_{vh}} \tag{A4.8b}$$

$$= \frac{1}{2}\gamma \tag{A4.8c}$$

where γ was also defined in (4.41c). Thus, the eigenvalues can be expressed as,

$$\lambda_1 = -jk_0 + \frac{1}{2}(P_{hh} + P_{vv} + \gamma) \tag{A4.9a}$$

$$= -j\left[k_0 + \frac{j}{2}(P_{hh} + P_{vv} + \gamma)\right] \tag{A4.9b}$$

Similarly,

$$\lambda_2 = -j\left[k_0 + \frac{j}{2}(P_{hh} + P_{vv} - \gamma)\right] \tag{A4.10}$$

The above three equations are the same as (4.41). The two eigenvectors $\vec{X}_1$ and $\vec{X}_2$ are obtained by substituting $\lambda_{1,2}$ in (A4.4b),

$$\lambda_1 \begin{bmatrix} X_{h_1} \\ X_{v_1} \end{bmatrix} = \begin{bmatrix} M_{11} & M_{12} \\ M_{21} & M_{22} \end{bmatrix} \begin{bmatrix} X_{h_1} \\ X_{v_1} \end{bmatrix} \tag{A4.11a}$$

$$\lambda_2 \begin{bmatrix} X_{h_2} \\ X_{v_2} \end{bmatrix} = \begin{bmatrix} M_{11} & M_{12} \\ M_{21} & M_{22} \end{bmatrix} \begin{bmatrix} X_{h_2} \\ X_{v_2} \end{bmatrix} \tag{A4.11b}$$

From (A4.11a),

$$\lambda_1 X_{h_1} = M_{11} X_{h_1} + M_{12} X_{v_1} \tag{A4.12a}$$

$$\lambda_1 X_{v_1} = M_{21} X_{h_1} + M_{22} X_{v_1} \tag{A4.12b}$$

or,

$$\frac{X_{h_1}}{X_{v_1}} = \frac{M_{12}}{\lambda_1 - M_{11}} = \frac{\lambda_1 - M_{22}}{M_{21}} \tag{A4.12c}$$

Similarly, from (A4.11b),

$$\frac{X_{h_2}}{X_{v_2}} = \frac{M_{12}}{\lambda_2 - M_{11}} = \frac{\lambda_2 - M_{22}}{M_{21}} \tag{A4.13}$$

Choosing $X_{h_1} = 1$, $X_{v_1} = (\lambda_1 - M_{11})/M_{12}$; and $X_{h_2} = 1$, $X_{v_2} = (\lambda_2 - M_{11})/M_{12}$ the 2×2 eigenmatrix $[\mathbf{X}_m]$ can be formed:

$$[\mathbf{X}_m] = \begin{bmatrix} X_{h_1} & X_{h_2} \\ X_{v_1} & X_{v_2} \end{bmatrix} \tag{A4.14a}$$

$$= \begin{bmatrix} 1 & 1 \\ \dfrac{\lambda_1 - M_{11}}{M_{12}} & \dfrac{\lambda_2 - M_{11}}{M_{12}} \end{bmatrix} \tag{A4.14b}$$

Next, let $\vec{E} = [\mathbf{X}_m]\vec{E}'$, so that $\vec{E}'$ satisfies,

$$[\mathbf{X}_m]\frac{d\vec{E}'}{dz} = (M)[\mathbf{X}_m]\vec{E}' \tag{A4.15a}$$

or,

$$\frac{d\vec{E}'}{dz} = [\mathbf{X}_m]^{-1}[\mathbf{M}][\mathbf{X}_m]\vec{E}' \tag{A4.15b}$$

$$= \begin{bmatrix} \lambda_1 & 0 \\ 0 & \lambda_2 \end{bmatrix} \vec{E}' \tag{A4.15c}$$

In the above, the diagonalization theorem has been used (e.g. Kreyszig 1988). Thus, the solution for $\vec{E}'$ (compare with (4.6)) is simply,

$$\begin{bmatrix} E'_h(z) \\ E'_v(z) \end{bmatrix} = \begin{bmatrix} e^{\lambda_1 z} & 0 \\ 0 & e^{\lambda_2 z} \end{bmatrix} \begin{bmatrix} E'_h(0) \\ E'_v(0) \end{bmatrix} \tag{A4.16}$$

and from $\vec{E}' = [\mathbf{X}_m]^{-1}\vec{E}$, the solution for $\vec{E}$ is,

$$\begin{bmatrix} E_h(z) \\ E_v(z) \end{bmatrix} = [\mathbf{X}_m] \begin{bmatrix} e^{\lambda_1 z} & 0 \\ 0 & e^{\lambda_2 z} \end{bmatrix} [\mathbf{X}_m]^{-1} \begin{bmatrix} E_h(0) \\ E_v(0) \end{bmatrix} \tag{A4.17}$$

where $\vec{E}(0)$ is the electric field value at $z = 0$. Substituting for $[\mathbf{X}_m]$ from (A4.14b) into (A4.17) gives,

$$
\begin{bmatrix} E_h(z) \\ E_v(z) \end{bmatrix} = \begin{bmatrix} 1 & 1 \\ \dfrac{\lambda_1 - M_{11}}{M_{12}} & \dfrac{\lambda_2 - M_{11}}{M_{12}} \end{bmatrix} \begin{bmatrix} e^{\lambda_1 z} & 0 \\ 0 & e^{\lambda_2 z} \end{bmatrix}
$$

$$
\begin{bmatrix} 1 & 1 \\ \dfrac{\lambda_1 - M_{11}}{M_{12}} & \dfrac{\lambda_2 - M_{11}}{M_{12}} \end{bmatrix}^{-1} \begin{bmatrix} E_h(0) \\ E_v(0) \end{bmatrix}
$$

(A4.18a)

$$
= \begin{bmatrix} \dfrac{(\lambda_1 - M_{11})}{(\lambda_1 - \lambda_2)} e^{\lambda_2 z} - \dfrac{(\lambda_2 - M_{11})}{(\lambda_1 - \lambda_2)} e^{\lambda_1 z} & \dfrac{M_{12}}{(\lambda_1 - \lambda_2)}(e^{\lambda_1 z} - e^{\lambda_2 z}) \\ \dfrac{M_{12}}{(\lambda_1 - \lambda_2)}(e^{\lambda_1 z} - e^{\lambda_2 z}) & -\dfrac{(\lambda_2 - M_{11})}{(\lambda_1 - \lambda_2)} e^{\lambda_2 z} + \dfrac{(\lambda_1 - M_{11})}{(\lambda_1 - \lambda_2)} e^{\lambda_1 z} \end{bmatrix}
$$

$$
\begin{bmatrix} E_h(0) \\ E_v(0) \end{bmatrix}
$$

(A4.18b)

From (A4.7) it follows that,

$$
\lambda_1 - \lambda_2 = \sqrt{(M_{11} - M_{22})^2 + 4M_{12}^2}
$$

(A4.19a)

$$
= \sqrt{(P_{hh} - P_{vv})^2 + 4P_{hv}^2}
$$

(A4.19b)

$$
\frac{\lambda_1 - M_{11}}{\lambda_1 - \lambda_2} = -\frac{(M_{11} - M_{22})}{2(\lambda_1 - \lambda_2)} + \frac{1}{2}
$$

(A4.19c)

$$
-\frac{\lambda_2 - M_{11}}{\lambda_1 - \lambda_2} = \frac{(M_{11} - M_{22})}{2(\lambda_1 - \lambda_2)} + \frac{1}{2}
$$

(A4.19d)

Now define,

$$
\tan 2\phi = \frac{2M_{12}}{M_{11} - M_{22}}
$$

(A4.20a)

$$
= \frac{2P_{hv}}{P_{hh} - P_{vv}}
$$

(A4.20b)

It follows that,

$$
\sin 2\phi = \frac{2M_{12}}{(\lambda_1 - \lambda_2)} = \frac{2P_{hv}}{(\lambda_1 - \lambda_2)}
$$

(A4.20c)

$$
\cos 2\phi = \frac{M_{11} - M_{22}}{(\lambda_1 - \lambda_2)} = \frac{P_{hh} - P_{vv}}{(\lambda_1 - \lambda_2)}
$$

(A4.20d)

Substituting in (A4.18b) yields (compare with (4.42)),

$$
\begin{bmatrix} E_h(z) \\ E_v(z) \end{bmatrix} = \begin{bmatrix} \sin^2 \phi \, e^{\lambda_2 z} + \cos^2 \phi \, e^{\lambda_1 z} & \dfrac{\sin 2\phi}{2}(e^{\lambda_1 z} - e^{\lambda_2 z}) \\ \dfrac{\sin 2\phi}{2}(e^{\lambda_1 z} - e^{\lambda_2 z}) & \cos^2 \phi \, e^{\lambda_2 z} + \sin^2 \phi \, e^{\lambda_1 z} \end{bmatrix} \begin{bmatrix} E_h(0) \\ E_v(0) \end{bmatrix}
$$

(A4.21)

Appendix 5
Formulas for variance computation of autocorrelation functions, their magnitude, and phase, and for estimators in the periodic block pulsing scheme

The computation of mean power, velocity, differential reflectivity, and differential phase involves the autocorrelation and cross-correlation functions. This appendix provides variance computations of the relevant quantities.

A5.1 Variance of $\hat{R}_{xx}[l]$

$\hat{R}_{xx}[l]$ is the estimate of the autocorrelation function at lag "l" from the signal samples $x[1], x[2], \ldots, x[n]$. The sample autocorrelation function estimate of complex signals is given by (5.130a) as,

$$\hat{R}_{xx}[l] = \frac{1}{N} \sum_{n=1}^{N-l} x[n+l]x^*[n]; \quad 0 \le l < N \tag{A5.1}$$

where N is the length of the signal. The expected value of $\hat{R}_{xx}[l]$ can be written as,

$$E[\hat{R}_{xx}[l]] = \frac{N - |l|}{N} R_{xx}[l] \tag{A5.2}$$

The variance of $\hat{R}_{xx}[l]$ can be written as,

$$\text{var}[\hat{R}_{xx}[l]] = E\left[|\hat{R}_{xx}[l]|^2\right] - \left|E(\hat{R}_{xx}[l])\right|^2 \tag{A5.3}$$

The first term on the right-hand side of (A5.3) can be expressed for positive "l" as,

$$E\left(|\hat{R}_{xx}[l]|^2\right) = \frac{1}{N^2} \sum_{n_1=1}^{N-l} \sum_{n_2=1}^{N-l} E\left[x(n_1+l)x^*(n_1)x(n_2)x^*(n_2+l)\right] \tag{A5.4}$$

Equation (A5.4) involves computation of the fourth moment of a random process. For complex Gaussian processes the fourth-order moment can be reduced to products of second-order moments (Reed 1962). Specifically, if x_1, x_2, x_3, x_4 are complex Gaussian, then,

$$E\left(x_1 x_2^* x_3 x_4^*\right) = E\left(x_1 x_2^*\right) E\left(x_3 x_4^*\right) + E\left(x_1 x_4^*\right) E\left(x_3 x_2^*\right) \tag{A5.5}$$

Using (A5.5), (A5.4) reduces to,

$$E\left(|\hat{R}_{xx}[l]|^2\right) = \frac{1}{N^2} \sum_{n_1=0}^{N-l} \sum_{n_2=0}^{N-l} \left(R_{xx}[l]R_{xx}[-l] + R_{xx}[n_1 - n_2]R_{xx}[n_2 - n_1]\right) \tag{A5.6}$$

$$= \frac{1}{N^2} \sum_{n_1=1}^{N-|l|} \sum_{n_2=1}^{N-|l|} \left(|R_{xx}[l]|^2 + |R_{xx}[n_1 - n_2]|^2\right) \tag{A5.7}$$

$$= \left(\frac{N - |l|}{N}\right)^2 |R_{xx}[l]|^2 + \frac{1}{N^2} \sum_{n_1=1}^{N-|l|} \sum_{n_2=1}^{N-|l|} \left(|R_{xx}[n_1 - n_2]|^2\right) \tag{A5.8}$$

Substituting (A5.7) and (A5.2) into (A5.3), yields,

$$\text{var}(R_{xx}[l]) = \frac{1}{N^2} \sum_{n_1=1}^{N-l} \sum_{n_2=1}^{N-l} |R_{xx}[n_1 - n_2]|^2 \tag{A5.9}$$

This can be further simplified by letting $n = (n_1 - n_2)$ and collecting terms of the same lag (and extending to negative values of l),

$$\text{var}(R_{xx}[l]) = \left(\frac{1}{N}\right)^2 \sum_{n=-(N-|l|)}^{(N-|l|)} (N - |l| - |n|)|R_{xx}[n]|^2 \tag{A5.10}$$

$\hat{R}_{xx}[1]$ is extensively used in the context of mean velocity estimation. $\text{var}[\hat{R}_{xx}[1]]$ can be readily obtained from (A5.10) by substituting $l = 1$.

A5.2   Variance of $\hat{R}_{xy}[l]$

$\hat{R}_{xy}[l]$ represents the computational cross-correlation function of two generic signals $x[n]$, $y[n]$, $n = 1, \ldots, N$, defined as,

$$\hat{R}_{xy}[l] = \frac{1}{N} \sum_{n=1}^{N-l} x[n + l]y^*[n]; \quad 0 \leq l < N \tag{A5.11a}$$

$$\hat{R}_{xy}[l] = \frac{1}{N} \sum_{n=0}^{N-|l|} x[n]y^*[n + |l|]; \quad -N < l < 0 \tag{A5.11b}$$

The cross-correlation function is not symmetric, but satisfies the property,

$$R_{xy}[l] = R_{yx}[-l] \tag{A5.12}$$

The variance of $R_{xy}[l]$ can be computed following steps very similar to (A5.3)–(A5.10) and is given as,

$$\text{var}(R_{xy}[l]) = \frac{1}{N^2} \sum_{n=-(N-|l|)}^{(N-|l|)} (N - |l| - |n|) R_{xx}(n) R_{yy}^*(n) \tag{A5.13}$$

$R_{xy}(0)$ is used in the estimation of ρ_{co} in the hybrid mode. It is also used in the estimation of ρ_{cx}^h in the alternating as well as in the periodic block pulsing scheme. For this specific case $\text{var}[R_{xy}(0)]$ can be readily obtained from (A5.13) as,

$$\text{var}[R_{xy}(0)] = \frac{1}{N^2} \sum_{n=-N}^{N} (N - |n|) R_{xx}(n) R_{yy}^*(n) \tag{A5.14}$$

A5.3   **Variance of** $\arg \hat{R}[l]$

The argument of $\hat{R}[l]$ is used in (5.200) and (5.208) for the estimation of mean velocity. The most commonly used estimate of mean velocity is computed from $\arg[\hat{R}[1]]$. The variance of the argument of $\hat{R}[l]$ can be computed from perturbation analysis in (5.197) as follows.

Let $\hat{\theta}$ be the argument of $\hat{R}[l]$, then $\tan \theta = y/x$ where x and y are the real and imaginary parts of $\hat{R}[l]$. Using two-dimensional Taylor series expansion of $\arg \hat{R}[l]$, the variance of $\hat{\theta}$ can be expressed as,

$$\text{var}[\hat{\theta}] \approx \text{var}[x] \left[\frac{\partial \arg[\hat{R}[l]]}{\partial x} \bigg|_{\bar{x},\bar{y}} \right]^2 + \text{var}[y] \left[\frac{\partial \arg[\hat{R}[l]]}{\partial y} \bigg|_{\bar{x},\bar{y}} \right]^2$$

$$+2\,\text{cov}[x, y] \left[\frac{\partial \arg[\hat{R}[l]]}{\partial x} \bigg|_{\bar{x},\bar{y}} , \frac{\partial \arg[\hat{R}[l]]}{\partial y} \bigg|_{\bar{x},\bar{y}} \right] \tag{A5.15}$$

(A5.15) can be simplified taking derivatives as,

$$\text{var}[\hat{\theta}] \approx E \left[\frac{1}{|R[l]|^4} \{\bar{x}^2 \, \text{var}(y) + \bar{y}^2 \, \text{var}(x) - 2\bar{x}\bar{y} \, \text{cov}(x, y)\} \right] \tag{A5.16}$$

where $\bar{x}, \bar{y}$ are the means of x and y, respectively. If x and y can be expressed in terms of their perturbation as,

$$x = \bar{x} + \delta x, \quad y = \bar{y} + \delta y \tag{A5.17}$$

then (A5.16) can be written as,

$$\text{var}[\hat{\theta}] \simeq E \left[\frac{1}{|R[l]|^4} (\bar{x}^2 \delta_y^2 - 2\bar{x}\bar{y}\delta_x\delta_y + \bar{y}^2 \delta_x^2) \right] \tag{A5.18a}$$

$$= E \left\{ \frac{1}{|R[l]|^4} \text{Im}^2(R^*[l])\delta \right\} \tag{A5.18b}$$

where $\delta = \delta_x + j\delta_y$. Substituting $\hat{R}[l] - R[l]$ for δ, (A5.18b) reduces to,

$$\text{var}[\hat{\theta}] \approx E\left\{\text{Im}^2\left[\frac{R^*[l]\hat{R}[l]}{R^*[l]R[l]}\right]\right\} \tag{A5.19}$$

$$= E\left\{\text{Im}^2\left(\frac{\hat{R}[l]}{R[l]}\right)\right\} \tag{A5.20}$$

$$= \frac{1}{2}E\left\{\text{Re}\left[\left|\frac{\hat{R}[l]}{R[l]}\right|^2 - \left(\frac{\hat{R}[l]}{R[l]}\right)^2\right]\right\} \tag{A5.21}$$

(A5.21) forms the basic equation for computing the variance of $\arg[\hat{R}[l]]$ (see also Miller and Rochwarger 1972).

It was shown in (A5.6) (and from (A5.7)) that,

$$E[|\hat{R}[l]|^2] = \left[\frac{1}{N^2}\sum_{n=-(N-|l|)}^{(N-|l|)}(N - |l| - |n|)|R[l]|^2\right] + |E(R[n])|^2 \tag{A5.22}$$

Using the moment theorem of (A5.5) and principles similar to the derivation of $E[|\hat{R}[l]|^2]$, it follows that,

$$E[(R[l])^2] = \frac{1}{N^2}\sum_{n=-(N-|l|)}^{(N-|l|)}(N - |l| - |n|)R[n + l]R^*[n - l] + [E(R[l])]^2 \tag{A5.23}$$

Substituting (A5.23) and (A5.22) into (A5.21), $\text{var}[\hat{\theta}]$ can be expressed as,

$$\text{var}[\hat{\theta}] \simeq \frac{1}{2}\left\{\sum_{n=-(N-|l|)}^{(N-|l|)}(N - |l| - |n|)\left[\left|\frac{R[n]}{R[l]}\right|^2 - \left(\frac{R[n+l]R^*[n-l]}{R^2[l]}\right)\right]\right\} \tag{A5.24}$$

If the spectrum of the signal is of Gaussian shape then the autocorrelation function at lag n is given by (5.194). Substituting for $R[n]$ in (A5.24) yields,

$$\text{var}[\hat{\theta}] \simeq \frac{1}{2N^2}\frac{1 - |\rho[l]|^2}{|\rho[l]|^2}\sum_{n=-(N-|l|)}^{(N-|l|)}(N - |l| - |n|)|\rho[n]|^2 \tag{A5.25}$$

where $|\rho[n]|$ is the magnitude of the correlation coefficient function at lag n.

A5.4   Variance of magnitude of $\rho_{xy}[l]$

The computation of the variance of $|\rho[l]|$ is very useful in the variance computation of $|\rho_{co}|$ and $|\rho_{cx}^h|$ in all modes of operation.

The estimation $|\hat{\rho}_{xy}[l]|$ can be expressed as,

$$|\hat{\rho}_{xy}[l]| = \frac{\hat{R}_{xy}[l]}{[\hat{R}_{xx}(0)\hat{R}_{yy}(0)]^{1/2}} \tag{A5.26}$$

Using perturbation techniques similar to that applied in (A5.15)–(E.18) the terms in the numerator and denominator on the right-hand side of (A5.26) can be expressed (see also Liu et al. 1994) as,

$$\hat{R}_{xy}[l] = |R_{xy}[l] + \delta| \simeq |R_{xy}[l]| \left\{ 1 + \mathrm{Re}\left[\frac{\delta}{R_{xy}(l)} \right] \right\} \tag{A5.27a}$$

$$\frac{1}{[\hat{R}_{xx}[0]]^{1/2}} = \frac{1}{[R_{xx}[0] + \delta_x]^{1/2}} = \frac{1}{[R_{xx}[0]]^{1/2}} \left[1 - \frac{1}{2}\frac{\delta_x}{R_{xx}[0]} \right] \tag{A5.27b}$$

$$\frac{1}{[\hat{R}_{yy}[0]]^{1/2}} = \frac{1}{[R_{yy}[0] + \delta_y]^{1/2}} = \frac{1}{[R_{yy}[0]]^{1/2}} \left[1 - \frac{1}{2}\frac{\delta_y}{R_{yy}[0]} \right] \tag{A5.27c}$$

where δ, δ_x, δ_y are the perturbations of $R_{xy}[l]$, $R_{xx}[0]$ and $R_{yy}[0]$, respectively. Note that δ is complex, whereas δ_x and δ_y are real. Similarly, the perturbation of $|\rho_{xy}[l]|$ can be expressed as,

$$|\hat{\rho}_{xy}[l]| = |\rho_{xy}[l]| + \delta_\rho \tag{A5.28}$$

Substituting (A5.27a–c) in (A5.26) yields,

$$\delta_\rho = |\rho_{xy}[l]| \left[\mathrm{Re}\left(\frac{\delta}{R_{xy}[l]} \right) - \frac{1}{2}\left(\frac{\delta_x}{R_{xx}[0]} + \frac{\delta_y}{R_{yy}[0]} \right) \right] \tag{A5.29}$$

Therefore,

$$\frac{\mathrm{var}[|\rho_{xy}[l]|]}{|\rho_{xy}[l]|^2} = E\left[\mathrm{Re}\left(\frac{\delta}{R_{xx}[l]} \right) - \frac{1}{2}\left(\frac{\delta_x}{R_{xx}[0]} + \frac{\delta_y}{R_{yy}[0]} \right) \right]^2 \tag{A5.30}$$

$$= E\left[\mathrm{Re}\left(\frac{\delta}{R_{xy}[l]} \right) \right]^2 - E\left[\frac{\delta_x}{R_{xx}[0]}\,\mathrm{Re}\left(\frac{\delta}{R_{xy}[l]} \right) \right]$$

$$- E\left[\frac{\delta_y}{R_{yy}[0]}\,\mathrm{Re}\left(\frac{\delta}{R_{xy}[l]} \right) \right]$$

$$+ \frac{1}{4}\left[\frac{\mathrm{var}[R_{xx}[0]]}{(R_{xx}[0])^2} + \frac{\mathrm{var}[R_{yy}[0]]}{(R_{yy}[0])^2} + \frac{2\,\mathrm{cov}[R_{xx}[0], R_{yy}[0]]}{R_{xx}[0]R_{yy}[0]} \right] \tag{A5.31}$$

Further simplification of the terms in (A5.31) yields,

$$E\left[\mathrm{Re}\left(\frac{\delta}{R_{xy}[l]} \right) \right]^2 = \frac{1}{2}\,\mathrm{Re}\left[E\left(\frac{\delta}{R_{xy}[l]} \right)^2 + E\left| \frac{\delta}{R_{xy}[l]} \right|^2 \right] \tag{A5.32}$$

$$E\left[\frac{\delta_x}{R_{xx}[0]}\,\mathrm{Re}\left(\frac{\delta}{R_{xy}[l]} \right) \right] = \mathrm{Re}\left[\mathrm{cov}\left(\frac{\hat{R}_{xx}[0]}{R_{xx}[0]}, \frac{\hat{R}_{xy}[l]}{R_{xy}[l]} \right) \right] \tag{A5.33a}$$

$$E\left[\frac{\delta_y}{R_{yy}[0]}\,\mathrm{Re}\left(\frac{\delta}{R_{xy}[l]} \right) \right] = \mathrm{Re}\left[\mathrm{cov}\left(\frac{\hat{R}_{yy}[0]}{R_{yy}[0]}, \frac{\hat{R}_{xy}[l]}{R_{xy}[l]} \right) \right] \tag{A5.33b}$$

Thus, it can be seen that all the terms of (A5.31) have been reduced to terms involving variance and covariance of $R_{xy}[l]$, $R_{xx}[l]$ and $R_{yy}[l]$. $E[(R_{xy}[l])^2]$, and $E[|R_{xy}[l]|^2]$

have been derived in Sections A5.1–3. Further simplification of (A5.32) and (A5.33a, b) yields,

$$
E\left[\mathrm{Re}\left(\frac{\delta}{R_{xy}[l]}\right)^2\right] = \frac{1}{2N^2}\sum_{n=-(N-|l|)}^{(N-|l|)}(N-|l|-|n|)\frac{R_{xx}[n]R_{yy}^*[n]}{|R_{xy}[l]|^2}
$$

$$
+ \mathrm{Re}\left\{\frac{1}{2N^2}\sum_{n=-(N-|l|)}^{(N-|l|)}(N-|l|-|n|)\frac{R_{xx}[n+l]R_{yy}^*[n-l]}{(R_{xy}[l])^2}\right\}
$$

$$
\tag{A5.34}
$$

$$
\mathrm{cov}(\hat{R}_{xx}[0],\hat{R}_{xy}[l]) = \frac{1}{N^2}\sum(N-|l|-|n|)R_{xy}[n+l]R_{xx}^*[n-l] \tag{A5.35a}
$$

$$
\mathrm{cov}(\hat{R}_{yy}[0],\hat{R}_{xy}[l]) = \frac{1}{N^2}\sum(N-|l|-|n|)R_{xy}^*[n+l]R_{yy}^*[n-l] \tag{A5.35b}
$$

The equations in Section A5.4 provide the building blocks for deriving the variance of any correlation coefficient function. For example, if x and y were two copolar signals such as in the hybrid mode, and if $\rho_{xy}[l]$ can be written as $\rho_{xy}[l] = \rho_{xy}[0]\rho_{xx}[l]$, and if x and y have identical autocorrelation functions then $\mathrm{var}[|\rho_{xy}[0]|]$ can be computed from (A5.31) as,

$$
\frac{\mathrm{var}[|\rho_{xy}[0]|]}{|\rho_{xy}[0]|^2} = \frac{(1-|\rho_{co}^{hv}|^2)^2}{2N^2|\rho_{co}^{hv}|^2}\sum_{n=-N}^{N}(N-|n|)|\rho_{xx}(n)|^2 \tag{A5.36}
$$

A5.5   Variance of estimates under periodic block pulsing scheme

In the periodic block pulsing mode the estimates of the various auto and cross-covariances are obtained from regularly spaced signals multiplied by their corresponding indicator functions. The estimators described in Section 6.4.4 can be written in vector notation as,

$$
\hat{P}_{co}^h = \frac{\mathbf{p}_h^t\mathbf{P}_{hh}}{\displaystyle\sum_{n=1}^{N}p_h[n]} \tag{A5.37}
$$

where $\mathbf{p}_h^t$ is the transpose of the vector of indicator function elements, and $\mathbf{P}_{hh}$ is the vector of power samples, $|V_{hh}[n]|^2$, $n = 1,\ldots,N$. Similarly $\hat{P}_{co}^v$, $\hat{P}_{cx}$, $\hat{R}_{cx}^h$, and $\hat{R}_{cx}^v$ can be written as,

$$
\hat{P}_{co}^v = \frac{\mathbf{p}_v^t\mathbf{P}_{vv}}{\displaystyle\sum_{n=1}^{N}p_v[n]} \tag{A5.38}
$$

$$\hat{P}_{cx} = \frac{\mathbf{p}_h^t \mathbf{P}_{vh}}{\sum\limits_{n=1}^{N} p_h[n]} \tag{A5.39}$$

$$\hat{R}_{cx}^h = \frac{\mathbf{p}_h^t \mathbf{K}_{hh,vh}}{\sum\limits_{n=1}^{N} p_h[n]} \tag{A5.40}$$

$$\hat{R}_{cx}^v = \frac{\mathbf{p}_v^t \mathbf{K}_{vv,hv}}{\sum\limits_{n=1}^{N} p_v[n]} \tag{A5.41}$$

where $\mathbf{p}_v^t$ is the transpose of the vector of indicator function elements; $\mathbf{P}_{vv}, \mathbf{P}_{vh}$ are the vectors of power samples, $|V_{vv}[n]|^2, |V_{vh}[n]|^2$, respectively; $\mathbf{K}_{hh,vh}$ and $\mathbf{K}_{vv,hv}$ are the vectors of complex products $V_{hh}[n](V_{vh}[n])^*$ and $V_{vv}[n](V_{hv}[n])^*$, respectively.

(i) Variance of $\hat{P}_{co}^h$, $\hat{P}_{co}^v$, and Z_{dr}

The variance of mean power estimates defined above can be written as,

$$\text{var}[\hat{P}_{co}^h] = (C_h)^2 \mathbf{p}_h^t \boldsymbol{\gamma}_{hh} \mathbf{p}_h \tag{A5.42}$$

where C_h is the normalizing constant $1/(\sum_{i=1}^{N} p_n[n])$, and $\boldsymbol{\gamma}_{hh}$ is the autocorrelation matrix of the power samples defined as,

$$\boldsymbol{\gamma}_{hh} = E[\mathbf{P}_{hh}\mathbf{P}_{hh}^t] = (\mathbf{P}_{co}^h)^2 \begin{bmatrix} \rho_p(0) & \rho_p(-1) & \cdots & \rho_p(-N+1) \\ \rho_p(1) & \rho_p(0) & \cdots & \rho_p(-N+2) \\ \cdots & \cdots & \cdots & \cdots \\ \rho_p(N-1) & \rho_p(N-2) & \cdots & \rho_p(0) \end{bmatrix} \tag{A5.43}$$

where ρ_p is defined in (5.190). Similarly,

$$\text{var}[\hat{P}_{co}^v] = (C_v)^2 \mathbf{p}_v^t \boldsymbol{\gamma}_{vv} \mathbf{p}_v \tag{A5.44}$$

where C_v is the normalizing constant $(1/\sum_{n=1}^{N} p_v[n])$ and $\boldsymbol{\gamma}_{vv} = E[\mathbf{P}_{vv}\mathbf{P}_{vv}^t]$. Also,

$$\text{var}[\hat{P}_{cx}] = (C_h)^2 \mathbf{p}_h^t \boldsymbol{\gamma}_{vh} \mathbf{P}_h \tag{A5.45}$$

where $\boldsymbol{\gamma}_{vh} = E[\mathbf{P}_{vh}\mathbf{P}_{vh}^t]$. The variance of Z_{dr} can be obtained from (6.116),

$$\text{var}(\hat{z}_{dr}) = 2\left(\frac{P_{co}^h}{P_{co}^v}\right)^2 \left[\frac{\text{var}[\hat{P}_{co}^h]}{(P_{co}^h)^2} + \frac{\text{var}[\hat{P}_{co}^v]}{(P_{co}^v)^2} - \frac{2\,\text{cov}(\hat{P}_{co}^h, \hat{P}_{co}^v)}{P_{co}^h P_{co}^v}\right] \tag{A5.46}$$

In the above, $\text{cov}(\hat{P}_{co}^h, \hat{P}_{co}^v)$ is the only new term to be evaluated which is obtained as,

$$\text{cov}(\hat{P}_{co}^h, \hat{P}_{co}^v) = (C_h C_v)\mathbf{p}_h^t \boldsymbol{\gamma}_{hh,vv} \mathbf{p}_v^t \tag{A5.47}$$

where $\boldsymbol{\gamma}_{hh,vv}$ is the correlation matrix defined as $E[\mathbf{P}_{hh}\mathbf{P}_{vv}^t]$. Substituting (A5.42), (A5.44), and (A5.47) in (A5.46), the variance of $\hat{z}_{dr}$ can be calculated. Variance of Z_{dr} (dB) can be obtained using the formula given in (6.115).

(ii) Variance of $\arg(\hat{R}_{cx}^h)$ and $|\hat{\rho}_{cx}^h|$

The variance of $|\hat{\rho}_{cx}^h|$ and $\arg(\hat{R}_{cx}^h)$ can be readily calculated from (6.134) through (6.138) by substituting for $\hat{P}_{co}^h$, $\hat{P}_{cx}$, and $\hat{R}_{cx}^h$ in terms of the indicator functions. The behavior of the variance depends on the exact form of the indicator function.

(iii) Variance of Ψ_{dp} and $|\rho_{co}|$

The estimator of Ψ_{dp} and $|\rho_{co}|$ in periodic block pulsing mode is similar to the alternating mode, but slightly simpler. The variance of $\hat{\Psi}_{dp}$ is obtained from (6.104) which can be written equivalently as,

$$\Psi_{dp} = \arg\left[\hat{R}_{co}^{vh}[1]\right] - \arg\left[\hat{R}[1]\right] \tag{A5.48a}$$

Using the variance expansion similar to that given in (6.123), the variance of $\hat{\Psi}_{dp}$ can be obtained as,

$$\mathrm{var}(\hat{\Psi}_{dp}) = \mathrm{var}\,(\Psi_2) + \mathrm{var}\left[\arg\hat{\rho}[1]\right] - 2\,\mathrm{cov}\left[\Psi_2, \arg\hat{\rho}[1]\right] \tag{A5.48b}$$

where Ψ_2 is defined in (6.83b). The variance and covariance depend on the pulsing mode and can be computed using expressions equivalent to (6.124) and (6.125). The expression for $\hat{\rho}_{co}$ in periodic block pulsing mode is given in (6.103). Similar to the variance of $|\hat{\rho}_{co}|$ given in (6.127), the expression for variance of $|\hat{\rho}_{co}|$ for the periodic block pulsing mode becomes,

$$\frac{\mathrm{var}[|\hat{\rho}_{co}|]}{|\rho_{co}|^2} = \frac{\mathrm{var}[|\hat{\rho}_{hh,vv}[1]|]^2}{|\rho_{hh,vv}[1]|^2} + \frac{\mathrm{var}[|\hat{\rho}[1]|]}{|\rho[1]|^2} \\ - \frac{2\,\mathrm{cov}[|\hat{\rho}_{hh,vv}[1]|, |\hat{\rho}[1]|]}{|\rho_{hh,vv}[1]|\cdot|\rho[1]|} \tag{A5.49}$$

Each term on the right-hand side of the above equation can be expanded similar to (6.128a–c). The exact variance value depends on the indicator function.

References

Abou-El-Magd, A.M., Chandrasekar, V., Bringi, V.N., and Strapp, W. Multiparameter radar and in situ aircraft observation of graupel and hail. *IEEE Trans. Geosci. Remote Sens.*, **38**(1): 570–578, 2000.

Agrawal, A.P. and Boerner, W.M. Redevelopment of Kennaugh's target characteristic polarization state theory using the polarization transformation ratio formalism for the coherent case. *IEEE Trans. Geosci. Remote Sens.*, **27**: 2–14, 1989.

Al-Jumily, K.J., Charlton, R.B., and Humphries, R.G. Identification of rain and hail with circular polarization radar. *J. Appl. Meteor.*, **30**: 1075–1087, 1991.

Anagnostou, E., Krajewski, W.F., and Smith, J. Uncertainty quantification of mean areal radar rainfall estimates. *J. Atmos. Oceanic Technol.*, **16**: 206–215, 1999.

Andsager, K., Beard, K.V., and Laird, N.F. Laboratory measurements of axis ratios for large raindrops. *J. Atmos. Sci.*, **56**: 2673–2683, 1999.

Antar, Y.M.M. and Hendry, A. Correlation techniques in two-channel linearly polarized radar systems. *Electromagnetics*, **7**: 17–27, 1987.

Atlas, D. Advances in radar meteorology. In *Advances in Geophysics*. Landsberg and Mieghem, Eds, pp. 317–478. New York, Academic Press, 1964.

Atlas, D. Ed., *Radar in Meteorology*. Boston, MA, American Meteorological Society, 1990.

Atlas, D. and Ulbrich, C.W. Path- and area-integrated rainfall measurement by microwave attenuation in 1–3 cm band. *J. Appl. Meteor.*, **16**: 1322–1331, 1977.

Atlas, D. and Ulbrich, C.W. Early foundations of the measurement of rainfall by radar. In *Radar in Meteorology*. D. Atlas, Ed., American Meteorological Society, Boston, pp. 86–97, 1990.

Atlas, D., Kerker, M., and Hitschfeld, W. Scattering and attenuation by non-spherical atmospheric particles. *J. Atmos. Terrestrial Phys.*, **3**: 108–119, 1953.

Atlas, D., Rosenfeld, D., and Short, D.A. The estimation of convective rainfall by area integral: 1. The theoretical and empirical basis. *J. Geophy. Res.*, **95**: 2153–2160, 1990.

Atlas, D., Srivastava, R.C., and Sekhon, R.S. Doppler radar characteristics of precipitation at vertical incidence. *Rev. Geophys. Space Phys.*, **2**: 1–35, 1973.

Atlas, D., Ulbrich, C.W., Marks, F.D., Amitai, E., and Williams, C.R. Systematic variation of drop size and radar-rainfall relations. *J. Geophys. Res.*, **104**: 6155–6169, 1999.

Aunon, J.I. and Chandrasekar, V. *Introduction to Probability and Random Processes*. New York, McGraw-Hill, 1997.

Austin, P.M. Relation between measured radar reflectivity and surface rainfall. *Monthly Weather Rev.*, **115**: 1053–1070, 1987.

Aydin, K. and Tang, C. Relationship between IWC and polarimetric radar measurands at 94 and 220 GHz for hexagonal columns and plates. *J. Atmos. Oceanic Technol.*, **14**: 1055–1063, 1997.

Aydin, K. and Walsh, T.M. Separation of millimeter-wave radar reflectivities of aggregates and pristine ice crystals in a cloud. In *Proceedings International Geoscience and Remote Sensing Symposium (IGARSS)*, July, 6–10, Seattle, WA, pp. 440–443, Piscataway, NJ, IEEE, 1998.

Aydin, K., Bringi, V.N., and Liu, L. Rainrate estimation in the presence of hail using S-band specific differential phase and other radar parameters. *J. Appl. Meteor.*, **34**: 404–410, 1995.

Aydin, K., Direskeneli, H., and Seliga, T.A. Dual-polarization radar estimation of rainfall parameters compared with ground-based disdrometer measurements: October 29, 1982, Central Illinois experiment. *IEEE Trans. Geosci. Remote Sens.*, **GE-25**: 834–844, 1987.

Aydin, K., Seliga, T.A., and Balaji, V. Remote sensing of hail with a dual-linear polarization radar. *J. Climate Appl. Meteor.*, **25**: 1475–1484, 1986.

Aydin, K., Seliga, T.A., Cato, C.P., and Arai, M. Comparison of measured X-band reflectivity factors with those derived from S-band measurements at horizontal and vertical polarizations. In *Preprints, 21st Conference on Radar Meteorology*, September, 19–23, Edmonton, pp. 513–517. Boston, MA, American Meteorological Society, 1983.

Aydin, K., Zhao, Y., and Seliga, T.A. Rain-induced attenuation effects on C-band dual-polarization meteorological radars. *IEEE Trans. Geosci. Remote Sens.*, **27**: 57–66, 1989.

Aydin, K., Zhao, Y., and Seliga, T.A. A differential reflectivity radar hail measurement technique: Observations during the Denver hailstorm of 13 June 1984. *J. Atmos. Oceanic Technol.*, **7**: 104–113, 1990.

Azzam, R.M.A. and Bashara, N.M. *Ellipsometry and Polarized Light.* New York, North-Holland, 1989.

Bader, M.J., Clough, S.A., and Cox, G.P. Aircraft and dual polarization radar observations of hydrometeors in light stratiform precipitation. *Quart. J. Roy. Meteor. Soc.*, **103**: 269–280, 1987.

Balakrishnan, N. and Zrnić, D.S. Estimation of rain and hail rates in mixed-phase precipitation. *J. Atmos. Sci.*, **47**: 565–583, 1990a.

Balakrishnan, N. and Zrnić, D.S. Use of polarization to characterize precipitation and discriminate large hail. *J. Atmos. Sci.*, **47**: 1525–1540, 1990b.

Barber, P. and Yeh, C. Scattering of electromagnetic waves by arbitrarily shaped dielectric bodies. *Appl. Optics*, **14**: 2684–2872, 1975.

Barber, P.W. and Hill, S.C. *Light Scattering by Particles: Computational Methods.* New Jersey, World Scientific, 1990.

Barge, B.L. Polarization measurements of precipitation backscatter in Alberta. *J. Rech. Atmos.*, **8**: 163–173, 1974.

Battan, L.J. *Radar Observations of the Atmosphere.* Illinois, The University of Chicago Press, 1973.

Battan, L.J. and Theiss, J.B. Depolarization of microwaves by hydrometeors in a thunderstorm. *J. Atmos. Sci.*, **27**: 974–977, 1970.

Baumgardner, D. and Colpitt, A. Monster drops and rain gushes: Unusual precipitation phenomena in Florida marine cumulus. In *Preprints, Conference on Cloud Physics*, January, 15–20, Dallas, TX, pp. 344–349. Boston, MA, American Meteorological Society, 1995.

Beard, K.V. Oscillation modes for predicting raindrop axis and backscatter ratios. *Radio Sci.*, **19**: 67–74, 1984.

Beard, K.V. Simple altitude adjustments to raindrop velocities for Doppler radar analysis. *J. Atmos. Oceanic Technol.*, **2**: 468–471, 1985.

Beard, K.V. and Chuang, C. A new model for the equilibrium shape of raindrops. *J. Atmos. Sci.*, **44**: 1509–1524, 1987.

Beard, K.V. and Jameson, A.R. Raindrop canting. *J. Atmos. Sci.*, **40**: 448–454, 1983.

Beard, K.V. and Johnson, D.B. Raindrop axial and backscatter ratios using a collisional probability model. *Geophys. Res. Lett.*, **11**: 65–68, 1984.

Beard, K.V. and Kubesh, R.J. Laboratory measurements of small raindrop distortion. Part 2: Oscillation frequencies and modes. *J. Atmos. Sci.*, **48**: 2245–2264, 1991.

Beard, K.V., Kubesh, R.J., and Ochs, H.T. Laboratory measurements of small raindrop distortion. Part I: Axis ratios and fall behavior. *J. Atmos. Sci.*, **48**: 698–710, 1991.

Beard, K.V., Johnson, D.B., and Baumgardner, D. Aircraft observations of large raindrops in warm shallow convective clouds. *Geophys. Res. Lett.*, **13**: 991–994, 1986.

Beard, K.V., Ochs, H.T., and Kubesh, R.J. Natural oscillations of small raindrops. *Nature*, **342**: 408–410, 1989.

Beaver, J. and Bringi, V.N. The application of S-band polarimetric radar measurements to Ka-band attenuation prediction. *Proc. IEEE*, **85**: 893–909, 1997.

Beaver, J., Bringi, V.N., and Huang, G. Modelling of measured polarimetric radar parameters and associated propagation effects. In *Preprints, 29th International Conference on Radar Meteorology*, Montreal, pp. 281–284. Boston, MA, American Meteorological Society, 1999.

Bebbington, D.H.O., McGuinness, R., and Holt, A.R. Correction of propagation effects in S-band circular polarization diversity radars. *Proc. IEE*, **134**: 431–437, 1987.

Boerner, W.M., Yan, W.L., Xi, A.Q., and Yamaguchi, Y. On the basic principles of radar polarimetry: The target characteristic polarization state theory of Kennaugh, Huynen's polarization fork concept, and its extension to the partially polarized case. *Proc. IEEE*, **79**: 1538–1550, 1991.

Boerner, W.M., Yan, W.L., Xi, A.Q., and Yamaguchi, Y. Basic concepts of radar polarimetry, in *Direct and Inverse Methods in Radar Polarimetry*. Part I, volume 350 of NATO ASI C. W.M. Boerner, et al., Eds, Dordrecht, The Netherlands, Kluwer Academic, 1992.

Bohren, C.F. and Battan, L.J. Radar backscattering by inhomogeneous precipitation particles. *J. Atmos. Sci.*, **37**: 1821–1827, 1980.

Bohren, C.F. and Battan, L.J. Radar backscattering of microwaves by spongy ice spheres. *J. Atmos. Sci.*, **39**: 2623–2628, 1982.

Bohren, C.F. and Huffman, D.R. *Absorption and Scattering of Light by Small Particles*. New York, John Wiley, 1983.

Bohren, C.F. and Singham, S.B. Backscattering by non-spherical particles: A review of methods and suggested new approaches. *J. Geophys. Res.*, **96**: 5269–5277, 1991.

Bolen, S., Bringi, V.N., and Chandrasekar, V. A new approach to compare polarimetric radar data to surface measurements. In *Preprints, 28th International Conference on Radar Meteorology*, September, 7–12, Austin, TX, pp. 121–122. Boston, MA, American Meteorological Society, 1997.

Bolen, S., Bringi, V.N., and Chandrasekar, V. An optimal area approach to intercomparing polarimetric radar rainrate algorithms with gauge data. *J. Atmos. Oceanic Technol.*, **15**: 605–623, 1998.

Born, M. and Wolf, E. *Principles of Optics*. 5th edition, New York, Pergamon Press, 1975.

Brandes, E.A. and Vivekanandan, J. An exploratory study in hail detection with polarimetric radar. In *Preprints, 14th International Conference on Interactive Information and Processing Systems for Meteorology, Oceanography, and Hydrology*. Boston, MA, American Meteorological Society, 1998.

Brandes, E.A., Vivekanandan, J., and Wilson, J.W. Radar rainfall estimates of the Buffalo Creek flash flood using WSR-88D and polarimetric radar data. In *Preprints, 28th International Conference on Radar Meteorology*, September, 7–12, Austin, TX, pp. 123–124. Boston, MA, American Meteorological Society, 1997.

Bringi, V.N. and Hendry, A. Technology of polarization diversity radars for meteorology. In *Radar in Meteorology*. D. Atlas, Ed., pp. 153–190, Boston, MA, American Meteorological Society, 1990.

Bringi, V.N., Burrows, D.A., and Menon, S.M. Multiparameter radar and aircraft study of raindrop spectral evolution in warm-based clouds. *J. Appl. Meteor.*, **30**: 853–880, 1991.

Bringi, V.N., Chandrasekar, V., and Xiao, R. Raindrop axis ratios and size distributions in Florida rainshafts: An assessment of multiparameter radar algorithms. *IEEE Trans. Geosci. Remote Sens.*, **36**: 703–715, 1998.

Bringi, V.N., Chandrasekar, V., Balakrishnan, N., and Zrnić, D.S. An examination of propagation effects in rainfall on radar measurements at microwave frequencies. *J. Atmos. Oceanic Technol.*, **7**: 829–840, 1990.

Bringi, V.N., Cherry, S.M., Hall, M.P.M., and Seliga, T.A. A new accuracy in determining rainfall rates and attenuation due to rain by means of dual-polarization measurements. *IEE Conference Publ.*, **169**(2): 120–124, 1978.

Bringi, V.N., Knupp, K., Detwiler, A., Liu, L., Caylor, I.J., and Black, R.A. Evolution of a Florida thunderstorm during the convection and precipitation/electrification experiment: The case study of 9 August 1991. *Monthly Weather Rev.*, **125**: 2131–2160, 1997.

Bringi, V.N., Rasmussen, R.M., and Vivekanandan, J. Multiparameter radar measurements in Colorado convective storms. Part I: Graupel melting studies. *J. Atmos. Sci.*, **43**: 2545–2563, 1986a.

Bringi, V.N., Seliga, T.A., and Aydin, K. Hail detection with a differential reflectivity radar. *Science*, **225**: 1145–1147, 1984.

Bringi, V.N., Seliga, T.A., and Cherry, S.M. Statistical properties of the dual-polarization differential reflectivity (Z_{DR}) radar signal. *IEEE Trans. Geosci. Remote Sens.*, **21**: 215–220, 1983.

Bringi, V.N., Vivekanandan, J., and Tuttle, J.D. Multiparameter radar measurements in Colorado convective storms. Part II: Hail detection studies. *J. Atmos. Sci.*, **43**: 2564–2577, 1986b.

Brockwell, P.J. and Davis, R.A. *Time Series:Theory and Methods*. 2nd edition, New York, Springer-Verlag, 1991.

Browne, I.C. and Robinson, N.P. Cross polarization of the radar melting band. *Nature*, **170**: 1078–1079, 1952.

Browning, K. Organization and internal structure of synoptic and mesoscale precipitation systems in mid-latitudes. In *Radar in Meteorology*, D. Atlas, Ed., Boston, MA, American Meteorological Society, 1990.

Browning, K.A. and Beimers, J.G.D. The oblateness of large hailstones. *J. Appl. Meteor.*, **6**: 1075–1081, 1967.

Brunkow, D., Bringi, V.N., Kennedy, P.C., Rutledge, S.A., Chandrasekar, V., Mueller, E.A., and Bowie, R.K. A description of the CSU-CHILL National Radar Facility. *J. Atmos. Oceanic Technol.*, **17**: 1596–1608, 2000.

Brunkow, D.A., Kennedy, P.C., Rutledge, S.A., Bringi, V.N., and Chandrasekar, V. CSU-CHILL radar status and comparison of available operating modes. In *Preprints, 28th International Conference on Radar Meteorology*, September, 7–12, Austin, TX, pp. 43–44. Boston, MA, American Meteorological Society, 1997.

Brussaard, G. A meteorological model for rain-induced cross-polarization. *IEEE Trans. Antennas Propagation*, **24**, 5–11, 1976.

Bucci, N.J. and Urkowitz, H. Testing of Doppler tolerant range sidelobe suppression in pulse compression meteorological radar. In *Proceedings, IEEE National Radar Conference*, Piscataway, NJ, pp. 206–211, Boston, IEEE, 1993.

Burg, J.P. The relationship between maximum entropy spectra and maximum likelihood spectra. *Geophysics*, **37**: 375–376, 1972.

Byers, H.R. and Braham Jr, R.R. *The Thunderstorm*. pp. 287, Washington DC, U.S. Government Printing Office, 1949.

Calheiros, R.V. and Zawadzki, I. Reflectivity–rain rate relationships for hydrology in Brazil. *J. Climate Appl. Meteor.*, **26**: 118–132, 1987.

Campos, E.F. and Zawadzki, I. Z–R relations from independent measurements of raindrop size distributions. In *Preprints, 29th International Conference on Radar Meteorology*, Montreal, pp. 663–665. Boston, MA, American Meteorological Society, 1999.

Capon, J. High-resolution frequency–wavenumber spectrum analysis. *Proc. IEEE*, **57**: 119–127, 1969.

Carey, L.D. and Rutledge, S.A. Electrical and multiparameter radar observations of a severe hailstorm. *J. Geophys. Res.*, **103**: 13 979–14 000, 1998.

Carey, L.D., Rutledge, S.A., Ahijevych, D.A., and Keenan, T.D. Correcting propagation effects in C-band polarimetric radar observations of tropical convection using differential propagation phase. *J. Appl. Meteor.*, **39**: 1405–1433, 2000.

Carter, J.K., Sirmans, D., and Schmidt, J. Engineering description of the NSSL dual linear polarization Doppler weather radar. In *Preprints, 23rd Radar Meteorological Conference and Cloud Physics Conference*, September, 22–26, pp. 381–384, Snowmass, CO. Boston, MA, American Meteorological Society, 1986.

Caylor, I.J. and Chandrasekar, V. Time-varying ice crystal orientation in thunderstorms observed with multiparameter radar. *IEEE Trans. Geosci. Remote Sens.*, **34**: 847–858, 1996.

Caylor, I.J. and Illingworth, A.J. Radar observations and modeling of warm rain initiation. *Quart. J. Roy. Meteor. Soc.*, **113**: 1171–1191, 1987.

Caylor, I.J., Goddard, J.W.F., Hopper, S.E., and Illingworth, A.J. Bright band errors in radar estimates of rainfall: Identification and correction using polarization diversity. In *COST-73 International Seminar on Weather Radar Networking*, Collier, C.G., Ed., pp. 295–304. Brussels, XII/478/89-EN. Belgium, Commission of the European Communities, 1989.

Chandra, M., Hardaker, P.J., and Holt, A.R. Measurement of differential propagation phase and differential reflectivity using hybrid transmit–receive polarization bases. In *EUR 18567–COST-75–Advanced Weather Radar Systems*, p. 858, Luxembourg. C.G. Collier, Ed., European Communities, Luxembourg, 1999.

Chandra, M., Schroth, A., and Lueneburg, E. Analysis and application of weather radar S- and M-matrix measurements. In *Preprints, International Conference on Antennas and Propagation, ICAP-87*, number 274, pp. 328–333, York, IEE, 1987.

Chandrasekar, V. and Keeler, R.J. Antenna pattern analysis and measurements for multiparameter radars. *J. Atmos. Oceanic Technol.*, **10**: 674–683, 1993.

Chandrasekar, V., Cooper, W.A., and Bringi, V.N. Axis ratios and oscillations of raindrops. *J. Atmos. Sci.*, **45**: 1323–1333, 1988.

Chandrasekar, V., Gorgucci, E., and Scarchilli, G. Optimization of multiparameter radar estimates of rainfall. *J. Appl. Meteor.*, **12**: 1288–1293, 1993.

Cheng, L. and English, M. A relationship between hailstone concentration and size. *J. Atmos. Sci.*, **40**: 204–213, 1983.

Clift, R., Grace, J.R., and Weber, M.E. *Bubbles, Drops and Particles*. New York, Academic Press, 1978.

Collier, C.G. *Applications of Weather Radar Systems*. New York, Ellis Horwood, 1989.

Conway, J.W. and Zrnić, D.S. A study of embryo production and hail growth using dual-Doppler and multiparameter radars. *Monthly Weather Rev.*, **121**: 2511–2528, 1993.

Cotton, W.R. and Anthes, R.A. *Storm and Cloud Dynamics*. San Diego, CA, Academic Press, 1989.

Deschamps, G.A. and Mast, P.E. Poincaré sphere representation of partially polarized fields. *IEEE Trans. Antennas Propagation*, **21**: 474–478, 1973.

DeWolf, D.A., Russchenberg, H.W.J., and Ligthart, L.P. Modelling of particle distribution in the melting layer. *Proc. IEE*, **137**, Part H(6), 1990.

Dodge, J., Arnold, J., Wilson, G., Evans, J., and Fujita, T.T. Cooperative Huntsville Meteorological Experiment (COHMEX). *Bull. American Meteorological Society*, **67**: 417–419, 1986.

Doneaud, A.A., Ionescu-Niscov, S., Priegnitz, D.L., and Smith, P.L. The area–time integral as an indicator for convective rain volumes. *J. Climate Appl. Meteor.*, **23**: 555–561, 1984.

Dou, X., Testud, J., Amayenc, P., and Black, R. The concept of normalized gamma distribution to describe rain drop spectra and its use to parameterize rain relations. In *Preprints, 29th International Conference on Radar Meteorology*, Montreal, pp. 625–628. Boston, MA, American Meteorological Society, 1999.

Doviak, R.J. Bistatic radar detection of high altitude clear air atmospheric targets. *Radio Sci.*, **7**: 993–1003, 1972.

Doviak, R.J. and Zrnić, D.S. *Doppler Radar and Weather Observations*. 2nd edition, San Diego, CA, Academic Press, 1993.

Doviak, R.J., Bringi, V.N., Ryzhkov, A., Zahrai, A. and Zrnić, D.S. Polarimetric upgrades to operational WSR-88D radars. *J. Atmos. Oceanic Technol.*, **17**: 257–278, 2000.

Eccles, P.J. and Mueller, E.A. X-band attenuation and liquid water content estimation by a dual-wavelength radar. *J. Appl. Meteor.*, **10**: 1252–1259, 1973.

Fabry, F. and Zawadzki, I. Long-term radar observations of the melting layer of precipitation and their interpretation. *J. Atmos. Sci.*, **52**: 838–851, 1995.

Fabry, F., Bellon, A., and Zawadzki, I. *Long Term Observations of the Melting Layer Using Vertically Pointing Radars*. Technical Report MW-101, Cooperative Centre for Research in Mesometeorology, McGill University, Montreal, Canada, 1994.

Feingold, G. and Levin, Z. The lognormal fit to raindrop spectra from frontal convective clouds in Israel. *J. Climate Appl. Meteor.*, **25**: 1346–1364, 1986.

Fetter, R.W. Radar weather performance enhanced by pulse compression. In *Preprints, 14th Conference on Radar Meteorology*, pp. 413–418. Boston, MA, American Meteorological Society, 1970.

Fulton, R.A., Breidenbach, J.P., Seo, D.-J., Miller, D.A., and O'Bannon, T. The WSR-88D rainfall algorithm. *Weather and Forecasting*, **13**: 377–395, 1998.

Funahashi, K. On the approximate realization of continuous mappings by neural networks. *Neural Networks*, **2**: 183–192, 1989.

Gage, K.S., Carter, D.A., Cifelli, R., and Tokay, A. Use of Doppler radar profilers as a calibration tool in support of TRMM ground validation field campaigns. In *Preprints, 29th International Conference on Radar Meteorology*, Montreal, pp. 758–761. Boston, MA, American Meteorological Society, 1999.

Galloway, A., Pazmany, A., Mead, J., MacIntosh, R.E., Leon, D., Kelly, R., and Vali, G. Detection of ice hydrometeor alignment using an airborne W-band polarimetric radar. *J. Atmos. Oceanic Technol.*, **14**: 3–12, 1997.

Gingras, V., Torlaschi, E., and Zawadzki, I. A theoretical comparison between staggered and simultaneous H/V sampling in dual-polarization radar. In *Preprints, 28th International Conference on Radar Meteorology*, September, 7–12, Austin, TX, pp. 23–24. Boston, MA, American Meteorological Society, 1997.

Goddard, J.W.F. and Cherry, S.M. The ability of dual-polarization radar (copolar linear) to predict rainfall rate and microwave attenuation. *Radio Sci.*, **19**: 201–208, 1984.

Goddard, J.W.F., Cherry, S.M., and Bringi, V.N. Comparison of dual-polarization measurements of rain with ground-based distrometer measurements. *J. Appl. Meteor.*, **21**: 252–256, 1982.

Goedecke, G.H. and O'Brien, S.G. Scattering by irregular inhomogeneous particles via the digitized Green's function algorithm. *Appl. Optics*, **27**: 2431–2438, 1988.

Goldhirsch, J., Rowland, J., and Musiani, B. Rain measurement results derived from a two-polarization frequency-diversity S-band radar at Wallops Island, Virginia. *IEEE Trans. Geosci. Remote Sens.*, **GE-25**: 654–661, 1987.

Goldstein, H. *Classical Mechanics*. Reading, MA, Addison-Wesley, 1981.

Gorgucci, E. and Scarchilli, G. Intercomparison of multiparameter radar algorithms for estimating of rainfall rate. In *Preprints, 28th Conference on Radar Meteorology*. Boston, MA, American Meteorological Society, 1997.

Gorgucci, E., Chandrasekar, V., and Scarchilli, G. Radar and surface measurement of rainfall during CaPE. *J. Appl. Meteor.*, **34**: 1570–1577, 1995.

Gorgucci, E., Scarchilli, G., and Chandrasekar, V. A robust estimator of rainfall rate using differential reflectivity. *J. Atmos. Oceanic Technol.*, **1**: 586–592, 1994.

Gorgucci, E., Scarchilli, G., and Chandrasekar, V. Error structure of radar rainfall measurement at C-band frequencies with dual-polarization algorithm for attenuation correction. *J. Geophys. Res.*, **101**: 26 461–26 471, 1996.

Gorgucci, E., Scarchilli, G., and Chandrasekar, V. Specific differential phase shift estimation in the presence of non-uniform rainfall medium along the path. *J. Atmos. Oceanic Technol.*, **16**: 1690–1697, 1999a.

Gorgucci, E., Scarchilli, G., and Chandrasekar, V. A procedure to calibrate multiparameter weather radar using properties of the rain medium. *IEEE Trans. Geosci. Remote Sens.*, **17**: 269–276, 1999b.

Gorgucci, E., Scarchilli, G., Chandrasekar, V., and Bringi, V.N. Measurement of mean raindrop shape from polarimetric radar observations. *J. Atmos. Sci.*, **57**: 3406–3413.

Gossard, E.E. and Strauch, R.G. *Radar Observations of Clear Air and Clouds*. Amsterdam, Elsevier, 1983.

Graves, C. Radar polarization power scattering matrix. *Proc. IRE*, **44**: 248–252, 1956.

Green, A.W. An approximation for the shape of large raindrops. *J. Appl. Meteor.*, **14**: 1578–1583, 1975.

Groginsky, H.L. and Glover, K.M. Weather radar canceller design. In *Preprints, 19th Radar Meteorology Conference*, pp. 192–198. Boston, MA, American Meteorological Society, 1980.

Gunn, R. Mechanical resonance in freely falling drops. *J. Geophys. Res.*, **54**: 383–385, 1949.

Gunn, R. and Kinzer, G.D. The terminal velocity of fall for water droplets in stagnant air. *J. Meteor.*, **6**: 243–248, 1949.

Hall, M.P.M, Cherry, S.M., Goddard, J.W.F., and Kennedy, G.R. Raindrop sizes and rainfall rate measured by dual-polarization radar. *Nature*, **285**: 195–198, 1980.

Hall, M.P.M., Goddard, J.W.F., and Cherry, S.M. Identification of hydrometeors and other targets by dual-polarization radar. *Radio Sci.*, **19**: 132–140, 1984.

Haykin, S. *Neural Networks – A Comprehensive Foundation.* Prentice Hall, NJ, Upper Saddle River, 1999.

Haykin, S. and Van Veen, B. *Signals and Systems.* New York, John Wiley, 1999.

Hendry, A. and Antar, Y.M.M. Radar observations of polarization characteristics and lightning-induced realignment of atmospheric ice crystals. *Radio Sci.*, **17**: 1243–1250, 1982.

Hendry, A. and Antar, Y.M.M. Precipitation particle identification with centimeter wavelength dual-polarization radars. *Radio Sci.*, **19**: 115–122, 1984.

Hendry, A. and McCormick, G.C. Radar observation of the alignment of precipitation particles by electrostatic fields in thunderstorms. *J. Geophys. Res.*, **81**: 5353–5357, 1976.

Hendry, A., Antar, Y.M.M., and McCormick, G.C. On the relationship between the degree of preferred orientation in precipitation and dual polarization radar echo characteristics. *Radio Sci.*, **22**: 37–50, 1987.

Hendry, A., McCormick, G.C., and Antar, Y.M.M. Differential propagation constants on slant paths through snow and ice crystals as measured by 16.5 GHz polarization-diversity radar. *Ann. des Telecommunications*, **36**: 133–139, 1981.

Hendry, A., McCormick, G.C., and Barge, B.L. The degree of common orientation of hydrometeors observed by polarization diversity radars. *J. Appl. Meteor.*, **15**: 633–640, 1976.

Herrick, D.F. and Senior, T.B.A. Low frequency scattering by rectangular dielectric particles. *Appl. Phys.*, **13**: 175–183, 1977.

Herzegh, P.H. and Carbone, R.E. The influence of antenna illumination function characteristics on differential reflectivity measurements. In *Preprints, 22ⁿᵈ Radar Meteorology Conference*, Zürich, pp. 281–286. Boston, MA, American Meteorological Society, 1984.

Heske, T. and Heske, J. *Fuzzy Logic for Real World Design.* San Diego, CA, Annabooks, 1996.

Hildebrand, P.H., Lee, W., Walther, C.A., Frush, C., Randall, M., Loew, E., Neitzel, R., Parsons, R., Testud, J., Baudin, F., and Le Cornec, A. The ELDORA/ASTRAIA airborne Doppler weather radar: High resolution observations from Toga Coare. *Bull. Amer. Meteor. Soc.*, **77**: 213–232, 1996.

Hitschfeld, W.F. and Bordan, J. Errors inherent in the radar measurement of rainfall at attenuating wavelengths. *J. Meteor.*, **11**: 58–67, 1954.

Höller, H., Bringi, V.N., Hubbert, J., Hagen, M., and Meischner, P.F. Life cycle and precipitation formation in a hybrid-type hailstorm revealed by polarimetric and Doppler radar measurements. *J. Atmos. Sci.*, **51**: 2500–2522, 1994.

Holt, A.R. The scattering of electromagnetic waves by single hydrometeors. *Radio Sci.*, **17**: 929–965, 1982.

Holt, A.R. Some factors affecting the remote sensing of rain by polarization diversity radar in the 3- to 35-GHz range. *Radio Sci.*, **19**: 1399–1412, 1984.

Holt, A.R. Extraction of differential propagation phase from data from S-band circularly polarized radars. *Electron. Lett.*, **24**: 1241–1242, 1988.

Holt, A.R. and Santoso, B. A Fredholm integral equation method for scattering phase shifts. *J. Phys. B.*, **5**: 497–507, 1972.

Holt, A.R., Bringi, V.N., and Brunkow, D. A comparison between parameters obtained with the CSU-CHILL radar from simultaneous and switched transmission of vertical and horizontal polarization. In *Preprints, 29th International Conference on Radar Meteorology*, Montreal, pp. 214–217. Boston, MA, American Meteorological Society, 1999.

Houze, R.A. *Cloud Dynamics*. San Diego, CA, Academic Press, 1993.

Hu, Z. and Srivastava, R.C. Evolution of raindrop size distribution by coalescence, breakup and evaporation: Theory and observations. *J. Atmos. Sci.*, **52**: 1781–1783, 1995.

Huang, G., Bringi, V.N., and Beaver, J. Application of the polarization power matrix in analysis of convective storm data using the CSU-CHILL radar. In *Preprints, 28th International Conference on Radar Meteorology*, Austin, TX, pp. 48–49. Boston, MA, American Meteorological Society, 1997.

Hubbert, J. and Bringi, V.N. An iterative filtering technique for the analysis of copolar differential phase and dual-frequency radar measurements. *J. Atmos. Oceanic Technol.*, **12**: 643–648, 1995.

Hubbert, J. and Bringi, V.N. Specular null polarization theory: Applications to radar meteorology. *IEEE Trans. Geosci. Remote Sens.*, **34**: 859–873, 1996.

Hubbert, J., Bringi, V.N., Carey, L.D., and Bolen, S. CSU-CHILL polarimetric radar measurements in a severe hail storm in eastern Colorado. *J. Appl. Meteor.*, **37**: 749–775, 1998.

Hubbert, J., Chandrasekar, V., Bringi, V.N., and Meischner, P. Processing and interpretation of coherent dual polarized radar measurements. *J. Atmos. Oceanic Technol.*, **10**: 155–164, 1993.

Hubbert, J.C. A comparison of radar, optic and specular null polarization theories. *IEEE Trans. Geosci. Remote Sens.*, **32**: 658–671, 1994.

Hubbert, J.C., Bringi, V.N., and Schönhuber, M. 2D-Video distrometer measurements: Implications for rainrate and attenuation estimators. In *Preprints, 29th International Conference on Radar Meteorology*, Montreal. Boston, MA, American Meteorological Society, 1999.

Hubbert, J.C., Bringi, V.N., Hoeller, H., and Meischner, P. C-band polarimetric signatures from a convective storm during CLEOPATRA. In *Preprints, 27th International Conference on Radar Meteorology*, October, 9–23, Vail, CO, pp. 443–446. Boston, MA, American Meteorological Society, 1995.

Hudak, D., Currie, B., and Kochtubajda, B. Snow formation processes in winter storms over the Mackenzie river basin. In *Preprints, 29th International Conference on Radar Meteorology*, Montreal, pp. 197–200. Boston, MA, American Meteorological Society, 1999.

Humphries, R.G. *Depolarization Effects at 3 GHz due to Precipitation*. PhD dissertation, Stormy Weather Group, McGill University, Science Report MW-82, 1974.

Husson, D. and Pointin, Y. Quantitative estimation of hailfall intensity with a dual polarization radar and a hailpad network. In *Preprints, 24th International Conference on Radar Meteorology*, March, 27–31, Tallahassee, FL, pp. 318–325. Boston, MA, American Meteorological Society, 1989.

Huynen, J.R. *Phenomenological Theory of Radar Targets*. PhD thesis, Technical University, Delft, The Netherlands, 1970.

Illingworth, A.J. and Blackman, T.M. The need to normalize rsds based on the gamma rsd formulations and implications for interpreting polarimetric radar data. In *Preprints, 29th Radar Meteorology Conference*, Montreal, pp. 629–631. Boston, MA, American Meteorological Society, 1999.

Illingworth, A.J. and Caylor, I.J. Polarization radar estimates of raindrop size spectra and rainfall rates. *J. Atmos. Oceanic Technol.*, **6**: 939–949, 1989.

Illingworth, A.J. and Caylor, I.J. Correlation measurements of precipitation. In *Preprints, 25th International Conference on Radar Meteorology*, June, 24–28, Paris, pp. 650–653. Boston, MA, American Meteorological Society, 1991.

Illingworth, A.J., Goddard, J.W.F., and Cherry, S.M. Polarization radar studies of precipitation development in convective storms. *Quart. J. Roy. Meteor. Soc.*, **113**: 469–489, 1987.

Ishimaru, A. *Wave Propagation and Scattering in Random Media*, volume 1. New York, Academic Press, 1978.

Ishimaru, A. *Electromagnetic Wave Propagation, Radiation and Scattering*. Englewood Cliffs, NJ, Prentice Hall, 1991.

Ishimaru A. and Cheung, R.L.T. Multiple scattering effects on wave propagation due to rain. *Ann. des Telecommunications*, **35**: 373–379, 1980.

Ito, S., Oguchi, T., Kumagai, H., and Meneghini, R. Depolarization of radar signals due to multiple scattering in rain. *IEEE Trans. Geosci. Remote Sens.*, **33**: 1057–1062, 1995.

Jackson, J.D. *Classical Electrodynamics*. New York, John Wiley, 1975.

Jameson, A.R. Microphysical interpretation of multi-parameter radar measurements in rain. Part I: Interpretation of polarization measurements and estimation of raindrop shapes. *J. Atmos. Sci.*, **40**: 1792–1802, 1983.

Jameson, A.R. Microphysical interpretation of multi-parameter radar measurements in rain. Part III: Interpretation and measurement of propagation differential phase shift between orthogonal linear polarizations. *J. Atmos. Sci.*, **42**: 607–614, 1985.

Jameson, A.R. The effect of temperature on attenuation correction schemes in rain using polarization propagation differential phase shift. *J. Appl. Meteor.*, **31**: 1106–1118, 1992.

Jameson, A.R. and Caylor, I.J. A new approach to estimating rainwater content by radar using propagation differential phase shift. *J. Atmos. Oceanic Technol.*, **11**: 311–322, 1994.

Jameson, A.R. and Davé, J.H. An interpretation of circular polarization measurements affected by propagation differential phase shift. *J. Atmos. Oceanic Technol.*, **5**: 405–415, 1988.

Jameson, A.R. and Johnson, D.B. Cloud microphysics and radar. In *Radar in Meteorology*, pp. 323–347. D. Atlas, Ed., Boston, MA, American Meteorological Society, 1990.

Jones, R.C. A new calculus for the treatment of optical systems. *J. Opt. Soc. Amer.*, **31**: 488–493, 1941.

Joss, J. and Waldvogel, A. A raindrop spectrograph with automatic analysis. *Pure Appl. Geophys.*, **68**: 240–246, 1967.

Joss, J. and Waldvogel, A. Precipitation measurement and hydrology. In *Radar in Meteorology*. D. Atlas, Ed., pp. 577–606, Boston, MA, American Meteorological Society, 1990.

Keenan, T., Glasson, K., Cummings, F., Bird, T.S., Keeler, R.J., and Lutz, J. The BMRC/NCAR C-band polarimetric (C-Pol) radar system. *J. Atmos. Oceanic Technol.*, **15**: 871–886, 1998.

Kennaugh, E.M. Polarization Properties of Radar Reflections. MS thesis, Ohio State University, Columbus, OH, 1952.

Kennedy, P., Rutledge, S., Bringi, V.N., and Petersen, W. Polarimetric radar observations of hail formation. *J. Appl. Meteor.*, In Press.

Kerker, M. *The Scattering of Light and Other Electromagnetic Radiation.* New York, Academic Press, 1969.

King, R.W.P and Prasad, S. *Fundamental Electromagnetic Theory and Applications.* New Jersey, Prentice-Hall, 1986.

Kleinman, R.E. Low frequency electromagnetic scattering. In *Electromagnetic Scattering*, pp. 1–28, New York, Academic Press, 1978.

Knight, C.A. and Knight, N.C. The falling behavior of hailstones. *J. Atmos. Sci.*, **27**: 672–680, 1970.

Knight, C.A., Smith, P., and Wade, C. Storm types and some radar reflectivity characteristics in hailstorms of the Central High Plains. In *The National Hail Research Experiment*, volume 1, pp. 81–94. C.A. Knight and P. Squires, Eds, Boulder, CO. Association University Press, 1982. Available from NCAR, Boulder, CO 80307.

Knight, N.C. Hailstone shape factor and its relation to radar interpretation of hail. *J. Climate Appl. Meteor.*, **25**: 1956–1958, 1986.

Knollenberg, R.G. Techniques for probing cloud microstructure. In *Clouds: Their Formation, Optical Properties and Effects*, pp. 15–89. P.V. Hobbs and A. Deepak, Eds, New York, Academic Press, 1981.

Kostinski, A. Fluctuations of differential phase and radar measurements of precipitation. *J. Appl. Meteor.*, **33**: 1176–1181, 1994.

Kostinski, A. and Boerner, W.-M. On foundations of radar polarimetry. *IEEE Trans. Antennas Propagation*, **34**: 1395–1404, 1986.

Kozu, T., Nakamura, K., Meneghini, R., and Boncyk, W. Dual parameter rainfall measurement from space: A test result from an aircraft experiment. *IEEE Trans. Geosci. Remote Sens.*, **29**: 690–703, 1991.

Krehbiel, P., Chen, T., McCrary, S., Rison, W., Gray, G., and Brook, M. The use of dual channel circular-polarization radar observations for remotely sensing storm electrification. *J. Meteor. Atmos. Phys.*, **59**(1–2): 65–82, 1996.

Krehbiel, P.R. and Brook, M. A broad-band noise technique for fast-scanning radar observations of clouds and clutter targets. *IEEE Trans. Geosci. Electronics*, **GE-17**: 196–204, 1979.

Kreyszig, E. *Advanced Engineering Mathematics*. New York, Wiley, 1988.

Kry, P.R. and List, R. Angular motions of freely falling spheroidal hailstone models. *Phys. Fluids*, **17**: 1093–1102, 1974.

Kubesh, R.J. and Beard, K.V. Laboratory measurements of spontaneous oscillations for moderate-size raindrops. *J. Atmos. Sci*, **50**: 1089–1098, 1993.

Kwiatkowski, J.M., Kostinski, A.B., and Jameson, A.R. The use of optimal polarization for studying the microphysics of precipitation: Nonattenuating wavelength. *J. Atmos. Oceanic Technol.*, **12**: 96–114, 1995.

Lawson, R.P., Cormack, R.H., and Weaver, K.A. A new airborne precipitation spectrometer for atmospheric research. In *Preprints, 8th Symposium Meteorological Observations*, pp. 30–35. Boston, MA, American Meteorological Society, 1993.

Leonardi, R.M., Scarchilli, G., and Gorgucci, E. Polar C55: A C-band advanced meteorological radar developed for C.N.R. Italy. In *Preprints, 22nd Radar Meteorology Conference*, September, 10–13, Zürich, pp. 238–243. Boston, MA, American Meteorological Society, 1984.

List, R. A linear radar equation for steady tropical rain. *J. Atmos. Sci.*, **45**: 3564–3572, 1988.

Liu, H. and Chandrasekar, V. Classification of hydrometeors based on polarimetric radar measurements: Development of fuzzy logic and neuro-fuzzy systems and in-situ verification. *J. Atmos. Oceanic Technol.*, **17**: 140–164, 2000.

Liu, H., Chandrasekar, V., and Giang Xu An adaptive neural network scheme for radar rainfall estimation for WSR-88D observations. *J. Appl. Meteor.*, In Press.

Liu, L., Bringi, V.N., Chandrasekar V., Mueller, E.A., and Mudukutore, A. Analysis of the copolar correlation coefficient between horizontal and vertical polarizations. *J. Atmos. Oceanic Technol.*, **11**: 950–963, 1994.

Loney, M.L., Zrnić, D.S., Ryzhkov, A., and Straka, J.M. In-situ and multiparameter radar observations of an isolated Oklahoma supercell at far range. In *Preprints, 29th International Conference on Radar Meteorology*, Montreal, pp. 188–191. Boston, MA, American Meteorological Society, 1999.

Lorentz, H.A. *Theory of Electrons*. 2nd edition, New York, Dover, 1952.

Mardia, K.V. *Statistics of Directional Data*. New York, Academic Press, 1972.

Marks, F.D. Jr Radar observations of tropical weather systems. In *Radar in Meteorology*. D. Atlas, Ed., American Meteorological Society, Boston, MA, 1990.

Marshall, J.S. and Hitschfeld, W. Interpretation of the fluctuating echo from randomly distributed scatterers, part I. *Can. J. Phys.*, **31**: 962–994, 1953.

Marshall, J.S. and Palmer, W. McK. The distribution of raindrops with size. *J. Meteor.*, **5**: 165–166, 1948.

Marzoug, M. and Amayenc, P. A class of single and dual frequency algorithms for rain-rate profiling from a spaceborne radar. Part 1: Principles and tests from numerical simulations. *J. Atmos. Oceanic Technol.*, **11**: 1480–1506, 1994.

Mathur, P.M. and Mueller, E.A. Radar backscattering cross sections for non-spherical targets. *IRE Trans. Antennas Propagation*, **4**: 51–53, 1956.

Matrosov, S.Y. Prospects for the measurement of ice particle shape and orientation with elliptically polarized radar signals. *Radio Sci.*, **26**: 847–856, 1991.

Matrosov, S.Y., Reinking, R.F., Kropfli, R.A., and Bartram, B.W. Estimation of ice hydrometeor types and shapes from radar polarization measurements. *J. Atmos. Oceanic Technol.*, **13**: 85–96, 1996.

Maxwell-Garnet, J.C. Colors in metal glasses and in metallic films. *Phil. Trans. Roy. Soc.*, **A203**: 385–420, 1904.

May, P.T., Keenan, T.D., Zrnić, D.S., Carey, L.D., and Rutledge, S.A. Polarimetric radar measurements of tropical rain at a 5 cm wavelength. *J. Appl. Meteor.*, **38**: 750–765, 1999.

McCormick, G.C. Propagation through a precipitation medium. *IEEE Trans. Antennas Propagation*, **23**: 266–269, 1975.

McCormick, G.C. Relationship of differential reflectivity to correlation in dual-polarisation radar. *Electron. Lett.*, **15**: 265–266, 1979.

McCormick, G.C. Polarization errors in a two channel system. *Radio Sci.*, **16**: 67–75, 1981.

McCormick, G.C. On the completeness of polarization diversity measurements. *Radio Sci.*, **24**: 511–518, 1989.

McCormick, G.C. and Hendry, A. Method for measuring the anisotropy of precipitation media. *Electron. Lett.*, **9**: 216–218, 1973.

McCormick, G.C. and Hendry, A. Principles for the radar determination of the polarization properties of precipitation. *Radio Sci.*, **10**: 421–434, 1975.

McCormick, G.C. and Hendry, A. Radar measurements of precipitation-related depolarization in thunderstorms. *IEEE Trans. Geosci. Electron.*, **GE-17**: 142–150, 1979a.

McCormick, G.C. and Hendry, A. Techniques for the determination of the polarization properties of precipitation. *Radio Sci.*, **14**: 1027–1040, 1979b.

McCormick, G.C and Hendry, A. Optimal polarizations for partially polarized backscatter. *IEEE Trans. Antennas Propagation.*, **3**: 33–40, 1985.

McCormick, G.C., Hendry, A., and Barge, B.L. The anisotropy of precipitation media. *Nature*, **238**: 214–216, 1972.

Meischner, P., Bringi, V.N., Hagen, M., and Hoeller, H. Multiparameter radar characterization of a melting layer compared with insitu measurements. In *Preprints, 25th International Conference on Radar Meteorology*, June, 24–28, Paris, pp. 721–724. Boston, MA, American Meteorological Society, 1991a.

Meischner, P., Bringi, V.N., Heimann, D., and Hoeller, H. A squall line in southern Germany: Kinematics and precipitation formulation as deduced by advanced polarimetric and Doppler radar measurements. *Monthly Weather Rev.*, **119**: 678–701, 1991b.

Meneghini, R. and Kozu, T. *Spaceborne Weather Radar*. Boston, MA, Artech House, 1990.

Meneghini, R. and Liao, L. Comparison of cross sections for melting hydrometeors as derived from dielectric mixing formulas and a numerical method. *J. Appl. Meteor.*, **35**: 1658–1670, 1996.

Metcalf, J. Interpretation of the autocovariances and cross-covariances from a polarization diversity radar. *J. Atmos. Sci.*, **43**: 2479–2498, 1986.

Metcalf, J. Observation of the effects of electric fields on the orientation of hydrometeors in a thunderstorm. *Bull. Amer. Meteor. Soc.*, **74**: 1080–1083, 1993.

Metcalf, J. Radar observation of changing orientations of hydrometeors in thunderstorms. *J. Appl. Meteor.*, **34**: 757–772, 1995.

Metcalf, J. and Echard, J.D. Coherent polarization-diversity radar techniques in meteorology. *J. Atmos. Sci.*, **35**: 2010–2019, 1978.

Metcalf, J. and Ussailis, J.S. Radar sytem errors in polarization diversity measurements. *J. Atmos. Oceanic Technol.*, **1**: 105–114, 1984.

Middleton, D. *An Introduction to Statistical Communication Theory*. New York, McGraw-Hill, 1960.

Mieras, H. Comments on "Foundations of Radar Polarimetry". *IEEE Trans. Antennas Propagation*, **34**: 1395–1404, 1986.

Miller, K.S. and Rochwarger, M.C. A covariance approach to spectral moment estimation. *IRE Trans. Inf. Theory*, **IT-8**: 558–596, 1972.

Miller, L.J., Tuttle, J.D., and Foote, G.B. Precipitation production in a large Montana hailstorm: Airflow and particle growth trajectories. *J. Atmos. Sci.*, **47**: 1619–1646, 1990.

Miller, L.J., Tuttle, J.D., and Knight, C. Airflow and hailgrowth in a severe northern high plains supercell. *J. Atmos. Sci.*, **45**: 736–762, 1988.

Moffatt, D.L. and Garbacz, R.J., Eds, *Research Studies on the Polarization Properties of Radar Targets*, commemorative vols. I and II of Edward M. Kennaugh. Technical report, Electroscience Laboratory, The Ohio State University, Columbus, OH, 1984.

Morrison, J.A. and Cross, M.J. Scattering of a plane electromagnetic wave by axisymmetric raindrops. *Bell Syst. Tech. J.*, **53**: 955–1019, 1974.

Mott, H. *Antennas for Radar and Communications: A Polarimetric Approach*. New York, John Wiley, 1992.

Mudukutore, A.S., Chandrasekar, V., and Keeler, R.J. Pulse compression for weather radars. *IEEE Trans. Geosci. Remote Sens.*, **36**: 125–142, 1998.

Mudukutore, A.S., Chandrasekar, V., and Mueller, E.A. The differential phase pattern of the CSU-CHILL radar antenna. *J. Atmos. Oceanic Technol.*, **12**: 1120–1123, 1995.

Mueller, E.A. Calculation procedure for differential propagation phase shift. In *Preprints, 22nd Conference on Radar Meteorology*, September, 10–13, Zürich. Boston, MA, American Meteorological Society, 1984.

Nathanson, F.E. *Radar Design Principles*. New York, McGraw-Hill, 1969.

Nathanson, F.E. and Smith, P.L. A modified coefficient for the weather radar equation. In *Preprints, 15th International Radar Meteorology Conference*, pp. 228–230. Boston, MA, American Meteorological Society, 1972.

Nghiem, S.V., Yueh, S.H., Kwok, R., and Li, F.K. Symmetry properties in polarimetric remote sensing. *Radio Sci.*, **27**: 693–711, 1992.

Oguchi, T. Summary of studies on scattering of centimeter and millimeter waves due to rain and hail. *Ann. des Telecommunications*, **36**: 383–399, 1981.

Oguchi, T. Electromagnetic wave propagation and scattering in rain and other hydrometeors. *IEEE Proc.*, **71**: 1029–1078, 1983.

Olsen, R.L. A review of theories of coherent radio wave propagation through precipitation media of randomly oriented scatterers, and the role of multiple scattering. *Radio Sci.*, **17**: 913–928, 1982.

Papoulis, A. *Probability, Random Variables and Stochastic Processes*. New York, McGraw-Hill, 1965.

Passarelli, R.E. Jr and Siggia, A.D. The autocorrelation function and Doppler spectra moments: Geometric and asymptotic interpretation. *J. Climate Appl. Meteor.*, **22**: 1776–1787, 1983.

Petersen, W.A., Carey, L.W., Rutledge, S.A., Knievel, J.C., Doesken, N.J., Johnson, R.H., McKee, T.B., Vonder Haar, T., and Weaver, J.F. Mesoscale and radar observations of the Fort Collins flash flood of 28 July 1997. *Bull. Amer. Meteor. Soc.*, **80**: 191–216, 1999.

Poelman, A.J. and Guy, F.R.F. Multinotch logic–product polarization suppression filters. *Proc. IEEE*, **131**(F): 383–396, 1984.

Pointin, Y.C., Ramond, C., and Fournet-Fayard, J. Radar differential reflectivity: A real-case evaluation of errors induced by antenna characteristics. *J. Atmos. Oceanic Technol.*, **5**: 416–423, 1988.

Pozar, D.M. *Microwave Engineering*. 2nd edition, Reading, MA, Addison-Wesley, 1998.

Pratte, J.F. and Ferraro, D.G. Automated solar gain calibration. In *Preprints, 24th International Conference on Radar Meteorology*, March, 27–31, Tallahassee, FL, pp. 619–622. Boston, MA, American Meteorological Society, 1989.

Proakis, J.G. and Manolakis, D.G. *Introduction to Digital Signal Processing*. New York, Collier MacMillian, 1988.

Probert-Jones, J.R. The radar equation in meteorology. *Quart. J. Roy. Meteor. Soc.*, **88**: 485–495, 1962.

Pruppacher, H.R. and Beard, K.V. A wind tunnel investigation of the internal circulation and shape of water drops falling at terminal velocity in air. *Quart. J. Roy. Meteor. Soc.*, **96**: 247–256, 1970.

Pruppacher, H.R. and Klett, J.D. *Microphysics of Clouds and Precipitation*. 2nd edition, Dordrecht, Kluwer Academic Publishers, 1997.

Pruppacher, H.R. and Pitter, R.L. A semi-empirical determination of the shape of cloud and rain drops. *J. Atmos. Sci.*, **28**: 86–94, 1971.

Raghavan, R. and Chandrasekar, V. Multiparameter radar study of rainfall: Potential application to area time integral studies. *J. Appl. Meteor.*, **33**: 1636–1645, 1994.

Randall, M., Lutz, J., and Fox, J. S-POL's high isolation mechanical polarization switch. In *Preprints, 28th Conference on Radar Meteorology*, September, 7–12, Austin, TX, pp. 252–253. Boston, MA, American Meteorological Society, 1997.

Rasmussen, R.M. and Heymsfield, A.J. Melting and shedding of graupel and hail. Part I: Model physics. *J. Atmos. Sci.*, **44**: 2754–2763, 1987.

Rasmussen, R.M. and Pruppacher, H.R. A wind tunnel and theoretical study of the melting behavior of atmospheric ice particles. Part III: Experiment and theory for spherical ice particles of radius >500 μm. *J. Atmos. Sci.*, **41**: 381–388, 1984.

Rasmussen, R.M., Levizzani, V., and Pruppacher, H.R. A wind tunnel and theoretical study of the melting behavior of atmospheric ice particles. Part II: Theoretical study for frozen drops of radius <500 μm. *J. Atmos. Sci.*, **41**(3): 374–380, 1984.

Ray, P. Convective Dynamics. In *Radar in Meteorology*. D. Atlas, Ed., Boston, MA, American Meteorological Society, 1990.

Ray, P.S. Broadband complex refractive indices of ice and water. *Appl. Optics*, **11**: 1836–1844, 1972.

Reed, I.S. On a moment theorem for complex Gaussian processes. *IRE Trans. Inf. Theory*, **IT-8**: 194–195, 1962.

Reinking, R.F., Matrosov, S.Y., Bruintjes, R. T., and Martner, B.E. Identification of hydrometeors with elliptical and linear polarization Ka-band radar. *J. Appl. Meteor.*, **36**: 322–339, 1997.

Reinking, R.F., Matrosov, S.Y., Martner, B.E., Orr, B.W., Kropfli, R.A., and Bartram, B.W. Slant-linear polarization applied to detection of supercooled drizzle. In *Preprints, 29th International Conference on Radar Meteorology*, Montreal, pp. 285–288. Boston, MA, American Meteorological Society, 1999.

Rinehart, R.E. and Tuttle, J.D. Dual-wavelength processing: Some effects of mismatched antenna beam patterns. *Radio Sci.*, **19**: 121–131, 1984.

Rogers, R.R. The early years of Doppler radar in meteorology. In *Radar in Meteorology*. D. Atlas, Ed., pp. 122–129. Boston, MA, American Meteorological Society, 1990.

Rosenfeld, D., Atlas, D., and Short, D.A. The estimation of convective rainfall by area integrals: The height area threshold (HART) method. *J. Geophys. Res.*, **95**: 2161–2176, 1990.

Rosenfeld, D., Wolff, D.B., and Amitai, E. The window probability matching method for rainfall measurements with radar. *J. Appl. Meteor.*, **33**: 682–693, 1994.

Röttger, J. and Larsen, M.F. UHF/VHF radar techniques for atmospheric reasearch and wind profiler applications. In *Radar in Meteorology*. D. Atlas, Ed., pp. 235–281. Boston, MA, American Meteorological Society, 1990.

Russchenberg, H.W.J. *Ground-based Remote Sensing of Precipitation Using a Multi-polarized FM-CW Doppler Radar*. The Netherlands, Delft University Press, 1992.

Ryde, J.W. Attenuation of centimeter radio waves and the echo intensities resulting from atmospheric phenomena. *J. IEEE*, **93**(3A): 101–103, 1946.

Ryzhkov, A. and Zrnić, D.S. Beamwidth effects on the differential phase measurements of rain. *J. Atmos. Oceanic Technol.*, **15**: 624–634, 1998.

Ryzhkov, A. and Zrnić, D.S. Precipitation and attenuation measurements at 10 cm wavelength. *J. Appl. Meteor.*, **34**: 2121–2134, 1995a.

Ryzhkov, A. and Zrnić, D.S. Assessment of rainfall measurement that uses specific differential phase. *J. Appl. Meteor.*, **35**: 2080–2090, 1996.

Ryzhkov, A.V. and Zrnić, D.S. Comparison of dual-polarization radar estimators of rainfall. *J. Atmos. Oceanic Technol.*, **12**: 249–256, 1995b.

Ryzhkov, A.V., Zrnić, D.S., and Fulton, R. Areal rainfall estimates using differential phase. *J. Appl. Meteor.*, **39**: 263–268, 2000.

Ryzhkov, A., Zrnić, D.S., and Gordon, B.A. Polarimetric method for ice water content determination. *J. Appl. Meteor.*, **37**: 125–134, 1998.

Sachidananda, M. and Zrnić, D.S. Z_{DR} measurement considerations for a fast scan capability radar. *Radio Sci.*, **20**: 907–922, 1985.

Sachidananda, M. and Zrnić, D.S. Differential propagation phase shift and rainfall rate estimation. *Radio Sci.*, **21**: 235–247, 1986.

Sachidananda, M. and Zrnić, D.S. Rain rate estimates from differential polarization measurements. *J. Atmos. Oceanic Technol.*, **4**: 588–598, 1987.

Sachidananda, M. and Zrnić, D.S. Efficient processing of alternately polarized radar signals. *J. Atmos. Oceanic Technol.*, **4**: 1310–1318, 1988.

Sachidananda, M. and Zrnić, D.S. Systematic phase codes for resolving range overlaid signals in a Doppler weather radar. *J. Atmos. Oceanic Technol.*, **16**: 1351–1363, 1999.

Sauvageot, H. *Radar Meteorology*. Boston, MA, Artech House, 1992.

Scarchilli, G., Gorgucci, E., Chandrasekar, V., and Seliga, T.A. Rainfall estimation using polarimetric techniques at C-band frequencies. *J. Appl. Meteor.*, **32**: 1150–1160, 1993.

Schroth, A., Chandra, M., and Meischner, P. A C-band coherent polarimetric radar for propagation and cloud physics research. *J. Atmos. Oceanic Technol.*, **5**: 803–822, 1988.

Scott, R.D., Krehbiel, P.R., and Rison, W. The use of simultaneous horizontal and vertical transmissions for dual-polarization radar meteorological observations. *J. Atmos. Oceanic Technol.*, **18**: 629–648, 2001.

Sekhon, R.S. and Srivastava, R.C. Doppler radar observations of drop size distributions in a thunderstorm. *J. Atmos. Sci.*, **28**: 983–994, 1971.

Seliga, T.A. and Bringi, V.N. Potential use of radar differential reflectivity measurements at orthogonal polarizations for measuring precipitation. *J. Appl. Meteor.*, **15**: 69–76, 1976.

Seliga, T.A. and Bringi, V.N. Differential reflectivity and differential phase shift: Applications in radar meteorology. *Radio Sci.*, **13**: 271–275, 1978.

Seliga, T.A. and Mueller, E.A. Implementation of a fast-switching differential reflectivity dual-polarization capability on the CHILL radar: First observations. In *Preprints, URSI*, August, 23–27, Bournemouth, pp. 119–122. Washington, DC, American Geophysical Union, 1982.

Seliga, T.A., Bringi, V.N., and Al-Khatib, H.H. Differential reflectivity measurements in rain: First experiments. *IEEE Trans. Geosci. Electron*, **17**: 240–244, 1979.

Seliga, T.A., Bringi, V.N., and Al-Khatib, H.H. A preliminary study of comparative measurement of rainfall rate using the differential reflectivity radar technique and a rainguage network. *J. Appl. Meteor.*, **20**: 1362–1368, 1981.

Seliga, T.A., Humphries, R.G., and Metcalf, J.I. Polarization diversity in radar meteorology: Early developments. In *Radar in Meteorology*. D. Atlas, Ed., pp. 109–114. Boston, MA, American Meteorological Society, 1990.

Sempere-Torres, D., Porra, J.M., and Creutin, J.D. A general formulation for raindrop size distribution. *J. Appl. Meteor.*, **33**: 1494–1502, 1994.

Sempere-Torres, D., Sanchez-Diezma, R., Zawadzki, I., and Creutin, J.D. DSD identification following a pre-classification of rainfall type from radar analysis. In *Preprints, 29th International Conference on Radar Meteorology*, Montreal, pp. 632–635. Boston, MA, American Meteorological Society, 1999.

Senior, T.B.A. and Sarabandi, K. Scattering models for point targets. In *Radar Polarimetry for Geoscience Applications*, p. 364. F.T. Ulaby and C. Elachi, Eds, Norwood, MA, Artech House, 1990.

Senior, T.B.A. and Weil, H. On the validity of modeling Rayleigh scatterers by spheroids. *Appl. Phys. B*, **29**: 117–124, 1982.

Shannon, C.E. and Weaver, W. *The Mathematical Theory of Communications*. Urbana, IL, University of Illinois Press, 1963.

Sheppard, B.E. Effect of irregularities in the diameter classification of raindrops by the Joss-Waldvogel disdrometer. *J. Atmos. Oceanic Technol.*, **7**: 180–183, 1990.

Sihvola, A.H. and Kong, J.A. Effective permittivity of dielectric mixtures. *IEEE Trans. Geosci. Remote Sens.*, **26**: 420–429, 1988.

Sikdar, D.N., Schlesinger, R.E., and Anderson, C.E. Severe storm latent heat release: Comparison of radar estimate versus a numerical experiment. *Monthly Weather Rev.*, **102**: 455–465, 1974.

Sinclair, G. The transmission and reception of elliptically polarized waves. *Proc. IRE*, **38**: 148–151, 1950.

Skolnik, M.J. *Introduction to Radar Systems*. New York, McGraw-Hill, 1980.

Smith, P. Equivalent radar reflectivity factor for snow and ice particles. *J. Climate Appl. Meteor.*, **23**: 1258–1260, 1984.

Smith, P.L. On the sensitivity of weather radars. *J. Atmos. Oceanic Technol.*, **3**: 704–713, 1986.

Smith, P.L. Precipitation measurement and hydrology: Panel Report. In *Radar in Meteorology*. D. Atlas, Ed., Boston, MA, American Meteorological Society, 1990.

Smith, P.L. An update on weather radar system sensitivity. In *Preprints, 26th International Conference on Radar Meteorology*, May, 24–28, Norman, OK, pp. 384–386. Boston, MA, American Meteorological Society, 1993.

Smith, P.L., Musil, D.J., Detwiler, A.J., and Ramachandran, R. Observations of mixed-phase precipitation within a CaPE thunderstorm. *J. Appl. Meteor.*, **38**: 145–155, 1999.

Smyth, T.J. and Illingworth, A.J. Correction for attenuation of radar reflectivity using polarization data. *Quart. J. Roy. Meteor. Soc.*, **124**: 2393–2415, 1998.

Smyth, T.J., Blackman, T.M., and Illingworth, A.J. Observations of oblate hail using dual polarization radar and implications for hail-detection schemes. *Quart. J. Roy. Meteor. Soc.*, **125**: 993–1016, 1999.

Srivastava, R.C., Jameson, A.R., and Hildebrand, P.H. Time-domain computation of mean and variance of Doppler spectra. *J. Appl. Meteor.*, **18**: 189–194, 1979.

Steifvater, K.C., Brown, R.D., Wicks, M.C., and Vannicola, V.C. Performance evaluation for polarimetric algorithms. In *Direct and Inverse Methods in Radar Polarimetry*, W.M. Boerner et al., Eds., NATO ASI C: Mathematical and Physical Sciences, Dordrecht, The Netherlands, Kluwer Academic, 1992.

Stevenson, A.F. Solutions of electromagnetic scattering problems as power series in the ratio (dimension of scatterer to wavelength). *J. Appl. Phys.*, **24**: 1134–1142, 1953.

Straka, J.M., and Zrnić, D.S. An algorithm to deduce hydrometeor types and contents from multi-parameter radar data. In *Preprints, 26th Conference on Meteorology*, pp. 513–515. Boston, MA, American Meteorological Society, 1993.

Stratton, J.A. *Electromagnetic Theory*. New York, McGraw-Hill, 1941.

Tang, C. and Aydin, K. Scattering from ice crystals at 94 and 220 GHz millimeter wavelength frequencies. *IEEE Trans. Geosci. Remote Sens.*, **33**: 93–99, 1995.

Testud, J., Bouar, E.L., Obligis, E. and Ali-Mehenni, M. The rain profiling algorithm applied to polarimetric weather radar. *J. Atmos. Oceanic Technol.*, **17**: 322–356, 2000.

Tokay, A. and Beard, K.V. A field study of raindrop oscillations. Part I: Observation of size spectra and evaluation of oscillation causes. *J. Appl. Meteor.*, **35**: 1671–1687, 1996.

Tong, H., Chandrasekar, V., Knupp, K.R., and Stalker, J. Multiparameter radar observations of time evolution of convective storms: Evaluation of water budgets and latent heating rates. *J. Atmos. Oceanic Technol.*, **15**: 1097–1109, 1998.

Torlaschi, E. and Holt, A.R. Separation of propagation and backscattering effects in rain for circular polarization diversity S-band radars. *J. Atmos. Oceanic Technol.*, **10**: 465–477, 1993.

Tragl, K. Polarimetric radar backscattering from reciprocal random targets. *IEEE Trans. Geosci. Remote Sens.*, **8**: 856–864, 1990.

Tsang, L., Kong, J.A., and Shin, R.T. *Theory of Microwave Remote Sensing*. New York, Wiley, 1985.

Tuttle, J.D., Bringi, V.N., Orville, H.D., and Kopp, F.J. Multiparameter radar study of a microburst: Comparison with model results. *J. Atmos. Sci.*, **46**: 601–620, 1989.

Twersky, V. Coherent electromagnetic waves in pair-correlated random distribution of aligned scatterers. *J. Math Phys.*, **19**: 215–230, 1978.

Ulbrich, C.W. Natural variations in the analytical form of the raindrop-size distribution. *J. Climate Appl. Meteor.*, **22**: 1764–1775, 1983.

Ulbrich, C.W. and Atlas, D. Hail parameter relations: A comprehensive digest. *J. Appl. Meteor.*, **21**: 22–43, 1982.

Ulbrich, C.W. and Atlas, D. Rainfall microphysics and radar properties: Analysis methods for drop size spectra. *J. Appl. Meteor.*, **37**: 912–923, 1998.

Uzunoglu, N.K. and Evans, B.G. Multiple scattering effects in electromagnetic wave propagation through a medium containing precipitation. *J. Phys. A: Math. Gen.*, **11**: 767–776, 1978.

Van Bladel, J. *Electromagnetic Fields*. Revised Printing, New York, Hemisphere, 1985.

van de Hulst, H.C. *Light Scattering by Small Particles*. New York, Dover, 1981.

Van der Ziel, A. *Noise: Source, Characterization, Measurement*. New Jersey, Prentice-Hall, 1970.

van Zyl, J.J. and Ulaby, F.T. Scattering matrix representation for simple targets. In *Radar Polarimetry for Geoscience Applications*, F.T. Ulaby and C. Elachi, Eds, p. 364. Norwood, MA, Artech House, 1990.

van Zyl, J.J., Papas, C.H., and Elachi, C. On the optimum polarizations of incoherently reflected waves. *IEEE Trans. Antennas Propagation*, **35**: 818–825, 1987.

Vivekanandan, J., Bringi, V.N., and Raghavan, R. Multiparameter radar modeling and observations of melting ice. *J. Atmos. Sci.*, **47**: 549–563, 1990.

Vivekanandan, J., Bringi, V.N., Hagen, M., and Meischner, P. Polarimetric radar studies of atmospheric ice particles. *IEEE Trans. Geosci. Remote Sens.*, **32**: 1–10, 1994.

Vivekanandan, J., Zrnić, D.S., Ellis, S.M., Oye, R., Ryzhkov, A., and Straka, J. Cloud microphysics retrieval using S-band dual-polarization radar measurements. *Bull. Amer. Meteor. Soc.*, **80**: 381–388, 1999.

Wakimoto, R.M. and Bringi, V.N. Dual-polarization observations of microbursts associated with intense convection: The 20 July storm during the MIST project. *Monthly Weather Rev.*, **116**: 1521–1539, 1988.

Waldvogel, A., Werner, H., and Schmid, W. Raindrop size distributions and radar reflectivity profiles. In *Preprints, 27th International Conference on Radar Meteorology*, October, 9–13, Vail, CO, pp. 26–30. Boston, MA, American Meteorological Society, 1995.

Wallace, P.R. Intrepretation of the fluctuating echo from randomly distributed scatterers, Part II. *Can. J. Phys.*, **31**: 995–1009, 1953.

Wang, P.K., Greenwald, T.J., and Wang, J. A three parameter representation of the shape and size distributions of hailstones – a case study. *J. Atmos. Sci.*, **44**: 1062–1070, 1987.

Warner, C. and Rogers, R.R. *Polarization-diversity Radar: Two Theoretical Studies*. Technical Report MW-90, Stormy Weather Group, McGill University, Montreal, 1977.

Waterman, P.C. Symmetry, unitarity and geometry in electromagnetic scattering. *Phys. Rev. D*, **3**: 825–839, 1971.

Waterman, P.C. and Truell, R. Multiple scattering of waves. *J. Math. Phys.*, **2**: 512–535, 1961.

Wilson, D.R., Illingworth, A.J., and Blackman, T.M. Differential Doppler velocity: A radar parameter of characterizing hydrometeor size distributions. *J. Appl. Meteor.*, **36**: 649–663, 1997.

Wilson, J.W. and Brandes, E.A. Radar measurement of rainfall – A summary. *Bull. Amer. Meteor. Soc.*, **60**: 1048–1058, 1979.

Wiscombe, W.J. Improved Mie scattering algorithms. *Appl. Opt.*, **19**: 1505–1509, 1980.

Wurman, J. Vector winds from a single-transmitter bistatic dual-Doppler radar network. *Bull. Amer. Meteor. Soc.*, **75**: 983–994, 1994.

Xiao, R. and Chandrasekar, V. Development of a neural network based algorithm for rainfall estimation from radar measurements. *IEEE Trans. Geosci. Remote Sens.*, **35**: 160–171, 1997.

Zadeh, L.A. A computational approach to fuzzy quantifiers in natural languages. *Comput. Math. Appl.*, **9**: 149–184, 1983.

Zawadzki, I. On radar–raingage comparison. *J. Appl. Meteor.*, **14**: 1430–1436, 1975.

Zawadzki, I. Factors affecting the precision of radar measurements of rain. In *Preprints, 22nd Radar Meteorology Conference*, September, 10–13, Zürich, pp. 251–256. Boston, MA, American Meteorological Society, 1984.

Zeng, Z., Yuter, S.E., Houze Jr, R.A., and Kingsmill, D.E. Microphysics of the rapid development of heavy convective precipitation. *Monthly Weather Rev.*, In Press.

Zrnić, D.S. Spectral moment estimates from correlated pulse pairs. *IEEE Trans. Aerosp. Electron. Syst.*, **AES-14**: 344–354, 1977.

Zrnić, D.S. and Balakrishnan, N. Dependence of reflectivity factor–rainfall rate relationship on polarization. *J. Atmos. Oceanic Technol.*, **7**: 792–795, 1990.

Zrnić, D.S. and Doviak, R.J. Effect of drop oscillation on spectral moments and differential reflectivity measurements. *J. Atmos. Oceanic Technol.*, **6**: 532–536, 1989.

Zrnić, D.S. and Mahapatra, P. Two methods of ambiguity resolution in pulse Doppler weather radars. *IEEE Trans. Aerosp. Electron. Syst.*, **AES-21**: 470–483, 1985.

Zrnić, D.S. and Ryzhkov, A.V. Advantages of rain measurements using specific differential phase. *J. Atmos. Oceanic Technol.*, **13**: 454–464, 1996.

Zrnić, D.S., Bringi, V.N., Balakrishnan, N., Aydin, K., Chandrasekar, V., and Hubbert, J. Polarimetric measurements in a severe hailstorm. *Monthly Weather Rev.*, **121**: 2223–2238, 1993.

Index